GEOMECHANICS IN TROPICAL SOILS

2·ICOTS
1988 Singapore

PROCEEDINGS OF THE SECOND INTERNATIONAL CONFERENCE ON GEOMECHANICS IN TROPICAL SOILS
SINGAPORE / 12-14 DECEMBER 1988

GEOMECHANICS IN TROPICAL SOILS

Editors
PUBLICATIONS COMMITTEE OF 2 ICOTS
School of Civil and Structural Engineering, Nanyang Technological Institute

VOLUME 1

1 Characterization, identification and classification of tropical soils
2 Engineering properties of tropical soils
3 Stability of slopes and excavations in tropical soils
4 Foundations of buildings in tropical soils
5 Construction of dams, roads, airfields, harbours, land reclamation in or on tropical soils

co-sponsored by
Nanyang Technological Institute
International Society for Soil Mechanics and Foundation Engineering
South East Asian Geotechnical Society

A.A.BALKEMA / ROTTERDAM / BROOKFIELD / 1988

CONFERENCE ORGANISING COMMITTEE

Prof. B.B.Broms, Chairman
Dr A.Goh, General Secretary
Ass.Prof. K.S.Wong, Publications, Poster Sessions
Ass.Prof. V.Choa, Publications, Poster Sessions
Dr B.K.Low, Publications, Poster Sessions
Miss Janet Ng, Ladies Programme
Mr John S.Y.Tan, Registration, Tour, Social Programme

International Society for Soil Mechanics
Foundation Engineering

TECHNICAL COMMITTEE ON TROPICAL SOILS (TC 25)

Dr E.W.Brand (Chairman), Hong Kong
J.B.Massey (Secretary), Hong Kong
Prof. G.E.Blight, South Africa
A.Jayaputra, Indonesia
G.W.Nyland, Australia
J.B.Queiros de Carvalho, Brazil
Dr J.S.Nogami, Brazil
J.Nuques, Equador
M. Liautaud, France
M.Londez, France
Dr E.Wernick, FR Germany
Dr M.D.Gidigasu, Ghana
Dr A.V.Jalota, India
Prof. A.Sridharan, India
Dr L.Belloni, Italy
A.Shimazu, Japan
M.O.Adesunloye, Nigeria
J.E.Cocksedge, UK
W.B.Lum, USA
R.E.Barton, Zimbabwe

The texts of the various papers in this volume were set individually by typists under the supervision of each of the authors concerned.

Published by
A.A.Balkema, P.O.Box 1675, 3000 BR Rotterdam, Netherlands
A.A.Balkema Publishers, Old Post Road, Brookfield, VT 05036, USA

Complete set of two volumes: ISBN 90 6191 816 2
Volume 1: ISBN 90 6191 817 0 / Volume 2: ISBN 90 6191 818 9

Printed in the Netherlands

Foreword

The Technical Committee on Tropical and Residual Soils (TC 25) was set up in 1985 by the International Society for Soil Mechanics and Foundation Engineering with Dr E.W.Brand as chairman and Dr J.B.Massey as secretary. One task of this committee was to organise an international conference on tropical soils in 1988 or 1989 following the very successful one that was held in Brasilia, Brazil, 11-14 February 1985.

The committee was invited in 1986 by the Nanyang Technological Institute and the Southeast Asia Geotechnical Society to hold the Second Conference on Tropical Soil (2ICOTS) in Singapore in December 1988. The invitation was accepted by the committee. The main purpose of the conference was to review the progress that has been made since 1985 with respect to the geotechnical properties and classification of tropical and residual soils, soil exploration and testing, stability of slopes and excavations, use of tropical soils as construction material for e.g. roads, airfields and earth dams and as foundation for different types of structures.

The response to the invitation of the organizing committee for papers to the conference has been very encouraging showing that there is a need for a specialty conference entirely devoted to the special problems connected with tropical soils which affect almost all construction activities in such countries as Brazil, Columbia, Equador, Venezuela, Ghana, Nigeria, Zaire, Tanzania, Kenya, Thailand, Malaysia, Indonesia and the Philippines.

Problems with tropical and residual soils are often connected with the high annual rainfall in many countries in the tropics and the effect of the water on e.g. the erosion and the stability of slopes and the control of the water content during compaction. Of special interests are the effects of the negative pore water pressure and the cementation on the strength of residual soils and of weathered rocks, and the importance of the soil structure on the permeability and the drainage. Of interest are also the possible relationships between the Atterberg limits and the shear strength and compressibility of tropical and residual soils. Commonly used relationships are not applicable.

The papers presented at the conference will be published in two volumes, one volume before the conference with all the papers and the second volume after conference with the special lectures, general reports and discussions. It is expected that the proceedings of the conference about properties and special problems connected with construction and residual soils will be a very valuable source of information for many years to come.

Bengt B.Broms
President ISSMFE (1985-1989)

Contents

Session 1

Characterization, identification and classification of tropical soils

The field description and identification of tropical residual soils

J.R.COOK, Road Materials Technical Assistance Project, Institute of Road Engineering, Bandung, Indonesia
D.NEWILL, Transport and Road Research Laboratory, UK

ABSTRACT: Tropical soils often differ from those in temperate zones largely as a result of more intensive weathering. This is particulary reflected in residual soils where the effects of hydrolysis,chemical decomposition and re-deposition of some of the more soluble products can result in properties and behaviour which cannot easily be predicted or determined in normal soil engineering classification systems that apply to temperate zone soils. The United Kingdom Geological Society has set up a Working Party to prepare a report that should help civil engineers to obtain a better understanding of tropical residual soils. This paper describes a part of that report dealing with field identification and description. A classification system is used which relates to tropical residual soils but the major part of the paper describes briefly a methodology for fieldwork which will aid identification. The methodology deals separately with general site characteristics,and soil material and soil mass characteristics and can be varied according to the project being undertaken.

1.0 INTRODUCTION

This paper is derived from part of a United Kingdom Geological Society Working Party Report on tropical residual soils, the draft of which was issued for comment in March of this year. The authors of this paper had particular responsibility in this draft document for the preparation of the section on field description and identification. This aspect of tropical residual soil work is of particular interest, since an accurate and systematic field description process is of importance, not only for classification purposes, but also because it may be the first opportunity that a geologist or engineer has to assess the materials in question. A meaningfulassessment of tropical residual soil masses and materials at an early stage in a project can lead to a quicker and more beneficial programme of fieldwork that can take into account any identified potential problems Also the usefulness of a laboratory testing programme relies on a thorough logging,sampling and initial field description.

The objective of the proposed methodology outlined in this paper is to define a framework for the field description of tropical residual soils in a manner compatible with their classification and by using the parameters that have been perceived to be relevant with the respect to their development and maintenance.

Other authors within the Working Party Report have identified the factors governing the development and maintenance of tropical residual soil types and have recognised and classified these types largely by their physical characteristics. The field identification of these soils must therefore take cognisance both of the governing factors and classification characteristics.

2.0 CLASSIFICATION OF TROPICAL RESIDUAL SOILS.

The Working Party considered it important, in the light of the nature of the material being investigated, that the experience of soil scientists be called upon to form a sound basis for the classification. Similarly the contributions of geomorphology and geology are drawn together for the benefit of the practising civil engineer.

The view taken that the proposed classification should be based on scientific principles has meant that a different approach has been adopted for tropical residual soils to those adopted for temperate zone soils. The latter classification systems tend to depend largely on textural and physical composition and the response to index tests on remoulded soil (e.g. Atterberg Limits). This approach to classification has been successful when dealing with temperate zone materials, but generally less so when adopted for use with tropical materials. For certain tropical soils it is well known that drying or other forms of pre treatment can have marked effects on the results of laboratory tests which can give misleading information about the in situ behaviour of the soil. (Tuncer 1982).

The approach to classification adopted by the current Working Party for tropical residual soils is based upon weathering and pedogenic processes. This follows the work of Duchaufour (1982) who recognised three phases of residual soil development in tropical areas. If one adds to these phases the development of duricrusts, which form by the residual or positive accumulation of minerals precipitated out of solution to form indurated layers, the basis of a classification system emerges which can be adopted for engineering purposes. The main features of the classification are shown in table 1. A problem encountered in using a formal scienific approach to soilclassification such as that of Duchaufour is that it does not take into account fully all of the factors that may lead to the development of the soil profile. The effects of erosion and movement will often have a strong influence so that transported material can be associated with the residual soil in the same profile. The frequent occurence of stone lines is evidence of this. It emphasises the need to careful description especially for civil engineering purposes.

3.0 BASIS OF DESCRIPTION SYSTEM

Description methods have been long established for geotechnical practice which have been generally developed for dealing with temperate zone soils and their classification. It is logical that the adoption of a fresh classification system for tropical residual soils should lead to a re-appraisal of the field description and identification procedures for such materials.

It was recognised that a description framework for such soils must take cognisance of identified formation processes

Table 1 A Formal classification of residual soils

A MATURE SOIL

Subdivision	Classification Criteria Groups
Vertisols	a, Vertisols (Sensu Stricto) 1, Developed (Inc Black Cotton Soils) 2, Slightly developed b, Coloured Vertic Soils 1, Intergrades & Ferruginous 2, Vertic Tropical Entrophic Brown Soils
Fersiallitic Andosols	a, Vitrosols b, Andosols (Sensu Stricto) 1, Humic Andosols 2, Differentiated Andosols 3, Hydromorphic c, Andic Soils
Fersiallitic (Sensu Stricto)	a, Fersiallitic Brown Soils 1, Tropical Brown Eutrophic 2, Subtropical & Mediterranean b, Modal Fersiallitic Red Soils 1, Tropical Fersiallitc Red Soil 2, Subtropical & Mediterranean
Ferruginous (Sensu Stricto)	a, Argillic (Bt Horizon present) 1, Eutrophic Ferruginous 2, Oligotrophic Ferruginous 3, Hydromorphic Ferruginous b, Ferrisols (Bt horizon not essential) 1, Ferrisol, weak Bt 2, Ferrisol, weathered Bt 3, Hydromorphic 4, Humic
Ferrallitic	a, Ferrallitic 1, Ferrallitic (Kaolinitic) 2, Ferrallitic (Gibbsitic) b, Ferrallites 1, Ferrites 2, Allites c, Ferrallitic (with hydromorphic segregation of iron) 1, Hydromorphic 2, Plinthitic 3, Indurated

B DURICRUSTS

Subdivision	Classification Criteria Groups
Silcrete	a, Grain Supported Fabric b, Floating Fabric c, Matrix Fabric d, Conglomeratic Fabric
Calcrete	a, Calcified Sand b, Powder Calcrete c, Nodular Calcrete 1, Nodular 2, Concretionary/concentric d, Honeycomb 1, Coalesced Nodules 2, Cemented e, Hardpan Calcrete 1, Cemented Honeycomb 2, Cemented Powder 3, Recemented 4, Coalesced Horizontal Nodules 5, Case Hardened Calcic f, Laminar g, Boulder
Gypcrete	a, Desert Rose b, Mesocrystalline c, Indurated 1, Powder 2, Indurated Alabastine 3, Alabastine Cobbles d, Evaporitic 1, Laminated 2, Bedded e, Gypsum Rich Dune Sand
Ferricrete	a, Water Table Curiasses 1, Local 2, Plinthite 3, Petroplinthite b, Plateau Curiasses Or a, Pisolitic b, Scoraceous & Vesicular c, Petroplinthite
Alcrete (Alucrete)	a, Pisolitic in Plinthites b, Scoraceous & Vesicular c, Petroplinthite

and be closely allied to their classification.
Because of the more intense weathering that occurs in tropical climatic conditions, decomposition of rock is generally more rapid. Consequently the physico chemical and mineralogical properties of the resulting soils assume a greater importance in understanding their engineering behaviour. These properties, which influence, and are reflected in, the development of the texture, fabric and in situ behaviour of the resulting soil,must be adequately covered in any description system.
The adopted description framework involves the definition of parameters that have been identified as having control on the development and maintenance of tropical residual soils in addition to their physical and behavioural characteristics. These parameters have been grouped as follows:

Site Characteristics
Soil Material Characteristics
Soil Mass Characteristics

The main elements of this description system are summarised in Tables 2,3,and 4. These elements may be defined by a series of standared tables, as indicated in the Working Party Report. A logical development of this system could be the adoption of standard field sheets coded for ease of use and transfer to computerised data base systems.

4.0 SITE CHARACTERISTICS

4.1 Location
The location and extent of each site under consideration should be adequately defined. The subdivision of project locations will be governed by the size, stage and type of project together with naturally occuring boundaries within its area.
4.2 Landform
The inter relationship between tropical residual soil occurrence and landform has been well illustrated in the literature . It follows, therefore, that the identification and classification of landform should be an important element in the field description process. Although the scale and methods of defining landform may be a function of project size and type, the overall objective may best be achieved by use of currently utilised methods of terrain analysis (Mitchell,1973) in which, for instance, Land Systems may be defined in terms of their component facets and elements.
4.3 Geology
Bedrock type has a major influence on the development of residual soil type, and therefore the relevant geology should be defined, possibly in simple terms by use of tables adopted from BS5930. In conjunction with this bedrock identification the geological history, and particularly the tectonic maturity, of the site should be established.
4.4 Current Climate
Climate is of major importance in soil formation. In addition the influence of past climates should be borne in mind and how this may have influenced the development of the landscape, soil, etc. Rainfall, temperature and their seasonal and diurnal variations affect soil both directly and in conjunction with vegetation and hydrology (Fitzpatrick,1983). Climate should be described in terms both of the general governing conditions and the current, and the immediately recent, weather. The climatic description should be capable of quantifying the following:

Annual rainfall
Seasonal and diurnal rainfall variations
Seasonal and diurnal temperature variations
Humidity

For some projects it may prove useful to establish a local climatic index on the lines of that described by Weinert,(1980).

4.5 Hydrology
The natural drainage pattern should be defined and described in broad terms, although, in practice, parts of this may be covered within landform definition. Attention should be paid to the following items in particular:

Occurrence of streams/rivers/seepages
Seasonal variations
Stream/river bed maturity
Groundwater position and variablility
Infiltration capacity

4.6 Vegetation
Vegetation should be defined in terms of type, both natural and imposed, and percentage of cover.

TABLE 2: GENERAL SITE CHARACTERISTICS

Parameter	Classification
Location	1. Position 2. Boundaries/elevation 3. Site extent
Landform	1. Relative relief 2. Slope angle 3. Profile form 4. Plan for form
Geology	1. Bedrock type 2. Information source 3. Tectonic history
Current climate	1. Annual rainfall 2. Rainfall variations 3. Temperature variations 4. Humidity 5. Evapo/transpiration
Hydrology	1. Streams/rivers 2. Bed maturity 3. Water table
Vegetation	1. Type 2. Percentage cover

5.0 SOIL MATERIAL CHARACTERISTICS

It may be considered (Ferreira 1985) that characterisation of the decomposed rock and tropical pedogenic soils can be made on the basis of:

a. Morphological characteristics
b. Mineralogical characteristics
c. Identification tests
d. Engineering properties

Field identification techniques can only partly define some of the above, although indirect measurements or tests may help. The combination of field identification with selective laboratory work should enable a more complete characterisation, and hence classification to be achieved. Laboratory testing is outside the scope of this paper but for field identification of material characteristics it is necessary to concentrate on the description of visible elements and the recording of available material behaviour. Current material description standards, Unified and AASHTO (ASTM D2487 and D3282) or the British Soil Classification System incorporated in BS 5930, are dominated either by an emphasis on arnaceous material grain size distribution, or on plasticity, as measured by Atterberg Limits, for fine grained soils (Mitchell and Nicholas, 1982, Child, 1974). Such commonly used temperate zone classification systems can be difficult to apply in tropical soils because of their sensitivity to sample treatment. Thus to rely solely on such standards may be misleading in residual soil/rock environment where the undisturbed structure and fabric of the material particles may be contradicted by their disturbed or remoulded behaviour.

Although the above standard description elements may be significant within this environment it is considered that they should form only part of a more broadly based material description framework. This framework is outlined below and includes, where possible, parts from these currently recognised Standards.

5.1 Moisture Condition

Moisture condition may be used both as an aid to identification in some soil types and as an indication of potential behaviour. The general terms dry, slightly moist, moist, very moist and wet have been previously suggested (Jennings et al 1973). The apparent moisture condition will naturally relate to the soil texture and therefore must be assessed and interpreted with this in mind. In addition to a current moisture condition it may be of significance in some environments to assess any seasonal variation of this moisture condition which may be of significance for engineering design purposes.

5.2 Colour

Colour is given a much greater significance than is general with non tropical materials; colour may frequently be an index to chemical or mineralogical change reflecting as it may do the mineralology of the soil (Hodgson,1976). The use of colour charts in the field is therefore strongly recommended. Colour should be noted, where possible, both in a dry condition and after being wetted up.

5.3 Strength

Consistency/strength should be described using standardised definitions for both soil and rock materials (BS5930,& Weltman and Head 1983). The definitions to be used are those relevant to an engineering context. Other detailed strength definitions exist that are primarily of use in pedological work but may cause confusion in relation to existing geotechnical standards (Hodgson 1976).

5.4 Fabric

The description of material fabric is of great importance in residual material identification. Fabric is the physical constitution of the soil as expressed by the spatial arrangement of the soil particles and associated voids (Brewer 1968). In a residual soil/weathered rock environment the range in scale of these particles" could be great and range from discrete and compound grains to large core stones. Brewer proposed a system that dealt with the fabric analysis in pedological terminology which can be adapted to the wider context of this report. This system is based on the concept of units of organisation and layers of organisation, which may be too detailed for the majority of engineering geological or geotechnical projects. Therefore only an outline of his approach has been suggested, although it is possible that particular projects may require the more detailed study.

5.5 Texture

By texture is meant the size and shape of the material particles. It is suggested that they should be described using the established standard descriptions with respect to grain size and shape,(Anon,1979)

5.6 Bulk Density

The undisturbed bulk density of a soil, which takes into account the material particles and their structure, is being given increasing attention as a significant and distinctive characteristic of some soil types (Fitzpatrick 1983). The usefulness of a field estimation of this property could be enhanced by comparison with a remoulded density.

5.7 Apparent Behaviour

The behaviour of the material in question shouldbe described in respect to the natural or imposed conditions. These may be regarded in part as field index tests within the framework of field identification rather than definitions of engineering behaviour. The requirements of such field tests are that they should be rapid and simple, be relevant to the classification, and to the engineering properties, and be capable of discriminating between soil types.

An outline soil'behaviour' list that should be investigated during the field description process as follows:

- Soaking
- Drying
- Remoulding
- Sensitivity to disturbance
- Durability

The material behaviour pattern may be examined by the use of simple field tests in addition to those undertaken as part of any larger scale in situ field testing programme A list of such simple tests that could be considered is given below:

- Swell test
- Shrinkage test
- Fabric collapse
- Plasticity
- Schmidt hammer test
- Penetrometer/hand vane
- Quick absorption
- pH measurement
- Aggregate pliers test (Netterberg 1978)

5.8 Mineralology

If possible, a field estimation of the material mineralology should be undertaken with particular note being taken of residual minerals left as a consequence of the soil formation/weathering process. The ratio of these residual primary minerals to secondary minerals may be a significant defining characteristic (Lumb 1965). This ratio should

Table 3 : SOIL MATERIAL CHARACTERISTICS

Parameter	Classification
1, Moisture	1, Condition 2, Drainage 3, Wetness duration
2, Colour	Colour Chart
3, Strength	1, Cohesive 2, Non cohesive/cemented soil 3, Rock like/ indurated
4, Fabric	1, Origin 2, Voids 3, Orientation 4, Distribution 5, Fissuring
5, Texture	1, Particle size 2, Particle shape 3, Grading
6, Density	1, Bulk undisturbed 2, Remoulded
7, Apparent behaviour	1, Remoulded strength 2, Durability 3, Infiltration 4, Swell/shrink 5, Plasticity
8, Mineralogy	1, Primary 2, Secondary

always be considered in the light of the original nature of the parent rock.
The estimation of the occurrence and nature of any secondary minerals in the field may be aided by observation of material behaviour during the simple field tests listed above e.g. the presence of swelling clay minerals.

6.0 SOIL MASS CHARACTERISTICS

The soil mass, which may contain several material types, should be identified and described in terms compatible with the proposed classification system and relevant geotechnical properties

6.1 Composition

The material components of the soil mass must be identified and recorded in a systematic fashion. This may be achieved for instance by the construction of profiles or sections through the soil mass in question.

6.2 Structure

The description of geological structure is meant to include any naturally occurring boundaries and discontinuities, both relict and contemporary, within or surrounding the soil mass, e.g.

Joints or bedding
Slickensiding
Faulting
Folding/flexuring

The nature and occurrence of these boundaries or discontinuities should be fully defined using accepted standards (ISRM 1980). In addition the nature of boundaries between units being described should be recoreded.

Included within a description of the mass structure may be a definition of the spatial relationship of the material types forming the soil mass that could be complementary to the descriptions of composition.

6.3 Behaviour

The behaviour of the defined soil masses should be described with respect to the effect of natural or imposed conditions. These conditions may include the following:

Civil engineering structures
Cut slopes
Natural slopes
Erosion
Field tests

Field tests may include those undertaken as part of a ground investigation programme, and could include the following:

Standard penetration test
Vane shear test
Cone penetration test
Pressuremeter test
In situ CBR test
Plate bearing test

6.4 Nomenclature

From the field evidence the soil masses, composed of described material types, should be identified in line with the proposed tropical residual soil classifications. This identification will form the first judgement as to mass type which may be modified in the light of subsequent laboratory information.

7.0 USE OF DESCRIPTION SYSTEM

In addition to relating to the physical characteristics of the soil material and mass, an effective and geotechnically useful field description system must acknowledge the influence of project type. The scale and extent of fieldwork will be defined by the type and stage of project. For example a project could be a road, tunnel, or dam at stages that may vary from feasibility through to construction and maintenance The project could also be part of a more general resource study to provide information for future planning needs.

Table 4: SOIL MASS CHARACTERISTICS

Parameter	Classification
Composition	1, Materials 2, Geometry 3, Nature
Structure	1, Layering 2, Jointing 3, Faulting 4, Slickensiding 5, Contacts
Behaviour	1, Erodability 2, Slope angles 3, Slope failure 4, Strength

The fieldwork planning for the description and identification of tropical residual soils should be undertaken in the light of the project under consideration. The project definition may affect the approach to field description under a number of headings:

Aerial extent
Depth of soil/rock mass influenced
Extent of disturbance to soil/rock mass

Although the proposed classification and description systems have been developed from the sound pedalogical base of the soil scientist, the practical use of the system must recognise a difference in emphasis between the pedologist and the geotechnical engineer. Traditionally the soil scientist has been involved in potential of soils for agricultural purposes, with the consequent tendency to cover large surface areas to limited vertical extent. The geotechnical engineer in contrast will tend to be interested in more detailed, site specific, information with generally greater emphasis on the in situ soil/rock profile and its vertical and horizontal variability. It is not the purpose of this paper to be involved in the determination of engineering properties which will require testing in the laboratory but rather to refer to methods that can be used in the field to identify and describe tropical residual soils and thus lead to more meaningful laboratory test programmes.
In order to be at its most effective the above framework of field description should be organised so that the recovery of early data may be usefully fed back into the system. A typical organisation of a project into component parts (site, soil masses, soil material) is illustrated in Fig. 1 .
Adequate pre-fieldwork desk studies to collect information relating to the 'site' characteristics may enable early judgement as to the likely soil environments that should be anticipated in programming fieldwork.

8.0 DESK STUDIES

The initial desk studies for a project should make possible its preliminary organisation with respect to sites and soil masses. This may be aided by the use of remote sensing procedures allied to terrain evaluation principles (Dowling and Williams 1964 , Beaumont 1979). By this procedure it may prove possible to identify

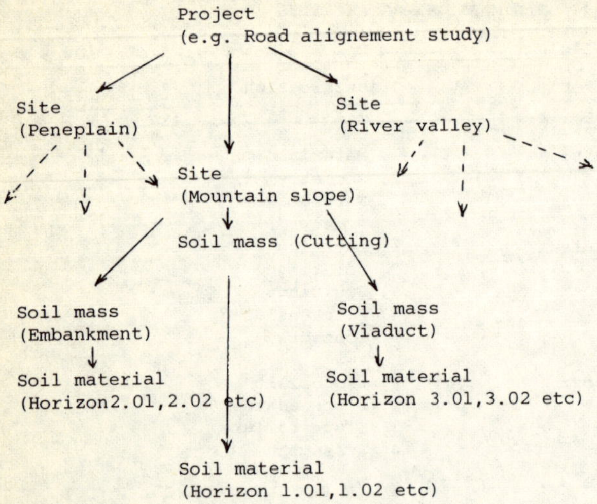

Fig 1. Example of possible site, mass, material relationships.

distinct project land systems together with typical land units or facets.

These initial desk studies should collect available information and identify available potential sources of further data. These further sources may vary from previous project work to information held in local ministries of work; typical of such sources are those identified for the Middle East by Eastaff et al (1978). The desk study should also be used to collect preliminary information with respect to site characterisation.

A vitally important part of desk studies and precursor to fieldwork is the effective planning of the ground investigations. The extent and style of ground investigations should be planned to take into account the type, stage and economic constraints of the project, together with its gemorphological and geological location (Cook and Younger, 1986).

9.0 Fieldwork

Ground investigation planning must take into account the type of information required by the project. The style of investigation, will be a function of economics, type of project, and the location of the project and will be influenced by the climatic , geotechnical and physical features outlined above. The location of projects within the broad terrain units in general terms, will influence the style, and hence the choice, of particular site investigation practices to be adopted.
The collection of information for the effective description of tropical residual material may utilise the whole range of current site investigation procedures. Although this section is primarily concerned with field descriptions, it is relevant to incorporate in this discussion the various desk activities that would be a necessary precursor to many field based projects.

A full description of site investigation procedures may be obtained from a number of established source references (BS 5930, Clayton et al 1982, Dumbleton and West 1974, Weltman and Head 1983); table 5 summarises briefly the main activities in the context of tropical residual soils.

10.0 FIELD ACTIVITIES

Field activities may be broadly divided into those undertaken primarily by the Geotechnical Engineer or Geologist, and those requiring the mobilisation of equipment and operatives. Although this section as a whole is principally concerned with the former, it is important that the Engineer or Geologist is involved in the effective organisation of the latter activities.
It is recognised that a range of disturbances may be imposed on a residual soil by different civil engineering activity It is essential, therefore, that the extent of remoulding or disturbance that occurs in sampling or testing the site materials should attempt to model the likely disturbance due to any proposed engineering activity.
Because of the sensitivity of tropical materials to disturbance this modelling can prove difficult to reproduce sensibly. The field description and classification of soil materials and masses by a geologist or geotechnical engineer can thus form a vital early link.

11.0 CONCLUSIONS

The more intensive weathering of rocks that is experienced in many tropical environments compared with temperate zone requires particular attention to be paid to the in-situ condition of the soil in its undisturbed state. Even samples being taken to the laboratory must be preserved in this state for more detailed examination. This especially applies to

TABLE 5 : EFFECTIVENESS OF FIELDWORK PROCEDURES

Procedure	Duricrust Environment
Boreholes	Techniques must take into account the presence of indurated layers/nodules stratified with weaker material. Sampling equipment must be of a 'robust' nature but still be capable of attempting to recover soil structure.
Test Pitting	May be very useful for low cost road projects or those in subdued terrain, but could be misleading in caprock/ layered situation. Very good for in situ examination of fabric and structure (e.g. calcrete classification Very useful for materials investigations.
Probing	Of limited usefulness in this environment.
Geophysics	Layered structure of any duricrust profiled unsuitable for effective geotechnical interpretation by these means.
In situ testing	May have to be of a robust nature (e.g. SPTs)
Mapping	Geomorphological and material mapping likely to be useful in establishing typical landform for local interpre - tation particularly in search for materials
Material logging	Emphasis for first geotechnical assessment on strength, grading, fabric and variability.
Remote sensing	Can identify ferricrete exposures in older landscapes e.g. Nigeria, or calcretes in dry areas e.g. pans or fossil drainage lines.

TABLE 5: EFFECTIVENESS OF FIELDWORK PROCEDURES Cont.

Procedure	Mature Residual Soil Environment
Boreholes	Techniques must take into account possible sensitivity of soil to sampling. This must be to a very high degree of quality,which even then may not be successful. High quality specialist rotary techniques may give best chances of effective sampling when undisturbed samples are required.
Test Pitting	Ideal for undisturbed sampling of sensitive soils. (e.g. Vertosols) May not be effective in terms of sampling the required depth of zone to be influenced by the project.
Probing	Can be of some use in preliminary assessment or of rock/soil Profile.
Geophysics	Can be used effectively to extrapolate soil profiles or to establish approximate depths to competent rock. Care needed in interpretation in what is likely to be a gradational rather than layered situation.
In Situ Testing	Should be used where possible, methods to be compatable with the disturbance to be effected by the proposed structure. Currently developed methods of pressuremeter testing might be applicable for instance.
Mapping	May be limited to general landform mapping.
Material Logging	Emphasis for first geotechnical assessment on mineralogy, behaviour tests, fabric, plasticity and strength.
Remote Sensing	Interperetive methods useful for detecting soil type, response to moisture etc.

residual soils where an understanding of the fabric, structure, chemical and mineralogical properties may be as important as the textural or particle size components in determining engineering behaviour.

In the paper a framework for the field description, examination and identification of soils is presented which links soil material characteristics within the soil profile to the soil mass characteristics. Established terrain evaluation techniques are one way both detailed and general needs can be related to suit the type and scale of the project. Field methods are identified which are taken from existing site investigation practices which if correctly used should not only enable an early detection of the presence of problem or unusual soils but which will also assist greatly in their classification.

ACKNOWLEDGEMENTS

The authors are grateful to members of the Working Party of the Geological Society, United Kingdom who are involved in the preparation of a report on tropical residual soils and advised in the preparation of this paper.

REFERENCES

Anon 1979 Classification of rocks and soils for engineering geological mapping Part 1: Rock and soil materials (IAEG commission of engineering geological mapping) Bull. Int Assoc Eng.Geol 19 365 371

Beaumont T.E. 1979 Remote sensing for the location and mapping of engineering construction materials in developing countries. Q.Jrnl. Eng. Geol. 12 147 158

BrewerR 1964 Fabric and mineral analysis of soils. John Wiley & Sons 470pp

British Standard Institution 1981 Code of Practice for site Investigations B.S. 5930 British Standards Institution London. 174pp.

Child G.H. 1984 Soil description quo vadis ? In Site Investigation Practice 20th Reg. Conf. Group Geol. Soc Guildford.

Clayton C.R. N.E. Simons and M.C. Matthews 1982 Site investigation A Handbook for Engineers. Granada Publishing. 423pp.

Cook J.R. and J.S. Younger 1986. The engineering geology of road projects in North Borneo. Proc.Eng.Group. Geol. Soc. Conf on Planning and Engineering geology.

Dowling J.W.F. and F.H.P. Williams 1964 The use of aerial photographs in material surveys and classification of landforms. Conf.Civ.Eng.Prob. Overseas ICE

Duchaufour P 1982 Pedology,pedogenisis and classification (English Edition Trans T.R.Paton) George Allen and Unwin, London.

Dumbleton M.J. and G.West 1974 Guidance on planning, directing and reporting site investigation. TRRL Report LR 625

Eastaff D.J. ,D.J.Beggs and M.D.Mc Elhenney 1978 Middle East geotechnical data collection. Q Inl.Eng.Geol. 9 37 55

Ferreira H.N. 1985 Characterisation, identification and classificaiton of tropical laterite and saprolite soils for geotechnical purposes.Int.Conf. on Tropical Soils,Brazil

Fitzpatrick E.A. 1983 Soils. Their formation, classification and description. Longman 253pp.

Hodgson N. 1973 Soilsurvey Handbook. Soil surveys of England and Wales. Rothamsted.

International Society of Rock Mechanics (ISMR) 1980 Rock characterisation, Testing and Monitoring: Suggested methods. Ed E T Brown, Pergamon Press 211pp.

Jennings J E , ABA Brink and A A B Williams 1973 Revised guide to soil profiling for civil engineering purposes in Southern Africa. The Civil Engineering in South Africa Jan 3.12

Lumb P 1965 The residual soils of Hong Kong Geotechnique 15 180 194

Mitchell C.W. 1973 Terrain Evaluation. Longman 221pp.

Mitchell J K and N Sitar 1982 Engineering properties of tropical residual soils. Proc. A S C E Conf. Engineering and Construction in Tropical and Residual soils 30 47

Netterberg 1978 Prospecting for calcrete road matrials in South and South West Africa. Civil Engineer in South Africa 20(1) 3 10.

Tuncer E R , R A Lohnes and T Demirel 1982 Desication of soils derived from volcanic ash. Proc ASCE Conf Eng. and Constn in tropical and residual soils. Hawaii 1982 A S C E New York.

Weiert H H 1974 A climate index of weathering and its application in road construction. Geotechnique 24 (4) 475 488

Weltman A J and J M Head 1983 Site investigation Manual CIRIA Spec Pub No 25.

Construction feature of southern China netted red clay

L.C.CUI, Changsha Communications Institute, People's Republic of China

ABSTRACT: Netted-red-clay is a kind of dark red clay or subclay with a netlike structure. It is located widely in the southern areas especially hills and urban areas in South China. For a long time it has been treated as a common redclay, and its special properties are not fully utilized. As a result, a lot of unnecessary waste has occured in base designing and constructing. Recently Chinese designers and constructers have determined that neted-red-clay is much better than common clay in physical and dynamic properties and in slope stability. It is waterproof and airproof, and it is hard for water to wear it down. This study adds comparative experiments to existing experiment achievements on neted-red-clay uses, and goes a step further in veryfying the existence of properties mentioned above by way of micro-texture and enviornment research. The author concludes that netlike clay can be considered as a special kind of redclay especially for uses in construction--houses, highways, bridges, harbors and airports in South China.

THE FORMATION AND DISTRIBUTION OF NETED-RED-CLAY

According to research made on soil by geologists and pedologists in China neted-red-clay is formed under the circumstemces of uncommonly high-temperature and torrential rains, for instance, subtropics. In such areas the average temperature is from 18-28 degrees centgrade. Dry seasons and rainly seasons are distinet. In Quaternary period several glacial stages and interglacial stages had taken place in southern China. It is in the interglacial stages with a high temperature, moist, hot climate that netted-red-clay was formed. During the forming process, under the influence of oxidation--reduction at the bottom, the ferric parts migrate along the crack, so red, yellow-white, greyish white nets are formed. The individual net size varies greatly, the fine is net-like, the coarsh is sausage-like or leopardskin-like, the longest streches out several metres, the shortest a few centmetres. From top to bottom the net gets bigger and bigger. The general orientation (direction) of nets may be sloping, horizontal, vertical which will indicate the record of the underground water movement. Neted-red-clay's formation is a long, slow historical process. In the light of the analyses of geological data, it needs at least 20,000 years.

Neted-red-clay distributes chiefly in such areas as Hunan, Hubei, Jiangxi, Anhuei, Shichuan, Yunnan, Gueizou, Jiangshu, Zejian and some parts of Guandon, Fujian, within a sea level of 500 meters and on the low hills, river banks, terrace and flood plain. However there are some exceptions. It has been found in Yunnan and Guizou high plateu with a sea level of 1000 meters, this is because of the vertical crustal movement; another example is found in Jiangxi Lushan mountain area, on top of the mountain with a sea level of 900 meters, there existes neted-red-clay. The discrepency between top area and the area around it is from 400 to 500 meters, but it is formed in the same surroundings. This shows that in this area the mountain plate elevated 400-500 meters. In some areas we find 4-6 layers of neted-red-clay on a cross-section and it also shows that during the course of it's formation the crustal plate raised discontinuously.

As shown above, the climate, landscape, underground water movement and crustal movement in Quaternary period had a profound influence upon the formation and distribution of the reted-red-clay in southern area.

THE ENGINEERING PROPERTIES OF NETED-RED-CLAY

The physical and dynamic characteristics are good. It is dense, high in strength, water-resistant, not easily erroded.

Data collected from four provinces give physical and dynamic properties as:

Indices of plasticity (I_p)	16.2-27.8
Liquil limit (W_l)	33.8-47.0
Water content (W)	18.85-25.31%
Bulk density (R)	1.93-2.07KN/m
Water viod (U)	0.55-0.53
Indices of liquidity (I_l)	0.15-0.1
Void (e)	0.70-0.75
Degree of saturation (S_r)	84-90%

Compressibity factor (a)	0.011-0.01 10 KPa
Angle of internal friction ( )	24.5-29.7
Cohesion (c)	0.34-0.76 10 KPa
Shear strength	2.52-2.15 10 KPa

Wang Yiou Ling did a good comparative expriment on neted-red-clay in Hubei. The result is shown as fig 1. The strength of neted-red-clay is much higher than other commonclay existing in the same area.

Fig 1.

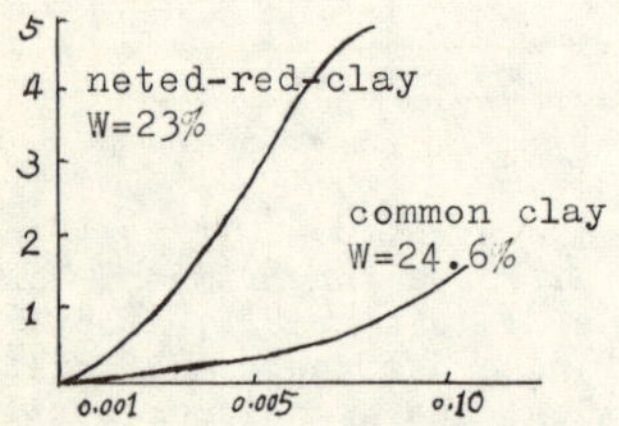

Fig 1. curve of unconfined compression strength of different clays

From statistics of 24 loading plate testing data, the proportional pressure (P) of neted-red-clay and ultimate pressure (P) and deformation modul E are shown as table I, fig 2.

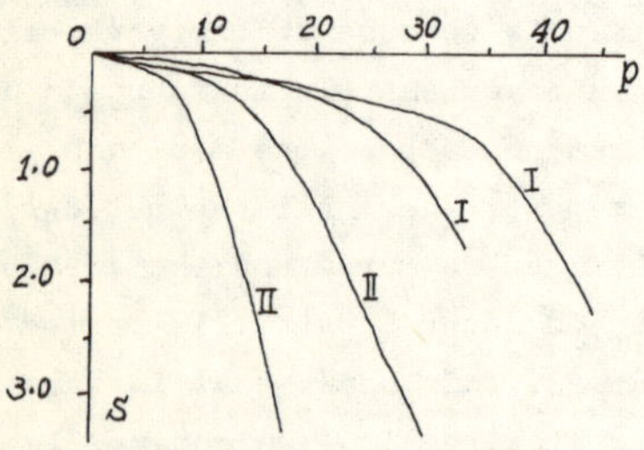

Fig 2. neted-red-clay s-p curve

Table I

type / target	I	II
proportional pressure (p_o)	11.3—12.2	6.9—9.3
ultimate pressure (p_u)	32.1—41.2	18.8—34.1
safety factor ($K=p_u:p_o$)	2.28—2.85	2.35—3.67
modulus of deformation (E_o)	1020—2532	880—1290

As shown above its special properties are that its P , E and P are higher than common clay, they are approximitely at the same level of half-dense graval or well-weathered hard rock. It has high P as well as P and E , so using P as allowable bearing capacity it is safe enough (safety factor P :P =2.2-37, mean 2.6; when P=P , the value of deformation under loading plates 0.04-0.011, mean 0.007). According to current Chinese code, we get a allowable bearing capacity that is approximately 2-4 times as "the ordinary clay" and "old clay". The correctness of loading plate testing data above has found proof in the long term sleuthing experiment. The data above show that it has brittle characteristics and has a high water--risistence structural strength. The existence of this structural strength is the chief reason that neted-red-clay has high strength and anti-water stability.

The properties shown above have been observed by the author on the bottom of some rivers in Shenzhen, China, around a bridge, the river bed composed of neted-red-clay. The author has observed it for several years but found no obvious evidence of errosion. The "U" shaped canel dug in this area composed of netted-red-clay, with no measures to protect against water, has not shown any obvious evidence of failure. Observing the extrior and testing the interior has shown its distinguishing properties.

According to surveying made by the author on highway slopes in three provinces, which have a slope of 1:0.75 in some areas, slopes (below 4 metres) can remain stable for a long time without any protection.

The discovery of MaWangDui first Han dynasty tomb (Han dynasty 2100 years ago) and the digging is a vare miracle in the world. The mummy had been kept very well. Inspite of good measures taken to protect it from corrosion at that time, it also has something to do with the geological conditions--net-red-clay being water-proof and air-proof.

THE FORMATION MECHANISM OF NETED-RED-CLAY PROPERTIES

In order to have some discussions about the formation mechanism of neted-red-clay properties, and on the basis of analysing the existing data, the writer again collected three samples in different rigions and at different depths. A series of comparative tests were made. The analysing methods include: "liquid migrating tube" granulanity analysis; Chemical analysis; X-ray differaction analysis; differential thermal analysis; energy spectrum anlysis; and high power electron microscope inspection. Besides analysing the whole specimen, some parts of brownbulk, yellow and white clay bulk were also tested.

I. grain size distribution

Table II is the results of grain size distribution analysis

Table II

analysing number	grain size (mm. %)					
	>0.25	0.25–0.1	0.1–0.05	0.05–0.01	0.01–0.005	<0.005
1146	0.98	13.21	29.71	15.6	10.0	30.5
1146-1	2.2	13.85	27.85	20.0	7.6	28.5
1146-2	0.55	8.97	29.18	17.6	6.0	37.7
1147	6.36	10.61	22.53	24.8	10.4	25.3
1148	10.62	19.57	25.71	8.8	6.4	28.9

Table III

analysing number	** tolal Fe (%)	disasociated Fe (%)	montmorillonite (%)	exchanging capacity (me/100g)	exchanging cation (me/100g)				sult-base tatal content (me/100g)	exchangeable acid (me/100g)		disasociated oxide (%)		
					K^+	Na^+	Ca^{++}	Ma^{++}		H^+	Al^{+++}	SiO_2	Al_2O_3	Fe_2O_3
1146	17.68	35.52	6.38	8.29	0.20	0.44	3.26	2.55	6.45	0.41	2.98	0.77	0.85	6.28
*1146-1	20.69	39.63	4.22	8.45	0.20	0.44	3.30	2.15	6.09	0.41	2.01	0.80	1.21	8.20
*1146-2	5.56	21.58	7.49	13.03	0.22	0.44	3.37	1.70	5.75	1.24	4.06	0.72	1.28	1.20
1147	12.60	42.78	7.13	9.70	0.20	0.37	3.07	1.36	5.00	0.99	1.66	0.62	0.90	5.39
*1147-1	16.42	74.67	3.13	9.28	0.15	0.31	2.90	1.63	4.99	0.83	1.66	1.09	0.92	12.26
*1147-2	5.89	19.53	6.84	9.36	0.20	0.37	3.75	2.25	6.57	1.55	2.07	0.61	0.96	1.15
1148	17.16	36.71	6.72	9.31	0.15	0.44	3.68	3.30	7.57	0.83	1.25	0.57	1.20	6.30
*1148-1	29.09	52.14	2.86	9.09	0.20	0.44	2.79	3.67	7.10	0.41	0.61	0.74	0.96	15.48
*1148-2	3.80	45.26	8.03	13.65	0.27	2.19	4.80	3.71	10.95	0.99	2.76	0.97	0.98	1.72

* 1146-1. 1147-1. 1148-1 —red bulk part, 1146-2. 1147-2. 1148-2—greyish white part.

** tolal Fe the tolal content of all kinds of Fe by means of soil fully Chemical analysis. inthes article it means the tolal content of crystallized or no-crystallized.

II. the main chemical composition

Table III shows the main chemical composition: The content of disasociate oxide in clay is high and ferric oxide (5.39-6.30%) is the highest. Due to the difference in composition and structure of the netted-red-clay, the disasociate oxide distributes itself mainly in the part of red bulk. The content is 8.20-15.48% in the yellow and white part. However, the content of ferric oxide is only 1.15-1.72%, 7-9 times less than the disassociate oxide. In different parts of clay the content of disasociate Al O and disasociate SiO are almost the same, (SiO : 0.57-1.09%, Al O : 0.85-1.28%). The cation exchange capacity of three specimens is commonly lower (8.20-9.70 meg/100g). So the specific surface is also small. The physical and chemical activity are weak and the PH value is lower (4.28-5.85), the clay is typical acidic clay. The acid value is determined by the existing of large amounts of exchangable Al and H . As the acidic clay is the salt-base unsaturated one, so in the composition of salt-base exchange the content of exchangable Na and K are both lower. Its value commonly does not exceed 0.44 meg/100g. The exchangable cation is mainly Ca (2.79-4.80 meg/100g) and secondly Mg (1.36-3.71 meg/100g). The composition of cation as shown above makes the colloidal partical attraction larger than the repulsion, so the coagulation structure combination is formed.

III. The main mineral composition

The clay mineral grain is fine and its propertes are complex. To indentify them we used more than four methods to do comparative testing.

The results of differential thermal analysis are shown in Fig. In the differential thermal curve of three specimens, near 560-565 C, nonsymmetric and most developed mid-temperature-heat-absorbing-valleys are existing: this is due to the thermal absorbing reaction of Kaolinite for lost crystal water (OH). But near 976-990 C, there exists an obvious thermal-releasing peak. This indicates the exothermic reaction of distruct of Kaolinite, and that the clay is chiefly composed of Kaolinite, near 310-350 C, a week thermal absorbing valley exists. This indicates the endothermic reaction of recrystallization of disasociate ferric oxide.

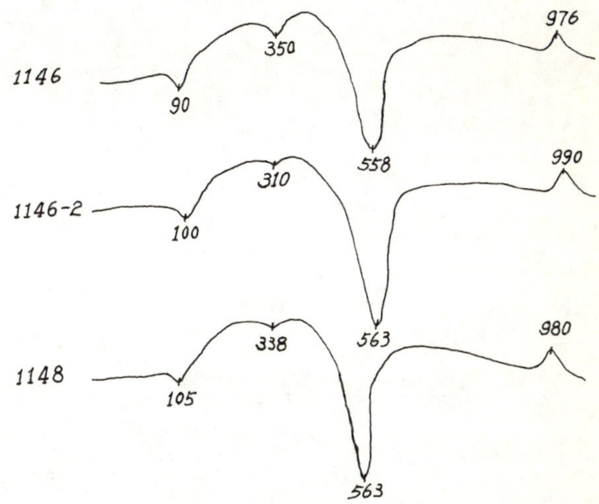

Fig 3. curve of differential thermal analysis

Under the circumstance that the amplification of scanning electro-microscope is more than 3000 times, a large number of highly dispersed kaolinite grain (d 1um) can be seen in the silt part of the clay. Besides the quartz fragments can be clearly seen (photo 1.2.). The self-growth quartz which formed from the red-clayization and the corrugated-pice-shaped montmorillonit accumulation (photo 3.) in the yellow and white part can also be seen.

As the the result of energy-spectrum analysis shows; besides Si and Al (the main composition of kaolinite) elements. Such as, Fe, K and Ti are also commonly distributed. It is noted that a small amount of disasociate ferric oxide and illite are also distributed throughout netted-red-clay. The content of disasociate ferric oxide is higher in the red bulk part than in the yellow and white parts. (fig 4.) This can also be verified by X-ray diffraction spectrum analysis.

To sum up, the composition is as follows:

The red bulk: K (kaolinite), Fe O (disasociate ferric oxide, hematie), SiO (quartz), I (illite), M (montmorillonite). The greyish white part: K, Ti, M, mica, Fe O.

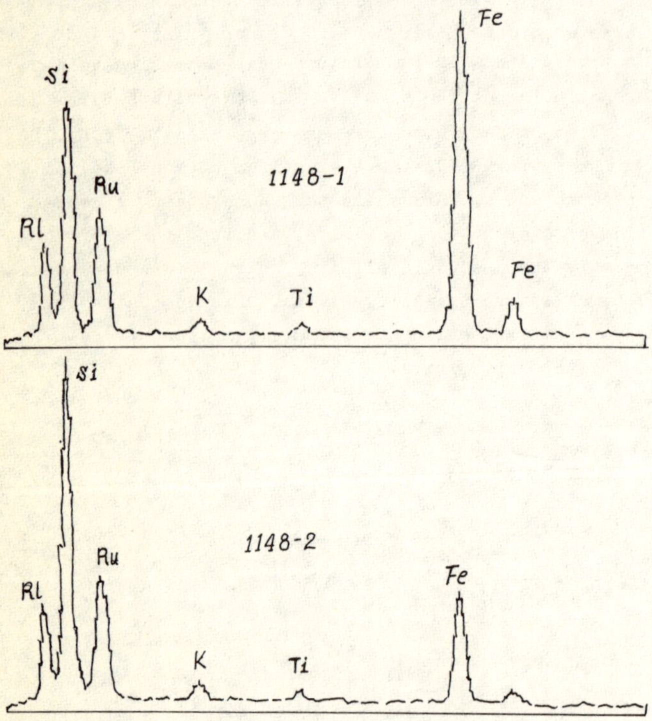

Fig 4. result of energy-spectrum

IV. The strudure properties of neted-red-clay

Macrostructure features

The neted-red-clay is composed of two parts, one part is red clay bulk, the other yellow and white clay bulk. They are different not only in color, but also in composition and physical property. For instance, in the red bulk the content of disasociate ferric oxide is high and the red bulk is hard and tough. In the yellow and white part the moisture content is high, it is soft, and it contains a lot of sand and mica fragments. So the strength of netted-red-clay is determined by the quantity and distribution of red bulk. The advantageous properties of netted-red-clay is governed by the bulk, mainly in that the brown bulk plays an important role in bearing the load.

Microstructure analysis

The microstructure property is mainly determind by the interparticle cohesion and the character of arrangments of touching particles. The netted-red-clay is mainly composed of Kaolinite. The Kaolinite has the weak effect of electric double layer. As the cation exchange capacity and specific surface is small, the thermodynamic potential of double layer is lower. The anions are Al and Ca , Mg etc, which have a constraining reaction to double layer. Since the Ph value is lower, the thickness of double layer is thiner and the electrok-inetic potential is lower. So the repulsion of interparticle force is weak, the attraction force is easily dominant, and the hard structure is formed (photo 5.8.)

The "cohesion" of disassociate ferric oxide plays an important role in the formation process of netted-red-clay structure. When the PH value is low, the unreversible cementation is formed between the ferric oxide colloidal particle which charges cation and kaolinite which charges the anions. This cementation gets much stronger as drying and ageing proceed. So the disassociate ferric oxide often forms the "contractile skin" of clay minerals. Between the integrities and between the integrity and mineral fragements, there exists a mosaic-type structure. This assembly feature has much to do with the cementation of the disassociate ferric oxide.

Under the SEM the quartz which was formed by recrystallization in the red bulk part and collective zone, mineral fragments can be seen. Within the yellow and white part, pores are distributs throughout.. These are formed by the dissoving and leaching process of ground water.

It should be pointed out that besides the reason explained above, the most important reason for which the fine engineering property of netted-red-clay is formed is that the clay has experinced a longer geological process than other clays. To some extent, it has been rockized, there crystallization of mineral is a mark of this rockization phenomenon.

Conclusion

The netted-red-clay is a special kind of clay that has very engineering properties it's engineering properties are determind by the formation history, composition and structure. The construction of cities, industrial project and trasportation construction in southern China has much to do with the distribution of netted-red-clay. So in classfication of soil, it should be devided independently,

Netted-red-clay should be considered as a unified foundation bearing-stratum, it's bearing capacity is much higher than the common clay. Making full use of it's engineering property will save a lot of capital. Now it has come into practice, economic efficiency and safty are excellent. But merely 5 years has elapsed since people studied the netted-red-clay in view of engineering, there is a lot of problems to be solved, the cooperation of various fields should be expected.

Acknowledgement

The author appriciates the work that has been done by Zou XinHua (Changsha Communication College) in experimentation data collection, editing and traslation. Also appriciated the cooperation of associate professor Zhou YinYin (Changsha Communication College) Qiu YunXin, Li YuRui and Xiu XiaoLan (Chinese acadamy of science).

REFERENCES

Xu Xin 1983. Discussion on the Vermiculated in the Iniddle and Lower Reaches of the Changjiang (Yangtge) River. Contribibution to the quaternary glaciology and geology. (I), P104.

Xiang Chunyao 1985. No.3. Enginneering--Geological Properties and Foundation of the Netly-Strealy Red Clay. Hydrogeology and Engineering Geology.

Plot I. The cementing of clay minerals such as kaolinite by highly dispersed disasociate Fe O layer quartz in colloidal masis and pores.

Plot II. fragments and recrystallized quartz in brown bulk parts.

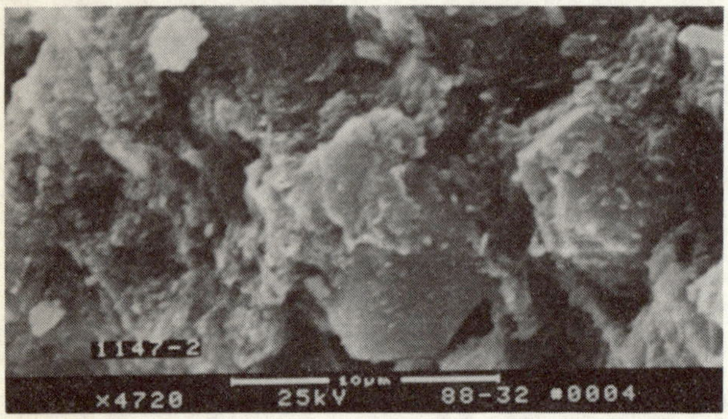

Plot III. Cementing of fragments by disasociate oxide and kaolinite.

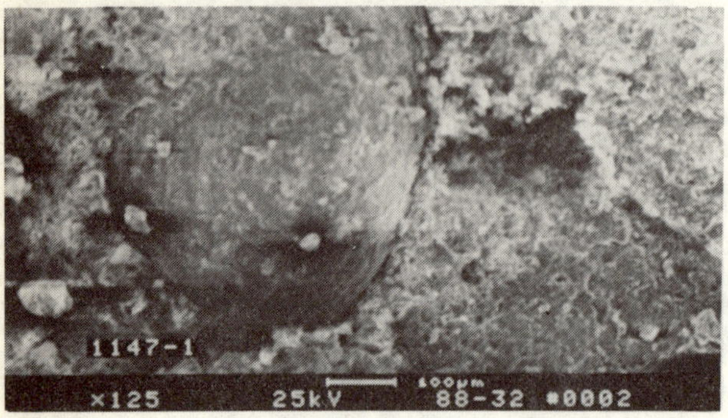

Plot IV. Ferric concretion in brown bulk parts, elements content sequence in energy spectrum analysing: Si, Fe, Al, Ti, K.

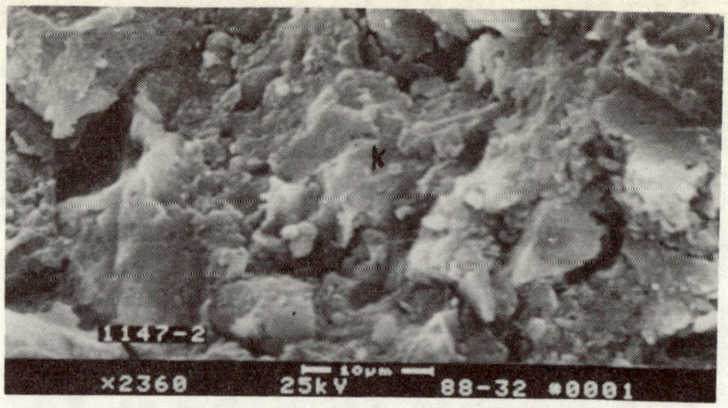

Plot V. disasociate oxide and kaolinite form the colloidal mass in yellow and white parts and its enclosing and covering to fragments.

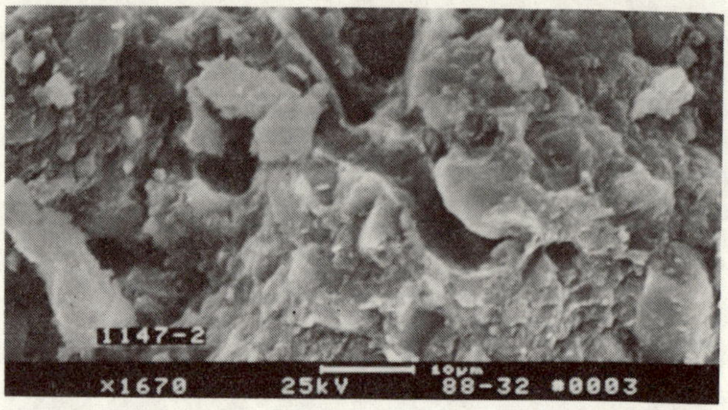

Plot VI. dissolved pores in yellow and white parts.

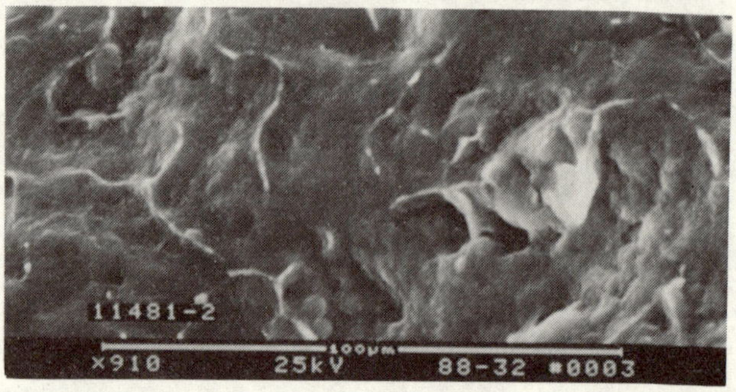

Plot VII. The shap characteristies of corrugated-pice-shaped montmorillonite in yellow and white bulk parts.

Plot VIII. water risistence coagulating mass formed by disasociate Fe O and clay mineral such as kaolinite in brown bulk parts

Potential application of engineering pedology in shallow foundation engineering on tropical residual soils

M.D.GIDIGASU, Council for Scientific and Industrial Research (CSIR), Building and Road Research Institute, Kumasi, Ghana

ABSTRACT. The paper has reviewed tropical soil foundation problems and highlighted the limitations of classisical soil mechanics in solving some shallow foundation problems in tropical residual soils.

It is emphasised that Engineering Geology has enhanced knowledge in foundation engineering and has become an indispensable tool in site investigation practices. Soil profile as a whole rather than a textural soil type is proposed here for characterising tropical shallow foundation condition . Because profile development processes involve all aspects of weathering, soil formation, including horizonisation and is also influenced by environmental factors, these facts are very useful to shallow foundation engineers in the tropics. For example, terrain classification and characterisation based upon profile features is useful for settlement developments. Engineering pedology which embraces the above stands out as a very useful tool for solving most shallow foundation problems, and has been used with considerable success in developing shallow foundation codes of practices in many tropical developing countries.

The only apparent limit in the practical application of engineering pedology in solving shallow foundation problems is the capacity and willingness of the geotechnical and foundation engineers to learn to use this branch of applied geoscience.

1. INTRODUCTION

The limitations of classical rock and soil mechanics in solving many foundation problems in some tropical environments has long been recognised (e.g. Donaldson, 1962). For example, the unique behaviour of the black cotton soils of India have prompted the development of special foundation types for light structures founded on these soils. It has been noted that the bearing capacity is sometimes not the main problem in these soils but rather swelling or shrinkage forces develped with wetting and drying of these soils. Wetting and drying cause swelling and shrinkage which respectively induce uplift and settlement of light structures. Some of the experiences acquired in India have been summarised by Mohan, Jain and Sharma (1967).

South Africa has three main types of foundation problem soils. These are, in expansive soils, the collapsing soils and the dolomitic rocks subject to sinkhole formation. Methods developed to deal with these problem soils as foundation materials have been succinctly reviewed by Donaldson (1962), Jennings (1965) and Jennings and Brink (1978).

Shallow foundation problems in Australia are associated with dispersive soils and highly compressive weak soils. Heiman (1967) has provided a comprehensive treatise on the solution of foundation problems associated with soft and compressible soils.

The relationship between the soil profile characteristics (morphology) and the performance of different types of foundations including standard strip footings and piles have also been studied in Australia for over 20 years and these have been reported elsewhere (Aitchison, 1953, 1974).

Studies relating to foundation problems on clay soils in the semi-arid zone of Ghana have been reported by small and White (1962); Gidigasu (1985); Gidigasu and Andoh (1980). Small and White have established that strip footings are generally not recommended unless located below the zone of seasonal moisture variations. Buildings on piles with bases located below the zone of soil moisture and volumetric changes are recommended for areas underlain by these soils.

Ola (1980) slso reported the distribution of foundation problems with soil types in Nigeria and has suggested specific recommendations for overcoming the problems in terms of selection of foundation types and details of building designs.

In a recent review by Hammond (1984), it has been shown that earth buildings founded on various tropical soils have to be given specific type of foundation types if the life assigned to the structures are to be achieved.

It is clear that various soil conditions and environmental factors require specific foundation types and construction techniques for specific types of buildings concerned to ensure their long term stability.

For example, some of the significant conclusions reported from Australian studies on foundations in structurally unstable soils (Heiman, 1967) has emphasized that the choice of foundation type should also be related to:

a. The type of construction that will be used, and the permissible foundation movements that can be tolerated without damage to the structure.

b. The use to which the building will be put. In some buildings minor deformations of structural members and architectural finishes will not be critical, but in other buildings cracking and differential movements of members will be unacceptable.

c. If more than one type of footing is satisfactory structurally, the final choice may be made on the basis of cost.

d. Sometimes footings which are satisfactory in other respects may have to be rejected because of construction difficulties.

There are a wide variety of piles available for use as footings (Small and White, 1962) and the decision as to which type of pile is most suitable for a particular situation will be influenced by the nature of the subsoil conditions, ease of construction, cost and the durability of the pile. For example, the nature of the foundation will determine whether the piles are to carry the load in end bearing, skin friction, or adhesion or partly in end bearing and partly in friction or adhesion.

It is known that raft foundations are sometimes used when buildings are placed on highly compressible foundation soils characterised by low bearing capacity and potential high settlement. The influence of the raft loading may extend down to the deeper foundations and also to adjacent buildings. It is therefore, important that the preliminary soil investigations be carried out over sufficient areas and depth to enable the likely behaviour of the raft to be determined.

Frequently, there are difficulties associated with the design of a raft, and some of the simplified methods of design that have been used in the past can lead to uneconomical construction or to serious errors in the assessment of the likely performance of the footing (e.g. Baker, 1957). However, the more accurate design methods are laborious and still depend for their accuracy on the availability of sufficient experimental data on soil parameters.

The construction of raft footings may also be costly, especially if the site is wet, and the cost may be further increased if it is necessary to waterproof part of the building below ground level (e.g. Bell, 1978).

2. THE LIMITATIONS OF CLASSICAL SOIL MECHANICS IN TROPICAL GEOTECHNICAL ENGINEERING PRACTICE

Considerable amount of research work has been carried out over the years on the classification and characterisation of intact rocks for civil engineering purposes (e.g. Deere, 1969; Miller, 1965; Stagg and Zienkiewicz, 1969; and Peck et al., 1973). For example, Peck et al. (1973) have proposed engineering characterisation of unweathered major rocks types and gave rough assessment of the suitability or otherwise of rocks types in their unweathered state for foundation engineering purposes. Jennings and Brink (1978) and Duncan (1970) have also suggested a classification system for rocks and soils on the basis of the unconfined compressive strength and saturation moisture content for general civil engineering purposes.

There are other rock and soil characterisation and classification systems developed for application in earthworks and for foundations for various temperate zone earthen materials. These have been discussed in detail in numerous books, journal papers, special reports and conference proceedings (e.g. Wagner, 1957; Miller, 1965; Deere, 1969; Stagg and Zienkiewicz, 1969). However, because it is not economical to place foundations sometimes on rocks found at considerable depths, studies have been undertaken to classify and characterise soils overlying various rocks for civil engineering purposes. For example, the standard textural classification chart, the Casagrande plasticity classification chart (Casagrande, 1948) and the numerous similar soil classification systems suggested in different countries have formed the basis of various temperate zone soil rating systems developed for earthworks and foundation engineering purposes (e.g. Wagner, 1957).

The most valuable and well known soil classification system available to soil engineers is perhaps the so-called Unified Soil Classification system which is the most elaborate form of the soil classification system developed by Casagrande (e.g. Casagrande, 1948).

The application of this internationally accepted soil classification and rating system for earth works, dams, pavement and foundation engineering for temperate zone environments has been succinctly reviewed by Wagner (Wagner, 1957).

The main limitation of the classical rock and soil classification and rating systems for foundation engineering in some tropical and residual materials and soils lies in the fact that the rate of chemical weathering in the tropics is very rapid and the end products of these processes differ in many respects from temperate zone earthen materials.

In particular, the composition of tropical and residual materials and soils is very erratic and complex. Moreover, these materials and soils respond in diverse ways to the controlling factors of the environment, especially climatic conditions. Similarly, results of studies (e.g. Little, 1969; Chandler, 1969, 1972; Attewell and Farmer, 1976; Dapples, 1959; Lornes et al., 1971, 1976; Hamrol, 1961; Hencher et al., 1982; Weinert and Clauss, 1962; Weinert, 1965; Dearman, 1974; Gidigasu, 1974, 1976, 1980) have thrown significant light on the influence of the degree of weathering of rocks and mode of soil formation on geotechnical properties and engineering behaviour of many tropically weathered materials and residual soils. However, the definition of degree of weathering and the geotechnical rating of partly weathered rocks is a very serious problem for engineering geologists and soil engineers. Lumb (1962); Hamrol (1961), Little (1969) and Duncan (1970) have made significant contributions to solving this problem through the use of the physical and physico-chemical properties including the so-called saturation moisture content while weinert (1965) proposed a very useful criterion based upon percentage secondary minerals content in the material compared to those in the parent rocks for defining the degree of weathering of rocks and their suitability or otherwise for earthworks and foundation engineering purposes in relation to the climatic, geologic, geomorphologic and drainage conditions.

Indeed, Lumb (1952) and Hencher et al. (1982) have attempted to quantify the degree of weathering of igneous rocks in terms of textural characteristics and profile features of overlying soils. The belief that the standard soil particle size analysis and plasticity test procedures are applicable to all types of tropical and residual soil types has been disproved by some studies which showed, for example, that index properties of halloysitic soils may change considerably with minor changes in the degree of pre-test drying, preparation and testing procedures (e.g. Terzaghi, 1958; Sherwood, 1967; Gidigasu, 1976).

The above considerations have called for caution in the universal application of the standard rock and soil testing procedures, characterisation and rating of some tropical and residual soils for earthworks and foundation engineering purposes.

A major contribution relating to the anticipation of geotechnical engineering problems from the knowledge of the parent rock type, mode of formation and the soil types was reported by Jennings and Brink (1978). Studies on the degree of weathering of rocks, the texture of the soils, their mineralogy, chemistry and the physico-chemical characteristics which have helped to interprete unusual soil behaviour fall within the field of engineering geology, whose importance in tropical and residual soils engineering practice should not be underestimated (e.g. Terzaghi, 1958; Loughnan, 1969; and Gidigasu, 1976, 1980, 1987).

3. APPROACH TO TROPICAL RESIDUAL SOILS ENGINEERING

Whereas the engineering behaviour of the temperate zone soils as foundation and construction materials have been the subject of countless and detailed studies reported in various publications and books there has been very little concern for the rather less predictable and or structurally unstable tropical and sub-tropical residual materials and soils.

The belief in some developing countries that the duty to develop tropical geotechnical engineering specifications and professional expertise lies in Europe and America should be discarded immediately because every country is concerned first and foremost with her problems. There is the need therefore, to develop tropical earthworks and foundation engineering expertise through organised laboratory, field construction and performance studies under tropical conditions for a number of years. This approach is necessary because it is known that numerous failures of earthworks, and foundations in tropical and residual soils could not be explained using the classical rock and soil mechanics principles developed for relatively stable temperate zone earthen materials.

The European and North American engineers unfamiliar with some tropical soils would blame most failures on wrong application of accepted international geotechnical specifications tailored for temperate zone soils with which they are familiar. For example, in many parts of the world it may not be the safe bearing pressure which should control the design of foundations but the reverse one of soil swelling pressure. Indeed, it is probable that on a world-wide basis swelling pressure would appear to be one of the most important of all the soil properties. The European and North American engineers may be familiar with the implications of the various water relationships and appreciate the reasons for setting the control limits at such values as given in the standard specifications of their homelands. When they are faced with working under tropical conditions they may be lost. This is not because there is anything wrong with the specifications with which they are familiar but because the soils for which these specifications were tailored are just not available. They have to realise that temperate zone standards are framed to suit temperate materials and soils and that a new form of tailoring may be necessary for the tropical and sub-tropical residual materials and soils. Apparently, the problem has to be tackled from the first principles to develop local geotechnical and foundations engineering standards and specifications. For example, it has been emphasised (e.g. Gidigasu, 1976) that special attention has to be paid to the contributions of the consequences of tropical physico-chemical and chemical soil forming processes to the deviations of engineering behaviour of some tropical pedogenic rocks and soils from the expectations of conventional rock and soil mechanics tailored for the temperate zone sedimentary soils and environments. Development of geotechnical engineering specifications for non-traditional materials and soils formed under tropical environments should involve a complete evaluation of the environmental factors and physico-chemical processes that affect and define their trends of formation.

For example, useful contributions to developing earthworks and foundation engineering standards for non-traditional and structurally unstable tropical and sub-tropical soils have been reported elsewhere (e.g. Wooltorton, 1936, 1955, 1966; Lumb, 1962, 1983; Aitchison, 1953, 1972, 1974; Gidigasu, 1976; American Society of Civil Engineers (ASCE), 1982; Brand and Phillipson, 1985).

One major point firmly established is that over-generalisation should be avoided in tropical soils engineering because of the complex inter-relationships between the parent rock type, the environmental factors, the different chemical weathering processes, the profile development patterns and their morphological implications, particularly, lithological, chemical and mineralogical composition on the one hand and the physico-chemical and geotechnical characteristics of pedogenic materials and residual soils on the other. Because of these complicated interrelationships between the soil forming processes and the nature of the soils, new procedures would have to be developed for testing and evaluating them. Similarly, criteria for predicting their behaviour both as foundations and construction materials, will have to be developed over the years (e.g. Aitchison, 1972; Gidigasu, 1976; Mitchell, 1976.

For example, in the case of residual soils the whole profile's response to the dynamic environmental factors affect the performance of shallow foundations a fact that should be borne in mind by designers and contractors operating in the tropical and sub-tropical environments (e.g. Aitchison, 1973).

Apparently, before any sense can be made of the behaviour of these soils one must comprehend the whole process of tropical soils formation, from initial physical and chemical decomposition of the rock to the final stage of the soil development and this would entail some knowledge of almost all aspects of earth science: geochemistry, geomorphology, hydrology, mineralogy, pedology and soil mechanics (e.g. Lumb, 1983). Particular attention will have to be paid to clay mineralogy, since allophane, halloysite and montmorillonite, the dominant minerals in tropical soils, behave differently from the temperate zone illites and kaolinites (e.g. Willis,1946; Terzaghi, 1958; Mitchell, 1976; and Kenney, 1967).

Unfortunately, ad-hoc tests developed for temperate-zone soils are too frequently simply transferred to tropical soils with disappointing results giving most tropical soils the undeserved reputation of being treacherous. The treachery lies in the test rather than the soils, for the most part, since irreversible changes in clay minerals and breakdown of the friable coarse fractions during pretreatment prior to American or British standard tests leads to misrepresentation of the soil behaviour in a natural state (e.g. Sherwood, 1969). There should be tests developed in the tropics for the non-traditional tropical soils; including shrinkage and absorption limits, abrasion and degradation tests. It has been shown (e.g. Mitchell, 1976) that these would prove more meaningful than the usual battery of tests developed for temperate zone materials and soils on the basis of which the imported geotechnical and foundation engineering Codes of Practices and specifications have been developed.

The first and fundamental task is to recognise the importance of the inter-relationships between the soil forming factors, namely, the climate, vegetation, the parent rock type (parent material), drainage condition, on the one hand and the processes of soil formation on the other all of which influence the profile features, the nature of the soils, the depth of weathering, physical, chemical, mineralogical and physico-chemical composition and how these affect the soil characteristics and behaviour in earthworks. Though these inter-relationships may not be obvious they do emphasise the need for detailed soil identification and recognition at both the macro and micro levels, using all available geoscience techniques and principles. Jennings and Brink (1978) have given some information on the relation between the parent rock type, the geological origin of the soils (residual or transported), textural characteristics and the potential shallow foundation problems to expect. For example, it is shown that residual yellow silty clays formed over igneous rocks in South Africa are potentially expansive while loose red silty sands or wind blown origin or some residual granitic soils are collapsible when wetted and under certain magnitude of applied loads (jennings and Brink, 1978; Knight, 1963).

4. PRINCIPLES OF ENGINEERING PEDOLOGY

Engineering pedology is the science of investigating engineering soil problems making use of all the natural laws known to be operative under a particular and specific conditions prevailing (Wooltorton, 1954, 1976). The engineering pedology begins with a knowledge of the rocks (parent materials) and weathering conditions and processes from which it deduces logically the nature of the end products, especially the morphology, lithology (texture), chemistry and mineralogy, etc. A comprehensive pedological soil report would be expected to contain the following information: Land region, geological age, rock group, rock type, climatic condition (rainfall, temperature, relative humidity, evapotranspiration), topography, slope angle, soil forming process, vegetation and landuse, type of material (e.g. gravel deposit), drainage conditions, approximate thickness of overburden diagram and description of profile morphology, including structure and fabric, clay mineralogy (predominant), organic matter content, cation exchange capacity and exchangeable ions, pH value, airdry moisture content, and particle size distribution (e.g. Brammer, 1952, Ahn, 1970).

It has been shown (e.g. Gidigasu, 1976) that the properties and the interpretation of test data for tropical non-traditional materials and soils will depend upon whether the weathering system and drainage conditions are productive of the kaolinite, halloysite or montmorillonite type mineral and whether allophane or some other lesser known secondary mineral may be present.

In pedology is to be found an understanding of the secrets of soil properties as those of cohesion, of their resistance to stress, their moisture relationships, their susceptibility to volume change and their reaction to the various kinds of additives incorporated for the purpose of moisture or strength stabilisation.

In conventional soil mechanics that understanding may be missing though that restricted outlook is rapidly changing gradually. For example, inferences deducted by ingles and Metcalf (1972) based upon field identification and recognition of soils using pedological principles have been reported.

The major potential merit of engineering pedology, however, is that through inter-relationships between the soil forming factors, pedogenesis, morphology, chemistry and mineralogy on one hand and the geotechnical properties and engineering behaviour in relation to the environmental controls, on the other, the expensive and time-consuming laboratory and field testing programmes can be reduced in most cases to only isolated confirmatory studies; and engineering experience gained for given soil series in one region can be transferred to similar soil series identified under similar environmental conditions elsewhere.

5. IMPORTANCE OF GEOMORPHOLOGY, ENGINEERING GEOLOGY, GEOCHEMISTRY AND CLAY MINERALOGY

Geomorphology is concerned with landform materials, and their related processes pertinent in all aspects of infrastructural development, (e.g. Cooke and Doornkamp, 1974). Significant advances in the mapping of a land's geomorphological characteristics have added an invaluable tool in assessing the potential of land for development, and in providing a valid basis for land-use planning. Geomorphological processes do not operate in isolation. They are generally a part of a whole system of inter-racting phenomena. In addition, their significance to environmental problems has to be assessed in the light of the socio-economic and regional development. Environmental geomorphology knows no traditional subject boundaries and embraces parts of such subjects as hydrology, engineering geology, soil mechanics, pedology, and geography. The potential for applying geomorphological knowledge depends not only on the problem but also on the willingness of the development agencies to appreciate the value of this knowledge. Apparently, the purpose of geomorphological maps is to portray the forms of the surfaces, the nature and properties of the materials of which these are composed, and to indicate the kind and magnitude of the processes in operation. As such they provide an integrated and comprehensive statement of land form and drainage. They contain information of considerable potential value in landuse and development planning, hydrological, civil, foundation, and highway engineering.

The most useful geomorphological maps, in environmental development terms, are those which are compiled in the context of a specified problem. Such maps may be based on a general geomorphological map, but in any event they should be explicit, clear and preferably in colour. No mapping exercise is ever complete in itself, however, and its results have to be considered in terms of the dynamic process systems of the kind described in the foregoing (Cooke and Doornkamp, 1974).

Engineering geology is concerned with many factors of importance in foundation engineering; the major ones being geological age, type of rock and petrography, structure, fabric and lithology, discontinuities such as joints, folds, faults, bedding and foliation, etc.; direction and angles of bedding, etc.; chemical and mineralogical composition of rocks (which determine the rapidity and the degree of physical and chemical weathering), secondary mineralogy of the soil, groundwater conditions - depth of water table and its seasonal variation, direction of flow of groundwater, chemical composition (such as sulphate content and the acidity pH). It is known (Kenny, 1967; Yatsu, et al., 1975 and Mitchell, 1976) that petrographic factors which influence engineering properties of soils include mineralogical composition of solid particles, grain size distribution, the shape, water and gas in the voids, chemical composition of soil water, etc. Related to this is also importance of geochemistry which deals with all the geological processes and phenomena related to chemistry; and is concerned with establishing abundance relationships of elements, distribution of elements in minerals and rocks, and laws governing the abundance relationships and distribution of elements.

Clay mineralogy which concerns all the laboratory techniques of identification and classification of clay minerals also should play significant role in geotechnical engineering. For example, the soils engineer needs information on the composition of clay soils to help him explain the fundamental causes of the properties, and, therefore, makes conclusions based upon empirical test data more certain. Information on mineralogical composition of clay soils also serve to warn the civil engineer if there is anything unusual about the clay mineral that needs special treatment. Experience has shown (e.g. Terzaghi, 1958; and Mitchell, 1976) that physical data alone may sometimes be very misleading in interpreting soil test results.

6. TERRAIN CLASSIFICATION AND ASSESSMENT OF SUBSOIL CONDITIONS

Terrain evaluation and classification are the complete appraisal of the capabilities and limitations of an area of ground in relation to a particular kind of landuse, in this case for purposes of shallow foundations. Most terrain evaluation surveys have been based on the technique of land classification because a natural breakdown of the land provides an ideal set of units for the storage of terrain data.

The classification is based on units of landscape defined by their geology, topography and water regime. A geographical reference system has several advantages over an abstract system of storing data, one of the most important being its flexibility for use at both general and detailed levels. The processes of weathering and erosion collectively constitute land forming processes. They control both the shape of the land surface and the soils formed on it. The area of unconsolidated sand, being permeable, is gently undulating and has a few indistinct streams. In the clay area all the surface water remains on the surface, giving an intricate network of small streams. A change of land forming process occurs at the steep escarpment where the streams flow more quickly and have cut deep valleys.

The foregoing illustrates the principle that as long as the geology, climate and land forming process remain uniform, the soils and landscape characteristics remain constant. Each feature of the landscape has a characteristic soil type with its own particular advantages and limitations for the engineer and for the farmer. The landscape can therefore be classified into individual features each of which occurs many times in a consistent pattern. Wherever a similar feature recurs there are the same opportunities and limitations both for agriculture and environmental engineering. The non-specialist recognizes this pattern as a landscape, a tract of land having certain obvious and recurring characteristics peculiar to itself and different in some ways from surrounding areas.

This pattern of landscape is called a land system and its component features are called land facets. These are the two main units of the terrain mapping method used in the United Kingdom (Beaven and Lawrence, 1972). A land system may be defined as an area of uniform geology (or a geological sequence) having a characteristic arrangement of topography and soils, repeated over a wide area. It is developed within a single climatic zone. Land systems are normally mappable at scales of about 1:½ million to 1:1 million, depending on the complexity of the terrain (Beaven and Lawrence, 1972).

The land facet is the basic unit of mapping. It is also defined by its geology, water regime and topography but in a much more restricted way than the land system. A land facet has a simple form (such as the steep scarp slope, a fairly homogeneous parent materials and a single water regime - both surface water and groundwater). The soils developed upon it are naturally fairly uniform, such that an engineer would accept a single design specification for specific foundation built on it. Land facets are normally mappable at scales of 1:10,000 to 1:50,000 depending on their size and arrangement (Beaven and Lawrence, 1972).

Land system maps are normally compiled using aerial photography to identify the land units. Because land systems form patterns of terrain, these patterns can be delineated in aerial photographs as preliminary land system boundaries. The individual photographs are examined under the stereoscope using geological maps and as much other ground information as possible to define the land facets in each pattern can be obtained.

The land system boundary can then be drawn more accurately, the complement of land facets indicating its precise extent. The procedure, compared with other mapping methods, is extremely rapid. The land facet boundaries are then checked in the field, to confirm the photo-interpretation and complete the mapping. At this stage it is also usual to collect samples so that the properties of the soils may be defined visually and in the laboratory.

In the land system method of terrain mapping the land facet and land system are the most important units. A land region is a group of land systems linked and related with an overall similarity of landforms. They are usually shown on a map at a scale of 1:1 million/1:3 million. A land province is larger still, representing a group of land regions united in a major lithological sequence or a major geological structure. Mapping scales up to 1:10 million are appropriate for this unit. Land region and land province maps can be compiled very quickly if adequate information on the geology and topography can be obtained. Apparently, this interplay of the various branches of geo-science could provide a sound basis for making progress in geo-technical engineering practice related to specific environment characterised by specific geologic, topographic, morphologic, soil, climatic and groundwater conditions.

7. SOILS CONDITIONS SURVEYS FOR SHALLOW FOUNDATIONS

Engineers, engineering geologists and pedologists now make considerable use of aerial photographs and satellite images of project sites to provide detailed information than from a map.

For example, in addition to giving information on general conditions and surface features, air photographs could be used to draw conclusions on terrain conditions such as soil type, drainage conditions, and the presence of marshy areas; unstable ground and outcrops, etc.

Geological maps provide first informations, but are unfortunately insufficient and rarely provide any information on the overburden sub-soil which is of more interest to the foundation engineer.

The usefulness of a geological map for foundation engineering purposes will depend in part on the date of mapping, the purpose for which the map was prepared, and the scale of the map. Recent geological methods, combined with techniques available through the use of pedological concepts, airphoto interpretations and engineering investigations, provide means of obtaining highly useful information for shallow foundation engineering purposes.

The engineer has relied and continues to rely more on geomorphologic and topographic than any other map for site selection for shallow foundations. This has been true chiefly because until fairly recent time, economics dictated that the most suitable foundation and the easiest method of construction should be selected.

Experience and firm background in the engineering significance of terrain features, good geomorphological and topographic maps provide a very effective tool in estimating the quality of a given areas and for making tentative selection of suitable foundation type. Where such maps are available they should be the first bit of data to be studied in an endeavour to develop preliminary information about the nature of the subsoil and the suitability or otherwise of the site for the project.

As in the case of geology, the science of pedology makes available a wealth of information that can be used for a general estimation of the entineering properties of soils. With emphasis on speed of construction these days, the engineer will find a good soil survey report of inestimable value for reconnaissance, site selection, and a preliminary indication of the engineering properties of the soil he will encounter in connection with building foundation projects. The pedological soil classification being of a qualitative rather than of a quantitative character, would of course, not give all the data which an engineer may like to have, such as density, moisture content and bearing capacity, etc., however, once these additional properties have been determined for a known pedological soil type through engineering tests, they continue to be of value for any future project on the same soil type in a similar environment.

A truly splendid example of the engineering uses and application of the pedological soil classification system for shallow foundation purpose has been reported by Aitchison (1953, 1974).

The field manual of foundation soil engineering if prepared based upon pedological information could be the best guide for practical application of pedology in shallow foundation. For example, in Australia it has been possible to establish correlations between the shallow foundation engineering practice and pedological soils series characterised by profile features, organic matter content, pH, airdry moisture content and grading of the soil in the profiles. Pedological soil maps also sometimes provide data on surface and internal drainage conditions as well as the depth of water table and its probable seasonal variations.

The Australian experience has demonstrated that the only limit in the practical uses of the pedological soil classification system for shallow foundation engineering is the capacity and willingness of the geotechnical engineers to learn and use this branch of applied geoscience.

8. SEASONAL SUBSOIL MOISTURE VARIATIONS IN PROFILES AND SHALLOW FOUNDATION ENGINEERING IMPLICATIONS

The significant operative factors that affect the performance of shallow foundations are features of the soil profile, extent and depth of seasonal moisture variation and groundwater. Field studies should establish the depth of moisture variation in the top soil which may take many forms. In foundation engineering, the vertical variation of moisture is very critical because the strength and compressibility of finegrained soils depend upon the moisture content at a given time. For example, the effect of moisture content on strength characteristics of rocks and soils have been reported by de Brujn (1967) and Duncan (1970). Similarly, the effect of the compositional factor, desication, and vertical seasonal variation of water content on the soil strength in profiles have been discussed by Aitchison (1952). Long term experiences in shallow foundation engineering in temperate and tropical climates have established that the base of shallow foundations should always be located:

a. Below the depth of structurally unstable strata or those susceptible to high volumetric changes as revealed by the identification properties.

b. Below the depth subjected to a cycle of severe wetting and drying.

c. Below the depth of artificially induced or naturally increased temperature.

d. On a soil layer having adequate bearing capacity and moderate or low compressibility.

In tropical and sub-tropical regions there could be great variations in seasonal influences of moisture content in profiles. The depth of seasonal variations of moisture content, physico-chemical characteristics, shear strength caused by severe drying and wetting may vary from as shallow depths as 3 metres to as deep as 10 metres (e.g. Aitchison, 1952; Gidigasu, 1980). Generally, the soil formed in any locality tends to be in a state of dynamic equilibrium with its climatic environment. Thus, many of the morphological features of a soil throughout the whole of the depth affected by seasonal climatic influences (that is, including the whole soil profile) are subjected to the cyclic changes of moisture regime (and perhaps of temperature). Apart from textural differentiation, perhaps the most notable feature of the soil is the structure whose development processes depend on the environment. Soil structure is developed as a result of processes of aggregation of primary and secondary particles. It has been shown that the processes at work in the creation of soil structure may be physical, chemical or biological in nature (e.g. Mitchell, 1976).

In the upper portion of the profile, the processes are mainly commentation of particle as a consequence of microbiological activity in the soil. In the deeper portions of the profile the formation of the structure may be mainly physical. The determination of the shear strength of soils at shallow depths generally pose two practical problems (Aitchison, 1952).

In the first place, it is necessary to ensure that the soil is in a moisture state which correctly represents the condition beneath the building, and secondly, it is essential that a proper technique, based upon soil propertics (including structure) be used in strength and compressibility measurements.

Long term studies at a permanent site to establish mean stable subsoil moisture condition for both the dry and the wet seasons are sometimes required to establish stable subsoil moisture content (e.g. Gidigasu and Appeagyei, 1985).

The definition of vertical variations of significant geotechnical properties for particular soil series is perhaps the key contribution of engineering pedology to geotechnical and shallow foundation engineering.

8.1 Relationship between the soil profile and the proposed engineering superstructure

Jennings and Brink (1978) have emphasized that while the general principles of soil profiling apply in all cases, the engineer responsible for observation in a trial hole will produce more useful work if he continually bears in mind the requirements of the proposed engineering structure. For example, if he is profiling for an earth dam he will be on the lookout for water-bearing strata and for materials which will affect foundation conditions, in this case, usually related to slip circles through the base of the dam. On the other hand, if the proposed work is a massive building of rigid construction, he will be more concerned with problems of compressibility.

It is important also that a sufficient number of profiles be examined to embrace the lateral extent of the structure. For small buildings, a single profile may be sufficient but for an extensive structure, such as a road, many profile may be required. The completed profile sheets should be laid side by side and careful consideration given to soil variations in a lateral direction.

In all cases the information provided on the profile sheets should extend to a depth which is equal to or greater than the depth which is significantly stressed by the structural loads. This depth is accepted to be of the order of 1.5B where B is the minimum plan dimension of the structure (e.g. Leonards, 1962).

A general note to observers of soil profiles is that it is advisable to group similar soil strata and to ignore minor changes in the profile. This may be achieved during the work of drawing up the profiles in the office - i.e. in the field it is advisable to record all observed facts in the notebook but when the profiles are finally prepared, these notes should be simplified, combining the strata of generally similar characteristics and avoiding unnecessary complicating details (Jennings and Brink, 1978).

9. SUMMARY OF CONCLUSIONS

Classical soil mechanics principles have failed in answering all problems relating to earthworks and foundation engineering for some tropical and residual soils formed under sub-tropical and tropical environments.

The limitations of wholesale importation of specifications developed for temperate zone materials have also been highlighted.

Consideration of the soil forming processes, the soil profiles characteristics (morphology), the soil compositional factors and how the whole profile under foundations responds to the influences of the environmental factors are vital for evolving realistic shallow foundation codes of practice for unstable tropical and residual soils.

Engineering pedology which virtually embraces all aspects of applied geoscience stands out as a useful tool for developing shallow foundation specifications for tropical environments.

Specific success stories of this approach have also been highlighted.

10. REFERENCES

Ahn, P., 1970. West African Soils. Oxford University Press, London, 3rd edition, 332pp.

Aitchison, G.D., 1952. The physical condition of the soil as a modifying factor in the measurement and interpretation of shear strength. Proc. 2nd Aust.-N.Z.Conf. Soil Mech.Found.Eng., Parkville, Australia, pp.15-29.

Aitchison,G.D., 1953. Soil morphology and foundation engineering. Proc.Intern.Conf.Soil Mech.Found.Eng., 3rd, Zurich, Vol.1, pp.3-7.

Aitchison, G.D., 1972. Problems of soil mechanics and construction on soft clays and structurally unstable soils. Proc.Intern.Conf.Soil Mech.Found.Eng., 8th, Moscow, Vol.3, pp.161-190.

Aitchison,G.D., 1974. Twenty-five years of application of soil survey principles in the practice of foundation engineering. Geoderma, Amsterdam, 10:99-112.

American Society of Civil Engineers (ASCE), 1982. Engineering and construction in tropical and residual soils. Am.Soc.Civ.Engrs., New York, 746pp.

Attewell, P.B. and Farmer, I.W., 1976. Principles of engineering geology. Chapman and Hill Publishers, London, 1045pp.

Baker, A.L.L., 1957. Raft foundations. Concrete Publications Limited, London, 148pp.

Beaven,P.J. and Lawrence, C.J., 1972. The application of terrain evaluation to the Trans-African Highway Project, P.I.A.R.C., Bulletin 206, 3rd Quarter, 1972, p.1-21.

Bell, F.G., 1978. Foundation engineering in difficult ground. Butterworths, London, 598pp.

Brammer, H., 1952. Ghana Soils. In: J.B.Wills (editor): Agriculture and land use in Ghana. Oxford University Press, London, pp.88-126.

Brand, E.W. and Philipson, H.B., 1985. Sampling and testing of residual soils - A review of international practice. Scorpion Press, Hong Kong, 194pp.

British Standards Institution (BS), 1972. Code of Practice for foundations, CP.2004, Brit.Std., London, 158pp.

Brujn, de C.M.A., 1967. Annual vertical movements and soil moisture redistribution at the Onderstepoort test site. Proc.Reg.Conf.Africa Soil Mech.Found.Eng., 4th, Cape Town, Vol.1, pp: 235-242.

Casagrande, A., 1948. Classification and identification of soils. Transactions, Am.Soc.Civ.Engrs., (with discussions), Vol.113, pp.901-991.

Chandler, R.J., 1969. The effect of weathering on the shear strength properties of Kenper Marl, Geotechnique, London, 19(3): 321-334.

Cooke, R.V., and Doornkamp, J.C., 1974. Geomorphology in environmental management. Claredon Press, Oxford, 413pp.

Dapples,E.C., 1959. Basic geology for science and engineering. John Wiley & Sons, New York, N.Y., 609pp.

Dearman, W.R., 1974. Weathering classification in the characterisation of rock for engineering purposes in British practice. Bull.Intern.Assoc.Eng., Geol., No.9, pp.33-42.

Deere,D.U., 1969. Geological considerations. In: Stagg and Zienkiewicz (editors): Rock mechanics in engineering practice. John Wiley & Sons, New York, N.Y., pp.1-20.

Donaldson, G.W., 1962. Foundation problems experienced in Africa. The CCTA Inter-African Housing Secretariat, Commission for Technical Co-operation in Africa South of the Sahara, Publication No.81; 17pp.

Duncan, N., 1970. Engineering geology and rock mechanics. Leonard Hill Publishers, London, Vol.1, 252pp.

Gidigasu, M.D., 1972. Mode of formation and geotechnical characteristics of laterite soils of Ghana in relation to soil forming factors. Eng.Geol., Amsterdam, 6:79-150.

Gidigasu, M.D., 1974. Degree of weathering in the identification of laterite materials for engineering purposes - A Review. Eng.Geol., Amsterdam, 8:213-266.

Gidigasu, M.D., 1976. Laterite soil engineering. Elsevier Scientific Publ. Company, Amsterdam, 554pp.

Gidigasu, M.D., 1980. Some contribution to tropical soils engineering in Ghana. Proc.Reg.Conf.Africa Soil Mech. Found.Eng., 7th, Accra, Vol.2, pp.509-626.

Gidigasu, M.D., 1985. Shallow foundation problems. Building Research and Practice, Paris, May-June, 1985, Vol.13, No.3, pp.153-160.

Gidigasu, M.D. and Andoh, M.B., 1980. Broad regional evaluation of expansive soil conditions for shallow foundations in Ghana. Proc.Reg. Conf.Africa Soil Mech. Found.Eng., 7th, Accra, Vol.1, pp.235-251.

Gidigasu, M.D. and Appeagyei, E., 1984. Shallow foundation conditions in tropical residual soils in relation to the environment. Proc.Reg.Conf.Africa Soil Mech.Found.Eng., 8th, Harare, Vol.1, pp.195-201.

Hammond, A.A., 1984. Foundations for earth buildings in the tropics. Building Research and Practice, Jan/Feb., 1984, Vol.12, No.1, pp.30-36.

Heiman, J.L., 1967. Footings for problem soils. Depart. of Works, Commonwealth Experimental Building Station, Sydney, Technical Study No.44, 59pp.

Hamrol, A.A., 1961. Quantitative classification of the weathering and weatherability of rocks. Proc.Intern. Conf. Soil Mech. Found.Eng., 5th, Paris, Vol.2, pp.771-774.

Hencher,S.R. and Martin, R.P., 1982. The description and classification of weathered rocks in Hong Kong for engineering purposes. Proc. 7th, S.E.Asian Geot.Conf. Hong Kong, Vol.1, pp.125-142.

Ingles, O.G. and Metcalf, J.B., 1972. Soil stabilisation: Principles and Practice. Butterworths, Sydney, 374pp.

Jennings, J.E., 1965. The theory and practice of construction on partly saturated soils as applied to South African conditions. Proc.Intern.Res. & Eng. Conf. on expansive clay soils. A & M Press, Texas, pp.345-363.

Jennings,J.E. and Brink, A.B.A., 1978. Application of geotechnics to the solution of engineering problems - essential preliminary steps to relate the structure to the soil which provides its support. Proc. Inst. of Engrs., London, Part 1, 64:571-589.

Kenney, T.C., 1967. The influence of mineralogical composition on the residual strength of natural soils. Proc. Geotech. Conference on shear strength properties of natural soils and rocks, Oslo, Sweden, Vol.1,pp.123-129.

Knight, K., 1963. The origin and occurrence of collapsing soils. Proc.Reg.Conf.Africa Soil Mech.Found.Eng., 3rd, Salisbury, Vol.1, pp.127-130.

Loughman, F.C., 1969. Chemical weathering of the silicate minerals. Elsevier, New York, N.Y., 154pp.

Lohnes, R.A., Fish, R.O., and Demiral,T., 1971. Geotechnical properties of selected Puerto Rican soils in relation to climate and parent rock. Geol.Soc.Am., Bull.82: 2617-2624.

Lohnes, R.A., Demirel, T., and Tuncer,E.R., 1976. Strength and structure of laterites and lateritic soils. Engineering Res.Inst., Univ.of Iowa,Rept. ISU-ERI-Ames - 77021; 83pp

Leonards, G.A., 1962. Foundation engineering. McGraw-Hill Book Company, New York, 1136pp.

Little, A.L., 1969. The engineering classification of residual tropical soils. Proc.Spec. Sess.Laterite Soils. Proc.Intern.Conf.Soil Mech.Found.Eng., 7th, Mexico, Vol.1, pp.1-10.

Lumb, P., 1962. The properties of decomposed granite. Geotechnique, London, Vol.12(3):226-243.

Lumb, P., 1983. Engineering properties of fresh and decomposed igneous rocks from Hong Kong, Eng.Geol., Amsterdam, Vol.27, pp.287-298.
Miller, R.P., 1965. Engineering classification and index properties for intact rock. Ph.D. Thesis, Univ. Illinois (unpublished).
Mitchell, J.K., 1976. Fundamentals of soil behaviour. John Wiley and Sons, New York, N.Y., 422pp.
Mohan, D., Jain, G.S. and Sharma, D., 1967. Bearing capacity of multiple under-reamed bored piles. Proc. Asian Reg.Conf.Soil Mech.Found.Eng., 3rd, Haifa, Vol.1, pp.103-106.
Ola, S.A., 1980. Mineralogical properties of some Nigerian residual soils in relation with building problems. Eng. Geol., Amsterdam, Vol.15, pp.1-15.
Peck, R.B., Hanson, W.R., Thornburn, T.H., 1973. Foundation engineering. 2nd edition, John Wiley & Sons, New York, 514pp.
Sherwood, P.T., 1967. Classification tests on African red clays and Keuper Marl. Q.J. Eng. Geol., London, 1(1): 47-55.
Simonson,R.G., 1974. Non-agricultural applications of soil surveys. Developments in Soil Science, Vol.4, Elsevier, Amsterda, 178pp.
Small, I. and White, B.G., 1962. A comparison of costs and methods of construction of house foundations on desiccated clays in Accra. West African Building Research Institute (WABRI), Research Note 8: 7pp.
Stagg, K.G. and Zienkiewicz, D.G., 1969. Rock mechanics in engineering practice. John Wiley & Sons, New York, N.Y., 442pp.
Terzaghi, K., 1958. Design and construction of the Sasumua Dam,. Proc.Brit.Inst.of Civil Engrs., London, 9:369-394.
Wagner, A.A., 1957. The use of the unified soil classification system by the Bureau of Reclamation. Proc.Intern. Conf.Soil Mech.Found.Eng., 5th, Paris, Vol.1, pp.125-132.
Weinert, H.H., 1963. Climatic factors affecting the weathering of igneous rocks. Agric. Meteorol., 2: 27-48.
Weinert, H.H. and Clauss, K.A., 1962. Weathering and climatic environment of a basic lava of the Lebombo, near Moamba. Province de Mozambique. Serv. Geol. Minas. Prov. Mozambique, Bull.29, pp.5-15.
Willis,E.A.,1946. Discussion on: A study of laterite soils, by Frukauf. Proc.Highw.Res. Board, 26:589-590.
Wooltorton,F.L.D., 1936. A preliminary investigation into the subject of foundations in Black Cotton and Kyatti soils of Mandalay District, Burma. Proc.Intern.Conf. Soil Mech.Found.Eng., 1st, Cambridge, Mass., Vol.3, pp.242-257.
Wooltorton,F.L.D., 1955. Engineering pedology and soil stabilisation. Highw.Res.Board, Washington, D.C., Bulletin 108, pp.29-57.
Wooltorton,F.L.D., 1966. Review of road making materials in Northern Borneo, by K.E.Clare and P.J. Beaven, Eng. Geol., Amsterdam, 1(3): 253-256.
Wooltorton, F.L.D., 1976. Foreword to "Laterite Soil Engineering", by M.D.Gidigasu, Elsevier, Amsterdam, 554pp.
Yatsu, E., Ward, A.J. and Adams, F., 1975. Mass wasting. Proc. Symposium on Geomorphology, 4th, Guelph, pp.69-78.

Fabric variability and index testing of a granitic saprolite

T.Y.IRFAN, Geotechnical Control Office, Hong Kong

ABSTRACT: Residual and saprolitic soils are complex and heterogeneous materials, and their engineering properties are generally influenced by the microfabrics developed from the weathering and other alteration processes. This paper describes the fabric variability observed both on site and in the test specimens of a granitic saprolitic soil and the effect of various soil microfabrics on the shear behaviour under direct shear test conditions and index properties, particularly grain size, void ratio and density. A brief review of semi-quantitative and quantitative index tests useful in describing and classifying weathered material types is also presented.

1 INTRODUCTION

There is worldwide diversity in the geotechnical engineering field over the terms used in describing and classifying the various components of the weathering profile and the weathering products. The terms residual soil, saprolitic soil and lateritic soil have sometimes been loosely and interchangeably used. For example, residual soil has been used to describe all the weathered rock zones above the "bedrock" (Brand and Phillipson 1985) while some argue that the term should be restricted to the topmost soil zone of the weathering profile which has lost its original rock fabric (Dearman and Turk 1985). In this paper, the term residual soil is confined to the latter meaning and saprolite is used as a general term to describe that portion of the weathering profile which retains its original rock fabric and structure but which can be considered a soil in terms of strength and consistency.

Saprolitic and residual soils (including laterites) formed from weathering of rocks in mainly tropical and subtropical environments present problems in geotechnical engineering because of their heterogeneity and complexity. Characterization and testing of such soils is correspondingly difficult. Many of the conventional soil mechanics methods and the relationships developed for transported soils in temperate climates do not readily apply to these soils (Vaughan 1985). The material behaviour of saprolitic and residual soils appears to be dominated by the microfabric of the soil which is related to the original rock fabric and weathering history. Many weathering indices of mainly geological nature have been proposed to quantify the degree of weathering, but few attempts have been made to relate them to the engineering properties (see, for instance, Lumb 1962, Baynes and Dearman 1978a for granitic soils). For the 'rock' range of the weathering spectrum, successful relationships have been derived between quantitative weathering indices, physical index properties and engineering properties (see, for instance, Irfan and Dearman 1979, Dearman and Irfan 1978 for weathered granite).

In Hong Kong, a six-fold classification scheme based on Moye (1955) has been traditionally used to characterize the weathering state of the rock material (Hencher and Martin 1982, GCO 1984). In this scheme, the term "decomposition" is used to describe the weathering state of rock material; grades I, II, III (fresh, slightly decomposed and moderately decomposed respectively) are used for rock and grades IV, V and VI (highly decomposed, completely decomposed and residual soil) for soil and soil-like materials. Various index tests have been suggested to assist in the identification of completely and highly decomposed materials.

Thick saprolitic soil mantles occur in Hong Kong, giving rise to slope stability problems due to a combination of high intensity rainfall, steep natural terrain and the large number of very steep cut slopes. The stability of both existing and new 'soil' cut slopes are conventionally analysed by the well-known limit equilibrium methods (e.g. Janbu 1973, Morgenstern and Price 1965) using shear strength parameters determined in the laboratory, generally by triaxial testing and more recently, to a limited extent, by direct shear testing. Direct shear testing is becoming increasingly used for the strength assessment of these soils because of its simplicity and because it is considered by some researchers to reproduce the field conditions more closely than the triaxial test. In the mountainous terrain of Hong Kong true residual soils are generally very thin (less than 3-4 m) or absent. Research by the Geotechnical Control Office (GCO) of the Hong Kong Government has therefore concentrated on determining methods applicable to a realistic assessment of the shear strength properties of the commonly occurring granitic and volcanic saprolites.

This paper describes the field and laboratory studies undertaken as part of this research programme to characterize the degree of weathering and microfabrics of a granitic soil in relation to its shear strength properties measured in direct shear. A brief review of semi-quantitative and quantitative index tests useful in describing and classifying weathered material types is also presented, and the early findings on the relationship between the microfabric of the granitic soils and the engineering properties are discussed. The theoretical aspects of shear strength determination and the evaluation of various laboratory strength testing methods are outside the scope of this paper. A brief review of shear strength testing in Hong Kong is given by Brand (1985) and a programme of direct shear testing undertaken by the GCO is described in a companion paper presented at this conference by Cheung, Greenway and Massey (1988).

Although the saprolitic zone is generally considered to behave as a uniformly distributed soil, both the strength of relict discontinuities and macro- and micro-structures formed as a result of weathering, as well as the soil matrix strength affects the behaviour of these soils in slope stability. Failures dominated by mass structure are common in this zone. This aspect of saprolitic soils is covered by Irfan and Woods (1988) in this conference.

2 SEMI-QUANTITATIVE AND QUANTITATIVE METHODS FOR EVALUATION OF DEGREE OF WEATHERING

There have been several attempts to quantify the degree of weathering and the weathered rock fabrics by using chemical, petrographical, X-Ray diffraction and electron microscope methods, or by simple field and laboratory mechanical and physical index tests. Although in most cases the quantitative and semi-quantitative weathering indices have been derived for the purpose of quantifying the change in purely geological properties, some quantitative weathering indices have been used to relate and to indirectly determine the engineering properties of weathered rocks (e.g. Dearman and Irfan 1978). Many of these weathering indices have been specifically developed for characterizing coarse-grained igneous rocks, particularly granites, because weathered granite has been encountered in numerous civil engineering schemes. Some of the quantitative weathering indices have also been applied to fine-grained igneous rocks (e.g. Weinert 1964, Dearman et al 1987). A review of various index test methods used to characterize the rock range of the weathering spectrum is given by Martin (1986).

Many different laboratory mechanical tests have been used primarily for rapidly and economically estimating material design parameters (strength and deformability) and for characterizing the degree of weathering. Good correlations have been reported between test indices and engineering design parameters of weathered granite (Hamrol 1961, Onodera et al 1974, Irfan and Dearman 1978a). Schmidt hammer, hand penetrometer and slake test have been used for classification purposes to broadly distinguish between classes of soil, for example between highly and completely decomposed granite (Hencher and Martin 1982). However, these tests are insensitive to the detection of fabric variations within each class of soil.

The chemical indices, such as weathering potential index (Reiche 1943), ignition loss (Sueka et al 1985), and pH and specific gravity of feldspars (Matsuo and Nishida 1968), are a relatively accurate measure of degree of chemical alteration. However, the engineering properties of saprolitic soils are also partly dependent on the remnant fabric of the parent rock. For example, granite and rhyolite in Hong Kong have similar chemical and modal compositions, but the physical and engineering properties are significantly different at similar degrees of weathering (Lumb 1965).

The relationships which have been successfully developed for transported soils relating the results of soil classification tests such as plasticity and grading parameters and various engineering properties do not seem to hold for saprolitic and lateritic soils (Gidigasu 1975, Mitchell and Sitar 1982). This partly stems from the difficulty of determining the 'exact' grain size of saprolitic soil which has relict fabric, including a weak bonding either retained from the parent rock or regenerated during weathering. The relict fabric generally determines the strength of these soils (Baynes and Dearman 1978a, Vaughan 1985), independent of stress history (Vaughan 1985).

It is known that particle size distribution of saprolitic and residual soils is sensitive to the test procedures used, e.g. use of pestle and mortar to crush soil, type of dispersing agent, amount of agitation, oven-drying, etc. (Mitchell and Sitar 1982, Sandroni 1985). The effect of pretreatment may also have a pronounced effect on normal plasticity parameters (Gidigasu 1975). Variations in grain size also result from variability in the initial grain size and mineral composition of the original rock.

Sowers (1985) considered that void ratios and degree of saturation are probably the only reliable simple indices that can be used to assess the degree of weathering of these soils. Specific gravity of feldspars in combination with void ratios have been used by Nishida and Anyama (1985) to classify weathered soil into four types for slope stability purposes. There is scope for developing special laboratory classification tests, but these can only be useful when combined with fabric observations.

3 MICROFABRIC STUDIES

The objective of any microfabric study should be to classify and quantify the various fabric aspects in order that these can be correlated with engineering properties. The term "microfabric" is used here to mean the geometrical arrangement of particles, including particle size and porosity distribution which are studied by optical and electron microscopy. Although the microfabric of any geological material exerts a fundamental influence on its engineering properties, the relationship is difficult to interpret because of the level of complexity of many microfabrics and the intuitive nature of their interpretations (Collins and McGown 1974, Baynes and Dearman 1978a). The latter authors reported a number of important observations related to the nature and engineering significance of weathered granite microfabrics (Baynes and Dearman 1978a,b). They concluded that microfabric reflects the duration and intensity of weathering, and that it is common to find very variable microfabrics within the same specimen, indicating the marked variability of weathering microenvironments. In the case of granites, they related the microfabric to the degree to which feldspars have been weathered, to the proportion of clay produced during the decomposition process, and also to the extent to which particles have been eluviated from the system.

The finer microfabric features which are beyond the resolving power of the ordinary microscope can be studied by scanning electron microscopy (Smart and Tovey 1982, Baynes and Dearman 1978c). Various microfabric configurations have been proposed to explain observed engineering properties. For example, Baynes and Dearman (1978a) superimposed two simple fabric models on a void ratio-degree of decomposition index (X_d) plot. Collins (1985a) proposed a microfabric characterization scheme for use in engineering studies, particularly for the saprolitic and residual soils, to provide a rational and solid framework which can be extended and added to, as the need arises. These studies, supplemented by data obtained using other identification methods such as XRD and DTA, can help in understanding the engineering behaviour of saprolitic and residual soils.

Standard petrographic modal analysis and other techniques have been used for the whole spectrum of weathered igneous rocks to quantitatively evaluate the successive mineralogical and textural changes brought about by weathering. These include microcrack determinations (Dixon 1969, Onodera et al 1974, Irfan and Dearman 1978b), modal analysis of the percentage of decomposed minerals (Weinert 1964, Mello Mendes et al 1966, Irfan and Dearman 1978b) and weight percentages of quartz and feldspars in both the weathered and parent rock (Lumb 1962). Quantitative petrographic indices derived from these methods have been directly correlated with a variety of physical and mechanical properties of weathered granites (Onodera et al 1974, Dearman and Irfan 1978).

The micropetrographic index, I_p, described by Irfan and Dearman (1978a) may be used to characterize altered rocks whether the alteration results from hydrothermal or weathering processes, in terms of its physical and chemical state. The other advantage of this technique is that it gives insight into the changing mineralogical and microcrack regimes involved in progressive weathering. The degree of decomposition index, X_d, described by Lumb (1962) gives an indication of the degree of decomposition of feldspars, but structural fabric elements (e.g. microcracks and voids) and decomposition state of other mineral constituents (e.g. biotite) are not taken into account.

4 SITE DESCRIPTION AND CLASSIFICATION OF SOIL TYPES

The sampling site is situated on a relatively flat hilltop (elevation 125 mPD) above a 10 to 15 m high cut slope on Hong Kong Island. The rock in this area is a typical granite with an average grain size of 1 to 4 mm, and occasional feldspars as large as 10 mm. In the fresh state the granite is light pinkish grey in colour and consists of quartz (35 to 40% by weight), alkali feldspars and plagioclase (55 to 60%), and biotite (less than 5%). The granite on site is weathered to a crumbly saprolitic soil to considerable depth (over 30 m), with occasional less weathered corestones and a thin 1 to 2 m, reddish brown residual soil developed over most of the hilltop area. The granite can generally be described as completely decomposed granite (CDG or grade V material of GCO, 1984) with considerable lateral and vertical variations in degree of weathering, both in terms of chemical decomposition and physical disintegration occurring throughout the site (Figure 1). In addition, the granite is hydrothermally altered, particularly adjacent to quartz veins, with all the feldspars altered to white kaolinite pseudomorphs. It is considered that hydrothermal alteration played a significant part during the late stage of the intrusive activity, with weathering effects under a hot and humid subtropical climate superimposed at a later stage when the rock was exposed to atmospheric agents. Within this general matrix of chemically decomposed granite, there are "corestones", and areas where feldspars show very little decomposition but where the rock is weathered to the condition of friable granitic soil (gravelly sand) as a result of mechanical disintegration.

The transition from saprolite to residual soil is gradual (Figure 1). Nearer to the ground surface the original granitic fabric is completely destroyed and the soil is somewhat compact due to collapse. The intensity of feldspar decomposition increases upwards in the weathering sequence with a related increase in silt-clay content. The final product is a clayey silty sand with no feldspar pseudomorphs recognizable near the surface.

Because of the variability observed in the weathered soil occurring throughout the site, an attempt was made to describe and classify the various saprolitic and residual soil materials on the basis of visually recognisable and easily determinable fabric properties, and in terms of degree and type of decomposition of mineral constituents, disintegration (fracturing, grain boundary opening), loss of parent rock fabric, etc. This allowed the identification of six types of soil (Table I). The terms used for the description of degree of change of the various weathering components are based on IAEG (1981). It should be noted that these soil types are useful for description and general classification, but transitions between the types occur frequently.

5 SAMPLING AND TESTING

5.1 Sampling

Eight hand-cut block samples were obtained from trial pits for subsequent direct shear testing. Care was taken during sampling to avoid major discontinuities, quartz veins and hydrothermally altered zones. Detailed examination in the laboratory showed that, although all samples could be classified as grade V granite (GCO 1984), there were differences between the samples in terms of degree of decomposition and disintegration of mineral constituents, type of alteration and other visually observable fabric properties. Four of the blocks (SH8, SH6, SH1 and SH2) were of type 3 material (Table II) or "completely decomposed" granite, block SH7 showed slight loss of granitic fabric, block SH3 showed partial loss of fabric throughout (i.e. type 2 material) and block SH5 showed effects of hydrothermal alteration.

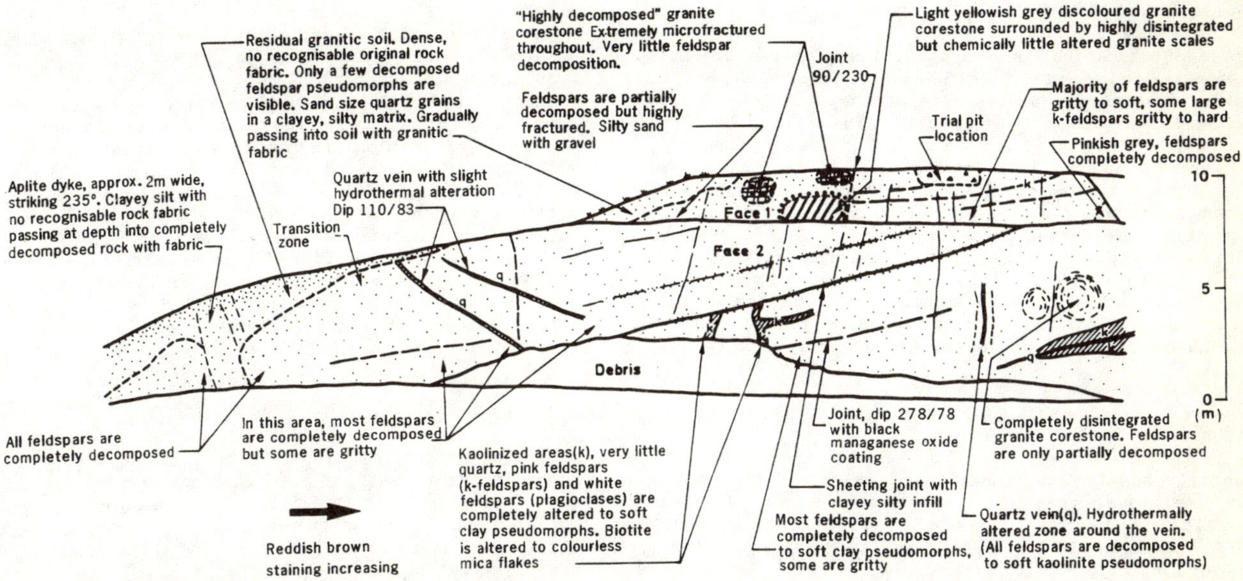

Figure 1. A sketch showing the variability of weathering characteristics at Shouson Hill

Table I. A material classification of soils formed from weathering and alteration of granite at Shouson Hill

Material Type	Material Weathering Grade (GCO, 1984)	Degree of Decomposition / Disintegration *	State of Decomposition / Disintegration of Major Mineral Constituents			Original Rock Fabric	Colour	Grain Size of Soil	Other Properties
			Feldspars *	Quartz	Micas				
1. Residual Granitic Soil	VI	Extreme decomposition	Extremely to completely decomposed; feldspar pseudomorphs unrecognisable	Reduced in size; loss of lustre	Biotite is completely decomposed to colourless hydrous mica	Not present	Light yellowish grey to reddish brown (leached)	Clayey silty sand	Slakes readily
2. Transition Granitic Soil	V/VI	Extreme to high decomposition	Extremely decomposed to powdery-to-soft grains	Fractured; some loss of lustre	Biotite is extremely decomposed	Partly lost	Light yellowish grey to reddish brown (leached)	Clayey silty sand to silty gravelly sand	Slakes readily
3. Decomposed Granitic Soil	V	High to extreme decomposition; complete disintegration	Highly to extremely decomposed to gritty-to-soft grains	Fractured; slight loss of lustre	Biotite is highly decomposed to yellowish grey flakes	Present	Light yellowish grey (may be leached)	Silty gravelly sand with little clay	Slakes readily
4. Altered Granitic Soil	*V	Extreme to complete alteration	Extremely to completely decomposed (altered) to soft clayey kaolinitic pseudomorphs	Greatly reduced in size & amount due to solution	Biotite is completely altered to colourless muscovite & chlorite (?)	Changed	Mottled light pink and creamy white	Clayey sandy silt	Slakes after agitation
5. Disintegrated Granitic Soil	V	Complete disintegration, slight to moderate decomposition	Slightly to highly decomposed to hard-to-powdery grains	Fractured; slight loss of lustre	Biotite is slightly to moderately decomposed to greenish black-yellowish green flakes	Present	Light yellowish grey	Sandy gravel to gravelly sand	Slakes readily
6. Highly Decomposed Granite	IV	Moderate to high decomposition; partial disintegration	Slightly to moderately decomposed to hard-to-gritty grains	Slight loss of lustre	Biotite is slightly decomposed to greenish black flakes; yellowish brown staining around grains	Present	Yellowish grey	Gravelly sand to sandy gravel	Slakes very little, very weak rock - dense soil; a series of very closely spaced macrocracks may be present
Fresh Granite	I	None	Plagioclase and potash feldspars are hard and sound	Lustreous; high quartz content (35 - 40 %)	Biotite is black, no muscovite	Equigranular; granitic	Light pinkish grey to light grey	Medium to coarse-grained (1 to 4mm)	

Blocks SH8, SH6, SH7 SH3 and SH5 were chosen for detailed micropetrographic examination. In addition to these samples, disturbed specimens were collected from an intensely altered (kaolinized) and completely decomposed granite (type 4), and completely disintegrated granite (type 5), to carry out some simple index tests, mainly Lumb's X_d index and particle size analysis.

5.2 Field Index Testing

A series of mechanical and physical index tests advocated by Hencher and Martin (1982) to classify weathered material were carried out on site at the location of each block sample within the trial pit. These included hand penetrometer, Schmidt hammer, slake, feldspar strength, quartz angularity and infiltration rate tests. All these tests showed that the material from where the blocks were obtained was grade V granite. Although generally useful in broadly differentiating between grade IV and grade V granitic materials, neither the individual test results nor the groups of results clearly highlighted the differences noted between blocks by detailed visual examination and by using fewer, simpler physical index tests, such as the feldspar grittiness index test (Table III).

5.3 Laboratory Soil Index Testing

Mechanical and physical index tests on the block samples and on individual shear test specimens included particle size distribution, degree of saturation, moisture content, density and specific gravity. All the tests were conducted generally in accordance with BS 1377 : 1975, except particle size determination. This was determined by a modified BS method routinely employed at the Public Works Central Laboratory. The method involved oven drying at 110°C followed by crumbling by hand, soaking with dispersant, hand stirring and wet sieving. In addition to the routine method, finger crushing of soft feldspars which were not already dispersed during pretreatment was undertaken on selected samples to determine a more representative degree of chemical breakdown of the mineral constituents. The particle size determined using this technique is termed "ultimate grain size"

The mean results of the soil index tests are summarised in Table II and the results on selected individual shear test specimens are given in Table IV.

Block samples SH1, SH2, SH6 and SH8 showed generally lower dry densities, lower moisture contents, lower degree of saturation and higher void ratios in comparison with the more weathered samples SH3 and SH7. The altered portion of sample SH5 showed the highest dry density and moisture content and the lowest void ratio.

The clay content was low for all samples (4 to 9%). Although clay content showed a general slight increase in ultimate grain size analysis, the silt content increased drastically to 15%, particularly for SH3 and SH5. The hand specimen HS4 taken from disintegrated granitic soil showed the least amount of clay (1%).

5.4 Direct Shear Testing

Consolidated drained direct shear tests were performed on soaked "undisturbed" samples within a normal stress range of 20 to 250 kPa. The bulk of the testing was carried out on 60 mm square x 20 mm thick, 100 mm square x 44 mm thick and 60 mm diameter x 36 mm thick specimens, as described by Cheung et al (1988).

Shear stress versus displacement curves of selected specimens from each test block tested at a normal stress

Table II. Results of laboratory index tests (SH = block sample, HS = hand sample, n.d. = not determined)

Block/ Sample No.	Description	Soil Type	Soil Index Properties: Initial Dry Density kg/m^{3}x10^3	Initial Moisture Content %	Degree of Saturation %	Initial Void Ratio	Specific Gravity kg/m^{3}x10^3	Grain Size Routine: G	Sa	Si %	C	Si+C	No. of Test	Grain Size Ultimate: G	Sa	Si %	C	Si+C	Degree of Decomp. $X_{d(mod)}$
HS4	Disintegrated Granitic Soil	5			n.d.			55	41	3	1	4	1	50	43	5	2	7	0.35
SH2	Decomposed Granitic Soil	3	1.21	14.6	33.5	1.17	2.63	40	33	17	10	11	6			n.d.			n.d.
SH8	Decomposed Granitic Soil	3	1.21	18.7	41.7	1.17	2.63	41	45	11	3	14	17	38	44	13	5	18	0.62
SH6	Decomposed Granitic Soil	3	1.21	18.6	42.0	1.17	2.63	42	40	14	4	18	14	40	42	14	4	18	0.73
SH7	Decomposed Granitic Soil with Slight Fabric Loss	3	1.21	17.1	38.6	1.16	2.62	40	45	11	4	15	18	24	38	26	12	38	0.89
SH3	Transition Granitic Soil with Partial Fabric Loss	2	1.30	18.7	45.0	1.09	2.63	31	42	18	9	27	1	31	39	20	10	30	0.92
SH5(w)	Decomposed Granitic Soil with Slight Alteration	3	1.29	20.6	52.0	1.05	2.63	35	38	22	5	27	33	26	38	30	6	36	0.94
SH5(a)	Altered Granitic Soil	4	1.31	20.6	54.0	1.01	2.62	34	34	26	6	32	5	26	31	34	9	43	0.97
HS5	Altered Granitic Soil	4			n.d.			6	30	56	8	64	1	5	26	60	9	69	0.94

Table III. Degree of decomposition of feldspars by grittiness test

Degree of Decomposition	Grittiness Term	Scale	Field Recognition
Fresh	Hard	1	Cannot be cut by knife; cannot be grooved with a pin
Moderately	Gritty	2	Can be cut by knife or grooved with a pin under heavy pressure
Highly to Extremely	Powdery	3	Can be crushed to silt size fragments by finger pressure
Completely	Soft	4	Can be moulded very easily with finger pressure

σ_n of 40 kPa using the 100 mm x 44 mm shear box are given in Figure 2. The shear stress curve of specimen tested at σ_n = 150 kPa is also shown in the figure. Table IV summarizes the laboratory test data for the same specimens. Markedly lower shear strengths are obtained for the transitional soil and altered soil types.

6. QUANTITATIVE AND OBSERVATIONAL FABRIC STUDIES

Thin sections (24 mm x 30 mm) were prepared from selected test blocks by first impregnating them with epoxy resin mixed with organic blue dye to differentiate between in-situ pores present in the soil sample and those produced by the thin sectioning process. Various fabric elements examined under an optical microscope included the mineral composition, the degree and type of chemical alteration of mineral constituents, the grain boundary relationships, the nature and type of microcracks and voids. The finer microfabric features such as form, packing and type of clay minerals and grain surface features produced by weathering, which are beyond th capabilities of optical microscope, were not determined.

6.1 Micropetrographic Index

The point counting technique described by Irfan an Dearman (1978a) was employed to determine the moda percentages of the unaltered and altered minera constituents, voids and microcracks. Micropetrographi index, I_p, was calculated from the modal analysis usin the following formula :

$$I_p = \frac{\%\ \text{Sound constituents}}{\%\ \text{Unsound constituents}}$$

$$= \frac{\%\ (\text{quartz + unaltered feldspars + unaltered biotite})}{\%\ (\text{altered minerals + voids + microcracks})}$$

A high magnification (100 times) and a minimum of thre thin sections per sample were used, with a minimum of 9C points counted in each case.

The results of micropetrographic analysis are given : Table V. A general agreement between the visual: observed fabric elements and the results (micropetrographic analysis is evident from the tab: indicating that this technique can be used 1 characterize and quantify some of the fabric elements (saprolitic soils. A wide range of values were obtain on the four thin sections taken from SH5, depending the location of the thin section within the block relation to the kaolinized area.

6.2 Modified Degree of Decomposition Index

Lumb's (1962) degree of decomposition index, X_d, w determined for the test blocks and other selected sampl collected from the site, using the following formula :

$$X_d = \frac{Nq - Nqo}{1 - Nqo}$$

Where Nq and Nqo are the weight ratio of quartz to qua: plus feldspar in the soil and in the original ro(respectively. Separating grains by hand under a

Table IV. Results of direct shear tests on selected individual specimens (Sample size : 100 □ x 44 mm; τ_c : Area corrected shear strength. Numbers in brackets are mean values for test block)

Specimen No.	Description	Dry Density ρ_d $X10^3$ kg/m^3	Initial Void Ratio e	Particle Size G %	Sa %	Si %	C %	Normal Stress σ_n kPa	Shear Strength τ_c kPa
SH8-1	Decomposed Granitic Soil	1.23 (1.21)	1.14 (1.17)	40	45	12	3	40	50
SH6-1	Decomposed Granitic Soil	1.21 (1.21)	1.16 (1.17)	40	48	9	3	40	46
SH6-2	Decomposed Granitic Soil	1.17 (1.21)	1.26 (1.17)	45	38	14	3	40	49
SH5-1	Decomposed Granitic Soil	1.28 (1.29)	1.05 (1.05)	36	34	25	5	40	48
SH5-2	Decomposed Granitic Soil	1.27 (1.29)	1.07 (1.05)	34	38	22	6	40	44
SH5-a	Altered Granitic Soil	1.31 (1.31)	1.01 (1.01)	31	37	25	7	40	37
SH3-1	Transition Granitic Soil	1.30 (1.30)	1.09 (1.09)	31	42	18	9	40	30
SH8-2	Decomposed Granitic Soil	1.21 (1.21)	1.17 (1.11)	36	47	13	4	150	140

reflecting microscope using the fraction retained on a BS 100 sieve (150 microns) was considered impractical for the present study, and a BS 25 sieve (500 microns) was used instead. Even with the use of the large sieve the technique was found to be time consuming and cumbersome. Petrographic examination revealed that feldspars are in various states of decomposition, and variations occur even within individual grains, with altered (clay minerals) and unaltered portions occurring together. In the normal method these are usually counted, unless broken down by fingers, as feldspar, giving an erroneous count of the true unaltered feldspar content in the soil. Iron oxide cementation of wholly altered feldspar pseudomorphs may also affect the results. With this in mind, a modified technique of finger crushing of partially decomposed feldspars was employed and $X_{d(mod)}$ values were determined (Table II). This gave a more accurate indication of unaltered feldspar content of the soil and resulted in relatively higher X_d values (Figure 3) than those reported by Lumb.

7 GRANITIC SOIL MICROFABRICS AND ENGINEERING PROPERTIES : SOME CONSIDERATIONS

7.1 Weathered/Altered Granitic Soil Microfabrics

Petrographic examination revealed that very variable soil microfabrics are produced, even within one test specimen, somewhat related to the initial fabric of parent rock and alteration and weathering history. The microfabric produced by decomposition of various feldspars vary from dense, highly packed aggregates to porous networks. Plagioclase feldspars which comprise about 20% of the original rock have completely altered to generally densely packed clay mineral aggregates, but still retain the original grain outline (pseudomorphs) with little or no voids. Alkali feldspars have a porous honeycombed structure with evidence of little chemical decomposition (Figure 4). This may have resulted from leaching out of fine clay minerals and colloids as they are formed during the course of weathering. The porous structure may also

Table V. Results of micropetrographic modal analysis

Block/ Sample No.	Description	Unaltered Feldspars %	Altered Feldspars %	Quartz %	Biotite %	Altered Biotite %	Other %	Microcracks and Voids %	Sound Minerals %	Altered Minerals %	Total Unsound %	I_p	$X_{d(mod)}$
SH8	Decomposed Granitic Soil	10.2	40.5	24.3	0.1	1.9	0.1	22.9	34.7	42.4	65.3	0.53	0.62
SH6	Decomposed Granitic Soil	9.4	41.4	21.5	0.8	3.0	0.0	23.7	31.7	44.4	68.1	0.47	0.73
SH3	Transition Granitic Soil	2.7	55.6	25.2	0.0	3.0	0.0	13.5	27.9	58.6	72.1	0.39	0.92
SH5-2w	Decomposed Granitic Soil	10.2	28.9	32.9	0.0	4.2	0.0	23.6	43.1	33.1	56.7	0.76	0.94
SH5-4a	Altered Granitic Soil	9.4	46.2	18.7	0.2	1.6	0.0	24.0	28.3	47.8	71.8	0.39	-
-1a	Altered Granitic Soil	7.0	46.2	22.7	0.0	4.4	0.0	19.5	29.7	50.6	70.1	0.42	-
-3a	Altered Granitic Soil	5.5	47.2	22.9	0.0	0.9	0.0	23.5	28.4	48.1	71.6	0.40	-
		7.3	46.5	21.4	0.1	2.3	0.0	22.3	28.7	48.8	71.2	0.40	0.97
	"Fresh Granite"	68.2	2.0	28.0	0.7	0.3	0.1	0.6	97.0	2.3	2.9	33.4	-

be the result of extensive etch-pit solution of feldspars (Berner and Holdren 1977) without forming clay minerals. Dearman and Baynes (1979) reported that direct solution etching is a dominant process in the early stages of weathering, with the formation of structurally controlled prismatic etch-pits and prismatic etch-trenches in the chemically weathered granites of S.W. England. Prismatic etch-trenches may extend uniformly through a crystal, producing a very porous regular structure with

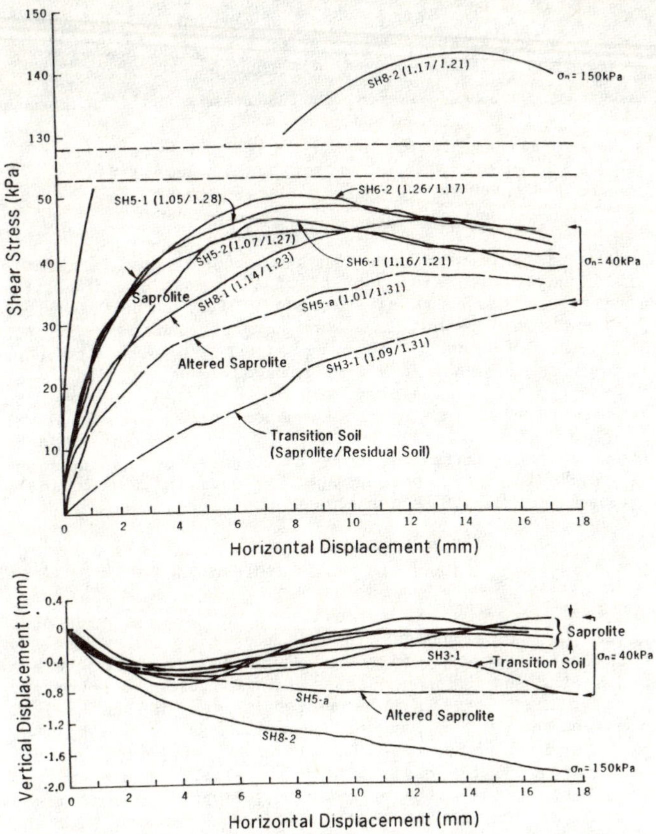

Figure 2. Shear stress-displacement curves of selected 100 x 44 mm size specimens at $_n$ = 40 kPa and 150 kPa (Shear rate : 0.06 mm/min.)

irregularly scattered clay mineral flakes. Where solution has been active, open honeycomb-like structures are produced in advanced stages of weathering, while in more stagnant areas, feldspars are decomposed to clay minerals with little leaching out and void formation. XRD results are not yet available for Shouson Hill samples. It is therefore not known whether gibbsite, a characteristic product of weathering of feldspars in the granites of S.W. England, is present in addition to kaolinite and other clay minerals. However, its presence is suspected due to leaching resulting from the well drained nature of the site and intense seasonal rainfall.

In addition to the highly porous honeycombed fabric of alkali feldspars, the soil, in general, has a very porous structure with numerous open microcracks and macrocracks (over 1 mm in width) through quartz grains. A great majority of grain boundaries, except some quartz-quartz boundaries, are open. Collapse of highly porous unstable feldspar structure may occur in advanced stages of weathering resulting in an aggregate of angular solution remnants with possible association of clay mineral aggregates and forming residual soils. Partially collapsed areas were observed in the transition soil samples SH7 and SH3. The upper horizon at Shouson Hill is in fact a dense residual soil. Biotite showed various types and degrees of alteration even within the same thin section. It is generally highly to completely altered to either mixtures of muscovite, clay minerals and iron oxides, or left as iron oxide skeletons. Most grains show open cleavage planes, and sometimes disturbed, broken up structures.

The effects of hydrothermal alteration are widespread at Shouson Hill, generally decreasing away from the quartz-kaolinite veins. The altered portion of test block SH5, which had the lowest void ratio, also had a less porous feldspar structure with significantly high clay mineral content (Table V). It is probable that the early hydrothermal processes chemically weakened the rock and altered in particular the plagioclase feldspars (sericitization), not just adjacent to the lines of structural weakness but also throughout the rock mass. Growth of sericite flakes were observed in what appeared to be a fresh rock in hand specimen (Table V).

7.2 Void Ratios, Metastability and Micropetrographic Characterization Plot

Laboratory determined void ratios of the test samples have been plotted against $X_{d(mod)}$ values in Figure 3. This chart, termed by Collins (1985b) "microstructure characterization plot", shows that a range of fabrics is possible for any given e value, depending on degree and type of weathering. A range of fabrics is also possible

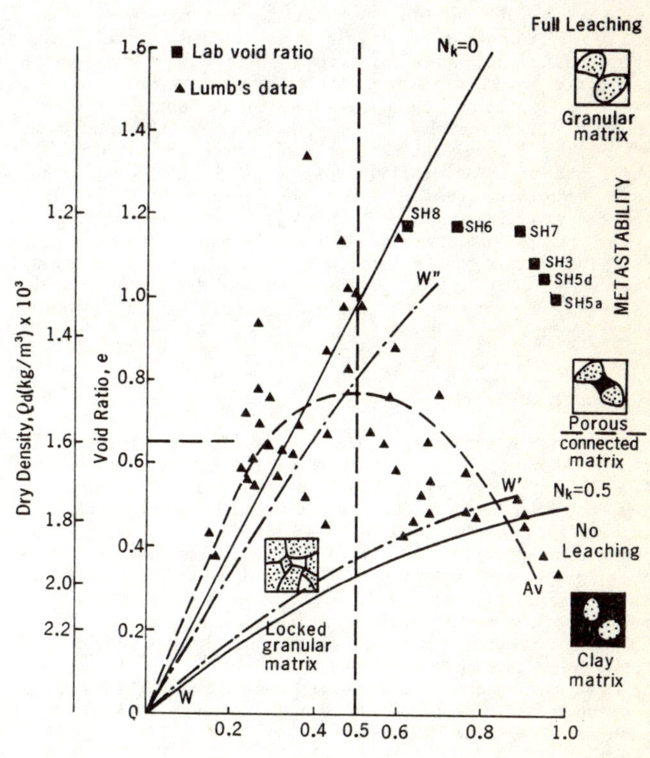

Figure 3. Microstructure characterization plot (after Lumb 1962, Baynes and Dearman 1978a, and Collins 1985b).

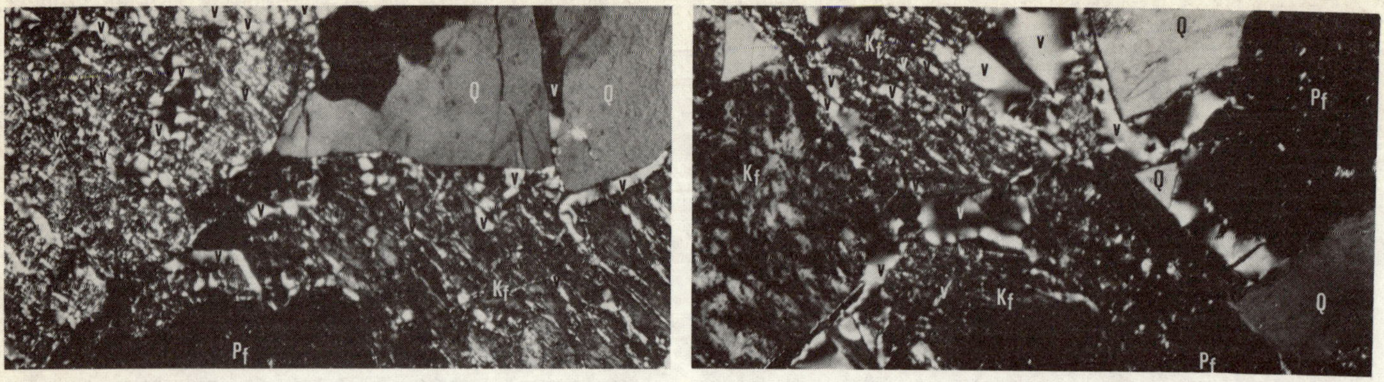

Figure 4. Photomicrographs of granitic saprolite (block no. SH8) illustrating the very porous, partially weathered alkali feldspar (K_f) microstructure (white areas marked V in K_f are voids and microcracks), completely decomposed, non-porous or little porous plagioclase (marked P_f) and microfractured quartz (light grey mineral marked Q). Crossed polarised light.

for any X_d value, depending on degree of leaching. The void ratios are generally higher than those reported by Lumb (1962) for granitic soils in Hong Kong. The X_d versus e plot given in Figure 3 should, however, be viewed in the light of the comments made in this paper on X_d determination. Due to their low densities, all the samples fall in the potentially metastable zone proposed by Baynes and Dearman (1978a) and all plot near the "full leaching" line somewhere between the "granular matrix" and "porous connected matrix" of Collins (1985b). Figure 3 shows that the critical void ratio for metastability is about 1.2 for Shouson Hill saprolite. This value is significantly higher than that indicated by Lumb's hypothetical curve for Hong Kong saprolitic and residual soils. Petrographic examination indicates that beyond this value collapse of porous structure occurs and saprolitic soil gradually transforms into residual soil near the surface. The upper surface at Shouson Hill has, in fact, transformed into residual soil with no evidence of relict fabric. This also agrees well with field observations of Brink and Kantey (1961) that a metastable soil structure develops near the surface as a result of intense leaching under well-drained site conditions.

The representativeness of this graph is questionable for lateritic soils and laterites where X_d is almost equal to 1.0 (i.e. all feldspars are decomposed) and e values can start to increase again in a leaching environment.

7.3 Specific Gravity

The specific gravity of the test samples (2.62 to 2.64 g/cm^3) show a slight change from the specific gravity of fresh granites (2.65 to 2.68 g/cm^3). The specific gravity of kaolinite and other clay minerals are generally much lower than 2.65 g/cm^3. The very small change in specific gravity of the test samples from fresh rock may indicate that small amounts of clay minerals have been formed. Although this was confirmed by petrographic studies, specific gravity is not considered to be a reliable indicator of degree of weathering since the values can be considerably influenced by the size of the sample, the preparation technique, the testing procedure and the presence of some unusual clay minerals such as halloysite (Gidigasu 1975).

7.4 Particle Size

A change in degree of weathering of samples is reflected in the particle size determination results given in Table II. However, the clay content of test samples, including altered specimens, rarely exceeded 10%, even when various pretreatment techniques were used. The petrographic analysis, however, indicated that the clay content (assuming all the altered minerals are of clay size fragments) is appreciably higher than that obtained in sieve analysis. The same phenomenon was also observed by Brenner et al (1978) in Thai granites and by Ruddock (1967). The former authors also observed an apparent decrease in clay content upon oven-drying, with soils having a considerable content of free iron oxide.

7.5 Behaviour of Granitic Soil in Direct Shear

As noted previously, saprolitic granitic soils have a relict primary bonding (interlocking of primary grains) related to original rock fabric and history and degree of weathering and other processes. In addition, a secondary bonding may develop at advanced stages of weathering, particularly in lateritic soils, as a result of continuous recrystallization of minerals during weathering and by precipitation of salts from pore water. It is this bonding (both relict and secondary) that gives a component of strength and stiffness to saprolitic and residual soils, particularly at initial stages of shearing, in addition to that due to the properties (including strength and porosity) of the particles composing the soil and their packing. As can be observed in detail by petrographic methods, the particles present in a saprolitic soil are individual grains or agglomerations of primary minerals of the original rock in various stages of decomposition and weakening, together with individual grains or agglomerations of secondary minerals (e.g. clay minerals) formed by weathering and/or other alteration processes. In the case of granites, quartz is the major soil constituent, significantly influencing the strength of the soil, as it survives as strong, relatively unweathered grains or agglomerations even in the overlying residual soil. In less weathered samples, for example highly decomposed granite (GCO 1984) or weakly cemented granite (Irfan and Dearman 1978a), the relict bonding is generally strong enough to give an appreciable cohesion value at low applied stresses.

Some loss of relict bonding is inevitable, even with high quality sampling, particularly in the weakly bonded saprolitic samples. Generally, expansion takes place which may cause loss of bond strength and hence subsequent loss of shear strength, yield strength and stiffness, and an underestimation of these properties in situ (Sandroni 1985, Vaughan et al 1988). Further destruction of the bonding and fabric disturbance is likely to occur during preparation particularly of small-sized samples, for testing in the laboratory. It is therefore desirable to use large specimens to be able to measure the component of strength arising from bonding.

A range of sample sizes was used in the shear strength testing of Shouson Hill saprolitic soil. The test results on fairly large 100 mm x 44 mm samples (Table IV) have been selected to illustrate the effects of soil microfabric and bonding on the shear strength behaviour of various soil types from this locality. As seen in Figure 3 and Table IV, neither the initial void ratios or dry densities, nor the grain size distributions can alone account for the variation in shear strength properties determined at a normal stress of 40 kPa (approximately equivalent to overburden pressure). Much lower peak shear strengths are obtained for the transition soil sample and altered samples.

The differences in strength values can, however, be explained when the microfabrics of these soil types, as discussed previously, are considered. For example, the limited results presented in Table V indicate a good correlation between various quantitative fabric elements such as I_p or percentage of altered mineral and shear strengths. Clearly defined peaks were generally obtained at low horizontal displacements for samples which showed no or very little fabric loss, but not for transition soil or hydrothermally altered soil samples. The compressive and dilatant behaviour of these soils during shear (Figure 3), can also be explained by again considering the microfabric of samples. Generally compressive (volume reduction) behaviour shown by the least weathered specimens during the initial phase of shearing under lower normal stress is attributed to gradual compression of the soil fabric with breaking up of relict primary and secondary bonding, and shearing and subsequent gradual collapse of porous honeycombed feldspars as the shearing progresses. This is followed by a "dilatant" phase dominated by mainly sand-size quartz grains and to a small extent by unaltered but broken up feldspar fragments, overriding each other and possibly rotating in the silty clayey compacted fabric along the already-formed shear plane. As the clay mineral content increases (i.e. for samples with clay microfabrics) and the void ratio decreases, this dilatant behaviour is expected to be weaker as was the case with transition soil sample SH3 and the altered sample SH5(a). At very large displacements, alignment of some clay minerals and smearing along the already formed shear surface may occur.

The contribution of weak bonding to the strength in the early stages of shearing cannot be assessed with any certainty because of the nature of the direct shear testing technique. However, the component of strength arising from relict primary bonding is considered to be small in the Shouson Hill soil. Wetting under load for 24 hours before testing can also soften the clay bridges (i.e. decomposed feldspars) between quartz grains and hence reduce the effect of bonding.

At high applied stresses, some collapse of the fragile fabric could have occurred throughout the sample during the soaking and consolidation stages, particularly for samples with high void ratios and high clay contents. The non-dilatant behaviour observed in Figure 3 indicates that a gradual but continuous volume reduction is taking place with further breakdown of fabric elements.

8 SUMMARY AND CONCLUSIONS

Saprolitic and residual soils are complex and heterogeneous materials and their characterization and testing present problems in geotechnical engineering. Field and laboratory studies undertaken in relation to a direct shear testing programme showed that significant variations in engineering and other soil index properties can occur in granitic saprolitic soils, not only between block samples taken within a few metres of each other, but also within the same specimen. This is mainly due to variations in microfabric including bonding of these soils related to both the initial fabric of the parent rock and to the history and microenvironment of weathering and other alteration processes.

Based on degree of decomposition of mineral constituents, disintegration, type of alteration and presence or absence of original granitic rock fabric, six broad "soil" types were recognised in the Shouson Hill saprolite with rapid variations between the types, in what is generally described as completely decomposed granite or grade V granite in Hong Kong.

Petrographic examination of soil sections revealed that the decomposed saprolitic soil is generally composed of partially weathered (mainly by etch-pit solution type weathering), very porous alkali feldspars with honeycombed microstructure, plagioclase feldspars completely altered to dense clay mineral aggregates and microfractured quartz and quartz agglomerations, some with tight grain boundaries, in a generally porous matrix. The degree of decomposition is more intense in samples affected by hydrothermal alteration, whereas in the transition soil samples a less porous microfabric with structureless, partially collapsed fabric areas are present.

Amongst the various chemical, mechanical and petrographical indices suggested to characterize weathering state of granitic soil, the micropetrographic index, I_p, of Irfan and Dearman (1978a) was found particularly useful in quantifying the degree of weathering and certain components of weathered soil microfabric. The technique also gives insight into changing mineralogical and microcrack regimes of weathered and altered rocks. In comparison the degree of decomposition index, X_d, which has been used as a basis for microfabric characterization was found to be affected by degree and type of pretreatment and iron-oxide cementation. In addition, structural components of microfabric such as cracks and voids, and decomposition state of other mineral constituents which can influence the strength behaviour, are not accounted for in X_d determination. I_p and other quantitative petrographical methods can be used to relate the microfabric of saprolitic soils, as well as the less weathered rock, to index properties and hence assist in understanding their behaviour. More work needs to be done on this aspect however, for the soil range of the weathering spectrum.

The relationships which have been developed for transported soils relating the results of soil classification tests and engineering properties do not seem to hold for saprolitic and residual soils. This partly stems from the difficulty of determining a meaningful grain size for these soils. The clay content of the test samples were determined to be less than 10% using routine laboratory procedures, whereas microscopic examination showed that the "clay" content is appreciably higher than that obtained by sieve analysis. Some of the clay size particles are locked-in in partially altered primary minerals and hence cannot be dispersed unless mechanically vigorously broken down. Iron oxide cementation of decomposition products may also affect the results.

The variations in shear strength cannot be entirely explained by variations in natural void ratio, density or grain size, which are themselves a function of weathering and alteration, without considering soil microfabric and bonding. The strength of saprolitic soil, particularly in initial stages of shearing at low applied stresses is a function of a weak relict and secondary bonding and the properties (including strength and porosity) of both the primary and secondary particles composing the soil and their packing. Quartz being the major soil constituent surviving as strong, relatively unweathered grains, influences the strenght of granitic soil significantly. The reported friction values determined by both triaxial and direct shear tests, of saturated samples, are generally similar to those of very sandy materials, i.e. in the range of 30° to 40°.

The component of strength arising from relict bonding is considered to be small in the Shouson Hill saprolite. Some loss of bonding is considered inevitable even with high quality sampling and during sample preparation for testing, particularly in weakly bonded soil samples, resulting in underestimation of strength properties insitu. It is therefore desirable to use large specimens to be able to measure the component of strength arising from bonding. Vaughan (1985) considered that bedding errors may also affect the laboratory determined values.

Significantly lower peak strengths, by as much as 60%, were obtained on samples showing the effects of hydrothermal alteration and also in transition soil samples which were broadly described as completely decomposed granite. Hydrothermal alteration may itself locally or areally intensify the effects of subsequent weathering. Many areas in commonly occurring granitic rocks in Hong Kong and elsewhere show both effects. An understanding of the relationship between the structural control exerted by the granite and the distribution of chemical and physical weathering effects and alteration would undoubtedly assist in the planning of site investigation and laboratory testing programme and interpretation of the results.

ACKNOWLEDGEMENTS

This paper is published with the permission of the Director of Civil Engineering Services of the Hong Kong Government. Direct shear testing and preliminary analysis of results was carried out by C.K. Cheung. The author acknowledges valuable comments made by J.B. Massey, A. Cipullo and R.P. Martin of the Geotechnical Control Office in the preparation of this paper.

REFERENCES

Baynes, F.J. & Dearman, W.R. (1978a). The relationship between the microfabric and the engineering properties of weathered granite. Bulletin of the International Association of Engineering Geology, No. 18, p. 191-197.

Baynes, F.J. & Dearman, W.R. (1978b). The microfabric of a chemically weathered granite. Bulletin of the International Association of Engineering Geology, No. 18, p. 91-100.

Baynes, F.J. & Dearman, W.R. (1978c). Scanning electron microscope studies of weathered rock : a review of nomenclature and methods. Bulletin of the International Association of Engineering Geology, No. 18, p. 199-204.

Berner, R.A. & Holdren, G.R. Jr. (1977). Mechanism of feldspar weathering: some observational evidence. Geology, vol. 5, p. 369-372.

Brand, E.W. (1985). Predicting the performance of residual soil slopes (Theme Lecture). Proceedings of the Eleventh International Conference on Soil Mechanics and Foundation Engineering, San Fransisco, vol. 5, in press.

Brand, E.W. & Phillipson, H.B. (1985). Review of international practice for the sampling and testing of residual soils. Sampling and Testing of Residual Soils : A Review of International Practice (ed. E.W. Brand and H.B. Phillipson), Scorpion Press, Hong KOng, p. 7-21.

Brenner, R.P., Nutalaya, P. & Bergado, D.T. (1978). Weathering effects on some engineering properties of a granite residual soil in Northern Thailand. Proceedings of the Third International Congress, International Association of Engineering Geology, Madrid, Section II, p. 23-36.

Brink, A.B.A. & Kantey, B.A. (1961). Collapsible grain structure in residual soils in Southern Africa. Proceedings of the Fifth International Conference on Soil Mechanics and Foundation Engineering, Paris, vol. I, p. 611-614.

British Standards Institution (1975). Methods of Test for Soils for Civil Engineering Purposes (BS 1377 : 1975). British Standards Institution, London, 143 p.

Cheung, C.K., Greenway, D.R. & Massey, J.B. (1988). Direct shear testing of a completely decomposed granite. Proceedings of the Second International Conference on Geomechanics in Tropical Soils, Singapore.

Collins, K. (1985a). Towards characterization of tropical soil microstructure. Proceedings of the First International Conference on Geomechanics in Tropical Lateritic and Saprolitic Soils, Brasilia, vol. 1, p. 85-96.

Collins, K. (1985b). Discussion. Proceedings of the First International Conference on Geomechanics in Tropical Lateritic and Saprolitic Soils, Brasilia, vol. 3, p. 207-209.

Collins, K. & McGown, A. (1974). The form and function of microfabric features in a variety of natural soils. Geotechnique, vol. 24, p. 223-254.

Dearman, W.R. & Baynes, F.J. (1979). Etch-pit weathering of feldspars. Proceedings of the Ussher Society, vol. 4, p. 390-401.

Dearman, W.R. & Irfan, T.Y. (1978). Assessment of the degree of weathering in granite using petrographic and physical index tests. Proceedings of the International Symposium on Deterioration and Protection of Stone Monuments, Paris, paper 2.3, 35 p.

Dearman, W.R. & Turk, N. (1985). Sampling and testing of residual soils in the United Kingdom. In Sampling and Testing of Residual Soils : A Review of International Practice (ed. E.W. Brand and H.B. Phillipson), Scorpion Press, Hong Kong, p. 175-181.

Dearman, W.R., Turk, N., Irfan, T.Y. & Rowshanei, H. (1987). Detection of rock material variation by sonic velocity zoning. Bulletin of the International Association of Engineering Geology, No. 35, p. 3-8.

Dixon, H.W. (1969). Decomposition products of rock substances. Proposed engineering geological classification. Rock Mechanics Symposium, Stephen Roberts Theatre, University of Sydney, p. 39-44.

Geotechnical Control Office (1984). Geotechnical Manual for Slopes (second edition). Geotechnical Control Office, Hong Kong, 295 p.

Gidigasu, M.D. (1975). Laterite Soil Engineering. Elsevier, Amsterdam, 554 p.

Hamrol, A. (1961). A quantitative classification of the weathering and weatherability of rocks. Proceedings of the Fifth International Conference on Soil Mechanics and Foundation Engineering, Paris, vol. 2, p. 771-774.

Hencher, S.R. & Martin, R.P. (1982). The description and classification of weathered rocks in Hong Kong for engineering purposes. Proceedings of the Seventh South-east Asian Geotechnical Conference, Hong Kong, vol. 1, p. 125-142.

IAEG (1981). Rock and soil description and classification for engineering geological mapping. Bulletin of the International Association of Engineering Geology, No. 24, p. 235-274.

Irfan, T.Y. & Dearman, W.R. (1978a). Engineering classification and index properties of a weathered granite. Bulletin of the International Association of Engineering Geology, No. 17, p. 79-90.

Irfan, T.Y. & Dearman, W.R. (1978b). The engineering petrography of a weathered granite in Cornwall, England. Quarterly Journal of Engineering Geology, vol. 11, p. 233-244.

Irfan, T.Y. & Dearman, W.R. (1979). Micropetrographic and engineering characterization of a weathered granite. Annals de la Societe Geologique de Belgique, vol. 101, p. 71-77.
Irfan, T.Y. & Woods, N.W. (1988). The influence of relict discontinuities on slope stability in saprolitic soils. Proceedings of the Second International Conference on Geomechanics in Tropical Soils, Singapore.
Janbu, N. (1973). Slope stability computations. In Embankment Dam Engineering (Casagrande Volume), (ed. R.C. Hirschfield and S.J. Poulos), Wiley, New York, p. 47-107.
Lumb, P. (1962). The properties of decomposed granite. Geotechnique, vol. 12, p. 226-243.
Lumb, P. (1965). The residual soils of Hong Kong. Geotechnique, vol. 15, p. 180-194.
Martin, R.P. (1986). Use of index tests for engineering assessment of weathered rocks. Proceedings of the Fifth International Congress, International Association of Engineering Geology, Buenos Aires, p. 433-450.
Matsuo, S. & Nishida, K. (1968). Physical and chemical properties of decomposed granite soils. Soils and Foundations, vol. 10, p. 1-9.
Mello-Mendes, F., Aires-Barros, L. & Rodrigues, F.P. (1966). The use of modal analysis in the mechanical characterization of rock masses. Proceedings of the First International Congress, International Society of Rock Mechanics, Lisbon, vol. 1, p. 217-223.
Mitchell, J.K. & Sitar, N. (1982). Engineering properties of tropical residual soils. Proceedings of the ASCE Speciality Conference on Engineering and Construction in Tropical and Residual Soils, Honolulu, p. 30-57.
Morgenstern, N.R. & Price, V.E. (1965). The analysis of the stability of general slip surfaces. Geotechnique, vol. 15, p. 79-83.
Moye, D.G. (1955). Engineering geology for the Snowy Mountains Scheme. Journal of the Institution of Engineers Australia, vol. 27, p. 281-299.
Nishida, K. & Anyama, C. (1985). Physical properties and shear strength of decomposed granite soil. Proceedings of the First International Conference on Geomechanics in Tropical Lateritic and Saprolitic Soils, Brasilia, vol. 1, p. 371-382.
Onodera, T.F., Yashinaka, R. & Oda, M. (1974). Weathering and its relation to mechanical properties of granite. Proceedings of the Third International Congress, International Society of Rock Mechanics, Denver, vol. 2, p. 71-78.
Reiche, P. (1943). Graphic representation of chemical weathering. Journal of Sedimentary Petrology, vol. 13, p. 58-68.
Ruddock, E.C. (1967). Residual soils of the Kumasi district. Geotechnique, vol. 17, p. 359-377.
Sandroni, S.S. (1985). Sampling and testing of residual soils in Brasil. In Sampling and Testing of Residual Soils : A Review of International Practice (ed. E.W. Brand and H.B. Phillipson), Scorpion Press, Hong Kong, p. 31-50.
Smart, P. & Tovey, N.K. (1982). Electron Microscopy of Soils and Sediments - Techniques. Oxford University Press, 264 p.
Sowers, G.F. (1985). Residual soils in the United States. In Sampling and Testing of Residual Soils : A Review of International Practice (ed. E.W. Brand and H.B. Phillipson), Scorpion Press, Hong Kong, p. 183-191.
Sueka, T., Lee, I.K., Muramatsu, M. & Imamura, S. (1985). Geomechanical properties and engineering classification for decomposed granite soils in Kaduna district, Nigeria. Proceedings of the First International Conference on Geomechanics in Tropical Lateritic and Saprolitic Soils, Brasilia, vol. 1, p. 175-186.
Vaughan, P.R. (1985). Mechanical and hydraulic properties of in-situ residual soils. General Report, Session 2, Proceedings of the First International Conference on Geomechanics in Tropical Lateritic and Saprolitic Soils, Brasilia, vol. 3, p. 231-263.
Vaughan, P.R., Maccarini, M. & Mokhtar, S.M. (1988). Indexing the engineering properties of residual soil. Quarterly Journal of Engineering Geology, vol. 21, p. 69-84.
Weinert, H.H. (1964). Basic igneous rocks in road foundations. C.S.I.R. Inf 218, National Institution for Road Research Bulletin, Pretoria, vol. 5, p. 1-47.

Physical characterization of weathering profile of clastic metasediments in Peninsular Malaysia

I.KOMOO, University Kebangsaan, Malaysia
S.N.MOGANA, University Kebangsaan, Malaysia

ABSTRACT: Characterization and classification of weathering profiles of clastic metasediments are generally more difficult compared to homogenous igneous rocks. This is due to several factors, such as stratification of various lithologies, different degree of lithologic resistance towards weathering, and complexicity of its geologic structures. However, field observations indicate that to a certain extent, there exist some uniformities in the weathering characteristics can be effectively described and classified.

The pattern and thickness of weathering profile has close relationship with geomorphic conditions. Five morphological classifications are recognised to illustrate this relationship. The weathering profile in metasediments seldom form corestones, therefore weathering classification based on rock:soil ratios is not practical to be used. This paper propose some modification on material and mass descriptions for weathering classification. For accurate field classification, physical characterization of weathearing profile has been studied and discussed with some examples. The scheme presented is a first attempt on systematic characterazition and require finer improvement.

1 INTRODUCTION

Weathering profile classifications are derived from the classification for engineering purpose used by Moye (1955) on the Snowy Mountains Scheme, Australia, and the geological classification developed by Ruxton and Berry (1957) for the Hong Kong granite. A decade later, Little (1967, 1969) made some modification and introduced a standard format in weathering classification. The weathering profile is graded from grade I (fresh rock) to grade VI (residual soil). The grading is based mainly on assessment of degree of discolouration of rock material, the rock : soil ratios, and the survival or destruction of the original texture of the granite in soil produced by weathering. This form the general basis for a recent weathering classification of rock mass for engineering purposes. Little's classification was refined by Dearman (1974, 1976), and finally with minor modification adopted by IAEG (1981) and ISRM (1981) as their proposed standard weathering classification of rock mass.

This proposed standard weathering classification is practical and suitable for most igneous rocks and perhap massive homogenous relatively hard rocks. However, attempt to use it for sedimentary and metasedimentary rocks indicate that the classification require further modification. The main reason being that sediments and metasediments weathered continously to form residual soils. In most cases it is difficult to accurately identify boundary between rock and soil. Therefore, the concept of rock : soil ratios cannot be used in weathering classification. A typical example of weathering profile on sedimentary or metasedimentary rocks has been discussed by Ibrahim Komoo (1985).

This paper will highlight the proposed modification of weathering classification for clastic metasediments, and observations in Peninsular Malaysia.

2 GENERAL GEOLOGY

The rock formations in Peninsular Malaysia can broadly be classified into five main groups, namely unconsolidated surficial deposits, Triassic or younger sedimentary rock, metamorphic or metasedimentary rock, marble and limestones, and igneous intrusion (Figure 1).

Metamorphic rock or most commonly called clastic metasedimentary rock consist of phyllite, schist and quartzite. In some places phyllite is interbedded with slightly metamorphosed shale and sandstone.

3. WEATHERING AND GEOMORPHIC CONDITIONS

Due to the tropical climate, most metasediments in Peninsular Malaysia experience extensive chemical weathering. The thickness of weathered prifile (slightly weathered rock to residual soil) varies from few metres up to more than 100m. The variations in thickness are mainly due to the type of lithology, mass structure and morphological conditions. In many places, the lithology and structure of rock mass are generally reflected in its geomorphic conditions. Therefore, the pattern and thickness of weathering profile has a close relationship with the morphological classification.

To illustrate the close relationship for clastic metasediments, an east-wast cross-section at 50 km south of Kuala Lumpur is shown in Figure 2. From this cross-section a complete morphological or terrain conditions can be described. They are as follows:

Type A -- low flat landform, in most areas covered by thin layer of surficial deposits

Type B -- low undulating landform, generally consist of phyllite, slate and metamorphosed shale

Type C -- low rounded hill, generally consist of interbedded phyllite and quartzite

Type D -- ridge-like hill, predominantly quartzite

Type E -- isolated hill, predominantly schist

Table 1 Classification of rock mass weathering profile (after ISRM, 1981).

Weathering Classification		Description
Term	Zone	
Residual Soil	VI	All rock material is converted to soil. The mass structure and material fabric are destroyed. The material has not been significantly transported.
Completely Weathered	V	All rock material is decomposed and/or disintegrated to soil. The original mass structure is still largely intact.
Highly Weathered	IV	More than half of the rock material is decomposed and/or disintegrated to soil. Fresh or discoloured rock is present either as a discontinuous framework or corestones.
Moderately Weathered	III	Less than half of the rock material is decomposed and/or disintegrated to soil. Fresh or discoloured rock is present either as a discontinuous framework or corestones.
Slightly Weathered	II	Discolouration indicates weathering of rock material and discontinuity surfaces. All the material may be discoloured by weathering and may be somewhat weaker than in its fresh condition.
Fresh Rock	I	No visible sign of rock material weathering. Some discolouration on major discontinuity surfaces.

As shown in Figure 2, morphological type A and B is characterised by thick layers of completely weathered and residual soils (more than 10m), and followed by considerably thick layers of highly to moderately weathered rocks (generally more than 40m). Fresh and slightly weathered rocks are seldom exposed. For morphological type C and D, the weathering profile is characterised by relatively thin layers of completely weathered and residual soils (about 5m), and followed by considerably thick layers of weathered rock materials. Slightly weathered rock may be present at depth of more than 15m, however the fresh rock is seldom exposed. For morphological type E, and in the same place type D (with steeper slope), the profile is characterised by very thin layers of completely weathered and residual soils (generally less than 2m), and followed by relatively thin layers of highly to moderately weathered rocks (less than 5m). Slightly weathered rock can be of several metres thick, and fresh rock is often exposed at the depth of less than 10m. Ibrahim Komoo (1986) had described the relationship between metasediments in Kuala Lumpur area as being in the form of highly, moderately and slightly weathered zones.

4 WEATHERING CLASSIFICATION

A completely weathering profile will indicate the gradual changes from fresh rock to residual soil. Much has been discussed about weathering profile, however one commonly accepted classification of rock mass weathering profile is as suggested by ISRM Commission on Classification of Rock and Rock Masses (1981), and is shown in Table 1.

Table 2 Proposed classification of metasedimentary rock mass weathering profile in Peninsular Malaysia.

Weathering Classification		Description
Term	Zone	
Residual Soil	VI	All rock material is concerted to soil. The mass structure and material fabric (texture) are destroyed. The material shows homogenous colour and has not been significantly transported.
Completely Weathered	V	All rock material is decomposed to soil. The mass structure and material texture are partially preserved. The material is friable if soak in water or squeeze by hand. The material commonly contain iron-rich concretions, and non-homogenous or mottled in colour.
Highly Weathered	IV	The rock material is in the transitional stage to form soil. More than 50% mass structure and total material texture are preserved. The material formed smaller fragments or blocks if soak in water or squeeze by hand.
Moderately Weathered	III	Discolouration indicates some of rock material is affected by weathering. The mass structure and material texture are preserved. Material fragment or block corners can be chipped by hand. Discontinuity commonly filled by iron-rich material.
Slightly Weathered	II	Discolouration along discontinuity and may be part of rock material. The mass structure and material texture are preserved. Material is generally weaker but fragment or block corners cannot be chipped by hand.
Fresh Rock	I	No visible sign of rock material weathering. Some discolouration on major discontinuity surfaces.

Weathering profile on metasediments is often far more complicated to be classified. This is, amongst other factors, due to the inhomogeneity (layering), and lithologic and textural (grain sizes) variations. However, field observation on several completely weathering profiles, particularly at massive and thickly bedded formations, to a certain extent indicate that there exist uniformities in the characteristics of the profiles. Table 2 shows the proposed classification of metasedimentary rock mass weathering profile based on field experience in Peninsular Malaysia.

For zones I, II (engineering rock) and V, IV (engineering soil), the proposed classification does not differ very much from the standard classification by ISRM (1981), except for some additional information. However, for zones III and IV the basic difference is that the proposed classification did not use the concept of rock : soil ratios. It was practically impossible to use rock : soil ratios mainly because the weathered materials does not form corestones and its strength varies considerably. Therefore, it was easier to classify zones III and IV based on material texture and mass structure.

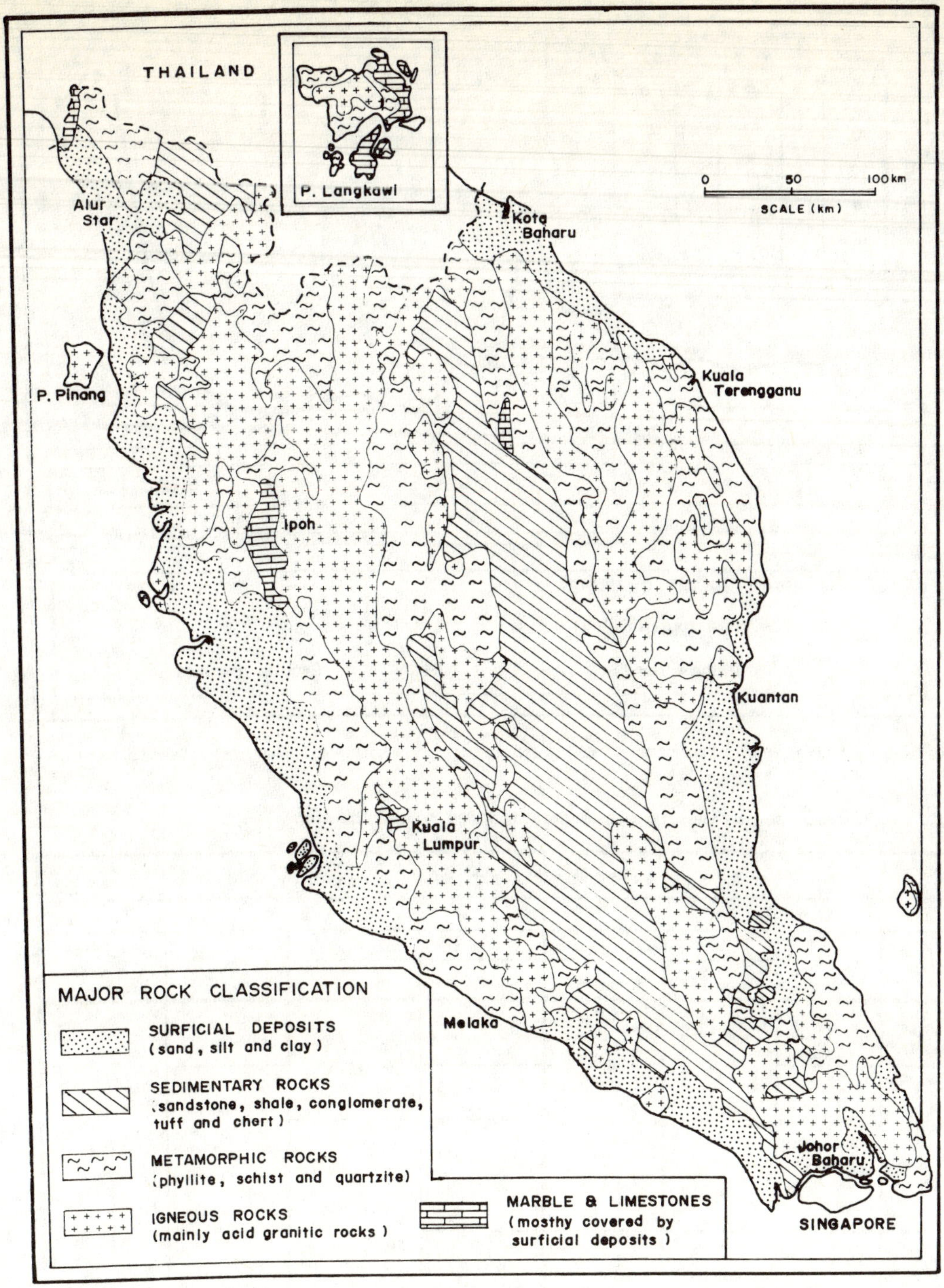

Figure 1 General geology of Peninsular Malaysia.

5 PHYSICAL CHARACTERIZATION

For accurate weathering classification in the field one should be able to recognise several physical characters or diagnostic features of weathering profile. This approach has already been adopted and discussed by Anon (1972), but developed from granitic terrains. The characterization should be based on two main properties, namely rock material (mineralogy and its textual features), and rock mass (structure and strength).

To establish the weathering classification on metasediments, the physical characterization scheme used is as shown in Table 3. This scheme is a first attempt on systematic characterization and has been tested on several profiles. Further work is still needed to perfect the scheme. Basically, the rock material is divided into 3 main properties:

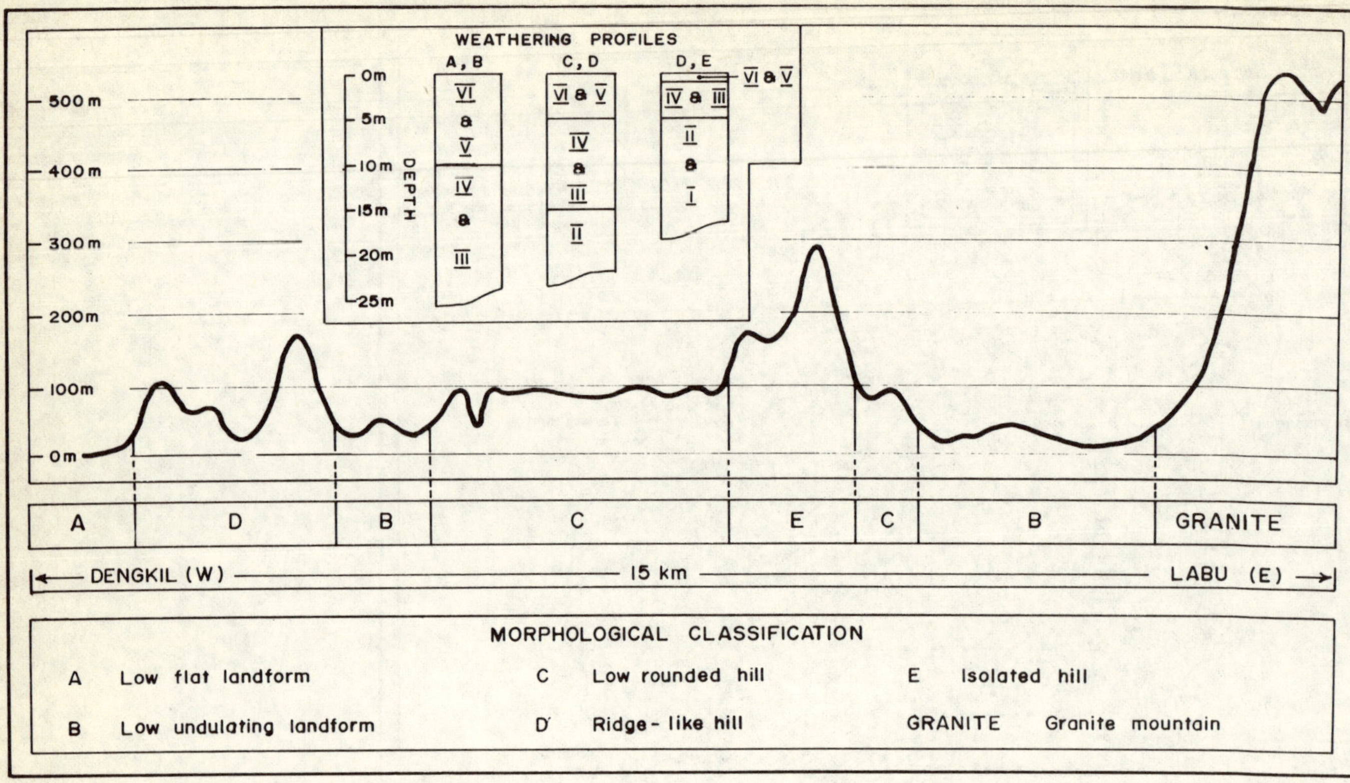

Figure 2 Cross-section on metasediments terrain to illustrate the relationship between morphological classification and weathering profile.

Table 3 Physical characterization scheme of the weathering classification for metasediments.

WEATHERING			PHYSICAL CHARACTERIZATION							
			MATERIAL				MASS			
TERM	ZONE		Discolouration	Texture	Friability in water	Friability by hand	Structure % preserved	Structure changes	Iron-rich layer	Strength (Schmidt hammer)
Residual Soil	VI		total changes (homogenous)	destroyed	friable	friable	100% destroyed	total changes	absent	nil
Completely Weathered	V	c	total changes (non-homogenous or mottled)	partially preserved			< 25% preserved		commonly present	
		b					25 - 50% preserved	iron-rich filling along discontinuity plane		
		a							may be present	
Highly Weathered	IV	b	completely discolouration	preserved	form smaller fragments or blocks	form smaller fragments or blocks	50 - 75% preserved		absent	
		a					> 75% preserved			less than 25
Moderately Weathered	III		some discolouration		remain as single block	fragment corners can be chipped	100% preserved			
Slightly Weathered	II		no changes			fragment corners cannot be chipped		discolouration along discontinuity		more than 25
Fresh Rock	I							no changes		

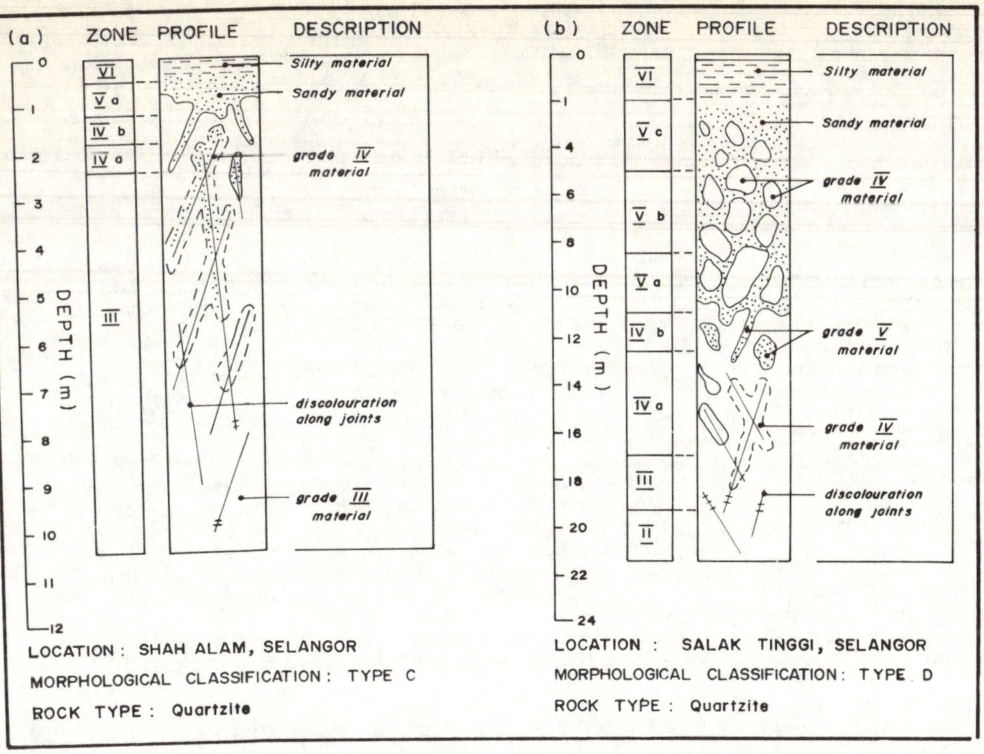

Figure 3 Examples of weathering profiles on massive quartizite rocks. a) morphological type C at Shah Alam, Selangor, b) morphological type D at Salak Tinggi, Selangor.

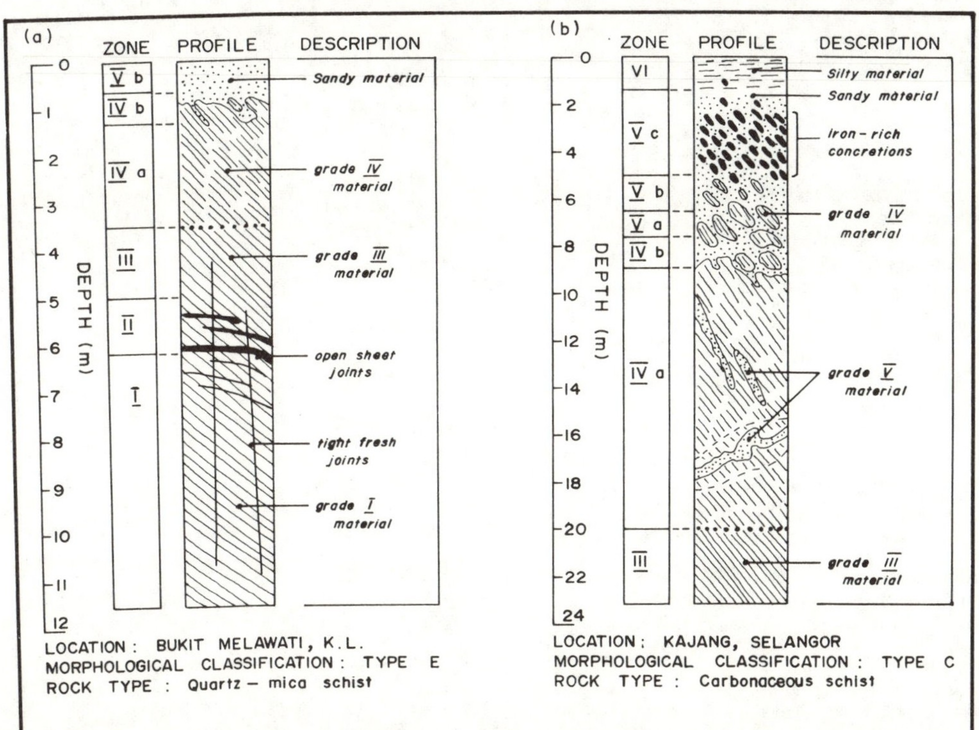

Figure 4 Examples of weathering profiles on massive schist formations. a) morphological type E at Bukit Melawati, Kuala Lumpur, b) morphological type C at Kajang, Selangor.

discolouration, texture and friability; and for rock mass the criteria are based on the degree of preservation and type of changes of the structure, and its estimated strength. Since most fine-grained metasediments in Peninsular Malaysia consists of layers of iron-rich concretions, this structure can also be used as one of the criterion.

6 SOME EXAMPLES

The proposed weathering classification and physical characterization scheme have been used to establish some weathering profiles on cut-slopes at several places. Four of the profiles will be discussed here.

Figures 3a and b show examples of weathering profiles of massive quartzite on morphological types C and D, respectively. Please note that zones VI and V are easily distinguished because they show textural difference (silty and sandy) in the material. However, zones IV, III and II can only be distinguished by a combination of several diagnostic features. Amongst important criteria to differentiate these zones are discolouration, friability, changes along discontinuity and mass strength.

Figures 4a and b show examples of weathering profiles of massive schist on morphological types E and C, respectively. Figure 4a shows a typical schist weathering profile on isolated hilly morphology. Zones VI and V normally are very thin, in many cases less than a meter, then followed by relatively thin layer of zone IV (less than 3m). A thin layer of zone III can only be recognised by its material friability and mass strength. Zones II and I show no textural and structural changes, except at zone II the sheet joints are opened and slightly discoloured. Figure 4b represents schist weathering profile on low rounded hilly terrain. The thickness of weathered materials is generally more than 40m. A layer or iron-rich concretions develop at the top part of zone V. Zones IV and III are usually of great thickness, and these zones can only be differentiated by its friability, discolouration, and up to a certain extent, its mass strength.

7 CONCLUSION

Geomorphic conditions have close relationship with the pattern and thickness of weathering profile on metasediments, and therefore can be used to estimate the depth of weathered materials. For practical weathering classification, it is important to recognise the physical characters of rock material (degree of discolouration, texture and friability) and rock mass (structure, strength and occurence of iron-rich layer) by using the proposed physical characterization scheme.

REFERENCES

Anon. 1972. The preparation of maps and plans in terms of engineering geology. Q.Jl. Engng Geol. 5: 295-382.

Dearman, W.R. 1974. Weathering classification in the characterisation of rock for engineering purpose in British Practice. Bull. Int. Assoc. Engng. Geol. 9: 33-42.

Dearman, W.R. 1974. Weathering classification in the characterisation of rock: A revision. Bull. Int. Asspc. Engng. Geol. 13: 123-127.

IAEG. 1981. Rock and soil description and classification for engineering geological mapping. IAEG Commission on Engineering Geological Mapping Report. Bull. Int. Assoc. Engng. Geol. 24: 235-274

ISRM. 1981. Basic geotechnical description fo rock massas. ISRM Commission on Classification of Rocks and Rock Masses. Int. J. Rock Mech. Min. Sci & Geomech Abstr. 18: 85-110.

Ibrahim Komoo 1985. Engineering properties of weathered rock profiles in Peninsular Malaysia. Proc. Eight Southeast Asian Geotech. Conf. Kuala Lumpur. 1: 3-81 - 3-86.

Ibrahim Komoo 1986. Engineering geological aspects of clastic metasediments in Kuala Lumpur area, Malaysia. Bull. Geol. Soc. Malaysia. 19: 597-612.

Little, A.L. 1967. Laterites. proc. Asian Reg. Conf. Soil Mech. Found. Engng. Haifa. 2: 61-71.

Little, A.L. 1969. The engineering classification of residual tropical soils. Proc. Int. Conf. Soil Mech. & Found. Engng. Mexico. 1: 1-10.

Moye, D.G. 1955. Engineering geology of the Snowy Mountains Scheme. JI. Inst. Eng. Australia. 27: 281-299.

Ruxton, B.P. amd Nerry, L. 1957. Weathering of granite and associated erosional features in Hong Kong. bull. Geol. Soc.

The role of ironoxide in tropical soil properties

S.M.RAO, Indian Institute of Science, India
A.SRIDHARAN, Indian Institute of Science, India
S.CHANDRAKARAN, Indian Institute of Science, India

ABSTRACT : The paper investigates the role of iron oxide in the plasticity and compressibility characteristics of a kaolinite rich red soil and a montmorillonite dominant black soil. Results indicated that the iron oxide present in the kaolinitic soil binds individual soil particles into coarser aggregates and contributes to the development of a random soil structure. The open structural units of the flocculated fabric facilitates greater water entrapment at liquid limit. The random structure also maintains the soil at a higher equilibrium void ratio and undergoes greater particle re-adjustment under a given load leading to enhanced compressibility.

With the montmorillonite dominant black clay, the aggregates formed by iron oxide bonding reduces the fines content and the effective surface area for development of diffuse double layer. As a consequence there results a suppression in the plasticity and compressibility characteristics of the soil.

1 INTRODUCTION

Tropical soils are characterized by the presence of iron and aluminium oxides and hydroxides which accumulate as a consequence of a chain of chemical and leaching processes. The presence of the sesquioxides, especially iron oxide is considered to play a key role in cementation and aggregate formation in tropical soils and exert an important influence on their engineering behaviour (Townsend et al. 1969, Townsend et al. 1971, Townsend 1985, Vargas 1973, Foss 1973, Wesley 1974). Newill (1961), Rao and Raymahashay (1981) and Townsend (1987) by chemically extracting iron oxide from lateritic soils demonstrated the sesquioxide's association with aggregate formation. Interestingly the workers (Rao and Raymahashay 1981, Townsend 1987) observed a decrease in liquid limit of the soils, despite an increase in fines content on iron oxide removal. The plausible reasons for such variations in liquid limit of the lateritic soils on sesquioxide removal are not understood. In addition, corresponding studies highlighting the influence of accumulated sesquioxides on the other major soil properties of compressibility, permeability and strength are unavailable.

The present paper attempts to understand the role of iron oxide and the mechanisms by which the sesquioxide influences the index and compressibility characteristics of a kaolinite rich red soil and a montmorillonite dominant black soil based on results obtained with remolded saturated specimens tested in the untreated and iron oxide removed (decemented) conditions.

2 EXPERIMENTAL PROGRAMME

2.1 Description of Soils Used

The soils used in the study are from Karnataka State, India. Sufficient amounts of disturbed soils in the desiccated state were collected from Bangalore district (kaolinite rich red soil) and from Dharwar district (montmorillonite dominant black clay) from depths of 0.5m to 3m. Information about the in-situ void ratio, moisture content, density and degree of saturation conditions have been provided elsewhere (Sridharan and Allam 1982, Sridharan et al. 1988b) and the ranges of values encountered are summarized in Table 1.

After removal of gravel fraction (by passing the soils through a 4.75 mm sieve), the soils were air-dried and pulverized to pass a 425 μm (ASTM Standard No.40) sieve. Procedural details for mineralogical and chemical (soil pH, exchangeable cation content) determinations have been provided in earlier papers (Sridharan and co-workers 1986, 1988a,b) and the test results for the disturbed soil samples are presented in Table 2.

Table 3 presents information regarding the specific gravities, grain size distributions, liquid, plastic and shrinkage limits of the kaolinitic and montmorillonitic soils. The procedures adopted for determining the physical properties have been detailed in earlier papers (Sridharan et al. 1986, 1988a).

Tropical soils, in particular, lateritic soils are known to be susceptible to property changes on drying (Terzaghi 1958, Newill 1961,

Townsend 1985). Oven-drying the kaolinitic and montmorillonitic soils from their field moisture contents in the present study however had no effect on their liquid limit values indicating that these soils are not susceptible to drying effects.

2.2 Methods

Chemical extraction of iron oxide from the kaolinitic and montmorillonitic soils were accomplished by a single treatment of the soils with dithionite-citrate-bicarbonate reagents (Mehra and Jackson, 1960). On completion of the treatment, the excess chemical reagents were discarded by centrifugation and the specimen washed once with distilled water. The amounts of iron oxide chemically extracted from the kaolinitic and montmorillonitic soils corresponded approximately to 2% (by weights of the soils).

The resulting decemented specimens in the moist state were tested for liquid, plastic and shrinkage limits by standard procedures. Specific gravity and grain size distributions were obtained for the decemented specimens in the air-dried state. Air-drying of the decemented specimens from the moist state was verified to have no effect on their plasticity characteristics. It is hence reasonable to expect that air-drying would also have no influence on the other physical properties of the specimens.

Conventional oedometer tests were conducted with the untreated and decemented kaolinitic and montmorillonitic soils in standard fixed

Table 1 Physical properties of undisturbed soils

Soil type	In-situ water content %	In-situ dry density Mg/m^3	In-situ void ratio	Degree of saturation %
Kaolinitic red soil	19-32	1.48-1.54	0.77-0.87	63-100
Montmoril--lonitic black soil	23-28	1.19-1.64	0.61-1.20	76-100

Table 2 Physico-chemical properties of the tropical soils

Soil type	Soil pH 1.2 suspension	Exchangeable cation content meq. 100 g^{-1} Ca	Mg	Na	K	Total	X-ray diffraction mineralogy
Kaolinitic red soil	7.3	4.8	2.9	5.7	0.6	14.0	Kaolinite, montmorillonite muscovite, quartz
Montmoril-lonitic black soil	8.6	45.8	24.1	8.3	1.6	79.8	Montmorillonite, quartz

Table 3 Index properties of the untreated and decemented tropical soils

Sl. No.	Soil type	Specific gravity	Grain size distribution Sand %	Silt %	Clay %	Liquid limit %	Plastic limit %	Plasticity index %	Shrinkage limit %
1	Kaolinitic red soil	2.57	14.0	19.5	66.5	80	35.3	44.7	21.9
2	Kaolinitic red soil on iron oxide removal	2.50	10.0	11.5	78.5	66	31.3	34.7	16.7
3	Montmorill-onitic black soil	2.79	35.0	17.0	48.0	132	41.0	91.0	11.8
4	Montmorill-onitic black soil on iron oxide removal	2.71	23.0	16.0	61.0	145	52.7	92.3	9.0

ring consolidometers using brass rings 76 mm in diameter and 25 mm height. The inside of the rings were coated with silicone grease to minimize side friction between the ring and the soil specimen.

The untreated and decemented kaolinitic and montmorillonitic soils were mixed into soft easily remoldable states at moisture contents slightly above (5% - 10%) individual liquid limit values. The remolding water contents ensured complete saturation of the soil specimens. The moist soil specimens were hand remolded into the consolidation rings (specimen thickness = 25 mm) and any air entrapped was displaced by water under vacuum. Using a load increment ratio of unity, consolidation tests were performed conventionally, the duration of each load being 48 hours. Each soil specimen was loaded to a maximum of 400 kPa and unloaded to the initial pressure of 6.25 kPa. Time-deformation curves obtained at each load increment were used to determine the coefficient of consolidation by the rectangular hyperbola method (Sridharan et al., 1987). The coefficient of permeability was computed from the value of coefficient of consolidation.

3 RESULTS AND DISCUSSION

3.1 Index Properties

Table 3 presents the index properties of the untreated and iron oxide removed, kaolinitic and montmorillonitic soil specimens. On extraction of iron oxide, the physical properties of the kaolinitic and montmorillonitic soils are observed to be perceptibly altered.

In recent studies (Sridharan et al. 1986, 1988a,b) the role of soil compositional factors and associated inter-particle forces in governing the index and compressibility characteristics of kaolinite rich red soils and montmorillonite dominant black clays have been highlighted. Studies with kaolinitic soils (Sridharan et al., 1988a) have indicated that the inter-particle attraction and repulsion forces have an important bearing on the liquid limit of these soils. The Coulombic forces determine the particle arrangement and the latter regulates the liquid limit values. The presence of di- and tri-valent cations or a low soil pH favors electrostatic attraction between positive edges and negative clay faces leading to a flocculated particle arrangement. The latter arrangement results in a higher liquid limit because of greater water entrapment in the large void spaces of its structural unit. Conversely, the presence of monovalent sodium exchangeable cations or a high soil pH results in weak inter-particle repulsion leading to a lesser degree of edge-face flocculation and a lower liquid limit.

Examination of results in Table 3 indicates a decrease in liquid and plastic limits of the kaolinite rich red soil on iron oxide removal inspite of an increase in the fines content. Similar observations were earlier made by Rao and Raymahashay (1981) and Townsend (1987). Also the shrinkage limit of the kaolinitic soil shows a decrease on sesquioxide removal.

The possible reasons for variations in the index properties of the kaolinitic soil on iron oxide removal are examined.

It has been earlier observed (Lambe 1958, Yong and Warkentin 1966) that the amount of shrinkage upon drying can be used as a measure of average particle orientation as a soil with a more oriented particle arrangement undergoes greater volume reduction upon drying than the same soil with its particles in a random/flocculant array. Evidently, then the lower shrinkage limit of the decemented kaolinitic soil (Table 1) is indicative of a less random soil structure apparently brought about by the breakdown of soil aggregates. The less porous structure in turn encloses smaller void spaces for water entrapment accounting for the lower liquid limit of the decemented specimen. The results also emphasize that in case of these tropical soils, besides the conventional Coulombic edge-face bonds, the sesquioxides accumulated by leaching effects contribute to the development of a random structure (Alexander and Cady 1962, Townsend et al. 1971).

In comparison, iron oxide removal from the montmorillonitic soil leads to an increase in liquid and plastic limits of the soil specimen (Table 3). The liquid limit of montmorillonitic soil is recognized to be a function of the electrical diffuse double layer forces (Yong and Warkentin 1966, Sridharan and Venkatappa Rao 1975, Sridharan et al. 1986). Possibly, the increase in fines content on sesquioxide removal provides a greater surface area for the development of diffuse double layer explaining the higher liquid limit of the decemented montmorillonitic soil specimen. The increase in diffuse double layer forces of the montmorillonitic soil on iron oxide removal also promotes greater inter-particle repulsion and a more oriented particle arrangement (Yong and Warkentin, 1966). The latter arrangement facilitates a relatively easier movement of particles and particle groups during drying, accounting for the lower shrinkage limit of the decemented montmorillonitic specimen.

3.2 Oedometer Results

Fig. 1 presents the void ratio-pressure curves for the kaolinite rich red soil and its iron oxide extracted counterpart. At any given pressure, the untreated soil exhibits a higher equilibrium void ratio in comparison to the decemented specimen. Recent studies (Sridharan and co-workers, 1988b) with remolded kaolinitic red soils have indicated that the soil structure exerts an important influence on their compressibility behaviour; a kaolinitic soil with a more random structure (inferred from its higher shrinkage limit) sustains the applied pressure at a higher void ratio in relation to a soil with a less flocculant array. Removal of iron oxide from the kaolinitic soil was observed to result in de-aggregation of particles and a decrease in extent of particle flocculation (indicated by its lower shrinkage limit). Apparently, then the lower void ratio sustained by the decemented kaolinitic soil is a consequence of its less random structure in comparison to the untreated soil.

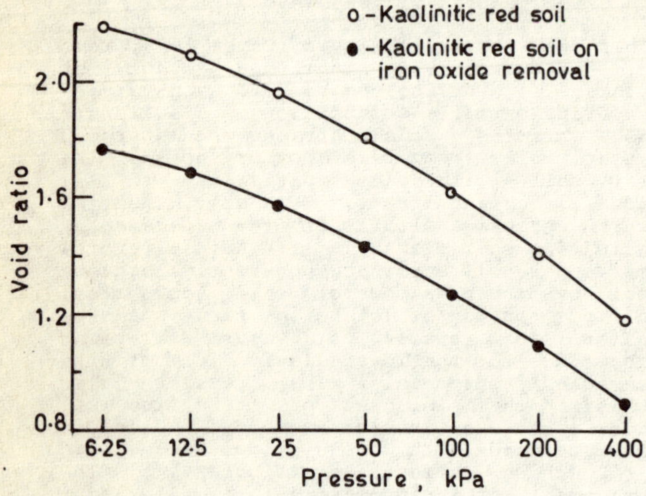

Fig. 1 Void ratio - pressure plots for the untreated and decemented kaolinitic soils

Fig. 2 plots Δe/Δlog p versus pressure for the kaolinitic soil in the untreated and decemented conditions. At any given pressure, the untreated soil exhibits a higher Δe/Δlog p value indicating greater compressibility. Application of consolidation pressure tends to re-orient particles into a parallel array (Leonards and Altschaeffl, 1964). Presumably, the higher compressibility of the untreated kaolinitic soil is a consequence of its more flocculated structure which entails greater particle re-adjustment under a given load.

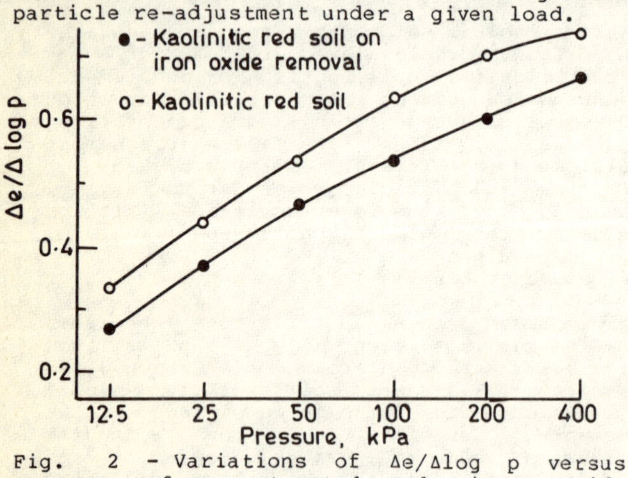

Fig. 2 - Variations of Δe/Δlog p versus pressure for untreated and iron oxide extracted kaolinitic soils

Fig. 3 plots the void ratio-pressure curves for the montmorillonitic soil in the untreated and iron oxide extracted conditions. At any pressure, the decemented soil shows a higher equilibrium void-ratio in comparison to the untreated specimen. The equilibrium void-ratios sustained by montmorillonitic soils at a given pressure are governed by physico-chemical repulsion forces arising from interaction of adjacent diffuse double layers (Bolt 1956, Olson and Mesri 1970, Sridharan and Venkatappa Rao 1973). Extraction of iron oxide from the montmorillonitic soil leads to an increased clay sized content (Table 3) and hence to a greater surface area. The consequent increase in diffuse double layer thickness is presumably responsible for the higher void-ratio sustained by the decemented soil.

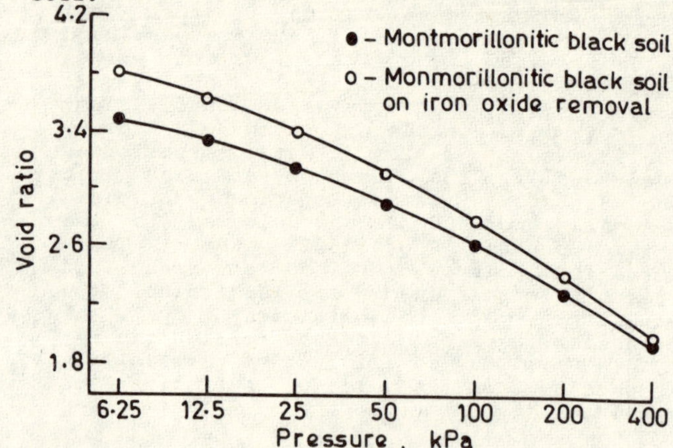

Fig. 3 - Void ratio-pressure plots for untreated and iron oxide extracted montmorillonitic soils

Fig. 4 presents the Δe/Δlog p versus pressure plots for the untreated and iron oxide extracted montmorillonitic soil specimens. At a given pressure, the decemented soil exhibits perceptibly higher Δe/Δlog p values; apparently, the enhanced diffuse double layer thickness resulting on iron oxide removal being responsible for the higher compressibility of the soil.

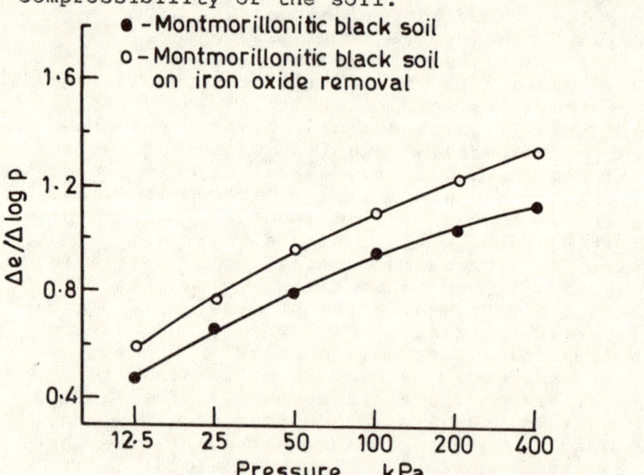

Fig. 4 - Variations of Δe/Δlog p versus pressure for the untreated and decemented montmorillonitic soils

Fig. 5 and 6 respectively present the coefficient of permeability (calculated from consolidation data) versus void-ratio plots for the kaolinitic and montmorillonitic soils in the untreated and decemented conditions. Conceptually, the permeability coefficients of the kaolinitic red soil and of the montmorillonitic black soil should decrease on iron oxide extraction; owing to breakdown of aggregates and a decrease in the extent of particle flocculation in the first instance and an increase in diffuse double layer thickness and reduction in effective pore space (Mesri and Olson, 1971) in the latter case. Experimental results (Fig. 5 and 6) show that the montmorillonitic soil exhibits a small but distinguishable reduction in permeability coefficient at a given void-ratio on iron oxide removal, however the

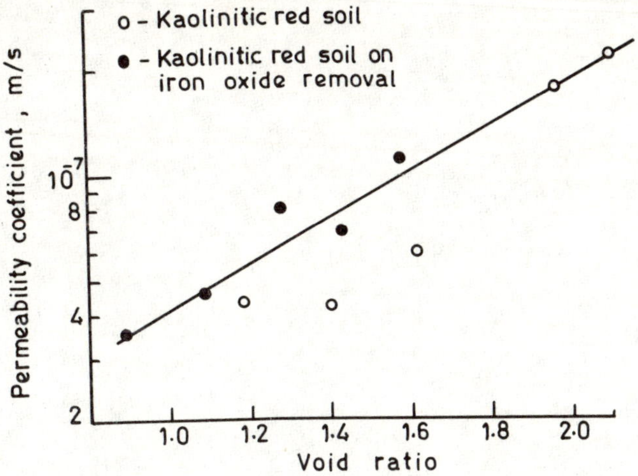

Fig. 5 - Coefficient of permeability versus void ratio plots for the untreated and iron oxide removed kaolinitic soils

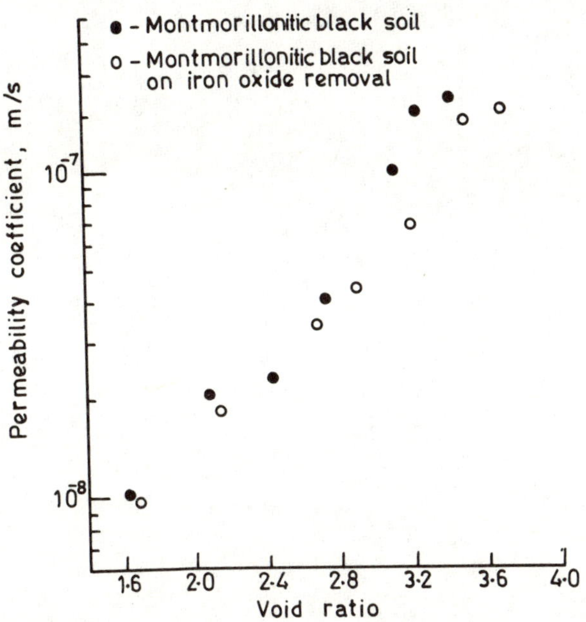

Fig. 6 - Coefficient of permeability versus void ratio plots for the untreated and iron oxide removed montmorillonitic soils

permeability coefficient of the kaolinite rich soil appears to be unaffected on the sesquioxide removal.

4 CONCLUSIONS

The presence of iron oxide has a significant influence on the index and compressibility characteristics of tropical soils. The mechanisms by which the sesquioxide influences the geotechnical properties is a function of the clay mineralogical composition of the soil. In case of the kaolinite rich red soil, the accumulated iron oxide binds indigenous soil particles into coarser aggregates and aids the development of a random soil structure. The flocculated structure in turn exerts an important influence on the liquid limit and compressibility behaviour of the kaolinitic soil.

In case of the montmorillonite dominant black clay, the aggregates formed by iron oxide bonding reduce the fines content and in effect the surface area available for development of diffuse double layer. The depression in diffuse double layer thickness in turn leads to a suppression of the plasticity and compressibility characteristics of the soil.

5 REFERENCES

Alexander, L.T. and Cady, J.G., 1962. Genesis and hardening of laterite in soils. US Dept. Agric. Tech. Bull., 1282 : 90pp.

Bolt, G.H., 1956. Physico-chemical analysis of the compressibility of pure clays. Geotechnique, 6 : 86-93.

Foss, I., 1973. Red soil from Kenya as a foundation material. Proc. Eighth International Conference on Soil Mechanics and Foundation Engineering, Vol. 2.2, Moscow : 73-80.

Lambe, T.W., 1958. The structure of compacted clay. J. Soil Mech. Fdns. Div. Am. Soc. Civ. Engrs., 84, SM2 : 1654-1 to 1654-34.

Leonards, G.A. and Altshaeffl, A.G., 1964. Compressibility of clays. J. Soil Mech. Fdns. Div. Am. Soc. Civ. Engrs., 90 : 133-155.

Mehra, O.P. and Jackson, M.L., 1960. Iron oxide removal from soils and clays by a dithionite citrate system buffered with sodium bicarbonate, Clays and Clay Minerals, 7 : 317-327.

Newill, D., 1961. A laboratory investigation of two red clays from Kenya. Geotechnique, 11 : 303-318.

Olson, R.E. and Mesri, G., 1970. Mechanisms controlling compressibility of clays. J. Soil Mech. Fdns. Div. Am. Soc. Civ. Engrs., 96 : 1863-1878.

Rao, S.K. and Raymahashay, B.C., 1981. Influence of clay minerals and iron oxide on selected properties of two laterite soils. Indian Geotechnical J., 11 : 255-266.

Sridharan, A., and Allam, M.M., 1982. Volume change behaviour of desiccated soils. J. Geotechnical Engg. Proc. Am. Soc. Civ. Engrs., 108 : 1057-1071.

Sridharan, A. Murthy, N.S. and Prakash, K., 1987. Rectangular hyperbola method of consolidation analysis. Geotechnique, 37 : 355-368.

Sridharan, A., Rao, S.M. and Murthy, N.S., 1986. Liquid limit of montmorillonite soils. ASTM Geotechnical Testing J., 9 : 198-203.
Sridharan, A., Rao, S.M. and Murthy, N.S., 1988a. Liquid limit of kaolinitic soils, Geotechnique, 38(2) : (in press).
Sridharan, A., Rao, S.M. and Murthy, N.S., 1988b. Compressibility behaviour of residual soils. J. Geotechnical Engg. Proc. Am. Soc. Civ. Engrs., (provisionally accepted).
Sridharan, A. and Venkatappa Rao, G. 1973. Mechanisms controlling volume change of saturated clays and the role of effective stress concept. Geotechnique, 23 : 359-382.
Sridharan, A. and Venkatappa Rao, G., 1975. Mechanisms controlling liquid limit of clays. Proc. Istanbul Conf. on Soil Mech. Found. Engg., Vol. 1, Istanbul : 75 - 84.
Terzaghi, K., 1958. Design and performance of the Sasumua dam. Proc. Brit. Inst. Civ. Engrs., London, 9 : 369-394.
Townsend, F.C., Manke, G.P. and Parcher, J.V., 1969. Effect of remolding on the properties of a laterite soil. Highway Res. Board Rec., 284 : 76-84.
Townsend, F.C., Manke, G.P. and Parcher, J.V., 1971. The influence of sesquioxides on laterite soil properties. Highway Res. Board Rec., 374 : 80-92.
Townsend, F.C., 1985. Geothechnical characteristics of residual soils. Geotechnical Engg. Proc. Am. Soc. Civ. Engrs., 111 : 77-93.
Townsend, F.C., 1987. Closure to discussion. J. Geotechnical Engg. Proc. Am. Soc. Civ. Engrs., 113(4) : 398-399.
Vargas, M., 1973. Structurally unstable soils in Southern Brazil. Proc. Eighth International Conference on Soil Mechnanics and Foundation Engineering, Vol. 2.2, Moscow : 239-246.
Wesley, L.D., 1974. Tjipanundjan dam in South West Java, Indonesia. J. Geotechnical Engg. Div. Proc. Am. Soc. Civ. Engrs., 100 : 503-522.
Yong, R.N. and Warkentin, B.P., 1966. Introduction to Soil Behaviour, New York : Macmillan.

Influence of climate on the properties of tropical soils

K.S.SUBBA RAO, Indian Institute of Science, India
P.V.SIVAPULLAIAH, Indian Institute of Science, India
J.R.PADMANABHA, Indian Institute of Science, India

ABSTRACT

Climate has a profound influence on the properties of soils, particularly, of tropical regions. Climate is usually characterized by relative humidity and temperature. The present paper attempts to study the effect of humidity and temperature on the index properties of four tropical soils : Na, Ca-montmorillonite, black cotton soil, kaolinite and red earth. The changes in relative humidity and temperature cause significant changes in the liquid limits of montmorillonitic clays but kaolinite and kaolinitic clays are unaffected by climatic changes. Plastic limit does not vary with relative humidity for any soil and the variations in the plasticity index follow the same trend as that of liquid limit. However, the earlier empirical correlations between liquid limit and plasticity index do not hold good for the humidity equilibrated soils. While relative humidity does not have any influence on the liquid limit of kaolinite, shrinkage limit decreases with increase in relative humidity.

1. INTRODUCTION

The consistency limits of clays are known to play a dominant role in affecting their engineering properties. Many useful correlations have also been established over the years, like for example, between the liquid limit and compression index, plasticity index and swelling potential, etc. Soils in equilibrium with different climate will have different thickness of adsorbed water layers and so will have different effective particle sizes. Added to this, there will also be factors like aggregation and soil fabric. It is interesting, therefore, to study how the index and other basic properties change with different climatic influences. Climate is usually characterised by relative humidity and temperature. Barring a few attempts, no detailed studies have yet been conducted to investigate the manner in which the climate affects the properties. The present paper attempts to study the effect of humidity and temperature on the index properties of tropical soils.

2. EXPERIMENTAL

Four soils: Na,Ca-montmorillonite, black cotton soil, kaolinite and red earth have been chosen for the study. The properties of these soils are presented in Table I.

The soil powders were taken in beakers, and kept in desiccators containing various salt solutions for maintenance of different relative humidities. An equilibrium time of 90 days was allowed under all relative humidities. Table II gives the saturated salt solutions used and the humidities produced.

The liquid limit experiments were conducted using the Casagrande apparatus as per ASTM D423-66 (1972). Water was thoroughly mixed ith the soil samples and the liquid limit test conducted without any specific moisture equilibration time being allowed. In the case of Na, Ca-montmorillonite, two more variations were also considered : (i) water was thoroughly mixed with water and 72 hours of equilibration time was allowed before the liquid limit tests, (ii) soil was mixed roughly 4 times the liquid limit water content and thoroughly churned in a stirrer at 3000 rpm for 10 minutes and left for evaporation for 10 days before the liquid limit test.

For the series of tests on temperature equilibrated soils, the powdered soils were kept in the temperature controlled oven for 45 days. The controlled temperatures were 35°C, 45°C and 55°C. Soil samples taken out from the ovens were cooled at room temperature in an empty desiccator for 2 hours before adding water. Water was thoroughly mixed with soil and the liquid limit test was conducted without any specific maturation time.

The plastic limits of humidity and temperature treated soils were determined as per ASTM D424-59 (1971).

The shrinkage limits of soils were determined as per ASTM D247-61 (1974).

3. RESULTS AND DISCUSSION

Liquid Limit. Table III gives the results of liquid limit tests conducted on humidity and temperature treated soils. Figure 1 shows graphically the liquid limit variation for Na, Ca-montmorillonite. The liquid limit varies from 463% to 503% for Na, Ca-montmorillonite, the lowest value being at 32% relative humidity. As the relative humidity increases from 2.5% to 96.5% relative humidity, the liquid limit value is seen to decrease from

TABLE I - PROPERTIES OF SOILS

Sl No	Properties	Kaolinite	Na, Ca-montmorillonite	Black Cotton Soil	Red earth
1	Specific gravity	2.64	2.83	2.72	2.78
2	Liquid limit, %	49	503	99	48
3	Plastic limit, %	29	44	33	28
4	Shrinkage limit, %	33	8	20	20
5	Surface area, m^2/gm	18	64	31	20
6	Sand fraction, %	38.2	-	31.0	65.2
7	Silt fraction, %	51.1	-	42.0	26.2
8	Clay fraction, %	10.7	100	27.0	8.5

TABLE II - SATURATED SALT SOLUTIONS FOR RELATIVE HUMIDITY CONTROL

Saturated Solution	% Relative Humidity
Calcium Chloride, $CaCl_2$	02.5
Magnesium Chloride, $MgCl_2$	32
Sodium Dichromate, Na_2CrO_3	52
Sodium Chloride, $NaCl_2$	75
Potassium Sulphate, K_2SO_4	96.5

486% at 2.5% relative humidity to 463% at 32% relative humidity and then steadily increase to 503% at 96.5% relative humidity.

Even at 2.5% relative humidity, monolayer of water adsorption is almost completed and second layer is fully developed only at about 75% relative humidity. Table IV shows how the humidity affects the hydration water layers. So between 2.5% and 75% relative humidity, the second layer of water is being formed on the montmorillonite particle. During this process, it is giving relatively less liquid limit. At 2.5% and above 75% relative humidity the particle attractions will be relatively less and consequently the formations of clusters will be less which explains the higher liquid limit values. This trend continues even when the liquid limits were determined after 72 hours of maturation time. The values as expected are higher. It is seen to vary from 519% to 563%. This shows that the interparticle force system caused by the humidity is not destroyed during the maturation time. The same trend continues even when the liquid limits were determined after churning for 10 minutes and keeping it for 10 days. This shows that even keeping for 10 days of maturation cannot destroy the fabric arrangements caused by adsorption of water with various humidities.

Referring to Table III, it is seen that humidity does not affect the liquid limit of kaolinite, which remains unchanged at 48%. The reason for this is not difficult to find. In kaolinite successive unit layers are so arranged that oxygen atoms and OH groups of adjacent layers approach one another in pairs. Hence the units are held together fairly tightly by hydrogen bonding between the layers. So the interlayer water cannot exist for

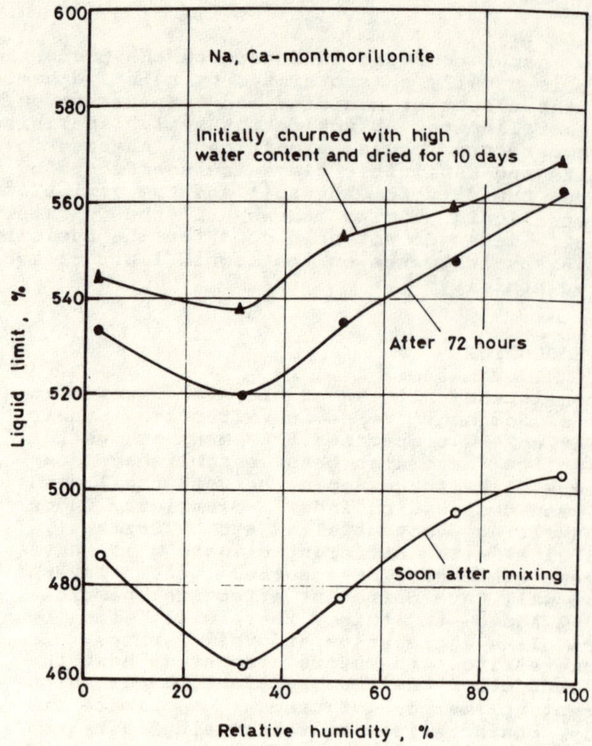

Figure 1 - Variation of liquid limit with humidity of Na, Ca-montmorillonite

kaolinite minerals. In other words, the d-spacing remains constant, which is seen from Table IV. Hence the exposure of kaolinite to different relative humidities will not alter the interparticle forces. Consequently the liquid limit remains constant at all relative humidities. The same explanation holds good for the behaviour of red earth in which the major clay mineral is kaolinite.

For black cotton soil, liquid limit decreases as relative humidity increases, showing a steady decrease from 101% at 2.5% relative humidity to 91% at 96.5% relative humidity. For black cotton soil, d_{001} reflections in the X-ray diffraction patterns are not sharp and hence the completion of water layers could not be established. The montmorillonite present in the black cotton soil does not show sufficient crystallinity and so the

TABLE III - LIQUID LIMIT VALUES - PER CENTAGE

Relative humidity %	Na,Ca-montmorillonite Immediately %	After 72 hours %	After 10 days with churning %	Kaolinite %	B.C.Soil Amb.Temp. %	35°C %	45°C %	55°C %	Red earth %
2.5	486	533	544	49	101	111	97	88	47
32	463	519	538	48	100	97	95	87	47
52	477	535	553	48	96	104	97	81	48
75	495	548	559	48	93	103	95	79	49
96.5	503	563	569	48	91	89	89	73	50

TABLE IV - X-RAY DIFFRACTION DATA OF Na,Ca-MONTMORILLONITE AND KAOLINITE

Relative humidity %	Na,Ca-montmorillonite d-spacing A°	Hydration water layers	Kaolinite d-spacing A°
2.5	12.62	1.04	7.2
32	12.99	1.17	7.2
52	13.59	1.38	7.2
75	15.23	1.94	7.2
96.5	15.77	2.13	7.2

basal reflections are not sharp. However, a similar trend of completion of water layers found for Na, Ca-montmorillonite can be expected for black cotton soil also. The continuous decrease in the liquid limit of black cotton soil indicates, that the inter mineral interaction is higher at higher humidities, causing aggregation leading to decrease in liquid limit.

Figure 2 shows the variation of liquid limits of humidity - treated black cotton soil with temperature. As the temperature increases the liquid limit decreases at any particular relative humidity. At 2.5% relative humidity, the liquid limit varies from 111% to 88% for a temperature change of 35°C to 55°C. At 75% relative humidity, the corresponding change is from 103% to 79%. The variation of liquid limit with increase in temperature is minimum at 32% relative humidity. This is because at 32% relative humidity, the soil particles are already in aggregated state due to humidity as seen in the case of Na, Ca-montmorillonite and the additional aggregation effect of temperature is not further felt markedly. At other humidity levels, particularly at 2.5% relative humidity and above 75% relative humidity, the effect of temperature is felt more markedly.

<u>Plastic Limit</u>. The plastic limits of the soils do not show any significant variation with relative humidity and temperature.

<u>Plasticity Index</u>. Table V and Figure 3 brings out the variations of plasticity index with relative humidity and temperature. The plasticity index shows qualitatively a similar

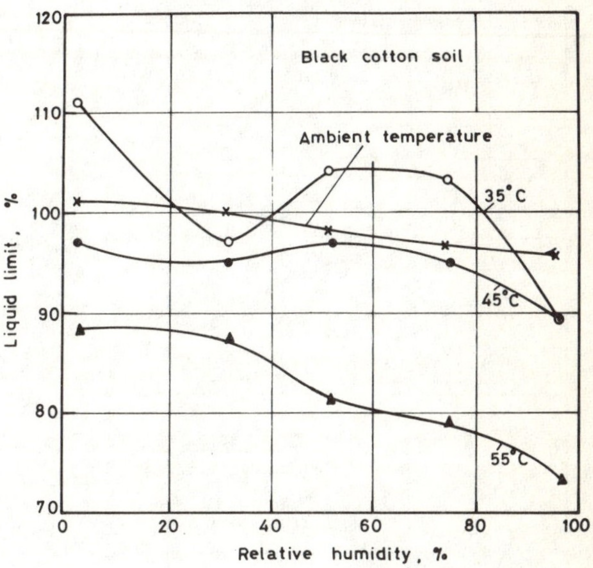

Figure 2 - Variation of liquid limit with humidity and temperature of black cotton soil

variation with relative humidity as that of liquid limit.

Figure 4 shows the relationship between plasticity index and the liquid limit. The relationship for these humidity treated soils can be expressed as

$$I_p = 0.96\ (w_L - 39.2)$$

Even the published data [Sivapullaiah and Sridharan (1985)] on bentonite-sand mixtures fall on this line. It will be interesting to compare the relationship with those suggested by others. Thomas Paal and Post (1984) obtained the equation as

$$I_p = 0.825(w_L - 17.87)$$

and Nagaraj and Jayadeva (1984) as

$$I_p = 0.74\ (w_L - 8)$$

Figure 4 shows the relationships between plasticity index, I_p, and liquid limit, w_L along with Casagrande's 'A' line. It is clear

TABLE V - PLASTIC LIMIT AND PLASTICITY INDEX VALUES

Relative humidity %	Na,Ca-montmorillonite		Kaolinite		Black cotton soil						Red earth	
	Plastic limit %	Plasticity index %	Plastic index %	Plasticity index %	35°C		45°C		55°C			
					PL%	PI%	PL%	PI%	PL%	PI%	PL%	PI%
2.5	52	434	42	7	39	72	43	54	42	46	28	19
32	58	405	37	11	41	56	44	51	41	46	28	19
52	56	421	40	8	40	64	40	57	45	36	28	20
75	61	434	40	8	39	64	42	53	46	33	28	21
96.5	58	445	39	9	39	50	42	47	40	33	28	22

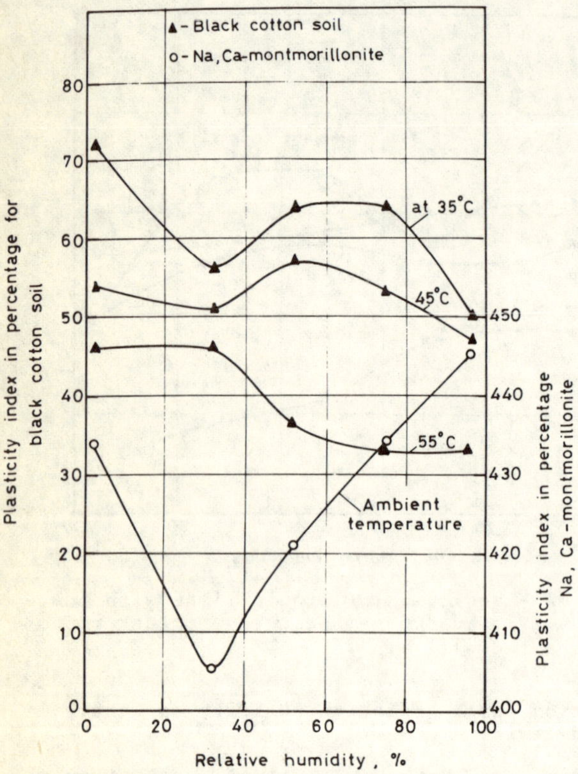

Figure 3 - Variation of plasticity index with humidity and temperature of Na, Ca-montmorillonite and black cotton soil

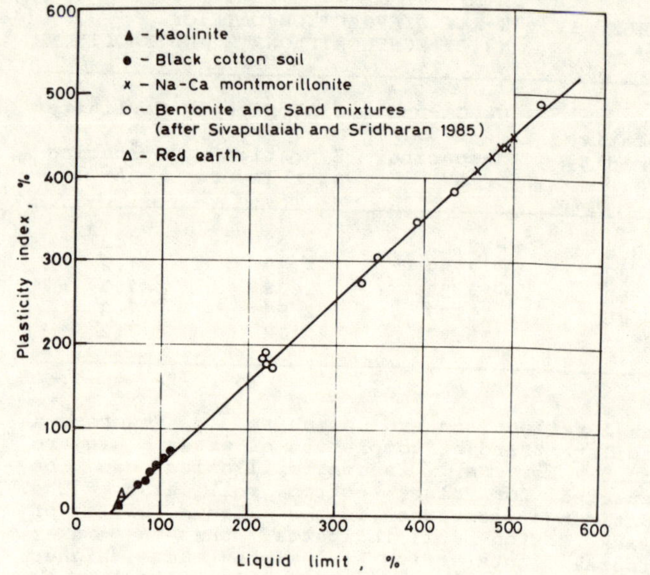

Figure 4 - Liquid limit vs. plasticity index for different soils

that there is considerable variation among the different approaches particularly in the low and high regions of liquid limit.

<u>Shrinkage Limit</u>. As seen from Table VI and Figure 5, the shrinkage limit values of Na, Ca-montmorillonite varies from 9.8 to 12.5, the lowest being at 96.5% relative humidity and the highest at 32% relative humidity. The shrinkage limit increases initially and then decreases steadily, as the relative humidity changes from 2.5% to 96.5%. This behaviour is opposite to that of liquid limit versus relative humidity. This actually corroborates the explanations given for the variation of liquid limit with variations in relative humidity, i.e., when the second layer of water is being formed on the montmorillonite clay particle, the interparticle attractions are more and the particles will be aggregated. At 2.5% and 75% and above relative humidities either monolayer or two layers of water are fully developed and the interparticle forces will be less and the aggregation will be minimum. So it follows that shrinkage limit decreases when the aggregation is less and vice versa.

For kaolinite, the shrinkage limit steadily decreases with increase in relative humidity. It is interesting to observe while the effect of relative humidity is negligible on the liquid limit of kaolinite particle there is noticeable effect on the shrinkage limit of kaolinite mineral. The reason for this is the change of fabric. At lower humidity levels, the edge to face flocculations are stronger thus showing the higher shrinking limits. With increase in humidity the adsorbed water amount increases on the edges with a consequent dispersion of particles. The adsorbed water in kaolinite can only be present at the edges since there is no interlayer water. Figure 6 shows schematically, arrangements of particles

TABLE VI - SHRINKAGE LIMIT VALUES

Relative humidity %	Black cotton soil %		Na, Ca-montmorillonite %	Kaolinite %	Red earth %
	35°C	45°C			
2.5	7.1	11.0	11.5	39.8	21.5
32	8.0	10.2	12.5	37.0	21.0
52	10.2	10.0	10.5	36.7	20.0
75	11.0	11.0	11.0	36.3	20.0
96.5	12.2	10.3	9.8	34.5	20.0

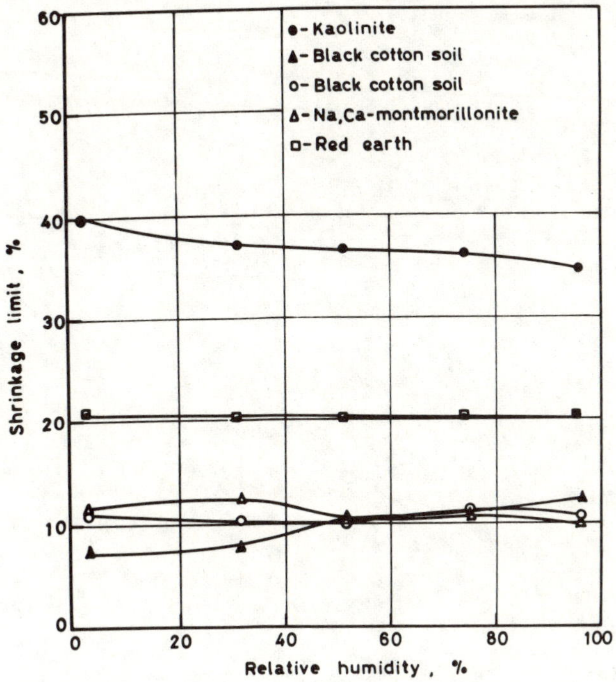

Figure 5 - Variation of shrinkage limit with humidity and temperature of Na, Ca-montmorillonite, kaolinite and black cotton soil

at low and high humidities. The soil initially at higher humidity permits a higher shrinkage and so a lower shrinkage limit. As the amount of water adsorbed increases on the edges with increase in humidity, dispersion of particles takes place. When the shrinkage limit test is conducted on this soil, the particles orient themselves parallel and reduce the shrinkage limit.

The shrinkage limit of red earth remains unaffected by humidity and is fairly constant at 20% (Figure 5). This limit is seen to be much lower than that of kaolinite probably on account of the collapse of flocculant structure owing to the presence of non-clay minerals. humidity does not seem to effect any further changes.

For black cotton soil, the shrinkage limit values at 35°C steadily increase from 7.1 to

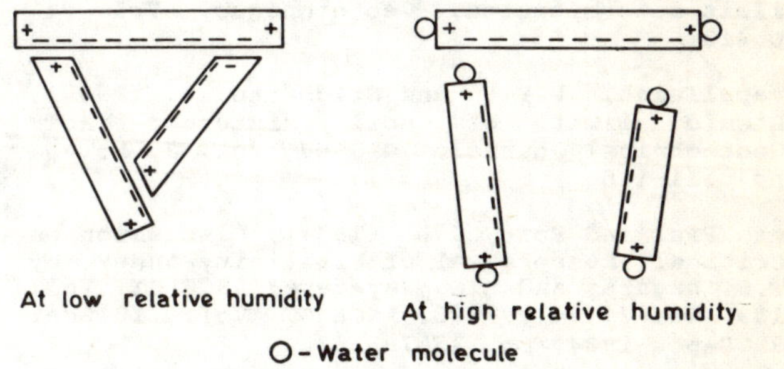

Figure 6 - Particle associations of kaolinite

12.2 as the humidity changes from 2.5% to 96.5%. Here also it is noticed that while the liquid limit values decrease continuously with increase in relative humidity shrinkage limit increasing with increasing humidity. However at 45°C the shrinkage limit values remain fairly constat at about 11%. This is probably because at 45°C the particles are already aggregated and the effect of humidity is not seen.

4. CONCLUSIONS

1. The liquid limit of Na, Ca-montmorillonite is significantly affected by humidity changes owing to changes in the adsorbed water layers.

2. The liquid limits of black cotton soil, kaolinite and red earth are unaffected by humidity changes. However, the liquid limit of black cotton soil decreases with increasing temperature due to particle aggregation.

3. The plastic limit is unaffected by humidity and temperature.

4. The empirical relationship between the liquid limit and plasticity index of humidity treated soils is

$$I_p = 0.96\ (w_L - 39.2)$$

   This relationship is different from the earlier empirical correlations.

5. While the shrinkage limit of kaolinite decreases with increase in humidity, the shrinkage limit of black cotton soil increases with humidity.

REFERENCES

ASTM Designation D423-66 (1972). Test for liquid limit of soils. Annual book of ASTM Standards 19, Philadelphia.

ASTM Designation D424-59 (1971). Test for plastic limit and plasticity index of soils: Annual book of ASTM Standards, 19, Philadelphia.

ASTM Designation D4247-61 (Reapproved 1974). Test for shrinkage factors of soils. Annual book of ASTM Standards, 19, Philadelphia.

Nagaraj, T.S., and Jayadeva, M.S. (1981), Re-examination of one-point method of liquid limit determination, Geotechnique, Vol. 31, p.413.

Sivapullaiah, P.V., and Sridharan, A. (1985), Liquid limit of soil mixtures; ASTM Geotechnical Journal, GTJODJ, Vol. 8, No. 3, pp. 111-116.

Tamas Paal and Post, J.L. (1984), Discussion on Critical Reappraisal of Plasticity index by T.S. Nagaraj and M.S. Jayadeva; ASCE GT. Vol. 109, No. 7, July 1983. ASCE GT. Vol. 110, No. 9, Sept. 1984, pp. 1367-1368.

Identification and classification of granitic residual soils using chemical weathering index

T.SUEOKA, Technical Research Centre of Taisei, Japan

Abstract: The degree of chemical weathering is chosen to identify and classify the granitic residual soils including lateritic and laterite in several different places in different climate in the world. Sesquioxide ($R_2O_3=Al_2O_3+Fe_2O_3+(TiO_2)$) and loss on ignition ($=H_2O(\pm)$) in molecular percentage of the granites and the residual soils were chosen to represent widespread chemical changes because of several reasons. Using the new proposed chemical weathering index method, the tendencies of full weathering process of granite to residual soil, lateritic soil and laterite, quantitative comparison between the residual soils derived from different places and climate can be expressed. Engineering classification of granite and weathered residue can be also exhibited by this method, and it is seen that the new chemical weathering index method could be applicable for both geological and engineering purposes.

1. Introduction

Evaluating the degree of weathering of residual soils is crucial not only from the viewpoint of engineering geology but also has a very significant meaning with a veiw to disaster prevention as the failures of decomposed granite soil slopes prove. In line with the investigation of residual soils and construction of civil engineering structures which have recently been taking place frequently by engineering geologists and geotechnical engineers who have worked in overseas, it has become increasingly necessary to establish the globally common geological and geotechnical engineering definitions and standards of residual and decomposed rocks in tropical and temperate regions. However, soils intrinsically involve local problems, their names varying by districts, for example, in case of granitic residual soils, it is called Masado (decomposed granitic residual soil) in Japan, grus or granular disintegration sand in Australia and U.S.A., thus variously defined. Furthermore, in tropical zones, further weathered reddish soil is called lateritic soil, and its weathered residue is called laterite. It is hardly to say that these soils are sufficiently defined.

In this paper, the chemical weathering of granitic residual soils extensively distributed in both tropical zones and temperate zones are assumed as a residue and integration of paticular chemical components, and chemical weathering index (CWI) as a general and quantitative weathering index is proposed, thereby defining residual soils at their diversified weathering stages. This paper also touches on their engineering properties through comparison with the conventional methods for rock and residual soil classification and categorization.

2. Investigated sites and used data

In order to investigate the chemical weathering extents of granite and its residual soils, investigations were carried out in tropical zones and temperate zones shown on Table-1. For reference, data on the site 4 through 6 is derived from available presentations. In Kaduna site, boring log samples were used for their chemical analysis and in Darling Range and Tottori site, excavated slope block samples were used for their chemical analysis.

Table 1 Zones where granites and their residual soils were sampled

No.	Mark	Sampling location	Latitude	Type of granite
①	◐	Darling Range, Australia	33°S	Precambrian granite
②	○	Kaduna, Nigeria	10°N	Precambrian granite
③	△	Tottori city, Japan	35.5°N	Tertiary granite
④	⊗	Mikawa Plain, Japan 1)	35.3°N	Cretaceous and Tertiary granites
⑤	□	Kitsugi, Shimane 2) prefecture, Japan	35.3°N	Tertiary granitic diorite
⑥	ϕ	Gotsu, Shimane prefecture, Japan 3)	35°N	Tertiary granitic diorite

3. Proposal on new chemical weathering index

In order to satisfy the following criteria for chemical weathering index, the new chemical weathering index for multi-purpose usage such as geotechnical engineering, engineering geology and agriculture will be proposed.

1. it can express the whole chemical weathering process,
2. it is suitable for engineering purposes,
3. it is a relatively simple (empirical) expression.

The new chemical weathering index is as follows:-

(1) According to the chemical composition change during the weathering tested in Table-1, the general and main trend of the weathering is that, at first stage, alkali and alkaline earths components such as K_2O, CaO, Na_2O, MgO, MnO, FeO, are leached. Secondly, silica (SiO_2) is leached gradually and sesquioxides such as Fe_2O_3, Al_2O_3 and TiO_2 are accumulated, and loss on ignition ($H_2O(+)$+$H_2O(-)$) increases rapidly. Therefore, chemical components can be classified into three groups. 1 Alkali and alkaline earths 2 Silica and 3 sesquioxides and loss on ignition ($H_2O(+)+H_2O(-)$). Fig-1 shows the method proposed.

(2) There are two methods to express the chemical components, by weight (percentage) or by molecular ratios. Chemical analyses are usually reported in weight percentage of the oxides of the elements. This form is useful if one is concerned with tonnages as in mining work. It is also satisfactory if the data are used for direct comparison of the quantities of each component in unaltered and altered rock; that is, in the form of ratios involving only individual substances. In all other cases, however, weathering studies are clarified if weight percentages are changed into molecular ratios[4] (Reiche, 1943). This is accomplished by the sum of the atomic weights which they represent. For lighter substances such as alkali and alkaline earths, it is more accurate to use molecular ratios in preference to weight percentage.

(3) If the chemical component groups (all alkali and alkaline earths, SiO_2, $R_2O_3+H_2O(\pm)$) of the granite are expressed in a triangular graph, it is very interesting that the groups of the granite derived from six different areas can be located in almost same line from the viewpoint of $R_2O_3+H_2O(\pm)$ component (Fig-1).

(4) Thornton et al[5] (1960) and Aramaki et al[6] (1972) investigated the chemical components of granite using five thousand pieces of data from studies in the United States and one thousand two hundred items of data from Japan, respectively. In their studies, DI (Differential Index) defined as the sum of the normative percentage of quartz, orthoclase, albite, nepheline, leaucile and kalsilite were adopted to compare the each chemical component. According to their studies, the most typical granite (DI=80) and granodiorite (DI=67) consist of 14%-16% of Al_2O_3 content and 1.5%-2.5% of Fe_2O_3 content in the United States and Japan.

According to the facts mentioned in (1) through (4), it is apparent that the variations of sesquioxides (Al_2O_3, Fe_2O_3) are very small compared with the other chemical components such as alkali and alkaline earths and silica, and if the molecular ratios are adopted, the variations of Al_2O_3 and Fe_2O_3 become smaller in the range of granite and granodiorite (DI=65-85) because these sesquioxide atomic weights are relatively heavier than other chemical components. This means that the molecular ratio of the sesquioxides (Al_2O_3, Fe_2O_3) of granite and granodiorite can be seen as steady in all regions on the earth. Of course the molecular ratios of loss on ignition ($H_2O(\pm)$) of the fresh granite and granodiorite are very small, and the molecular ratio of $H_2O(\pm)$ increases rapidly when the rock is weathered because of the atomic weight of $H_2O(\pm)$ is the smallest in the chemical components of the rocks. These facts mentioned above can prove the validity of the molecular ratio of sesquioxides and loss on ignition ($R_2O_3+H_2O(\pm)$) as the chemical

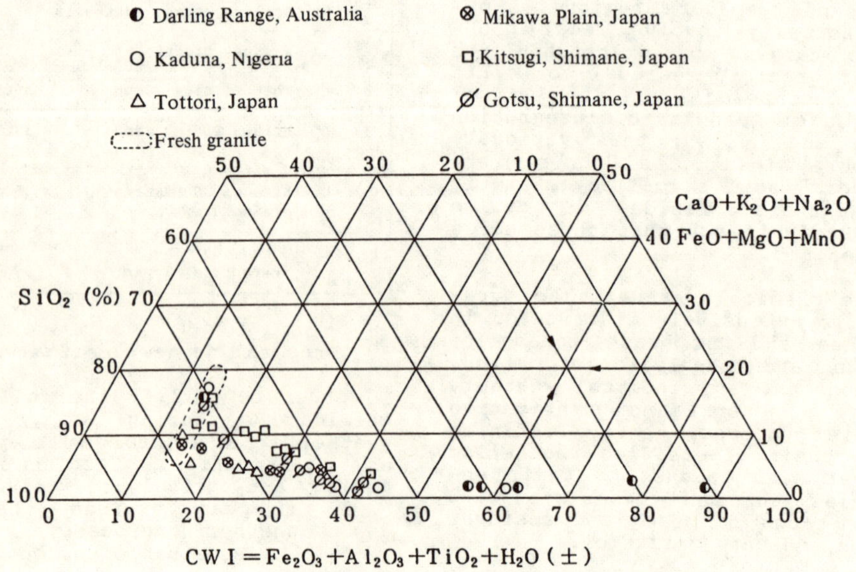

Fig. 1 Chemical Weathering Tendencies of Granitic Residual Soils in Various Area (molecular %)

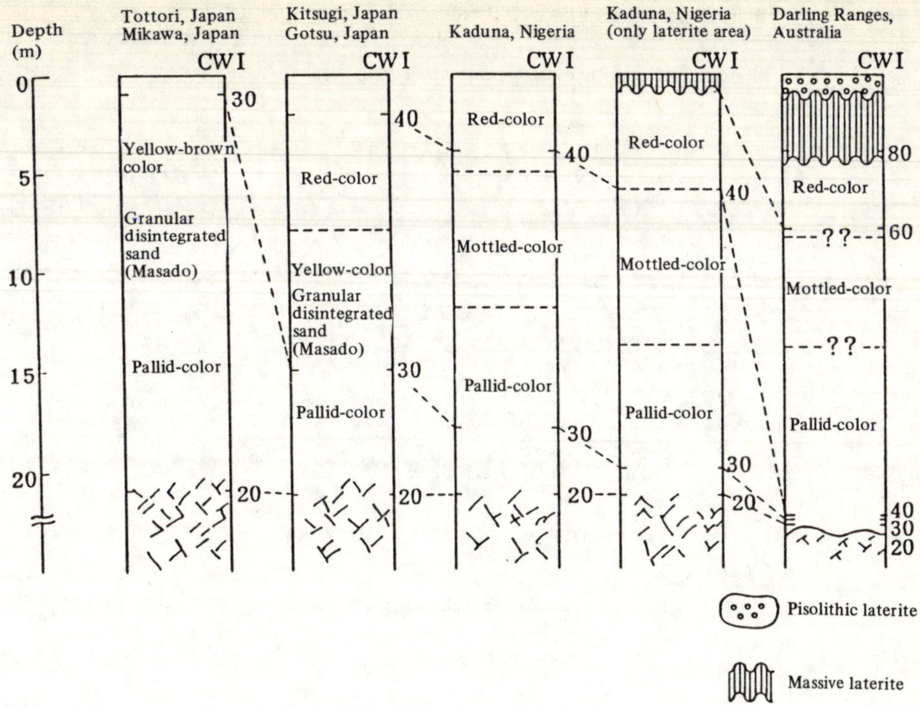

Fig. 2 Soil Profiles in Different Countries

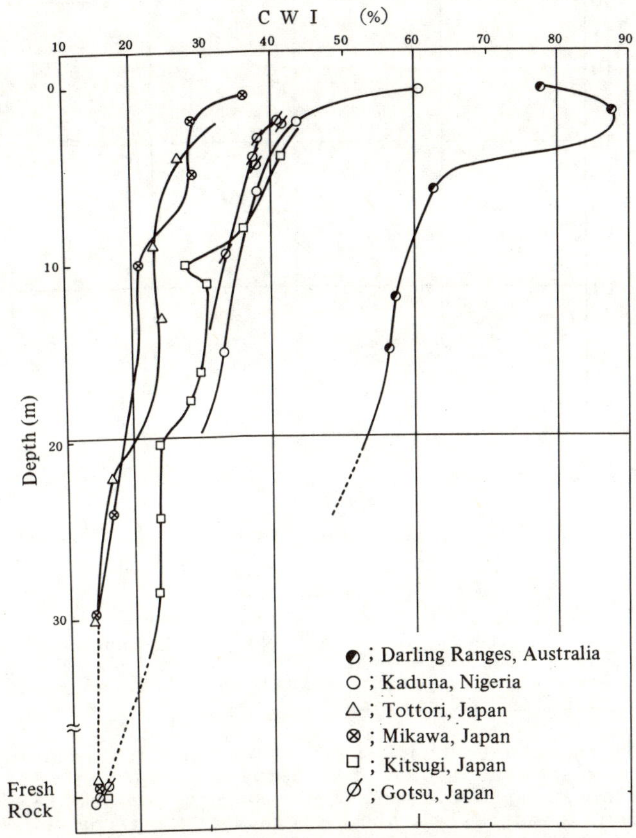

Fig. 3 The Relationship between Depth and New Checmial Weathering Index (molecular %)

weathering index for granitic rock and the residual soils. Therefore, chemical weathering index defined by the following percentage can express the weathering trend of the granitic rock and the residual soils, the degree of weathering of weathering of them and other geotechnical engineering properties.

$$CWI = \left(\frac{Al_2O_3+Fe_2O_3+TiO_2+H_2O(\pm)}{\text{All chemical compornents}}\right)_{mole} \times 100(\%)$$

4. Chemical weathering index and geotechnical properties of the residual soils

Several geotechnical properties can be seen by using the new proposed chemical weathering index. In this section the relationship between CWI and geotechnical properties of residual soils will be discussed.

4.1 CWI and soil layer sections of residual soils

Fig-2 shows the cross sections of granitic residual soils distributed in tropical zones and temperate zones. In this figure and in Fig-3, the CWI values in respect to depths are represented. It can be seen from the figures that the colors of the residual soils become reddish with the progress of chemical weathering, but the colors of the soils are not always in agreement with the extents of chemical weathering index CWI. As compared with the residual soils in Nigeria and Australia, the residual soils in Tottori city and Mikawa Plain, Japan are less chemically weathered, but those in Kitsugi and Gotsu have approximately same CWI values in their residual soil cross

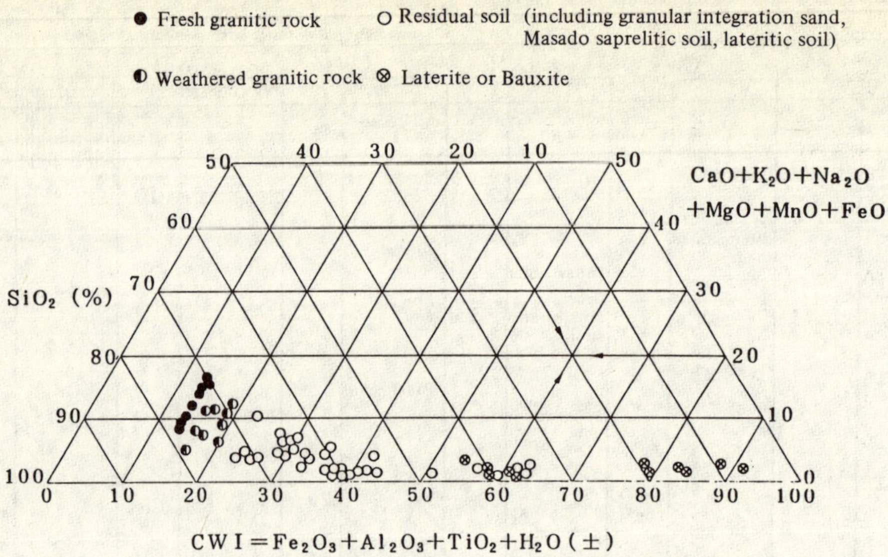

Fig. 4 Classification of the Decomposed Materials Derived from Granitic Rocks (Molecular %)

sections to those in Kaduna, Nigeria, despite dissimilar weather conditions between the two districts, which is very interesting. This phenomenon is likely to result from the fact that the topographies in Gotsu and Kitsugi are gently sloped, and affected by warm palaeclimate, thus the weathered residual soils are preserved in good condition.

4.2 CWI and physical conditions of weathered products

In order to investigate the degree of weathering of granitic rock and the residual soils, it is basically important what is the physical condition of the weathered materials. Fig-4 shows the relationship between CWI and the physical state of granitic weathered residual soils. The weathered products having dissimilar physical conditions such as fresh granite, weathered granite, residual soil (which is otherwise called saprolite, Masado, grus, or granular disintegration sand), laterite (as cemented) are all approximately in correspondence to CWI, intimating that CWI is effective as an index to express the physical conditions of weathered products. Namely, rocks can be largely classified from the magnitudes of CWI, such as fresh granite (CWI=13-15%), weathered granite (CWI=15-20%), residual soil (CWI=20-60%) and laterite (CWI=60- %).

In addition, in case of CWI, the same residual soil can be quantitatively classified into different chemically weathered extents, thus CWI being convenient for engineering geologists and geotechnical engineers who simultaneously handle the residual soils distributed in tropical zones, subtropical zones and temperate zones. Fig-4 also shows that a range of change in the CWI of weathered granite is slight, that is in agreement with the degree of weathering of weathered granite is slight, which was

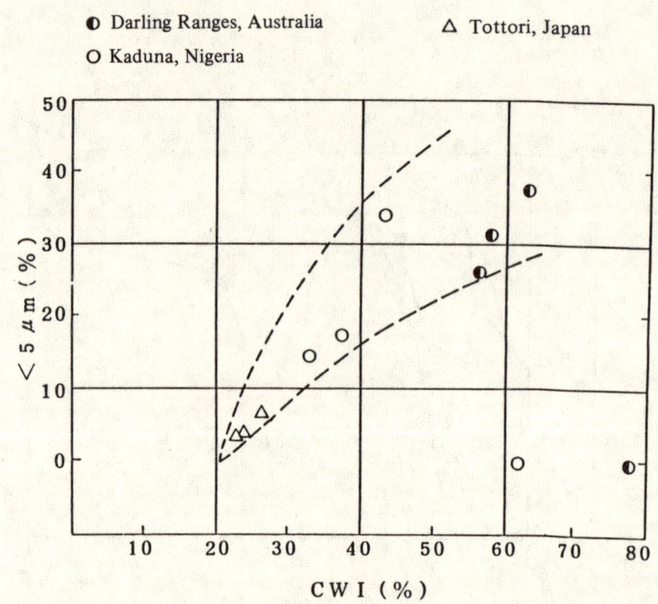

Fig. 5 The Relationship between Fines (< 5μm) and CWI.

pointed out by Onodera, Yoshinaka and Oda[7]), largely increase the porosity in weathered granite, decreases in alkali and other chemical components are slight.

On the other hand, Fig-5 shows the relationship between the CWI of residual soil in its weathering stage and fine content smaller than 5μm. Generally, the percentage of fires contained in residual soil increases as CWI increases, which is considered as a reasonable result from the fact that weathered products

Table 2 Relationship between CWI and rock classification and engineering characteristics

CWI (%)	Classification*	Extent of weathering	Classification of weathered granite	Cohesion and porosity
13 - 15	I	Fresh rock	Fresh rock	increase
	II	Slightly weathered		
15 - 20	III	Moderately weathered	Weathered granite	Cohesion
	IV	Highly weathered	Granular disintegration sand	
20 - 40	V	Completely weathered	Masa soil Saprolite	Porosity
40 - 60	VI	Residual soil	Lateritic soil	
60 - 90		Weathered hard pan (as cemented)	Laterite or bauxite	

* Dearman, et al (1978)

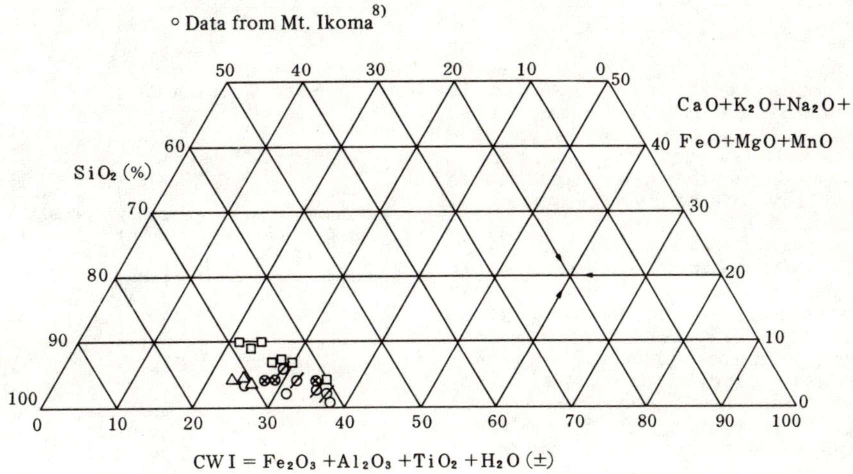

Fig. 6 Degree of chemical weathering of Masado in Japan (mol %)

become more finely granulated with the progress of chemical weathering. However, it seems that with the increasing of CWI, the clay contents smaller than 5μm more greatly vary.

In Japan, decomposed granite soil or weathered residual soil derived from granite is called Masado, and Fig-6 shows the degree of weathering and the tendencies of weathering of Masado[8]. It is apparent that the degree of weathering of Masado is almost limited between 20 to 40% of CWI, and the chemical weathering tendency is same as the other granitic weathering products shown in Fig-1 and Fig-4.

4.3 CWI and engineering classification and idnetification of residual soils

Versatile proposals have been presented regarding the rock classification of granite. In this section, initially, those rock classification will be examined how the new chemical weathering index CWI corresponds to it. Employing Dearman's method as a rock classification method[9], the relationship between CWI and rock classification is shown on Table-2. Dearman's rock classifications II, III and IV are approximately equal 15 to 20% in CWI, and V is approximately equal to 20-40% in CWI. That is, rock classification is a classification

Table 3 Relationships between CWI and climatic province and types of weathered products

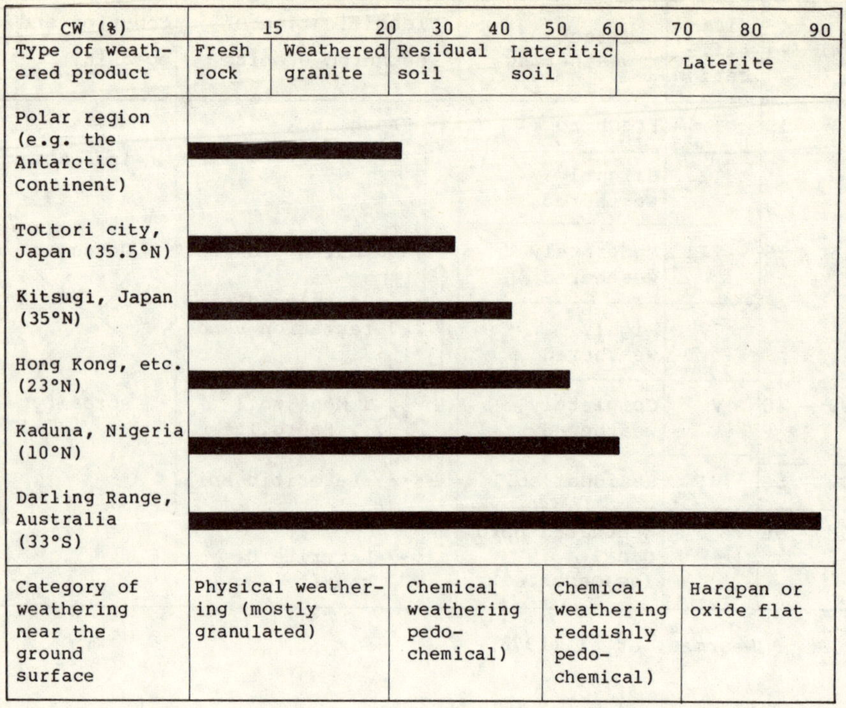

method centering on rocks, and the range of such changes in chemical weathering extents is small. On the other hand, it seems that engineering parameters such as cohesion and porosity change as CWI changes in the table. These parameters are crucial ones useful in examining the stability of slopes composed of granitic residual soil.

Now then, how will CWI relate to the climatic province on the earth? Table-3 shows their conceptual relationship. In the polar regions such as the Antarctic Continent, rocks have been granulated mostly due to physical weathering, and not much soil development is sighted. In temperate zone such as Japan, there are reddish residual soils remaining in those areas where the topographies are preserved owing to palaeoclime. On the contrary, Masado (sandy and silty) is the typical residual soil in mountainous configurations with sharp slopes. In tropical zones and subtropical zones, reddish residual soils (lateritic soil) are predominant as a whole except laterite flats. In addition, in the periphery of Darling Range, Australia, which belongs to the temperate zone at the present day, but because it was under tropical climate in the Tertiary period and the Australian Continent has been a stable continent, weathered residues have remained for a long period, thereby building up bauxite (laterite).

As mentioned in the introduction, the definition and identification of many kinds of residual soils has not been defined because these residual soils are the products in local regions. Using the chemical weathering index CWI and the proposals done by several researchers

Table 4 Definitions of granitic residual soils using CWI

Type of residual soil	Definition of term
Granular disintegration sand	Sandy residual soil derived from granite having smaller than 30% CWI
Masado	Sandy and silty residual soil derived from granite, having smaller than 40% CWI
Saprolitic soil [9)]	Residual soil which exhibits clear inherited structure features and lead to easy identification of the parent rock, in case of granite origin, having more than 20% CWI
Lateritic Soil [9)]	It belongs to horizons A or B of well drained profiles developed under humid tropical climates, its clay fraction is constituted essentially of minerals of the kaolinite group and of iron or aluminum hydrated oxides. It is a soil in a physical sense and has more than 40% CWI
Laterite	It belongs to horizons A or B of well drained profiles developed under tropical (sometime temperate) climates, it is constituted essentially of minerals of sesquioxide of iron and aluminum. It is cemented or pisolithic and having more than 60% CWI

and committee, new definitions of several kinds of granitic residual soils can be shown in Table-4. The definitions and identifications of the granitic residual soils can be more correct because quantitative values are also added in the definitions.

5. Conclusion

The new proposed CWI can express the weathering tendencies from freshrock to residual, lateritic soils and laterite fo granite origin, and it can also exhibit the degree of weathering quantitatively. Therefore, it will become possible for granitic residual soils from different places to be compared quantitatively. In these days common scale and common definition of geotechnical terms such as residual, lateritic, saprolite, laterite are needed for geotechnical engineer or engineering geologist in the world, so if the new proposal can help it, it is a great pleasure for the authors.

Reference

1) Kimiya K. 1975 Tensile strength as a physical scale of weathering in granitic rocks. Chishitsugaku-Zasshi Vol. 81, No. 6 pp. 349-364 (In Japanese)

2) Miura K. 1973, 74, 75 Weathering in plutonic rocks I, II, III. Engineering Geology Vol. 14 No. 3, Vol. 15 No. 1, Vol. 16 No. 2 (In Japanese)

3) Miura K. 1980 Weathering of granites and their mechanical properties Tsuchi-to-Kiso (Soil and Foundation) Vol. 28, No. 7 pp. 11-20

4) Reiche P. 1943 Graphic representation of chemical weathering, Journal of Sedimentary Petrology, Vol. 13, No. 2 pp. 58-68

5) Thornton C.P. and Tuttle O.F. 1960 Chemistry of igneous rock I Differential index American Journal of Science Vol. 258 pp. 662-684

6) Aramaki S., Hirayama K. and Nozawa T. 1972 Chemical composition of Japanese granites Part 2 Variation trend and average composition of 1200 analysis Journal of the Geological Society of Japan Vol. 78 No. 1 pp. 39-49

7) Onodera T., Yoshinaka R. and Oda M. 1974 Weathering and its relation to mechanical properties of granite, Advances in Rock Mechanics, National Academy of Science, Washington D.C. pp. 71-78

8) Nishida K. Sasaki and Kuboi 1984 Amorphous material of decomposed granite soil (Masado) and its properties Soil and Foundation Vol. 24 No. 2 pp. 180-190

9) Infan T.Y. and Dearman W.R. 1978 Engineering classification and properties of weathered granite Bulletin of Int. Assoc. of Engineering Geology No. 17 pp. 79-90

Pore size distribution characteristics of tropical soils in relation to engineering properties

E.R.TUNCER, Associate Professor of Civil Engineering, Jordan University of Science and Technology, Irbid, Jordan

ABSTRACT: To predict the engineering behavior of tropical soils in terms of index properties usually fails. Recent studies have emphasized that the engineering behavior of tropical soils is controled, to a large extent, by their structure. Pore-size distribution characteristics give significant information about the fabric and enable one to quantify pore structure. In this study, nine soils derived from basalt and five soils derived from volcanic ash, all from Hawaii, were subjected to pore-size distribution measurements, shear strength tests, engineering index tests, and mineralogical analyses. Various parameters were generated from the pore-size distribution curves to characterize the pore structure of each soil series. The test results indicate that the pore structure parameters have significant correlations with degree of weathering, mineralogy, and strength properties of the basalt derived soils. In volcanic ash derived soils, on the other hand, shrinkage seems to be the most significant behavior having strong correlations with pore structure characteristics.

INTRODUCTION

The engineering behavior of highly weathered tropical soils in the undisturbed state is mainly controled by their structure. For this reason, regular classifications which are based on engineering index tests using disturbed samples, fail to predict the field behavior of tropical soils (Gidigasu, 1971; Lohnes and Demirel, 1973; Tuncer and Lohnes, 1977). In an effort to better understand the structure of such soils in relation to various engineering properties, the pore-size distribution of nine soils derived from basalt and of five soils derived from volcanic ash was measured by mercury injection porosimetry tests. The usefulness of mercury injection porosimetry to determine pore-size distribution characteristics of soils was introduced by Diamond (1970). Since then several studies have been performed on soils providing detail descriptions of the method. The mercury injection technique differs from microscopic methods which are utilized to study the structure of soils, because it enables one to quantify the pore structure of soils and to search various relationships between the soil structure and other soil properties, quantitatively.

The objective of this study is to relate engineering properties to pore-size distribution characteristics of tropical soils.

LOCATION OF SAMPLE SITES

The soils investigated in this study were collected from the islands of Oahu, Kauai, and Hawaii. The islands of the state Hawaii are located in the central part of the Pacific Ocean.

The soil series from the island of Oahu are Molokai, Lahaina, Wahiawa, Manana, and Paaloa. All were derived from the Koolau basalt series on the leeward side of the Koolau mountain range. According to Wentworth and Winchell (1947), this basalt series is very uniform chemically and petrologically, so it may be assumed that the soils were derived from essentially the same parent rock.

The soil series from the island of Kauai are Lihue, Puhi, Kapaa, and Halii. All were derived from Koloa basalt series on the windward side of the island. Beinroth et al. (1974) state that these soil series can be considered as derived from the same parent material.

The five soil series from the island of Hawaii are Kilohana Waimea, Kukaiau, Honokaa, and Hilo, all derived from volcanic ash. The Kilohana and Waimea weathered on the leeward side of the island and will be considered as the dry soils; the Kukaiau, Honokaa, and Hilo formed under the higher rainfall of the windward side of the island and will be grouped together as the wet soils. The parent material for all five series is the Pahala ash of Pleistocene age (MacDonald, 1949).

METHODS OF STUDY AND TEST RESULTS

Thin-walled Shelby tubes were utilized to collect undisturbed samples. Two or three borings were drilled at each location. All the samples were extracted from B-horizon, since it is influenced by the weathering process more than the other horizons and can form a good basis for the comparison of different soils.

Although standard engineering index tests do not accurately predict the field behavior of tropical soils, they were performed for completeness, and for any possible correlation that might show up. Wet densities were determined by bulk measurements. Water contents were measured and used together with wet densities to compute dry densities. In the determination of specific

Table I. Results of the physical and engineering index tests.

Soil Series	Water Content (%)	Dry Density (kg/m³)	Specific Gravity	Void Ratio	Sand (%)	Silt (%)	Clay (%)	Shrinkage Limit (%)	Liquid Limit (%)	Plastic Limit (%)	Plasticity Index (%)
Molokai	23.03	1410	2.946	1.088	28.7	51.0	20.3	24.6	45.5	38.7	6.8
Lahaina	22.82	1350	2.937	1.174	31.0	38.7	30.3	32.6	49.1	38.5	10.6
Wahiawa	27.28	1480	2.989	1.020	10.7	45.5	43.8	26.0	51.3	36.0	15.3
Manana	33.85	1400	2.991	1.137	69.3	16.7	14.0	39.1	67.1	50.4	16.7
Paaloa	39.56	1320	3.098	1.347	27.8	39.2	33.0	23.8	57.1	39.3	17.8
Lihue	35.54	1330	3.112	1.342	12.4	28.8	58.8	34.2	72.6	54.9	17.7
Puhi	41.33	1290	3.326	1.577	24.5	39.9	35.6	36.2	44.0	40.3	3.7
(shallow)	43.09	1260	3.297	1.660	-	-	-	-	-	-	-
(deep)	39.14	1340	3.351	1.463	-	-	-	-	-	-	-
Kapaa	45.79	1170	3.395	1.899	24.0	41.7	34.3	31.0	52.0	42.0	10.0
Halii	43.74	1230	3.391	1.755	45.1	34.4	20.5	29.8	40.0	35.4	4.6
(soil matrix)	52.53	1170	3.493	1.905	-	-	-	-	-	-	-
(concretion)	24.77	1660	2.985	0.802	-	-	-	-	-	-	-
Kilohana	39.20	920	2.904	2.155	69.0	24.8	6.2	44.7	-	-	-
Waimea	45.79	810	3.084	2.802	33.0	57.7	9.3	35.0	66.8	63.0	3.8
Kukaiau	164.13	440	3.071	5.993	89.0	10.4	0.6	47.9	93.3	88.6	4.7
Honokaa	244.92	300	2.980	8.901	89.5	9.0	1.5	50.5	-	-	-
Hilo	234.90	350	2.842	7.130	87.5	7.4	5.1	56.3	61.4	-	-

gravities, Atterberg limits, and grain-size distributions, the trimmings from thin-walled tube samples were mixed and quartered with a sample splitter, in order to prepare representative samples for each soil series. Porosity and void ratio values were also calculated. In laboratory testing, it was observed that Puhi gives different characteristics for samples from shallow and deep depths, and Halii has different properties for soil matrix and concretions it contains. So, each of these two soil series was treated as being two different soils, and whenever it was possible the soil properties of each were determined separately. Table I summarizes the results of the physical and engineering index tests.

In the qualitative mineralogical analysis, representative samples of each soil series were subjected to x-ray diffraction and differential thermal analysis (DTA), after oven drying and sieving them through No. 200 sieve. In x-ray diffraction, either copper-K_α or molybdenum-K_α radiation was used. Randomly oriented powder samples were examined oven-dried, glycolated, and heated. Analysis showed that kaolinite and gibbsite are the main clay minerals present in the basalt derived soils. Hematite is the most abundant iron compound detected. Volcanic ash derived soils have significantly different mineralogy. The driest soil has high contents of primary minerals, especially plagioclase feldspar; some hydrated halloysite is also present in that soil. The second driest soil contains higher halloysite content, but the three wet soils contain none. Allophane and gibbsite contents are, in general, higher in the wetter soils. Using DTA, kaolinite, gibbsite, halloysite, and allophane were identified readily. Hematite, on the other hand, does not have any particular reaction within the temperature range (20° - 1000°C) covered.

In the second step of mineralogical analysis, the objective was to determine the contents of predominant minerals in these soils, i.e., kaolinite, gibbsite, allophane, and iron oxides. Determination of the kaolinite and gibbsite contents was achieved by using thermal gravimetric analysis (TGA) and DTA as suggested by Wendlandt (1974). Hematite was determined quantitatively by x-ray diffraction using the method developed by Tuncer et al. (1976). The basis of this method is a comparison of x-ray diffraction peak intensities of the mineral in the sample with the intensities of the same peak in specimens which are a mixture of the natural soil sample and known additional amounts of the mineral in question. A quantitative estimate of allophane in the soils was the most difficult determination, because of its amorphous nature and difficulty to obtain pure allophane. The allophane content was determined indirectly by selecting the Hilo soil, which has the fewest mineral species in it, as a standard and assuming that the only minerals present are oxides of iron, gibbsite, and allophane. Once the gibbsite and iron oxide contents are determined, the allophane content is calculated by subtraction. For the remaining four soils derived from volcanic ash, the allophane content was determined by comparing the weight loss for the most characteristic reaction for allophane in each soil with the weight loss for the reaction in the Hilo. The total amount of iron oxides occuring in these soils was determined by first measuring the amount of total iron present, and then calculating total iron oxides on the basis of total iron content. To achieve this, a quantitative x-ray fluorescence analysis was utilized as suggested by Tuncer et al. (1977). In addition, replicates of each soil sample were analyzed for iron content with an atomic absorption spectrophotometer. Organic matter contents were determined by the dichromate method. The mineral and chemical contents of basalt derived soils are shown in Table II, and similar data for the ash derived soils are shown in Table III.

Strength characteristics of the soils were analyzed by Paulson (1975). Direct shear testing

Table II. Mineral and chemical contents of basalt derived soils.

Soil Series	Kaolinite Content[(a)] (%)	Gibbsite Content[(b)] (%)	Hematite Content[(b)] (%)	Total Iron Content by x-ray fluor. (%)	Total Iron Content by atomic absorption spec. (%)	Total Fe_2O_3 Content[(c)] (%)	Organic Matter (mg/g)
Molokai	52	9	12.6	11.4	13.5	16.3	4.28
Lahaina	60	9	10.3	11.9	12.7	17.0	4.14
Wahiawa	54	5	13.3	12.4	14.3	17.7	3.81
Manana	43	16	6.4	16.3	17.1	23.3	8.43
Paaloa	30	33	20.1	17.3	18.9	24.7	8.67
Lihue	51	10	3.8	14.3	14.3	20.4	8.91
Puhi	11	47	15.0	17.7	25.1	25.3	11.30
(shallow)	10	42	12.1	14.2	26.0	20.3	12.10
(deep)	13	52	8.6	19.3	24.1	27.6	10.50
Kappa	10	48	6.6	24.8	26.5	35.4	12.12
Halii	1	53	12.1	28.1	30.4	40.1	15.40
(soil matrix)	1	52	16.2	18.6	28.5	26.6	13.21
(concretion)	0	76	13.6	19.2	32.2	27.4	17.58

(a) By thermal gravimetric analysis, (b) By x-ray diffraction, (c) From x-ray fluorescence.

Table III. Mineral and chemical contents of ash derived soils.

Soil Series	Allophane Content (%)	Gibbsite Content (%)	Iron Oxide Content (%)	Organic Matter (%)	Relative Contents Halloysite	Relative Contents Plagioclase
Kilohana	26	0	10.0	4.6	X[(a)]	XX[(b)]
Waimea	32	8	16.4	2.2	XX	0
Kukaiau	42	27	17.7	5.3	0	0
Honokaa	55	22	19.6	9.2	0	0
Hilo	42	32	26.0	4.2	0	0

(a) X- small amount, (b) XX- large amount

was used to determine laboratory undisturbed soil strength. It is very difficult to ascertain drainage conditions during direct shear testing; therefore, the results are used only as empirical indicators of the strengths of the soils studied. Table IV outlines the representative strength parameters of the soils. In general, the ash derived soils have much lower strength than those of basalt derived soils.

Table IV. Strength parameters of soils studied.

Soil Series	Cohesion (kPa)	Internal Friction Angle (degree)
Molokai	280	34
Lahaina	350	52
Wahiawa	175	27
Manana	210	57
Paaloa	184	38
Lihue	119	41
Puhi	84	52
Kapaa	70	44
Halii	49	35
Kilohana	(Too friable to test)	
Waimea	(Too friable to test)	
Kukaiau	70	37
Honokaa	70	33
Hilo	42	33

Volcanic ash derived soils were subjected to shrinkage by allowing them to air dry in the laboratory at a relative humidity of 20% ± 3%. The water content versus time curves for this experiment for each soil series are shown in Figure 1. The theory of drying as developed in chemical engineering indicates that the form of the drying curves depends on the structure and composition of the soil and the mechanism by which the water moves within the soil.

The soils were subjected to mercury porosimetry tests in order to obtain their pore-size distribution characteristics. The instrument used in this study has a pressure range from 7 kPa to 350 MPa. Using a mercury surface tension of 474 dynes/cm and a contact angle of 140°, it is possible to measure pore radii between 105.4 μm and 0.0021 μm. Prior to testing with a mercury porosimeter, soil samples must be dried. To evaluate the influence of drying on the pore size distribution of all soil series, two or three portions of each soil from the same boring and depth were subjected to freeze drying, oven drying and/or air drying. Among those three techniques, freeze drying accepted as the best to keep the disturbance at a minimum level. To freeze dry the soil samples, the procedure suggested by Tovey and Wong (1973) was used. The results showed that other than five soil series, Puhi, Halii, Kukaiau, Honokaa, and Hilo, all

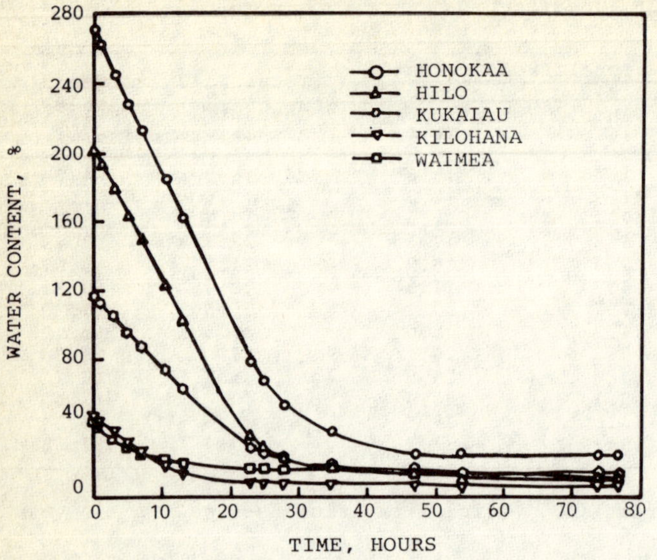

Figure 1. Drying curves for each ash derived soil.

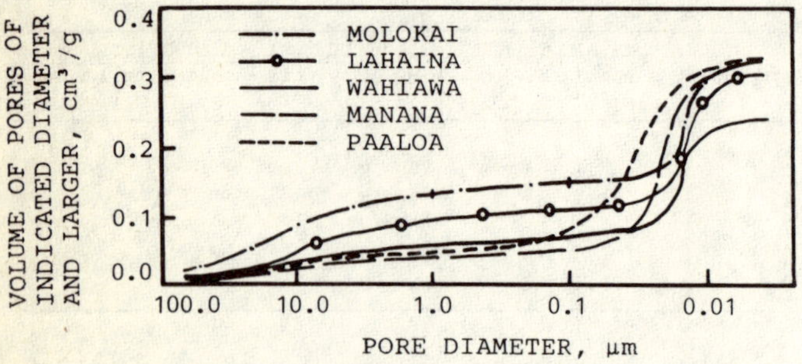

Figure 2. Pore size distribution curves for soils from the island of Oahu.

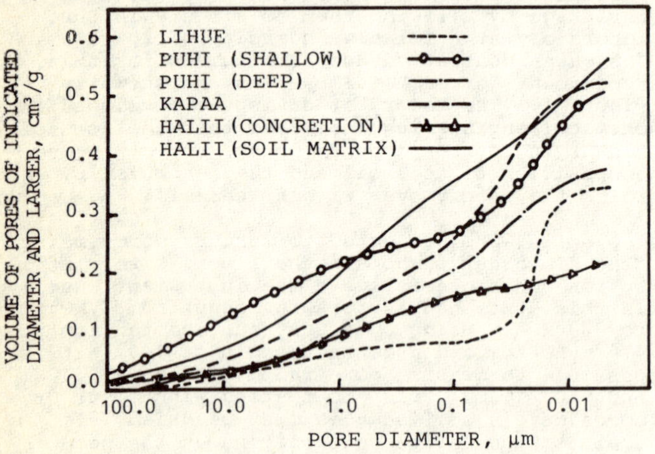

Figure 3. Pore size distribution curves for the soils from the island of Kauai.

give practically the same pore-size distribution curve for freeze dried, oven dried, and air dried samples. The last three soils, all derived from ash, exhibit volume changes of the order of 75%. For each series, 4 to 10 samples from various depths and borings were subjected to pore-size determinations. Figures 2 and 3 show the

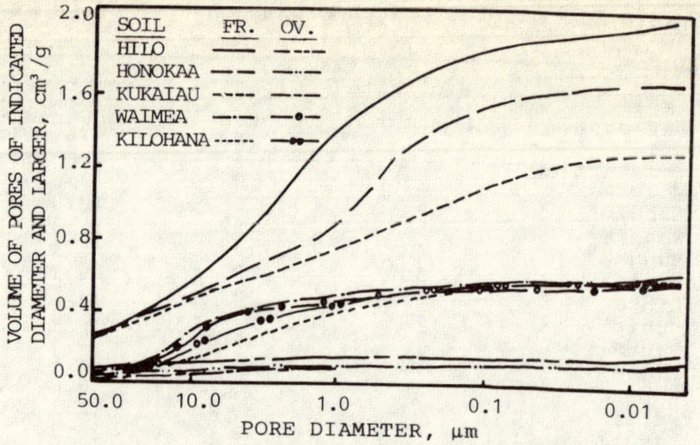

Figure 4. Pore size distribution curves for soils from the island of Hawaii.

representative pore-size distribution curves for the basalt derived soils. Figure 4 shows pore-size distribution curves for freeze dried and oven-dried ash derived soils.
From these curves, it is possible to generate a set of parameters to characterize the pore structure of each series. The parameters generated are the median (50%) and small mode pore diameters, the uniformity coefficient of pores, and the void ratios computed from the mercury injection data as well as from bulk measurements. The uniformity coefficient of pores is defined, in analogy with the uniformity coefficient calculated from particle-size distribution curves, as the ratio of the minimum pore diameters corresponding to 40% and 80% of the total pore volume penetrable by mercury, respectively. Table V outlines all the parameters generated from mercury porosimetry results.

DISCUSSION OF TEST RESULTS

The specific gravity of tropical soils appears to be a very useful parameter in relating, especially basalt derived tropical soils, to the degree of weathering. By definition, the specific gravity of solids in a soil sample is the weighted average of the specific gravities of the minerals which comprise the soil. According to Alexander and Cady (1962), primary feldspars are reduced to kaolinite and primary ferromagnesian minerals are converted to sesquioxides; even the kaolinite is converted to gibbsite, as weathering proceeds. Higher specific gravity indicates the presence of relatively larger amounts of high specific gravity minerals in the soil and suggests a more weathered soil. The Hawaiian basalt derived soils investigated in this study exhibit a good correlation between specific gravity and total iron content, measured by x-ray fluorescence analysis, as shown in Figure 5. Figure 6 shows a median pore diameter versus specific gravity plot, indicating that pore sizes are getting larger as the soils get more weathered in basalt derived tropical soils. The very same thing is indicated in a small mode pore diameter versus specific gravity plot, shown in Figure 7. The uniformity coefficient of pores tends to increase with increasing specific gravity implying that in the cases of more severely weathered soils the distribution gets less uni-

Table V. Pore size parameters generated from mercury porosimetry data.

Soil Series	Void Ratio by Bulk Measurement	Void Ratio by Hg-injection	Percent of Total Pore Volume Intruded by Hg	Median Pore Diameter (μm)	Small Mode Pore Diameter (μm)	Uniformity Coefficient of Pores d_{40}/d_{80}
Molokai	1.088	0.960	88.5	0.037	0.015	107.9
Lahaina	1.174	0.919	78.3	0.019	0.014	5.2
Wahiawa	1.020	0.723	70.9	0.017	0.015	1.8
Manana	1.137	0.978	86.0	0.023	0.022	1.6
Paaloa	1.347	1.029	76.4	0.038	0.031	2.2
Lihue	1.342	1.075	80.1	0.021	0.020	1.6
Puhi (deep)	1.463	1.255	85.8	0.135	0.032	13.9
Puhi (shallow)	1.660	1.660	100.0	0.190	0.025	82.0
Kapaa	1.899	1.739	91.6	0.102	0.040	9.0
Halii (soil matrix)	1.905	1.902	99.8	0.380	-	41.2
Halii (concretion)	0.802	0.609	75.9	0.590	-	27.4
Kilohana	2.155	1.765	81.9	2.710	-	14.3
Waimea	2.802	2.022	72.2	8.200	-	22.1
Kukaiau	5.993	4.049	67.6	2.000	-	42.1
Honokaa	8.901	5.074	57.0	1.380	0.400	14.3
Hilo	7.130	5.801	81.4	2.120	-	8.5

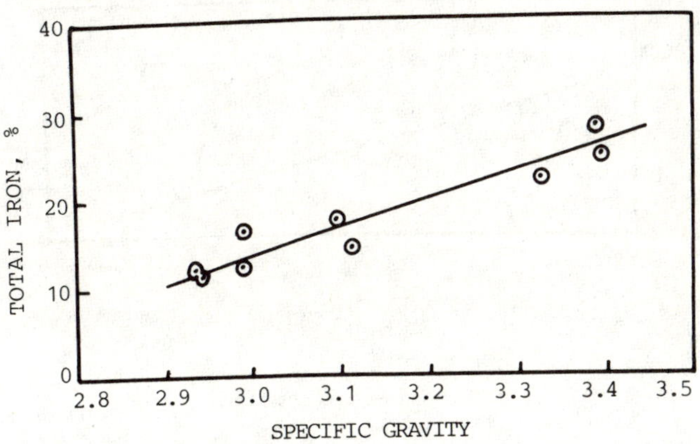

Figure 5. Relationship between total iron content and specific gravity for basalt derived soils.

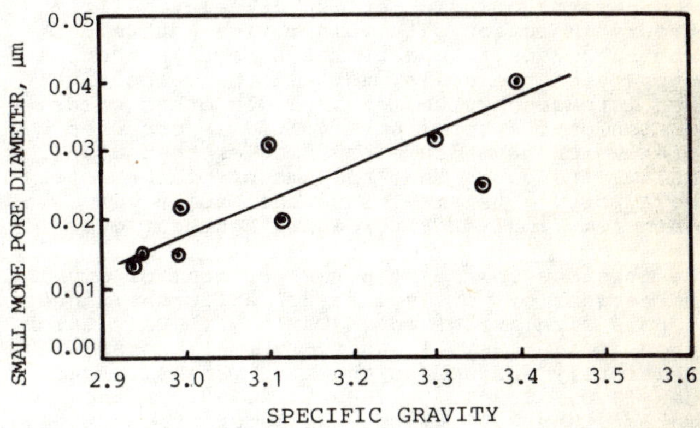

Figure 7. Relationship between small mode pore diameter and specific gravity for basalt derived soils.

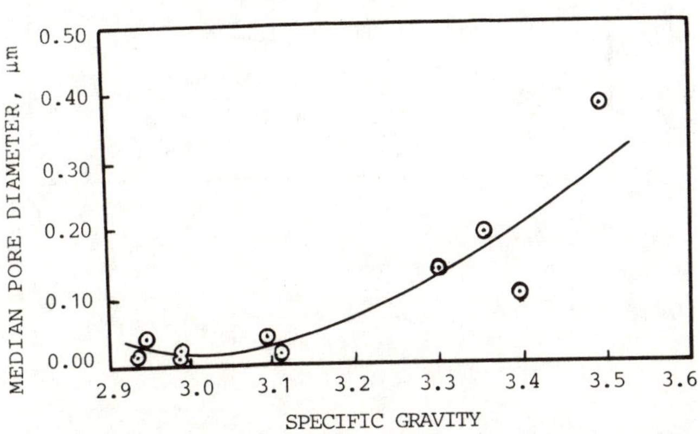

Figure 6. Relationship between median pore diameter and specific gravity for basalt derived soils.

form, in other words, there are all sizes of pores distributed more evenly within the soil domain.

The occurrence of large size pores in more weathered soils can be interpreted as implying larger soil aggregates in such soils. A soil composed of fine grains with large size pores among them is not very likely. In the course of weathering at one stage, kaolinite is the major mineral component in the soil. In the further stages, kaolinite starts to decrease while sesquioxides of iron and aluminum increase. Kaolinite is known to be very inert as a binding agent. Sesquioxides, on the other hand, have the capability of cementing the small grains together and forming larger, water stable soil aggregates (Baver et al., 1972). As weathering proceeds further, sesquioxide content keeps on increasing, leading to the formation of larger soil aggregates. It might be said that toward the last stages of weathering, soils possess a granular quality due to the aggregation taking place.

Table VI. Amounts of pores in different size ranges for the soils studied.

Soil Series	Very Large Pores (cm³/g)	Very Large Pores (%)	Large Pores (cm³/g)	Large Pores (%)	Medium Pores (cm³/g)	Medium Pores (%)	Small Pores (cm³/g)	Small Pores (%)	Very Small Pores (cm³/g)	Very Small Pores (%)
Molokai	0.087	23.6	0.046	12.5	0.016	4.3	0.148	40.1	0.072	19.5
Lahaina	0.046	11.5	0.049	12.3	0.015	3.7	0.159	39.8	0.131	32.7
Wahiawa	0.036	10.6	0.026	7.6	0.010	2.9	0.149	43.7	0.120	35.2
Manana	0.026	6.8	0.015	3.9	0.010	2.7	0.249	65.5	0.080	21.1
Paaloa	0.031	7.1	0.016	3.7	0.038	8.7	0.228	52.0	0.125	28.5
Lihue	0.022	5.1	0.041	9.6	0.017	3.9	0.245	56.8	0.106	24.6
Puhi (shallow)	0.119	23.7	0.095	18.9	0.055	10.9	0.200	39.8	0.034	6.7
Puhi (deep)	0.035	8.0	0.068	15.6	0.100	22.9	0.152	34.7	0.082	18.8
Kapaa	0.050	9.0	0.085	15.2	0.118	21.1	0.237	42.4	0.069	12.3
Halii (soil matrix)	0.070	12.8	0.122	22.4	0.061	11.2	0.235	43.1	0.057	10.5
Halii (concretion)	0.030	11.1	0.060	22.3	0.069	25.7	0.036	13.4	0.074	27.5
Kilohana	0.134	18.1	0.260	35.0	0.145	19.5	0.170	22.9	0.033	4.5
Waimea	0.288	31.7	0.212	23.3	0.072	7.9	0.037	4.1	0.300	33.0
Kukaiau	0.459	23.5	0.288	14.8	0.347	17.8	0.203	10.4	0.654	33.5
Honokaa	0.484	16.2	0.407	13.6	0.647	21.7	0.146	4.9	1.303	43.6
Hilo	0.547	21.8	0.789	31.4	0.516	20.6	0.132	5.3	0.525	20.9

A comparison of void ratios calculated from mercury injection with void ratios determined from bulk density measurements reveals that, in majority of the soils, not all the voids have been intruded by the mercury. In other words, a portion of the total pore volume is occupied by pores which have diameters smaller than 0.0042 µm. Table V indicates that the amount of these very small pores gets smaller as the soils get more weathered, implying increasing aggregation.

The absolute volume of pores per gram of dry soil and percent of pore volume for different ranges of pore size are given in Table VI. Very large, large, medium, small, and very small pores are arbitrarily defined as the size ranges larger than 10.0, 10.0-1.0, 1.0-0.1, 0.1-0.01, and smaller than 0.01 µm pore diameter, respectively. Table VI indicates that smaller size pores decrease while medium, large, and very large size pores increase in percent with increasing weathering, again due to increasing aggregation as weathering proceeds.

Strength parameters, cohesion and internal friction angle, are plotted against median pore diameter and small mode pore diameter, respectively, as shown in Figures 8 and 9. These figures show that as pore size increases internal friction angle also increases, while cohesion decreases. Increasing pore size indicates a soil with more aggregates in it, causing a granular soil type behavior.

Tropical soils, weathered from volcanic ash, usually have high natural water contents, and they exhibit large volume changes upon drying. The shrinkage limit on disturbed representative samples of each volcanic ash derived soil was determined and the volume change was calculated by dividing the difference between the volumes of the soil at the natural water content and at the shrinkage limit by the volume of the soil at the natural water content. The shrinkage limits and volume changes for the five ash derived soils are shown in Table VII. Pore-size distribution tests performed on oven dried and freeze dried specimens enable one to calculate void ratio changes and consequently volume changes resulting from drying as given in Table VII. The calculations for volume change based on both shrinkage limit and pore-size distributions

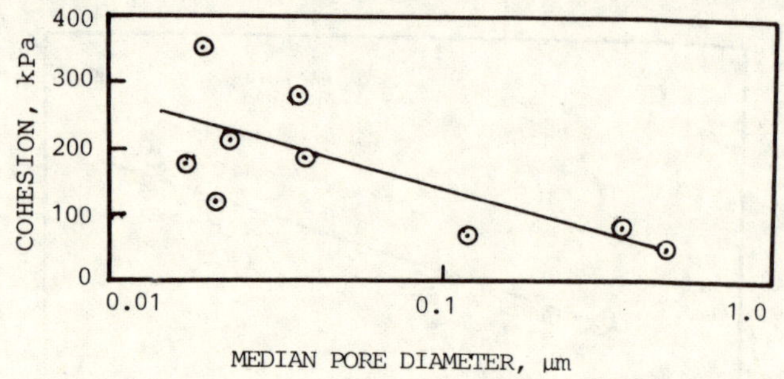

Figure 8. Relationship between cohesion and median pore diameter for basalt derived soils.

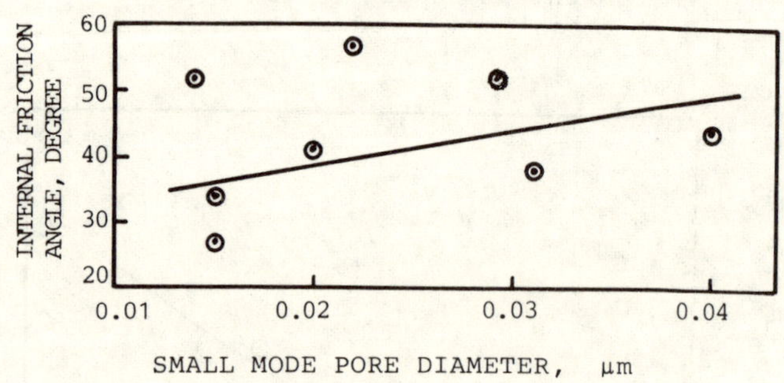

Figure 9. Small mode pore diameter versus internal friction angle relationship for basalt derived soils.

Table VII. Volume change characteristics of ash derived soils.

Soil Series	Shrinkage Limit, SL (%)	Volume Change from SL (%)	Void Ratio Change from Porosimetry	Volume Change from Porosimetry (%)	Critical Water Content (%)
Kilohana	44.7	0[a]	0.110	3.5	27
Waimea	35.0	8.3	0.539	14.2	38
Kukaiau	47.9	51.1	5.156	73.7	50
Honokaa	50.5	58.2	8.057	81.4	80
Hilo	56.3	62.5	6.109	75.1	56

[a] The field water content is below the shrinkage limit.

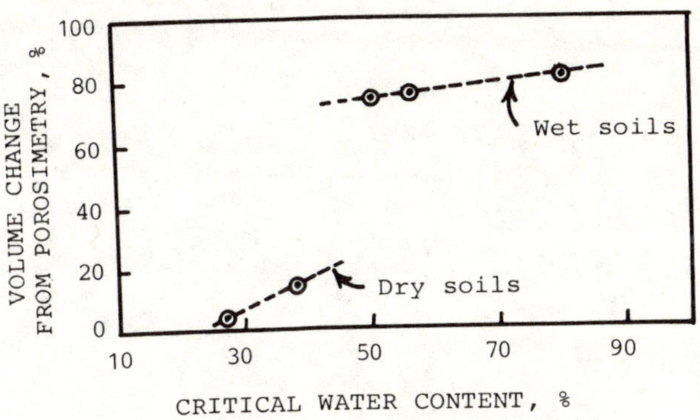

Figure 10. Critical water content versus volume change relationship for ash derived soils.

demonstrate that the dry soils show negligible volume changes, whereas the wet soils exhibit very large volume changes upon drying. The void ratio change from porosimetry measurements should be the most accurate measures of the shrinkage since these measurements reflect the behavior of the undisturbed soil.

By studying the drying curves of volcanic ash derived soils, it is possible to define the critical water content as the water content at which the curve deviates from linearity while drying. The critical water contents are shown in Table VII. In all soils, except Honokaa, the critical water contents and the shrinkage limits are nearly equal. The good correlation shown in Figure 10 between the critical water contents and the volume changes as measured by mercury porosimeter suggests that the critical water content might be considered as a parameter in predicting relative shrinkage in these soils.

CONCLUSIONS

Based upon the data from the Hawaiian soils studied here, the following conclusions may be drawn regarding the relationships between engineering properties and pore-size distribution characteristics.

1. Mineralogy has a very good correlation with the degree of weathering. In basalt derived tropical soils, specific gravity of solids, which depends on the minerals comprise the soil, is a reliable parameter that reflects the degree of weathering. Pore-size distribution parameters versus specific gravity relationships reveal that soil aggregation increases with increasing weathering since sesquioxides act as binding agents, in advanced weathering stages.

2. Strength parameters, cohesion and internal friction angle, have systematic trends with increasing degree of weathering. These trends can be shown further by the relationships obtained between the pore-size distribution characteristics and strength parameters.

3. In volcanic ash derived soils, the magnitude of volume change upon drying as obtained by using the void ratio change from porosimetry measurements seems to be more reliable than the one computed on the basis of shrinkage limit measurements.

4. The critical water content as determined from drying curves of volcanic ash derived soils may be a valuable parameter in predicting relative shrinkage in these soils, since this parameter exhibits a good correlation with the volume change as calculated from the mercury porosimetry measurements.

5. Pore structure parameters suggested here seem to have a great potential in quantifying soil fabric and studying the influence of fabric on various properties of soils, quantitatively.

ACKNOWLEDGEMENTS

This research was conducted with funds from the U.S. Army Research Office and the Engineering Research Institute of Iowa State University. Oran Bailey, State Soils Scientist of Hawaii, helped select sample sites and provided pedologic data. Prof. R.A. Lohnes and Prof. T. Demirel of the Iowa State University, Civil Engineering Department, contributed to the interpretations of experimental results. The author wishes to thank all the above for their contributions to this work.

REFERENCES

Alexander, L.T. and J.G. Cady, 1962, Genesis and hardening of laterite in soils, U.S. Department of Agriculture Tech. Bull. 1282.

Baver, L.D. et al., 1972, Soil physics, 4th ed. John Wiley and Sons, Inc., New York.

Beinroth, F.H. et al., 1974, Geomorphic relationships of oxisols and ultisols on Kauai, Hawaii, Soil Science of America 38: 128-131.

Diamond, S., 1970, Pore size distribution in clays, Clays and Clay Minerals 18: 7-23.

Gidigasu, M.D., 1971, The importance of soil genesis in the engineering classification of Ghana soils, Engineering Geology 5: 117-161.

Lohnes, R.A. and T. Demirel, 1973, Strength and structure of laterites and lateritic soils, Engineering Geology 7: 13-33.

MacDonald, G.A., 1949, Petrography of the island of Hawaii, U.S. Geological Survey, Professional paper 214-D: 51-96.

Paulson, S.K., 1975, The strength, structure, and mineralogy of selected Hawaiian lateritic soils, Unpublished M.S. thesis, Iowa State University, Ames, Iowa.

Tovey, N.K. and K.Y. Wong, 1973, The preparation of soils and other geological materials for the SEM, Proceedings of the International Symposium on Soil Structure, Gothenburg, 59-67.

Tuncer, E.R. et al., 1976, A method for quantitative determination of soil minerals by x-ray diffraction, Transportation Research Board Record 560: 75-80.

Tuncer, E.R. et al., 1977, Quantitative analysis of elements in sediments and soils by x-ray fluorescence, Clay and Clay Minerals 25 (No. 2): 73-77.

Tuncer, E.R. and R.A. Lohnes, 1977, An Engineering classification for certain basalt-derived lateritic soils, Engineering Geology 11: 319-339.

Wendlandt, W.W., 1974, Thermal methods of analysis, John Wiley and Sons, Inc., New York.

Wentworth, C.J. and H. Winchell, 1947, Koolau basalt series, Oahu, Hawaii, Bulletin of the Geological Society of America 58: 46-78.

Characterization, identification and classification of tropical soils

M.VARGAS, Themag Engenharia Ltd, São Paulo, Brazil

ABSTRACT: A discussion about the advantages and disadvantages of the pedological, geological and geotechnical approaches for the characterization, identification and classification of tropical soils is presented. The discussion concludes with the statement of the necessity to use geotechnological index properties when identifying tropical soils for engineering purposes, although pedological and geological information would be very usefull for characterization of such soils. Examples of tropical soils occurring in Brazil are given to illustrate the idea that besides grain-size distribution and plasticity, mineralogical nature not only of clay-fraction but also of silt-fraction is necessary to identify and classify such soils for engineering purposes, if appropriate test methods are adopted. A simple method to achieve that purpose, within the possibilities of field identification is presented. However design parameters of tropical soils cannot in general, be inferred from those index properties only, because they depend not only on the nature of such soils but mainly on the states they occur in the field. So, it is to keep in mind that it is incorrect to introduce, in a general soil classification, any criterion based on their occasional state properties.

1 INTRODUCTION

Since the publication of Casagrande's paper on classification and identification [1] of soils, in 1948, questions have arisen about the pertinence to consider, in the classification of soils, the mineralogical nature of the clay fraction and thus, the necessity "to add new major sub-divisions, more or less arbitrarily, as soils become better known", as the author had the opportunity to state in his discussion of Casagrande's paper. At that time it had already become clear that kaolin-type clays, that occurred frequently in Brazil, could not be univocally classified in the A.C. system. They could not be distinguished, for instance, from the silts and silty clays of ML and MH groups, if only visualized by their plots in the plasticity chart. So it was proposed, in the discussion, to subdivide silt and clay groups, according to mineralogical composition [2], denoting kaolin-type clays by the symbols KL and KH. According with Casagrande, such clays could be easyly identified by learning their "feel".

Many new systems of classification have been proposed to substitute the A.C. system based on the argument that Casagrande's classification was unable to classify tropical soils. But most of those proposals were systematically wrong, because they confused index properties, which are indices related to the nature of soils, with mechanical properties, which are related to the state the soil has acquired, depending on how it occurred in the field.

It is a principle in the logic of classification that any classification system of natural materials should consider their nature and not the eventual state how they may be found in a certain circumstance. For instance: water is identified by its chemical composition: H_2O; and not by its eventual states: solid, as ice, liquid or gaseous, as vapor. So, index-property of a certain material is what is only related to its nature, necessary and sufficient to classify it for the purpose in view.

Casagrande based his system in grain-size distribution and plasticity. But he, immediately became aware that it was necessary to introduce two new indices: the fibrous nature of peats and the organic matter in organic silts and clays. This, and the fact that he mentioned specifically kaolin-type clays, had opened consideration of the mineralogical composition of clay fractions as another index.

2 GENETIC CHARACTERIZATION OF TROPICAL SOILS

The apparent incapability of Casagrande's system to properly identify and classify tropical soils has led many Brazilian engineers and geologist to evoke geological and pedological approaches for the genetic characterization of tropical soils. For instance: identifying soils, simply by sight, as: residual soil, from such and such rock, or colluvium, alluvium, or lateritic soils or whatever pedological origin they may have. But it became quite obvious that such identification if sufficient for geologists or pedologists, could not give the necessary information needed by engineers, to forecast deign parameters. Only regional correlations between geological and pedological classifications and the engineering properties of soils have been established; but, in general, they are not universally valid. However it became a tradition in Brazilian Geotechnology to rely much more or geological than on pedological investigations, to identify soils for engineering purposes.

Taking into account these geological and pedological considerations it was proposed, in the First Conference on Tropical Soils, held in

Brasilia in 1984, to consider tropical soils as lateritic or saprolitic. The lateritic soils would be those belonging to well drained surface profiles, developed under humid climates. In a report distributed during the Conference [3] it is stated: "Its clay-fraction is constituted essentially of minerals of the kaolinite group and of iron or aluminum hydrated oxides", assembled in peculiar "porous", hardened or aggregate structures. On the other hand, soils would be considered saprolitic if evidently residual showing relict structures from the parent rock.

In spite of being undoubtedly usefull for engineering use, the separation of the tropical soils in lateritic and saprolitic cannot be considered a basic grouping of tropical soils for three reasons. First: it is hopeless to expect to judge mechanical properties, as compressibility or resistance - that is, at the end, main scope of engineers - only by identifying a soil as lateritic or saprolitic. Second: dividing tropical soils strictly into two denominations, leaves no place for other possible types of soils that eventually can be found in the tropics. As, for instance, the surface expansive soils that cannot be nominated lateritic or saprolitic. Third: there are, in the tropics, clay or sand terciary deep beds that do not show any relict structure, overlain by surface lateritic horizons. It is impossible to call those deep beds saprolitic but they are neither lateritic, according to the above mentioned definition.

So the geological or pedological classification systems are not sufficiently developed for pratical geotechnical purposes. Their terminology can and must be used for complementary elucidation but not as basic information for engineers.

3 IDENTIFICATION OF TROPICAL SOILS

The identification of a soil is the act of assigning a proper name to the soil that would individualise it among others. This would not necessarily refer to the eventual state how it is found in the field. For instance: a person's name is the same during his youth or old age. So the name of a soil should be related to its nature and not to its state.

Nature of a soil is that which is maintained whatever be the state or circumstance how it is found. On the other hand, its state: soft, hard, porous, concretionated, collapsible, expansive, etc, is something acquired at a certain circumstance. In general, a soil can be found, in the field, in any of these states without avoiding to be the same type of soil. According to Casagrande's classification system the index--properties sufficient to classify a soil are: grain-size distribution, plasticity and organic matter content. But after almost fifty years of soil mechanics experience in soil classification for engineering purposes, it can be stated than it is necessary to add to such indices not only the mineralogy of clay-fraction but also the mineralogical nature of the silt size grains (as ferrous-oxides, mica cristals or big kaolin cristals).

The grain size distribution and plasticity of the soil must be determined in samples from which all the structure or aggregation of grains has been destroyed. In the case of tropical soils this is a most important point, as in those soils the tendency for aggregation of grains to form clusters is often manifested. As in dry tropical regions, samples are likely to dry out, during storage or transportation to the laboratory, careful re-hydration of the samples must be performed before tests. This is a process that should be investigated in every tropical region and carefully included in the test-method specifications. The simple translation of test methods for grain-size distribution or Atterberg limits, prepared in temperate climate countries, lead to questionable results of tests carried out in the tropics. It is possible that the incomplete disaggregation of grain clusters, in tropical soils, is the main source of criticism against the adoption of grain-size and Atterberg's tests as basic for tropical soils identification and classification for engineering purposes. So appropriate test methods must be adopted in order that classification tests are made on samples of which all traces of sctructure and aggregation are destroyed.

Mineralogy of clay-fraction is something that is not liable to be made by means of simple tests run by engineers and technicians engaged in practical works, as are often the case of soils identification labs. "It should be kept in mind, however" . as remarked by Casagrande - "that any expansion (of A.C. Classification) should not be beyond the possibilities of field identification" [1]. Having in mind this remark many attempts have been made to find out a simple test that would identify the clay minerals present in the clay-fraction of soils, within field possibilities. In such circumstance an approximate characterization of these minerals can be obtained by means of Skempton's "activity index" [4]. Low activity corresponds to kaolin type clays and high activity to montmorilonite and other complex structured clays. It seems that to divide clays, according to the activity index, into low activity ($I_a < 0,75$) and high activity ($I_a > 1,25$) would be sufficient for identification of clay minerals for engineering purposes.

That is the reason why I proposed to use a combination of Casagrande's plasticity and Skempton's activity charts to identify fine grained tropical soils, as shown in figure 1. On the right side chart identification is made according to region where sample plots (above or below A-line and on the right or left side of B-line). But the identification must be completed by the plot of the same sample on the left side chart (above or below D or E-lines and to the right or left of C-line). Combination of the two charts will allow identification of the fine grained soil by means of three index-properties: grain size, plasticity and clay-minerals.

Figure 1 show plots of a lateritic soils from gneiss, along or below A-line and along or below E-line. This is a typical plot of kaolin type soil. A saprolite soil from micaschist plots below A-line and above or below D-line, on the right of C-line. This is the typical plot of a micaceous residual soil.

Figure 1 also shows the plot of a saprolite soil from sandstone that plots above A-line, and above D-line, but with a very low clay content. This is due to a very active clay in its clay-fraction.

However a fourth and very important index property of the tropical soil must be investigated: it is the mineral nature of silt size grains. Unfortunately there is no standardized test for this purpose. But the identification of the

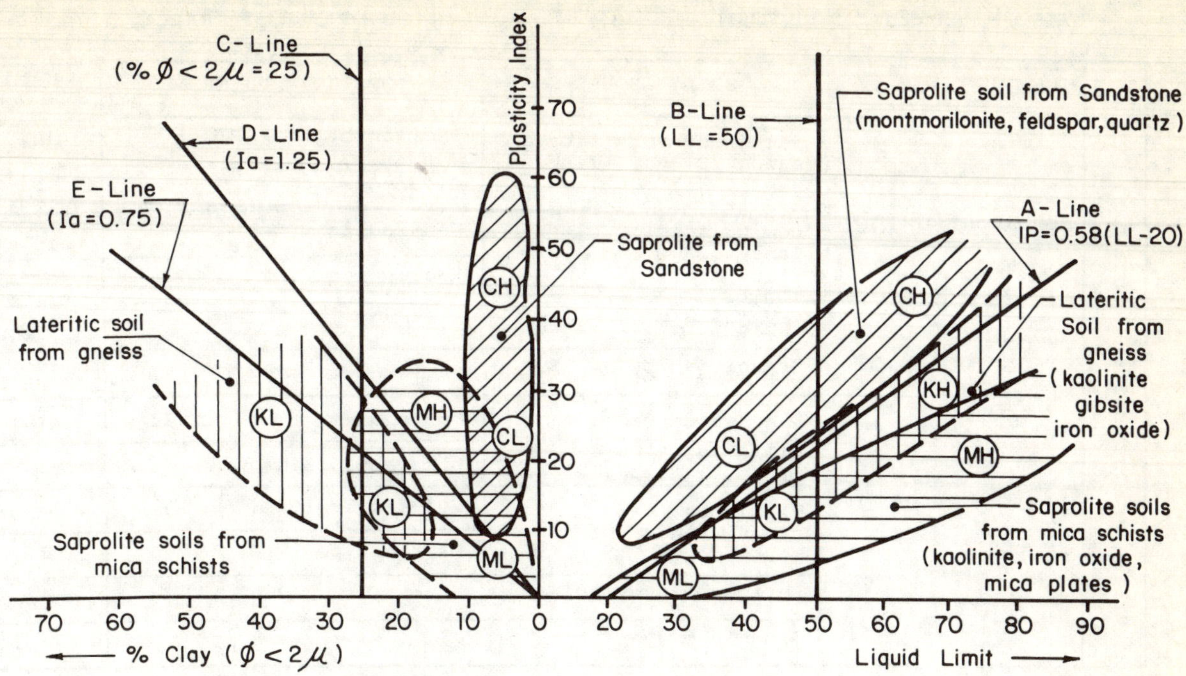

FIG. 1 - ASSOCIATED PLASTICITY AND ACTIVITY CHARTS

nature of iron oxide or mica particles can be made by visual inspection with a magnifying glass, of a partion of pulverized dry soil sample, spread on a white sheet of paper. The micaceous soils shall be classified in special MLm or MHm groups and rich ferrous oxide grain content soils, in special KL_f and KH_f groups.

4. CLASSIFICATION OF TROPICAL SOILS

Taking into account all the above mentioned considerations, classification of tropical soils can be made, in principle, according to AC classification system. The coarse grained soils would be classified, according to grain size distribution, exactly as proposed by Casagrande in one of the eight soil groups with its typical manes. Perhaps a future expansion of AC classification would be by taking into consideration not only the grain sizes but also the form or grains: as for instance: round, angular or flake grains.

The fine-grained soils, containing little or no coarse grained material, shall be divided into two major division: low compressibility (liquid limit less than 50) and high compressibility (kiquid limit more than 50). That is: respectively on the left and the right side of B-line. But now, these two major divisions should be subdivided into silts and clays; and those grop, subdivided - in order to include tropical soils - according to its clay minerals and mineralogical nature of silt grains as, for instance, according to the Table I.

A certain content of mica flakes in the silt fraction of micaceous soils tends to increase its apparent activity index. Consequently in figure 1, samples of saprolite silty soil, from micaschist, that contains mica particles in its silty fraction, plot above the D-line; but have to be classified in the ML and MH groups. Soils from a lateritic surface layer plot along and below A-line and below D-line. However a lateritic soil of group KL_f or LH_f may plot along A-line but above E-line, due to iron oxide grains in its silty fraction. A saprolite soil from sandstone decomposition plots usually along A-line and on the right side of B-line. But if it is a soil with a low content of an active clay, that confers to the sandy-clay soil the behaviour of a pure clay, then it will plot as in figure 1. That is why the sandy or silty clays, showing high plasticity due to the high activity of their clay fraction, were also included in the second group CH or KH, of high plasticity clays.

There is a research going on at the soils laboratory at the University of São Paulo, Brazil, to find out the effects of dehydration of tropical soils samples and the best way to re-hydrate them for classification tests purposes [5]. Only by preparing samples for identification tests free from aggregation of grains, due to eventual dehydration, can one be sure that the index-properties, obtained as results of tests, corresponds to the nature of soils and not to eventual states the soil has acquired in the field or in the laboratory.

The Brazilian official method, for preparation of samples for identification tests, specifies that samples shall be air dried and then pulverized passing standard sieve nº 10. The proccess used to destroy all aggregation of grains was to soak completely the samples and then stir it in an almost liquid paste for at least one hour.

Figure 2 shows the variation of Atterberg limits and percentage of clay-fraction when complete disaggregation of grains is obtained. The liquid limits tends to increase along A-line and clay content, along E and D-lines, with dehydration. In figure 2 there are indications of the geological origins, clay minerals and silica-sesquioxides ratio of each sample.

TABLE I - PROPOSED CLASSIFICATION OF FINE GRAINED SOILS IN ORDER TO INCLUDE TROPICAL SOILS

	MAJOR DIVISION	SOIL GROUPS	PROPOSED SUB-GROUPS	GROUP SYMBOLS
Fine grained soils, containing little or no coarse grained material	Low Compressibility $L_\omega < 50$	Silts and very fine sands. Clayey fine sands. Silty clays	Inorganic. Rock Flour	ML
			Micaceous soils	ML_m
		Low plasticity clays Sandy clays Silty clays	Non-kaolin type	CL
			Kaolin type	KL
		Ferrous silty soils		KL_f
		Organic silty soils		OL
	High Compressibility $L_\omega > 50$	Fine sandy or silty soils	Diatomaceous. Elastic soils	MH
			Micaceous soils	MH_m
		High plasticity clays, sandy or silty clays	Non-kaolin type	CH
			Kaolin type	KH
		Ferrous clayey soils		KH_f
		Organic clayey soils		OH

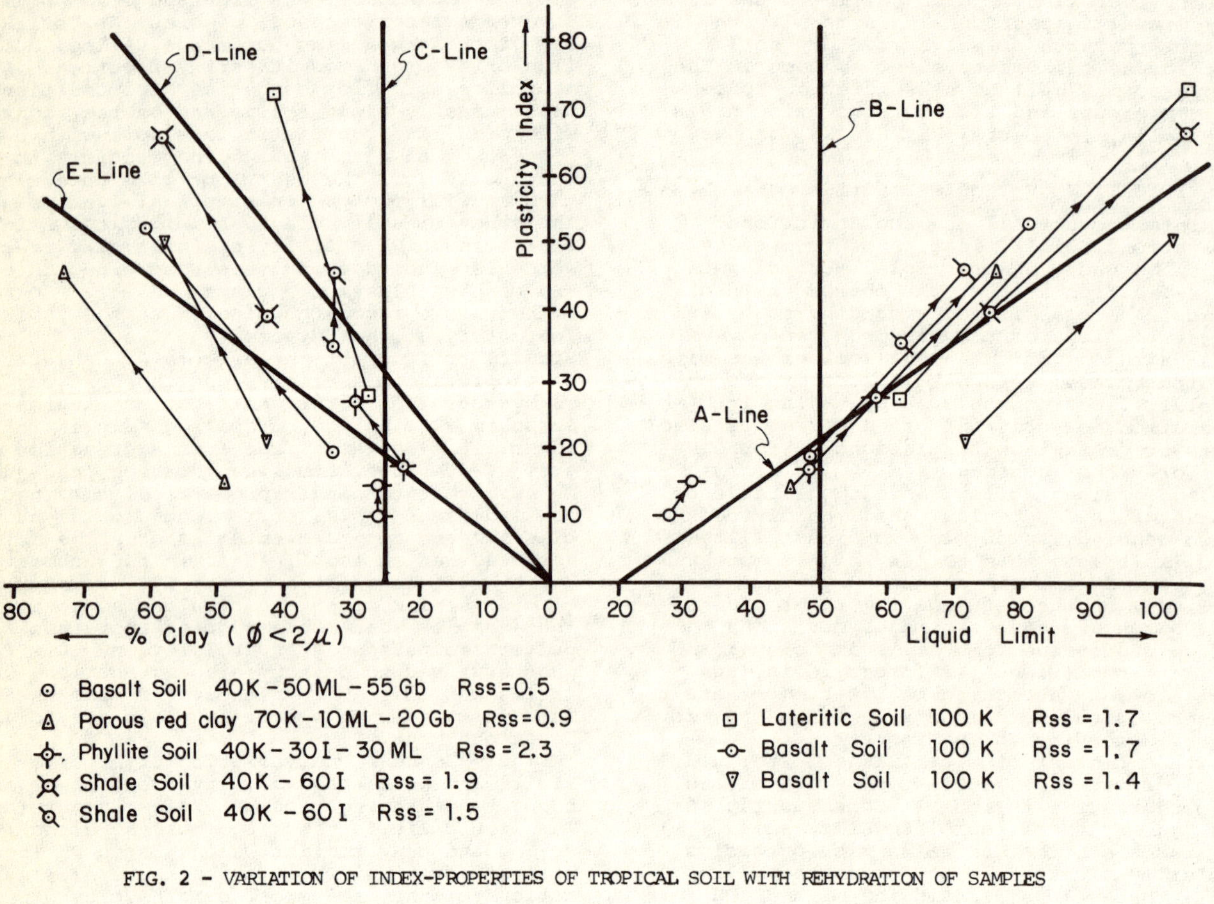

FIG. 2 - VARIATION OF INDEX-PROPERTIES OF TROPICAL SOIL WITH REHYDRATION OF SAMPLES

Clay minerals are indicated by the percentages of kaolinite K; gibbsite Gb; illite I, and mixed layers minerals ML. Molar rations between silica and sexquioxides of iron and aluminium are denoted by R_{ss}.

5 CONCLUSIONS

Considering that soil identification and classification must be done on basis of index-properties only (to mantain their values whatever be the state how the soil occurs in the field) Casagrande's AC classification system can be used to classify tropical soils, provided the fine grained soil groups are sub-divided, in order to include clay mineral and mineralogical nature of silt-fraction also as index-properties besides grain size distribution and plasticity.

Preparation of tropical soils samples, for identification test must be done with appropriete methods in order to destroy completely any aggregation of their grains or any structure imposed to their fabric by any eventual state that occurred to the soil.

Thus, classification of soils cannot be directly related to their geotechnical state properties - such as consistance, compaction or structure - because these obviously depend on the occasional states they occur in a certain circumstance that can be changed as they are removed, compacted or used differently. The purpose of a classification system is only to group a sample of soil in a class with commom index-properties. It can be inferred then that the soil, thus classified, will probably behave, in identical circumstances and same states, as the other soils of its own class.

6 REFERENCES

Casagrande, A. - "Classification and Identification of Soils" - Transactions A.S.C.E., Vol.113, 1948.

Vargas, M. - "Discussion" - Transactions A.S.C.E., Vol. 113, 1948.

Nogami, J. - "Theme 1: Characterization, Identification and Classification of Tropical Soils" Progress Report Committee on Tropical Soil of the I.C.S.M.F.E. 1st International Conference on Geomechanics in Tropical Lateritic and Saprolitic Soils - Brasília, 1948.

Skempton, A.W. - The Colloidal Activity of Clays" - Proc. 3rd I.C.S.M.F.E. - Zurich, 1953.

Ignatius, S.C. - Umpublished Master's Thesis - University of São Paulo - Brazil.

Engineering classification of residual soils

L.D.WESLEY, Senior Lecturer in Civil Engineering, University of Auckland, New Zealand

ABSTRACT: A brief review is given of classification systems in use for residual soils. An alternative classification or grouping system is proposed based on mineralogical composition and in situ structure. The purpose of this grouping system is not to provide a systematic classification method but to identify residual soil groups which can be expected to have similar engineering properties.

INTRODUCTION

A number of attempts have been made over the years to devise methods for the description or classification of residual soils. However, despite these attempts, no generally accepted methods have been established. This is not at all surprising, in view of the very diverse nature of residual soils, and it is unlikely that a universal scheme is a practical possibility. The purpose of this paper is to examine the types of classification or descriptive schemes which have been used to date, and to propose a different grouping or division of residual soils, not so much in order to make possible the systematic classification of residual soils, as to enable engineers to identify residual soil types which belong together, and which can be expected to have similar engineering properties.

PRESENT CLASSIFICATION SYSTEMS

Methods at present in use for grouping or classifying residual soils appear to fall into three broad types as follows:

(a) Methods based on the weathering profile.
(b) Methods based on pedalogical classification.
(c) Methods intended for local use on specific soil type only.

These methods are considered and discussed in the following sections.

Methods Based on Weathering Profile

Based on the work of Moye (1955), Little (1969) presented a typical example of a method for classifying residual tropical soils on the basis of the degree of weathering evident in a typical profile of such soils. The profile concept and the six categories of material type are shown in Fig. 1. Similar systems have been proposed by a number of authors, sometimes on a general basis and sometimes in relation to a particular formation or locality. Saunders and Fookes (1970) review and describe some of these systems.

Little (1969) stated that the classification system he proposed was intened to cover the

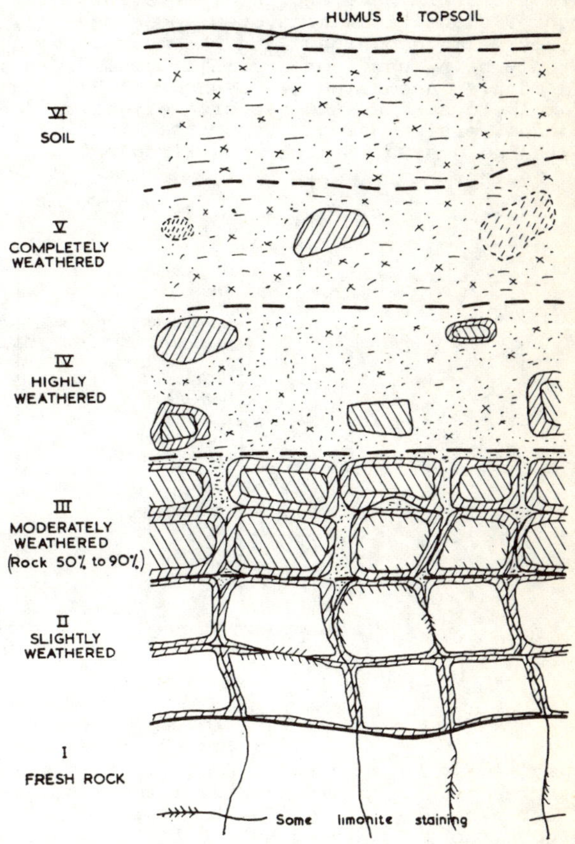

Fig. 1 : Classification of residual soils by degree of weathering (after Little, 1969)

residue resulting from the weathering of igneous rocks in the humid tropics. This is a significant and important limitation to the usefulness of such methods. A more important limitation in systems such as this, however, lies in the fact that they provide no comparative information on the nature of the top horizon, that is the true soil layer, which is often of greatest interest to the engineer. These systems are useful therefore on a localised basis, where an engineer is seeking to describe or classify the properties of a particular formation, but are not useful for comparing the properties of totally different formations. For example, they will not assist an engineer to come to terms with the differences in properties between, say, the weathered granite soils of Malaysia or Hong Kong, and the black cotton soils of India.

These methods are better regarded as systems for describing weathered rock profiles than as classification systems for true residual soils. The methods provide information on the in situ state of the soil only; they provide no information on the actual composition of the soil. They are presumably intended to be used as supplements to systems which describe the composition or nature of the soil itself such as the Unified Soil Classification System.

Both the usefulness and the limitations of this type of classification system appear to be fairly well recognised, and the method is rightly being used for specific local situations and adapted accordingly. Pender (1971) for example, describes the use of a slightly amended version used in the classification of weathered greywacke rock in Wellington, New Zealand.

in Java, Indonesia. Andosols, or volcanic ash soils, may also show a rather abrupt boundary between soil and the underlying rock, in the form of basaltic lahar or basalt rock. Townsend (1985) emphasises this point with the comment, "one major difference between the basic and acidic rocks is that most pedologists suggest that basic rocks weather rapidly into soils providing a sharp contact zone with the weathering of minerals occurring within a layer of only a few millimetres. Conversely, the zone of alteration in acidic quartz-rich rocks appears to be quick thick".

With sedimentary rocks, the picture is even more complex, and the soil profile above the rock is likely to reflect both the weathering sequence and the differences in the parent rock. Where the parent rock consists of interbedded sandstone and mudstone, for example, this may be reflected in interbedded silt and clay layers in the residual soils in the Auckland area of New Zealand derived from weathering of a sandstone and mudstone formation known as Waitemata series. Figure 2 shows three types of weathering profile for residual materials.

<u>Methods Based on Pedological Groups</u>

Geotechnical engineers have made use of pedological terms to designate various soil groups for many years. The term laterite or lateritic soil was one of the first such uses and goes back at least to Bee (1948) Ranganathan (1961) made one of the earliest uses of the term black cotton clay. Wesley (1974) used the terms

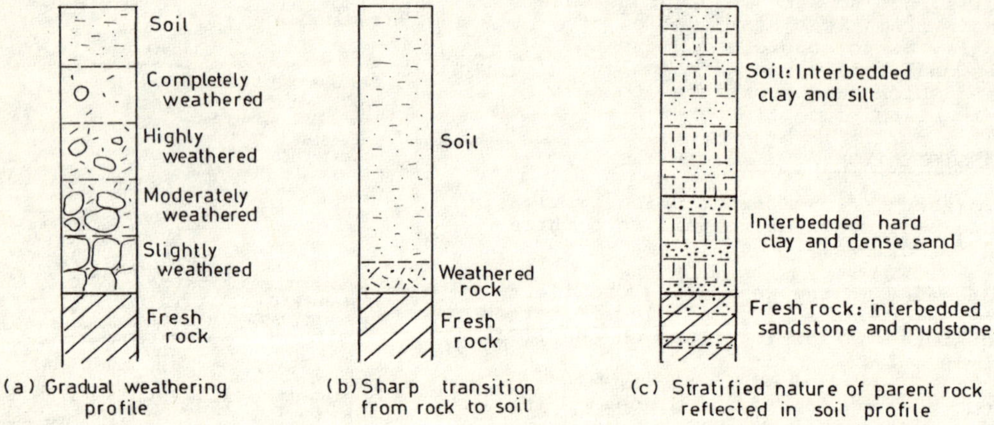

Fig. 2 : Variations in residual soil profiles

It should perhaps be pointed out at this stage that the profile shown in Fig. 1 is relevant only to certain types of residual soils, such as the igneous rocks mentioned by Little. As pictured in Fig. 1, the profile consists of a series of zones of not greatly differing thicknesses. With some residual soils, however, this is not the case at all, and the boundary between soil and rock is very abrupt, with only a very thin zone of transition material. This is frequently the case with the weathered basalt soils found in the North Island of New Zealand, and also the case with red clays derived from basaltic rock

latsol and andosol to designate two soil groups in Indonesia. This usage was taken directly from the pedological classification system being used at that time by Indonesian soil scientists (Junas Dai and Driessen, 1972, Lenvain et al, 1972). Lohnes and Tuncer (1977) also used the term andosol in describing volcanic ash soils in Hawaii.

Various additional terms have since been added to the geotechnical literature, and different terms are used for the above groups by different countries. The terms oxisols, andepts and vertisols are in common use for lateritic soils (latosols), andosols and black cotton soils respectively. Mitchel and Sitar (1982) present a table showing the variety of names used by three pedological systems, namely the French, F.A.O. and the U.S. Taxonomy. Uehara (1982) gives a useful account of the various pedological groups and their associated properties. The three soil types mentioned above, however, remain the three most distinctive tropical soil types, and the types of most interest to the engineer. Table 1 summarises the various names used for these groups.

Table 1 : Distinctive tropical soil groups of interest to geotechnical engineers

Commonly used Names	Rigorous Pedological Names			Dominant clay minerals	Important characteristics
	FAO	US Soil Taxonomy	French		
Lateritic soils Latosols Red clays	Ferralsols	Oxisols	Ferralitic soils	Halloysite Kaolinite Gibbsite Geothite	Very large group with wide variation in characteristics
Volcanic ash soils Andosols	Andosols	Andepts	Eutropic brown soils of tropical regions on volcanic ash	Allophane minor Halloysite	Characterised by very high water content and irreversible changes when dried
Black cotton soils Black clays Tropical black earths Grumusols	Vertisols	Vertisols	Vertisols	Smectite (montmorillinite)	Problem soils, high shrinkage and swell, low strength

Table 1 also shows the predominant clay minerals associated with each group, and suggests that mineralogical composition is a strong influence on the properties of each group.

The use of pedological names has not been done with the intention of establishing rigorous classification systems along the lines of those in use by soil scientists. The names have simply been borrowed as a convenient way of identifying particular soil groups. Unfortunately, some confusion has been created by the rather unsystematic use of these terms. Mitchell and Sitar (1982) for example include the red soils of Foss (1973) and basalt-derived lateritic soils of Tuncer and Lohnes (1977) in the group andosols (Table 7 of Mitchell and Sitar, 1982), although there is no suggestion by the original authors that the soils belong to this group, and no evidence that they contain allophane, the characteristic mineral of andosols. Morin (182) also includes andosols as part of the tropical red clay group. In the writer's view, andosols are a very distinctive group and although in the tropics they may be associated with red clays to some extent, their composition and properties are different, and for geotechnical engineering purposes they should not be confused with red clays. Andosols in countries like Japan and New Zealand are not associated with red clays at all.

Methods Intended for Specific Local Use

In view of the complexity of residual soils, and the almost total lack of any common features among some residual soil groups, (for example black cotton soils and weathered granite soils) it is not surprising that descriptive or classification methods have been developed for local use in particular formations. Tuncer and Lohnes (1977) for example, describe a system suggested for use with lateritic soils from Hawaii and Puerto Rico. Pender (1971, 1981) describes empirical correlations for the weathered greywacke of Wellington, New Zealand. Wirth and Zeigler (1982) describe a system specifically developed for use on the Baltimore subway project.

In the writer's view these methods represent a very promising development, as such methods are highly desirable for dealing in a systematic way with particular formations, and it is likely that the profession will see an increasing use of such systems in the future. A word of warning should however be exercised to those seeking to develop such systems, as there is a danger that existing systems will merely be modified in some way, or correlations valid for one group of residual soils will be assumed to provide a basis for correlations within another group. For example correlations between strength and void ratio may be valid for some soils (Pender 1981, Tuncer and Lohnes 1977) but attempting to find such a correlation for volcanic ash soils would be likely to be a futile exercise. Each soil type must be evaluated on its merits.

A COMPREHENSIVE APPROACH TO GROUPING OR CLASSIFYING RESIDUAL SOILS

Basis

The specific characteristics of residual soils which distinguish them from sedimentary soils can generally be attributed to either the presence of particular structural effects, such as the presence of unweathered or partially weathered rock, planes of weakness, interparticle bonds and the like. These influences can be grouped under the general headings of composition and structure.

Composition includes the particle size, shape and especially the mineralogical composition.

Structure can be subdivided into two main categories as follows:

(a) Macro-structure or discernible structure
This includes all features discernible to the naked eye, such as layering, planes of weakness, fissures, pores, presence of unweathered or partially weathered rock and relic structures.

(b) Micro-structure or non-discernible structure
This includes interparticle bonding or cementation, clustering of particles etc.

The first classification method discussed above, which subdivides the weathering profile of the rock, is based essentially on the macro-structure of the soil. By visual examination of the macro-structure, it is possible to divide the weathering profile up into a series of layers.

The second classification method (i.e. the pedological systems), on the other hand, appears to be much more closely related to mineralogical composition than to either macro-structure or micro-structure. The distinctive engineering characteristics of the three soil groups listed in Table 1 arise primarily from their distinctive mineralogical composition.

Thus we really have two parallel systems which describe different features of the soil. The two systems come together essentially in the top layer of the weathering profile, that is, the true soil layer. However, with the three soil types listed in Table 1, the engineer may not be interested in any underlying partially weathered material, either because the soil layer is so thick that the material below is irrelevant to the geotechnical works be is concerned with, or because the intermediate zone is too thin to be of engineering importance.

Methods of classification and description which are in current use with sedimentary soils consist of two distinct parts. The first part is the classification of the material itself, and makes no reference to the undisturbed state in which the soil exists in the ground. Particle size and Atterberg limit measurements form the bases of such classification; the Unified System is probably the best known of these methods. The second part is the description of the soil in its undisturbed state, and covers such aspects as stiffness (for cohesive soils), relative density (for granular soils) and structural features such as bedding planes, faults and joints, or other discontinuities.

A Proposed Grouping of Residual Soils

Following on from the above consideration a useful first step in the grouping of residual soils is to divide them into groups on the basis of composition alone, without reference to their undisturbed state. The following three groups are suggested:

Group A: Soils without a strong mineralogical influence

Group B: Soils with a strong mineralogical influence coming from conventional clay minerals normally found in sedimentary soils

Group C: Soils with a strong mineralogical influence coming from special clay minerals not found in sedimentary soils

Group A : Residual Soils

By eliminating those soils which are strongly influenced by particlar clay minerals, there is some possibility of identifying a group of soils which can be expected to have similar properties. Soils which come into this group will generally be those of a fairly coarse nature, with a relatively low clay fraction. The residual soils of Hong Kong appear to be a good example of soils which belong in this group. They are derived from the weathering of granites or volcanics (rhyolite), and particle size data given by Lumb (1965) shows the clay fraction to average only about 20%. Even the completely weathered soil layer, classed as "red earth" by Lumb has a maximum clay fraction of only 40%. Lumb and Lee (1975) show that the principal clay mineral is halloysite but believe that this has little influence on the engineering performance of the soils and suggest that behaviour is "controlled essentially by the coarser silt and sand fractions".

In general, soils which have a weathering profile of the type illustrated in Fig. 1 will come within this group. In relatively rare instances, the top layer (i.e. the soil layer) may be sufficiently advanced in weathering to become a true clay with properties strongly influenced by distinctive clay minerals.

Group A soils can be subdivided further on the basis of the extent and manner in which their behaviour is influenced by structural effects. It is convenient to separate structural effects into the two broad groups mentioned earlier, namely macro-structure and micro-structure.

Group A can therefore be divided into two main sub-groups:

Sub-group (a)

These are soils in which macro-structure plays an important role in the engineering behaviour of the soil. The lower horizons of the soils which weather according to the pattern shown in Fig. 1 fall into this category.

Sub-group (b)

These are soils without macro-structure, but with a strong influence from micro-structure. The most important form of micro-structure is inter-particle bonding or cementation, and although this cannot be identified by visual inspection it can be inferred from fairly basic aspects of soil behaviour. Sensitivity in particular is a very good measure of micro-structure, as high sensitivity results from the presence of a distinctive structure (involving some form of bonds) which is destroyed on remoulding. In the writer's view, more attention should be paid to the value of sensitivity with residual soils, as sensitivity is closely related to liquidity index which is in turn closely related to structural effects such as interparticle bonding. Residual soils which have a high liquidity index (or exist in a state analogous to this concept) are also those which show pronounced bonding or structural effects. It is the presence of specific bonding or similar structural effects which enables the soil to exist in a state close to the liquid limit.

As Vaughan (1985) has pointed out, the presence of considerable coarse material in many residual soils makes the determination of Atterberg limits and the liquidity index somewhat problematical. Focussing on the sensitivity of the soil avoids this difficulty.

Sub-group (c)

Soils which are not greatly influenced by macro or micro-structural effects are included here as a third sub-group.

However, this sub-group is a very minor group, as very few residual soils of Group A will fall into this category.

Group B : Residual Soils

This group is made up of soils which are strongly influenced by conventional clay minerals such as those normally found in sedimentary soils. The most significant member of this group is the black cotton soils or "vertisols", whose characteristic properties are high shrink and swell potential, high compressibility and low strength. These characteristics are directly related to their predominant mineralogical constituent which is montmorillinite or similar minerals of the smectite group.

Information in the literature suggests that not many other residual soils belong to this group, although there are some soils derived from sedimentary rocks (sandstones and mudstones) which have properties which are fairly strongly influenced by mineralogical composition. Soils in the Auckland area of New Zealand derived from the weathering of Waitemata sandstone come into this category. Some of the soils are of high shrink and swell characteristics due to the presence of montmorillinite; others show the opposite characteristics and mineralogical composition appears to be a very minor influence on their behaviour.

Group C : Residual Soils

These are the soils which are strongly influenced by the presence of clay minerals which are not found in sedimentary clays. The two most important minerals involved here are the silicate clay minerals halloysite and allophane. Halloysite is a lattice (crystalline) mineral of tubular form and is normally associated with the same group as kaolinite. Allophane is an amorphous (none-lattice) mineral. In addition to these silicate minerals, tropical soils may contain non-silicate minerals (or "oxide" minerals), in particular the hydrated forms of aluminium and iron oxide (the sesquioxides) gibbsite and geothite.

The influence of halloysite and allophane on the soil properties is fairly clear from the case studies recorded in the literature. The influence of the sesquioxides is less well documented. It is convenient however to subdivide this group into three sub-groups:

(a) Halloysitic Soils

The principal influence of halloysite appears to be that the engineering properties of the soil are good, despite a high clay fraction, and fairly high values of natural water content and Atterberg limits. The good engineering properties appear to be the direct result of their mineralogical bonding or cementation arising from the presence of the sesquioxides (Wesley 1973). Terzaghi (1958), Matyas (1969) and Wesley (1974) have given accounts of the good engineering performance of these soils.

(b) Allophanic Soils

The influence which allophane has on soil behaviour has been described elsewhere (e.g. Lohnes and Tuncer 1977, Wesley 1974, 1977) and will not be repeated here. It should be understood however that the influence of allophane is both dramatic and puzzling, in that it results in soils having water contents ranging from about 80% to 250%, but which still perform very satisfactorily as engineering materials. They are frequently much superior to soils with water contents only a fraction of the above values.

(c) Soils Influenced by the Presence of Sesquioxides

The principal role of the sesquioxides appears to be to act as cementing agents which bind the other mineral constituents into clusters or aggregations. With sufficient concentration of sesquioxides, the hard concretionary materials commonly known as laterite are formed. The silica/alumina ratio (SiO_2/Al_2O_3) and the silica/sequioxide ratio have both been used as indicators of degree of laterisation. This sub-group could perhaps be termed the lateritic group, but the term laterite is generally used very loosely, sometimes to include both halloysite and allophane clays which contain only a trace of sesquioxides, and whose behaviour is not significantly influenced by the sesquioxides.

The above groups, especially the halloysite and allophane groups can be further subdivided on the basis of structure. Allophanic soils (which appear to be always associated with volcanic ash as parent material) show considerable variation in their structural influence. Allophane soils in Indonesia are generally of low to moderate sensitivity, while those in Japan are likely to be of moderate to very high sensitivity (Kuno et al 1978), indicating a strong strucutral component in their undisturbed state. The halloysitic soils of Java, Indonesia, with which the writer is familiar do not appear to have significant micro-structure or macro-structure. They are of very low sensitivity and their behaviour in the remoulded and undisturbed state is often almost identical. This is not necessarily true of other halloysite soils.

Table 2 : Grouping of residual soils which can be expected to have similar engineering properties

MAJOR DIVISION	SUB — GROUPS	COMMENTS
GROUP A Soils without a strong mineralogical influence	(a) Strong macro — structure influence	Nature of macro — structure needs definition — stratification — fractures, fissures, voids etc
	(b) Strong micro - structure influence.	Remoulding likely to strongly influence behaviour — sensitivity should be a useful indicator
	(c) Little or no structure influence	Probably a rather minor sub — group
GROUP B Soils strongly influenced by 'normal' clay minerals	(a) Smectite (montmorillinite) group	Problem soils, characterised by low strength, high compressibility, high shrink swell behaviour (similar characteristics to any montmorillinite soil)
	(b) other minerals ?	? ?
GROUP C Soils strongly influenced by clay minerals found only in residual soils	(a) Halloysite	Low activity soils, good engineering properties
	(b) Allophane	Low activity soils, with good engineering properties, characterised by very high water contents, and large irreversible changes on drying.
	(c) Sesquioxide (lateritic)	Extremely variable group, ranging from silty clay to gravel

This suggested grouping system is summarised in Table 2.

DISCUSSION

General

As mentioned earlier the above grouping of residual soils is intended to provide a basis for identifying groups of residual soils which can be expected to have similar engineering properties. It appears to provide a better basis for doing this than using pedalogical terms.

The terms red clay, or lateritic soil, for example are used to cover such a wide range of materials as to be almost useless to the engineer. The predominant clay minerals in some red clays (or lateritic soils) are kaolinite and montmorillinite; these soils plot above the A-line on the plasticity chart. The predominate clay mineral in other red clays, as already mentioned, is halloysite; these soils plot below the A-line and have quite distinct engineering properties.

The disadvantage of using mineralogical composition as a classification basis is readily apparent; geotechnical engineers seldom have ready access to the facilities needed for mineral identification. However, most countries have institutions which undertake mineralogical studies, and cooperation between geotechnical engineers and these institutions ought not to be difficult and hopefully should be to the benefit of both parties.

The Place of Conventional Classification Tests

The usefulness of conventional classification systems such as the Unified System for residual soils has been questioned on a number of occasions (e.g. De Graft-Johnson, 1969) primarily because of the dependence of particle size and plasticity measurements on the method of sample preparation. It is argued also that the in situ character of the soil is so destroyed in preparing the soils for testing that the results do not give an indication of the properties of the undisturbed soil.

There is some truth in these arguments; at the same time it is of interest to note that there is a great deal more data on particle size and Atterberg limits for residual soils than there is data on shear strength, compressibility or other properties of direct interest to the engineer. It is the writer's view that the conventional particle size and Atterberg limit tests are very useful with residual soils, and knowledge of these properties is much more useful than for example the knowledge that the soil is a red clay or lateritic soil. The argument that the results are influenced by drying is not an argument for rejecting the tests since there is no difficulty in avoiding drying the soil. Frost (1967) in drawing attention to this question, rightly called for correct pre-testing procedures rather than rejection of the tests.

It has been the author's experience that with residual soils the position which a soil occupies on the conventional plasticity chart provides a good indication of soil properties, possibly just as good as with sedimentary soils. Soils which plot well below the A-line behave as silts while those which plot well above the A-line behave as clays.

Problems arise when attempts are made to relate specific soil properties, or classification boundaries to one or other of the liquid and plastic limits. For example, the British classification system (BS 5903:1981) divides soils into a number of categories based on the liquid limit. Such a division is not very relevant to residual soils. It is the position above or below the A-line which is of most significance, especially with tropical residual soils.

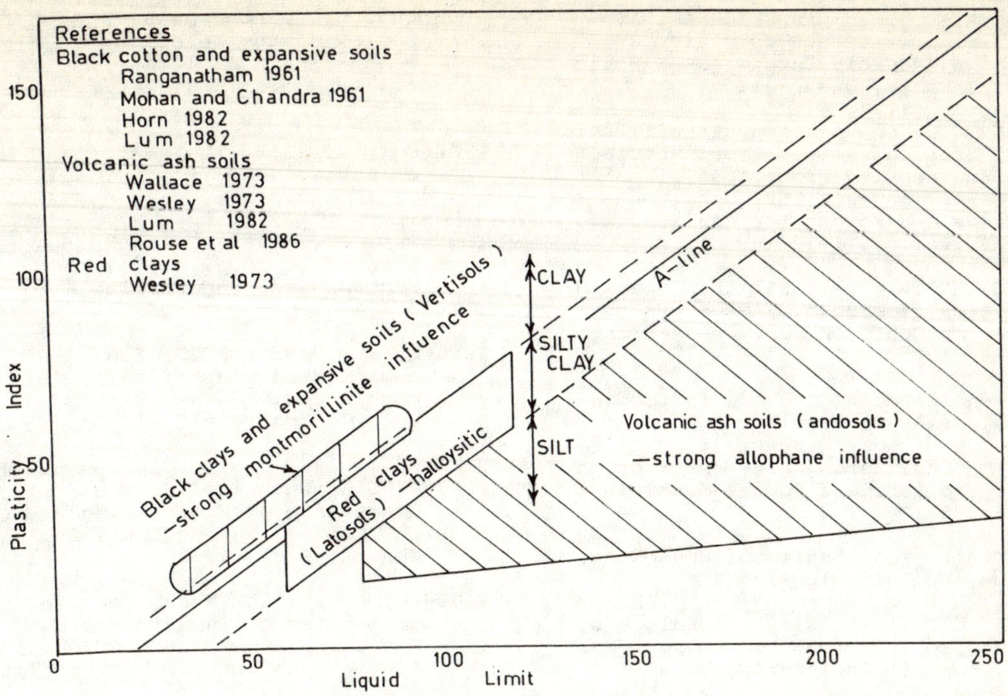

Fig. 3 : Possible division of plasticity chart for use with residual soils

Rather than a subdivision based on the liquid limit, a subdivision along the lines shown in Fig. 3 would be more relevant to residual soils. The lines drawn parallel to the A-line divide soils into three types labelled clay, silty clay, and silt. Many residual soils behave as silty clays for engineering purposes, and rightly fall into the category of silty clay on this chart. The more distinctive residual soil types, such as vertisols (Black Cotton soils) and andosols, would rightly be classified as clays and silts respectively. The relative positions of vertisols, latosols (red clays) and andosols is shown in Fig. 3.

It should be noted that the influence of increased mixing (or even drying) of the soil on the Atterberg limits is to move the point on the plasticity chart parallel to the A-line (Morin and Todor 1975); hence if we use distance above or below the A-line as our main criteria for evaluating soil this movement is not of great significance.

CONCLUSION

As indicated in the introduction it is not the purpose of this paper to provide a systematic classification procedure for residual soils; the intention has been to help geotechnical engineers understand the place of the systems which have been proposed in the past, and to propose a basis for dividing residual soils into groups which can be expected to have similar geotechnical properties. The scheme proposed is believed to fulfil this objective but suffers from the draw-back that mineralogical composition is one of its components and information of this nature is not a normal point of geotechnical investigations.

REFERENCES

BEE, R.J. 1948 Some notes on laterite - a soil of engineering importance in the tropics, Conf. on Civil Engineering Problems in the Colonies, Inst. Civil Engineers, London, p 191

DeGRAFT-JOHNSON, J.W.S. and BHATIA, H.S. 1969 General report - engineering characteristics, Proc. Specialty Session on Engineering Properties of Lateritic Soils, Seventh Int. Conf. on Soil Mechanics and Foundation Engineering, Mexico, 1969, Vol. 2, 13-43

FOSS, I. 1967 Red soil from Kenya as a foundation material, Proc. Eighth Int. Conf. on Soil Mechanics and Foundation Engineering, Moscow, Vol. 2.2, 73-80

FROST, R.J. 1967 Importance of correct pre-testing preparation of some tropical soils, Proc. First Southeast Asian Regional Conf. on Soil Engineering, Bangkok, 43-44

HORN, A. 1982 Swell and creep properties of an African black clay, ASCE Geotechnical Engineering and Construction in Tropical and Residual Soils, Honolulu, Hawaii, 1982, 199-215

JUNUS Dai and DRIESSEN, P.M. 1972 A general orientation preceeding the resumption of clay mineralogical studies at the soil research institute, Bogor, Indonesia, Second ASEAN Soil Conference, Jakarta, Indonesia

LENVAIN, J., MULYADI, D. and ABDURACHMAN, A. 1972 Artificial structure formation and aggregation in andosol, Second ASEAN Soil Conference, Jakarta, Indonesia

LITTLE, A.L. 1969 The engineering classification of residual tropical soils, Proc. Specialty Session on Engineering Properties of Lateritic Soils, Seventh Int. Conf. on Soil Mechanics and Foundation Engineering, Mexico, 1969, Vol. 1, 1-10

LOHNES, R.A. and TUNCER, E.R. 1977 Engineering characteristics of andosols, Proc. Fifth Southeast Asian Conf. on Soil Engineering, Bangkok, 305-312

LUM, W.B. 1982 Engineering problems in tropical and residual soils in Hawaii, ASCE Geotechnical Engineering Specialty Conference on Engineering and Construction in Tropical and Residual Soils, Honolulu, Hawaii, 1982, 1-12

LUMB, P. 1965 The residual soils of Hong Kong, Geotechnique, Vol. 15, 180-194

LUMB, P. and LEE, C.F. 1975 Clay mineralogy of the Hong Kong soils, Proc. Fourth Southeast Asian Conf. on Soil Engineering, Kuala Lumpur, Vol. 1, 41-50

MATYAS, E.L. 1969 Some engineering properties of Sasamua clay, Proc. Specialty Session on Engineering Properties of Lateritic Soils, Seventh Int. Conf. on Soil Mechanics and Foundation Engineering, Mexico, 1969, Vol. 1, 143-151

MITCHELL, J.K. and SITAR , N. 1982 Engineering properties of tropical residual soils, ASCE Geotech. Engng. Specialty Conference on Engineering and Construction in Tropical and Residual Soils, Hawaii 1982, 30-57

MOHAN, D. and CHANDRA, S. 1961 Frictional resistance of bored piles in expansive clays, Geotechnique 11.4, pp 294-301

MORIN, W.J. and TODOR, P.C. 1975 Laterite and lateritic soils and other problem soils of the tropics, USAID 3682, Lyon Associates, Baltimore, U.S.A.

MOYE, D.G. 1955 Engineering geology for the Snowy Mountains Scheme, Institution of Engineers, Australia, Vol. 27, p 287

PENDER, M.J. 1971 Some properties of weathered greywacke, First Aust.-N.Z. Conf. on Geomechanics, Melbourne, Vol. 1, 423-429

PENDER, M.J. 1981 Friction and cohesion parameters for highly and completely weathered Wellington Greywacke, Wellington, Vol. 1, 171-175

RANGANATHAM 1961 Soil structure and consolidation characteristics of black cotton clay, Geotechnique 11.4, 333-338

ROUSE, W.C., READING, A.J. and WALSH, R.P.D. 1986 Volcanic soil properties in Dominica, West Indies, Eng. Geology, Vol. 23, 1-28

SAUNDERS, M.K. and FOOKES, P.G. (1970) A review of the relationship of rock weathering and climate and its significance to Foundation Engineering Geology, Vol. 4, 289-325

TERZAGHI, K. 1958 The design and performance of Sasomua dam, Proc. Institution of Civil Engineers, Vol. 9, 369-394

TOWNSEND, F.C. 1985 Geotechnical characteristics of residual soils, ASCE Journal of Geotechnical Engineering, Vol. 3, No. 1, 77-94

TUNCER, E.R. and LOHNES, R.A. 1977 An engineering classification for certain basalt-derived lateritic soils, Eng. Geology, Vol. 11, No. 4, 319-339

UEHARA, G. 1982 Soil science for the tropics, ASCE Geotechnical Engineering Specialty Conference on Engineering and Construction in Tropical and Residual Soils, Honolulu, Hawaii, 1982, 13-29

VAUGHAN, P.R. 1985 Mechanical and hydraulic properties of in situ residual soils, First Int. Conf. on Geomechanics in Tropical Lateritic and Saprolitic Soils, Brazil, 1-33

WALLACE, K.B. 1973 Structural behaviour of residual soils of the continually wet highlands of Papua New Guinea, Geotechnique, Vol. 23, No. 2, 203-18

WESLEY, L.D. 1973 Cluster hypothesis and the shear strength of a tropical red clay, Geotechnique, Vol. 23, No. 1, 109-113

WESLEY, L.D. (1973) Some basic engineering properties of halloysite and allophane clays in Java, Indonesia, Geotechnique, Vol. 23, No. 4, 471-494

WESLEY, L.D. 1977 Shear strength properties of halloysite and allophane clays in Java, Indonesia, Geotechnique, Vol. 27, No. 2, 125-136

WIRTH, J.L. and ZEIGLER, E.J. 1982 Residual soils experience on the Baltimere Subway, ASCE Geotechnical Engineering Specialty Conference on Engineering and Construction in Tropical and Residual Soils, Honolulu, Hawaii, 1982, 557-577

Identification and classification of tropical soils

B.C.R.YUDHBIR, Indian Institute of Technology, India
B.K.SAHU, Indian Institute of Technology, India

ABSTRACT: A soil classification chart incorporating the Casagrande A-line & Skempton's activity concepts has been developed from data on mineral mixture studies. The proposed chart has been checked against natural soils. From the trends between the three indices used in the classification chart, it can be demonstrated that soils of same mineralogical composition will lie on or parallel to the A-line. It is suggested that such a classification chart meets the necessary conditions for proper soil (tropical) identification as advocated by Vargas (1985).

1 INTRODUCTION

Casagrande's soil classification system takes into account only grain size distribution, plasticity of the fines, and organic matter content. The mineralogical nature of the clay-fraction of soils though very important, is not reflected in this classification scheme. The fact that more than one clay mineral may be present in clay fraction, and also that even small fraction of more active minerals may control soil behaviour, makes it more difficult to include mineral composition into a classification scheme. Skempton's activity index, (Skempton 1953) for a soil, however, reflects the nature of predominant clay mineral and according to Skempton's scheme, inorganic soils are divided into inactive, normal and active soils. Clearly in addition to plasticity index, and liquid limit (used in Casagrande scheme) the activity index is one more classification parameter of soils.

Vargas (1985) has advocated the use of these indices for classification of tropical soils. Sahu (1977), working with mineral mixtures to produce a large range of soils examined large number of natural soils with known mineral composition and proposed a composite plot for soil classification along these same lines. Using Kondner's (Kondner et al 1964) two mineral mixture law, all the classification index data for mineral admixtures (Sahu 1977 & Seed et al 1964) has been interpreted to correlate W_L, PI and Activity with the ratio of two constituent minerals. A classification plot based on these interpretations has been checked against values for natural soils of known mineral composition. It is proposed that such a classification chart meets the necessary conditions for proper soil identification as advocated by Vargas (1985).

2 CLASSIFICATION SYSTEMS AND SOIL MINERALOGY

The plasticity chart for identifying fine grained soils, as suggested by Casagrande is shown in Figure 1. This classification system is essentially based on plasticity index, PI and liquid limit, W_L and does not take into account the mineralogy of the clay fraction. Skempton (1953) introduced the concept of soil activity to remedy this situation. Left hand portion of Figure 1 shows Skempton's classification scheme based on soil activity. The activity of the clay fraction reflects the nature of the predominant clay mineral, but this predominance may not be due to quantity alone. Based on the study of clay mineral admixtures (Seed et al 1964, Sahu 1977), it has been shown that in case of Kaolinite - montomorillonite admixtures, activity of kaolinite increases rapidly with the increase of montmorillonite content and even for a small quantity of montmorillonite (<5%) the activity of kaolinite increases from 0.4 to 1.0. However, in case of illite-montmorillonite/bentonite admixtures, even upto 20% montmorillonite the activity does not change at all, and only for montomorillonite content greater than 70%, significant increase in activity is obtained. A similar trend is observed in case of natural soils (Sahu 1977) except that activity is suppressed upto 40% montmorillonite in the presence of illite. The suppression of any change in activity by montmorillonite in the presence of illite was also brought out by White (1955) and Lambe & Martin (1957); the percentage of montmorillonite upto which illite suppresses activity is however likely to depend on the exchangable cations of montmorillonite and the nature of illite (crystalline or amorphous).

Vargas (1985) has proposed the use of a composite classification chart shown in Figure 1 for classification of tropical soils. The data point plotted in Figure 1, is an expansive residual soil from Bahia shown in Figure 21 of the paper by Vargas. This chart will identify this soil as an inactive clay of high plasticity whereas the A-line method alone would have suggested a clay of high plasticity supposedly having all the attributes of clays of this category. In case of tropical soils, depending on the environmental conditions controlling drainage & the intensity of leaching, the actual weathering profile will undergo changes in mineralogy and thus also the nature of soil. Vargas (1985) has used Figure 1 to classify a variety of residual soils. Use of natural soils to develop a composite classification

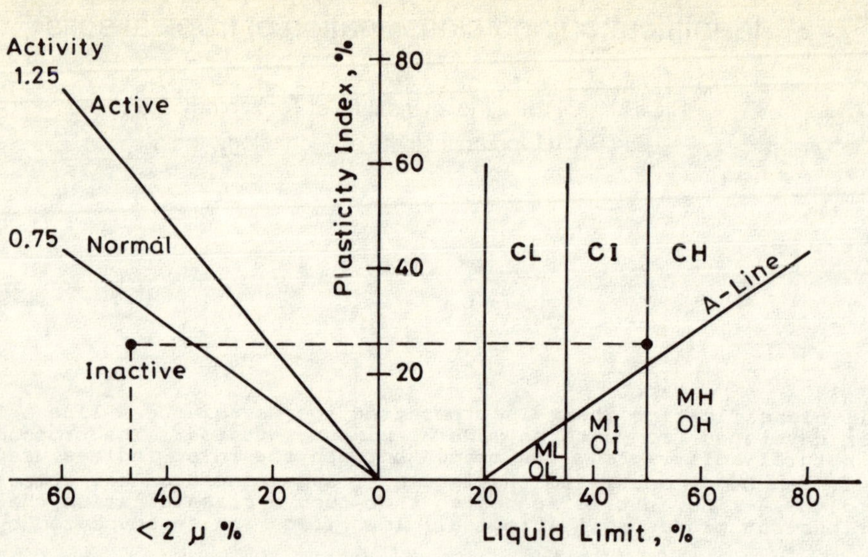

Figure 1: Classification chart proposed by Vargas (1985)

chart incorporating soil Atterberg limits & activity poses certain problems. In order to develop consistent trends over a wide range of soils, one needs data on large number of soils exhibiting a wide range of Atterberg limits & activity values. Differences in mineral composition and particle gradation have usually produced large scatter in data. However, results from mineral admixture studies afford an opportunity to develop a unified classified system incorporating the features of Casagrande's A-Line & Skempton's activity.

From the mineral mixture studies reported by Seed et al (1964) & Sahu (1977), a relationship between activity, plasticity index and liquid limit can be developed. The interaction between clay minerals in a two mineral mixture system can be quantified by Kondner's two mineral mixture law (Kondner et al 1964). The indices; liquid limit, plasticity index, and soil activity can be expressed in the following general form:

$$I_j = (I_x - I_y)\, e^{-AP} + I_y$$

where I_j is the value of an index, i.e. Liquid Limit, plasticity index or activity for a soil comprised of two minerals x & y,

I_x is the value of the same index of mineral component x

I_y is the value of the same index of mineral component y, $I_y < I_x$.

P is the proportion of P_y/P_x

A is the two mineral interaction constant and has a unique value for each index upto a certain proportion of P_y.

In case of illite-monmorillonite mixture, for example, the values of A are 1.56, 1.02 & 0.91 in case of the three indices; liquid limit, plasticity index and activity respectively. In case of liquid limit, for example I_x is the liquid limit of montomorillonite investigated (=315%) & I_y is the liquid limit of illite investigated (=107%). In case of liquid limit, the interaction constant A=1.56 is applicable for a proportion of illite less than 35%. Details for other indices & other mineral mixtures are given by Sahu (1977).

Using these interaction relations for clay mineral mixtures & sand-clay mineral mixtures investigated by Seed et al (1964) & Sahu (1977), a soil classification chart is developed as shown in Figure 2. Sahu (1977) checked this plot against sixty natural soils for which Atterberg limits, activity & mineral composition were known.

3 DISCUSSION

The proposed classification chart, as shown in Figure 2, has been checked for soils of wide variety as lited in Table I. The numbers on each data point, in brackets, are values of liquid limit. There is reasonable agreement between the value of liquid limit shown in brackets and the value that would be obtained from the proposed contours of liquid limit. This agreement should be viewed in the light of the fact that in general natural soils are multimineral mixtures where as the proposed chart utilises 2 clay mineral mixtures & Quartz-clay mineral mixtures. This classification chart gives a complete description of the soil compared to the A-line classification. The proposed chart not only achieves the objectives for soil identification, as stated by Vargas (1985), it also provides a much wider range of index values over which some patterns can be observed. Attempts to use natural soils & employ Figure 1 for classification purposes, as advocated by Vargas (1985), cannot allow generalisations of expected pattern over a wide range because the extreme variability in gradation & mineral composition produces large scatter.

It is interesting to make some observations based on the trends depicted in Figure 2. For example, if two soils on the A-line have

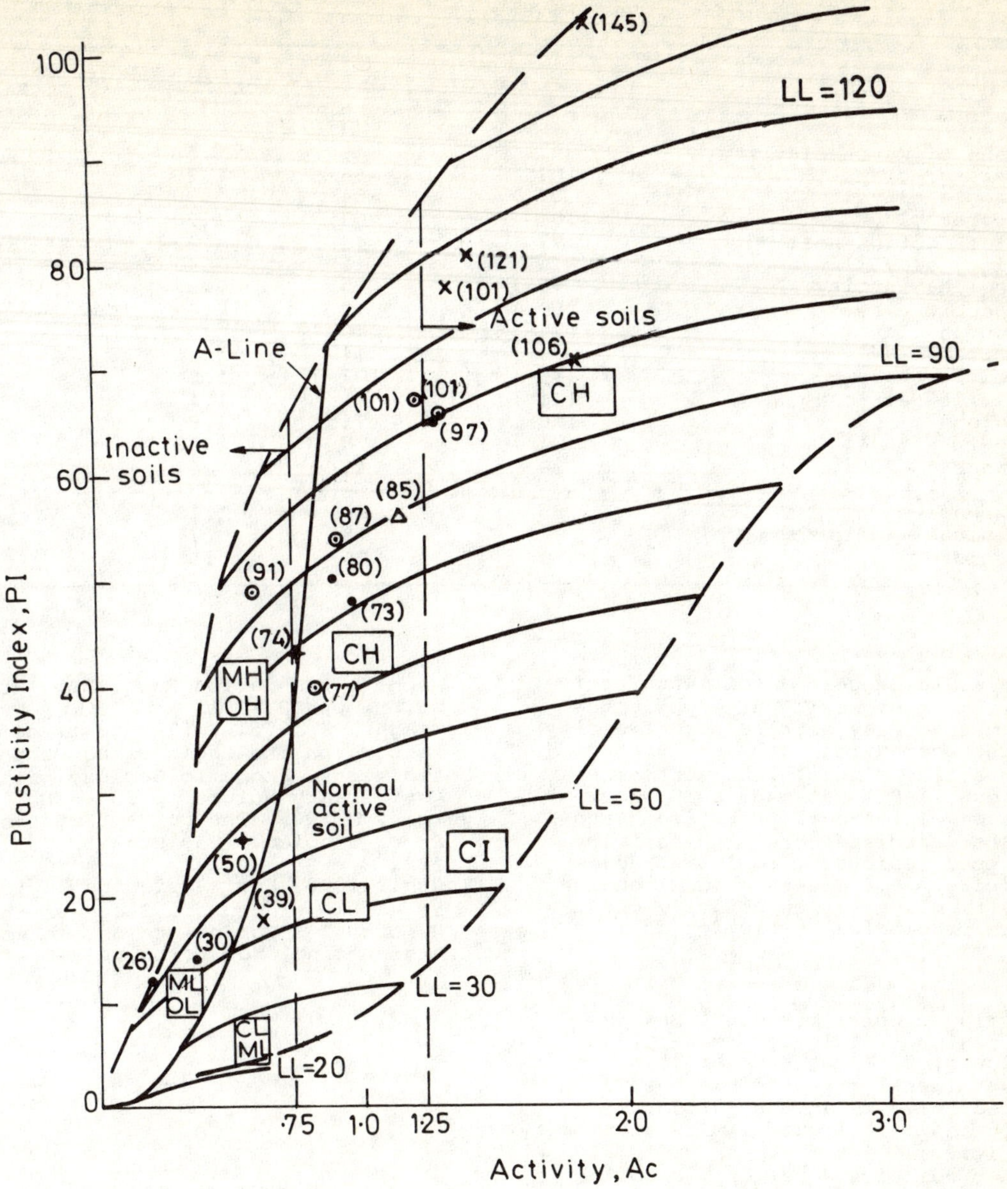

Figure 2: Proposed classification chart

liquid limit values of 50% & 100% & plasticity index values of 22 & 58%, the chart in Figure 2 will indicate that activity changes slightly from 0.6 to 0.82 giving corresponding clay contents of 36% & 71% respectively. The doubling of liquid limit values along A-line is practically due to doubling up of clay content ($<2\mu$), which is quite an expected result. Also the activity value is within ± 0.15 of 0.75, the border between inactive & normal soils. This result tends to support the conclusion (Terzaghi, 1955, Yudhbir 1978) that lines parallel to A-line represent soils of same geologic origin, the lower lines (below A-line) represent low plastic micaceous residual soils derived from Granite, Schist etc. and the upper lines (above A-line) represent highly plastic expansive soils derived from basalts, clay shales etc.

Another observations, in the light of the conclusion stated above, is also quite interesting. Two soil with same liquid limit of 50%; one on A-line (PI=22%) & the other above A-line (PI=32%) will show activity values of 0.6 & 1.25 respectively on the basis of Figure 2. The former would have a clay content of 36% while the latter would have a clay content of 26% only. Obviously the

TABLE I

Symbol	Soil	W_L	PI	Activity	Reference
⊙	Rann of Kutch clay	91	49	.58	Sahu (1977)
⊙	Paravur clay	101	66	1.29	
⊙	Kandla clay	87	54	.90	
⊙	Gujrat clay	101	67	1.20	
⊙	Black cotton soil	77	40	.82	
⊙	Silty clay	26	12	.40	
●	Shellhaven clay	97	65	1.27	
●	Gosport clay	80	50	.89	
●	London clay	73	48	.96	
●	Horten clay	30	14	.36	
×	Peirre shale	145	103	1.84	
×	Little Belt clay	121	81	1.40	
×	Vajont I	106	71	1.80	
×	Bearpaw Shale	101	78	1.32	
×	Pasabache clay	39	18	.62	
△	Expansive	85	56	1.13	Vargas (1985)
✦	Residual soils from	74	43	.75	
✦	Bahia	50	26	.54	

latter result would suggest a soil with active clay mineral like montomorillonite while the former would mean a clay with less active clay minerals or the one having much smaller percentage of a clay mineral like montmorillonite. Vargas (1985) has made such observations in cases of residual soils from sandstone where the saprolites are high activity soils & the upper residual & colluvial soils have low activity. He suggests gradual change of high activity clays to low activity ones, as the surface pedological evolution goes on.

4 CONCLUSIONS

Based on two clay mineral mixture & Quartz-clay mineral mixture studies the mineral interaction law proposed by Kondner et al (1964) for two mineral mixtures has been used to develop predictive relations for three classification indices: liquid limit, plasticity index and activity.

Using these predictive equations, a soil identification chart, incorporating Casagrande's A-Line & Skempton's activity concepts, has been proposed (Figure 2). The chart has been checked against natural soils of known mineralogy & gradation. The overall patterns between the three indices, over a wide range of values have been shown to predict trends which bring out the influence of geologic origin and ongoing pedological processes on the nature of soils.

REFERENCES

Kondner, R.L. & J.R. Vendrell 1964. Consolidation coefficient: cohesive soil mixtures Proc. Jl Soil Mech. Foundn. Eng. Div. ASCE, Vol.90, Sept. pp 31-42.

Sahu, B.K. 1977. Basic studies on mineral mixtures in relation to engineering behaviour of natural soils, Ph.D. Thesis, IIT Kanpur.

Seed, H.B. et al. 1964. Clay mineralogical aspects of the Atterberg limits, J. soil Mech. & Foundation Engg. Div. ASCE, Vol.90 No.SM4 pp 107-131.

Terzaghi, K 1955. Influence of geological factors on the engineering properties of sidiments. Economic Geology, Fiftieth Anniversary Volume pp 557-618.

Vargas, M 1985. The concept of tropical soils, Proc. Tropicals'85, Brazil, Vol.3, pp 101-134.

Yudhbir 1978, Engineering geological aspects of earth structures, State-of-the-Art Report, IGS Conference on geotechnical engineering, New Delhi pp 9-21.

Session 2
Engineering properties of tropical soils

Soil-lime reactions: The behaviour of a Brazilian saprolitic soil

N.BARELLI, Instituto de Química, UNESP, Brazil
M.H.PONTIERI, Instituto de Química, UNESP, Brazil
C.B.MELIOS, Instituto de Química, UNESP, Brazil

SYNOPSIS

In tropical and subtropical regions- where strongly weathered soils predominate - there is lack of information about soil-lime treatment, particularly concerning the chemical reactions involved and/or the products formed.

In the present work a saprolitic soil, situated under the stone line, was selected to react with varying amounts of commercial hydrated lime.

Specimens compacted at optimum moisture content were submitted to unconfined compressive strength tests, chemical analysis (reacted lime), X-ray diffraction and scanning electron microscopy (SEM) to ascertain the effects of lime on the soil. The improvement of bearing strength developed by lime addition to the soil, X-ray diffraction and SEM data were compared with the results associated with other soils from Brazil.

INTRODUCTION

According to the literature lime has been extensively used as stabilizing agent for many years and the great majority of the investigations were related to soils of temperate regions (Eades et al., 1962: Ormsby et al., 1973; Perret, 1977; Carvalho, 1979).

There is lack of information about soil-lime treatment on red tropical soils. Lime stabilization of soils for use in construction of pavements can often be beneficially and economically utilized. In most cases, however, sufficient knowledge is not yet available for evaluating the probable effects of lime stabilization of a soil without extensive testing of the individual soil. This situation is particularly prevalent in tropical and subtropical regions where soil stabilization research has been quite limited. Investigations carried out in Brazil (Ziolkowsky, 1977; Narde, 1979; Chauvel et al., 1980 and Lima, 1981) dealt mainly with the effects of lime on the compression strength properties of the soils but almost nothing has been reported concerning the chemical reactions involved and products formed. Available data concerning reactivity of advanced weathered soils are conflicting and indicate a lack of systematic investigation of the significant factors influencing lime stabilization of such soils. The investigation of expedient and economical means of construction utilizing indigenous materials in the tropics and subtropics, where strongly weathered soils predominate, is of importance from the stand point of development of highways systems, airfields, and other facilities required for the operation of transport and supply systems.

In the present work a saprolitic soil, situated under the stone line, and located in the Km. 280 of the SP 310 "Washington Luis" highway, in São Paulo State, Brazil, was selected to react with a commercial lime, compacted at optimum moisture content by employing the Proctor Method (Proctor, 1933), with the main purpose of investigating how it reacts with lime and which products are formed.

MATERIALS AND METHODS

Soil

In wet tropical regions the weathering affect simultaneosly all the weathered minerals that desappear in the surrounding horizon. In this condition the neoformation process is clearly dominant and will lead to the formation of a level where only secondary minerals will be present.

In some cases the weathered profile is composed by two levels: the upper one, which is the pedologic soil and another, more deep, formed by alterations, i.e., the saprolitic soil.

The resulting "in situ" altered material from the mother rock is considered saprolitic soil (Vargas, 1969), more or less affected by the laterization process. It is dense, generally heterogeneous with a major silt fraction and a high hydration degree. It is an easy-working soil, even above the optimum moisture content, but the final compacted products is more heterogeneous as compared with the lateritic soils.

In some cases there exists a discontinuity between lateritic and saprolitic soils, marked by a stone line. In this case the lateritic soil has been formed from transported material which is different from the deeper saprolitic one.

Generally the saprolitic soil is tick, and this is a function of the mother rock, topographic position and local conditions. It has a variety of colors with red, yellow, brown and green shades.

From the mineralogical point of view there is no major differences between saprolitic and lateritic soils; the main difference resides in the components organization. In the lateritic soils the mineralogical components are associated (plasma clay minerals - iron minerals) forming structural units. In the saprolitic one the componentes are isolated without particular organization beyond that inherent to the mother rock. So, the plasma in the saprolitic soil is constituted by a clay mineral formed direcly by transformation or neoformation, from a primary mineral (pseudomorphose). Another important characteristic of the saprolitic soil is the low quantity of fines (clay minerals) and the large proportion of coarse fraction (sand and gravel).

The field observation must be complemented by laboratory work, with granulometric data,mineralogy and grain format. Saprolitic soils with mica and kaolinite display low plasticity but have a relativelly large expansion index and loss of support when moist.

The soil used in this work has a mineralogical composition of Kaolinite,Gibbsite, Hematite and Quartz (16.5 % SiO_2, 2.2 % Al_2O_3, 15.6 % Fe_2O_3) and Kr = 2.3. It was mixed with commercial hydrated lime (I.L. 26.7 %;CaO 50.8%; SiO_2 2.0 %; MgO 17.9 %; I.R. 2.5 %; R_2O_3 1.9 %) in amounts of 3, 5, 7 and 9 g./100 g. soil. From de microstructural point of view (figure 1) it is characterized by a microgranular mass with a great apparent porosity.

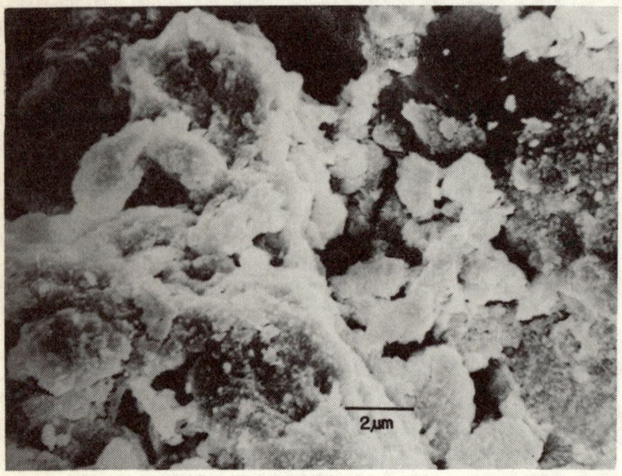

Figure 1. Scanning electron micrograph of the natural saprolitic soil.

SPECIMEN PREPARATION

The moisture -dry density relationship for the untreated soil and soil-lime mixtures (after passing through the 2.00 mm sieve), were determined by using a special equipment (Hilt <u>et al</u>., 1960; Pinto, 1965). The determinations were carried out for specimens compacted in molds and in a single layer. Each mixture was given five blows from the hammer assembled with the apparatus; the mold was inverted and another five blows were applied. Specimens of 50 x 50 mm were obtained and the compactation energy achieved was very nearly equivalent to the one given by the standard Proctor Method. Specimens were sealed in plastic bags to avoid lime carbonatation and loss of moisture.

PREPARATION OF MIXTURES

A predetermined amount of lime was added to an accurately weighed air-dried sample of the soil and the mixture was homogenized in a plastic bag (Ormsby <u>et al</u>., 1973). Water was added in appropriate amounts; the resulting system was homogenized inside the plastic bag, to prevent lime carbonatation and moisture loss. Final homogenization was established by visual inspection and hand contact.

UNCONFINED COMPRESSIVE STRENGTH

This test was used to show the effects of hydrated lime on the strength behavior of the soil. The vertical load was applied at a rate of 1.27 cm./min..Taking into account the results of dry density - moisture studies, sixteen replicates for each fixed amount of lime added were prepared and divided in four sets of four specimens each. These specimens were subjected to the compression test. Values of rupture strength are highly dependent on the test and specimen conditions (Willoughby <u>et al</u>. 1968). The compression data were therefore used only as an indication of the relative effects on strength rather than as an indication of absolute values pertaining to particular specimens.

DETERMINATION OF RESIDUAL LIME

At the end of the curing period, the LeDuc Method (Perret, 1977) was employed for determining unreacted lime: an accurately weighed sample (<u>ca</u>. 1.5 g.) is stirred in a tightly closed erlenmeyer for 30 min. with 20 ml. of a 20 % (w/v) aqueous solution of saccharose. As soon as most of the precipitate has settled, the supernatant liquid is decanted through a filter paper, leaving as much of the precipitate as possible in the erlenmeyer. The entire operation is repeated once again using a fresh 20 ml. portion of the saccharose solution. The precipitate is then transferred to the filter and washed several times with small volumes of water. The filtrate and washings are combined and the final volume is completed to 100 ml. with water. An aliquot of the resulting solution is titrated with 0.0100N H_2SO_4; the end point is located by using a mixed indicator (0.1 % bromocresol green and 0.5 % phenolphtalein, both in ethanol, and mixed in 1:2 volumetric proportion). The reacted lime is obtained by difference.

X-RAY DIFFRACTION ANALYSIS

Specimens intended for X-ray diffraction analysis were taken randomly from four samples previously subjected to the rupture process at the end of the curing period. The diffraction patterns were obtained by the diffractometric method in the Bragg-Brentano arrangement using CoK_α radiation (λ = 1.7902 Å) with 38 Kv and

20 mA, with proporcional counter and pulse - height discriminator. The $d's_{hkl}$ obtained were compared with standard ASTM (1967). Mixtures of all compositions were X-rayed.

SCANNING ELECTRON MICROSCOPY (SEM)

The scanning electron microscope proves to be ideally suited for studying the morphology, texture and fabric of clays in soils treated with lime, including also products formed in the reactions. The outstanding characteristic of these images is their remarkable three - dimensional quality. It is also seen that certain samples are more easily examined by SEM because of their larger crystallite size,better crystalinity and open texture (Barelli, 1982).

Specimens for SEM examination were taken as in the X-ray diffraction analyses and the material was glued to a SEM specimen stub with enamel and coated sucessively with carbon and gold in a vacuum evaporator. After this coating the microstructure and morphology of the clay and reaction products were studied in a "Jeol" model JSM-P15 scanning electron microscope.

EXPERIMENTS AND CONCLUSIONS

The approach used for this investigation consisted in making chemical, X-ray diffraction and compressive strength analyses as well as SEM examinations of the untreated soil so that their individual peculiarities could be noted before it is treated. The second step was to treat the soil with varying amounts of lime which was followed by curing periods ranging from 3 to 31 days at 23-28°C. The effects of the given treatment were evaluated through the following studies:

a) dry density as a function of moisture content (fig. 2).
Note that there is a variation in the optimum moisture content; this results is in disagreement with some literature reports (e.g. Perret, 1977; Carvalho, 1979).

b) lime consumption vs. amount of lime added, for all curing periods.

c) lime consumption vs. curing period, for all fixed lime contents (fig. 3).

d) compression strength vs. amount of lime content, for all curing periods.

e) compression strength vs. curing period, for all fixed lime contents (fig. 4). Note that for 7 g. lime/100 g. soil, the compressive strength is at lower values as compared with the other mixtures.

f) compression strength vs. percent reacted lime, for all fixed lime contents.

The following conclusions could be drawn from the body of experimental evidence:

1) The soil reacts very well with hydrated lime, with consumption within 71-86 % (fig. 3).
2) The improvement of bearing strength developed by lime addition was substantially lower, as compared with some lateritic soils (Barelli et al., 1985,1986).
3) From the X-ray diffration pattern (Table I) we see the formation of calcium silicate hydrate as a reaction product without a discrete morphology (fig. 5), but resembling a snow field. According to Taylor (1950) the heating of this product at 800°C should lead to the formation of wollastonite. However, this last compound has not been detected for the presently investigated system. Instead, amorphization of the entire complex was observed
4) The gathered data convey strong additional support to previous reports (Cabrera et al., 1979; Carvalho, 1981) which point out that the clay minerals found in the soils are the sole materials responsible for stabilization; the high pH system created by addition of lime promotes the breakdown of the crystalline structure of the clay minerals leading to the generation of pozzolanic reaction products.

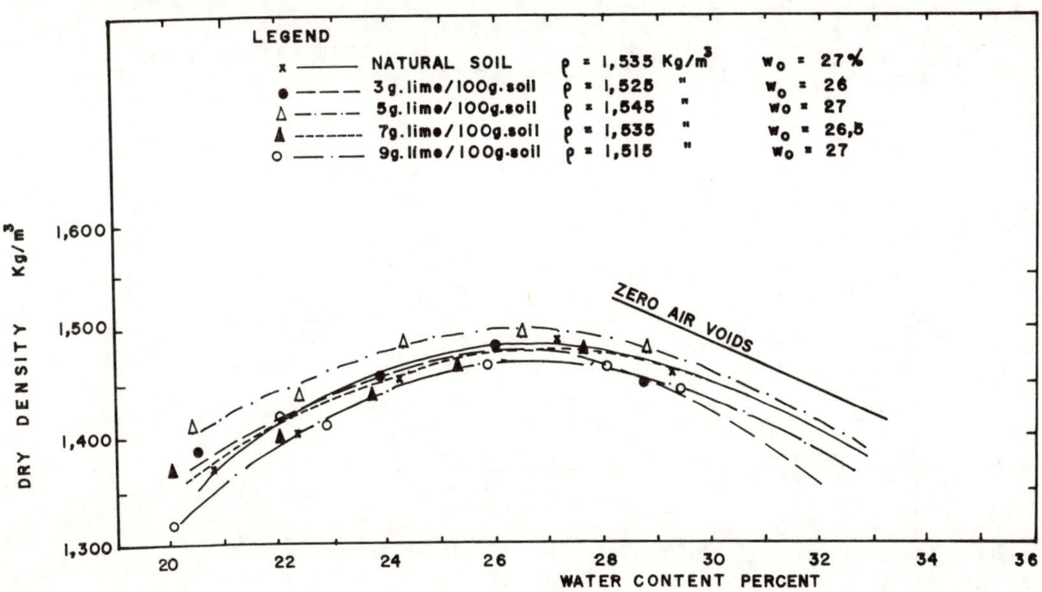

Figure 2. Dry density as a function of moisture content.

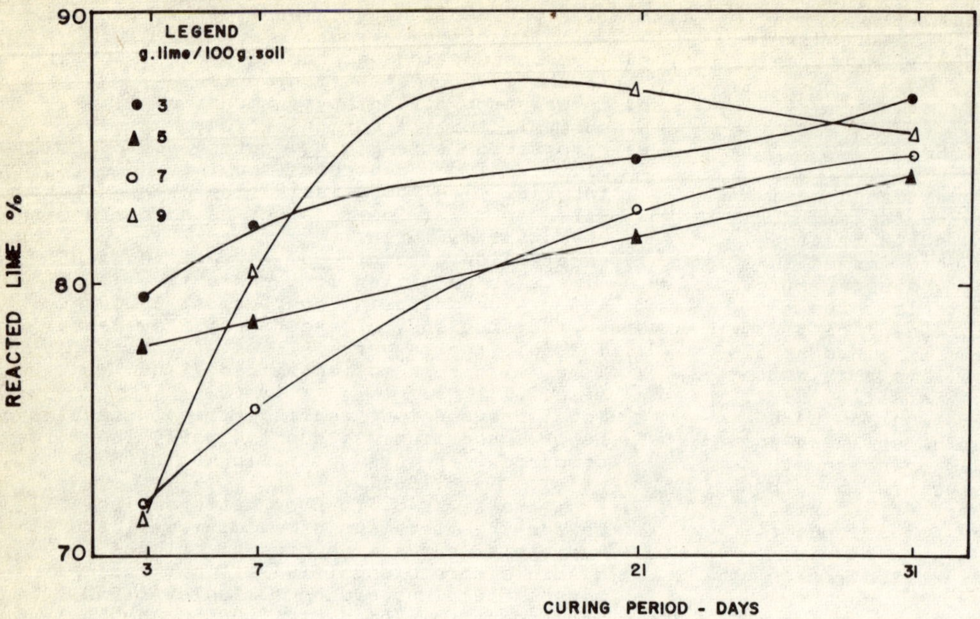

Figure 3. Reacted lime (%) vs. curing period.

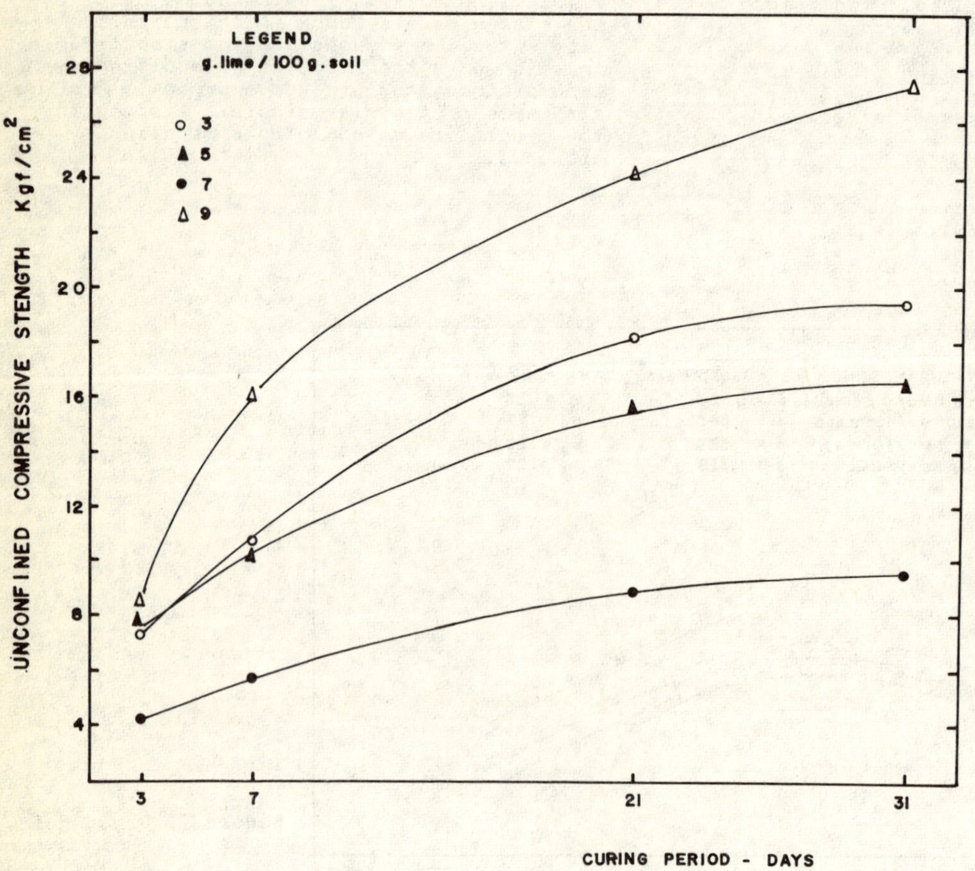

Figure 4. Unconfined Compressive strength vs. curing period.

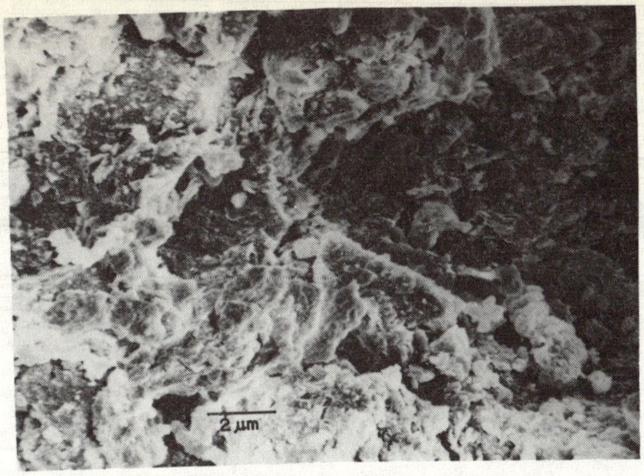

Figure 5. Scanning electron micrograph of the soil-lime reaction product.

TABLE I - X-Ray diffraction pattern data of the untreated soil superimposed to the corresponding soil-lime mixtures.

Line	I/I_o*	2θ	d_{nkl}(Å)	Materials
1	S	13.80	7.44	Calcium Silicate Hydrate
2	S	14.50	7.08	Kaolinite
3	M	21.50	4.80	Gibbsite
4	M	23.82	4.33	Kaolinite
5	M	24.80	4.16	Kaolinite
6	W	28.30	3.85	Kaolinite
7	S	20.70	3.49	Gibbsite
8	W	31.25	3.31	Quartz
9	M	34.50	3.09	Calcium Silicate Hydrate
10	S	38.85	2.68	Hematite
11	W	41.20	2.54	Kaolinite
12	W	41.90	2.39	Kaolinite
13	W	45.20	2.39	Kaolinite
14	W	76.40	1.44	Kaolinite

*S = strong M = medium W = weak

ACKNOWLEDGMENT

The financial support given by CNPq (Brazilian National Research Council) through Fellowships (Procs. 305632/78 and 300575/86) is gratefully acknowledged.

REFERENCES

ASTM (1967): Joint Committee on Powder Diffraction Standards. The Powder Diffraction File, Philadelphia, Pa.

BARELLI, N. (1982): Microscopia Eletrônica de Varredura da Parte Argilosa de Solos Usados como Base de Pavimentos Rodoviários (Scanning Electron Microscopy of Clays in Soils Used as Road Foundation) Ecl. Quím.,7, pp. 25-29.

BARELLI, N. and MELIOS, C.B. (1985): Soil-Lime Reactions: A Comparison Study Between Two Brazilian Red Soils. First International Conference on Geomechanics in Tropical Lateritic and Saprolitic Soils, Brazil - February 11-14, vol. 2, Technical Sessions, pp. 271-281.

BARELLI, N.; MELIOS, C.B. and R.C.F. da SILVA (1986): Soil-Lime Reactions: The Behaviour of a Brazilian Red Soil, Proc. 13th. Meet. Internat. Mineral. Assoc., Varna, Bulgaria, pp. 305-311.

CABRERA, I.G. and NWAKANMA, C. (1979):Pozzolanic Activity and Mechanism of Reaction of Red Tropical Soil-Lime Systems., Transportation Rearch Record 702, National Academy of Sciences - Washington D.C., Low Volume Roads: Second National Conference, pp.199-207.

CARVALHO, J.B.Q. (1979): Soil Properties Affecting the Lime Stabilization of Red Tropical Soils from North East Brazil, PhD. Thesis, Department of Civil Engineering, The University of Leeds, England, 344 pp.

CARVALHO, J.B.Q. (1981): Amorphous Materials and Lime Stabilized Soils. Proc. 10 th. Internat. Conf. Soil Mechanics and Foundations Engineering, Stockholm, Sweden, Session 12, pp. 761-763.

CHAUVEL, A. and NOBREGA, M.T. (1980): Comportamento dos solos Argilosos Desenvolvidos sobre Basaltos do Brasil Meridional Após Tratamento com Cal (Behaviour of Clayed Soils Developed on Southern Brazilian Basalts

Following Lime Treatment) Boletim nº 12, IV Reunião Aberta da Indústria da Cal., pp. 122-172.
EADES, J.L.; NICHOLS, F.P. and GRIM, R.E.(1962): Formation of New Minerals with Lime Stabilization as proven by Field Experiments in Virginia. Highway Research Board, Bull. nº 335, pp. 31-39.
HILT, H.G. and DAVIDSON, D.T. (1960): Lime Fixation in Clayey Soils. Highway Research Board, Bull. nº 262, pp. 20-32.
LIMA, D.C. (1981): Algumas Considerações Relativas a Aspectos da Estabilização dos Solos, em Particular à Estabilização Solo-Cal (Some Considerations Concerning Certain Aspects of Soil Stabilization, Particularly Soil-Lime Stabilization), M.Sc. Dissertation, Escola de Engenharia de São Carlos, USP, São Carlos, Brazil, 171 pp.
NARDE, J.V. (1979): Introdução à Estabilização de Solos com Cinzas Volantes e Cal (Introduction to Soil Stabilization with Fly Ash and Lime), Publicação Técnica, Departamento de Estradas de Rodagem de Santa Catarina, Florianópolis, SC, Brazil.
ORMSBY, W.C. and KINTER, E.B. (1973): Strength Development and Reaction Products in Lime - Montmorillonite - Water System. Public Roads, 37(4) pp. 136-148.
PERRET, P. (1977): Contribuition a L'étude de la Stabilization des Sols Fins par la Chaux: Étude Globale du Phénomene et Applications. Thése de Docteur - Engenieur, Institute Nationale des Sciences Apliquées des Rennes, France, 180 p.
PINTO, S.C. (1965): Equipamento Reduzido para Moldagem de Corpos de Prova de Solo Aditivo (Reduced Equipment for molding Additive Soil Specimens) IPR, Trabalho Apresentado no I Simpósio sobre Pesquisas Rodoviárias, Rio de Janeiro, 16-21 agosto.
PROCTOR, R.R. (1933): Waterway Experiment Station 1945.
TAYLOR, H.F.W. (1950): Hydrated Calcium Silicates. Part I. Compound Formation at Ordinary Temperature.J. Chem. Soc. pp.3682-3690.
VARGAS, M. (1969): "Residual Soil Sampling.Proc. Spec. Session nº 1 on Soil Sampling, 7th. Int. Group on Soil Sampling, pp. 50-53, Melbourne, Australia.
WILLOUGHBY, D.R.; GROSS, K.A.; INGLES, O.G.; SILVA, S.R. and SPIERS, V.M. (1968):Strength Development and Reaction Products of Alkali-Stabilized Clays by Electron Microscopy and Electron Diffraction, paper nº 430.
Australian Road Research Board Proceedings, Vol. 04, Part 2, pp. 1386-1408.
ZIOLKOWSKY, J.C. (1977): Estabilização de Solos Lateriticos com Cal (Lime Stabilization of Lateritic Soils) Publicação Técnica, Instituto Militar de Engenharia, Rio de Janeiro, Brazil.

In-situ testing of residual soils in Singapore

M.F.CHANG, Nanyang Technological Institute, Singapore

ABSTRACT : This paper discusses some common in-situ tests currently used for site investigation in residual soils in Singapore. Tests were carried out at five sites, two involving the residual soil of the sedimentary Jurong Formation and the other three the residual granite soil. Some approximate correlations of test results were established. A comparison of various in-situ tests was made to explore the applicability of these tests to residual soils in terms of material variability and design parameter evaluations. A proper way of selecting the in-situ test methods for a well-balanced site investigation work was recommended.

1. INTRODUCTION

Tropical residual soils cover almost two-thirds of Singapore Island. Construction activities involving residual soils have increased due to the establishment of several public housing estates and the construction of two major public transport works, the Singapore Mass Transit System and the Bukit Timah Expressway, in the heart of these originally relatively undeveloped land areas. There is an economic incentive to look for reliable and efficient site investigation methods which can provide geologic data and engineering characteristics of subsurface materials at project sites covered by residual soils in Singapore.

Residual soils of Singapore, similar to those in many other parts of the world, are generally heterogeneous and highly textured in nature. Quite often, they are partially saturated and relatively stiff. As a result, "undisturbed" sampling in these materials is very difficult and samples recovered are often not representative. In-situ testing is often the only reliable means for the determination of engineering parameters, including the subsurface profile and the engineering characteristics, of the geological materials for geotechnical designs in residual soils.

This paper presents some experience on the use of in-situ tests in site investigation for the evaluation of the stratification and the stress-deformation and strength characteristics of residual soils at five test sites in Singapore. The in-situ tests discussed include the standard penetration test (SPT), the Swedish ram sounding test (RST), Swedish weight sounding test (WST), the cone penetration test (CPT), the flat dilatometer test (DMT), the pressuremeter test (PMT) and the plate load test (PLT). Some test results are presented and compared. Correlations between the results of various test methods are established whenever possible. The selection of in-situ tests for site investigation in residual soils is discussed.

2. RESIDUAL SOILS OF SINGAPORE

In geotechnical engineering, residual soils are generally considered to encompass Grade V and IV materials (completely and highly weathered rocks) in addition to Grade VI (completely decomposed rocks). The residual soils of Singapore are derived mainly from weathering of two main rock formations, namely the Bukit Timah granite and the sedimentary Jurong Formation. Each of these two residual soil types covers approximately one-third of the land area of Singapore Island. They are unsaturated above the ground water table, which is commonly encountered at great depths.

The residual granite soil varies between clayey silt or clayey and silty sand and sandy/silty clay. It is of medium to high plasticity depending primarily on the degree of weathering. Its thickness generally varies between 10 and 35 metres. According to Poh et al (1985), the residual granite soil has totally lost its rock texture and can be classified as Grade IV material based on the classification proposed by Little (1969). The materials are generally of medium stiff to very stiff or hard consistency.

The residual soil of the sedimentary Jurong Formation is usually present in the form of interbedded layers of predominantly medium plasticity clayey silt, sandy clay and clayey to silty sand materials. Due to the severe folding of parent rock structures and the high frequency of joints and fractures in the parent rock, the residual soil of the sedimentary origin is spatially heterogeneous and highly variable. Its depth varies from a few centimeters to some ten meters. The consistency of the materials generally ranges from stiff to hard.

3. COMMON IN-SITU TEST METHODS

The in-situ test methods that are commonly used in Singapore for the determination of subsurface stratification and the stress-deformation and strength characteristics of residual soils fall into the following three categories:

(i) Dynamic penetration tests
(ii) Quasi-static penetration tests
(iii) Stress-deformation tests

3.1 Dynamic Penetration Tests

Tests in this category include the standard penetration test (SPT), the most common in-situ test in residual soils in Singapore, and the Swedish ram sounding test (RST). The MacKintosh probe which is popular in Malaysia is not commonly used here.

The standard penetration test, normally conducted at the bottom of a borehole, involves driving a standard split spoon sampler (35 mm I.D. and 50 mm O.D.), with a split barrel of approximately 457 mm long, into the soil by means of a 63.5 kgf hammer dropping from a height of 760 mm and counting the number of blows, N, required for 300 mm penetration of the sampler. The N-value is defined as the standard penetration resistance. The variation of N with depth is used as an aid to soil profile description and weathering classification and as a measurement of strength and its variability with depth. The test is particularly suitable for hard-driving materials in which other in-situ testing tools fail to penetrate. One major problem related to the test is the variable energy delivered to the sampler which can significantly affect the measured N-value. This variation in input energy is either apparatus dependent or operator dependent or both.

The Swedish ram sounding test which employs an automatic ram sounding apparatus (Dahberg and Bergdahl, 1974) aims at minimizing the variation in the input energy delivered to the sampler in each blow. The test involves driving a cone tip, approximately 90 mm long with a diameter of 45 mm and an apex angle of 90°, directly into the ground by means of a 63.5 kgf hammer free falling at a height of 500 mm. The penetration resistance is recorded at 200 mm intervals and the resistance required for every 200 mm penetration is taken as the ram sounding resistance, N_{20}. In this test, no borehole is required and no sample is recovered. The resulting resistance profile is almost continuous with depth and it reflects the vertical strength variability of the subsurface materials. In residual soils, its penetration capacity is not very different from or slightly better than the SPT (Chang and Wong, 1986).

3.2 Quasi-Static Penetration Test

Tests in this category include the cone penetrometer test (CPT) and the Swedish weight sounding test (WST).

In the cone penetrometer test, a standard cone tip with a base area of 10 cm^2 and an apex angle of 60° is hydraulically pushed into the subsurface materials and the point resistance, q_c, and the unit shaft friction, f_s, can be recorded independently or separated by calculation. The main limitation of the test is that its penetration capacity is limited by the capacity of the rig and the amount of reaction one can provide. In general, CPT's penetration capacity and the maximum sounding depth in residual soils are limited. The measurements, which are normally taken at 200 mm intervals in a mechanical cone test and almost continuously in an electric cone test, provide relative strength profiles for the subsurface materials and allow the materials to be classified reasonably accurately without recovery of samples.

In Singapore, both the classical manual weight penetrometer and the modern engine driven type of weight sounding machine (Broms and Bergdahl, 1982) have been used in the Swedish weight sounding test. The weight penetrometer consists of a screw shaped point, with a maximum diameter of 35 mm, connected to 22 mm diameter extension rods. In the classical WST, the number of half turns required for every 200 mm of penetration of the penetrometer when the penetrometer is subjected to a total weight of 100 kgf is recorded. The weight sounding resistance is normally taken as the total number of half turns per meter of penetration, N_w*. In the modern WST, the rod subject to a normal force of 100 kgf rotates at a constant speed and the number of half turns required for a penetration of 200 mm of the penetrometer is recorded and taken as the weight sounding resistance, N_w. The N_w-profile provides an assessment of the relative strength of the subsurface material and its variation with depth. The apparatus is light, the operation is simple, and the test is fast and easy to conduct.

3.3 Stress-Deformation Tests

Tests normally carried out mainly for the determination of stress-deformation characteristics include the flat dilatometer test (DMT), the pressuremeter test (PMT) and the plate load test (PLT).

The flat dilatometer test is a newly developed in-situ test method (Marchetti, 1980). The dilatometer tip is a 95 mm wide and 14 mm thick stainless steel blade mounted with a thin expandable steel membrane (60 mm in diameter) on one side. The tip is connected to a gas pressure and flow control box by a pneumatic-electric tubing. During the test, the dilatometer tip is normally pushed into the ground by a CPT rig. The penetration is temporarily stopped at every 200 mm intervals. Two readings, one corresponding to the lift-off pressure of the membrane and the other the pressure required to cause a prefixed deflection of approximately 1 mm at the center of the membrane, are taken. These two reading, after correcting for membrane stiffness are denoted as p_0 and p_1, respectively. The p_0-value is a measure of the in-situ horizontal stress or the stress history, whereas Δp, which equals to $p_1 - p_0$, is a measure of the stiffness of soil according to Marchetti (1980). These two pressures, p_0 and p_1, are used to calculate three diameter indices, namely the material index, I_D, the horizontal stress index, K_D, and the dilatometer modulus E_D, which can be correlated with various engineering soil properties (see Marchetti, 1980). The E_D-value was found to equal to 38.2 $(p_1 - p_0)$ based on the theory of elasticity. The E_D-value can also be related to the elastic modulus E and Poisson's ratio ν by $E_D = E/(1 - \nu^2)$.

The DMT is strictly speaking not a direct stress-deformation test. It relies on empirical correlations for the interpretation of design parameters. However, it has the practical advantage of a penetration test. Similar to the mechanical CPT, it gives practically continuous profiles of soil properties. The depth of penetration, however, is also limited by the capacity of the pushing equipment and the reaction available.

Pressuremeter tests commonly carried out in residual soils in Singapore involve two different types of apparatus : the Menard type pressuremeter and the OYO Elastmeter. The Menard pressuremeter consists of a measuring cell and two guard cells. In the test, water is used to pressurize the measuring cell in steps and the volume change of the measuring cell in each step is recorded as a measurement of the volumetric or radial strain of the soil. A pressure-volume change (p vs v) curve is plotted based on the measurements after corrections to account for rubber restitution and stiffness are made. The OYO Elastmeter is a single cell pressuremeter. The change in borehole radius or the radial strain of the soil is directly monitored by a pair of calipers located in the middle section of the cell. Either gas or water can be used as the agent in the application of pressure. The result is presented in a form of pressure versus borehole radius (p vs R) curve.

The pressure-volume change curve or the pressure-borehole radius curve normally consists of three parts: the initial recompression part, the pseudo-elastic part, and the plastic part. The slope of the pseduo-elastic part gives an indication of the elastic modulus of soil. A pressuremeter modulus E_p is normally calculated from this slope using an assumed Poisson's ratio for the soil. The pressure corresponding to a volumetric strain of 100% (or a radial strain of 50%) is arbitrarily taken as the limit pressure, p_L. The limit pressure is an indirect measure of shear strength of soil. The yield pressure, p_y, which represents the upper pressure limit of the pseudo-elastic range is also often recorded. Occasionally, the at-rest pressure, p_0, can also be estimated from the intersection of the recompression and the pseudo-elastic parts of the curve.

One of the main advantages of PMT is its suitability for materials with a very wide range of consistency and stiffness. As long as a reasonably stable and undisturbed borehole is formed in the material, the stress-deformation response of a reasonably large volume of in-situ material next to the borehole can be measured by the pressuremeter. Other advantages also include the speed of the test and its capability of testing soils that are difficult to sample or to penetrate with other in-situ test devices.

The plate load test (PLT) is a classical type of stress-deformation test. In the test, a square or circular bearing plate is set up at a prepared base, either on grade or at the bottom of a test pit or shaft. Vertical loads are applied in increments and the corresponding settlement of the plate is recorded after each load increment. A load settlement $(p - \delta)$ curve is then plotted. The linear portion of the curve is used to calculate the modulus of subgrade reaction k_s, based on $k_s = p/\delta$. The elastic modulus of the underlying soil, E, can then be calculated from the k_s-value using $E = \pi/4\ (1 - \nu^2)\ Bk_s$ for plates resting on an open surface, where ν is Poisson's ratio and B is the plate diameter.

The test is only suitable for homogeneous soil conditions and it requires very careful preparations. Due to the reaction required and the necessary preparation works, the test is rather expensive, especially when it is carried out at a great depth.

4. TEST SITES AND TEST PROGRAMS

A number of in-situ tests were carried out at five test sites in Singapore. Table 1 lists the general soil conditions and types of tests carried out at various sites.

Site S-1 and S-2, located on the campus of the Nanyang Technological Institute, involved the residual soil of the sedimentary Jurong Formation. Site S-1 was situated on a high ground where the water table was below the depth of investigation. Site S-2 was located on a very gentle sloping ground where the water table was 2 to 3 meters below the ground surface at the time of investigation.

Sites G-1, G-2 and G-3 involved the residual soil of the Bukit Timah Granite. Site G-1 was a fresh cut site occasionally with random boulders. No ground water was encountered during the investigation. Site G-2 was located near the toe of a slope and the ground water table was at around 4.5 m depth. Site G-3 was on top of a low hill. The location of the ground water table was around 9 meters below the ground surface.

During the investigation, in-situ test points were spaced at typically 1.5 to 3 meters apart. This was to avoid interaction among various tests on one hand and to minimize the possibility of large variations in surface materials on the other hand. It should also be pointed out that a classical weight penetrometer was used in the WST at Site G-3 and an engine driven apparatus was used for WSTs at other test sites. For PMTs, an OYO Elastmeter was used at Sites S-1 and G-1, whereas a Menard pressuremeter was used at Sites G-2 and G-3.

5. CORRELATIONS OF TEST RESULTS

In-situ tests can be classified into indirect and direct tests. Indirect tests are those simple tests which produce measurements that

Table 1 Soil conditions and types of tests carried out at various test sites

Site No.	Residual Soil Types	General Soil Conditions	Types of Tests
S-1	Jurong Formation	Clayey Silt (CL) % Fines = 60 - 95 PI = 10 - 20	SPT, RST, PMT, (DMT), (CPT)
S-2	Jurong Formation	Sandy Clayey Silt (MH) % Fines = 40 - 60 PI = 21 - 27	SPT, RST, WST, CPT, DMT
G-1	Bukit Timah Granite	Sandy Clayey Silt (MH) % Fines = 50 - 75 PI = 22 - 27	PMT, PLT, (SPT), (RST), (WST)
G-2	Bukit Timah Granite	Clayey Silt (MH) % Fines = 60 - 67 PI = 20 - 27	SPT, PMT, DMT
G-3	Bukit Timah Granite	Sandy Clayey Silt (MH) % Fines = 52 - 95 PI = 16 - 34	SPT, CPT, WST, DMT, (PMT)

( ) indicates minor tests or tests been reported elsewhere

Table 2 Comparison of various in-situ test methods

Test	Nature	Cost*	Practical Advantages/ Limitations	Applicability to Residual Soils +	
				Variability Evaluation	Parameter Evaluation
SPT	Indirect Dynamic strength	M	o providing samples	A	B
RST	Indirect Dynamic strength	L-M	o high mobility and speed	A	B
CPT	Indirect Static strength	M	o moderate penetrability	B	B
WST	Indirect Static strength	L	o limited penetrability o high mobility and speed	C	B
DMT	Indirect Stress-deformation	M	o moderate penetrability	B	B
PMT	Direct Stress-deformation	H	o high suitability to variable soil condition	C	A
PLT	Direct Stress-deformation	VH	o applicable to homogeneous soils	-	A

* L = low, M = medium, H = high, VH = very high
\+ A = high, B = moderate, C = limited, - = not applicable

can be empirically related to engineering properties of soils using empirical correlations, whereas direct tests are those generally more sophisticated tests which produce measurements that can be theoretically interpreted to give the intended engineering properties without a need to resort to empirical correlations. The SPT, the RST, the CPT, the WST and the DMT can be classified as indirect tests and the PMT and the PLT can be classified as direct tests.

Correlations of test results between the more-expensive direct tests and the less-expensive indirect tests are important for an optimum use of site investigation methods. Correlations between the well-estabilished tests and the less well-estabilished tests are also useful.

This section explores possible useful correlations between various test results for the two types of residual soils investigated. Some important findings that are useful for practical applications in the residual soils of Singapore are presented.

5.1 RST versus SPT

The correlation between the SPT resistance, N, and the RST resistance, N_{20}, is useful. The cheaper, faster and more standardised RST can be a supplement or even a possible substitute for the well-estabilished SPT in major projects involving a large number of test points if correlations between N and N_{20} for similar subsurface materials are available.

Figures 1 and 2 shows the relationships between N and N_{20} reported by Chang and Wong (1986) for the residual granite soil and the residual soil of the sedimentary Jurong Formation. A correlation of N_{20} = 0.5 N was found to be generally valid for both types of residual soils. For the residual soils of the Jurong Formation, the correction is backed-up by additional data collected from Sites S-1 and S-2 in this investigation. Nevertheless, for N > 30, N_{20} = 0.67 N seems to be a better fit, as was also indicated by Chang and Wong (1986).

Although previous studies, such as Muromachi and Kobayashi (1982), and a theoretical consideration of the energy delivered to the soils seem to suggest a correlation of N_{20} = N between these two tests, this cannot be verified in the residual soils of Singapore. Possible explanations for the RST being a more effective penetrometer in residual soils as compared to the SPT, as given by Chang and Wong (1986), include the lower rod energy ratio (partly due to larger height of fall and heavier anvil) and the larger side friction (as a result of the longer penetrometer) in the SPT, the difference in penetrometer shape and the high in-situ horizontal stress in residual soils which is believed to have a significant contribution to the large difference on side friction on the two different penetrometers. One possible reason for the N_{20}/N ratio to increase as N gets larger than 30 could be that, under hard driving conditions, a strong plug is likely to form in the split spoon sampler and causes the sampler to behave like a close end penetrometer with reduced resistance to driving (Randolph, 1987).

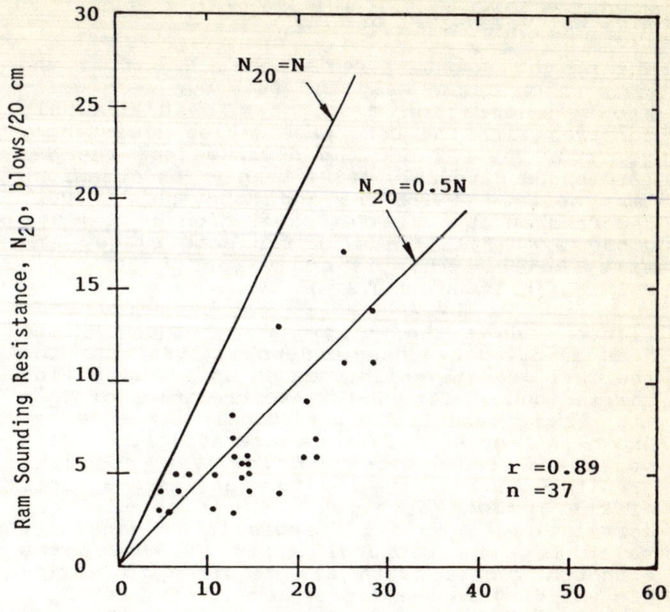

Figure 1 Relationship between N_{20} and N for residual granite soil (after Chang and Wong, 1986)

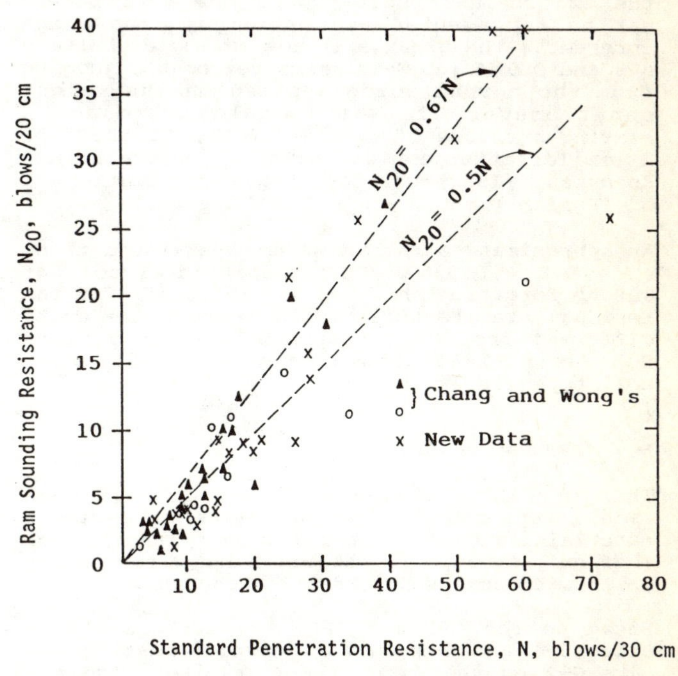

Figure 2 Relationship between N_{20} and N for residual soil of the Jurong Formation (after Chang and Wong, 1986)

5.2 WST versus CPT or SPT

The weight sounding resistance, N_w, from the WST, which is an easy and fast but less well-known penetration test, has been commonly compared with the cone penetration resistance, q_c, from the CPT and the N-value from the SPT (Broms and Bergdahl, 1982) mainly for cohesionless soils. Muromachi and Kobayashi, (1982) reported an approximate correlation of N = 3 + 0.050 N_w* (half turns/m) for some clayey and silty soils. This is equivalent to N = 3 + 0.25 N_w (half turns/0.2 m).

Figure 3 shows the relationship between N_w and N as reported by Chang and Wong (1986) for the residual granite soil based on an investigation carried out at Site G-1. A correlation of N_w = 1.5 N was found for $N_w < 30$ half turns/20 cm. Shown in Figure 3 are also some additional data collected from Site G-3 for the residual granite soil and Site S-2 for the residual soil of the Jurong Formation. The simple linear correlation: N_w = 1.5 N seems to be generally valid for the residual soils of Singapore, although further data are required to verify its validity in the Jurong Formation. For $N_w >$ 30 half turn/20 cm, the data points scatter and approach the correlation reported by Dahlberg (1974) for sands. It seems that the accuracy of the N_w as a measurement of soil strength decreases considerably as the soil gets stiffer with the N_w-exceeding 30 half turns/20 cm.

Figure 4 shows the relationship between N_w and q_c for the two residual soil types as obtained from one test site in each rock formation. The correlation is quite different from one residual soil type to the other. For each of the two residual soil types, the data points can be represented by a straight line which intersects the q_c-axis at a q_c-value of between 0.8 and 0.9 MPa. This seems reasonable judging from the normal force applied on the weight penetrometer tip which could introduce a bearing pressure of around 1.0 MPa at the tip. A similar intercept was found by Muromachi and Kobayashi (1982) who reported a correlation of q_c (Kgf/cm^2) = 6.7 + N_w* (half turns/m).

An approximate correlation has been found to be q_c = 0.8 + 0.025 N_w for the residual soil of Jurong Formation and q_c = 0.9 + 0.07 N_w for the residual granite soil. This seems to be quite different from that reported by Muromachi and Kobayashi (1982) based on several different soil types in Japan.

5.3 CPT versus SPT

The correlations between N and q_c have been widely reported for various types of geologic materials and the commonly quoted range of q_c (kgf/cm^2)/N ratio is between 1 and 2 for clayey soils with some sand and silt contents.

Based on the CPT and SPT data collected from Sites S-1 and S-2 on the Jurong Formation and Site G-2 on the Bukit Timah Granite, Figure 5 was derived. An average correlation of q_c/N = 0.20 (with q_c in MPa) was found for the residual soils of Singapore, although the ratio is closer to 0.21 for the residual granite soil

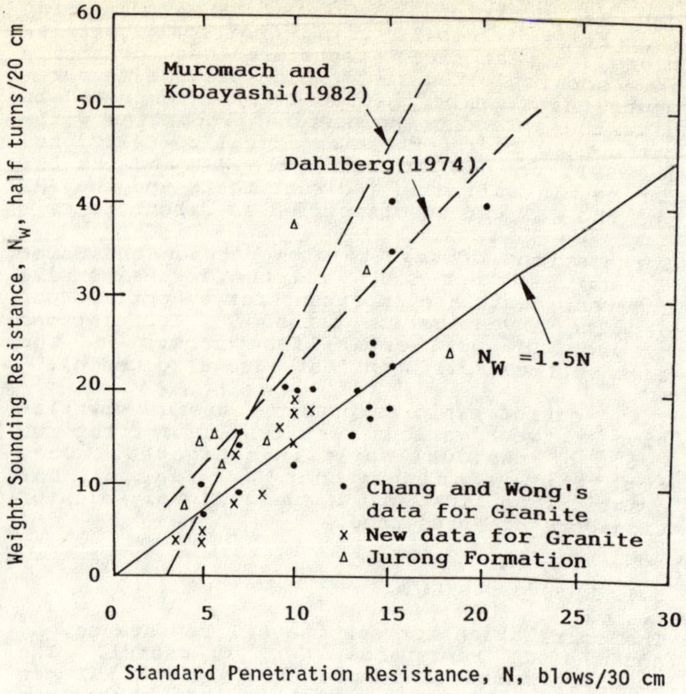

Figure 3 Relationship between N_w and N for residual soils of Singapore (modified from Chang and Wong, 1986)

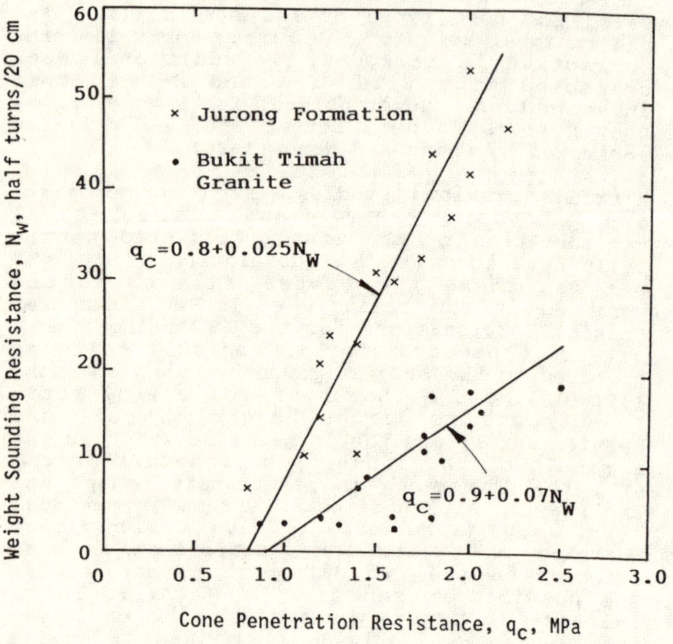

Figure 4 Relationship between N_w and q_c for two residual soil types.

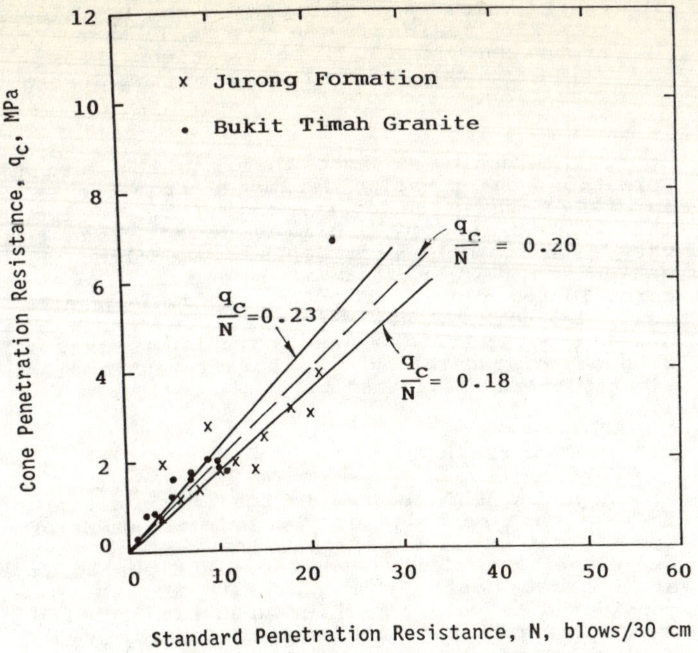

Figure 5 Relationship between q_c and N for residual soils of Singapore.

and 0.18 for the residual soil of the Jurong Formation. This value of $q_c/N = 0.20$ is the same as that reported by Poh et al (1985) for the residual granite soil.

5.4 DMT versus CPT

Although CPT is basically a shear strength test, the q_c-value has often been used as an estimate of the elastic modulus of soils. The ratio of E/q_c quoted by previous investigators for stiff sandy clays to clayey sand ranges from 1 to 6. Since the E_D-value from the DMT is a measure of elastic soil modulus, it is interesting to see whether a correlation exists between E_D and q_c. A very limited data collected in this investigation indicates an approximately average E_D/q_c ratio of 4.7 for the residual soil of the Jurong Formation and 6.9 for the residual granite soil. Considering $E = E_D (1 - \nu^2)$ and a ν-value of between 0.3 and 0.5, the deduced E/q_c ratio ranges from 3.5 to 4.3 for the Jurong Formation and 5.2 to 6.3 for the Granite. These values are believed to be slightly on the high side, judging from the fact that in the DMT the soils are prestrained by the advance of the dilatometer tip prior to the measurement.

5.5 PMT versus SPT and CPT

Pressuremeter tests are often carried out primarily for the assessment of the stress-deformation and strength characteristics of soils. However, the ratio of E_p/p_L from the PMT as well as correlations between p_L and N and p_L* and q_c* have been found to be closely related to soil types and their consistencies or compactness. According to Baguelin et al (1978), the E_p/p_L ratio for stiff to very stiff clay is between 10 and 20 and for soft to firm clay is beteen 8 and 10. Results of PMTs accumulated in this investigation indicate that E_p/p_L tends to increase with an increase in p_L. For the residual granite soil, the E_p/p_L ratio is between 12.5 and 25. For the residual soil of the Jurong Formation, the E_p/p_L ratio ranges from 7.5 to 20 in general. These ranges of E_p/p_L seem to compare reasonably well with those reported by Baguelin et al (1978) for clayey soils.

Figure 6 shows a comparison between the N and the p_L-value obtained from the PMT. An approximate correlation of $N/p_L = 0.0133$ (or $p_L/N = 75$), when p_L is expressed in kPa, seems obvious for the residual soils of Singapore. The same correlation also seems to apply when the limit pressure p_L is replaced by the net limit pressure p_L*. Interesting enough, if $c_u = 6$ N, where c_u is the undrained strength in kPa, as proposed by Stroud (1970) for low plasticity clays is considered, the resulting correlation between p_L* and the undrained strength, c_u is $c_u = p_L*/12.5$. This is in agreement with the commonly considered relationship between p_l* and c_u for stiff clays (e.g. Bagnelin et al, 1978).

An attempt was also made to correlate the q_c*-value with the p_L*-value based on a limited size of data. The q_c*/p_L* ratio was found to range from 2.3 to 3.8 and averaged 3.2 for the residual soil of the Jurong Formation. For the residual granite soil, the ratio ranged from 2.8 to 3.1 and averaged 2.95. These values compare relatively well with Baguelin's reported values of between 2.5 and 3.5 for firm to stiff clay and between 3 and 4 for very stiff to hard clay.

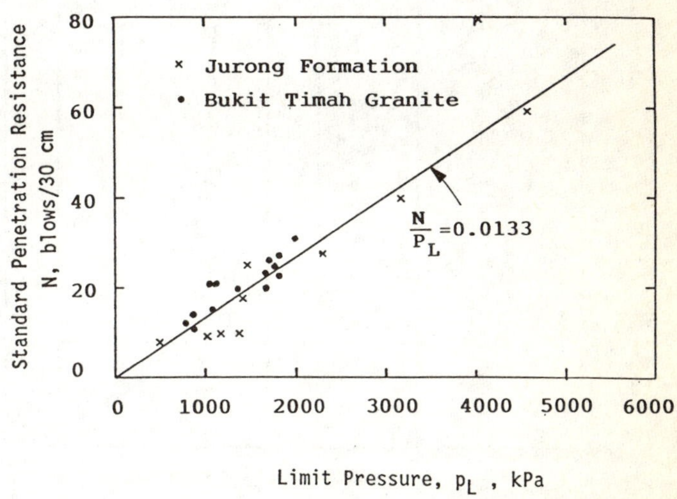

Figure 6 Relationship between N and p_L for residual soils of Singapore.

5.6 PMT versus DMT

The comparison of results based on the PMT and the DMT is particularly useful as both tests are stress-deformation tests. Since the PMT is more expensive to perform and the number of test that can be performed at a particular test location is very limited compared to the DMT which gives almost continuous profiles of soil characteristics, the DMT could be a good supplement to the PMT in large projects.

Figures 7 and 8 show the results of two dilatometer tests, one carried out in the residual granite soil at Site G-1 and the other in the residual soil of the Jurong Formation at Site S-1, together with the corresponding PMT results. The soil conditions as revealed by adjacent direct borings and by the DMT's material index (see Marchetti, 1980) are also presented in the figures. Three important points are noticeable:

(1) the DMT's p_0-value is higher than, in the residual granite soil, or approximately equal to, in the residual soil of the Jurong Formation, the yield pressure p_y as determined from the PMT,

(2) the DMT's p_1-value is very close to the p_L-value as determined from the PMT; the p_1/p_L ratio averages around 1.15 for the granite residual soil and 0.87 for the residual soil of the sedimentary Jurong Formation,

(3) the E_D-value is comparable to the E_p-value for the residual granite soil and is around one and a half to two times the E_p-value for the residual soil of the Jurong Formation.

It is interesting to note that, in the Jurong Formation, the p_o-value estimated from the PMT is higher than the corresponding σ_{vo}'-value. The DMT's p_0 and p_1 profiles do not seem to vary significantly with depth, resembling the unique characteristic of a heavily over-consolidated clay. For the residual granite soil, the PMT's P_0 is slightly less than σ_{vo}' and the P_0 and P_1 values increase gradually with depth, resembling the characteristic of a lightly overconsolidated clay.

5.7 PMT versus PLT

Although the PLT is one of the most suitable in-situ test methods for the determination of the elastic modulus of soil, the test is very expensive, especially when the elastic modulus at a great depth is required. The PMT is considered to be a possible substitute for the PLT. The fact that the soil is loaded in the horizontal direction in the PMT, whereas in the PLT the vertical response of the soil is measured, does not seem to be an issue. Shields and Bauer (1975) found that the modulus measured by the PLT was practically independent of the loading direction.

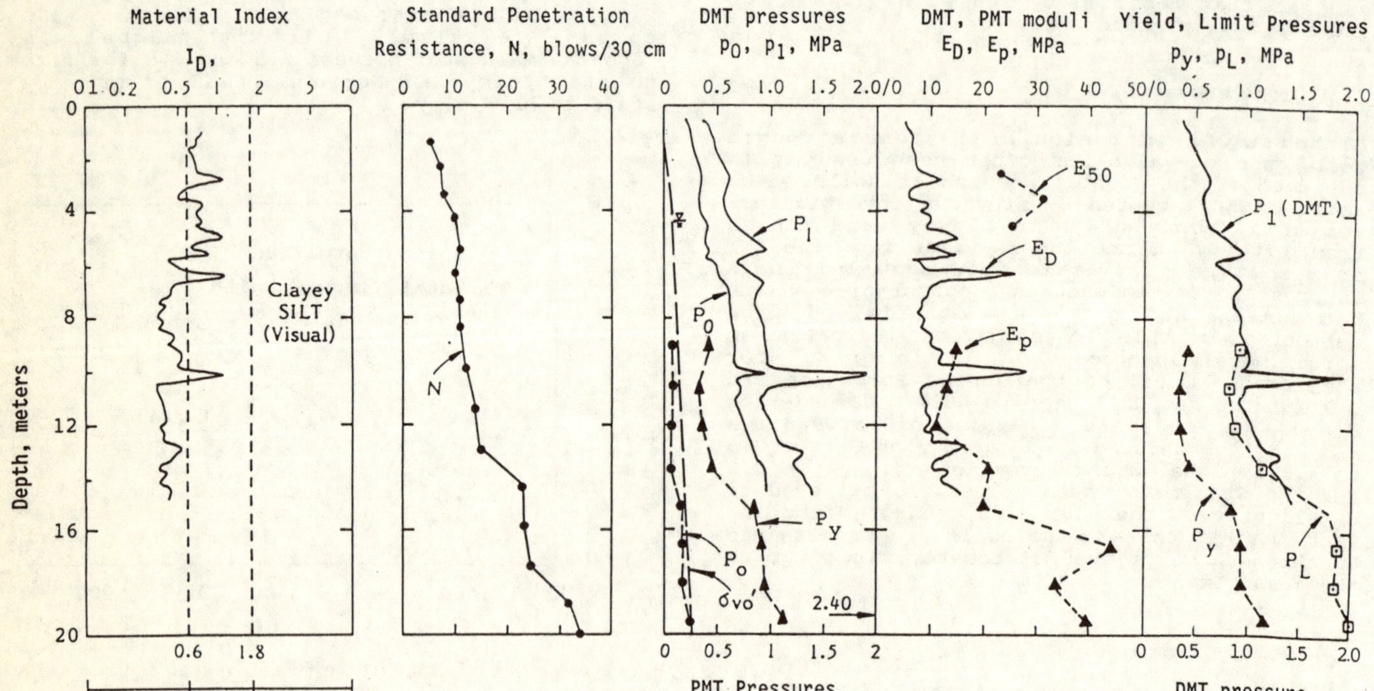

Figure 7 Comparison of DMT and PMT results in residual granite soil.

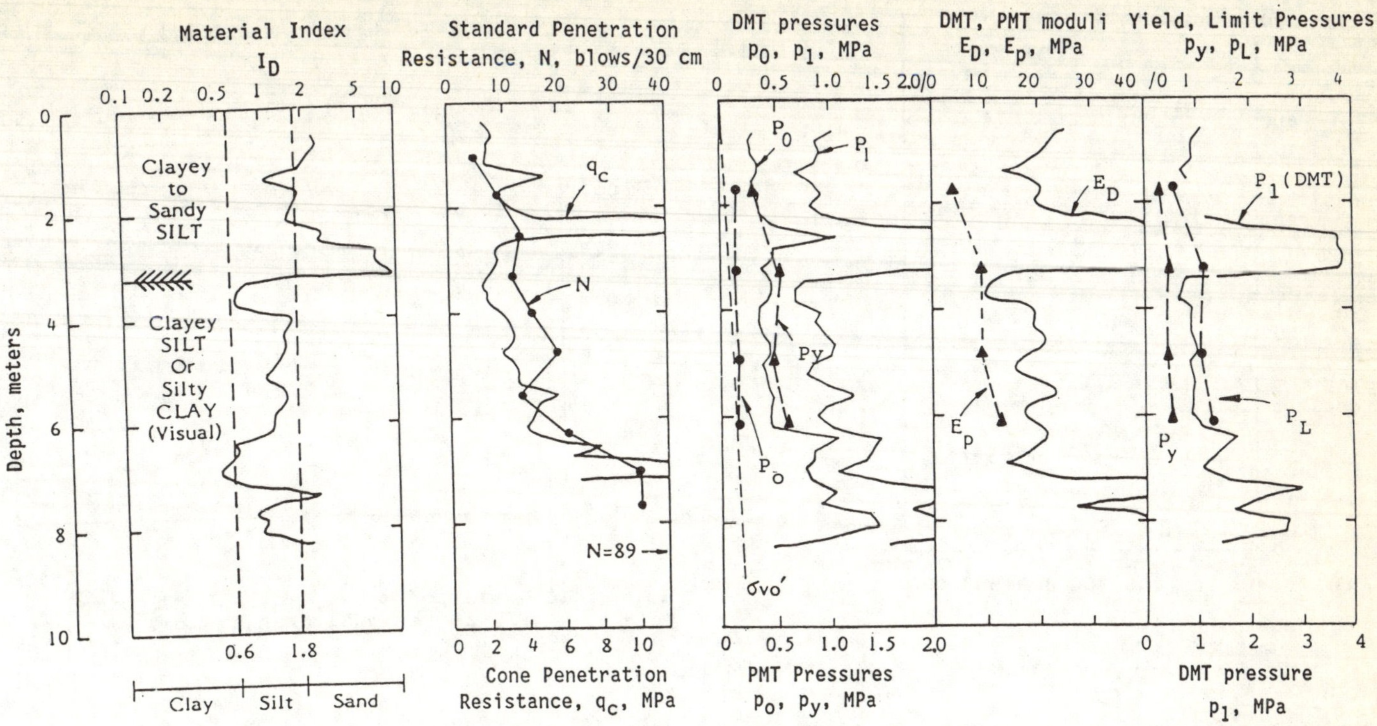

Figure 8 Comparison of DMT and PMT results in residual soil of the Jurong Formation.

A series of PLTs with different plate diameters were carried out at five test points on a roadway cut surface on the Bukit Timah Granite (Site G-1). Figure 9 shows the measured modulus of subgrade reaction, k_s, plotted against the plate diameter, B, for various test points. The k_s-value was found to decrease with an increase in plate diameter until B > 450 mm, except at Test Point 5 where the test was carried out at the bottom of a 0.5 m deep fresh cut and a constant k_s-value was observed. If one assumes that the stress bulb under the influence of the applied loads extends 2B below the plate and the average k_s-value over this influence zone can be taken as that corresponding to a depth of B, a variation of k_s with depth, as shown in Figure 10, can be derived. It should be noted that the k_s-value seems to be relatively high close to the ground surface, possibly due to surface compaction effect induced by heavy vehicles and dessication, except at Test Point 5 at which the compacted or dessicated crust has been removed. The k_s decreases with depth to a depth of around 0.4 m and below which it seems to assume a constant value of around 90 MN/m^3.

Based on the theory of elasticity, the elastic modulus, E, of a subsoil with a constant k_s-value can be calculated from $E = \pi/4\ (1 - \nu^2)\ B\ k_s$ as mentioned earlier. Applying k_s = 90 MN/m^3 and assuming $\nu = 0.3$, the calculated E-value varies linearly from 25.7 MN/m^2 at a depth of 0.4 m to 77.2 MN/m^2 at a depth of 1.2 m. Figure 11 shows the variation of E with depth so obtained from the PLTs. The E_p-values obtained from nearby pressuremeter tests are also shown in the figure. It should be noted that the E_p-value, averaged around 6 MN/m^2 within the depth influenced by the plate, is much smaller than the E-value calculated from the PLT. The E/E_p ratio is approximately between 5 and 11 in the depth range where the PLT results are available. It is interesting to note that the E_{50} (secant modulus corresponding to 50% failure stress) measured in UU triaxial tests on block samples recovered from the base level was around 8 MN/m^2, only slightly higher than the E_p-value. The average laboratory UU undrained strength of the soil was around 110 kN/m^2.

6. SELECTION OF IN-SITU TESTS

Since "undisturbed" sampling in residual soils, which are generally stiff, heterogeneous, and textured, is difficult, engineers have to rely heavily on in-situ tests for the determination of soil stratigraphy and engineering properties. A large number of in-situ test methods are available. In addition to those discussed in this paper, other in-situ tests such as field vane, borehole shear and field shear box tests, and geophysical methods, are occasionally used in residual soils. For the special purpose of determining permeability, borehole permeability test, packer test, and piezometer permeability test are also used. The selection of proper insitu test methods to

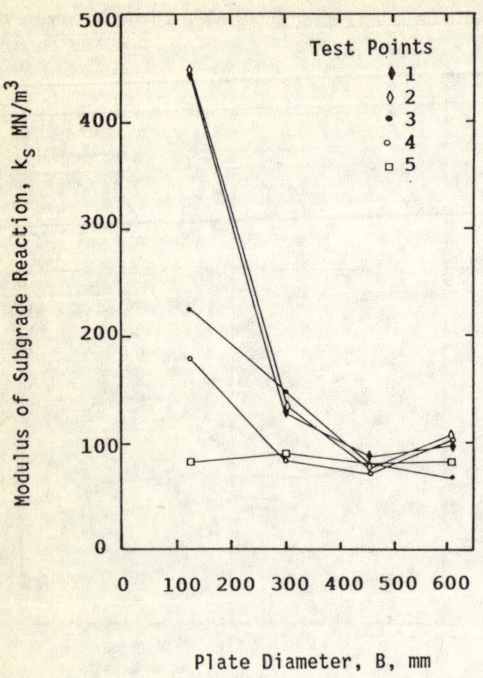

Figure 9 Variation of k_s with plate diameter based on PLT in residual granite soil.

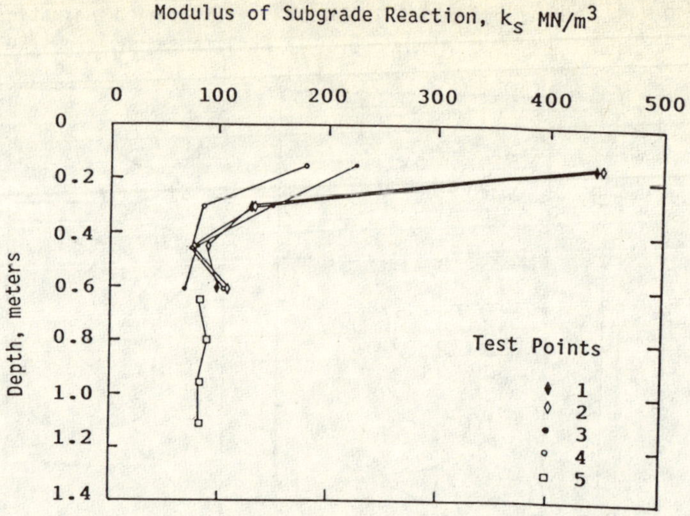

Figure 10 Variation of k_s with depth based on PLT in residual granite soil.

achieve a balance between economy and technical requirements becomes a major concern to geotechnical engineers.

There are a number of special features present in residual soils which may predominantly affect the selection of in-situ test methods. Residual soils are products of weathering of parent rocks. The material is often heterogeneous with its consistency or hardness often varying significantly. Due to differential weathering, the material gradually gets stronger as the depth gets larger, in general. An abrupt decrease in material resistance is seldom found, although a sudden increase in resistance is possible due to the presence of hard weathered rocks in the residual soil matrix or oxidized or altered layers such as iron veins which are very common in the residual soil of the sedimentary Jurong Formation. Residual soils could also be partially saturated, textured and highly anisotropic in nature. As a consequence, there are at least three important requirements that should be considered in the selection of in-situ test methods for site investigation in residual soils:

(1) The variability of the material, which may affect the performance of engineering works, needs to be carefully defined in terms of its extent. Possible minute changes in soil characteristics in the vertical direction and the detailed stratigraphy, however, are not crucial, in general.

(2) The wide range of hardness or consistency of the materials to be investigated necessitates tools or test methods which have the greatest adaptability.

(3) Interpretation of test results could be difficult even with the use of sophisticated direct test methods due to partial saturation, texture and anistropy.

Based on the above-mentioned considerations, the suitability of the seven in-situ tests discussed in this paper to site investigation in residual soils are compared. Table 2 summarizes the comparison of these tests methods in terms of their costs, practical advantages or limitations, and their applicability to residual soils for material variability and design parameter evaluations. In terms of variability evaluation, the SPT and the RST, which have similar penetration capability seem to standout. Comparing these two, the RST may offer slightly greater advantage than the SPT as it produces almost continuous operator independent resistance profile and the RST apparatus has a greater mobility, although no sample is recovered in the RST. The applicability of the other indirect tests to variability evaluation is often hindered by their limited penetrability. This is particularly the case for the WST which utilizes a light-weight machine that makes it extremely suitable for difficult-to-access site conditions such as along a steep slope. Generally, the WST is stopped when the N-value of the material exceeds 50 blows/30 cm.

In terms of design parameter evaluation, the direct tests, namely the PMT and the PLT, are more capable. Comparing these two, the very high cost involved in the PLT and the fact that it is generally applicable only to homogeneous

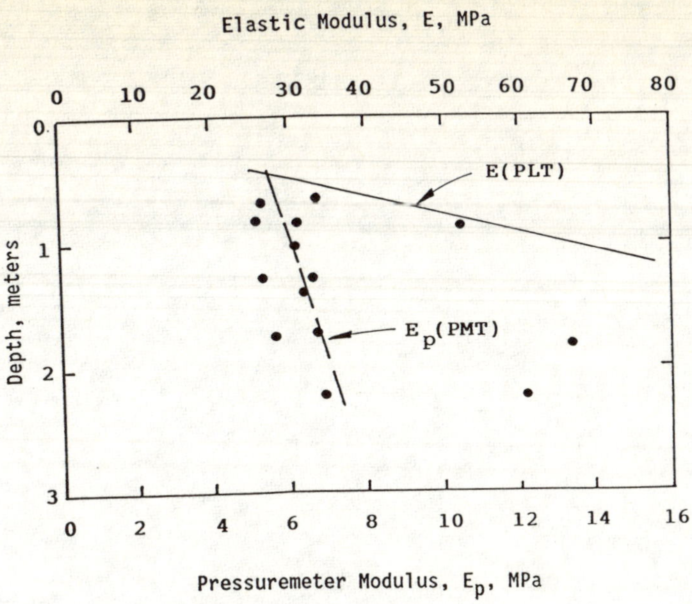

Figure 11 Comparison of E and E_p profiles in residual granite soil.

materials at shallow depths prevent it from being widely used. The PMT, on the other, seems to be a more practical and valuable test particularly for the assessment of the variation of stress-deformation properties with depth. The tests can be carried out at an interval as close as 1.5 m or so in a properly prepared borehole. The pressuremeter modulus E_p corresponding to the initial loading, which can be calculated directly from the PMT curve, can be used for direct evaluation of elastic modulus of soil, although an empirical factor has to be applied to account for borehole disturbance and discrepancy of theory used in the interpretation. This situation can be improved if a measurement of the reload modulus is included in the routine test. The reloaded modulus is known to be more representative of the field elastic modulus. Laboratory tests on "undisturbed" samples should also be carried out wherever possible to verify the usefulness of the pressuremeter results.

It should be noted that, while the direct tests are most suitable for the evaluation of design parameters, the number of tests that can be carried out is very limited due to their sophistication and high costs. On the contrary, the indirect tests are most useful for the evaluation of material variability mainly because of their low to moderate costs and high speed; but the interpretation of design parameters from the measurements requires validated empirical correlations. In planning a reasonable investigation program, it is therefore important that locally available in-situ test methods are properly combined after a careful consideration of the advantages and limitations of each method. This could include, for example, the use of less expensive indirect tests as supplements to other more expensive indirect tests or as sounding tools to find out where and what types of much more costly direct tests should be carried out. Only with a good prior understanding of the engineering problems and a careful selection of available site investigation techniques based on their technical and economic requirements can one achieve his goal of developing a well-balanced site investigation program.

7. CONCLUSIONS

The use of SPT, RST, CPT, WST, DMT, PMT and PLT in the residual soils of Singapore were discussed. Reasonable correlations were found between results of various tests. Some useful approximate correlations obtained are:

(i) for both residual soil types

$N_{20} = 0.5\ N$
$N_w = 1.5\ N \quad \text{for } N_w \leq 30$
$q_c\ (\text{MPa}) = 0.2\ N$
$p_L\ (\text{kPa}) = 75\ N$

(ii) for the residual soil of the Jurong Formation

$q_c\ (\text{MPa}) = 0.9 + 0.007\ N_w$

$q_c^* = 3.2\ p_L^*$

$p_1\ (\text{DMT}) = 0.87\ p_L$

(iii) for the residual granite soil

$q_c\ (\text{MPa}) = 0.8 + 0.025\ N_w$

$q_c^* = 2.95\ p_L^*$

$p_1\ (\text{DMT}) = 1.15\ p_L$

A comparison of various in-situ tests discussed based on their nature, costs, practical advantages and limitations and applicability in terms of material variability and design parameter evaluations clearly reveals that a well balanced site investigation program requires a combination of various in-situ test methods. For a major project site involving residual soils, this can be achieved by combining SPTs, supplemented by RSTs, with PMTs, backed-up by laboratory stress-deformation tests on "undisturbed" samples.

Further research should be devoted to the establishment and verification of correlations between test results obtained by various site investigation methods.

ACKNOWLEDGEMENT

The author would like to thank Nanyang Technological Institute in providing partial financial support for this study. He also wish to express his gratitude to those who have contributed the test data, in particular Kiso-Jiban Consultants (Singapore) and several former NTI students. Thanks are also due to the assistance of the technical staff in the NTI Geotechnics Laboratory. The manuscript was typed by Ms Jamillah Bte Haji Sa'adon.

REFERENCES

Baguelin, F. Jesequel, J.F. and Shields, D.H. 1978. The Pressuremeter and Foundation Engineering. Trans Tech Pub, Germany.

Broms, B.B. and Bergdahl, U. 1982. The weight sounding test (WST) - state-of-the-art Proc 2nd Europ Symp on Penetration Testing. Amsterdam, pp 203-214.

Chang, M.F. and Wong, I.H. 1986. Penetration testing in the residual soils of Singapore. Proc Spec Geomech Symp on Interpretation of Field Testing for Design Parameters. Univ of Adelaide, Australia.

Dahberg, R. 1974. A comparison between the results from Swedish penetrometers and standard penetration results in sand. Proc Europ Symp on Penetration Testing. Stockholm, Vol 2.2, pp 67-68.

Dahberg, R. and Bergdahl, U. 1974. Investigation of the Swedish ram sounding method. Proc Europ Symp on Penetration Testing. Stockholm, Vol 2.2, pp. 93-102.

Little, A.L. 1969. The engineering classification of residual tropical soils. Proc of Spec Session on Engineering Properties of Lateristic Soils. 7th ICMFE, Mexico, Vol 1, pp 1-10.

Marchetti, S. 1980. In situ tests by flat dilatometer, J. Geot Engrg Div. ASCE, Vol 106, GT3, pp 299-321.

Muromachi, T. and Kobayashi, S. 1982. Comparative study of static and dynamic penetration tests currently in use in Japan. Proc 2nd Europ Symp on Penetration Testing. Amsterdam, pp 297-302.

Poh, K.B., Chua, N.L. and Tan, S.S. 1985. Residual granite soil of Singapore. Proc. 8th SEAGC, Kuala Lumpur, p. 3-1 ≃ 3-9.

Randolph, M.F. 1987. Modelling of the soil plug response during pile driving, Proc. 9th SEAGC, Bangkok, pp 6-1 - 6-14.

Shields, D.H. and Bauer, G.E. 1975. Determination of the modulus of deformation of a sensitive clay using laboratory and in-situ tests, Proc. ASCE conf. on In-situ Measurements of Soil Properties, Raleigh, North Carolina, pp 395 - 421.

Stroud, M.A. 1974. The standard penetration test in insensitive clays and soft rocks. Proc 1st Europ Symp on Penetration Testing. Stockholm, Vol 2.2, pp 367-375.

Direct shear testing of a completely decomposed granite

C.K.CHEUNG, Geotechnical Control Office, Hong Kong
D.R.GREENWAY, Geotechnical Control Office, Hong Kong
J.B.MASSEY, Geotechnical Control Office, Hong Kong

ABSTRACT: The direct shear test offers some advantages over the triaxial test. However, before it can be applied with confidence to the routine measurement of shear strength of saprolites and residual soils, some of the practical constraints associated with the test need to be resolved. This paper describes the results of a series of tests on 'undisturbed' completely decomposed granite, aimed at examining the behaviour of the soil in direct shear, and exploring the influence of specimen size and rate of shearing on the measured strength of the soil.

1 INTRODUCTION

Steep terrain, heavy seasonal rainfall and deep tropical weathering contribute to numerous slope failures in Hong Kong every year. Very many of these failures are shallow, extending to depths of two to three metres. Intense hillslope development, with associated formation of large cuttings, requires that very substantial resources be devoted to shear strength testing of the saprolitic and residual soils for the purpose of safe and economic slope design.

Traditionally, the preferred methods of testing have been the consolidated drained triaxial test and the consolidated undrained triaxial test with measurement of pore pressure, conducted at low effective confining pressures which are relevant to the field conditions. Tests are normally carried out on specimens saturated by the application of high back pressure. However, not only are triaxial tests costly and time consuming to perform, but some doubts exist concerning specimen disturbance during saturation at low confining pressures, and the accurate estimation of specimen cross-sectional area, particularly at high axial strain.

The direct shear test is a much simpler and quicker test to perform, and appears to offer a solution to the concerns relating to the triaxial test. Theoretical objections to the use of the test remain, such as rotation of principal stresses, progressive failure of the specimen, and inability to measure pore pressures. However, the field conditions of low overburden pressure, and wetting of the soil by infiltration and soaking under low head, can be simulated more easily in the direct shear test, and the strain conditions are more representative of the majority of slope failures observed in the field. Rotation of principal stresses is of less concern where there is no pronounced soil anisotropy.

There are practical constraints associated with the use of the direct shear test for saprolitic and residual soils, which must be resolved before the test can be widely adopted as a routine investigation tool; these apply particularly to the decomposed granites, which may have a maximum particle size ranging up to 8 mm or more. This paper describes the results of a series of tests on 'undisturbed' completely decomposed granite (saprolitic soil), aimed at examining the behaviour of the soil in direct shear, and exploring the influence of specimen size and rate of shearing on the measured strength of the soil. Tests have been carried out using 60 mm and 100 mm square shear boxes with specimen thickness varying from 20 mm to 44 mm, and at shearing rates from 0.0072 mm/min. to 0.6 mm/min.

By resolving questions concerning these test parameters, it has been found that the direct shear test could be used satisfactorily for the routine measurement of shear strength of many of the saprolitic and residual soils. The paper presents the results of the tests carried out, and also briefly examines the effects of soil anisotropy.

2 REVIEW OF PREVIOUS WORK

2.1 Use of the Direct Shear Test

The direct shear test is a conceptually simple test that apparently was used for soil testing as early as 1776 by Coulomb (Lambe & Whitman, 1969), and which certainly featured prominently in the work of Collin in 1846 (Skempton, 1984). Because of the shape of the specimen used, the direct shear apparatus is particularly suited to consolidated drained testing, as the drainage paths within the specimen are short and excess pore pressures can rapidly dissipate.

Attributes of the direct shear test, and comparisons with the more refined triaxial test, have been discussed since at least 1939 (Hennes, 1952), and were systematically presented in Lambe's classical test on soil testing in 1951 (Lambe, 1951). Similar discussions can be found in separate British and American symposia on shear strength of soils held in 1950 and 1952, respectively (ICE, 1951; ASTM, 1952). During the intervening years, the direct shear test has proven to be useful for testing reinforced soils (Jewell & Wroth, 1987; Gray & Ohashi, 1983) and rocks (ISRM, 1981; Schneider, 1978), as well as a wide range of soil types.

Interpretation of direct shear data to obtain shear strength parameters has been briefly summarised by Morgenstern & Tchalenko, 1967. The most common interpretation of the drained test (considering the case with c' = 0) is:

$$\tau/\sigma = \tan \phi' \quad ....(1)$$

where τ = apparent shear stress and

σ = apparent normal stress.

The presumption inherent in equation 1 is that the shear strength is mobilized in the direction of the apparent shear stress (horizontally in the shear box) and that this direction coincides with the failure surface. As early as 1950, however, it was pointed out that the direction of the apparent shear stress is the direction of the maximum shear strain rate, and if the axes of the principal stress coincide with those of strain rate, then:

$$\tau/\sigma = \sin \phi' \quad ....(2)$$

Equation 2 is the appropriate interpretation if conditions of simple shear are assumed to exist within the specimen, and it has been supported by a number of researchers (Bishop, 1954; Morgenstern & Tchalenko, 1967). Potts et al (1987) have shown that, considering the angle of dilation of the material, ψ :

$$\tau/\sigma = \frac{\sin \phi' \cos\psi}{1-\sin \phi' \sin\psi} \quad ....(3)$$

which reduces to equation 2 when ψ = 0 and to equation 1 when $\psi = \phi'$.

In this paper, equation 1 is used for the interpretation of the direct shear test data.

Direct shear testing is further described by Akroyd (1957), Bowles (1970), Head (1982) and Vickers (1978). Procedures for consolidated drained testing are described in ASTM D 3080-72. Several particular aspects of direct shear testing are addressed in the following sections which are essential to the development of a standard test method for saprolitic and residual soils.

2.2 Shear Box Size

As reviewed by Lambe (1951), Taylor & Leps (1938) performed direct shear tests using both 75 mm and 300 mm square shear boxes on dry Ottawa sand (maximum grain diameter, d_m, approximately 1 mm). They found that the size of the shear box had a negligible effect on the results in this case.

Bishop (1948) carried out comparative direct shear tests using both 60 mm and 305 mm (12 inch) square shear boxes. He also reported a negligible scale or size effect for a medium sand (d_m = 0.6 mm). Using the larger shear box, which was 152 mm (6 inches) deep, he reported well behaved, smooth shear stress-displacement curves for a well graded gravel with d_m = 25 mm (1 inch), but 'jerky' curves for a uniform coarse beach gravel consisting of rounded pebbles about 25 mm in diameter. He therefore suggested that the pebbles probably represented an upper limit to the size of uniform material that could be tested accurately in the large shear box.

Skempton & Bishop (1950) noted that the shear box test normally utilised specimens either 60 mm square by 20 mm thick or 305 mm (12 inches) square by 152 mm (6 inches) thick. They considered the larger shear box to be capable of testing gravels with d_m up to 25 mm, or up to 38 mm (1½ inches) if they were well-graded. Bishop (1950) further considered that unless the thickness of the direct shear specimen was at least one quarter of its length, no unique measurement of strength could be obtained.

Hennes (1952) reported the testing of dry gravels and crushed aggregates with d_m up to 19 mm (3/4 inch) in a shear box 152 mm wide by 305 mm long (depth unspecified). They discontinued testing of stone larger than 19 mm because "the results were less consistent".

From Japan, Onodera et al (1976) reported shear tests on undisturbed decomposed granitic soil using 60 mm square by 20 mm thick specimens at natural moisture content. They encountered difficulty when d_m of the granitic soil was larger than 2 mm, as suitable strengths could not be determined for such small specimens. They used an advanced direct shear apparatus capable of maintaining either constant volume (undrained) conditions or constant normal stress drained conditions, and prepared specimens by setting them into gypsum plaster before shearing. However, this technique does not lend itself to control of specimen moisture content.

Brady et al (1983) performed comparative tests using 60 mm and 300 mm square shear boxes on sand (d_m = 2 mm approximately) at one normal stress. After discovering substantial errors in the large box due to fluctuations in applied normal load, they subsequently modified the large box to accept a load cell and then obtained results comparable to the small box. In fact, the results from the modified large box showed less scatter than those from the smaller box.

These references support the general conclusions that there is no significant scale effect in direct shear testing of cohesionless soils due to the size of the shear box itself, but that particle size is important in relation to shear box size, soils with larger particles obviously requiring larger shear boxes. ASTM (1985) actually gives two guidelines for direct shear testing:

(a) specimen thickness to be at least six times the maximum grain diameter of the soil;

(b) specimen diameter (or width) to be at least twice the thickness.

As many granitic soils in Hong Kong have maximum grain diameters of 8 mm or more, it is apparent that it may be difficult to simultaneously meet both ASTM guidelines for some local granitic soils using a shear box which is small enough to be used conveniently for routine investigations.

Extensive investigations on the shear box size effect in gravel testing have also been carried out in China. The Chinese Specification for Geotechnical Tests compiled for use in the hydroelectricity and hydraulic bureaus (HBPRC, 1980) gives two guidelines for testing sandy gravel in direct shear:

(a) specimen thickness to be between four and

(a) ... eight times the maximum grain diameter of the soil,

(b) specimen diameter to be between eight and twelve times the maximum grain diameter.

These guidelines are largely based on the result of a study using both 500 mm diameter by 300 mm thick and 500 mm diameter by 500 mm thick shear boxes on alluvium with maximum particle sizes from 20 to 120 mm.

2.3 Shear Rate

In the ordinary direct shear test on saturated soils, the rate of shear is the principle factor that determines whether the test is drained or undrained. It is impossible in theory to attain either a fully drained or fully undrained condition in a constant rate of shear test, but in practice it is often possible to select a shear rate so that the deviation from ideal conditions is not significant.

Gibson & Henkel (1954) derived an expression for the time to failure (t_f) in a drained direct shear test necessary to attain a specified degree of dissipation of pore pressure at the centre of the test specimen, namely:

$$t_f = \frac{H^2}{2C_b(1-U_c)} \qquad (4)$$

where H = half thickness of specimen

C_b = soil consolidation coefficient

U_c = degree of dissipation of pore pressure at the centre of the specimen.

Equation 4 is based on the assumption that the upper and lower surfaces of the direct shear specimen are allowed to drain freely, and that the rotation of principle stresses during the test can be neglected. Gibson & Henkel (1954) also reported direct shear tests on London clay where the shear rate computed by equation 4 with 95% pore pressure dissipation gave results virtually identical to the fully drained strength of the clay.

Based on equation 4, Head (1982) recommended for drained direct shear tests:

$$t_f = 12.7\ t_{100} \qquad (5)$$

where t_{100} is the time to 100% primary consolidation obtained by extrapolation of the linear portion of the compression - square root time plot obtained in the consolidation phase of the test. Equation 5 is also based on 95% pore pressure dissipation, but it has an advantage over equation 4 in that it utilises more readily available input data. Catering for 98% pore pressure dissipation, this expression would transform to:

$$t_f = 30\ t_{100} \qquad (6)$$

ASTM (1985) recommends:

$$t_f = 50\ t_{50} \qquad (7)$$

where t_{50} is the time to 50% primary consolidation. This equation gives essentially the same time to failure as equation 5, and would therefore represent 95% pore pressure dissipation.

After an appropriate time to failure has been computed using one of the equations above, the limiting (maximum permissible) rate of shear for a drained direct shear test can be estimated as follows:

$$\text{rate of shear} \leqslant \delta_f / t_f \qquad (8)$$

where δ_f is the horizontal displacement of the shear box at peak strength (failure). However, δ_f can only be estimated prior to the test itself.

The foregoing discussions all relate to pore pressure effects, which have a dominant influence on the appropriate rate of shear in drained tests on saturated specimens. Minor influences may also arise from the viscous component of shear strength of the specimen, or from inertial effects in the test apparatus. The 1938 data of Taylor & Leps (as reviewed by Lambe, 1951) showed that varying the shear rate from 0.15 mm/min to 2.5 mm/min in direct shear tests on dry Ottawa sand produced a variation in the measured friction angle of less than 2%. This would indicate that the viscous component of shear strength in a granular material is generally small. Similarly, inertial effects are considered negligible for modern direct shear equipment operating within the normal range of shear rates employed in drained testing of soils.

2.4 Strength Anisotropy

Sowers (1963) found that residual soils in America often had anisotropic strength properties because of the parallel orientation of their flaky grains, particularly micas and kaolinite clays. He anticipated that they were weaker in shear parallel to the grain faces than perpendicular to them, but concluded that more investigation of this should be required.

Onitsuka et al (1985) found that strength anisotropy existed in both undisturbed and compacted granitic soils in Japan. When tested in a shear box, the shear strength of vertical specimens was 1.1 to 1.5 times greater than that of horizontal specimens of undisturbed soils. For static and impact compacted (remoulded) soils, the corresponding strength anisotropy ratio was found to be 1.1 to 1.2.

3 SAMPLING, INDEX TESTS AND OEDOMETER TESTS

3.1 Sampling site

The sampling site for this testing programme was situated on a relatively flat hilltop above a cutting, in an area of Hong Kong known as Shouson Hill. The cutslope was formed during the early construction of an adjacent road, and varies in height from 10 to 15 m. Figure 1 shows the general location of the sampling site.

The rock type at the sampling site is classified as fine- to medium-grained granite (GCO, 1986). In the fresh state, this granite is light pinkish grey in colour, with grey quartz, light pink potash feldspar, milky white plagioclase feldspar and black biotite. Quartz comprises approximately 35 to 40% of the rock and generally occurs as subrounded grains 1 to 6 mm in diameter, although grains as large as 8 mm or more are not uncommon. The feldspars are typically 1 to 5 mm in diameter, but may be as large as 10 mm. Biotite comprises less than 5% of the rock.

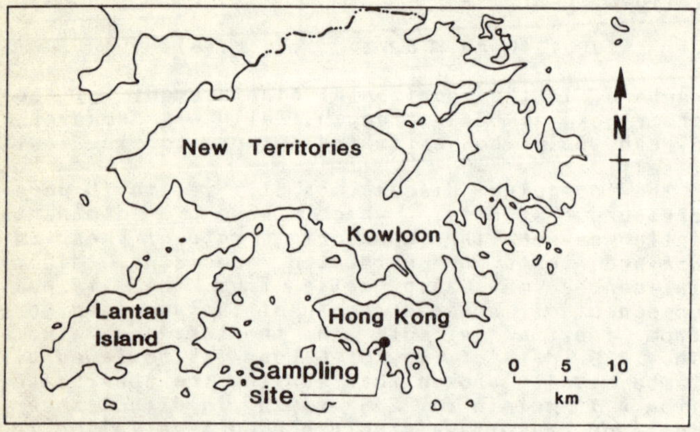

Figure 1. Location Plan of Sampling Site.

The material exposed in the adjacent road cutslope is predominantly completely decomposed (Grade V) granite (GCO, 1984), defined as "rock wholly decomposed but rock texture preserved", with a gradual transition towards residual soil in which all original rock texture has been destroyed. However, some hydrothermally altered granitic soils are found, particularly near quartz veins in the exposure. The broad descriptions of the two soil types sampled for testing are as follows:

(i) Decomposed Granitic Soil - Grade V granite with light yellowish brown colour that may be leached in some spots. Feldspars are decomposed to gritty-to-soft grains, the quartz grains are slightly fractured and the biotite is decomposed to yellowish grey flakes. However, variations occur within this broad class in terms of degree of decomposition of mineral constituents and disintegration, and two sub-types, termed "transition" and "disintegrated" granite soil respectively, were identified. This soil is a silty gravelly sand with a small clay fraction.

(ii) Altered Granitic Soil - Grade V granite with a mottled light pink and creamy white colour, hydrothermally altered. Feldspars are all decomposed to soft clayey kaolinite pseudomorphs, quartz grains are greatly reduced in size and biotite is altered to colourless muscovite. This soil is a clayey sandy silt, and is found mainly along the quartz veins. It is considered that hydrothermal alteration occurs as a result of rising hot aqueous silica-rich solutions from within the cooling magma during the late stages of intrusive activity.

Detailed descriptions of the site and the test specimens are given by Irfan (1988).

3.2 Sampling and Field Index Tests

Hand-cut block samples were obtained for this investigation in accordance with the recommendations of USBR (1974) and GCO (1987).

Insitu density tests were performed adjacent to each block. The mean dry unit weight from these field tests was 13.6 kN/m³, and the natural moisture content was about 18%.

During block sampling, a series of index tests including hand penetrometer, Schmidt hammer, slake, feldspar strength, quartz angularity and microfracture tests were carried out on each block, together with simple insitu constant head infiltration tests, following the procedures recommended by Hencher & Martin (1982). In general, hand penetrometer strength ranged from 110 to 220 kPa, Schmidt hammer values were zero in all tests, small lumps of soil slaked completely in water, feldspar grains could be easily grooved with a pin, fragments could be crumbled by hand, quartz grains were mostly sub-angular and minor microfractures could be identified. The infiltration rates from the constant head tests ranged from 1 to 8 x 10^{-4} m/s, indicating that the material is very permeable.

The field index tests proved useful for broadly classifying the soil, but not in detecting the significant variations in material apparent from visual examination.

3.3 Laboratory Index Tests

A range of laboratory index tests were performed on both block samples and individual shear specimens. Each specimen prepared for testing in direct shear was also tested for particle size distribution, specific gravity, unit weight and degree of saturation; void ratio could thereby be derived.

The overall mean dry unit weight of the specimens before shear was 12.4 kN/m³, which is considerably less than the field value of 13.6 kN/m³. This discrepancy is thought to arise from the relatively small specimen size and trimming procedure in the laboratory, as large quartz grains encountered by the cutter normally had to be totally removed (plucked away). This would lead to an underestimate of dry density, and to an overestimate of void ratio.

The initial moisture contents agree well with the natural moisture contents, indicating that there had been little drying out of the specimens. The specific gravity of solids was found to be 2.63 almost uniformly throughout.

The overall grading envelope for the granitic soil is shown in Figure 2. On average, the sand and gravel fractions in the soil were essentially equal, with the combined silt and clay fraction (i.e. fine fraction) generally comprising from 15 to 30%. The soil falls within the bounds of "average fine" and "extreme coarse" given by Lumb (1962).

3.4 Oedometer Tests

Vargas (1953), Vaughan (1985) and others have reported the deformation behaviour of saprolites and residual soils. They have noted that a 'virtual preconsolidation pressure' may be detected from the results of oedometer and isotropic stress tests, and that this is a useful parameter for interpreting shear behaviour.

Several oedometer tests were carried out on specimens of the granitic soil, and some typical results are plotted in e log p space in Figure 3. No pronounced change in stiffness is

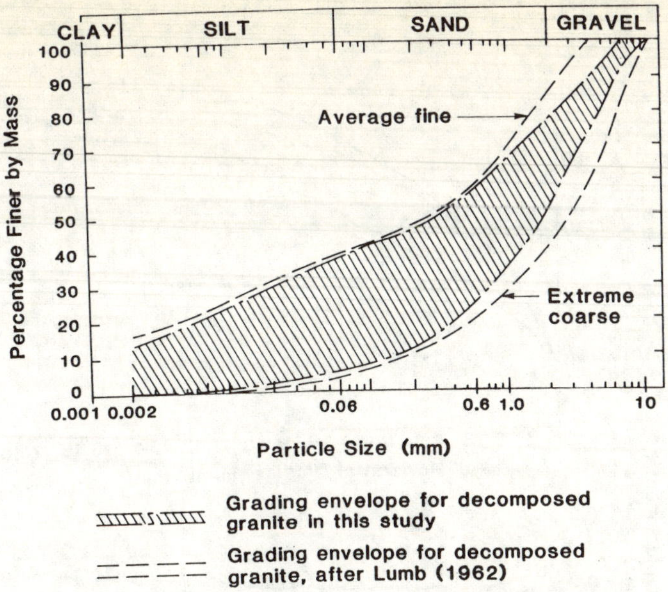

Figure 2. Grading Envelope of Granitic Soil.

apparent from these curves, and linear plots of these data (not shown here) as suggested by Vaughan (1985) do not indicate a clearly defined yield stress, despite the fact that observation of microstructure indicated the presence of weak bonding in varying amounts (Irfan, 1988). This may be due to disturbance during trimming of such small specimens, to the maximum grain size in relation to the specimen thickness, to bedding error (Vaughan, 1985), to yield occurring at such low stress levels that it is not detected, or a combination of some of these factors. Isotropic stress tests on large specimens would be a more suitable means of detecting yield in this material.

4 DIRECT SHEAR TESTS

4.1 Specimen Preparation

Direct shear specimens 100 mm or less in size were trimmed from the block samples using a knife and cutter. Large quartz grains at or near the boundary of the specimen were the main challenge during trimming.

Once the specimen had been fully assembled in the shear machine, a normal stress of 5 kPa was applied and then the specimen was flooded with water. The specimen was allowed to soak for 16 hours or longer before being consolidated under the chosen normal stress. A 16 hour soaking period could be readily accommodated in the test schedule by overnight soaking, and it was considered that this procedure would produce nearly complete saturation.

After soaking, consolidation to the required normal stress was carried out. When at least 98% consolidation had been achieved, the specimen was then sheared to failure.

4.2 Testing Programme

The testing programme was separated into several phases, designed to investigate the effects of varying specimen size and rate of shear, and to attempt to identify any soil anisotropy.

The 60 mm square by 20 mm and 100 mm square by 20 mm boxes were supplied as standard equipment from Wykeham Farrance Engineering Ltd., and were used in conjunction with their WF 25300 direct shear machine. A 60 mm square by 36 mm modified box was obtained by removing both the upper and lower porous stones and replacing them with filter papers. A 100 mm square by 44 mm modified box was obtained by removing the two porous stones and the lower platen. Filter papers were again used as backing for the perforated grid plates, and the lower filter paper was extended beneath the lower shear box as well.

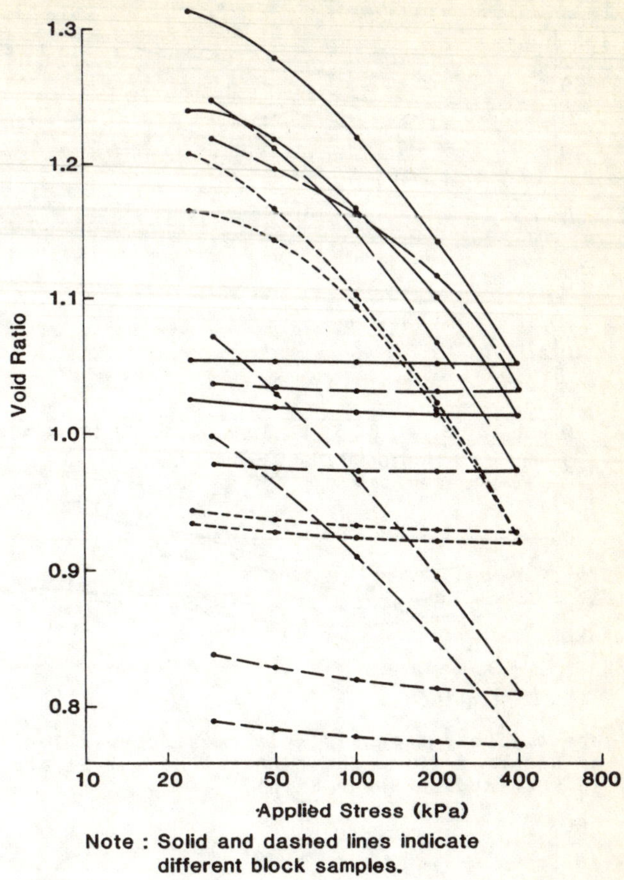

Figure 3. One Dimensional Consolidation Curves from 75 mm Diameter x 19 mm Oedometer.

4.3 Effect of Shear Box Size

Typical shear stress-displacement curves from the standard and modified 60 mm boxes are given in Figures 4 and 5, respectively, and from the standard and modified 100 mm boxes in Figures 6 and 7. The 60 mm square by 20 mm box produced reasonably smooth curves, although the curves did not always peak, even at high displacements. The 100 mm square by 44 mm box gave smooth curves with clearly defined peak shear stresses.

The rather large displacements at failure for

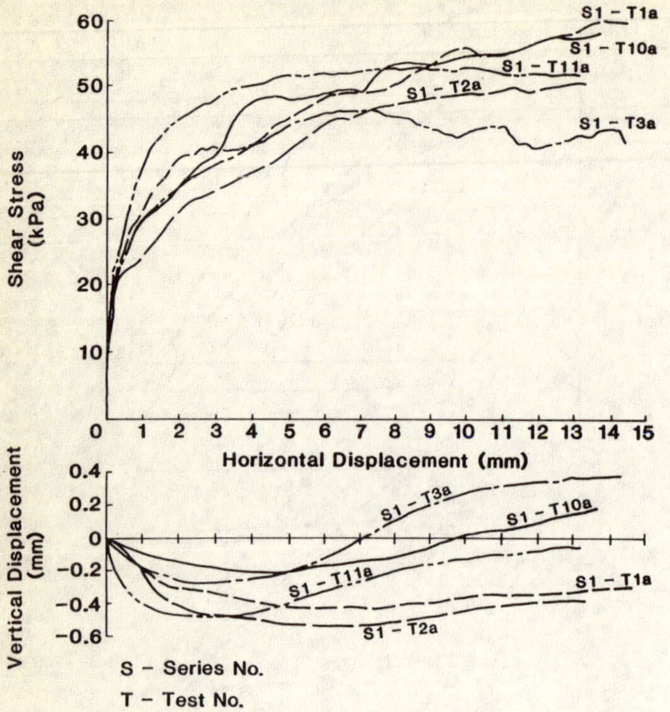

Figure 4. Shear Stress-Displacement Curves for 60 mm Square x 20 mm Shear Box at Normal Stress of 40 kPa (Medium Shear Rate).

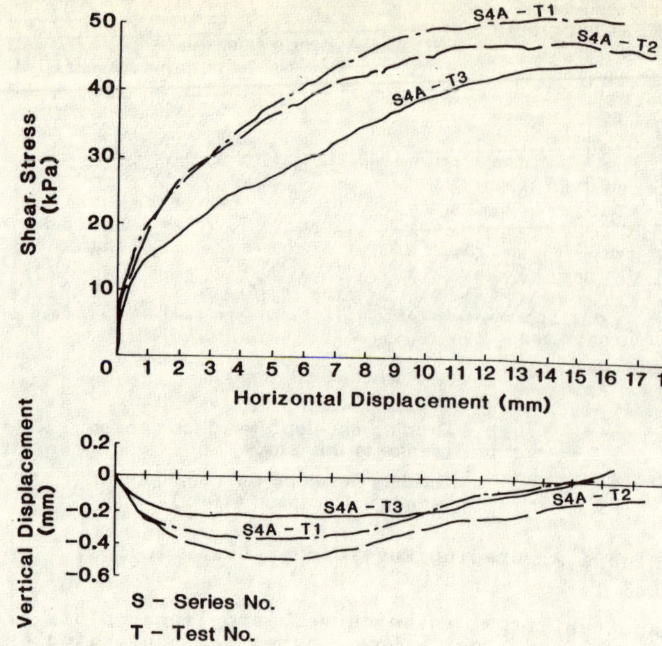

Figure 6. Shear Stress-Displacement Curves for 100 mm Square x 20 mm Shear Box at Normal Stress of 40 kPa (Medium Shear Rate).

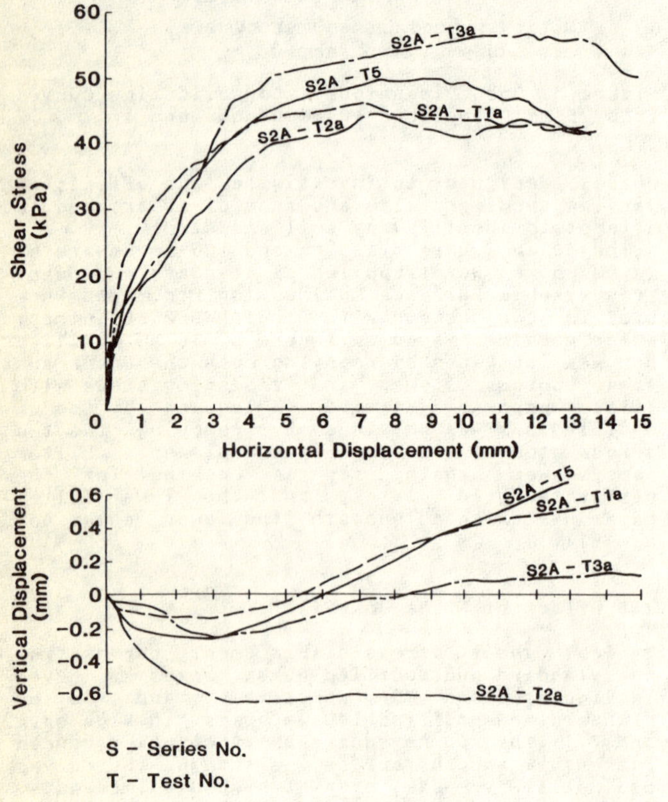

Figure 5. Shear Stress-Displacement Curves for 60 mm Square x 36 mm Shear Box at Normal Stress of 40 kPa (Medium Shear Rate).

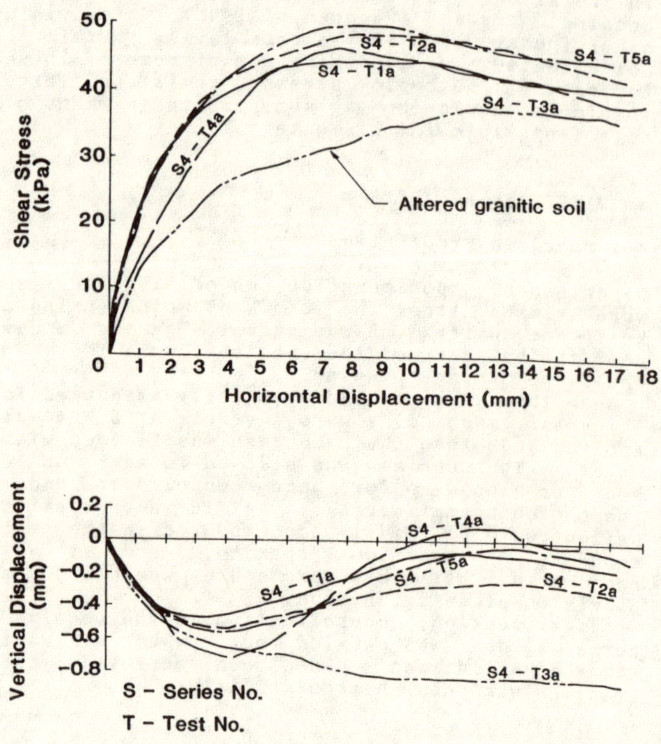

Figure 7. Shear Stress-Displacement Curves for 100 mm Square x 44 mm Shear Box at Normal Stress of 40 kPa (Medium Shear Rate).

this material might be explained by the generally high degree of decomposition and low dry density of the material. The volume change behaviour was always compressive for the first few mm of displacement, but thereafter showed dilatant tendencies in many instances under low normal stresses. At high normal stresses, volume change continued to be compressive to failure.

The measured shear strengths at 40 kPa normal stress in the various shear boxes are summarised in Table I. The 100 mm square by 44 mm box gave the lowest mean shear strength, with the least scatter in the data. The smallest shear box gave the highest shear strength and the most scattered results, indicating that the jerky shear stress-displacement curves were produced by particle crushing and rearrangement within the confines of the small box, and that this led to an overestimate of the shear strength. A similar result was found at 150 kPa normal stress (Table II).

It should be noted that even the 100 mm square by 44 mm box did not meet the ASTM guideline for thickness to maximum grain diameter ratio. This guideline would undoubtedly cater for poorly graded or uniform materials with a large proportion of particles near the maximum size, and it can perhaps be relaxed somewhat when only occasional particles are of the maximum size (as in granitic soil). This idea was also put forward by Skempton & Bishop (1950), as explained previously.

Shear Box Size	Dry Unit Weight (kN/m³)	Shear Strength, τ (kPa)	Mean Shear Strength (kPa)	Standard Deviation (kPa)
100□x44 mm	12.7 12.8 12.1 11.7 12.3 12.1	44 48 46 49 45 50	47.0	2.4
100□x20 mm	12.5 12.5 12.3	51 47 44	47.3	3.5
60□x36 mm	12.4 12.5 12.3 13.2	46 44 57 50	49.3	5.7
60□x20 mm	12.4 12.6 12.6 12.5 12.4	60 50 46 57 52	53.0	5.6

Table I. Effect of Shear Box Size on Shear Strength at 40 kPa Normal Stress (Medium Shear Rate).

Shear Box Size	Dry Unit Weight (kN/m³)	Shear Strength, τ (kPa)	Mean Shear Strength (kPa)	Standard Deviation (kPa)
100□x44mm	12.1 12.6 12.3 12.8	140 144 137 152	143.3	6.5
100□x20mm	12.0 11.8 11.8	136 145 152	144.3	8.0
60□x36mm	11.7 12.0 11.9 12.1 12.3	162 160 131 160 154	153.4	12.9
60□x20mm	13.5 13.7	150 159	154.5	–

Table II. Effect of Shear Box Size on Shear Strength at 150 kPa Normal Stress (Medium Shear Rate).

4.4 Effect of Shear Rate

The nomenclature adopted for shear rate is:

(a) slow - 0.0072 mm/min,
(b) medium - 0.06 mm/min,
(c) fast - 0.6 mm/min.

The time to failure in direct shear tests at the fast shear rate was less than half an hour, while tests at the medium shear rate took about four hours and slow tests took 30 to 40 hours. As shown by the tabulated values of 30 t_{100} given in Table III, even the fast tests easily satisfied equation 6. Furthermore, equation 6 is based on the rather stringent requirement of 98% pore pressure dissipation at the centre of the specimen.

Shear Rate (mm/min)	Consoli- dation $30t_{100}$ (min)	Dry Unit Weight (kN/m³)	Shear Strength, τ (kPa)	Mean Shear Strength (kPa)	Standard Deviation (kPa)
Fast (0.6)	2.8 4.8 2.8	12.1 12.4 11.4	48 43 46	45.7	2.5
Medium (0.06)	4.5 4.5 4.5 4.5 3.8 2.8	12.7 12.8 12.1 11.7 12.3 12.1	44 48 46 49 45 50	47.0	2.4
Slow (0.0072)	2.8 4.8 2.8	12.2 12.0 12.0	50 46 46	47.3	2.3

Table III. Effect of Shear Rate on Shear Strength in the 100 mm Square x 44 mm Shear Box at 40 kPa Normal Stress.

Shear test results at the three different shear rates using the 100 mm square by 44 mm box are summarised in Table III. No significant differences in measured shear strengths were noted. Clearly, tests lasting more than a few hours are not required to obtain reliable 'drained' results for this highly permeable soil.

4.5 Strength Envelope

Figure 8 shows the shear strength data derived from the 100 mm square by 44 mm shear box using the medium shear rate.

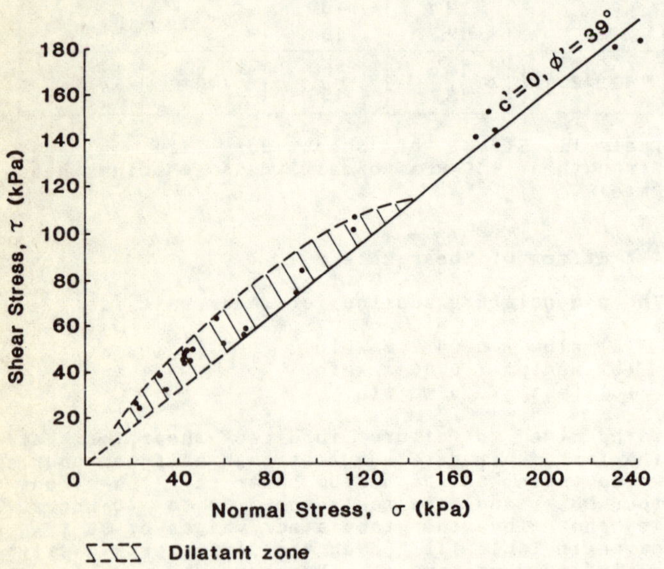

Figure 8. Shear Strength Results from 100 mm Square x 44 mm Shear Box.

The average shear strength envelope can be described by a straight line through the origin ($c' = 0$, $\phi' = 39°$), with dilatant behaviour, and possibly also the shearing of weak bonds at small horizontal displacements, increasing the shear strength within the low stress range by up to 20 kPa. The 'dilatant' zone intersects the straight line describing non-dilatant or compressive behaviour over a broad range of stresses from about 60 to about 160 kPa. This interpretation is generally supported by the vertical displacement measurements obtained during shearing. Specimens tested in the lower stress range generally dilated due to intergrain contact between quartz particles and unaltered feldspars, after an initial volume reduction at low strain due to progressive disturbance to the soil structure and collapse of individual porous feldspars (Irfan, 1988).

From Figure 7, markedly lower shear strength can be observed for the altered granitic soil than for the decomposed granitic soil. Specimens of the former were generally denser and contained more silt and clay than the latter. In the low stress range, dilation did not occur after the initial volume reduction.

4.6 Strength Anisotropy

Two specimens were trimmed so that the shear strength could be measured on a plane that had been oriented vertically insitu. The results of these two tests are shown in Table IV. As the results are very similar to those from specimens oriented horizontally, no pronounced strength anisotropy was found in this instance.

Orientation	Dry Unit Weight (kN/m^3)	Shear Strength, τ (kPa)	Mean Shear Strength (kPa)
Horizontal	12.7	44	47.0
	12.8	48	
	12.1	46	
	11.7	49	
	12.3	45	
	12.1	50	
Vertical	12.0	45	47.5
	11.8	50	

Table IV. Effect of Specimen Orientation on Shear Strength in the 100 mm Square x 44 mm Shear Box at 40 kPa Normal Stress (Medium Shear Rate).

5 CONCLUSIONS

A 100 mm square by 44 mm shear box was found to be adequate for testing granitic soils with a maximum grain diameter up to approximately 8 mm. Testing the granitic soil in smaller shear boxes generally produced irregular results and higher shear strengths, probably due to excessive particle crushing and rearrangement within the confines of the box.

Laboratory determined dry densities were found to be lower than the field determined values, and they probably underestimated the true values because of large quartz particles encountered and removed during trimming of the relatively small specimens.

No significant differences in shear strength were detected when the shear rate was varied from 0.007 to 0.6 mm/min. The limit of existing theory, as represented by equation 6, has not been tested due to the high permeability of the granitic soil.

The average shear strength of the completely decomposed granite tested in direct shear has been defined within the normal stress range of 20 to 200 kPa, and can be represented as shown in Figure 8. The strength in the upper stress range may be described by a linear strength envelope with $c' = 0$, while in the lower stress range the shear strength exceeds this envelope due to dilation, and possibly also weak bonding. It is of interest to note that even though the dry densities of the samples tested are within the range considered to be potentially metastable (Irfan, 1988), shearing the test specimens at low stress levels caused dilation, after the specimens experienced an initial gradual volume reduction at low strain (Figure 7) due to progressive disturbance to the soil structure and collapse of individual porous feldspars.

The strength of the hydrothermally altered granite was much lower than that of the decomposed granite. Care should therefore be taken to identify the effects of alteration on saprolitic soils during ground investigation.

In a very preliminary investigation, no pronounced strength anisotropy was apparent in the granite soil tested.

ACKNOWLEDGEMENTS. This paper is published with the permission of the Director of Civil Engineering Services, Hong Kong Government. The authors acknowledge the contribution of T.Y. Irfan of the Geotechnical Control Office to the testing programme, and thank him for his useful comments in the preparation of this paper.

REFERENCES

Akroyd, T.N.W. (1957). Laboratory Testing in Soil Engineering. Soil Mechanics Ltd., London, 249 p.

ASTM (1952). Symposium on Direct Shear Testing of Soils. STP 131, American Society for Testing and Materials, Philadelphia, 88 p.

ASTM (1985). Standard method for direct shear test of soils under consolidated drained conditions. 1985 Annual Book of ASTM Standards, Test D3080-72, American Society for Testing and Materials, Philadelphia, pp 514-518.

Bishop, A.W. (1948). A large shear box for testing sands and gravels. Proceedings of the Second International Conference on Soil Mechanics and Foundation Engineering, Rotterdam, vol. 1, pp 207-211.

Bishop, A.W. (1950). Discussion on "The measurement of the shear strength of soils" by A.W. Skempton & A.W. Bishop. Géotechnique, vol. 2, pp 113-116.

Bishop, A.W. (1954). Discussion on "Shear characteristics of a saturated silt, measured in triaxial compression" by A.D.M. Penman. Géotechnique, vol. 4, pp 43-45.

Bishop, A.W. (1966). The strength of soils as engineering materials. Sixth Rankine Lecture, Géotechnique, vol. 16, pp 91-130.

Bowles, J.E. (1978). Engineering Properties of Soil and Their Measurement. (Second Edition). McGraw-Hill Book Co., New York, 212 p.

Brady, K.C., Awcock, I. & Wightman, N.R. (1983). A Comparison of Shear Strength Measurements using Two Sizes of Shearbox. Laboratory Report 1105, Transport and Road Research Laboratory, U.K., 25 p.

BSI (1975). Methods of Test for Soils for Civil Engineering Purposes. BS 1377:1975, British Standards Institution, London, 143 p.

GCO (1987). Guide to Site Investigation. (Geoguide 2). Geotechnical Control Office, 362 p.

GCO (1986). Hong Kong and Kowloon: Solid and Superficial Geology. Sheet 11, 1:20 000 Geological Map HGM 20, Geotechnical Control Office, 1 sheet.

GCO (1984). Geotechnical Manual for Slopes. (Second edition). Geotechnical Control Office, Hong Kong, 295 p.

Gibson, R.E. & Henkel, D.J. (1954). Influence of duration of tests at constant rate of strain on measured 'drained' strength. Géotechnique, vol. 4, pp 6-15.

Gray, D.H. & Ohashi, H. (1983). Mechanics of fiber reinforcement in sand. ASCE Journal of Geotechnical Engineering, vol. 100, no. 3, pp 335-353.

HBPRC (1980). Specification for Geotechnical Testing. SDS01-79, Hydro-electricity Bureau of the People's Republic of China, vol. 2 (in Chinese), 486 p.

Head,K.H. (1982) Manual of Soil Laboratory Testing. Volume 2 : Permeability, Shear Strength and Compressibility Tests, Pentech Press, London, 747 p.

Hencher, S.R. & Martin, R.P. (1982). The description and classification of weathered rocks in Hong Kong for engineering purposes. Proceedings of the Seventh Southeast Asian Geotechnical Conference, Hong Kong, vol. 1, pp 125-142.

Hennes, R.G. (1952). The strength of gravel in direct shear. American Society for Testing and Materials, STP 131, pp 51-62.

ICE (1951). Proceedings of the Conference on the Measurement of Shear Strength of Soils in Relation to Practice. Géotechnique, vol. 2, pp 89-250.

Irfan, T.Y. (1988). Fabric variability and index testing of a granitic saprolite. Proceedings of the Second International Conference on Geomechanics in Tropical Soils, Singapore (in press).

ISRM (1981). Suggested methods for determining shear strength. In Rock Characterization Testing & Monitoring, E.T. Brown (Editor), Pergamon Press, Oxford, pp 129-140.

Jewell, R.A. & Wroth, C.P. (1987). Direct shear tests on reinforced sand. Géotechnique, vol. 37, pp 53-68.

Kjellman, W. (1951). Testing the shear strength of clay in Sweden. Géotechnique, vol. 2, pp 225-235.

Lambe, T.W. (1951). Soil Testing for Engineers. John Wiley & Sons, New York, 165 p.

Lambe, T.W. & Whitman, R.V. (1969). Soil Mechanics. John Wiley & Sons, New York, 553 p.

Lumb,P. (1962). The properties of decomposed granite. Géotechnique, vol. 12, pp 226-243.

Morgenstern, N.R. & Tchalenko, J.S. (1967). Microscopic structures in Kaolin subjected to direct shear. Géotechnique, vol. 17, pp 309-328.

Onitsuka, K., Yoshitake, S. & Nanri, M. (1985). Mechanical properties and strength anisotropy of decomposed granite soil. Soils and Foundations, vol. 16, no. 1, pp 17-26.

Potts, D.M., Dounias, G.T. & Vaughan, P.R. (1987). Finite element analysis of the direct shear box test. Géotechnique, vol. 37, pp 11-24.

Skempton, A.W. (1984). Selected Papers on Soil Mechanics. Thomas Telford Ltd., London, 281 p.

Skempton, A.W. & Bishop , A.W. (1950). The measurement of the shear strength of soils. Géotechnique, vol. 2, pp 90-116.

Sowers, G.F. (1963). Engineering properties of residual soils derived from igneous and metamorphic rocks. Proceedings of the Second Pan American Conference on Soil Mechanics and Foundation Engineering, Brazil, vol. 1, pp 39-62.

Taylor, D.W. & Leps, T.M. (1938). Shearing properties of Ottawa standard sand as determined by the M.I.T. strain control direct shearing machine. Record of Proceedings of Conference on Soils and Foundations, Corps of Engineers, Boston. Reviewed in Lambe (1951),

p. 93.
USBR (1974). Earth Manual. (Second Edition). United States Bureau of Reclamation, US Government Printer, Washington, 810 p.
Vargas, M. (1953). Some engineering properties of residual clays occurring in Southern Brazil. Proceedings of the Third International Conference on Soil Mechanics and Foundation Engineering, Zurich, vol. 1, pp 67-71.
Vaughan, P.R. (1985). Mechanical and Hydraulic Properties of In-situ Residual Soils. (General Report). Proceedings of the First International Conference on Geomechanics in Tropical Lateritic and Saprolitic Soil, Brasilia, vol. 3, pp 231 - 263.
Vickers, B. (1978). Laboratory Work in Civil Engineering - Soil Mechanics. Granada Publishing, London, 170 p.

In-situ direct shear tests for lateritic gravels in Taiwan

B.-L.CHU, Department of Civil Engineering, National Chung-Hsing University, Taichung, Taiwan
T.-W.HSU, Department of Civil Engineering, National Chung-Hsing University, Taichung, Taiwan
T.-C.LAI, Department of Civil Engineering, National Chung-Hsing University, Taichung, Taiwan

ABSTRACT: This paper first presented the geological surveyings on test site and the results of the preliminary laboratory tests on the gravelly deposits. And in essence, the primary objective of this project was to conduct the in-situ direct shear tests to determine the mechanical properties of lateritic gravels. Shear strength of this natural deposits were determined by performing direct shear tests on specimens of 1.5m x 1.5m sample size in the field. Geological observations show that gravels in Sanyi area have been strained into fragments due to geological plate tectonic stress. Grain size analysis indicates that the soil particles with sizes greater than 1.5 in. predominate the main components of this skip-graded stratum. The test results show that in Sanyi area, the angle of internal friction was ϕ_p=54.3° with cohesion intercept C_p=0.15kg/cm^2 at peak points, while at residual points, its friction angle was ϕ_r=44.9° with cohesion intercept C_r=0. Results also indicate that ϕ_p=46° with C_p=0.15kg/cm^2, and ϕ_r=34° with C_r=0 in Linkou area.

1 INTRODUCTION

Cobbles, gravels and sandy materials are extensively used in the highway pavement, railroad ballast, embankment, retaining wall and filling constituents of the dam. Also, these aggregates are the most essential in the various construction activity of the civil, hydraulic and architecture engineering. According to investigation, annual requirement for the aggregates in Taiwan has reached 100 million tons(2)(3), and its demand still increasing in the recent years for highly economic development.

Gravelly deposits are usually distributed in Taiwan in the following areas such as river bed, seashore and mountain ranges. Due to the traffic and economical limitation as well as the defense consideration, most exploitation for the gravelly deposits were constrained in the river area. Investigation (3) indicates that resourses of the gravelly deposits in the river area were estimated approximately 30 billion tons. However, this exploitative area was reduced because of the construction of the reservoir and check dam, as well as the consequence of overuse in recent years. According to the study of Economic Ministry, the available gravelly aggregates for exploitation in the river bed are only few billion tons left.

Therefore, it will become insufficient for the need of the engineering construction in the coming years. This situation are more serious in the northern part of Taiwan. In order to provide the necessary amount of gravelly aggregates for the national construction, it is getting vital for searching the alternative resource area besides the river bed. In addition, more investigation and research for seeking the high quality and great quantity of gravelly aggregates in the new resources come urgent.

Taiwan Bureau of Mining Started studying the possible places for the gravelly aggregate resources. From the preliminary reconnaissance, it was found that in the areas between Linkou and Sanyi, of northern part of Taiwan exhibit the desirable amount of gravelly aggregates. Before complete developing for exploitation in these areas, it become necessary to investigate the distribution of the gravel aggregates, the cost of the exploitation, the transportation fee for delivering the materials and the effect of environmental innovation. Additionally, important consideration for the engineering application must be contemplated prior to its development. Hence, pertinent experimental study of this site materials on engineering property and mechanical characteristics become essential. However, the particle sizes in the deposit are large and the structural arrangement of these particles can not be reduplicated in the laboratory, hence, in-situ direct shear tests are suggested to perform, and the results can be explicitly applied in the engineering practice(5)(6).

This study first adopted the sampling work in the field, grain size analysis, index property as well as material engineering property tests in the laboratory. Geological investigation of the quality and quantity on the gravelly deposits in Linkou and Sanyi sites were also explored. Secondly, in-situ direct shear tests were performed in this two areas to determine the shear strength parameters of gravelly deposits. The results can be employed in the foundation design and slope stability analysis of these exploitation sites.

2 GEOLOGICAL SETTING

There is a footheel slopeland belt distributing in NNE-direction between the central range and the western coast plain in Taiwan. In western edge of this belt there is a series of tableland and slopeland. This study is focus at two of them. The first located at Linkou tableland and the second at Sanyi slopeland. The geological conditions of the two sites are described in following paragraphs.

(a) Test site in Linkou

Linkou tableland locates 20km southwest of Taipei with height about 230m above sea level. The surface of the tableland is relatively rough because of erosion. The test site locates on the slope with dipping angle of 10° - 50°, which is the south eastern edge of the tableland. The formation overlies on the tableland is Linkou formation deposited in Pleistocene. The formation is nearly horizontal with dipping angle less than 10°. It is composed of conglomerate and some lenticular sand and silt pockets which can be seen obviously in level 75m and 175m. The top part is covered by several meter thick lateritic soil. According to the lithological property, the Linkou formation can be divided into two sections, the dense conglomerate and the loose conglomerate. The dense conglomerate locates at levels between 35m and 95m. It is composed of graywacke(60%), quartzite (25%) gravels and matrix (15%). The graywacke gravels are moderately weathered and brown. The quartzite gravels are relatively fresh and whitish brown. The grain size is commonly less than 10cm. The shape is round to subround. The sorting of this section is fair. Some sand lenses are interbedded in conglomerate, usually with thickness less than 1m and short extension.

The loose conglomerate overlies on the dense conglomerate conformly. The section contains more matrix than the first section. The percentage of gravels with diameter larger than 30cm is also high. By the way sorting in this section is relative poor. Sand lenses can be seen obviously especially near the level of 175m. The test site located in dense conglomerate.

(b) Test site in Sanyi, Miaoli
The second site locates in Sanyi, Miaoli, about 150km south of Taipei, capital city which is in the northern Taiwan. It is a slopeland with a narrow ridge striking north. Both sides of the ridge are composed of a series of narrow ridges and deeply cutting valleys. The formation deposited in Plio-Pleistocene, namely Toukoshan formation distributes in this area.

This formation is composed of two phases, namely Fouyenshan and Shianshan phase. The first one is a conglomerate phase and the second is composed of mudstone. This formation was compressed by a strong tectonic movement in early Pleistocene, so the formation is strongly folded and faulted. The study area is located at the east flank of Sanyi syncline. The bedding in this area is dipping to west with a dipping angle of 65°. The Toukoshan phase contains gravles, sand and silt. The percentage of gravels including quartzite and sandstone is about 80% in volume. The quartzite is about 70% and sandstone is about 10% of the total gravels. Usually quartzite is coarse grain fresh to lightly weathered and white to pale. It is commonly 1 cm to 30 cm diameter, subrounded to rounded shape. The gravels have fractures because of high compression during tectonic movement period in Pleistocene. The sandstone is fine to medium grain, subrounded to rounded shape and moderately sorted. The matrix contains sand (10%) and silt (5%). The bedding is not perfectly developed in conglomerate but the contact boundary of conglomerate and mudstone shows the attitude of the formation in N20° E65° NW. The total thickness of conglomerate is over 600m. On the west part of study area there are secondary deposited gravels and alluvial cones over the conglomerate. They are composed of gravels eroded and transported from the upper part of the slopeland. The thickness of the secondary deposit is about 50m. The lithology of this deposit is similar as the conglomerate formation but it is usually loose and poor cemented. The test is performed in original Fouyenshan phase.

3 LABORATORY AND IN-SITU MATERIAL PROPERTY TESTS ON THE GRAVELLY DEPOSITS

Four types of test were performed to study the property of gravelly deposit, (1) Sampling of the opening pit in the field (2) Grain size analysis (3) Soil index property test (4) Material engineering property test on the gravel particles.

3.1. Sampling of the opening pit in the fiels.

Select the typical spot of this two sites, surfacing the ground, dig an opening with dimensions 1.5m x 1.5m x 1.5m. Measure the dimensions of the opening to calculate the volume, weigh the gravelly deposit sample excavated from the opening to compute the weight. From the volume and weight, the bulk unit weight of the gravelly deposit in this opening could be estimated. For this investigation, thirty openings around each site were selected for study.

3.2. Grain size analysis

This test was performed by using 60cm x 40cm sieves to determine the grain size distribution. The samples excavated from the opening are grouped into three parts (a) Coarse aggregates coarser than 1" (b) Fine aggregates finer than 1" but coarser than 200# (c) Soil particles finer than 200#. The analyses are listed as follows: In the field, all the sample materials were shaken through a set of sieves, such as 12", 12" - 9", 9" - 6", 6" - 3", 3" - 3/2", 3/2" - 1", 1" - 3/4" and below 3/4". Weigh the sample retaining on each sieve separately and choose typicle sample 10 to 15 kg from each sieve for water content measurement.

(a) Coarse aggregates coarser than 1"; Dry sieve analysis was used. First, clean all the soil materials attached on the gravel particle, and measure the weight W_{w1} of the sample retaing on each seive of 1", 1½", 2", 3", 6", 9" and 12" separately. Divide each weight by the water content m_1 determined in the field to get the dry weight $W_{d1} = \dfrac{W_{w1}}{1 + m_1}$.

(b) Fine aggregates finer than 1" but coarser than 200#; wet sieve analysis was used. This is due to the large amount of adhesive clay particle fastened on the coarse particle, and it becomes impractical if dry sieve analysis is applied. This method was to measure the wet weight of the sample W_{w2} first, and get the corresponding dry weight W_{d2} by using the equation $W_{d2} = \dfrac{W_{w2}}{1 + m_2}$, in which m_2 is the water content of the sample determined in the field. Quatering method was applied to get the typical sample from the specimen of dry weight 25kg. Put these sample in the sieves and wash them by water to pass through 1", 3/4", 1/2", 3/8", 4#, 8#, 16#, 30#, 50#, 100# and 200# sieves, following the procedures of ASTM D2217. After having the dry retaining weight on each sieve, time the factor $\dfrac{W_{d2}(kg)}{25(kg)}$ for each dry weight to get the true dry weight of each sieve in the opening pit.

(c) Soil particles finer than 200#; By using the hydrometer analysis to determine the soil distribution finer than 200#. The deflocculating agent which was mixed into the soil specimen for dispersing the sample was sodium hexametaphosphate.

Combined the results obtained from tests (a), (b) and (c), the grain size distribution for this opening pit can be plotted.

3.3. Soil index property tests.

In order to understand the soil property of fine particle, and for the soil classification in engineering application, the following tests such as bulk unit weight, specific gravity and Atterberg limits determination were performed in the laboratory. The procedures follow the American Society of Testing Material and the book entitled "Soil Testing for Engineers" written by T.W. Lambe.

(a) Bulk unit weight; It indicates that the unit weight was measused from all the sample weight excavated from the opening divided by the opening volume determined in the field.
(b) Specific gravity: This test was aimed to determine the specific gravity of particles finer than 4#. The procedures follow ASTM D854-58.
(c) Atterberg limits; These tests were performed for the soil particles passing 40#. The procedures follow ASTM D423-66 to get the plastic limit and liquid limit, from these two limits, plastic index could be obtained.

3.4. Material engineering property test on the gravel particles:

In order to estimate the material property of the cobble and gravel in the gravelly deposits, experiments were performed to study the aggregates coarser than 4#. The tests are listed as follow:

(a) Absorption test; This test was to determine the absorption of coarse aggregates for two groups of particle sizes 4# - 1½" and 3" - 6" respectively. The procedures follow ASTM C127-80.

(b) Los Angeles abrasion test; The test was performed on two groups of particle sizes 1½" - 3" and 3/8" - 1½" separately.
For particle sizes 1½" - 3"; follow ASTM C535-69 to test the sample of E gradation of aggregates. And for particle sizes 3/8" - 1½"; follow ASTM C131-71 to test the sample of A-gradation of aggregates.

(c) Soundness test; Select the aggregates with the particle sizes of 4# - 1½" and 1½" - 3" from the opening pit, test procedures follow ASTM C88-76.

(d) Compressive strength test; Core the sample from the cobble and smooth it with the dimensions 5cm x 5cm x 10cm. Follow ASTM C12-68 and C39-72 to test the uniaxial compressive strength of the cobble.

4 GENERAL ENGINEERING PROPERTY OF LATERITIC GRAVEL DEPOSITS IN LINKOU AND SANYI SITES

The results of engineering property in these two site, such as bulk unit weight, specific gravity and Atterberg limits are listed in Tables 1 and 2. Also the graphs of grain size distribution in Figs. 1 and 2 show that the gravel deposit gradation was gaped and classified as poorly graded aggregate materials. Coarse aggregates seem to exhibit greater amount than fine aggregate. The mean grain sizes of the aggregates are approximately equal to 2" in 70% of the openings, and 90% of them are over 1½". Follow the unified soil classification system, it was grouped as "GP-GM", "GP-GC" or GP. It was also classified as A-1-a, A-1-b, A-2-4 or A-2-6 according to AASHTO classification system. Tables 1 and 2 also show the results of bulk unit weight, specific gravity, absorption, soundness, and abrasion for coarse aggregates (3/2" - 3") and fine aggregates (4#-3/2") as well as the compressive strength of the cobbles in these two sites. Generally, the coarse aggregates have slightly higher quality, comparing with the fine aggregates. However, the aggregates in these two sites could be used as the constituent materials in the dam as well as the base and subbase filling materials in the highways and airport runways according to AASHTO M57-64 code. In comparison, the aggregate in Linkou site has low quality and contains certain amount of lateritic clay, for this reason, it can not be used as the mixing aggregates in the concrete manufacturing. However, in Sanyi site, the aggregate has better quality and less clay mineral contamination, hence, it is recommended as the mixing materials in the asphalt or cement concretes.

TABLE 1

Site	$D_{10,mm}$	$D_{30,mm}$	$D_{50,mm}$	$D_{60,mm}$	$C_c=\frac{D_{30}^2}{D_{10}D_{60}}$	$Cu=\frac{D_{60}}{D_{10}}$	Dry unit weight T/M^3	P.I.
Sanyi	0.247	21.88	58.21	84.86	31.35	520.05	2.07	75% : NP 25% : <6
Linkou	0.256	19.78	54.67	80.5	18.94	313.72	1.81	8 - 13

TABLE 2

Site	Specific Gravity	absorption (%)	abrasion (%)		Soundness (%)	Compressive Strength(kg/cm^2)
Sanyi	2.60(<4#) 2.55(>4#)	1.30(<4#) 1.10(>4)	12.49* (100 turns) 9.28** (200 turns)	42.5* (500 turns) 25.51** (1000 turns)	3.2	1156
Linkou	2.58(<4#) 2.53(>4#)	1.96(<4#) 2.66(>4#)	19.4* (100 turns) 417.6** (200 turns)	48.93* (500 turns) 38.28** (1000 turns)	3.1	570

*A gradation **E gradation

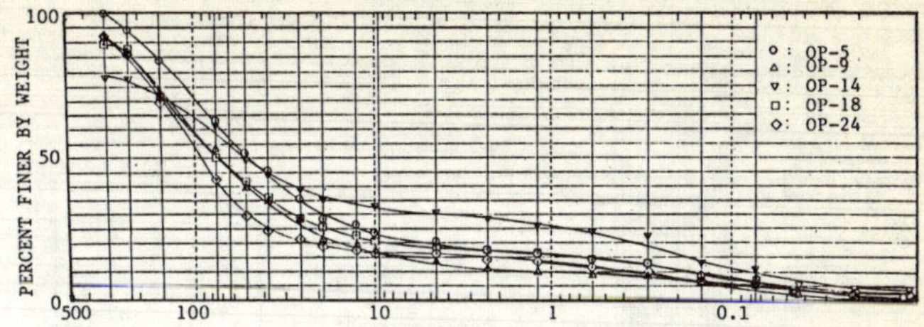

Fig. 1 Typical Grain Size Distribution of Opening Pits at Sanyi Site

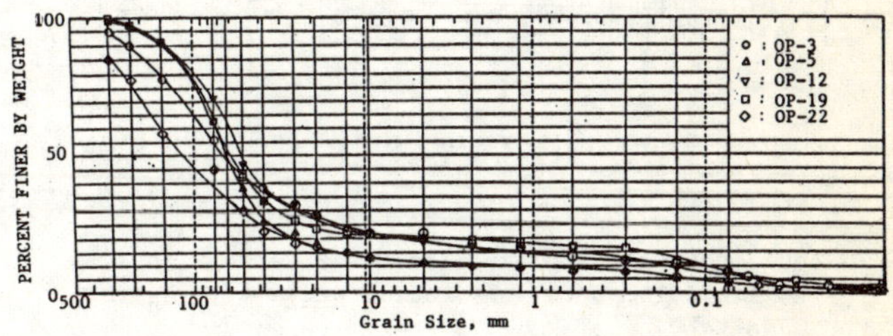

Fig. 2 Typical Grain Size Distribution of Opening Pits at Linkou Site

5 EQUIPMENT AND PROCEDURES OF IN-SITU DIRECT SHEAR TEST

The in-situ direct shear tests on undisturbed gravel deposits were performed in Linkou and Sanyi sites. In Sanyi, five specimens were applied with normal stresses; 0.41, 0.51, 0.76, 1.17 and 1.31 kg/cm^2 respectively. While in Linkou, five specimens were employed with stresses, 0.48, 1, 2, 4, and 6 kg/cm^2 separately. Except the different reacting system used in the hydraulic jack, the equipment and procedures of the in-situ direct shear tests were similar and described below:

5.1. The equipments for applying the normal load

(a) Two 100 T capacity hydraulic jacks with 20cm stroke.

(b) Reaction system to transfor normal load uniformly to the test blocks. In Sanyi, the reaction system with three combinations of dead weight are described below.
(1) Four concrete blocks with each weight 4.1 T were fabricated in the field, the concrete block dimensions are 1.8m x 1.2m x 0.8m with 210kg/cm^2 compressive strength.
(2) Not only for the dead weight, but for the excavation and the transportation purposes, Japan Mitsubishi MS 180 backhoe with weight of 18T was used.
(3) Grillage was formed with rails in the field. 19 rails with each weight 375kg and length 19m were combined in parallel, four other short rails with each 180kg weight and 5m length were 90° linked with the upper rails by bolts and nuts. This combination was approximately 7.5T in weight.

All the equipments and set-up were shown in Fig.3. In Linkou site grouted anchor tendons were used as the reaction system. The steel reaction beam with the dimensions 6.5m x 0.4m x 0.6m was fastened on the pair of the anchor tendons with each tendon prestressed 10T load. The set-up was depicted in Fig.4.

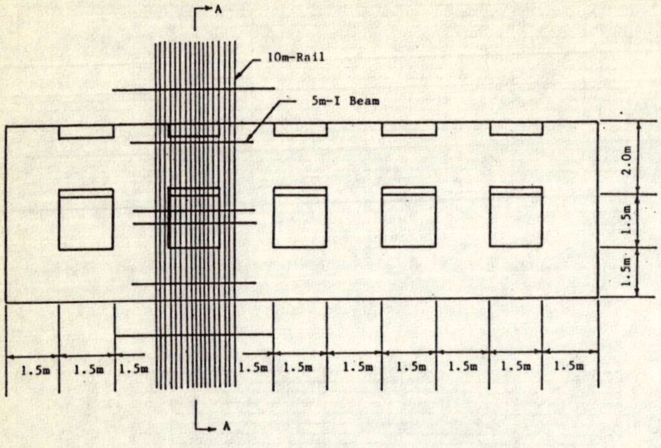

Fig. 3 Plane View of Setup for In-Situ Direct Shear Test at Sanyi Site

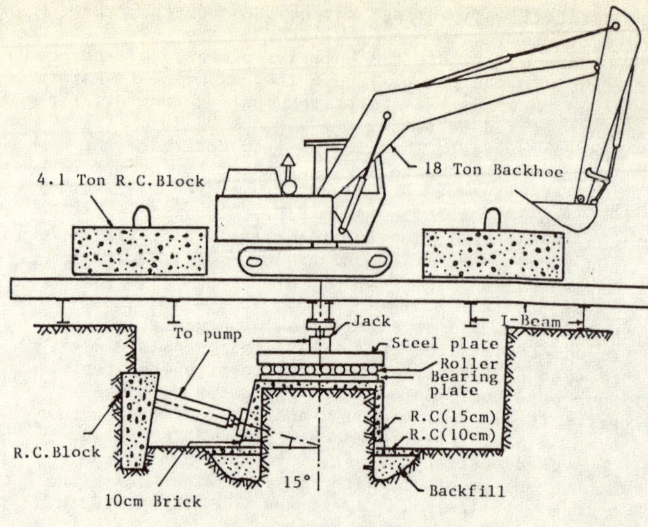

A-A Section

5.2. Equipments for applying the shear force.

(a) Two 50T capacity hydraulic jacks with 34cm stroke.

(b) Wedge shape concrete block in front of the wall which is excavated for the shear reaction system. The wedge concrete blocks with dimensions of 50cm x 20cm x 90cm and 50cm x 20cm x 700cm were poured in the place just adjacent to the wall. A 30cm x 30cm x 3.81cm steel plate was placed in the front face of the wedge concrete block. The ball bearing with 5cmϕ x 30cm was put in the between of the hydraulic jack and the steel plate.

(c) During the shear, the wall was used as the reaction system to transmit the shear force to the specimen. The shear force should be distributed uniformly on the face of specimen. Also, the resultant line of applied shear forces should pass through the center of the shear plane base, with an angular tolerance of ±3°. In this study, the resultant shear force was designed to have 15° with the horizontal plane as shown in Figs. 3 and 4.

5.3. Equipments for measuring the displacements:

(a) Displacements were measured by Transducer Type Dial Guage with an accuracy better than 0.01mm. The Gauges were installed at eight locations of the specimen block to measure the displacements on the vertical, shear and lateral direction movement.

(b) For the purposes of accuracy and expedition, all the gauges were recorded by Data logger.

5.4. Preparation work in the field

(a) Excavation work; First, use the backhoe to level the ground. In the designated place, excavate the trench of appropriate dimensions, as well as the earth block for the specimens as shown in Fig. 5.

(b) Cutting work of the specimen; Use the hammer and chisels to form the specimen with the dimensions 1.5m x 1.5m x 0.7m.

(c) Encapsulating work of the specimen; For providing sufficient strength and rigidity to prevent collapse or distortion of the specimen during the testing, encapsulate the specimen with 15cm thick of 210 kg/cm^2 reinforced concrete as shown in Fig.6.

5.5. Experimental procedures:

Refer to the procedures described in ISRM (1)(4), the outlines are listed below:

(a) Consolidation; The consolidation stage may be considered complete when the rate of change of

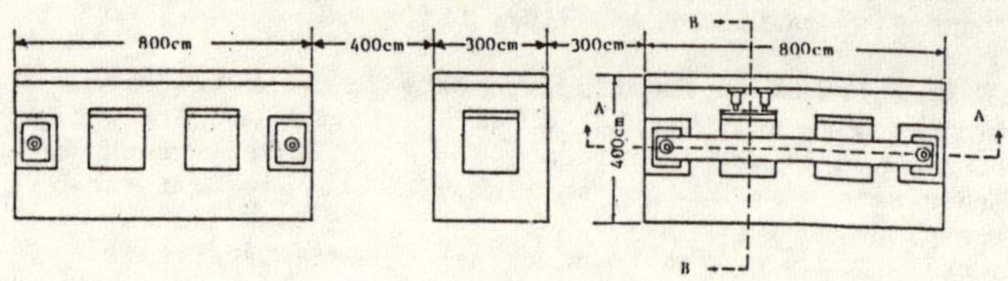

Fig. 4 . Plane View of Setup for In-Situ Direct Shear Test at Linkou Site

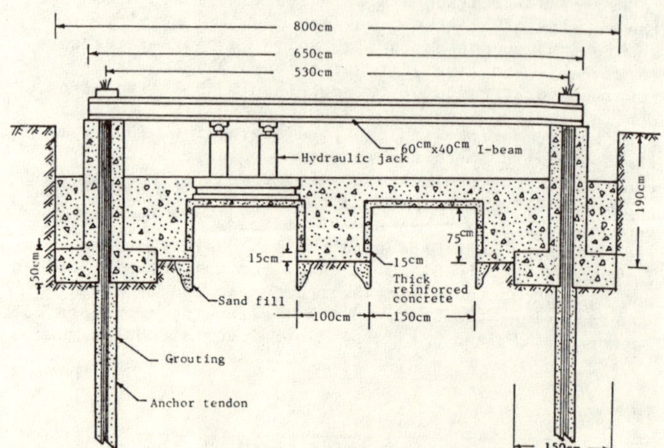

A-A SECTION

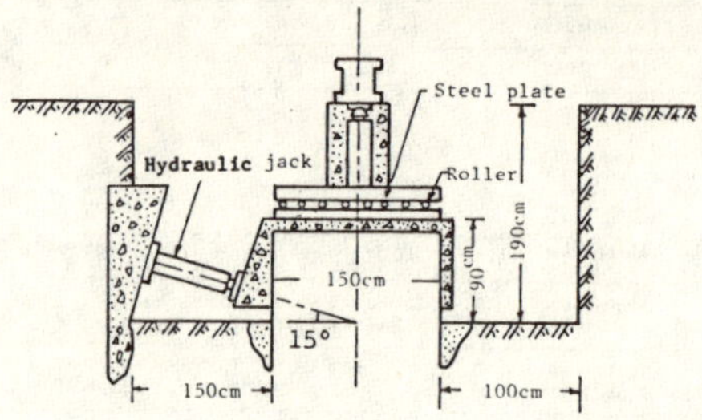

B-B SECTION

Fig.5. Schematic Picture of Earth Block for Specimen.

Fig.6. Schematic Picture of Encapsulating Concrete on the Specimen.

normal displacement is less than 0.05mm in 10 minutes. Shear loading may then be ready to apply.

(b) Shearing; Correction to the applied normal load should be required to hold the normal stress constant during shearing. The shear force is applied either in increments or continuously in such a way to control the rate of shear displacement. Approximately 10 sets of readings were taken before reaching peak strength. The rate of shear displacement should be less than 0.1mm/min. in the 10 minute period before taking a set of readings. This rate may be increased to not more than 0.5mm/min. between steps of readings provided that the peak strength is adequately recorded. After reaching peak strength, readings should be taken at increments of from 0.5 to 5mm shear displacement as required to adequately define the force-displacement curves. The rate of shear displacement should be 0.02-0.2mm/min. in the 10-minute period before a set of readings in taken. It may be possible to establish a residual strength value when the sample is sheared at constant normal stress and at least four consecutive sets of readings are obtained which show not more than 5% variation in shear stress over a shear displacement of 1cm. The specimen under shearing was depicted in Fig.7.

Fig.7. Schematic Picture of Specimen under Shearing.

(c) After the test, the specimen was inverted, photographed and fully described as shown in Fig.8, sample in the area of shear surface was taken for index testing.

6 RESULTS

(a) The consolidation curve for each normal stress is plotted during the consolidation stage of testing as shown in Fig.9.

(b) For each specimen test, graphs of shear stress versus shear displacement are plotted as shown in Figs.10, and 11. Values of peak and residual shear strength are obtained from the graph.

(c) Graphs of peak and residual shear strength versus normal stress are plotted from the combined results

Fig.8. Schematic Picture of the Shearing Surface on the Specimen.

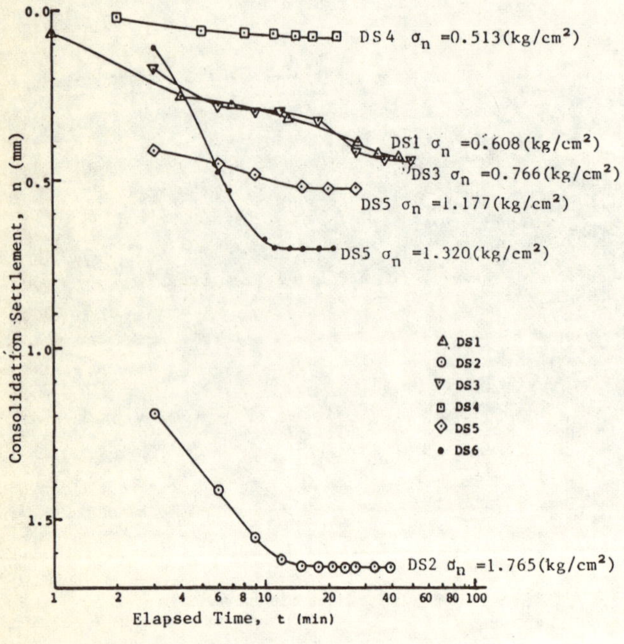

Fig.9 Consolidation Settlement Versus Time for the In-Situ Direct Shear Test at Sanyi Site

for all test specimens. The results according to Mohr's Coulumb failure criteria, indicate that in Sanyi area, the angle of internal friction was $\phi_p = 54.3°$ with cohesion intercept $C_p = 0.15$ kg/cm² at peak point, while at residual point, angle of internal friction was $\phi_r = 44.9°$ with cohesion intercept $C_r = 0$. In Linkou area, $\phi_p = 46°$ with $C_p = 0.15$ kg/cm², and $\phi_r = 34°$ with $C_r = 0$ as shown in Figs. 12 and 13.

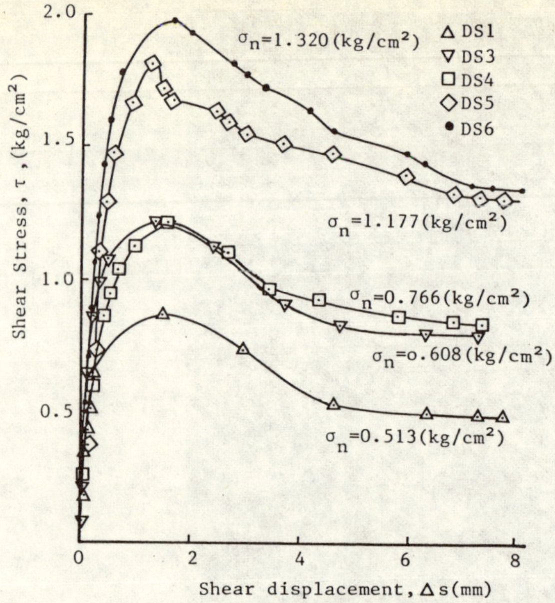

Fig.10. Shear Stress Versus Shear Displacement for the In-Situ Direct Shear Test at Sanyi Site.

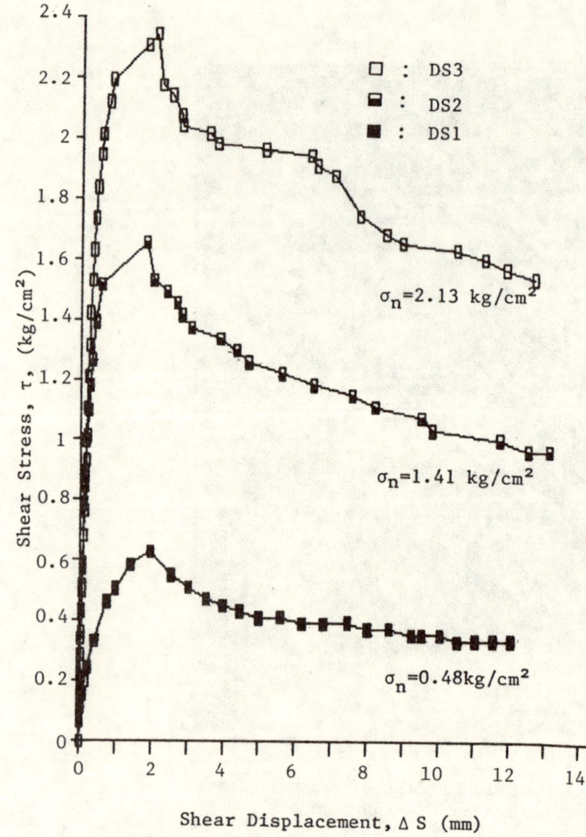

Fig.11. Shear Stress Versus Shear Displacement for the In-Situ Direct Shear Test at Linkou Site.

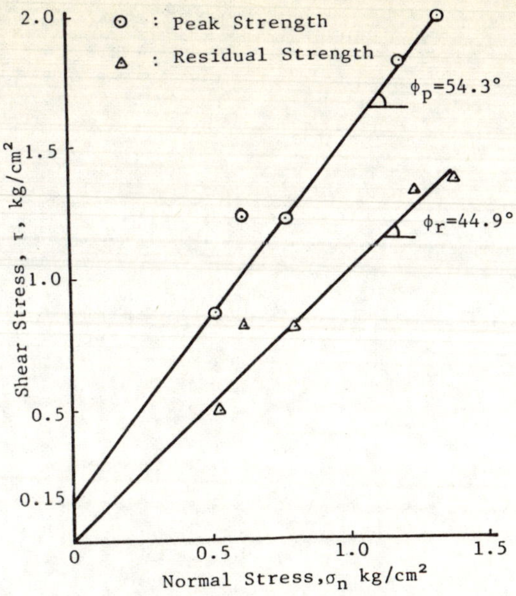

Fig.12.Shear Stress Versus Normal Stress for the In-Situ Direct Shear Test at Sanyi Site.

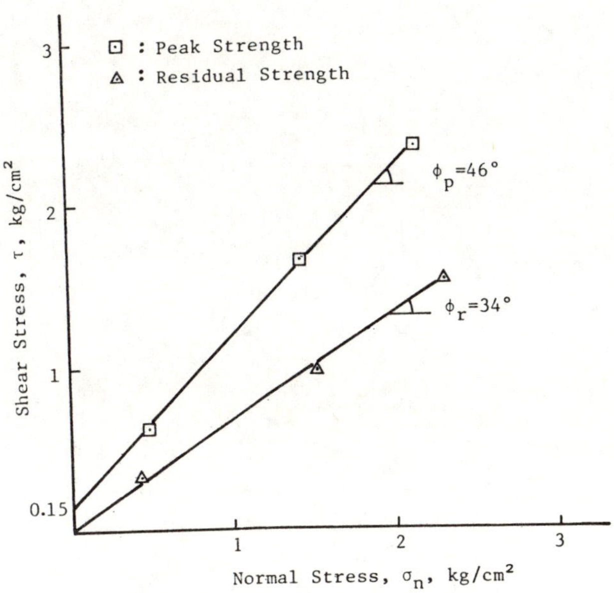

Fig.13. Shear Stress Versus Normal Stress for the In-Situ Direct Shear Test at Linkou Site.

7. CONCLUSION

(a) In northern Taiwan, the demanding for seeking the alternative resources of the aggregates becomes very urgent due to the economic development. Based upon the findings from this study, it was suggested that Sanyi and Linkou terrace deposits exhibit good quantity of aggregates for engineering usage, such as filling materials in the embankment, base and subbase of the highways and airport runways. In comparision, it was also recommended that the aggregates in Sanyi area could be used in the concrete manufacturing rather than those in Linkou area.

(b) Aggregates coarser than 1½" were the fundmental constituents in these two sites. However, grain sizes between 3/8" and 30# were gaped, for this reason, the deposits in two sites were classified as GP, GP-GM or GP-GC in the unified soil classification system, and grouped as A-1-a, A-1-b, A-2-4 or A-2-6 in the AASHTO classification system.

(c) From the consolidation results prior to shearing in the in-situ direct shear tests, the final consolidated settlement was only 1.33mm ($\varepsilon_v \doteqdot 0.20\%$) for normal stress of 1.75 kg/cm^2 at Sanyi site. It indicates that this gravel deposits exhibits high bearing capacity.

(d) The graphs of shear stress versus shear displacement at both Sanyi and Linkou sites showed that peak points were clearly demonstrated in each test. It indicates that in these areas, the gravel deposits were overconsolidated.

(e) The deposit in Sanyi area showed that the angle of internal friction was: $\phi_r = 44.9°$ with the cohesion intercept $C_r = 0$. It indicates that this site could stand stable at steep slope angle. In Linkou area, $\phi_p = 46°$ with $C_p = 0.15$ kg/cm^2, and $\phi_r = 34°$ with $C_r = 0$.

8 REFERENCES

1. Brown, E.T. 1981. "Rock Characterization Testing & Monitoring", ISRM Committee on Field Tests, pp.131-140.
2. Chu, B.L. 1982. "A Preliminary Study on Material Properties of Gravels from Foothills and Terraces in Taiwan", Journal of Civil and Hydraulic Engineering Vol. 9, No.2 -Paper No. 9-05 pp. 73-86.
3. Ho, S.C. 1975. "An Introduction to the Geology of Taiwan Explanatory Text of the Geologic Map of Taiwan". The Ministry of Economic Affairs, pp. 85-87.
4. Lama, R.D. & Vutukuri, V.S. 1978. "Handbook on Mechanical Properties of Rocks" Trans Tech Publications, Clausthal, Germany.
5. Prakash, S. & Ranjan, G. 1982. "In-situ Tests for Measurement of Soil Properties in Boulder Deposit", IUTAM Conference on Deformation and Failure of Granular Materials /Delft/ 31 Aug. 3 Sept. 615-623.
6. Surendra P.J. & Ramesh C.G. 1974. "In-situ Shear Test For Rock Fills," Journal of the Geotechnical Engineering Division, Proceedings of the ASCE, Vol. 100, No. GT9, pp.1031-1050.

Geotechnical properties of a compacted lateritic gravel from Cachoeira Porteira Dam

L.M.COSTA FILHO, Senior Geotechnical Engineer, ENGE-RIO Engenharia e Consultoria S.A., Brazil
T.M.P.DE CAMPOS, Associate Professor, Catholic University of Rio de Janeiro, Brazil
E.RITTER, Post graduation Student, Catholic University of Rio de Janeiro, Brazil
A.C.O.RIBEIRO, Geologist, ELETRONORTE, Centrais Elétricas do Norte do Brasil, Brazil

ABSTRACT: During the investigations for borrow areas of the Cachoeira Porteira hydroelectric development in the Amazon region of Brazil, a large volume of soils has been found containing lateritic concretions, with a matrix varying from clayey to sandy. A testing program has therefore been carried out in order to study its basic geotechnical characteristics. Due to the variation of lateritic gravel content in the natural soil, an investigation of the effect of percentage of concretions on the geotechnical properties has also been carried out.

This paper presents a general description of the material, including its classification (grain size distribution, plasticity) and compaction characteristics as well as some data about the concretions. The results of triaxial and permeability tests carried out on the three main types of lateritic gravel presented and compared with values obtained for the matrix soils. A series of triaxial and permeability tests has been performed on samples of a lateritic gravel with clayey matrix with 20, 40, 60 and 80% in weight of concretions (greater than nr. 4 sieve). The results are compared with those of the clayey matrix and it is concluded that the presence of concretions increases strength and stiffness of the compacted soil. However, small differences were observed between materials containing 20, 40 and 60% of concretions, whereas for 80% a marked increase in stiffness and strength was noted together with an increase in permeability.

1 INTRODUCTION

The Cachoeira Porteira hydroelectric development, currently under basic design by ENGE-RIO for the Centrais Elétricas do Norte do Brasil - ELETRONORTE, involves a complex of dam works located on the Trombetas river, state of Pará in the Amazon region of Brazil (Figure 1). The volume of compacted fill will be of the order of 13 millions of cubic meters. All the region is covered by dense tropical forest.

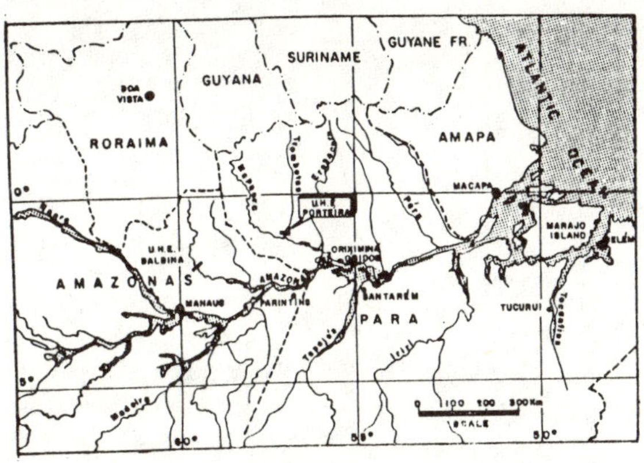

Figure 1. Site location

During the investigations of borrow areas, large volumes of soil with lateritic concretions (lateritic gravel) were encountered, with a matrix varying from clayey to sandy. Due to the better workability of these materials, observed during the construction of fills for several purposes in the region, it was decided to consider its use in the earth fills of the dam and dykes. Previous experiences with materials with similar granulometric distributions (terrace gravel, saprolites) indicated favourable geotechnical characteristics for use in compacted fills. Therefore, it was decided to carry out a series of laboratory tests to determine some geotechnical properties of the compacted lateritic gravel of the Cachoeira Porteira region.

This paper presents a summary of the general characteristics of the lateritic gravel from the proposed borrow areas and the results of triaxial and permeability tests on selected samples.

2 GENERAL DESCRIPTION OF THE LATERITIC GRAVEL

The occurrence of materials containing lateritic concretions is fairly common throughout the Amazon region. Besides its intense use as pavements and embankment for highways and airports in the region, lateritic gravels have already been used for construction of embankment for dams in the region but in much smaller scale than predicted for Cachoeira Porteira Dam. In Tucuruí Dam soils lateritic concretions were preferably used in the shoulders of the left wing embankment (Borges et al, 1983). In Balbina Dam, due to the smaller volume available of lateritic gravel, it was used only locally in the cut-off and under the upstream berm. In the dykes of Samuel hydroelectric development soil with lateritic concretions was employed in several stretches. In Pitinga Dam this material was used in the upper part of the dam, with the purpose of heightening it at a faster rate due to its better workability under very wet weather (Cruz, 1987).

The borrow areas investigated for Cachoeira Porteira Dam are located in the domain of sedimentary rocks (mainly sandstones and siltstones) of the Trombetas (Ordovician and Silurian ages) and Maecuru (Devonian) formations. Details about the regional geology are presented by Pupo et al, 1986.

Due to the presence of different lithologies, associated to mass movements and intense weathering and laterization processes, several types of soils were formed in the area. Their main characteristics and the criteria used for classification are presented by Machado Filho et al, 1986.

For specific purposes of the design, lateritic gravel has been defined as having a minimum of 20% by weight of material retained in the sieve number 4 (4,8 mm). It is considered that below 20%, its geotechnical behaviour is entirely controlled by the matrix. According to the grain size distributions and plasticity characteristics of the matrix, the lateritic gravels have been classified in 4 groups, related to the probable rocks of origin. These are shown in Table I. However, type LF is not found in the borrow areas for the chosen dam axis.

Throughout the region, the lateritic gravel normally occurs as a continuous layer following the topography with thicknesses varying from a few decimeters to a few meters. In a general way, it occurs associated to the more planar areas in the top of the hills and in more gentle slopes, disappearing in the talwegs with steeper slopes. This is illustrated in Figure 2

LATERITIC GRAVEL TYPE	MATRIX DESCRIPTION	PROBABLE ROCK OF ORIGIN
LA	Very plastic, slightly sandy (fine sand), slightly silty	Pure siltstone
LC	Medium plasticity, slightly silty, slightly sandy to sandy clay	Siltstone with sandstone interbeddings or vice-versa
LD	Slightly clayey, fine sand of low to medium plasticity	Sandstone with few siltstone interbeddings
LF	Fine sand with very little clay, frequently non-plastic	Almost pure sandstone

Table I - Lateritic gravel types in Cachoeira Porteira

Which shows for a borrow area curves of equal thickness of lateritic gravel, obtained from the investigations by pits and auger holes. In most cases it occurs immediately under the organic soil cover, some 0,5 to 1,0 m thick.

The lateritic concretions are formed mainly by iron oxides (goethite and hematite), with amorphous silica also found. They are of various shapes, but predominantly subrounded and platey. Visual and binocular glass observations have indicated that they show very rough surface, with many cavities, particularly those with more irregular shape. This is reflected in their apparent dry unit weight varying from 2,1 to 2,8 t/m^3 compared with a specific gravity of 3,1 to 3,6 for the material which composes the concretions. The matrix material has a specific

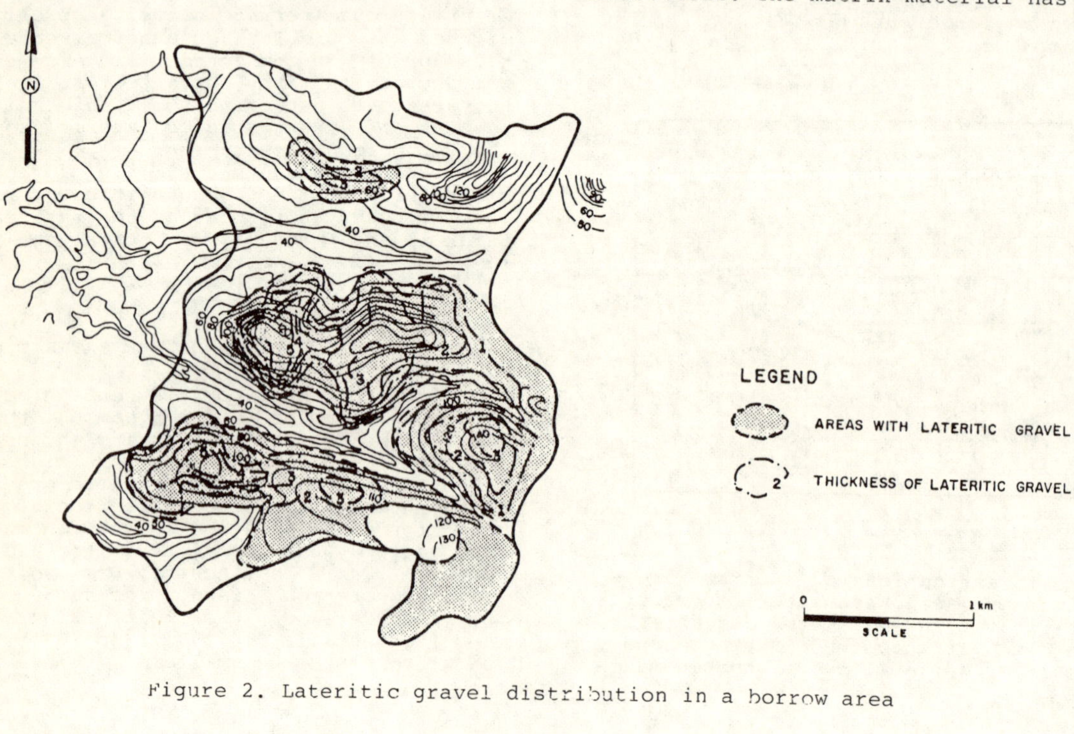

Figure 2. Lateritic gravel distribution in a borrow area

gravity varying from 2,65 to 2,90. In situ the natural moisture content of the concretions has been found to vary between 6 and 18%, with high degree of saturation. Also, due to their roughness, a significant amount of fines is normally found adhered on the surface.

As for the dimensions of the concretions, these are found to be fairly variable with a maximum size of about 7,5 cm but in several places blocks with dimensions of the order of 20 cm were observed. Locally cuirasses with thicknesses of up to 50 cm were noted, which presented some difficulties to the opening of pits.

The concretions are generally very resistant, mainly those with subrounded shape. The more platey ones sometimes can be broken by hand. During compaction in the laboratory the amount of breakage has been generally reduced.

3 GEOTECHNICAL CHARACTERIZATION

Figure 3a shows the average grain size distributions for the 3 types of lateritic gravel from a large number of tests together with those for their corresponding matrixes. Also presented in the same figure are their plasticity and compaction (standard Proctor) characteristics contours.

An analysis of about 400 grain size distributions in the 3 types of lateritic gravel indicated that the gravel content (material greater than nr. 4 sieve) is distributed uniformly between 20 and 60%, with a few tests giving up to 80%. The same occurs for each type of lateritic gravel considered separately.

The compaction results shown in Figure 3c correspond to those carried out in 15 cm diameter moulds with 2,54 cm maximum particle size. It was found that the conventional procedure of taking only the material passing in sieve nr. 4 and correcting the results by absorption tests in the gravel fraction gave erratic and in most cases erroneous results, particularly for higher gravel contents.

4 BASIC STUDIES OF SHEAR STRENGTH

For a preliminary evaluation of the shear strength and permeability characteristics of the compacted lateritic gravels, a series of laboratory tests was performed on materials LA, LC and LD. For this, selected samples were chosen for each of the materials, corresponding to average grain size distributions and plasticity, as shown in Figure 3. A detailed description of the testing program and result was presented by Costa Filho et al, 1986 and only a summary is presented here.

The compaction was carried out in 15 cm (6)

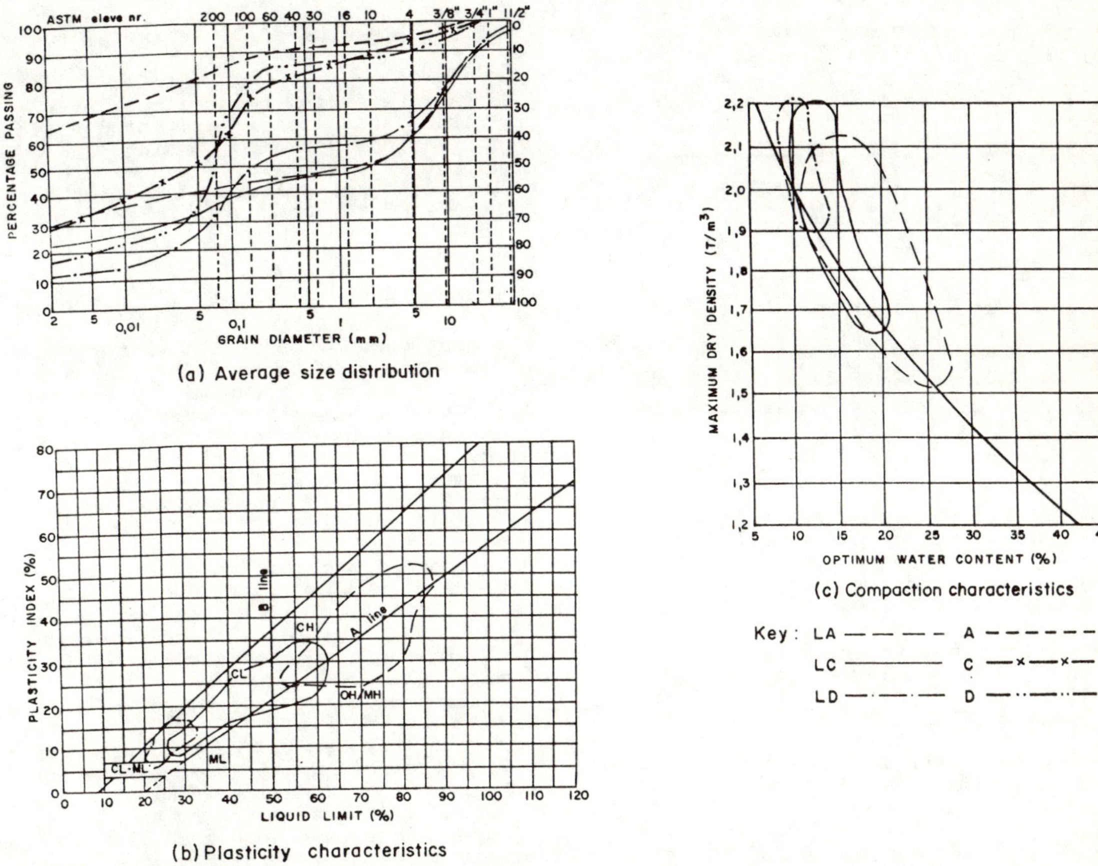

Figure 3. Geotechnical characteristics of lateritic gravels

diameter mould, without reuse or previous drying of the material, with a maximum particle size of 1,90 cm (3/4"). During preparation, the material retained in the 3/4" sieve was compensated by an equal weight of material with sizes between 3/4" and 4,8 mm (# 4).

The following testing program was executed on each lateritic gravel type:

- consolidated drained triaxial (CDsat) tests on saturated 15 cm diameter specimens;
- unconsolidated undrained triaxial (PN) tests with measurement of pore water pressure keeping a relationship σ_3/σ_1 constant and equal to 0,6, on 15 cm diameter specimens;
- permeability tests during the triaxial CDsat tests and in permeameters.

All specimens were compacted with degree of compaction 98%. For the CDsat triaxial and permeability tests the specimens were compacted at around optimum moisture content whereas for the PN tests optimum and optimum +2% were used.

A typical set of stress-strain curves obtained in the triaxial CDsat tests is shown in Figure 4. The material LA was found to be the least resistant and more compressible of the three types, as expected, reflecting its more clayey matrix. Nevertheless, a comparison between the lateritic gravels and their respective matrixes, compacted at about the same conditions, has indicated:

- a significant increase in effective strength parameters, as presented in Table II;
- a stiffer behaviour particularly under isotropic compression, but also during shearing.

STRENGTH PARAMETERS	LATERITIC GRAVEL			MATRIX		
	LA	LC	LD	A	C	D
C' (KPa)	30	46	30	15	27	27
Ø'	32°	37,5°	37°	28°	32°	32°

Table II - Effective strength parameters for lateritic gravels and matrixes

In relation to pore pressure development during PN tests, the results indicated values of the ratio $u/\sigma 1$ generally below 0,1 with the exception of material LA compacted at 3% above optimum which gave values of 0,2 to 0,3. It was noted that the presence of the concretions did not alter very much the pore pressure development in the PN tests since for the matrix soils it was also fairly low, particularly for soils C and D.

The coefficients of permeability obtained both in the triaxial tests and in the permeameters were fairly low for the three types of compacted lateritic gravels, being of the order of 10^{-6} to 10^{-7} cm/s. The only exception were values of $3x10^{-5}$ cm/s obtained in the triaxial tests on material LC at effective confining stresses of 0,5 and 1,0 kg/cm² but this value reduced to $3x10^{-7}$ cm/s at confining stress of 3,0 kg/cm².

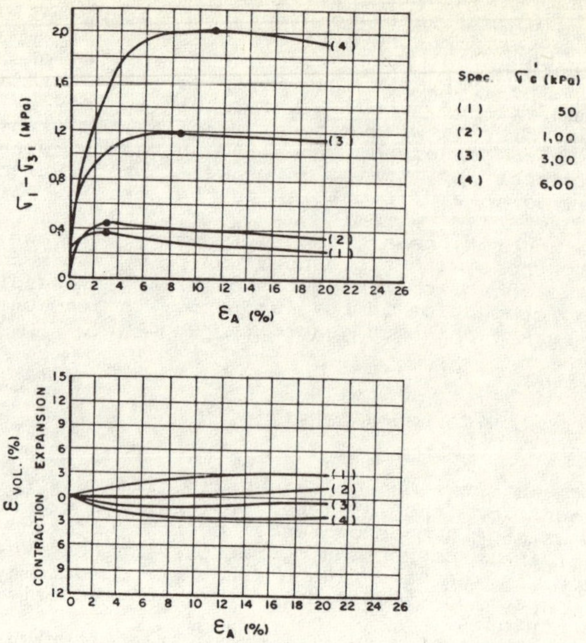

Figure 4. Results of CID triaxial tests on LC material

It is interesting to note that even for these low values of permeability, the coefficients of consolidation measured during the isotropic stage of the triaxial tests were of the order of $1x10^{-1}$ cm²/s, reflecting the relatively high stiffness of the compacted laterite gravels.

5 INFLUENCE OF THE LATERITIC GRAVEL CONTENT

As mentioned before, the lateritic gravel content was found to be fairly variable in the potential borrow areas of the Cachoeira Porteira Dam. It was, therefore, of interest to investigate the effects of the lateritic gravel content on the geomechanical properties of the compacted material. To do so, a project oriented research program was carried out at the Pontifical Catholic University of Rio de Janeiro.

The testing program comprized conventional isotropically consolidated undrained (CIU) and drained (CID) triaxial tests, refer to Ritter (1988), and flexible wall permeability tests on saturated, compacted samples of the LA type of material.

Mixtures of soil with nominal proportions of gravel of 0 (matrix), 20, 40, 60 and 80% were prepared. For that, the required proportions in weight of air dried, washed lateritic concretions with a maximum nominal dimension of 19.05 mm (3/4") was added to the matrix. The mixtures were then compacted with a degree of compaction of 98% (relative to the normal Proctor energy).

As a reference, the optimum compaction conditions of the mixtures are summarized in Table III.

LATERITIC GRAVEL CONTENT(%)	MAXIMUM DRY WEIGHT (kN/m^3)	MOISTURE CONTENT (%)
0	14,9	30,5
20	16,2	26,0
40	17,4	23,1
60	18,3	20,9
80	18,5	16,0

Table III - Optimum Normal Proctor Compaction Conditions for the Mixtures

For the triaxial and permeability tests on the matrix material, samples with 38.1 mm of diameter were used. Specimens with 101.6 mm of diameter were employed for the 20 to 80% gravel content mixtures.

A constant gradient of 10 was used in the permeability tests, which were performed on mixtures compacted at their respective optimum water contents.

Only results of CIU triaxial tests performed on samples compacted at optimum moisture content are presented in this work. Data from CIU tests run in specimens with other moisture contents and from the CID tests will be discussed elsewhere.

Consolidation and Permeability Results

Isotropic compressibility characteristics of the mixtures, for effective consolidation stresses, $\sigma c'$, between 50 kPa and 800 kPa are shown in Figure 5.

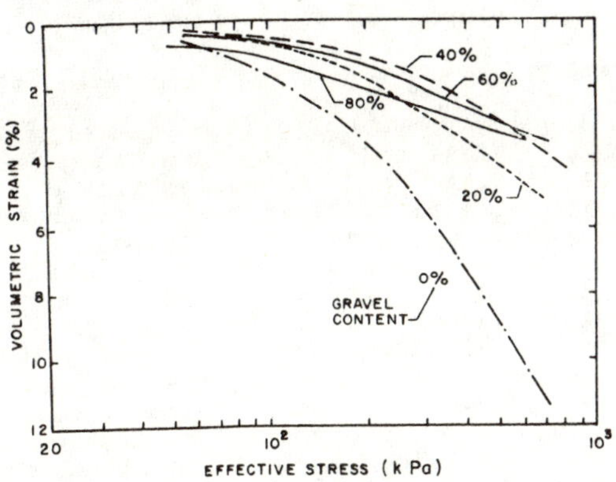

Figure 5. Isotropic compression curves of the mixtures

At the initial consolidation stress levels there was a trend for the volumetric strains to decrease with an increase in gravel content only up to 40%. An increase in volumetric strains with the gravel content was then observed, refer to Figure 5. On the other hand, considering only the relative inclination of the curves, there is a clear trend for a decrease in compressibility with the increase in gravel content; the matrix being much more compressible (bulk modulus of the order of 5 to 9 MPa) than the other mixtures (bulk modulus varying between 10 and 30 MPa).

Results of constant back pressure gradient permeability tests, for effective consolidation stresses of 50 and 600 kPa, are shown in Figure 6.

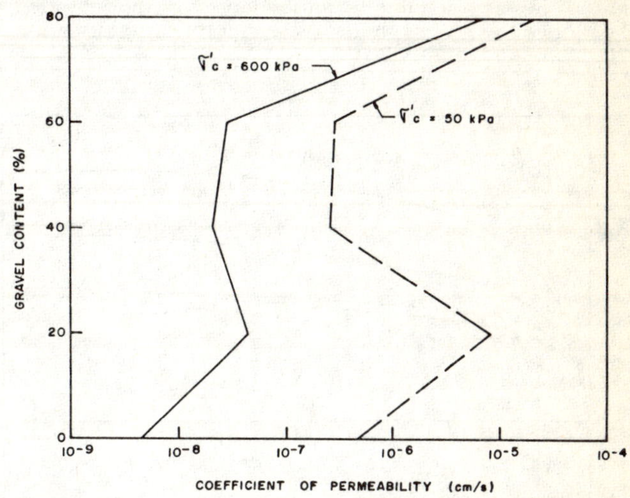

Figure 6. Permeability coefficients for the mixtures

As can be seen, the effective stress level has an important influence on the value of the permeability coefficient for the 0 to 60% mixtures. This influence decreases for the mixture of 80% of gravel content.

Although differences in the coefficient of permeability for the 40 and 60% mixtures are negligible, there is a trend of increase in this parameter for gravel contents above 40%. Largest differences in permeability, particularly at the higher level of effective consolidation stress, occurs between the matrix and the 80% lateritic gravel content materials.

It is interesting to notice that average values of coefficient of permeability, back calculated from the consolidation phases of the triaxial tests, showed the same trend of variation with gravel content observed in Figure 5. These infered data were more or less coincident with the direct measurements of permeability only for the effective consolidation stress of 50 kPa. Values about 10 times smaller than those shown in Figure 5 were calculated for the effective stress level of 600 kPa. Care should, therefore, be taken when interpreting results from these indirect measurements.

Consolidation coefficients of the order of 7.0×10^{-2} cm^2/s and of 1.5×10^{-1} cm^2/s were found, respectively, for the matrix and 20% mixture, and for the 40% and 60% gravel content samples in the consolidation phases of the triaxial tests. Higher C_V values, of the order of 1 cm^2/s were found for the material compacted with 80% of concretions. Considering that the compressibility of the samples with 40 to 80% of gravel did not differ much (see Figure 4), these higher C_V values are basically due to the higher permeability associated to the 80% mixture.

Stress - Strain - Strength Results

Figure 7 shows stress-strain and pore water pressure-strain curves obtained in CIU tests for a consolidation effective stress of 50 kPa. Curves with the same general shape were also obtained for the samples consolidated at 100, 300 and 600 kPa.

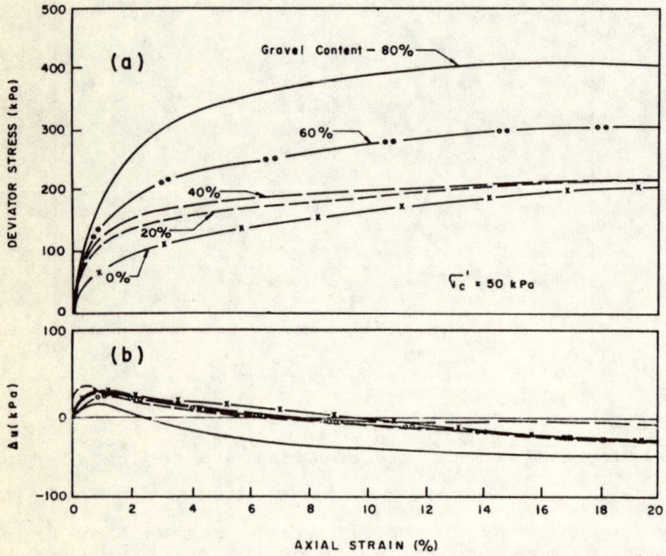

Figure 7. Results of triaxial CIU tests with σ'_C = 50 kPa for the mixtures

As exemplified in Figure 7, a continuous increase in deviator stress with the axial strains was observed in all samples. Yielding ocurred at relativelly small strains (bellow 2%). A trend for a decrease in the yielding strain level with an increase in gravel content was observed for the samples consolidated at 300 and 600 kPa.

Undrained secant moduli of deformation at axial strains of 0.5 and 1.0%, for effective consolidation stresses of 50 and 600 kPa are shown in Figure 8. As can be seen, the lower the gravel content, the higher is the deformability of the material. The largest relative changes in undrained modulus occur between the mixtures with 0 and 20% of gravel content and between those with 60 and 80%.

Considering the effective stress paths, relatively little variations were obtained for the mixtures with 0% to 60% of gravel, refer to Figure 9. Although presenting the same general shape, the mixture with 80% of concretions showed a stronger tendency for dilation, even at the highest levels of effective consolidation stresses.

The similarity between the effective stress-paths for 0 to 60% gravel content could

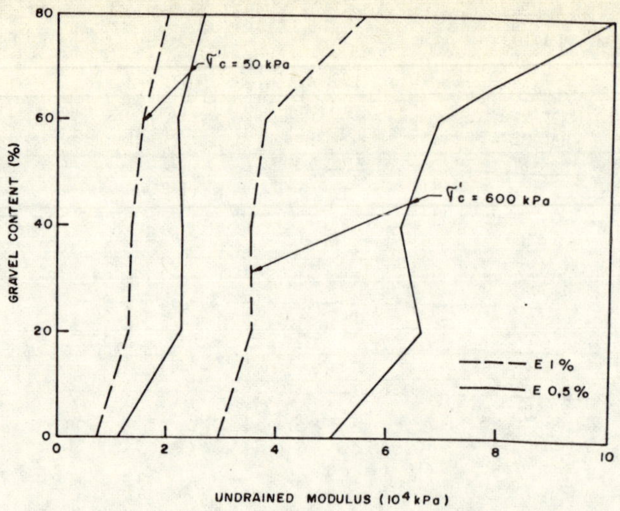

Figure 8. Undrained secant moduli for the mixtures

be due to the fact that the specimens were prepared at optimum moisture content of the mixture and in this condition the matrix will be on the wet side of the optimum, with the difference increasing with gravel content. Therefore a smaller tendency for pore pressure generation due to increase in stiffness of the mixture could be counterbalanced by a wetter matrix.

As can be noticed in Figure 7b, pore pressure peaks occur, for all mixtures, at axial strain levels around those corresponding to yielding.

Considering Figure 9, these peaks correspond basically to the points where the effective stress paths touch the strength envelope associated to each mixture. The observed trend for the effective stress paths to climb up the strength envelopes after touching them is compatible with the fact that none of the mixtures showed peak strengths, as shown in Figure 7a.

Assuming a maximum obliquity failure criteria (σ_1'/σ_3' maximum) for the undrained tests, and taking into account results from drained tests, the linearized effective strength envelopes shown in Table IV were obtained. A general trend for an increase in the effective angle of friction with the increase in the gravel content was observed, the largest relative variation in ϕ' occurring between the mixtures with 20% and 40% of concretions. If only CIU tests are considered, the envelopes for 20% gravel content are very similar for those for 40%. There are evidences that the strength envelopes may be curved at the lower levels of effective stresses (Ritter, 1988). If that is true, changes in the trend of variation of the cohesion intercept with the gravel content, which can be infered from the data shown Table IV, may occur.

From the effective stress paths of the triaxial CIU tests (Figure 8) it can be seen that the undrained strengths for the gravel contents from 0 to 40% are fairly similar for all consolidating stresses, with 60% presenting a

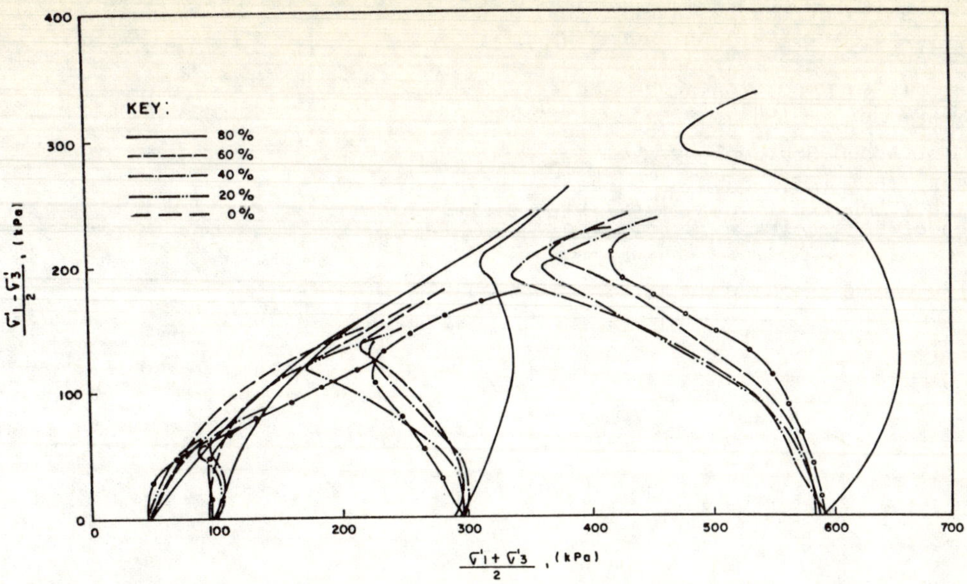

Figure 9. Effective stress-paths for the mixtures CIU triaxial tests

small increase. For 80% significantly higher values are obtained.

LATERITIC GRAVEL CONTENT	Ø'	C' (kPa)
0	29,0	19,0
20	30,0	25,0
40	35,5	14,0
60	35,5	24,0
80	37,0	31,0

Table IV - Effective strength parameters (CIU + CID tests)

6 CONCLUSIONS

Based on the studies carried out up to the present moment, one can arrive to the following main conclusions:

a - The compacted lateritic gravels of Cachoeira Porteira present geomechanical characteristics which are perfectly satisfactory for use as construction materials for embankment dams, confirming previous empirical experiences in highway fills;

b - Even the addition of only 20% by weight of lateritic concretions provided the materials with geomechanical characteristics significantly better than those of the matrix soils (clayey or sandy);

c - The results of studies of influence of porcentage of concretions greater than 4,8 mm for the laterite gravel with clayey matrix, indicate small variation in the geomechanical parameters for 20, 40 and 60%. While a change on behaviour seems to occur for a gravel content between 20 to 40%, a single geomechanical behaviour could be preliminarily assumed for engineering purposes. For 80% of concretions higher stiffness, shear strength and permeability was obtained. It would be interesting to extend these studies to other types of soils with lateritic concretions to confirm these facts as well as other moulding conditions;

d - The greater uncertainties in relation to the use of compacted lateritic gravels in fills for dams seem to be associated with their trafficability limits and tó compaction control techniques to these types of materials.

ACKNOWLEDGEMENTS

The authors are grateful to ELETRONORTE - Centrais Elétricas do Norte do Brasil - the support given for the development of the studies and to ENGE-RIO the support given for the preparation of this paper. Mrs. E. Ritter had the financial support of CAPES for her M.Sc. course.

REFERENCES

- BORGES, C.M. et al 1983. Considerations on the design of the earth and rockfill dams of Tucuruí. XV Brazilian Seminar of Large Dams, Theme IV, Vol. 1 (in Portuguese).
- COSTA FILHO, L.M; MACHADO FILHO, O.V.; AMARAL, M.A.B., 1986. Geotechnical characteristics of Cachoeira Porteira compacted laterite gravel. VIII Brazilian Congress of Soil Mech. and Found. Eng., Vol. 8 (in Portuguese).
- CRUZ, P.T. 1987. Personnal Communication.
- MACHADO, FILHO, O.V. 1985. Geotechnical in-

vestigations of soils of borrow areas for Cachoeira Porteira Dam. 5th. Int. Congress of Int. Assoc. Eng. Geology, Vol. 4.

- PUPO, G. et al. 1986. Conditioning geological factors on the selection of a site for the Cachoeira Porteira dam. 5th. Int. Congress of the Int.Assoc.Eng.Geology. Vol.1.
- RITTER, E. 1988. The influence of gravel content on the behaviour of a lateritic soil from Cachoeira Porteira - M.Sc. Thesis, PUC-RJ.

Influence of the moisture content in the shear strength of undisturbed lateritic soil

R.DAVISON DIAS, Reader in Soil Mechanics, Graduate Program in Civil Engineering, Federal University of Rio Grande do Sul, Brazil

SYNOPSIS

This paper presents results of a triaxial compresion test done in samples collected from B horizon with lateritic behavior localized in south of Brazil, near Porto Alegre city. The profiles were classified pedologically as a red yellow Podzolic originary from terciary sediments. Tests were done in undisturbed specimens.

The tests were performed at different saturation degrees, from natural moisture to complete saturation. The tests were Consolidated Isotropic Undrained (CIU) with pore water pressure measurement. In the samples with natural moisture the tests were drained.

Based on lateritic soils pedogenetic conditions, it was made an analysis of the moisture conditions in the engineering applications that better represent the field conditions.

The paper analyse the stress path found and the variation of the strenght parameters, the porewater pressure and the A parameter in accordance to the moisture content.

INTRODUCTION

The region situated in the tropics and subtropics present large red soils extensions, known as lateritic soil.

The soils with lateritic behavior is parcially saturated in natural state. However, by climatic variation the water content is not constant. In order to estimate the shear strength for a geotecnical project, it is necessary to know the conditions of saturation that the profile is going to present.

The shear strenght of a parcially saturated soil is complex. It depends on a negative pore pressure that decreases when the saturation degree increases, as shown by Fredlund(1978, 1985 and others).

When subjected to the water action, without oxigenation, the lateritic soil may present an iron oxid reduction process. The soil shows gray spots with gleyzation characteristics. This process happens under imperfect drainage conditions, drain impediments or complete inundation. Basically, lateritic soils are parcially saturated, the moisture content is bellow the plastic limit and its colours are reddish and homogeneous without spots. These characteristics do not evidence neither the occurence of the gleyzation process nor the complete saturation.

In the case of a C horizon with a low permeability material, in rainy times, the saturation degree may increase. The total or almost total saturation may happens when this material is present on the foundation of a dam.

Tests in different saturation degrees were performed in order to compare the shear strenght in the different conditions.

SOIL CHARACTERISTICS

The blocks were taken from the B horizon of a red yellow Podzolic with all the characteristics lateritic soils. This horizon is similar to B latossolic. There is just one texture difference between A and B horizon. A is sandiest then B. The C horizon begins in the depht of 3.0 meters. It is formed by clay (40%),silt (40%) and fine sand (20%). This grain size occurs up to 7.5 meters. The C horizon is poorly drained and shows tipics gray spots of gleyzation.

By the geology, the region is formed by terciary sediments of sandstone and claystone. In the local the samples were taken there were claystone.

The soil analysed is red and formed by sand (22%, in what 20% is fine sand), silt (32%) and clay (46%). The liquid limit is 47% and the plastic limit is 30%.

The average content, when the material was collected, was 27%. The soil is porous and the average physical properties are: unit weight: 16.5 kN/m^3, dry unit weight: 13.1 kN/m^3, void ratio: 1.1 and saturation degree: 68%.

This soil is permeable with coefficient of permeability 1.6×10^{-6}m/s for a 100 kN/m^2 consolidated pressure. The virtual overconsolidation pressure results in 140 kN/m^2 (natural moisture and 130 kN/m^2 (when the sample is soaked).Other characteristics of this soils are described in Davison Dias (1987).

METHOD

Triaxial compression tests were performed in samples with natural water content and tests in which the saturation degree was increase. In these ones the viewed conditions were: macropores saturations; almost total saturation and total saturation.

In the tests with macropores saturations two kind of tests were considered. The first one uses Skempton B parameter (Skempton, 1954) as 55% and the second one uses the Skempton B parameter as 57%. The macropores saturation were done using water percolation through two tubes during 48 hours untill steady flow was achieved. In the tests with B equal 55% the samples lost the natural water content, the new average value was 19.8%.

In the tests with almost total saturation the back pressure gave a B parameter equal 90%. In the samples completelly saturated the B parameter was equal 100%. The total saturation was obtained through the application of back pressure after the macropores saturation. In the tests the used techniques were the ones described by Bishop and Henkel (1962).

The tests performed in the samples at natural moisture were consolidated drained, once the soil is permeable, with a low saturation degree (near 68%) and a high coefficient of consolidation ($Cv = 8.5 \times 10^{-4}\ m^2/s$).

RESULTS AND ANALYSIS

The table I presents the average values of initial water content (w), determinated during moulding of the samples, of the water content after water percolation (w_p) and after the tests (w_f).

TAB I - Values of water content in the begin, after percolation and in the end of the tests.

Tests Conditions	w % initial	w % percol.	w % final
Water Natural Content	26.6	-	24.9
B = 55%	20.8	33.8	30.5
B = 57%	24.8	35.4	32.8
B = 90%	27.3	36.2	34.7
B =100%	25.9	35.4	43.2

The figure 1 show a suction average curve versus the water content. This test was performed in the suction plate. The samples were initially soaked and after was applied the suction pressure. The tests performed with the natural soil structure without overload.

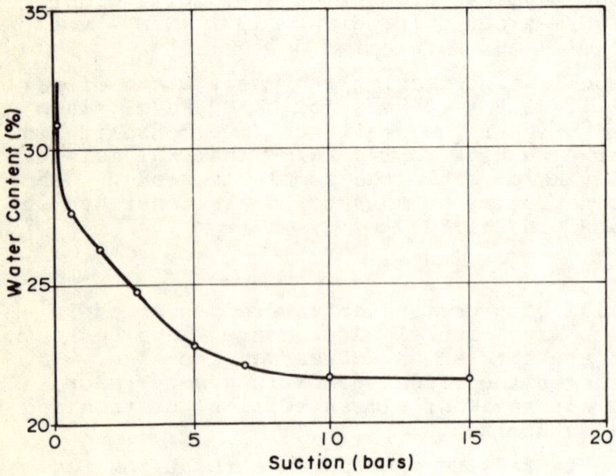

Fig. 1: Suction versus water content

Analysing the results of table I, the suction curve and considering that the percolation does not modify the soil structure, we have practically total reduction of suction when the water content is increased by percolation. If in field, in rainy days, the percolation water content was reached, greater than 33%. It is possible to have the conditions of percolation tests.

Water content measurements in this profile, in the depht that the samples were collected (1.5m), resulted in values between 22.0 and 27.0%. The determinations were made in seven differents ocasions. Two of them were in rainy days. In the depht 2.8m, near the C horizon, the water content variated between 25.0 and 36.0%, exceeding the percolation water content.

The knowledge of soil profile behavior must consider the specific application. For example: in the case of a shallow foundation, with 2.20m depht, subjected to an uplift, the loaded soils will be the ones adjacent to the foundations. In this case, the tests that better represents the fields conditions are the ones performed at natural water content as verified by Davison Dias (1987) in field uplift test in actual size footings.

Except in the foundations of dait is very unlikely to happen the saturation conditions in a lateritic soil profile. If the saturation occurs probably we will have a change in the pedogenesis and the soil do not present lateritic behavior any more. If often happens in the low part among elevations.

Figure 2 shows the deviator stress curves versus the vertical strain and the p versus q diagram results from CID tests done in samples at natural water content.

The samples with natural water content tested with consolidation pressure greater than the virtual overconsolidation pressure presented a shear strenght increase as the vertical strain increase. For pressures lower then overconsolidation pressure the stress level keeps constant after reaching the maximum value. It happens also for large strain.

There was a descontinuity in the failure envelope obtained for the vertical strain of 5%. For consolidation pressures smaller than the overconsolidation pressure the cohesion and angle of internal friction values were greater than for the higher pressures. For vertical strain of 10% and 15% the envelopes were linears.

Figures 3 to 6 show results of triaxial compression test CIU with water pressure measurements. In this figures are present the curves: deviator stress (σ_d), ratio between deviator stress and confining stress ($\sigma_d/\sigma_{c'}$), ratio between change of water pressure and confining stress (Δ_u/σ_d) and pore water pressure parameter A, all of them versus vertical strain.

The figure 3 contain the curves obtained from the tests with macroporos saturation throught percolation (B = 55%) and in figure 4 the curves with B = 57%. In the figure 5 it is represented the results of the tests done with almost total saturation (B = 90%) and in figure 6, the

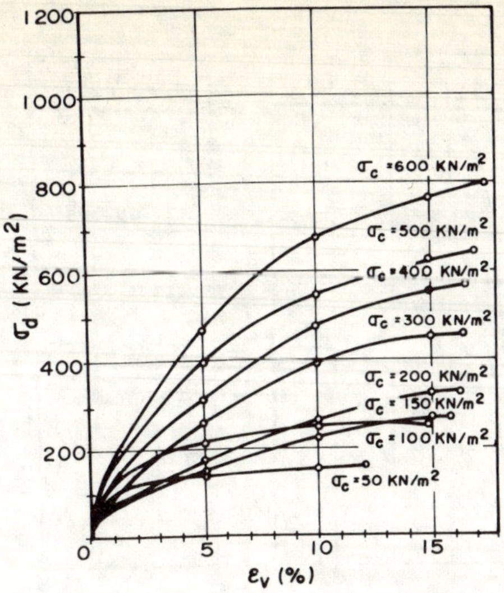

a) Deviator stress x vertical strain

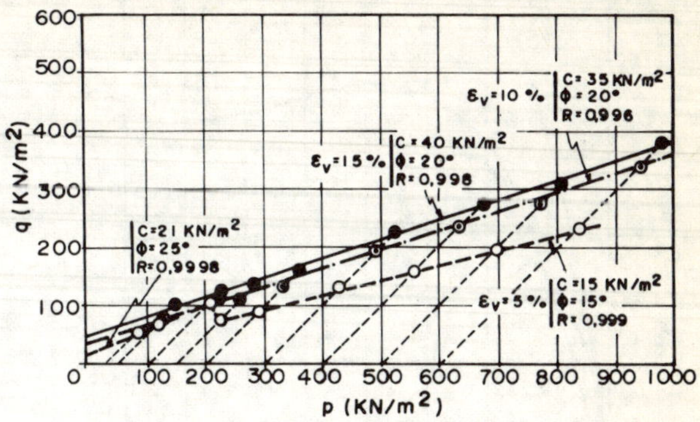

b) fix q diagram

Fig. 2: CID Test - Natural moisture

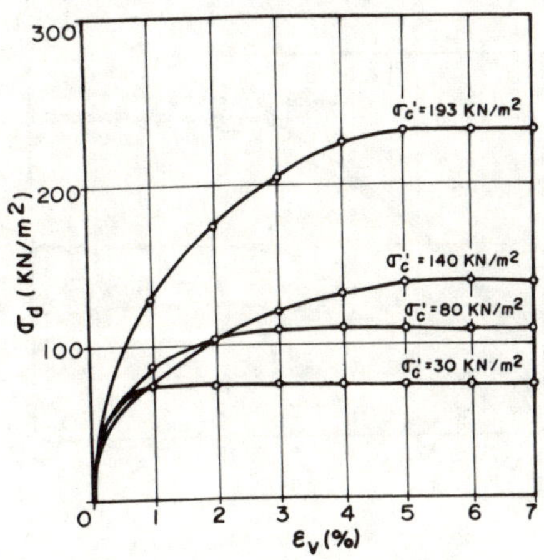

a) Deviator stress x vertical strain

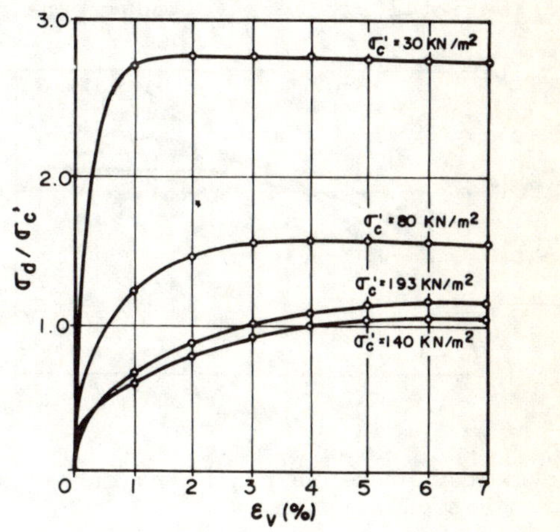

b) Deviator stress/confining stress x vertical strain

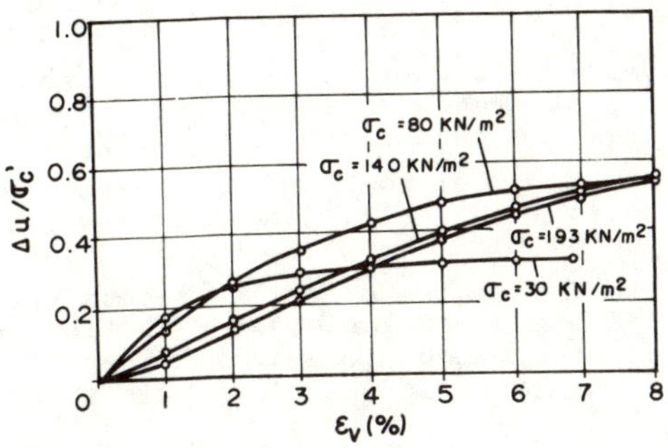

c) Water pressure/confining stress x vertical strain

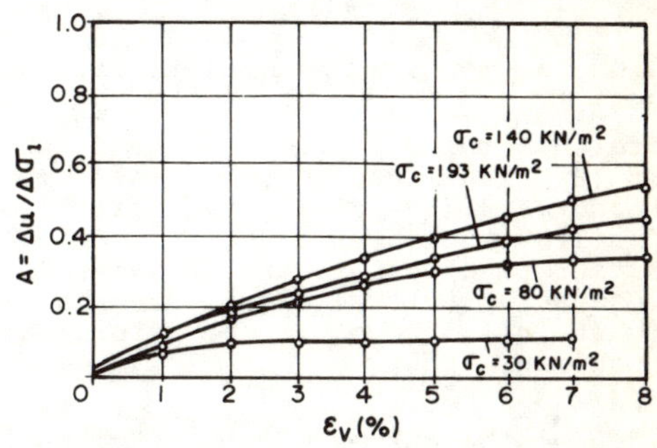

d) A parameter x vertical strain

Fig. 3: CIU Test with percolation - B = 55%

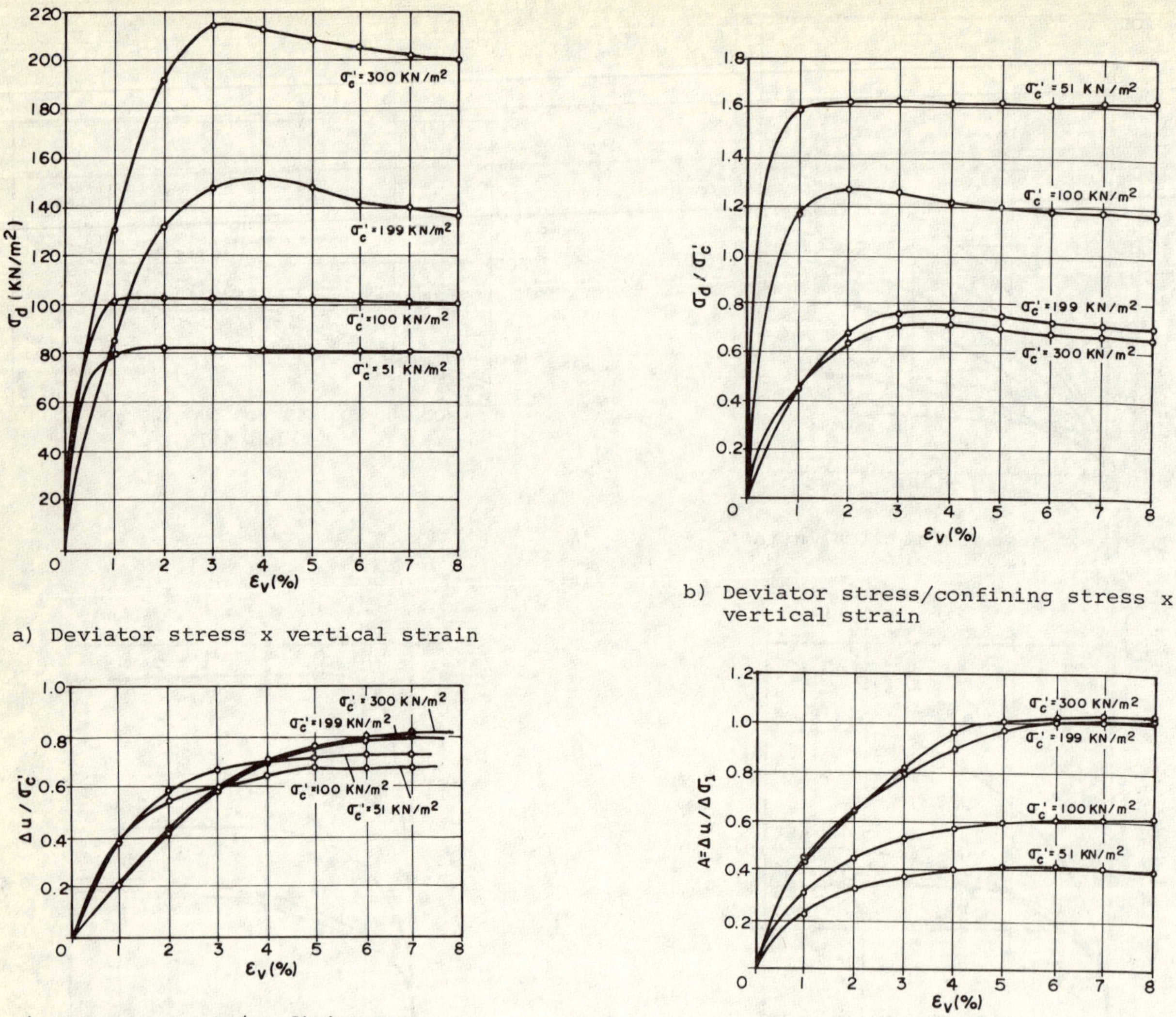

a) Deviator stress x vertical strain

b) Deviator stress/confining stress x vertical strain

c) Water pressure/confining stress x vertical strain

d) A parameter x vertical strain

Fig. 4: CIU test with percolation - B = 57%

results of the tests with complete saturation (B = 100%).

The difference that happened in the complete saturation (B = 100%) and in the almost complete saturation tests (B = 90%) were small, probably resulting from the material heterogeneity. It is important to emphasyse that the lateritic soil presents structural aggregates that have a significant influence on its behavior (Davison Dias e Gehling, 1985).

In the tests done with an increase in the saturation degree the curves of stress versus vertical strain present a small strenght decrease after the peak for consolidation pressures superiors than the overconsolidation pressure. The stress level kept constant after the maximum value for pressures inferior than overconsolidation pressure.

The curves resulted pratically normalized for consolidation pressures superior the overconsolidation pressure in all the CIU tests.

The strain in order to reach the maximum value in the tests were small, generally lower than 2%, except two of them that reached 3%.

The figure 7 show the effective stress path plotted in p' x q diagram. The effective stress path presented a behavior similar to the normally consolidated soils with the increase of the B parameter.

The variation of the relation between the maximum values of deviator stress, increase of poro pressure and A parameter with the confining stress is shown in the figure 8.

Considering Su equal q, it is possible to say that the undrained strenght decrease as the sa-

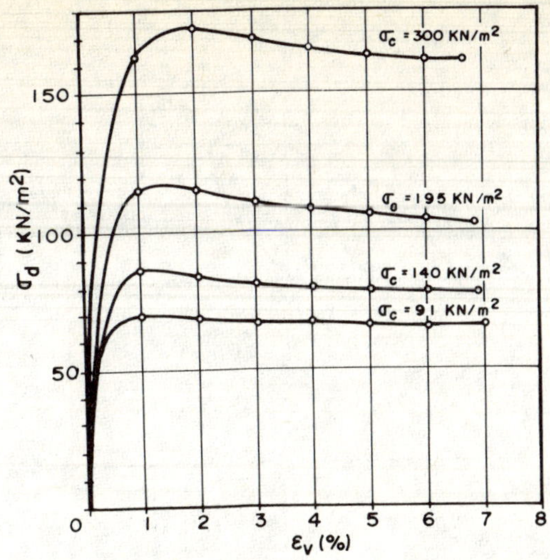

a) Deviator stress x vertical strain

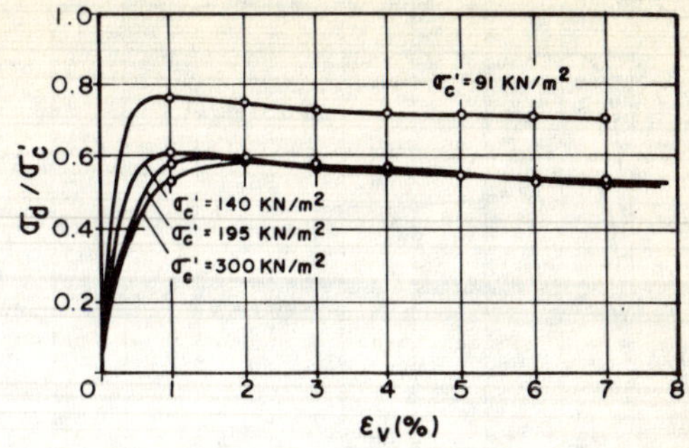

b) Deviator stress/confining stress x vertical strain

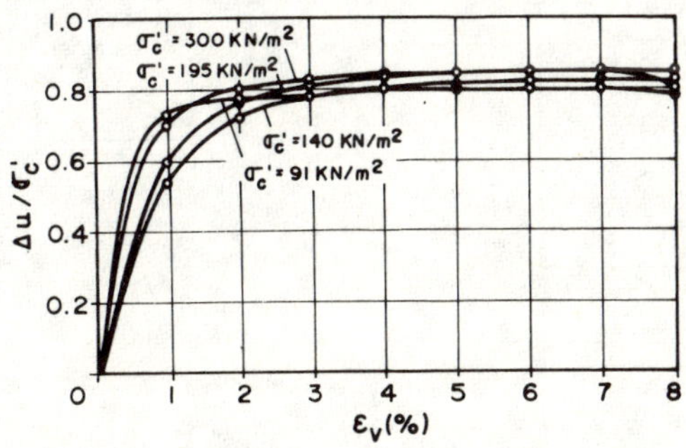

c) Water pressure/confining stress x vertical strain

Fig. 5: CIU Test with percolation - B = 90%

turation degree increase. There were just a little diference with full saturation (B = 100%) or almost full (B = 90%). But if the saturation is restrict to the macropores there is an strenght increase, eventhough the soil have lost almost all the suction. This difference is due to a little residual suction that may exist and of a probable disturbance in the soil structure with high back pressure levels.

For pressures higher than the overconsolidation pressure the relation between σ_d/σ_c' is almost equal to 0,57, when the soil is satured. When the macropores are satured the relation increase to 0,75 and, in soils at natural water content increases to 0.82. The relation between the water pressure at failure and confining pressure was about 0.7.

As expected the smallest values for maximum A parameter were found for the test when B=55%, and increased with the increase in the saturation degree.

The effective failure envelope and total failure envelope are presented in figure 9.In this picture it is also presented the values obtained from direct shear test, done in the some material at natural water content and soaked conditions.

The effective strenght parameters obtained from soaked tests presented small variation. The cohesion intercept resulted between 10.0 and 12.7 kN/m^2 and the angle of internal friction resulted between 24.3 and 27.8 degree. The average envelope for all the data present a cohesion intercept equal to 10.8 kN/m^2 and angle of internal friction equal to 26.6 degree.

In total stresses the highest envelope was obtained from the direct shear test at natural moisture. In sequence cames the test CIU with percolation (B = 57%), CIU with B = 90% and the one which shown the lowest envelope was the CIU with complete saturation (B = 100%), as expected.

All results of failure in a p' x q diagram for low and high confining pressure were plotted in figure 10. Analysing the figure, it is observed that the envelope is not a straight line, it is curve and have an increase in the cohesion intercept and a decrease in the angle of internal friction for pressures greather than 200 kN/m^2.

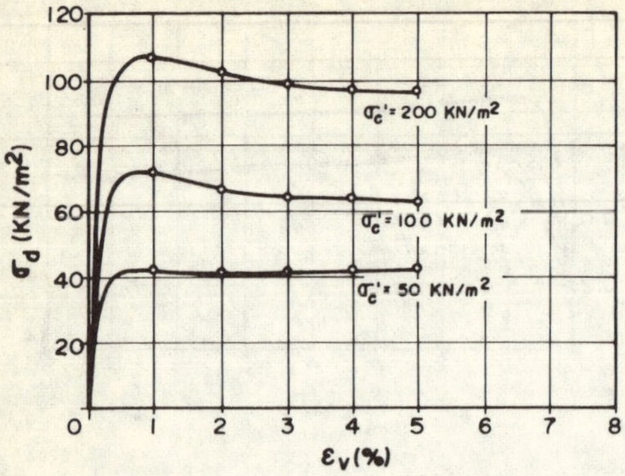

a) Deviator stress x vertical strain

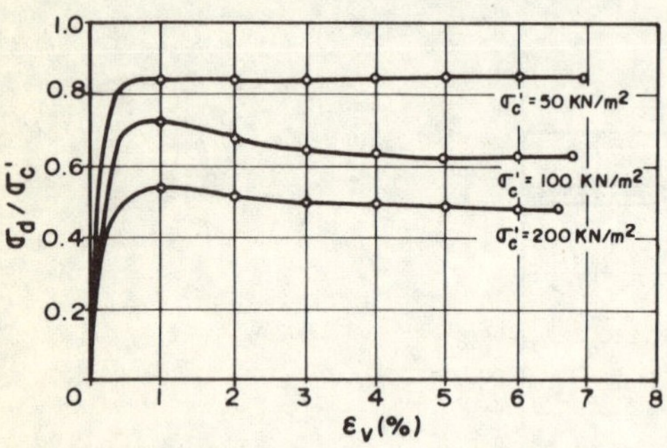

b) Deviator stress/confining stress x vertical strain

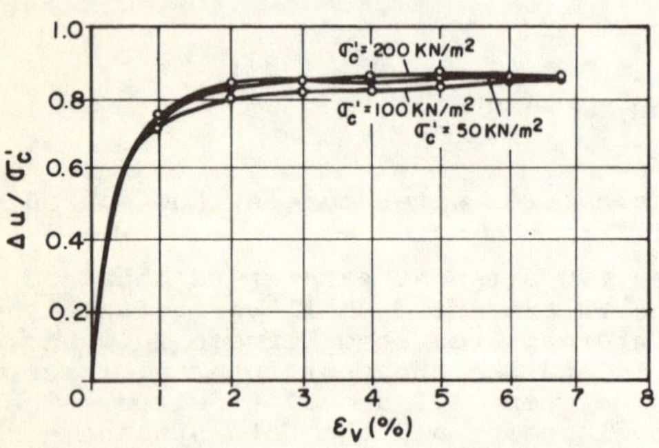

c) Water pressure/confining stress x vertical strain

Fig. 6: CIU Test - Full saturation - B = 100%

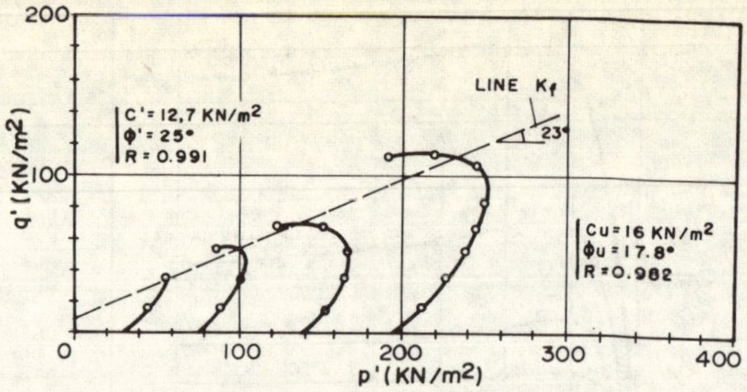

a) B = 55%

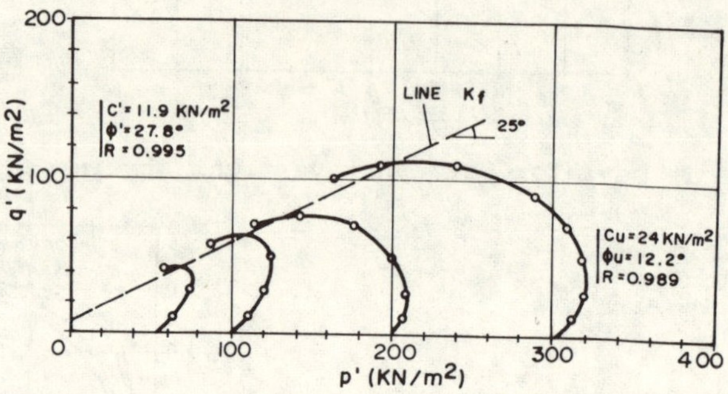

b) B = 57%

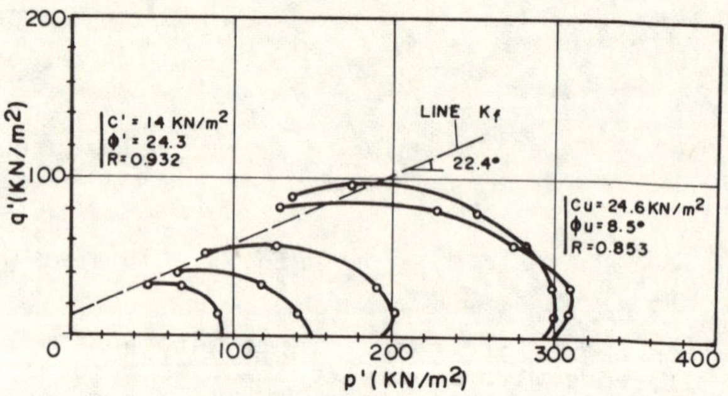

c) B = 90%

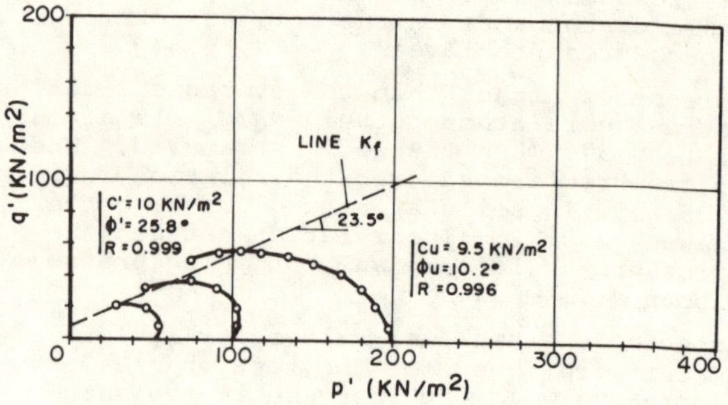

d) B = 100%

Fig. 7: Effective stress path in p' x q diagram

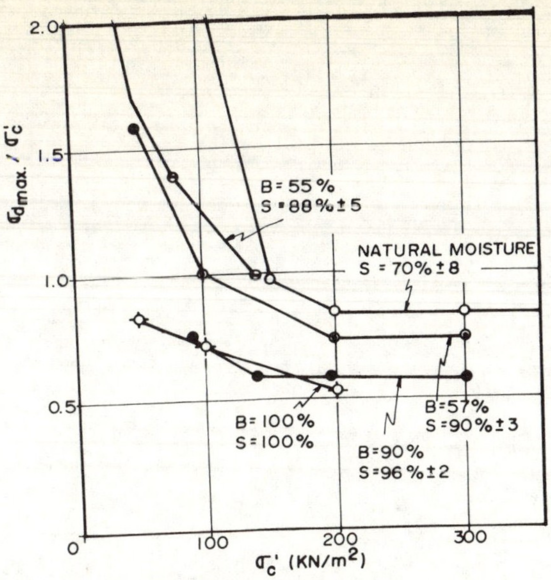

a) Maximum deviator stress/confining stress x confining stress

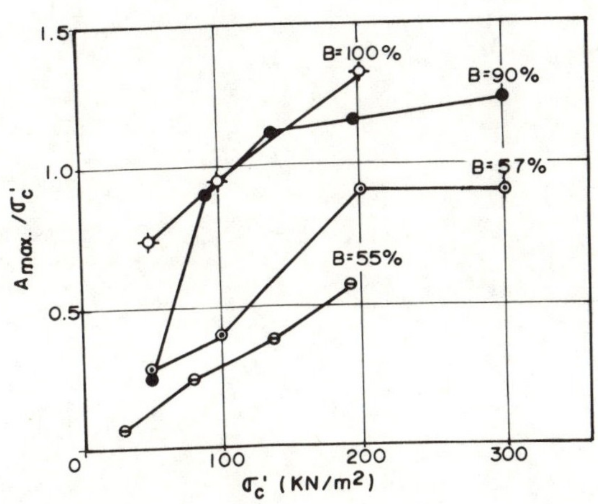

b) Maximum A parameter/confining stress x confining stress

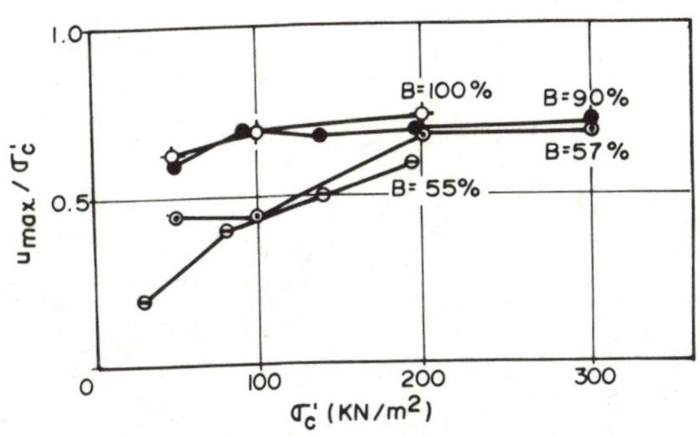

c) Maximum water pressure/confining stress x confining stress

Fig. 8: Parameters variation with confining stress

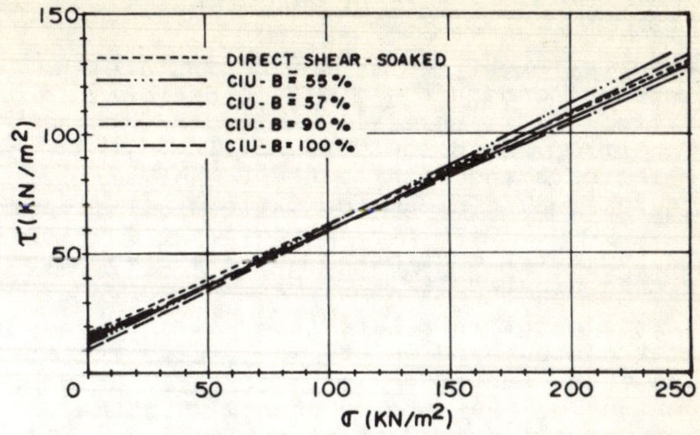

a) Effective failure envelopes

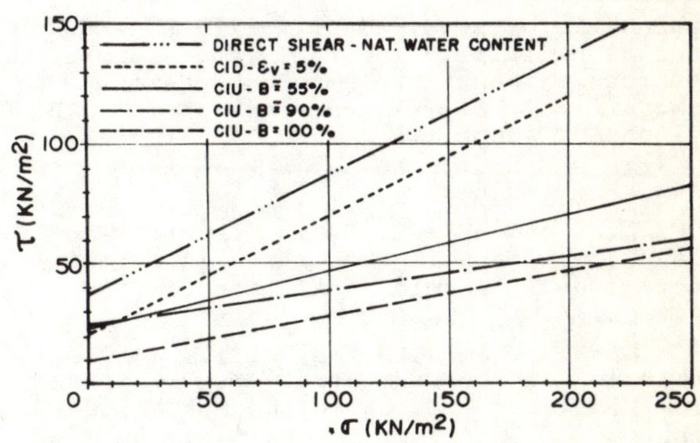

b) Total failure envelopes

Fig. 9: Failure envelopes

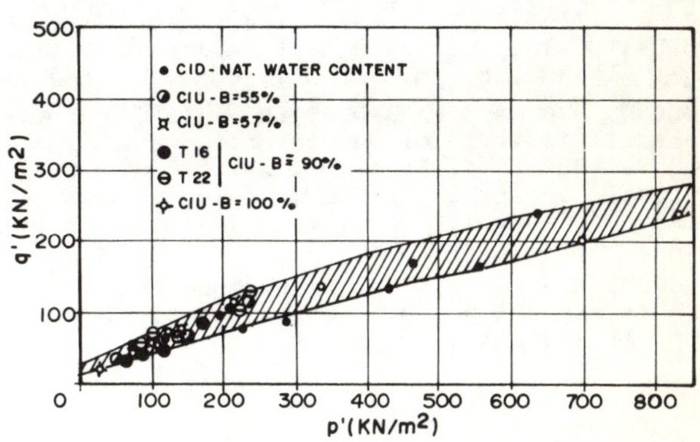

Fig. 10: All results of failure in a p' x q diagram

CONCLUSION

The shear strenght tests should simulate the real present and future field conditions. A lateritic soil is partially saturated. Near the C horizon, with low permeability, it is possible to reach the water content that saturate the macropores. In dam foundations, it is possible to obtain complete saturation. In other cases, the water content in the field is often lower than plastic limit.

The stress paths in a lateritic soils depends on water content and on stress level. The effective stress envelope is practically unique, independent of the water content, when the suction stress is reduced almost to zero. The total stress envelopes are lower as the saturation degree increases.

ACKNOWLEDGEMENT

The author is grateful to Dr. Willy Lacerda and Dr. Jarbas Milititsky for the suggestions and discussions. Thanks are also due Helena Tessele Nodari for helpful and FINEP and CNPq by the financial support.

REFERENCES

BISHOP, A.W. and HENKEL, D.J. (1962): The Measurement of Soil Properties in Triaxial Test. Edward Arnold Ltda., Londres, 2ª edition.

DAVISON DIAS, R. and GEHLING, W.Y.Y. (1985): Influence of the Macrostructure Originated from Pedologic Process on the Mechanical Behavior of Lateritic Porous Soils in South Brazil. First International Conference on Geomechanics in Lateritic and Saprolitic Soils, Vol. 1, pp 296-308, Brazil.

DAVISON DIAS, R. (1987): Pedology and Geotechnology Applied to Design of Foundations for Transmission Towers. Thesis of Doctor of Science, 361 p., COPPE/UFRJ, Brazil.

FREDLUND, D.G., MORGENSTERN, N.R. and WIDGER, R.A. (1978): The Shear Strength of Unsaturated Soils. Canadian Geotechnical Journal, Vol. 15, No. 3, pp. 313-321, Canada.

FREDLUND, D.G. and RAHARDDJO, H. (1985): Theoretical Context for Understanding Unsaturated Residual Soil Behavior. First International Conference on Geomechanics in Lateritic and Saprolitic Soils, Vol. 1, pp. 296-308, Brazil.

SKEMPTON, A.W. (1954): The Pore-Pressure Coefficient A and B. Geotechnique, V. 4, pp. 143-147.

Geotechnical properties of laterites of Goa

A.M.DESHMUKH, Department of Civil Engineering, College of Engineering, India

ABSTRACT Laterites are encountered extensively in India as indeed throughout the tropical countries of the World. The high degree of variability in these deposits poses unexpected geotechnical problems and as such an elaborate evaluation of their nature and engineering behaviour is imperative. This paper presents data which briefly delineates nature and describes physical and geotechnical properties of laterite samples collected from the territory of Goa in India. The nature was isolated by chemical analysis. The study of physical properties included determination of specific gravity, texture, water absorption and consistency limits. The geotechnical characteristics were assessed through compaction, CBR, impact and crushing strength tests. Chemical composition is found to have significant influence on these characteristics. The influence of weathering on nature and behaviour of laterites is also highlighted.

INTRODUCTION

Laterites are extensively encountered throughout the tropical zones of the World like Australasia, Africa and South America. Their study is of particular significance to countries like India which are situated in the tropics. In India laterites occur widely in the Western coastal region, North eastern region and along the Eastern coastal zone,see Fig.1. Laterites are considered as diverse and problematical geotechnical materials as these exhibits unpredictable behaviour when used either as a construction material or as a foundation supporting material. Due to inadequate data on their geotechnical characteristics and in-situ performance, many tropical countries are unable to utilize successfully this local material. As such, engineers will have to develop an understanding of this highly variable material. An indepth study of the nature and response of laterites to various stress systems is thus imperative. With this aim in view a systematic study of laterites of Goa in India was carried out and this paper presents typical data on their nature, physical properties and geotechnical characteristics.

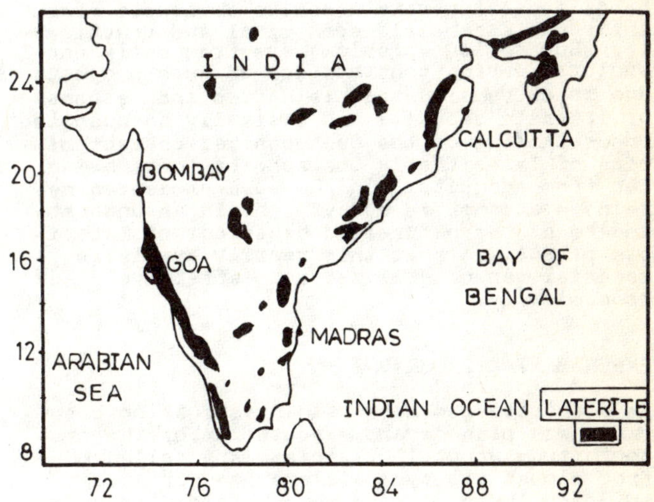

Fig.1. Distribution of Laterites in India

TERRITORY STUDIED

The territory of Goa, lies on the West coast of India, see Fig.1. It lies between the North latitudes of 14° 53' 57" and 15° 47' 59" and East longitudes of 73° 40' 54" and 74° 20' 11". It is bound by the Arabian sea on the West and by hill ranges on the North, East and South. The area of this territory is 3072 sq. km. It measures 105 km. from the North to the South and 60 km. from the East to the West. The climate is typical one with the mean annual precipitation ranging from 3000 mm to 3500 mm. The mean annual maximum temperatures are 32°C and 21.55°C respectively. The average relative humidity varies from 76 to 86 percent. Except for a narrow strip on the North Eastern border of Goa which is covered by the Deccan trap, the rest of the area is covered with pink Phyllites and varities of Quartzites and Graywackes. The territory has been divided into four physiographical subdivisions as Eastern hill ranges, Central rolling to undulating uplands, flood plains of two major rivers and Coastal planes. The typical climate, geology and topography have played a pro-

TABLE I

Chemical Analysis of Laterites of Goa

Sample No.	Description of sample	Fe_2O_3 %	SiO_2 %	Al_2O_3 %	$\frac{SiO_2}{Al_2O_3}$	$\frac{SiO_2}{Fe_2O_3+Al_2O_3}$
1	Plains	22.88	32.24	29.76	1.08	0.61
2	Plains	25.74	19.30	36.70	0.53	0.30
3	Quarry	28.60	30.60	24.10	1.27	0.58
4	Quarry	42.90	14.60	25.94	0.56	0.21
5	Coast	26.46	18.90	21.50	0.87	0.39
6	Coast	34.83	06.00	24.00	0.25	0.10
7	Hill side	30.03	25.20	26.90	0.84	0.44
8	Hill side	37.77	02.80	36.20	0.08	0.04
9	Mine	36.03	10.20	26.90	0.38	0.16
10	Mine	43.90	05.10	30.40	0.17	0.07

minent role in the formation of huge laterite deposits in Goa. The areas with pink Phyllites, appeared to be almost completely laterized. It is estimated that about 80 percent of total area of the territory is covered with laterite soils. At present in Goa various major projects involving construction of earth dams, highways, airport, bridges and high rise structures are being undertaken. The abundance of local laterite deposits obviously make their use essential, economical and practicable. But serious problems are frequently encountered during construction of such projects due to extremely variable nature and response of laterite deposits and basically because the understanding of the geotechnical characteristics of laterites in Goa appears to be being far from adequate. In the work presented herein an attempt is made to obtain an understanding of the nature and behaviour of laterites of Goa, so that this readily available material can be utilized more effectively and economically.

EXPERIMENTAL PROGRAMME

An elaborate geotechnical investigation programme was planned and executed. For the present study about 150 samples were collected from almost all typical laterite formations in Goa. These samples were obtained from different locations including hilly regions, river beds, stone quarries, road cuttings, plains and coastal areas. The understanding of these samples was achieved through a study of their nature, and evaluation of their physical and geotechnical properties and an assessment of the influence of nature on the geotechnical characteristics. The nature was studied by determining the chemical composition. The study of physical properties included determination of specific gravity, texture, consistency limits and water absorption. The geotechnical properties were assessed through compaction tests, California Bearing Ratio tests, impact strength and crushing strength tests. The experimental programme was adopted to match the nature of the sample and was conducted as per Indian Standard Specifications. About 1200 tests were conducted. The geotechnical investigations were mainly oriented towards assessing suitability of laterites particularly for design and construction of highway/airport pavements. The results of the study are summarized and presented hereafter.

NATURE

The fresh samples collected from laterite stone quarries were observed to be generally red to pinkin colour.Such freshly dug samples were initially observed to be soft but further these samples used to harden after exposure to climatic conditions. The exposed samples collected from hilly regions, plains and coastal zones were observed to be generally reddish to dark brown in colour and appeared hard. The samples obtained in the vicinity of iron ore mines were observed to be dark brown to blackish brown in colour and appeared very hard. The chemical analysis of typical laterite samples is presented in Table I. The chemical analysis reveals that for laterites of Goa the Silica-sesquioxide ratio is less than 1.33 and as such these laterites are true laterites. These laterites generally show a high degree of variation in their chemical composition. The iron oxide generally varies from 11.44 to 56.7 percent, the aluminium oxide from 22.26 to 49.40 percent, silica from 2.80 to 27.95 percent and the ignition loss varied from 9.25 to 15.0 percent. In Table I results of chemical analysis are presented in a particular sequence for the sake of isolating influence of weathering. For example sample numbers 1 and 2 represent the same site but sample number 1 shows less degree of weathering while sample number 2 exhibits high degree of weathering. It is as such observed from these results that as the weathering progresses (that is as the hardening of the material occurs) the Silica-sesquioxide ratio decreases, the percentage of iron oxide and aluminium oxide increases while the percentage of silica decreases. The nature of laterite soils thus appears to be considerably influenced by its degree of weathering.

PHYSICAL PROPERTIES

A very wide range of specific gravity from 2.55 to 3.57 was obtained. In Fig.2. variation in specific gravity with percentage of iron oxide is presented. The specific gravity

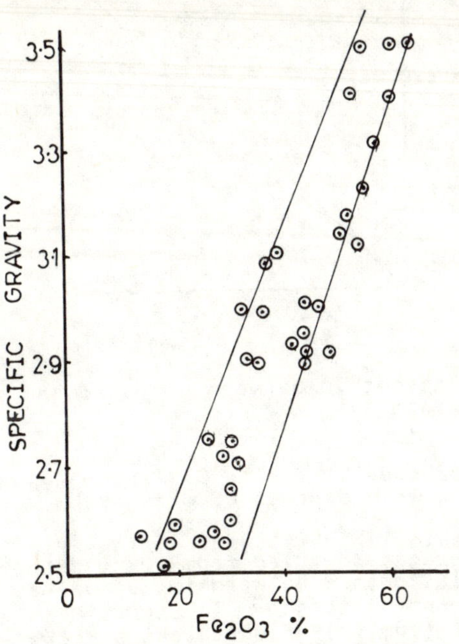

Fig.2. Variation of Specific Gravity with percentage of Iron Oxide.

was observed to increase with the increase in percentage of Iron Oxide and with the degree of weathering.

Typical particle size distribution curves are presented in Fig.3. The particle size analysis indicated that most of the soils are gravels and gravelly sands with little silt fraction. The coefficient of uniformity generally varied from 1.73 to 1000 while the coefficient of curvature varied from 1.0 to 750. The gradation of particle size of these soils varied from well graded to very non uniform. Texturally these soils are very variable and mostly contain all fractions of particles. In laterite soils difficulties were experienced in achieving consistent results. This was probably due to the vulnarability to degradation of some relatively weak materials which gave rise to erratic results. The fresh samples were generally observed to be initially soft which further hardened with time. Hence grain size distribution for same sample in fresh state differed considerably with that in weathered state.

Consistency limits of only a few laterite soils could be determined. The results of such tests are plotted on plasticity index chart

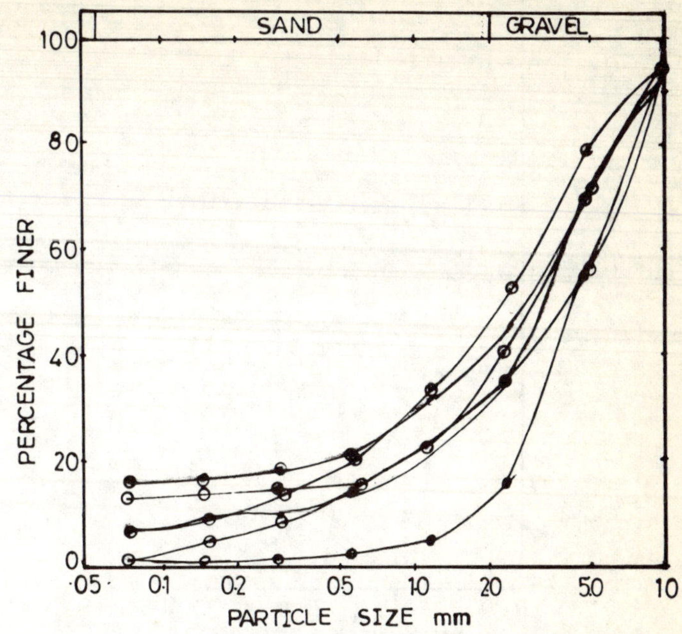

Fig.3. Particle Size Distribution Curves

in Fig.4, which shows that such fine grained soils are of inorganic nature with medium compressibility. The percentage of clays

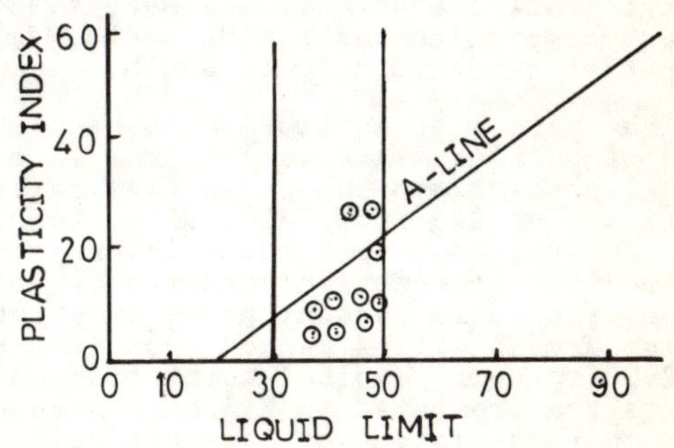

Fig.4. Plasticity Index Chart

was almost negligible in these soils and as such the range of plasticity was observed to be low.

The water absorption varied in a moderate range from 7 percent to 22 percent. The relationship between water absorption and specific gravity is presented in Fig.5. It is observed from this figure that samples with low specific gravity values absorb more moisture. It thus appears that weathered samples absorb less moisture.

The Physical Properties of laterite soils are as such appear to be influenced by the chemical composition i.e. percentage of Iron Oxide

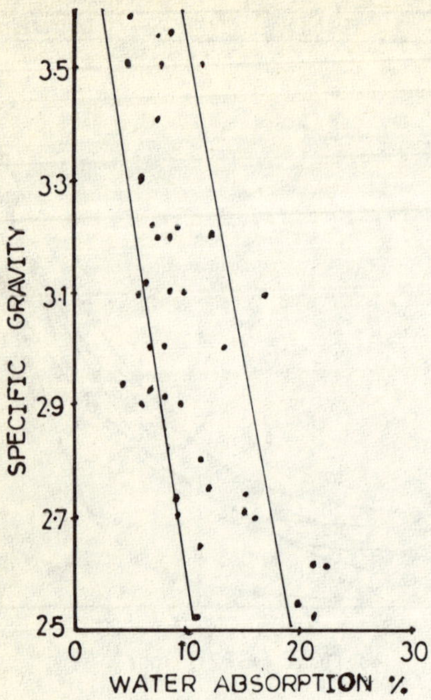

Fig.5. Variation of Specific Gravity with water absorption percentage.

and obviously by the degree of weathering.

GEOTECHNICAL CHARACTERISTICS

The geotechnical characteristics were evaluated through compaction tests, CBR tests, impact strength tests and crushing strength.

The optimum values of moisture contents from standard compaction tests varied from 12 percent to 23 percent and the respective maximum dry densities varied from 1750 kg/cum to 2200 kg/cum. The values of optimum moisture contents from modified compaction tests ranged between 8 percent to 15 percent and the respective maximum dry densities ranged from 1820 kg/cum to 2880 kg/cum. Typical compaction characteristics are presented in Fig.6. These compection charecteristics of laterite soil resemble remarkably with compaction charecteristics of conventional coarse grained soil.

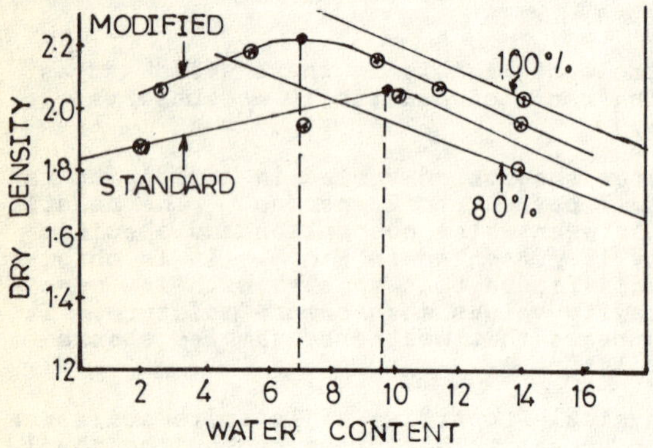

Fig.6. Typical Compaction Characteristics

However, degree of weathering of soil is likely to alter the behaviour under compaction. In Fig.7. a relationship between unsoaked and soaked CBR values of laterite soils is presented. All points are observed to fall below

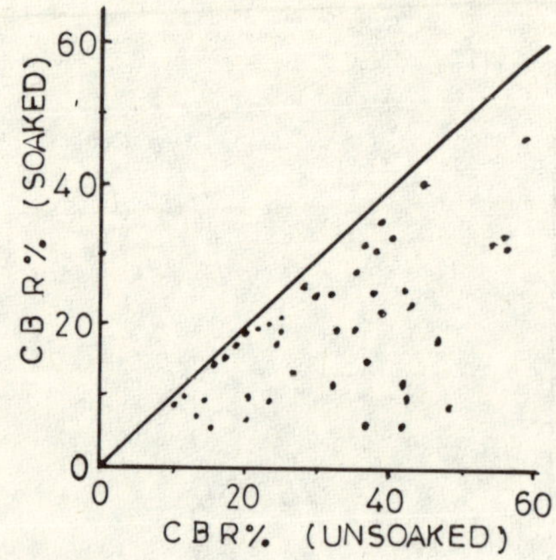

Fig.7. Relationship between unsoaked and soaked CBR values

the line of equality. The CBR was observed to increase with degrees of weathering.

The aggregate impact strength values for laterite aggregates varied from 22 to 56 percent. A relationship between impact values and water absorption is shown in Figure 8. It is evident from this figure that samples with low impact values absorb less moisture. In mine waste samples with high Iron Oxide percentage the impact strength values were observed to be comparatively lower than highly weathered samples of laterites which were collected from other sites. This may be due to the fact that the samples with high Iron Ore percentage usually get crushed easily under the force of impact and as such yield high impact value.

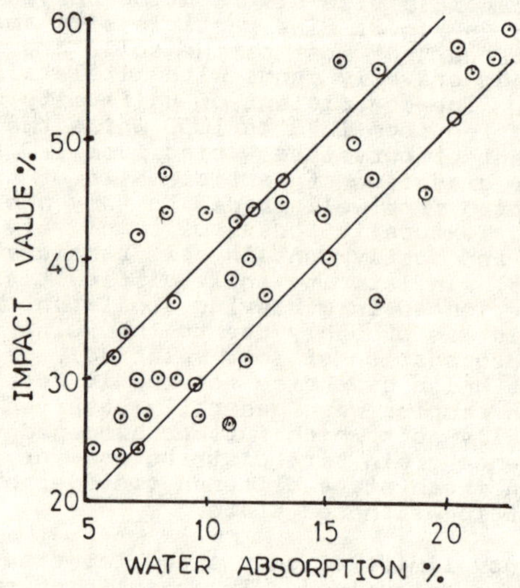

Fig.8. Impact Strength V/s Water Absorption

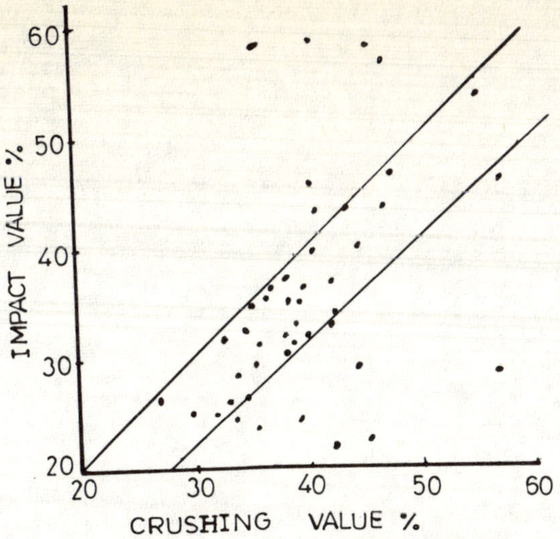

Fig.9. Relationship between Impact and Crushing Values.

A relationship between impact and crushing strength values is shown in Fig.9. The impact strength values are usually observed to be higher than the crushing strength value. However, in few fresh laterite samples crushing strength values were observed to be slightly higher than the impact strength values. The impact and crushing strengths are observed to be function of degree of weathering. These characteristics in general observed to decrease with degree of weathering. In case of laterite soils impact and crushing strength values are however observed to be higher as compared to conventional material used for pavement construction.

CONCLUSIONS

A study on the nature and engineering behaviour of laterites is presented. The main conclusions of study are :

1- Laterites of Goa are chemically true laterites. The chemical analysis reveales that as the weathering of laterites progresses, the sesquioxide ratio reduces, percentage of Iron and Aluminium Oxides increases and percentage of Silica reduces.

2- Specific Gravity varies widely. Higher values are generally associated with increasing percentage of Iron Oxide and degree of weathering. The water absorption varies in a moderate range. It generally reduces as the weathering progresses.

3- Texturally laterite soils are mainly coarse grained. The fine fractions in a few laterite samples exhibit low plasticity. The coefficients of uniformity and curvature show that these soils are mostly non uniform materials which vary between well graded to poorly graded range. The particle size distribution is influenced by weathering process.

4- These soils are improved by compaction in the same way as conventional soils. This material exhibits fairly high CBR values. Influence of soaking on CBR is observed to be less pronounced. The impact and crushing strength values are observed to be on higher side. The laterite aggregates are as such observed to be weaker than conventional aggregates generally used for construction of pavements.

5- Chemical composition and degree of weathering do influence the properties of laterites.

The results presented in this paper may help in the preliminary estimates of design and construction. Such data may enable engineers to employ laterites rationally and predict with confidence their field performance with a fairly good degree of accuracy.

REFERENCES

Deshmukh, A.M. (1980). A study of geotechnical properties of the laterites of Goa, Research report submitted to State RD&D Committee, Government of Goa, unpublished.

Deshmukh, A.M., Khanolkar, B.V. and Dharwad, R.C.(1987). Plastic cutoff wall for seepage control at Salauli Dam, VIII Asian Regional Conference on Soil Mech. Found. Engg. 449-452, Japan.

Gidigasu, M.D. (1976). Laterite Soil Engineering. Pedogenesis and Engineering principles. Elsevier Scientific Publishing Co. N.Y.

Oertel, G. (1957). Notes ebibliografia Ocerca de geologia de norte de Goa, Lisbon.

Preliminary consolidation and triaxial strength test results for some undisturbed tropical red clay soils from West Java, Indonesia

P.R.N.HOBBS, British Geological Survey, Keyworth, Nottingham, UK
M.G.CULSHAW, British Geological Survey, Keyworth, Nottingham, UK
K.J.NORTHMORE, British Geological Survey, Keyworth, Nottingham, UK
A.RACHLAN, Institute of Road Engineering, Bandung, Indonesia
D.C.ENTWISLE, British Geological Survey, Keyworth, Nottingham, UK

ABSTRACT: Many tropical red clay soils, or 'laterite' soils, are formed by intense weathering of volcanic and other igneous rocks in warm, humid conditions as part of the process of lateritisation. They are rich in iron sesquioxides and kaolinite clay minerals, including halloysite and/or allophane, and are characterised by high in situ void ratios maintained by an open aggregated ped-like structure. A laboratory study is underway to investigate the relationship between the engineering properties of tropical red clays from Kenya and West Java and their geological provenance and geographical distribution. Stress-path triaxial tests and oedometer consolidation tests have been carried out on four samples from West Java to investigate in situ stress-strain and settlement characteristics. Preliminary results have identified two distinct 'yield' points at approximately 100-200 KPa and >500 KPa which can be interpreted as failure or de-structuring of inter-ped and intra-ped bonding, respectively. Standard geotechnical index tests indicate the soils to be plastic clays. However, the pseudo-overconsolidated nature of these materials and measured permeability values, representative of fine-grained granular soils, suggest that their in situ engineering behaviour is more akin to weakly cemented silts.

1 INTRODUCTION

Tropical red clay soils (or 'laterite' soils as defined by Gidigasu [1976] and Malomo [1982]) are extensively developed throughout the tropics. Locally, they have been given a variety of names; for example, in East Africa they are termed Red Coffee Soils. They are usually considered to be residual soils, although Gidigasu (1976) included transported soils within his definition.

The work described in this paper is concerned with residual clay soils from West Java, Indonesia and forms part of a broader investigation that is examining the engineering geological and geotechnical characteristics of some of these tropical soils in relation to their geological provenance and the environmental controls on their formation, principally climate. The overall project is being carried out by the Engineering Geology and Reservoir Properties Research Group of the British Geological Survey (Nottingham) in collaboration with the Indonesian Institute of Road Engineering (Bandung) and the Kenyan Materials Branch of the Ministry of Transport and Communications (Nairobi).

Terzaghi (1958), Newill (1961) and Wesley (1973) all appreciated that the engineering properties of many of these soils were largely influenced by their unusual (compared with soils developed in temperate climates) clay mineralogy which markedly affects the results obtained from "standard" geotechnical classification tests. For example, moisture content determinations are affected by the alteration of halloysite, found in many tropical red clay soils, to metahalloysite when the soil is dried at the "standard" temperature of 105°C. This is caused by the irreversible loss of 'bound' water from the crystal structure. The transition of halloysite to metahalloysite is also controlled by climate, notably humidity, and is observed to take place naturally where halloysite-rich soils at higher and wetter elevations of, for example, the highland areas of the Aberdare Range in Central Kenya give way to metahalloysite-rich soils at lower, and drier, elevations near Nairobi (Newill 1961). The nature of this change is still not clear, though detailed mineralogical examination of a range of samples from different locations in Kenya and Indonesia is underway using X-ray diffraction, differential thermal analysis, scanning electron microscopy, transmission electron microscopy and other techniques. These results will be reported in the near future.

Little work has been reported in the literature on the shear strength and consolidation characteristics of these soils, partly because of the difficulty of obtaining high quality, undisturbed samples. Wesley (1977) carried out shear strength, consolidation and other tests on samples from West Java and observed that "undisturbed" samples for triaxial and consolidation testing could only be prepared by using push-in cylindrical tubes of the appropriate diameter and then extruding the samples from the tubes. This was because he found the soils very difficult to trim due to their "stickiness". Contrary to this experience, the authors found the trimming of samples in pits and in the laboratory to be relatively easy, if time consuming. The method of sampling in the field and the preparation of sub-samples for laboratory testing are described below. The only way to explain this difference, assuming that the soils described in this paper are similar to those described by Wesley (which is likely, as some of the sites are fairly close to those cited in his paper), is that whilst the samples described here were collected towards the end of the dry season (October/November 1987), Wesley's samples <u>may</u> have been obtained during or immediately following the wet season. This might account for their apparent stickiness.

The importance of sample disturbance is highlighted by Klem (pers comm) in relation to oedometer tests carried out to determine the collapse potential of Kenyan soil samples. He states that even small disturbance of the oedometer sample can affect the measured collapse potential significantly. However, whilst Klem took his field samples in a similar manner to the authors, he recovered the samples in the laboratory by hydraulic extrusion from the sample tubes which, in many cases resulted in disturbance of the in situ soil structure. To avoid this, the samples described here were removed from plastic sample tubes by carefully cutting away the tube with an electric saw. Whole samples (of 100mm diameter) were used for triaxial testing and sub-samples for oedometer testing prepared by trimming from the soil core. Sample extrusion was avoided at all stages of sample preparation.

Newill (1961) suggested that the clay particles in these soils form aggregations or peds, possibly with an iron oxide coating. Triaxial testing was carried out to determine strength parameters and to try to obtain a better understanding of how this aggregated ped-like structure might influence the strength characteristics.

The oedometer testing was primarily carried out to investigate the reported tendency of some of these soils to collapse on wetting under load. Collapse is here defined as the rapid reduction in voids ratio of a partially saturated sample when wetted under a load greater than the highest effective overburden pressure existing in situ.

Preliminary triaxial strength and consolidation data are available for four undisturbed pit samples and are presented here.

2 SAMPLE DESCRIPTION

The sample sites were all located in West Java (one near Bandung, one near Lembang, one close to Cikampek and the fourth between Padalarang and Purwakarta) and at elevations ranging from 80 to 1280 m (Table 1). Climatically, the sites ranged from warm temperate-humid (Pit 2) through warm-humid (Pit 1) to hot-humid (Pits 5 and 6) (Table 2). The underlying geological formations comprise basic and intermediate volcanic deposits (probably tuff) of Pleistocene age, though it is probable that the tuffs at pit sites 1 and 2 are younger than those at sites 5 and 6 which are much further away from the volcanic centres. The Soil Map of West Java (Anon 1966) indicates that, pedologically, the soils at nearly all of the sample sites are latosols, with the exception of site 2 where the soil is an andosol. Wesley (1973) discussed the transition from latosol to andosol, pointing out that the soil map arbitarily shows the boundary to be at an elevation of about 1000 m. Wesley (1973) also shows that the andosols are rich in allophane while the latosols consist predominantly of halloysite. Preliminary mineralogical examination of the four samples using X-ray diffraction and thermogravimetric analysis indicates that the samples consist mainly of disordered kaolin with significant amounts of goethite. At this stage it is not possible to comment on whether the disordered kaolin is allophane, halloysite or a mixture of the two. The results of the mineralogical analysis are presented in Table 3.

The sample from Pit 1 was a slightly friable, homogeneous, slightly sticky reddish brown clay (5YR 4/4 on the Munsell Soil Color Chart [Anon 1975]). At Pit 2 the soil consists of a friable brown to dark brown (10YR 4/3) clay with a few sand-cobble sized fragments of partially weathered crystalline tuff. The less completely weathered nature of this soil is probably indicative of its younger age. The Pit 5 soil consists of a friable, homogeneous, moist, dark reddish brown (2.5YR 3/4) clay and at Pit 6, a friable, homogeneous brown to dark brown (7.5YR 4/4) clay containing occassional angular crystalline fragments of sand size.

3 SAMPLING METHOD

The samples were taken at four locations from hand-dug pits, at a depth of 2.0 m. A 'pedestal' of soil was prepared at one end of the pit and protected from drying out by covering with plastic sheet during excavation. An undisturbed sample was then obtained from the pedestal by trimming around the cutting edge of a greased, sharpened 100 mm diameter x 200 mm plastic tube as it was gently eased into the soil. Penetration progressed in increments of 10 to 20 mm using the minimum of force. On filling the tube completely, the sample was cut away from below and the soil trimmed square with the tube ends at top and bottom. Plastic caps were placed on the tube ends and secured with waterproof tape. The complete tube was then sealed with paraffin wax, covered with a plastic bag, and placed in a foam-lined metal box for transportation. Duplicate tube samples were taken in each pit; one for preparing oedometer specimens and the other for triaxial testing. Samples selected for return by air-freight to the BGS UK laboratories were, on arrival, immediately placed in a temperature and humidity controlled store until the moment of testing.

4 TEST METHODS

4.1 Consolidation Testing

Two types of 1-dimensional oedometer consolidation test apparatus were used; firstly a conventional front-loading dead-weight lever-arm type and secondly a specially adapted high pressure (up to 24,000 KPa) 15 ton Denison dead-weight lever-arm type. The test samples for each apparatus comprised discs of 75mm diameter x 20mm thickness and 76mm diameter x 19mm thickness, respectively.

Four oedometer discs (specimens A-D) were prepared from each 100mm x 200mm tube with specimen 'A' uppermost. Weights were added or removed at 24 hour intervals and each cell flooded at a different axial stress; specimen 'A' was flooded after application of the first load increment (ranging from 20-40 KPa), specimen 'B' after application of the second increment and so on (Table 4). The

Table 1 Location details for pit sample sites

Pit No.	1:50000 Map No.	1:50000 Map name	Location Easting	Northing	Elevation m a.M.S.L	Location Description
1	4522 III	Bandung	788750	242000	1035	Approximately 8 km N of Bandung; 500 m E of track from Jalan Ciumbeluit to Lembang
2	4522 IV	Segala-herang	790200	244300	1280	Approximately 2.25 km SE of Lembang; 50 m W of track from Jalan Ciumbeluit to Lembang
5	4423 I	Tjikampek	773700	287400	80	Approximately 4 km SSW of Cikampek; 300 m NW of junction of Purwakarta to Cikampek road and new Jakarta to Semarang toll road
6	4422 I	Tjikalong -Wetan	768950	260050	590	Approximately 22 km S of Purwakarta and 27 km N of Padalarang. On W edge of Sawit village; 100 m W of Purwakarta to Padalarang road

Table 2 Terrain/climatic details for pit sample sites

Pit No.	Type of Topography	Slope Angle	Underlying Geology	Soil Type	Mean Annual Rainfall, mm	Mean Temperature Range, oC
1	Volcanic foot ridges -site on ridge top	8° down ridge line	Undiff old volcanic (Quaternary) deposits. Probably intermediate tuffs	Brown Latosol	1500-2000	16-30
2	Volcanic foot ridges -site on ridge top	12° down ridge line	Undiff old volcanic (Quaternary) deposits. Probably intermediate tuffs	Brown Andosol	c. 2000	11-26
5	Undulating low hills and valleys on lower-most volcano slopes	3°	Undiff Pleistocene volcanic deposits. Probably intermediate tuffs	Red Latosol	2000-2500	22-34*
6	Deeply incised volcanic foot ridges. Site on upper valley side	10° above 16° below into valley	Undiff Pleistocene volcanic deposits. Probably intermediate tuffs	Brown Latosol	3000-3500	21-29*

* estimated

Table 3 Summary mineralogical data obtained by X-ray diffraction and thermogravimetric analysis methods

Pit No.	Disordered Kaolin	Gibbsite	Goethite	Quartz	Other
1	68	nd	31	tr	Tridymite tr
2	65	3	10	**	Magnetite ** Cristobalite ** Tridymite * Plagioclase tr
5	72	~1	25	*	nd
6	68	8	17	tr	Tridymite * Cristobalite *

KEY
** major
* present
tr trace
nd not detected

Figures represent percentage concentration

Table 4 Stress at which specimen was flooded in the consolidation test

Pit No.	Depth, m	Specimen flooding stress, KPa					
		A	B	C	D	E	G
1	2.0	40	80	120	N/A	N/A	750
2	2.0	38	100	200	400	N/A	N/A
5	2.0	80	120	200	N/A	500	N/A
6	2.0	40	120	200	N/A	N/A	N/A

high-pressure tests were carried out in a similar manner but with minimum initial loadings of 125-250 KPa. Both types of test were monitored using Mitutoyo digital dial gauges and Hewlett-Packard HP85 computers running a GDS consolidation software package ('GDSDLS' version 1.1).

4.2 Triaxial Testing

Multi-stage, isotropically-consolidated, undrained (CIU) triaxial tests were carried out on single, 100mm diameter x 200mm, cylindrical specimens (sampled immediately adjacent to the oedometer test samples in each pit). The specimens were neither extruded nor trimmed, and side drains were not used. The apparatus used was a GDS stress-path system using a 100 mm Bishop-Wesley hydraulic cell with three GDS digital hydraulic actuators supplying cell pressure, back pressure and deviator stress and a GDS digital pressure interface (DPI) monitoring pore-pressure. Control and logging of the actuators and interface were effected by a Hewlett-Packard HP85 computer running a 'GDSTTS' (version 5.0) software package. All volume changes and displacements were monitored by the hydraulic actuators. Volume changes and displacements were measured to resolutions of 1 mm^3 and 0.5 KPa, respectively. The specimens were isotropically consolidated at effective confining stresses between 20 and 400 KPa in a fully saturated condition (B > 0.9) which required the use of back pressures ranging from 200 to 400 KPa. Permeability, by constant axial flow rate (Olsen et al 1985), and pore-pressure parameter 'B' determinations were made. Undrained compression was achieved using stress-rate control. The consolidation stresses used and stress rate are given in Table 5.

5 TEST RESULTS - OEDOMETER CONSOLIDATION

Voids ratio vs consolidation stress (e-logp) curves are given in Figures 1, 2, 3 and 4. Derived data and a summary of index test data are given in Table 6. Despite expectations to the contrary, 'collapse' was only observed in two cases, Pit 1-2.0m-test G and Pit 5-2.0m-test D, at axial stresses of 750 KPa and 200 KPa, respectively. The e-logp curves are almost identical for the test specimens A-D prepared from the individual sample tubes, and show

Table 5 Triaxial test conditions data

Pit No.	Depth, m	Consolidation stress, KPa / *Triaxial test stress rate, KPa/min				
		Test 1A	Test 1B	Test.2	Test 3	Test 4
1	2.0	25 / *10	25 / *15	100 / *20	200 / *30	N/A
2	2.0	25 / *10	N/A	100 / *50	300 / *50	N/A
5	2.0	35 / *80	N/A	100 / *30	193 / *50	400 / *100
6	2.0	20 / *50	N/A	100 / *60	200 / *80	400 / *100

pseudo-overconsolidated behaviour and large changes in voids ratio.

Initial voids ratios (e) lie in the range 1.6 to 2.2 and initial degrees of saturation [$S_{r(o)}$] in the range 60% to 89%. Compression indices (Cc = 0.33 to 0.75) are comparable with those from temperate clays of medium plasticity whilst swelling indices (Ce) are low at between 0.01 and 0.07 with the exception of high pressure specimen, Pit 1-2.0m-test G, which showed a Ce of 0.23. Values of the coefficient of volume compressibility (Mv) lie in the range 0.03 to 1.0 m^2/MN and for the high pressure tests between 0.01 and 0.43 m^2/MN. The coefficient of consolidation values (Cv) are extremely high (values greater than 5000 m^2/yr are not uncommon). This is due to rapid consolidation settlements which, in some cases, are so rapid that 80% of consolidation is completed within 10 seconds of loading. The secondary consolidation is generally small. The consolidation behaviour of these samples at normal stresses is more typical of a granular material.

A form of pseudo-overconsolidation (Perloff & Baron 1976) is clearly demonstrated by the e-logp curves of tests on sample Pit 2-2.0m where a sharp change (increase) of slope is seen at an axial stress of 200 KPa in each case. (The response of test A at 200 KPa is less marked but the sample has a low initial voids ratio.) Tests on samples Pit 6-2.0m and Pit 1-2.0m also exhibit distinct, but also less marked, slope changes, at between 40 and 100 KPa and between 200 and 600 KPa, respectively. Tests on sample Pit 5-2.0m show an interesting departure from the behaviour of

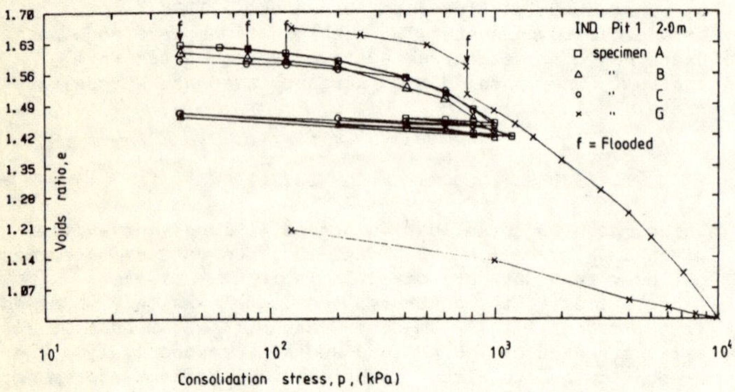

Figure 1 e vs logp curves for sample Pit 1-2.0 m

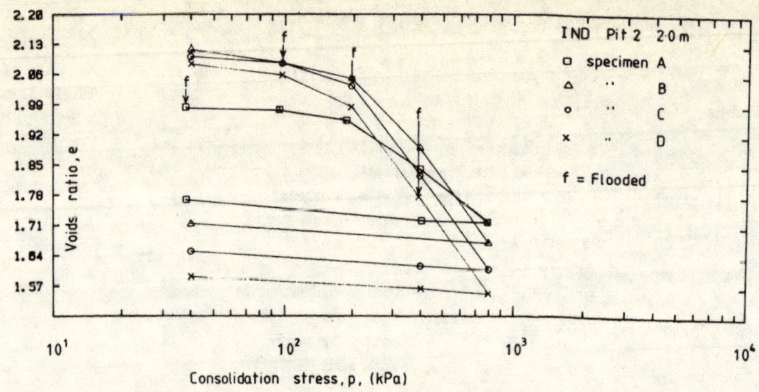

Figure 2 e vs logp curves for sample Pit 2-2.0 m

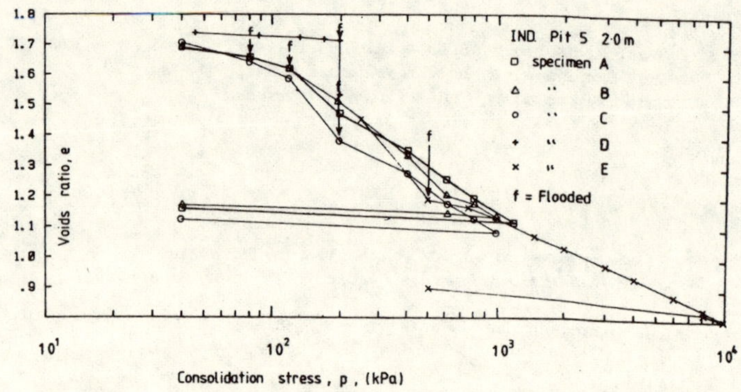

Figure 3 e vs logp curves for sample Pit 5-2.0 m

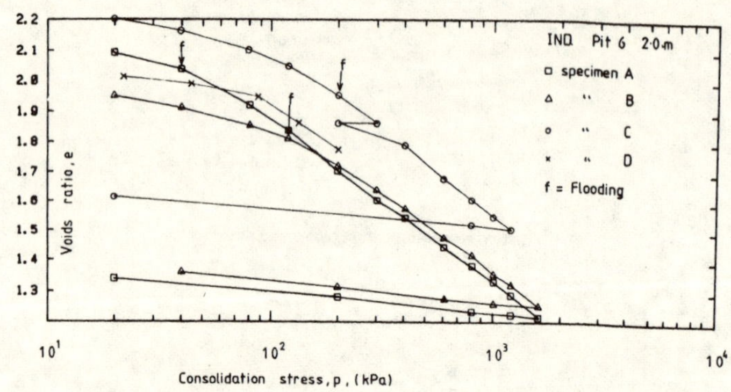

Figure 4 e vs logp curves for sample Pit 6-2.0 m

the other pit samples. In these tests a concave kink is seen in each case at axial stresses ranging from 200 KPa (test C) to 630 KPa (test B). This effect represents a form of microstructural collapse, being rapid and stress dependant, but apparently unrelated to flooding. The high pressure test (E) also shows this kink (at 500 KPa). In this case collapse has occurred prior to flooding.

In the case of high pressure test, Pit 1-2.0m-test G, a collapse of approximately 10% of total settlement has occurred on flooding at 750 KPa bringing the e-logp curve to match closely the duplicate test curves (A, B and C). This behaviour may be considered indicative of an initial state of sample disturbance or de-structuring (Vaughan et al 1988) for samples used in tests A, B and C, and arrival at this state, following collapse, in test G; hence the coincidence of all the curves. However, it is also possible that the sample used in test G had a higher in situ voids ratio than the others and its lower degree of saturation has resulted in the collapse on flooding. This possibility leads to consideration of the relationship between collapse potential and degree of saturation which, though not discussed here, forms part of the project's research programme. Nevertheless, despite rigorous efforts in the field to minimize sample disturbance, it may be possible that during penetration of the plastic sample tubes, in Pit 1, a localized collapse or 'de-structuring' has taken place at natural moisture content in the upper and central portions of the tube sample (specimen G is taken from the base) purely as a result of friction between the greased tube and soil sample. Relief of effective stress due to the inability of the capillaries to sustain it may also be a factor (Vaughan et al 1988) either prior to sampling as a result of pit excavation, or following removal of the protective tube in the laboratory.

The pseudo-overconsolidation behaviour referred to earlier and exhibited by all of the test specimens is probably due to a breaking of inter-ped bonding (as described by Vaughan et al 1988). In most cases this 'yield' point occurs once only. However, as seen earlier, tests on Pit 5-2.0m specimens each exhibit what may be interpreted as two yield points. Hence, two completely separate phases of bond yield are seen, each within a different axial stress range and apparently independent of the flooding process. These two components of bonding may be visualised as: i) bonding between peds (inter-ped bonding) which breaks down at a relatively low axial stress and ii) bonding within the ped itself (intra-ped bonding). In either case, yielding could involve a process whereby iron oxide-cemented inter-ped 'bridges' are broken and/or a process involving the re-arrangement or crushing of clay peds. Given the high void ratios of these soils it is possible that a localised form of partial-brittle, micro-shear failure is taking place despite confinement within the oedometer ring.

The unloading portion of the e-logp curves reveals small levels of rebound throughout the tests (low values of Ce). This pseudo-plastic behaviour is indicative of collapse of the soil structure or at least some form of irrecoverable compression. It is not until high axial stresses (greater than 1000 KPa) are reached that rebound on unloading becomes more marked (for example Pit 1 2.0 m test G and Pit 5 2.0 m test E).

6 TEST RESULTS - TRIAXIAL

The results of multi-stage CIU Triaxial tests are given in the form of stress-strain curves (Figures 5, 6, 7 and 8) and effective stress-path (s'-t) plots (Figures 9, 10, 11 and 12). The stress-strain curves show considerable stiffness in the case of samples Pit 2-2.0m and Pit 5-2.0m. Peak strength is achieved at net strains of between 1.5 and 3% (measured over the whole test). Values of tangential Young's Modulus, Ei, vary from 850 KPa to 1500 KPa.

The effective stress-path plots show the transition from pseudo-overconsolidated behaviour (dry of critical) to normally-consolidated behaviour (wet of critical) as effective stress increases with each stage of the test; the transition occurring at an average effective stress of between 50 KPa and 100 KPa. Based on the K_f line, values of Θ (the angle of the k_f line to the horizontal) were measured and values of effective cohesion (c') and angle of internal friction (ϕ') calculated (Table 6) using the relationships:

$$\sin \phi' = \tan \Theta \qquad (1)$$

$$c' = \frac{y}{\cos \phi'} \qquad (2)$$

where y is the y-axis intercept of the K_f line.

Values of c' range from 18 KPa to 46 KPa and ϕ' from 18° to 30°. Such high friction values are more indicative of silts or fine sands rather than of very plastic clays and reflect the bonded ped-like structure of the soil discussed earlier.

Isotropic consolidation of the triaxial specimens prior to undrained compression revealed the pseudo-collapsing nature of the soils similar to that observed in the one-dimensional oedometer consolidation tests. Volume decrease during early consolidation was very rapid, the primary stage having been completed in less than a minute in some cases (as also indicated by the oedometer results).

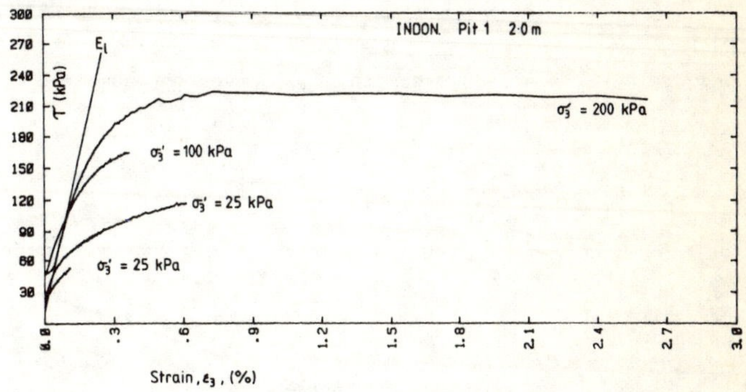

Figure 5 Stress (τ) vs strain (ε_3) curves for sample Pit 1-2.0 m

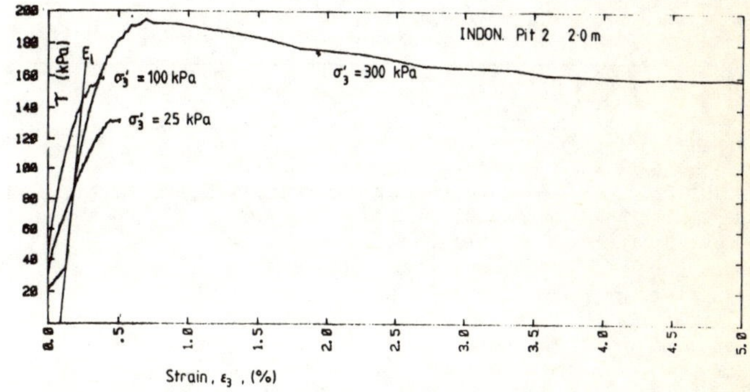

Figure 6 Stress (τ) vs strain (ε_3) curves for sample Pit 2-2.0 m

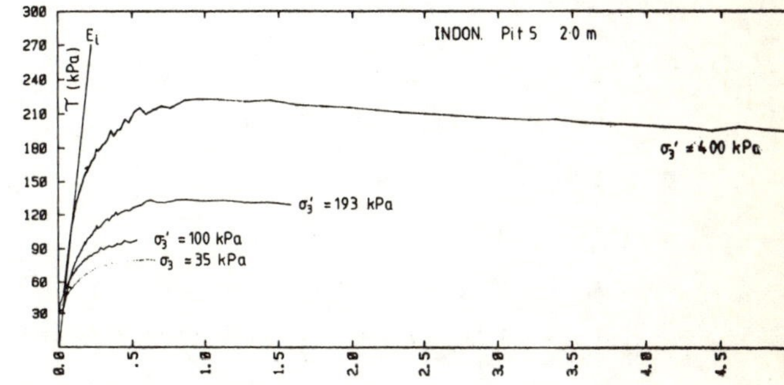

Figure 7 Stress (τ) vs strain (ε_3) curves for sample Pit 5-2.0 m

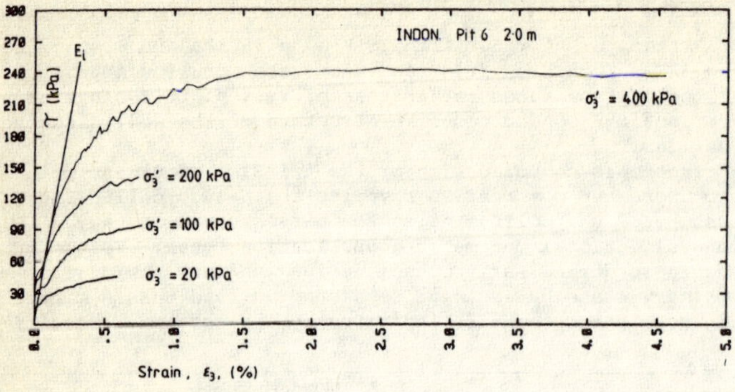

Figure 8 Stress (τ) vs strain (ε_3) curves for sample Pit 6-2.0 m

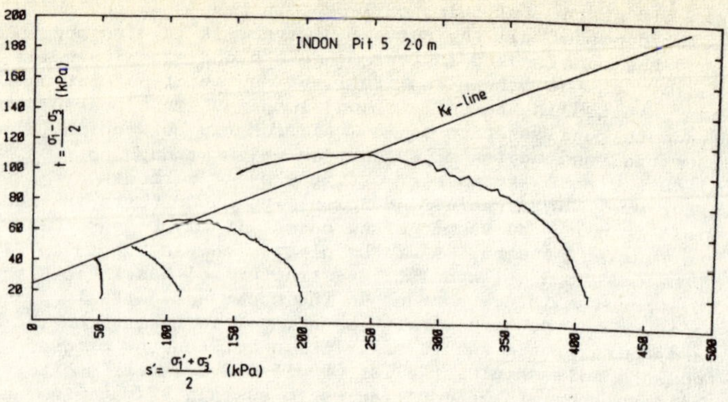

Figure 11 Stress path (s'-t) for multi-stage triaxial tests on sample Pit 5-2.0 m

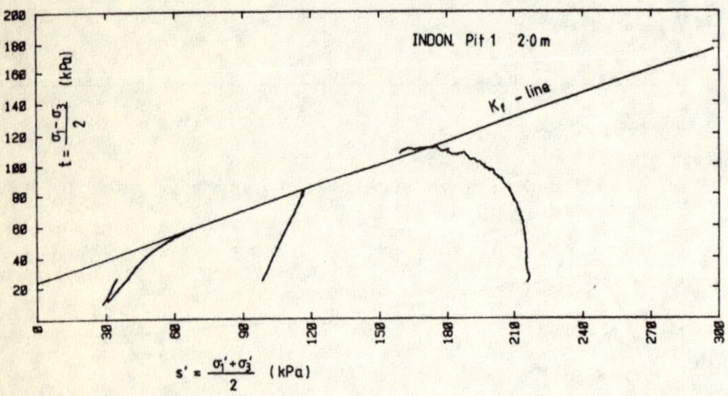

Figure 9 Stress path (s'-t) for multi-stage triaxial tests on sample Pit 1-2.0 m

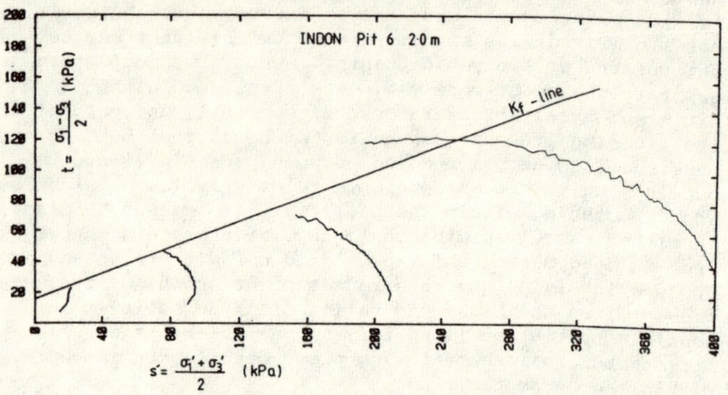

Figure 12 Stress path (s'-t) for multi-stage triaxial tests on sample Pit 6-2.0 m

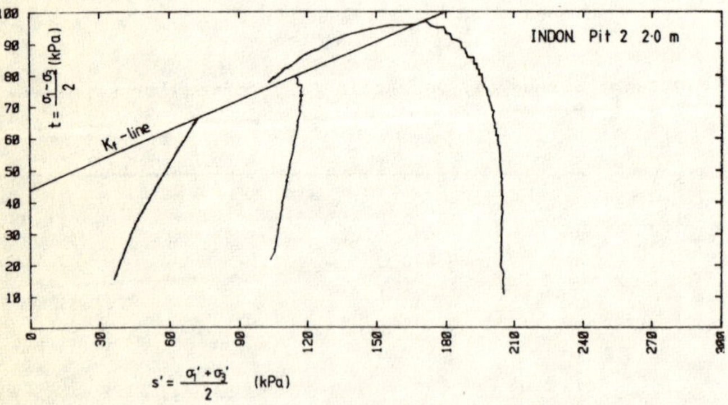

Figure 10 Stress path (s'-t) for multi-stage triaxial tests on sample Pit 2-2.0 m

Constant flow-rate permeability tests, carried out prior to each stage of triaxial compression, provided steady, consistent values of between 5.6 x 10^{-9} and 8.3 x 10^{-8} m/s (obtained using flow rates of between 0.5 and 2.0 mm^3/s). These permeability values are more representative of silts or fissured temperate clays (Head 1982) rather than texturally homogeneous clay soils with measured clay contents in excess of 75% (except for the sample from Pit 2 which had a clay content of 29%) and plasticity indices ranging from 21-45%. In general, permeability decreased with increasing effective confining pressure though an increase was observed in one instance (Pit 1-2.0m) following final stage shear failure (presumably due to the availability of shears or opened fissures providing channels for water flow within the test sample).

The mode of failure following final compression was found to be one of barrelling and multiple internal shear planes having both steep and shallow angles. Surface expression of these shear planes was usually poor even at large strains. Sample Pit 5-2.0m was found to contain a vertical, probably pre-existing, fissure which coalesced with angled shear planes at top and bottom of the specimen. At no point during setting up of the test specimen were negative pore-pressures observed.

Table 6 Results of consolidation, triaxial and index tests

Pit No.	Depth, m	Test type Low pr. +High pr.	mc %	Sr(o) %	eo	Mv (m2/MN)	Cc (Ce)	Test type	c' KPa	φ' degrees	Ei KPa	k m/s	WL (WP) [PI]	Specific Gravity	Clay content %
						Oedometer Tests				**Triaxial Tests**				**Index Tests**	
		1 2m A	52	87	1.64	0.135	0.33(0.03)	X4							
1	2.0	1 2m B	53	87	1.63	to	0.41(0.04)	multi	28.9	30.0	1110	2.2-4.4	94	2.77	86
		1 2m C	52	89	1.60	0.031	0.39(0.02)	stage			(St.4)	X10-8	(49)		
		+1 2m G	47	85	1.67	0.075 to 0.016	0.66(0.23)	CIU					[45]		
		2 2m A	63	85	2.14	0.358	0.39(0.01)	X3							
2	2.0	2 2m B	62	79	2.13	to	0.71(0.04)	multi	46.3	18.1	910	5.6X10-9	88	2.82	29
		2 2m C	59	76	2.21	0.084	0.74(0.03)	stage			(St.3)	to	(62)		
		2 2m D	58	75	2.17		0.72(0.03)	CIU				2.8X10-8	[26]		
		5 2m A	38	61	1.71	0.995	0.50(0.03)	X4							
5	2.0	5 2m B	38	61	1.73	to	0.52(0.02)	multi	25.7	20.9	1500	4.7-7.5	72	2.74	80
		5 2m C	38	60	1.76	0.101	0.54(0.03)	stage			(St.4)	X10-8	(51)		
		+5 2m E	36	60	1.80	0.429 to 0.010	0.36(0.06)	CIU					[21]		
		6 2m A	59	75	2.16	0.977	0.55(0.06)	X4							
6	2.0	6 2m B	57	79	2.00	to	0.53(0.07)	multi	18.1	27.9	857	8.3X10-8	99	2.78	75
		6 2m C	57	71	2.22	0.073	0.60(0.06)	stage			(St.4)		(61)		
		6 2m D	--	--	2.08		0.50(0.06)	CIU					[38]		

7 CONCLUSIONS

Though experimental work on this project is not yet complete, initial results from the undisturbed testing programme, comprising triaxial and oedometer consolidation tests, reinforce confidence in the sampling and laboratory preparation techniques used, and show that conventional test methods such as multi-stage triaxial incorporating sample saturation under back-pressure can be used successfully with partially-saturated sensitive tropical clay soils of the type described here. The permeability values (>10^{-9} m/s) measured directly during the triaxial tests using the GDS system, are more typical of fine-grained granular soils or fissured temperate clays and reflect the high in situ void ratios resulting from the open aggregated ped-like structure of these tropical clay-rich soils. Consolidation tests at natural moisture contents (i.e. unflooded) show that all the samples tested are pseudo-overconsolidated, almost certainly due to inter-ped bonding in the soil structure. Consolidation settlements are extremely rapid and although sensitive to applied normal stress, appear to be independent of flooding. Two 'yield' points have been identified as the soil structure deforms under different levels of effective stress. A first yield point generally occurs under effective loads of 100-200 KPa (interpreted as reflecting failure of inter-ped bonding as the soil de-structures) and a second yield point at higher stress levels of c. 500-600 KPa (interpreted as further de-structuring of the ped clusters due to failure of intra-ped bonds).

The oedometer test results have, to date, shown unexpectedly few instances of 'true' microstructural collapse; the relatively high degree of saturation of these samples compared with those described in the literature, being a possible explanation. It is possible that instances of collapse at low stresses, reported in the literature, are due, at least in part, to test specimens having dried out. Further work is currently underway on undisturbed samples from both Indonesia and Kenya, the results of which will be published in due course.

ACKNOWLEDGEMENTS

This paper is published with the permission of the Director of the British Geological Survey (N.E.R.C.) and the Director of the Institute of Road Engineering, Bandung, Indonesia.

REFERENCES

Anon, 1966. Soil map of the Province of West Java. Institute of Soil Research, Bogor, Indonesia.

Anon, 1975. Munsell soil color charts. Munsell Color Co. Inc., Baltimore, U.S.A.

Gidigasu, M. D., 1976. Laterite soil engineering. Elsevier Scientific Publishing Co., Amsterdam, Holland. 554p.

Head, K. H., 1982. Manual of soil laboratory testing. Volume 2: Permeability, shear strength and compressibility tests. Pentech Press, London. 747p.

Malomo, S. 1982. The use of the word "laterite" in engineering geology - a review. Proceedings of the 4th International Congress of the International Association of Engineering Geology, New Delhi, India, II, 201-208. A. A. Balkema, Rotterdam, Holland.

Newill, D. 1961. A laboratory investigation of two red clays from Kenya. Geotechnique, 11, 302-318.

Olsen, H. W., Nichols, R. W. and Rice, T. L. 1985. Low gradient permeability measurements in a triaxial system. Geotechnique, 35, 145-157.

Perloff, W. H. and Baron, W. 1976. Soil mechanics principles and applications. Ronald, New York, U.S.A.

Terzaghi, K. 1958. Design and performance of the Sasumua Dam. Proceedings of the Institution of Civil Engineers, 9, 369-393.

Wesley, L. D. 1973. Some basic engineering properties of halloysite and allophane clays in Java, Indonesia. Geotechnique, 23, 471-494.

Wesley, L. D. 1977. Shear strength properties of halloysite and allophane clays in Java, Indonesia. Geotechnique, 27, 123-136.

Vaughan, P. R., Maccarini , M. and Mokhtar, S. M. 1988. Indexing the engineering properties of residual soils. Quarterly Journal of Engineering Geology, 21, 69-84.

Engineering properties of weathered tropical soils from hill regions in Bombay area

P.JAGANNATHA RAO, Central Road Research Institute, India
K.A.JOHN, Central Road Research Institute, India
O.P.YADAV, Central Road Research Institute, India

ABSTRACTS: The paper presents data concerning engineering properties of soils for use as fill materials, from a number of quarries in Bombay Area. These are tropical residual soils derived from weathering of traps. A wide range was found in the index properties and compaction characteristics. The soils were divided into three groups for evaluation purposes. It was found that one group of soils have, low W_L with high I_p, low proctor density, high optimum water content and an ability to retain high water contents. In these characteristics, these soils showed a close resemblance to soils from Java, which also originated from weathering of volcanic parent materials in a tropical environment.

1 INTRODUCTION

Within 15-20 km of metropolitan Bombay region, yellowish brown soils of granular texture are found widely in the hillocks that dot the area. These soils are often referred to as "Moorum" and used as fill material for embankments constructed in the area. However, a number of cases of distress experienced by pavements in this area created a concern regarding the suitability of these soils as fill materials. In this paper, properties of fill materials from a number of sources in the area under discussion are presented. Peculiar properties of these soils are brought out and their suitability for use as fill materials is evaluated.

2 PROPERTIES OF FILL MATERIALS

Figure 1 shows the index properties of fill materials from a number of quarries around Bomby area. It may be observed that except for one group of samples which were found to belong to CH or CI groups, all other samples belong to MI or MH group i.e. inorganic silt or silty clays of medium to high compressibility.

Field experience has shown that in the case of fills made up of CI or CH groups of soils having maximum proctor density of the order of 1600-1700 kg/m^3 performance of the embankments was satisfactory. On the other hand, in one of the cases where the pavement has shown severe distress early in its life, investigation has shown that the embankment materials had a maximum proctor density of the order of 1300-1400 kg/m^3 only, with a relatively high optimum moisture content. As a result of these observations, a comprehensive study was conducted on the properties of fill materials from 12 selected quarries. This study has clearly brought out some of the peculiar characteristics of these borrow materials.

Table 1 summarises the relevant properties of the materials from the twelve quarries. Figures 2, 3 & 4 show the compaction charecteristics of the fill materials. On the basis of compaction characteristics it has been found convenient to divide the soils into three groups:

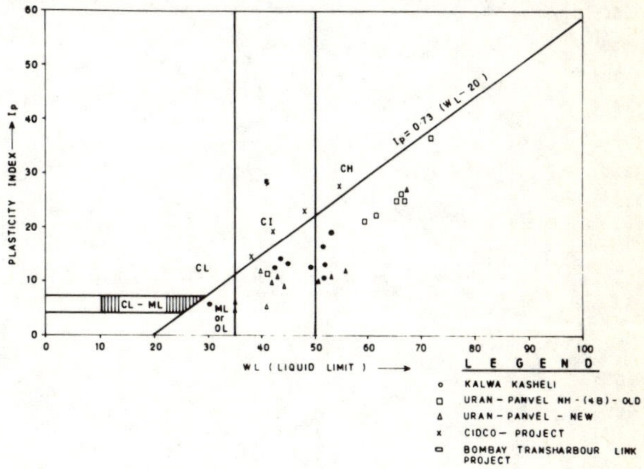

Figure 1. Properties of fill materials (Bombay area).

Group I	Sample Nos.	1,4,9	high density
Group II	Sample Nos.	3,5,6,10,12	medium density
Group III	Sample Nos.	2,7,8,11	low density

3 CHARACTERISTICS OF MATERIALS IN GROUP I

It was found that materials in group I are well graded and either non-plastic or have a low plasticity index. The maximum laboratory densities (Proctor) are high and range from 1800 kg/m^3 to 1900 kg/m^3. California Bearing Ratio values under saturated conditions were found to be in the range of 18-30. It was concluded that these materials would serve as satisfactory fill materials.

4 CHARACTERISTICS OF MATERIALS IN GROUP II

The materials under group II have proctor density

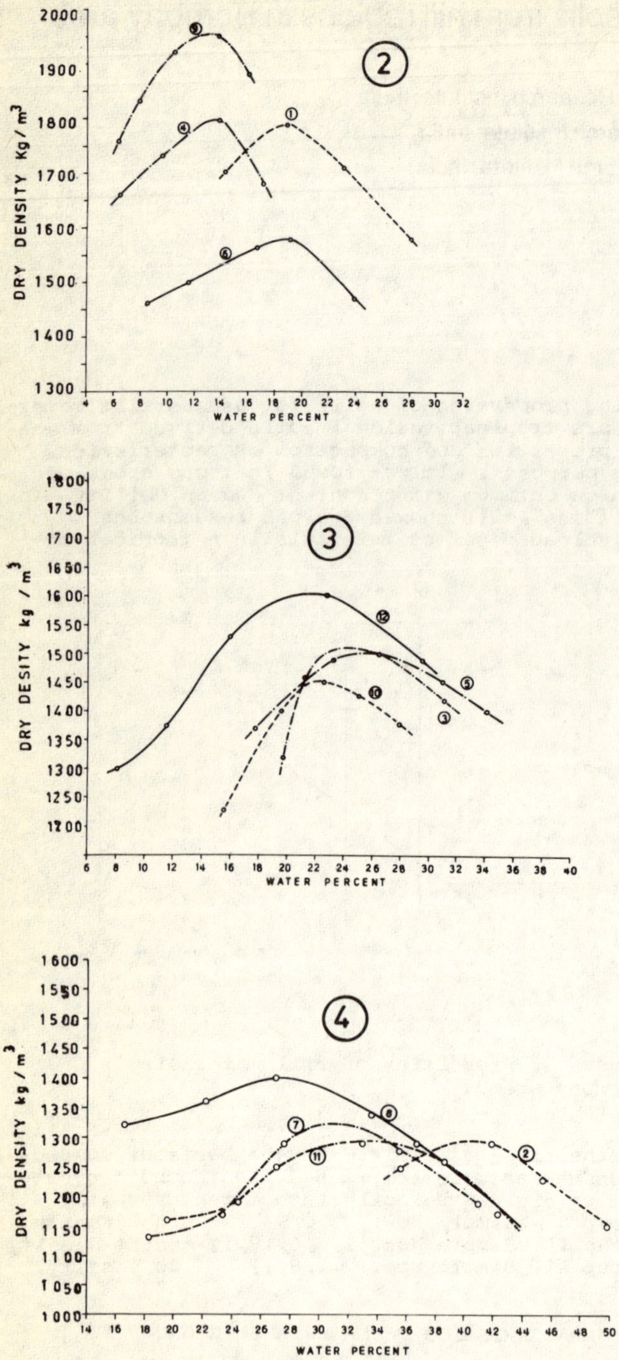

Figures 2, 3 and 4. Compaction characterics-fill material: Uran-Panvel Highway.

in the range 1450 kg/m^3 to 1650 kg/m^3. These were found to have higher liquid limits (W_L) and plasticity indices (I_P) than samples in group I. These materials are considered suitable for use as fill materials if proctor density of the materials are greater than 1520 kg/m^3. It was also proposed that the group II materials may not be used whenever materials conforming to characteristics of group I are available.

Table 1. Properties and classifications of fill materials.

a.

Quarry No.	W_L %	I_P %	Proctor density kg/m^3	OMC %	Classification
1.	Non-Plastic		1790	18.9	SM
2.	50	10	1290	40.0	ML
3.	42	10	1510	24.0	ML
4.	35	5	1800	13.2	SM
5.	67	27	1500	26.0	SM-MH
6.	41	5	1580	19.4	SM
7.	57	12	1290	33.4	MH
8.	44	9	1400	26.2	ML
9.	36	6	1950	13.1	SM
10.	43	11	1450	22.5	SM
11.	54	11	1300	31.0	SM-MH
12.	40	12	1600	22.0	SM

b.

Quarry No.	clay < 0.002 mm	Silt 0.002 to 0.075 mm	Sand 0.075 to 2.0 mm	Gravel > 2.0 mm
1.	2	37	59	2
2.	13	39	42	6
3.	16	34	36	14
4.	5	29	36	30
5.	17	21	8	54
6.	5	27	56	12
7.	14	37	46	3
8.	4	49	44	3
9.	3	12	47	38
10.	4	30	47	19
11.	10	26	56	8
12.	3	13	39	45

Specific gravity of particles: 2.90

5 CHARACTERISTICS OF MATERIALS IN GROUP-III

Group III materials represent fill materials having distinctly dissimilar properties from group I and group II materials. It was found that all the four samples in this group may be classified as inorganic silts or silty clays of medium to high compressibility, (ML to MH groups). These soils were found to be of non-swelling nature and may be termed as inactive to normal, on the basis of Skempton's activity ratio.

The proctor density of these materials was found to be very low, of the order of 1300 kg/m^3 only. The optimum moisture content (OMC) ranged from 30% to 40% which is in the range of plastic limit of these soils. Previous study of pavement failure in the area has shown that the embankment on which the pavement was built was made up of materials of similar characteristics. In these embankmentss, it was found that even in the relatively dry season, in-situ moisture content was high. In other words, the fill soils were found to have ability to a retain large quantity of water without undergoing volumetric expansion.

6 HYGROSCOPIC CHARACTERISTICS OF THE FILL MATERIALS FROM BOMBAY AREA

The non-swelling nature of the soils coupled with relatively high retained moisture content in the vicinity of plastic limit was observed in the field studies earlier. Hence a study was conducted on the hygroscopic moisture absorption characteristics of the materials and it was found that for the soils studied, the following linear relationships between the hygroscopic moisture content and different engineering properties exist;

i. Relationship between hygroscopic moisture an and Plasticity index

$H = 0.83\ P + 4.85$

ii. Relationship between hygroscopic moisture and optimum moisture content

$H = 0.30\ M + 4.80$

Where H = hygroscopic moisture (%)
P = Plasticity Index
M = Optimum moisture content

Figure 5 shows the above relationships graphically.

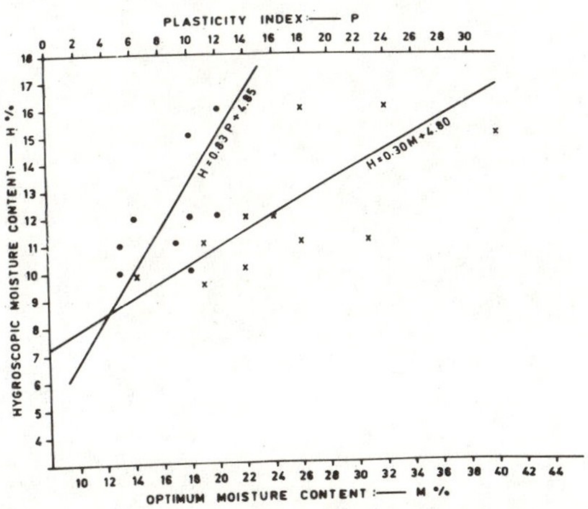

Figure 5. Correlation between Hygroscopic moisture content, OMC and Plasticity Index.

These relationships are valid for the residual tropical soils from Bombay area that were studied. It may be seen that as the Plasticity Index increases, the amount of hygroscopic moisture retained increases. This observation explains the presence of high retained water contents in fills made up of the above soils even during dry season.

Since the area is near sea coast, the atmospheric humidity remains high even in dry season and the peculiar nature of soils enables them to absorb water from atmosphere and retain it.

7 SUMMARY

Summing up, the peculiar properties of the borrow materials are:

a. Low I_p combined with relatively high W_L and W_p

b. Predominance of silt sizes in the particle composition

c. Low compacted density with high OMC

d. Ability to retain significantly high water content with negligible volume change.

Comparison of the properties with soils from other parts of the world shows that these soils properties fit in very well with those of tropical residual soils derived from weathering of volcanic materials, such as those found in Java(Wesely, 1973). Limited scanning Electron Microscope(SEM) studies carried out indicate the likely presence of allophane. Figures 6 shows SEM of the clay particles from sample No.8, magnified 10,000 times:

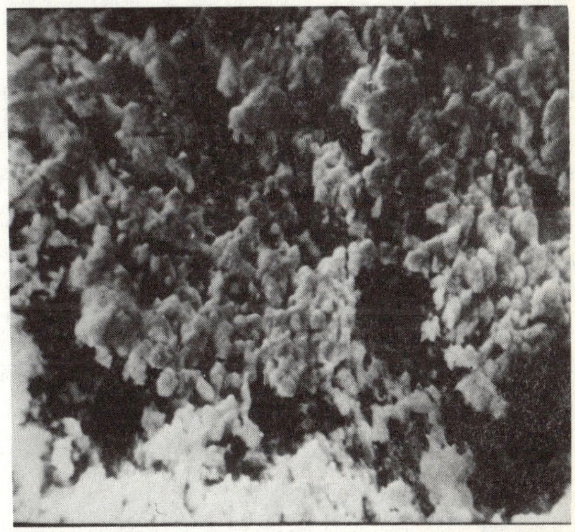

Figure.6. Scanning Electron Micrograph of clay particles.

8 CONCLUSIONS

The study brings out the need for due attention for comprehensive testing of the relevant engineering properties of borrow materials that are being considered for use as fills in highway embankments. Reliance on superficial features such as colour and grain size distribution alone can lead to problems in the service life of the fill. The origin and nature of the fill material has an important bearing on engineering properties of the fill and hence its performance.

ACKNOWLEDGEMENTS

The paper is published with kind permission of Dr. M.P. Dhir, Director, Central Road Research Institute, New Delhi. The authors are also grateful for his encouragement in carrying out the study.

REFERENCES

Wesely L.D. (1973) "Some basic Engineering Properties of Halloysite and Allophane clays, Geotechnique, Vol 23, No.4, 1973.

Some characteristics of a volcanic ash soil

P.J.MOORE, Reader in Civil Engineering, University of Melbourne, Australia
J.R.STYLES, Senior Lecturer in Civil Engineering, University of Melbourne, Australia

ABSTRACT: A volcanic ash with unusual characteristics was encountered during construction work in the New Guinea highlands. The volcanic ash has weathered to high plasticity silts and clays with high natural moisture contents. These soils were found to be amorphous to x-ray diffraction and they experienced a loss of plasticity following air drying. Compaction curves obtained under conditions of decreasing moisture content showed no resemblance to standard laboratory compaction curves. The soils were also found to be very sensitive with respect to strength characteristics.

1 INTRODUCTION

During the construction of a highway in the vicinity of Mt. Hagen in the New Guinea highlands, extensive deposits of a volcanic ash soil were encountered. Mr Hagen (Figure 1) is

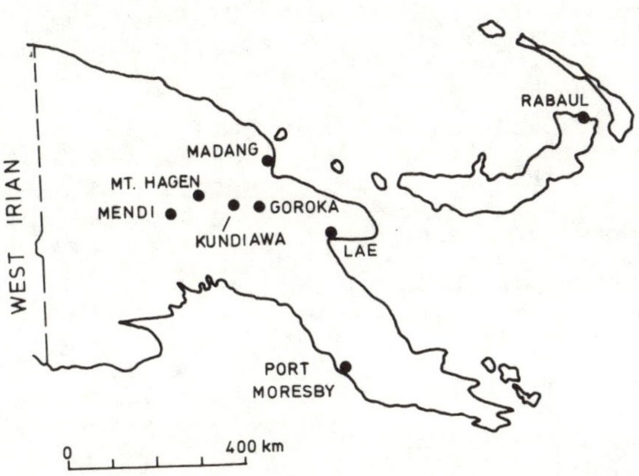

Figure 1 : New Guinea locality sketch

located about 6° south of the equator at an elevation of about 1700m and with an annual rainfall exceeding 2600mm. The volcanic ash was probably derived from the two extinct volcanoes, Mt. Giluwe and Mt. Ialibu located between Mt. Hagen and Mendi and in the immediate vicinity of the highway route. The volcanic ash has weathered into brown, white and light grey clayey silts and silty clays of high plasticity and high natural moisture content. These soils were far from ideal as construction materials and considerable difficulties were encountered when they were used for pavement or embankment construction.

2 MINERALOGY

Results of X-ray diffraction tests on samples of the volcanic ash soil showed that the samples were amorphous to X-ray diffraction and contained negligible distinctive clay minerals. The samples consisted of aluminium and silica dioxides and hydroxides with some ferric oxide. It was concluded that the mineral present was probably allophane.

Allophane has been described as having a structure of random gel-like fragments, linked randomly at a small number of locations (Wesley, 1973). The "fragments" were defined by Uehara (1982) as hollow spherical particles with a diameter of approximately 5×10^{-7} cm, and a corresponding specific surface of the order of 10^7 cm^2/gm : a characteristic which gives rise to very high retention of water.

Soils, in which allophane may be found, include those classed as Andosols (F.A.O.), Andepts (U.S. Soil Taxonomy) or Andisols (Uehara, 1982). Townsend (1985) refers to Andosols as allophanic soils - soils which contain allophane and hallosite, in addition to iron and alumina sesqui-oxides.

3 PLASTICITY CHARACTERISTICS

Atterberg limit test results carried out on samples of the volcanic ash without prior air drying shown in Figure 2. This shows that most of the soils would be classified as MH according

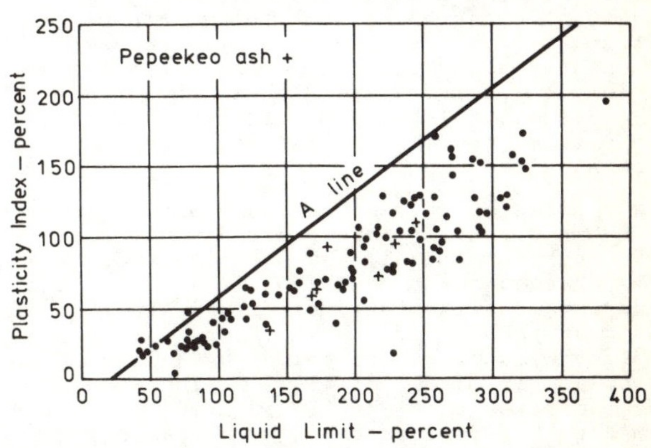

Figure 2 : Plasticity characteristics of volcanic ash

to the Unified Classification System. These soils are very similar to those described as Pepeekeo ash from the island of Hawaii and which was encountered during the construction of the Hawaii Belt Road. The characteristics of the Pepeekeo ash have been described by Hirashima (1948,1951), Terzaghi and Peck (1967) and Lum (1982).

Most of the volcanic ash soils are fully saturated with high natural moisture contents. As shown in Figure 3 these moisture contents are generally in the vicinity of the liquid limit although some soils possess moisture contents

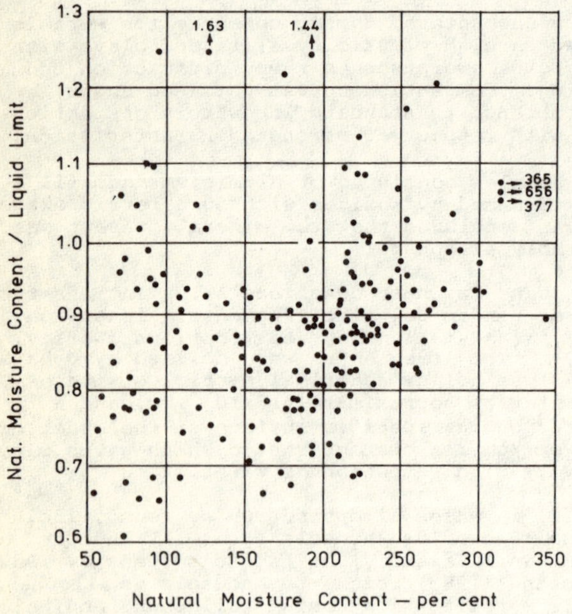

Figure 3 : Moisture content of volcanic ash

considerably higher than the liquid limit.

Significant irreversible changes occur to the volcanic ash soils following air drying. The majority of the soils tested became non plastic following air drying. These samples became sufficiently granular in behaviour that liquid and plastic limit tests could not be carried out. Frost (1967) has demonstrated that the grain size of the soil particles increases following air drying.

A few samples of volcanic ash did not entirely lose their plasticity after air drying. The way in which the liquid and plastic limit values changed for these samples as a result of air drying is illustrated in Figure 4. This shows that the "before" and "after" points on the plasticity chart are joined by a line sloping approximately at 45°. A line with this slope indicates that the liquid limit is the parameter that is decreasing with scarcely any change in the plastic limit.

Figure 4 also includes data compiled by Morin and Todor and reproduced by Townsend (1985). A feature of the volcanic ash encountered in this project was the generally much higher liquid limits than for other previously reported New Guinea andosols. In considering the plasticity characteristics of soils classed as andosols, the extremes of behaviour may be related to the predominant clay mineral. Table 1 gives Atterberg limit values for several soils classified as andosols.

The four examples cited in Table 1 may be more conveniently grouped by predominant clay mineral type. When this is done the first two soil samples could be grouped as latosols, with the second two as andosols. Latosols are typified by red/reddish brown soils and generally are a more advanced weathering product than the typically yellowish brown andosol group. From the description of the Sasumua clay given by Matyas (1969) it would appear that this clay more correctly belonged to the latosol group. Wesley (1973) notes that the predominantly halloysitic Tjpanundjang clay, whilst coming from an andosol area, showed the physical and mineralogical characteristics of a latosol.

The air drying mentioned above is a slow process under relatively low humidity laboratory conditions. One series of measurements (Figure 5) showed that it required about two weeks for the moisture content to decrease from 200% to 30%. Under the high rainfall, high humidity conditions in the field such air drying before compaction, as recommended by Mitchell and Sitar (1982) and Townsend (1985), is quite impracticable.

4 COMPACTION BEHAVIOUR

Typical specifications for field compaction require that the field dry density achieved be not less than a certain percentage (90% or 95%) of the maximum dry density as determined from a standard laboratory compaction test. For wet soils the laboratory procedure requires air drying. Apart from the impracticability of air drying the volcanic ash soil in the field in a tropical climate (as mentioned above), the use of the above mentioned typical specification requirement is quite inappropriate for these volcanic ash soils. This is because the field compaction behaviour of the volcanic ash can be significantly different from that in a standard laboratory test.

If laboratory compaction is carried out on the volcanic ash at natural moisture contents between about 100% and 300% (see Figure 3) the upper limit to the dry density that can be achieved (by laboratory or field compaction) is on the zero air voids line. This dry density is considerably lower than the maximum obtainable by a standard laboratory compaction test where the sample was initially air dried (Figure 6). If laboratory compaction is carried out on samples dried to moisture contents lower than the natural moisture content, compaction curves similar to those shown in Figure 6 are obtained. These curves exhibit a slowly increasing dry density as the moisture content decreases. Sometimes these curves peak but often they do not. The behaviour observed is very similar to compaction curves obtained by Wesley (1973) for allophane soils compacted at natural moisture contents. Both Frost (1967) and Wesley (1973) have shown that the peak in the compaction curve can be produced if the soil is rewetted following partial air drying. As shown in Figure 6 the

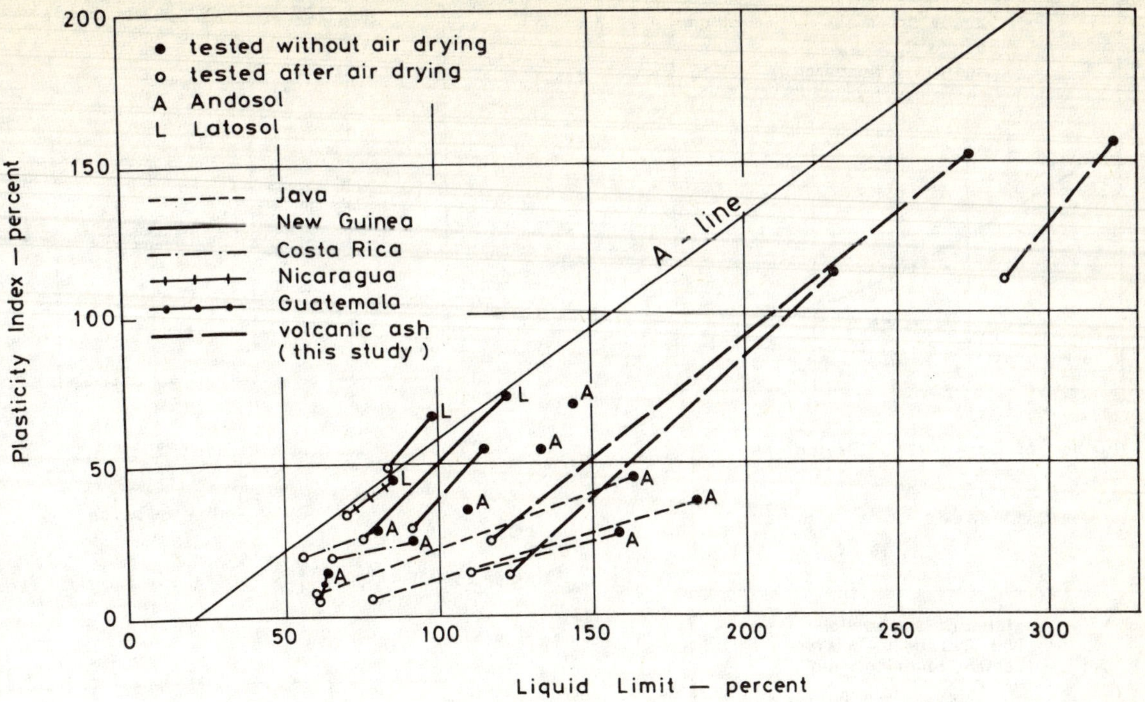

Figure 4 : Effect of air drying on plasticity characteristics

Table 1. Atterberg Limits of Andosols

Soil Type	w_n (%)	w_L (%)	w_p (%)	I_p (%)	Mineralogy
Sasumua clay:†					
natural	62	107	73	34	predominantly
oven dried	62	65	47	28	halloysite
Tjpanundjang clay:*					
natural	68	95	65	30	predominantly
oven dried	68	61	46	15	halloysite
Java clay:*					
natural	146	156	129	27	allophane
oven dried	146		non-plastic		
Tjpanundjang clay:*					
natural	128	165	119	46	predominantly
oven dried	128	44	42	2	allophane with 5-10% halloysite

† Matyas (1969) - classified by Mitchell & Sitar (1982)
* Wesley (1973)

compaction curves with decreasing moisture contents bear no resemblance to the standard compaction curve.

When field compaction is carried out on the volcanic ash at natural moisture content, typical dry densities achieved are as shown in Figure 7. The density ratio in this figure is defined as the ratio of the field dry density measured after compaction by the field roller equipment to the dry density obtained by laboratory compaction, both at the natural moisture content.

5 STRENGTH AND CBR MEASUREMENTS

The results of some unconfined compression tests on "undisturbed" and remoulded samples of the volcanic ash are shown in Figure 8. These results lead to values of sensitivity (ratio of undisturbed to remoulded strength) between 1.7 and 5.6, corresponding to an only moderately sensitive soil. However the "undisturbed" samples were really partially disturbed so the undisturbed strength values are likely to be greater than those shown in Figure 8. Reports that cuts in the volcanic ash can stand almost

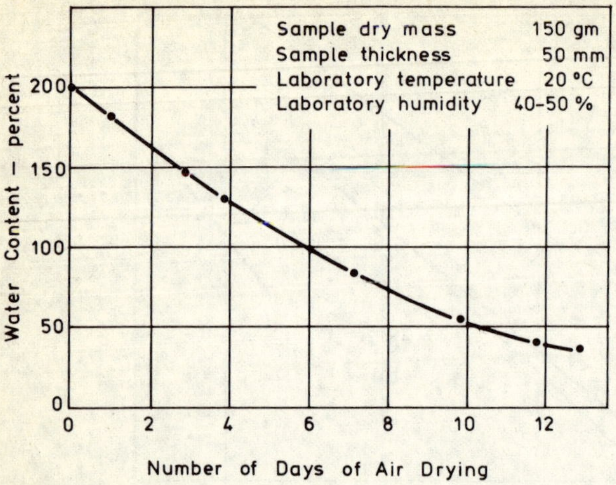

Figure 5 : Loss of water on air drying

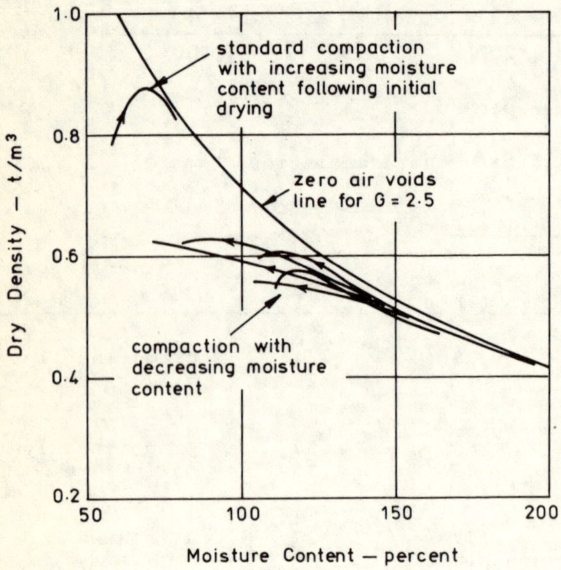

Figure 6 : Compaction curves for volcanic ash

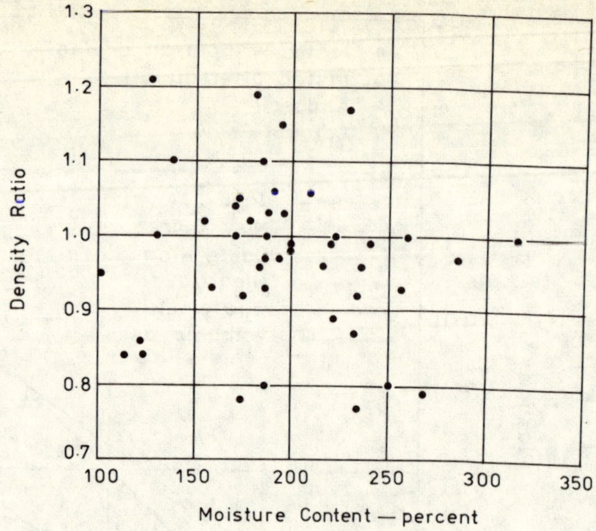

Figure 7 : Density ratios achieved by field compaction

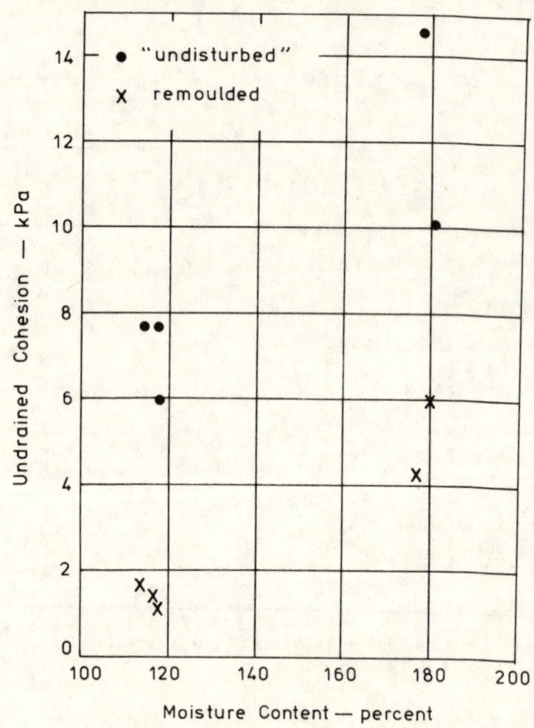

Figure 8 : Laboratory strength measurements

vertically to heights up to 10m supports the idea that the strength of the soil is several times that indicated in Figure 8. Further evidence comes from vane shear tests carried out in the field and which yielded values of undrained cohesion between 14 and 150 kPa. These figures suggest that the volcanic ash is likely to be very sensitive and difficult to handle by field compaction equipment.

Quite low values of CBR (around 1%) were measured on samples compacted at natural moisture content. A considerable improvement in the CBR is observed if the samples are partially dried out prior to compaction as shown in Figure 9. While this drying out may be a relatively simple procedure in the laboratory, as previously mentioned it is normally impracticable in a humid tropical environment.

6. CONCLUDING REMARKS

The high moisture content high plasticity soils encountered in the New Guinea highlands possessed some unusual characteristics, some of

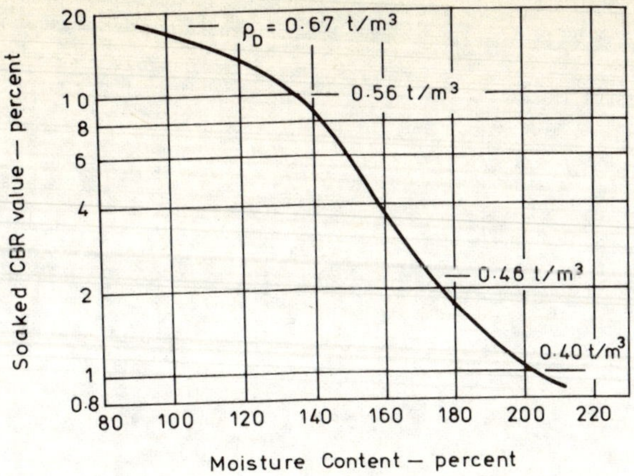

Figure 9 : CBR test results on samples with decreasing moisture content

which have been reported for other similar soils elsewhere. These characteristics include:

a) Soils amorphous to X-ray diffraction.
b) Loss of plasticity following air drying, most of the tested samples becoming non-plastic.
c) Compaction curves obtained on samples with decreasing moisture contents showed no resemblance to standard laboratory compaction curves on air dried and rewetted samples.
d) The soils are very sensitive with respect to strength behaviour.

7 REFERENCES

Frost, R.J. (1967), "Importance of Correct Pretesting Preparation of Some Tropical Soils", Proc. of S.E. Asian Regional Conf. on Soil Engineering, Bangkok, pp 43-53.

Hirashima, K.B. (1948), "Highway Experience with Thixotropic Volcanic Clay", Proc. Highway Res. Board, Vol. 28, pp 481-494.

Hirashima, K.B. (1951), "Highway Construction Problems Involving Plastic Volcanic Ash", Highway Res. Board Bulletin 44.

Lum, W.B. (1982), "Engineering Problems in Tropical and Residual Soils in Hawaii", Proc. Speciality Conf. on Engineering and Construction in Tropical and Residual Soils, ASCE Geotech. Eng. Div., Honolulu, pp 1-12.

Matyas, E.L. (1969), "Engineering Properties of Sasumua Clay", Proc. Speciality Session on Eng. Properties of Lateritic Soils, Vol. 1, 7th Int. Conf. Soil Mech. and Found. Eng., Mexico City, pp 143-152.

Mitchell, J.K. and Sitar, N. (1982), "Engineering Properties of Tropical Residual Soils",ASCE Spec. Conf. Engineering and Construction in Tropical and Residual Soils, Honolulu,pp 30-55.

Terzaghi, K. and Peck, R.B. (1967), "Soil Mechanics in Engineering Practice", Art. 51, John Wiley & Sons, New York.

Townsend, F.C. (1985), "Geotechnical Characteristics of Residual Soils", Jnl. Geot. Eng.Div., ASCE, Vol. 111, No. 1, pp 77-94.

Uehara, G. (1982), "Soil Science for the Tropics", ASCE Spec. Conf., Engineering and Construction in Tropical and Residual Soils, Honolulu, pp 13-26.

Wesley, L.D. (1973), "Some Basic Engineering Properties of Halloysite and Allophane Clays in Java, Indonesia", Geotechnique, Vol. 23, No. 4, pp 471-494.

An approach to predict tropical soil behaviour

T.S.NAGARAJ, Indian Institute of Science, India
N.S.PANDIAN, Indian Institute of Science, India
B.M.SHASHIKUMAR, Indian Institute of Science, India
G.L.SIVAKUMAR BABU, Indian Institute of Science, India

ABSTRACT : Most of the natural soils in tropical environment, arid and semiarid regions, are in partly saturated condition. Engineering behaviour of such soils are presently analysed with the help of modified effective stress and stress state variables approaches. In this paper the difficulties of practical application are identified. Based on the physico-chemical state of partly saturated soil, an integral state parameter $[(e/e_L)\sqrt{S_r}]$ has been identified to take into account pore water tension by $\sqrt{S_r}$ and nature of fine grained soil by void ratio, e_L, corresponding to liquid limit water content. It has been found that this state parameter bears a unique relation with external pressure, compressed both under constant water content and inundated condition.

1 INTRODUCTION

In a broad sense, classical geotechnical engineering has developed in regions of earth dominated by temperate climates. As a result, many of our concepts of soil behaviour have been conditioned by the earth materials found there. Over the last three decades or so, the need for engineering and construction has extended on a large scale into much less familiar areas such as tropics. This environment which encompasses many of the developing nations including India, has been of major concern due to the magnitude of the projects and the large populations affected.

The characteristics of tropical soils are such that they are formed under hot and humid climates mostly as residual deposits. With depth, the degree of rock weathering decreases and gradually weathering ceases with in situ rock unaltered. Most of the natural soils in such arid and semiarid regions and soils subjected to seasonal moisture deficits are in a partly saturated state. Further, the mechanics of partly saturated soil is a subject of considerable interest in all the areas of the world where weather conditions are such that there is an annual excess potential of evaporation over precipitation. It is apparent that far more than half of earth's surface is a moisture deficit area. Generally, an unsaturated soil is considered to be a three-phase system. However, on the basis of the definition of a phase, the air-water interphase has been considered as a fourth independent phase. The air-water interface has been referred to as contractile skin (Fredlund, 1979). The most distinctive property of contractile skin is its ability to exert a tensile pull. It behaves like an elastic membrane under tension interwoven throughout the soil structure.

The two approaches that have been pursued independently to understand and generalise partly saturated soil behaviour are :

i) modified effective stress equation to account for partial saturation,

ii) incorporation of independent stress state variables in the physical models.

2 EFFECTIVE STRESS CONSIDERATIONS

The conventional effective stress equation has been modified by many investigators (Bishop 1960, Aitchison 1960, Jennings 1961, Jennings and Burland 1962, Bishop and Blight 1963, Burland 1964, 1965, Sridharan 1968 and others) to account for partial saturation by including additional parameters. The one proposed by Bishop (1960) is the most generally accepted one and is in the form :

$$\sigma' = \sigma - u_a + \chi (u_a - u_w) \quad (1)$$

where σ' is the effective stress
σ is the total stress due to applied loads
u_a is the pore air pressure
u_w is the pore water pressure
and χ is the coefficient having a value between 0 and 1 and depends primarily on the degree of saturation and the soil type.

Critical reappraisal of effective stress equations developed for partly saturated soils (Gulhati 1975, Nagaraj and Srinivasa Murthy 1983) has revealed that analysis of partly saturated soil behaviour by this approach is not only a complex deficient exercise but also involves an accurate measurement of u_a, u_w and χ parameters.

3 STRESS STATE VARIABLE APPROACH

Fredlund and Morgenstern (1976, 1977) and Fredlund (1979) have developed basic concepts and technology required to relate the partially saturated soil behaviour analogous to saturated soil behaviour with the appropriate stress parameters of $[\sigma-u_a)$ and (u_a-u_w). By considering the contractile skin as an additional phase which brings the system into equilibrium along with the soil particles under an applied stress gradient and considering independent stress variables for the phases, it has been possible to analyse the strength and volume change behaviour of partly saturated soils (Fredlund 1985,1987). It has been rightly cited that the search for effective stress equations for analysis of partly saturated soil behaviour should be terminated and the illusive χ, β, ψ factors proposed in earlier effective stress equations should be dispensed with.

Still, with either of the above approaches for the analysis of the behaviour of partly saturated soil, it is very essential to measure pore water tension in soils under different test conditions. This is an arduous and the most unsought job by every field engineer. Hence it is desirable to develop an approach within the framework of a scientific base, with the input of minimum easily measurable soil parameters in routine field investigation, capable of interpreting, generalising and predicting partly saturated soil behaviour. The cardinal aim of this paper is to report the developemnt of generalised state parameter concept for partly saturated soil behaviour.

4 STATE PARAMETER APPROACH

State can be explained as a limited set of dynamical variables of the system, such as pressure, temperature, volume, etc. which are sufficient to describe or specify the state of the system completely (Fredlund 1987). Basically, the variables that can be considered pertain to stress state and physical state. The selection of the appropriate state reduces to the identification of

i) material state variables that can encompass a wide spectrum of variations in physical and chemical state of material.

ii) stress state variable as a non material variable required for the characterisation of a system.

For successful application of the state variables approach to practice, it is essential that the state variables can be easily evaluated or determined by simple tests and can be theoretically justified. Ultimately the above approaches have to be compliment to each other. This situation can be recognised for saturated normally consolidated clays. For a saturated clay the state parameter, void ratio can be uniquely linked with the effective stress mobilized in the system. If the state parameters can be generalized to encompass the type of soil and the past stress history effects can be identified and suitably taken care of, the link between the material state variable and stress state gets better established.

In the case of partly saturated soil, void ratio and degree of saturation are independent material state parameters, soil is another independent variable. It is necessary to identify a combined state parameter of all the above three variables. The relationship between this parameter and the effective stresses in partly saturated soil need be examined for practical exploitation. The effects of pore water tension need be accounted for in the generalization of partly saturated soil behaviour.

5 IDENTIFICATION OF APPROPRIATE STATE PARAMETERS

Recently, there have been serious attempts to identify state parameters for sands, saturated clays and partly saturated soils. Been and Jefferies (1985) have characterised the state parameters in terms of two variables.

i) a state parameter which combines the influence of void ratio and stress, and

ii) a farbic parameter which characterizes the arrangement of sand grains. It has been proposed that mean stress is a suitable measure for incorporation into the state parameter. The Kernel concept is that the steady state defines a reference state and the distance of a sand from this reference state in void ratio-stress space characterizes the state of sand. This measure of state is called the state parameter, ψ and the definition is illustrated in Fig. 1.

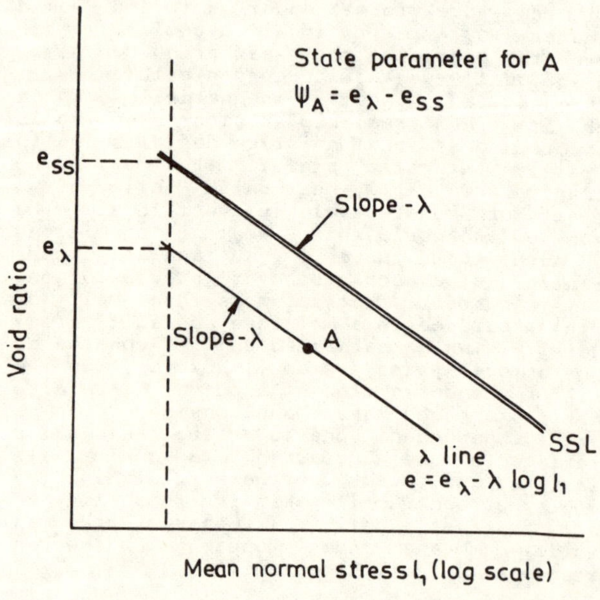

Fig. 1 State Line For Sand (Ref: Been and Jefferies, 1985)

It has been possible to develop a generalized state parameter approach for saturated clays (Nagaraj and Srinivasa Murthy 1985, 1986). From the micromechanistic considerations of clay soil behaviour it has been shown that the Gouy-Chapman diffuse double layer theory forms a basis to generalise the compressibility behaviour of saturated uncemented fine grained soils. The genaralized compressibility equations for normally and overconsolidated clays are given respectively by

$$e/e_L = 1.122 - 0.2343 \log p \quad (2)$$

and

$$e/e_L = 1.122 - 0.188 \log p_c - 0.0463 \log p \quad (3)$$

where p is the effective consolidation or overburden pressure in kPa, p_c is the preconsolidation pressure in kPa and e/e_L is the generalized soil state parameter and the void ratio at liquid limit, e_L ($=w_L G$). Equations (2) and (3) are the generalized stress-state equations representing mutual compatability between the terms. Fig. 2 shows the paths predicted by equations (2) and (3).

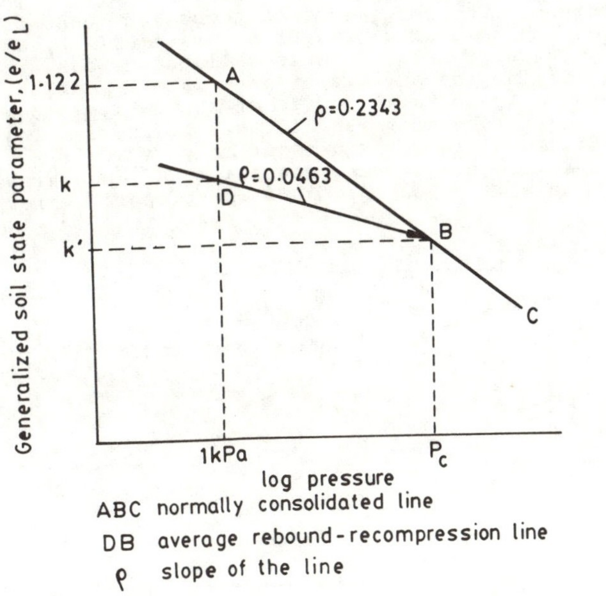

Fig. 2 State Lines For Normally Consolidated And Overconsolidated Saturated Uncemented Clays

In the case of partly saturated soils, such simple generalization is not quite apparent. This is beacuse that classification is necessary regarding;

i) the physico-chemical state of pore water

ii) the mode of combination of state parameters, void ratio and degree of saturation and physico-chemical potential, and

iii) separation of the effects of applied stress, pore water tension at any equilibrium condition.

The physico-chemical state of a partly saturated soil has been defined by Bolt and Bruggenwert (1976). In an aqueous environment, the thickness of the diffuse double layer extends to a distance at which the concentration of ions equal that of the bulk solution. As ions have to stay only in the liquid phase, the thickness of the diffuse double layer cannot exceed the thickness of the water film in contact with the charged surface. This results in readjustment of ion concentration. Such a double layer has been recognized as the truncated diffuse double layer (Fig. 3). An important feature of the truncated double layer due to partial saturation is to absorb water until the double layer develops compatible with the external loading.

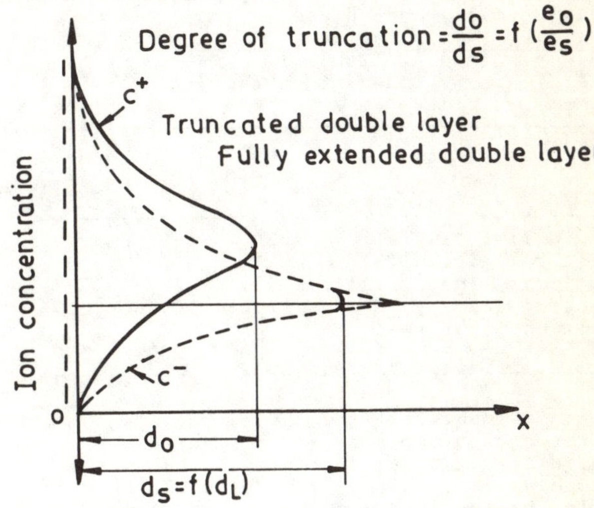

Fig. 3 Schematic Representation Of Truncated Double Layer

The mode of combination of state parameters to a generalized state parameter has already received some attention (Nagaraj and Srinivasa Murthy 1985). For the assumed physical model with two parallel interacting particles separated by a half space distance, d in equilibrium with change in pore water tension, Δu_w and changes in interparticle forces, $\Delta(R-A)$ being positive increments, it has been shown that the equilibrium equation can be represented by

$$\sigma - \Delta u_w = \Delta(R-A) \quad (4)$$

The change in surface tension is a function of the radius of minisci. This varies with the square root of the area of the water phase which reflects the degree of saturation. Thus the pore water tension can be accounted by $\sqrt{S_r}$. As in the case of saturated clays, the micro state parameter, d accounting for the influences of interparticle forces can be taken care of by the generalized state parameter (e/e_L). Hence [$(e/e_L)\sqrt{S_r}$] forms a combined generalized state parameter for partly saturated soils. For practical application the relationship between the state parameter and effective stress state need to be established. To determine the effective stress in partly saturated soil the pore water tension has to be determined. This again leads to the formidable task of measurement of pore water tension. The possibility of circumventing this predicament by development of unique relationship of state parameter with external load under equilibrium has been attempted by constant water content compression test from initial unstressed state.

6 EXPERIMENTAL INVESTIGATION

Index properties of soils used in this experimental programme are listed in Table I. The range of liquid limits considered (50 to 84) encompasses range of tropical fine grained soils normally encountered.

Table I
Index properties

	Red earth	Brown soil	Black cotton soil
Liquid limit, w_L(%)	50	60	84
Plastic limit, w_p(%)	27	31	47
Plasticity index, I_p(%)	23	29	37
Specific gravity, G	2.66	2.70	2.73
Fraction finer than 425μm	100	100	100

As is evident from the table, soils possess different potentials as reflected by the liquid limit water contents. This makes it possible to encompass the behaviour of other soils having intermediate values of potential automatically in the generalized relationships. Soils were excavated from the chosen sites, airdried, lightly pulverized and fractions finer than 425μm size were collected for liquid limit and compressibility tests. Liquid limit water contents were determined according to BS1377-1975 procedure and soil mixed with predetermined quantity of water was left for uniform moisture distribution before compressibility tests were conducted.

The influence of prestress effects on compressibility was eliminated by a datum state obtained by compression at 10 kPa from a very loose state. Compression tests at constant water content were conducted covering the range of pressure from 25 to 800 kPa. The loss of moisture during compression test was prevented by surrounding the soil sample with a snuggly fitting rubber membrane secured in place by O-rings. Different samples from the same datum were compressed at constant water content to different stress levels and inundated with water. The sudden compression experienced by each of the samples was monitored. The inundation condition was maintained till equilibrium condition was reached. Series of constant water content compression tests on samples with different initial water contents were determined and for each of the cases corresponding equilibrium conditions under inundated conditions were monitored.

7 TEST RESULTS AND DISCUSSION

Compressibility test data under constant water content and after inundation are illustrated in the form of e-log p plots - typical representation of which are in Fig. 4 to 8. The tests cover different initial water contents besides specifically referring to different liquid limits of soils. It is evident that along constant water content path, from a very low degree of saturation it increases as consolidation pressures are applied. Work is done by the soil and the applied stress on the menisci or water in displacing it from one position to other. This induces positive incremental changes in pore water pressure. Still pore water tension resists compression. The soils equilibriate at higher void ratios compared to the void ratios reached upon inundation under constant load. Irrespective of the soil, for the range of liquid limits considered in the pressure range of 25 to 800 kPa, there has been collapse upon inundation reaching lower void ratios. The magnitude of reduction depends upon the initial degree of saturation, S_r at any stress level. This collapse is due to nullifying the effects of pore water tension. e-log p paths with degree of saturation close to unity are shown in all the figures. The difference in pressure at any void ratio between constant water condition and inundated is indicative of the magnitude of pore water

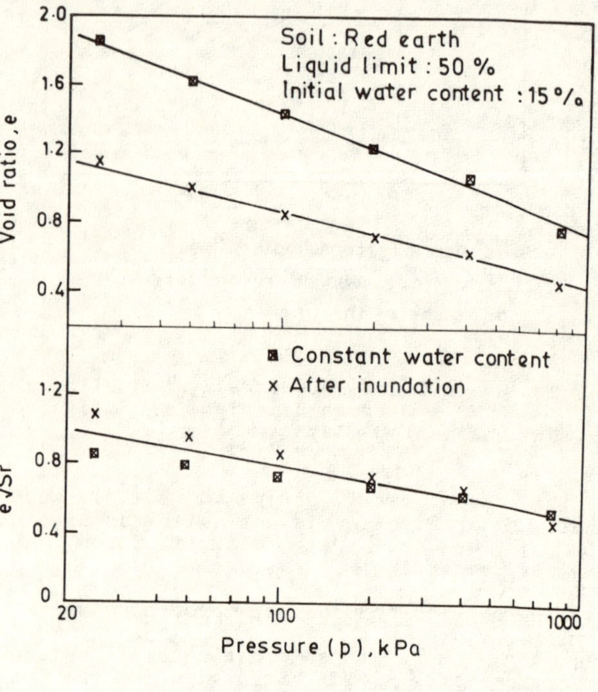

Fig. 4 Soil State - Pressure Relationship

tension. This inference can be drawn from the plots of $e\sqrt{S_r}$ versus log p at the bottom of each of Fig. 4 to 8. Since $\sqrt{S_r}$ directly accounts for pore water tension. State parameter, $e\sqrt{S_r}$ for a partly saturated soil bears a unique relationship with applied pressure at equilibrium condition. For monotonic increase in pressure, without any prestress effects, the pore water tension is directly due to external loading and hence unique state parameter $e\sqrt{S_r}$ versus pressure is possible. Upon normalization of $e\sqrt{S_r}$ by the void ratio of the soil at its liquid limit water content, e_L (=$w_L G$) all data points (Fig. 4 to 7) lie in a narrow band in the [(e/e_L) $\sqrt{S_r}$] versus log p plot yielding the

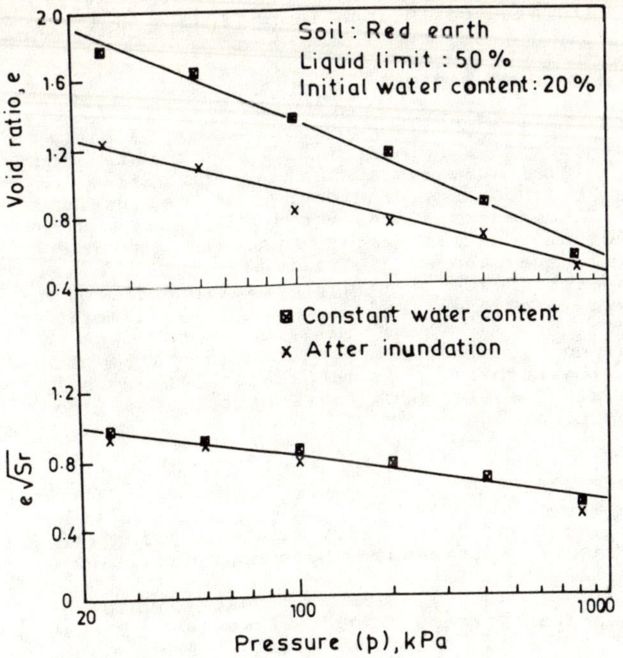

Fig. 5 Soil State - Pressure Relationship

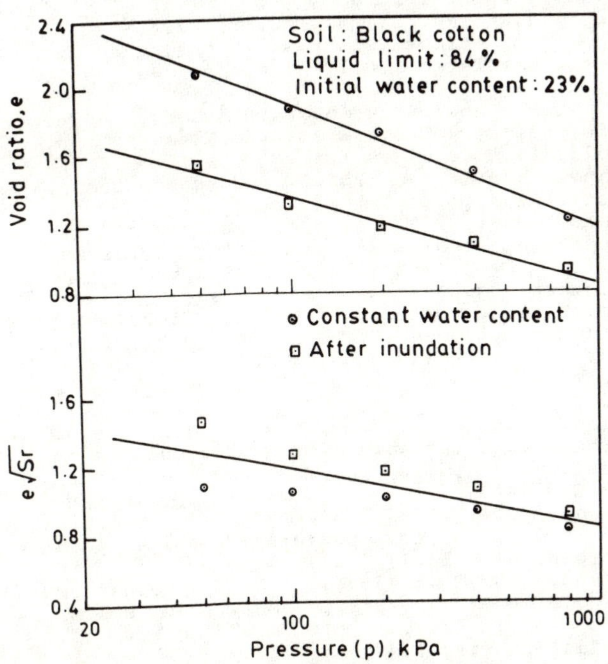

Fig. 6 Soil State - Pressure Relationship

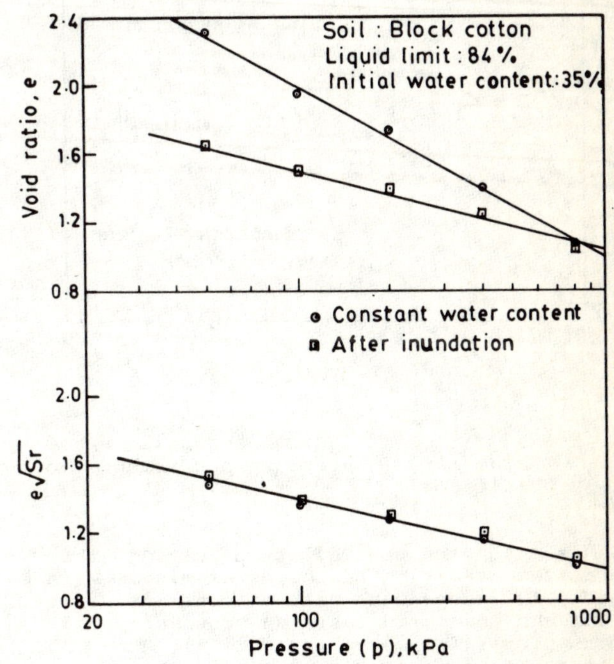

Fig. 7 Soil State - Pressure Relationship

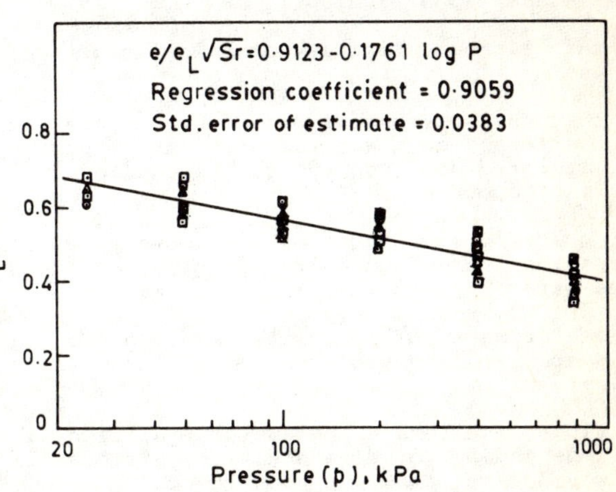

Fig. 8 Generalized Soil State Parameter - Pressure Relationship

generalized relationship

$$[(e/e_L)\sqrt{S_r}] = 0.9123 - 0.1761 \log p \quad (5)$$

with a regression correlation coefficient of 0.90. In the above relationship $[(e/e_L)\sqrt{S_r}]$ is the generalized state parameter for partly saturated soil and p is the overburden pressure in kPa.

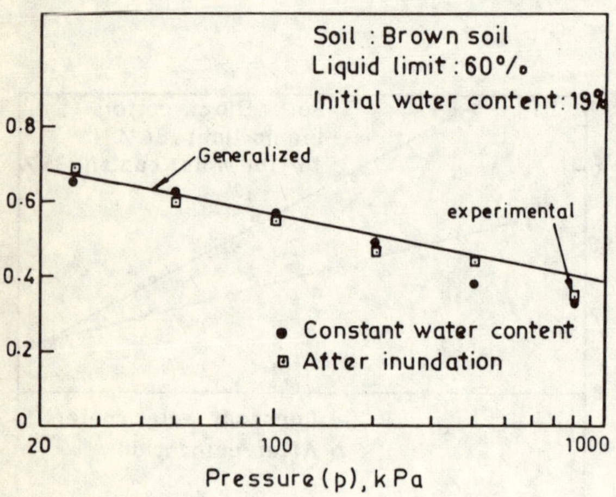

Fig. 9 Validation of Generalized Soil State Parameter - Pressure Approach To Field Soil

At this stage the possibility of having this generalized relation has been examined with respect to independent data of constant water content and inundated tests on brown clay soil (index properties cited in Table I) with an arbitrary initial water content of 19 percent. It can be seen in Fig. 9 that the data points of $[(e/e_L)\sqrt{S_r}]$ corresponding to different external loading for both the above conditions lie close to the generalised state line for partly saturated soils. With the following in built assumptions such as :

i) the size and rigidity of clay clusters, and

ii) the characteristics of fabric clay of clusters do not affect the partly saturated soil behaviour, the agreement is very good.

The specific advantages of the above mode of generalization by consideration of appropriate state parameter are that :

i) measurement of pore water tension can be overcome by use of only generalized state parameter,

ii) since the state parameter relation is independent of stress history effects it would be possible to identify other compatible states. Location of points $[(e/e_L)\sqrt{S_r}]$, p on the state space reflects different possibilities, points lying along the state line and also above due to partially saturated cemented experience collapse, whereas below the state line, upon inundation exhibit swelling response. Detailed investigations are envisaged to develop approaches based on generalized state parameter concept to predict swelling pressure, heave or collapse of partly saturated soils upon inundation.

8 CONCLUDING REMARKS

Based on the physico-chemical state of partly saturated soils, $[(e/e_L)\sqrt{S_r}]$ is a generalized soil state parameter accounting for nature of fine grained soil and degree of saturation. For uncemented fine grained soil with different initial water contents in loose state the above generalized soil state parameter bears a unique relationship with external pressure under equilibrium conditions. This relation provides a reference state for systematic identification of the other in situ states of partly saturated soils which heave or collapse upon inundation.

9 REFERENCES

Been, K. and Jefferies, M.G., 1985. State parameter for sands. Geotechnique. 35(2). 99-122.

Aitchison, G.D., 1960. Relationship of moisture stress and effective stress functions in unsaturated soil. Proc. Conf. on Pore Pressure and Suction in Soils. 47-52. Butterworth. London.

Bishop, A.W., 1960. The measurement of pore pressure in the triaxial test. Proc. Conf. on Pore Pressure and Suction in Soils. 38-46. Butterworth. London.

Bishop, A.W. and Blight, G.E., 1963. Some aspects of effective stress in saturated and partly saturated soil. Geotechnique. Vol. 13, No. 3, 177-197.

Bolt, G.H. and Bruggenwert, M.G.M., 1976. Soil chemistry Basic elements. Elsevier, Sc. Publication Co., New York.

Burland, J.B., 1964. Correspondence, Geotechnique, 14(1) : 64.

Burland, J.B., 1965. Some aspects of the mechanical behaviour of partly saturated soils, Symp. Moisture Equilibria and Moisture changes in Soils Beneath Covered Areas, Butterworths, Austrilia. 270-277.

Fredlund, D.G. and Morgenstern, N.R., 1976. Constitutive relations for volume changes in unsaturated soils, Can. Geotech. Jl., 13 : 261-276.

Fredlund, D.G. and Morgenstern, N.R., 1977. Stress state variables for volume changes in unsaturated soils, ASCE Jl. of GT. Dn., 103(GTS):447-466.

Fredlund, D.G., (1979). Appropriate concepts and technology for unsaturated soils, Can Geotech. Jl., 16(1) : 121-139.

Fredlund, D.G., 1985. Soil mechanics principles that embrace unsaturated soils. Proc of 11th Intl. Conf. on SM and FE, Vol. 2, San Francisco, 465-472.

Fredlund, D.G., 1987. Stress state for expansive soils. Lectures, additional papers, General reports, 6th Intl. Conf. on

Expansive Soils, New Delhi. 1-9.
Gulati, S., 1975. Shear behaviour of partially saturated soil. Proc. of 5th Asian Regional Conf. on SM and FE, Bangalore, Vol. 1 : 87-90.
Jennings, J.E.B. and Burland, J.B., 1962. Limitations of the use of effective stress in partly saturated soils, Geotechnique. 12(2) : 125-144.
Nagaraj, T.S. and Srinivasa Murthy, B.R., 1983. Rationalisation of Skempton's compressibility equations, Geotechnique. Vol. 33, No.4 : 433-443.
Nagaraj, T.S. and Srinivasa Murthy, B.R., 1985. Compressibility of partly saturated soils. Jl of ASCE GT. Dn. Vol. 111, No. 7 : 937-942.
Nagaraj, T.S. and Srinivasa Murthy, B.R., 1986. Prediction of compressibility of overconsolidated soil. ASCE GT. Dn. 112, 484-488.
Sridharan, A., 1968. Some studies on the strength of partly saturated soils. Ph.D. thesis, presented to Purdue University, Lafayette, Indiana.

Hydraulic conductivity of tropical residual soils

G.N.NNADI, Queen's University, Canada

ABSTRACT: Compacted tropical residual soils are used for the construction of earth-dams, embankments, waste disposal facilities, and other earth structures. The hydraulic conductivity of these soils is therefore, of great importance in engineering problems of seepage, settlement, stability, pollutant migration, and drainage. The hydraulic conductivity of compacted soils is known to be affected by three state variables: void ratio, degree of saturation, and the soil microstructure. The relationship between the hydraulic conductivity of compacted clays and void ratio as well as saturation is well known, but its relationship with soil microstructure, which is an important variable is not well understood. This study shows the importance of soil microstructure as a variable and develops relationships between conductivity and structural factors. The roles of other factors in controlling the conductivity of tropical residual soils are also further clarified.

1. INTRODUCTION

The hydraulic conductivity of a compacted soil is, in many cases, the controlling factor in seepage, settlement, stability and drainage problems of civil engineering construction. Unfortunately,it is the least investigated among soils' engineering properties. However, recently, significant contributions have been made to our understanding of the factors that control the hydraulic conductivity of compacted clays. Mitchell et. al (1965) has observed that the structure of compacted clay is the most important single variable influencing the hydraulic conductivity of compacted clays and that the structure changes as the molding water content is increased.

Olsen (1962), and Barden and Sides (1970) have shown that the variation of hydraulic conductivity of compacted clays with molding water content is controlled largely by flow through the intercluster pores of the clay. At low molding water content, the clusters have high resistance to deformation; thus creating large intercluster pores filled with air. As the molding water content is increased, the resistance to deformation from compactive effort is reduced and subsequently smaller intercluster pores are formed.

Considering the strength of compacted clays, Seed et. al (1960) observed that at the time of compaction the relative magnitude of the attractive and repulsive forces as well as the nature and magnitude of the compaction energy determine the structure of the soil. They also observed that the structure may range between two limiting conditions: (A) completely flocculated structure, and (B) completely dispersed structure. Lambe and Whitman (1969), found that the highest permeability will occur in samples which are in the most flocculated state and samples in the most dispersed state will have the minimum permeability.

Very little information is available in the literature on the hydraulic behaviour of tropical compacted residual soils (Ola, 1980; Nnadi,1987). Tropical residual soils occur in different forms and are used as construction materials in different parts of the world. In Nigeria, perhaps the largest group of residual soils occur as laterites.The presence of road-side quarries bears testimony of their economic value as construction materials. This investigation examines the hydraulic behaviour of compacted tropical soils, and it is the purpose herein to outline the experimental procedures, to summarize the results and to indicate the practical implications of these studies.

2. MATERIALS INVESTIGATED

The region under consideration; southeastern Nigeria shown in Figure 1, lies between latitude 5 and 7 north of the equator and between longitudes 6 and 9 east of the meridian. The climate is tropical with rainfall varying from 350cm in the south to 150cm to the north. The geological formation is mainly sedimentary rock predominantly Cenozoic in age.

The engineering properties of tropical residual soils have been shown to be influenced by the mode of formation (Gidigasu, 1971). The materials investigated are used as construction materials and their selection was based on certain considerations:(1) they were being considered as construction material for the lmo Airport runway.(2) there have been cases of failures where they were used for embankments.

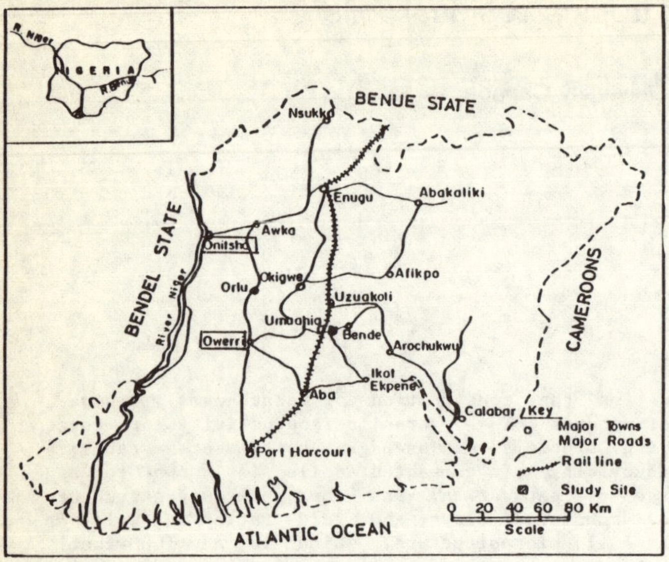

Figure 1. Map of the study areas.

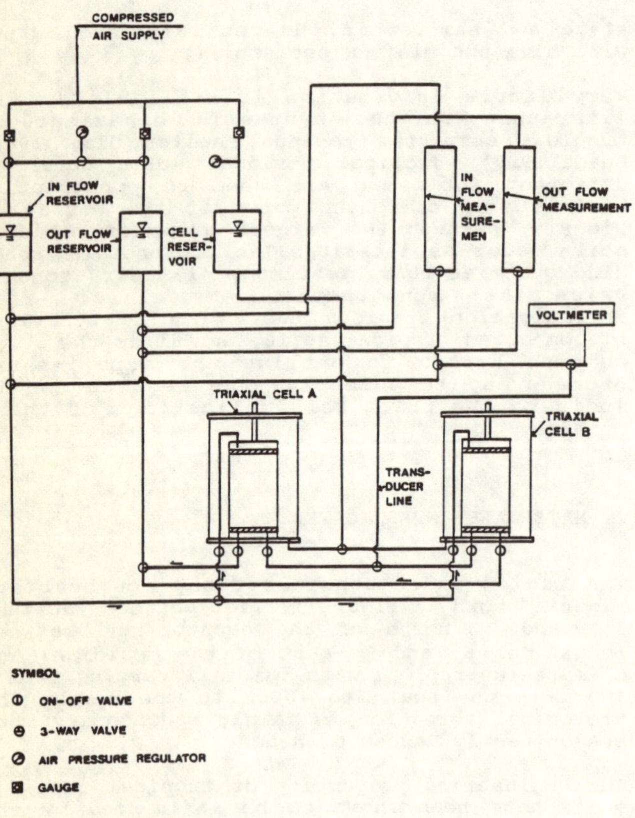

Figure 2. Schematic diagram of the two cells triaxial permeameter

Samples were collected from two locations: lmo-Airport site and Onitsha site. The locations are indicated in Figure 1. Samples from the lmo Airport site are designated in this report as IMOAP and Onitsha site samples as OS.Before samples were put into the air-tight polyethylene bags, soil "faces" were scrapped to remove exposed and oxidised materials so as to obtain fairly fresh samples.

3. EXPERIMENTAL PROCEDURES

A schematic diagram of the permeability test apparatus used is shown in Figure 2. lt allows the use of back pressure to saturate two specimens concurrently while permitting the flow measurements to be measured separately. Pressure gauge is used to adjust the applied pressure and hence the applied hydraulic gradient, but final adjustment of the hydraulic gradient is with the pressure transducer connected to a voltmeter. The transducer is also used to evaluate the pore pressure parameter "B" and the effective stress on the samples.

The cylinderical specimens are confined within flexible membranes on the sides with porous stones placed at the top and bottom. Leakages by osmosis and diffusion were reduced by enclosing the specimens with two flexible membranes separated by a thin layer of silicone fluid. Preferred flow between the specimen and the membranes was reduced by the application of a slightly high effective lateral stress of 10 kpa during saturation. To avoid over-consolidation of any portion of the specimens during saturation, each increase of back pressure was limited to 50 kpa and enough time allowed for pore pressure equilibration within the specimen. The hydraulic gradient used did not exceed 20 and this reduced the likelihood of particle migration.

Soil Materials:

The physical and chemical characteristics of the two soils as determined in this study are shown in Tables 1 and 2 respectively. Samples were carefully prepared to maxmize homogeneity and the results presented in Table 1 is a good representation of the properties of the compacted soils. Detailed results of the physical and chemical properties have been discussed by Nnadi (1988).

Although earlier investigators (Lambe,1958; Mitchell et.al,1965) advocated for the use of kneading compaction due to the fact that it reduces high shear strains on the specimen especially specimens compacted wet of optimum. Specimens for this study were however, compacted by static method because, values of laboratory hydraulic conductivity from static compaction quantitatively correlate better with field values than either kneading or impact compacted soils (Dunn and Mitchell, 1987). Morealso, static compaction gives better replicate samples more readily than any other laboratory method.

Specimens were carefully mixed and compacted to a tightly controlled density and water content. The dimensions were 50mm in diameter and 100mm

TABLE 1

Physical Properties of Test Soils

Property (1)	IMOAP (2)	OS (3)
Unified Soil Classification	SC-SM	SC-SM
Grain size distribution		
Percent Sand (%)	75	63
Percent Silt (%)	13	20
Percent Clay (%)	12	17
Atterbery Limits		
Liquid limit (%)	44	34
Plastic limit (%)	23	16.4
Specfic Gravity	2.64	2.68
Compaction Characteristics (ASTM D-1557)		
Maximum Dry Density(KN/m3)	17.6	18.8
Optimum Moisture Contents(%)	13.1	12.6

TABLE 2

Chemical Composition of Nigerian Residual Soils

Property (1)	IMOAP (2)	OS (3)
SiO_2	49.5	56.8
Fe_2O_3	8.98	10.45
Al_2O_3	12.3	17.6
CaO	0.01	0.01
MgO	0.015	0.106
K_2O	0.07	0.15
Na_2O	0.2	0.16
P_2O_5	0.22	0.18

high. By calculating the exact amount of soil and water required, any predetermined dry density and molding water content could be obtained. It was therefore possible to make sets of virtually identical specimens. Immediately following compaction, each specimen tube was covered with "parafilm M" at top and bottom and finally sealed with lucite cap and base for storage. Each specimen was stored for seven days in the humidity room before testing. Specimens were saturated with high back pressure. When the degree of saturation approached 90%, permeability was determined directly and also calculated using Cv values at each increment of back pressure. The coefficient of consolidation, Cv, was obtained from the pore pressure dissipation curves. The hydraulic conductivity, k, is related directly to the coefficient of consolidation, Cv, by the equation:

$$k = Cc r_w c_v$$

where
Cc = the slope of the $\delta V/Vo$ against the effective cell pressure at that point.

4. RESULTS AND ANALYSIS

INFLUENCE OF STRUCTURE:

To investigate the effect of soil structure on the hydraulic conductivity, the soils were compacted to the same dry density of 1700kg/m^3 at various molding water content. The results shown in Figure 3, indicate that the permeability decreases gradually with increase in molding water content up to an initial degree of saturation of about 90%. Above this level of saturation an abrupt decrease occurs. The sequence in the decrease in permeability illustrates the changes in the structure of the compacted residual soils from flocculated arrangement to dispersed orientation. The rate of decrease gives an indication of the structural sensitivity with respect to the molding water content.

It can be seen from Figure 3, that the permeability increased in IMOAP samples as the

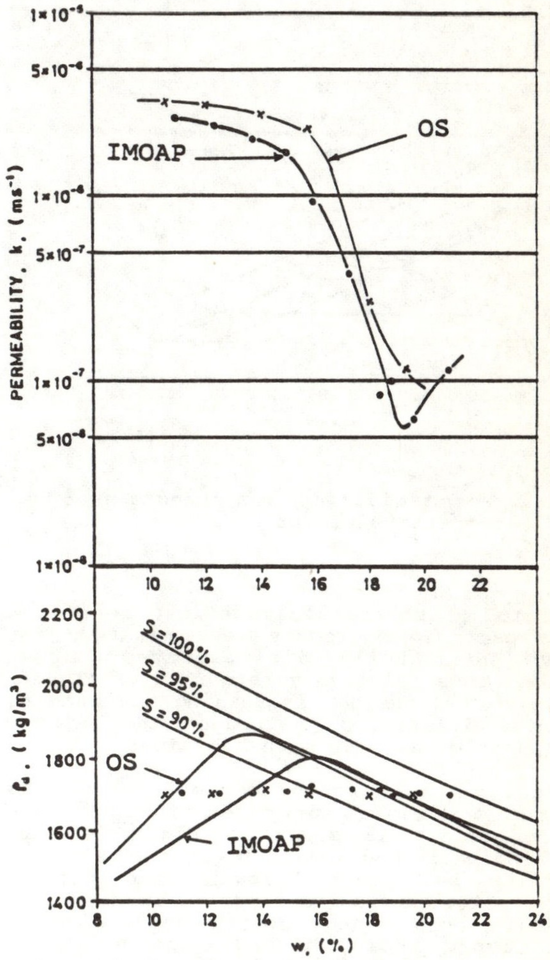

Figure 3. Influence of microstructure on the permeability of compacted tropical soils.

molding water content approaches 100%. Pervious investigators (Mitchell et. al, 1965; Ola, 1980), have shown that the compactive effort decreased with increasing molding water content. It might be argued that at molding water content

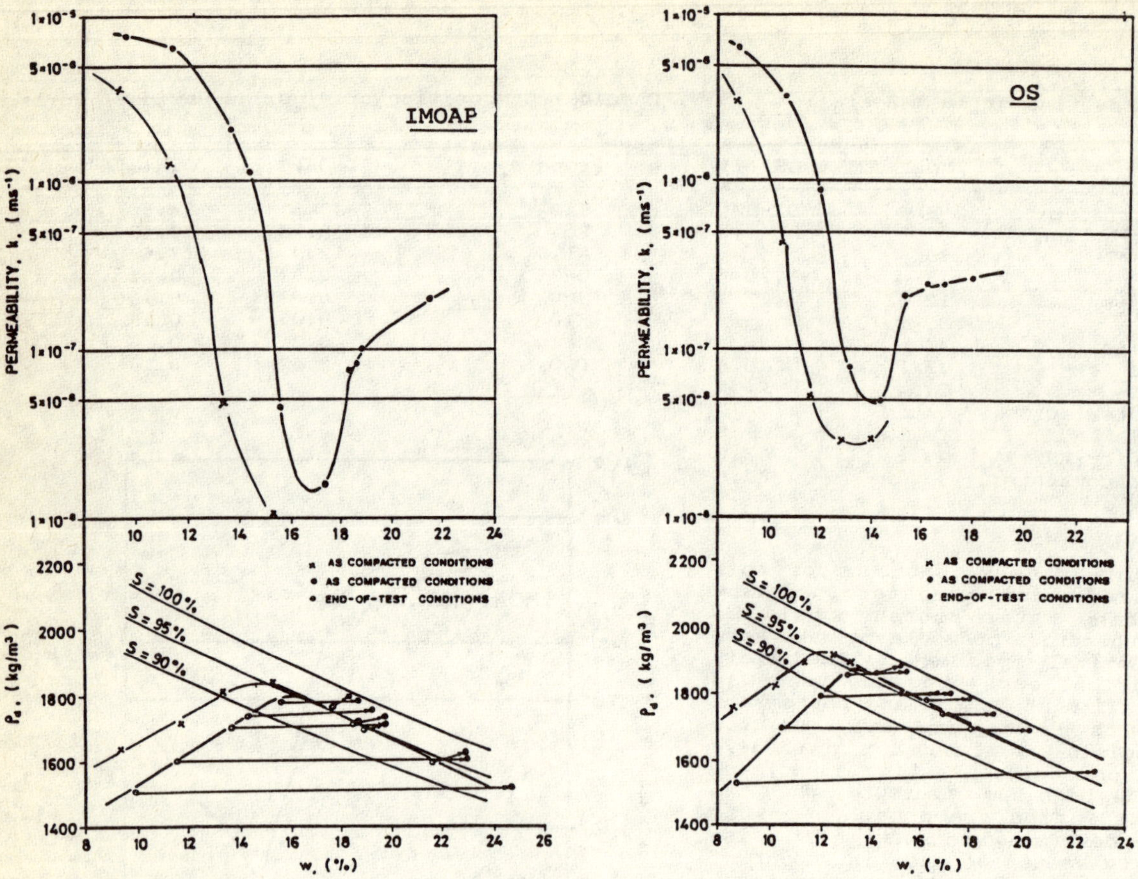

Figure 4. Permeability as a function of compactive effort.

close to 100%, the compactive effort is so low that the particle aggregates and clusters may not have been fully disintegrated during compaction. Thus relatively larger inter-cluster voids are formed thereby increasing the rate of flow. This explanation, although tenable, may not be the sole cause of such behaviour.

The effect of molding water content on the structure of samples prepared at the same compactive effort is shown in Figure 4. The results indicate the same abrupt decrease in permeability, but at a much lower water content. ln both Figures 3 and 4, the minimum permeability occurs at slightly above optimum water content, after which the permeability starts to increase. The implication of this behaviour is masked by the effect of the microstructure of the compacted soils because the samples are at various densities and void ratios.

Considering Hogentogler's theory of compaction, the moisture in a compacted soil is found as films surrounding the particles forming the soil. These films are adhesive in low water content and the adhesive force decreases as the water content is increased, until the limit of lubrication is reached. This limit is the same as the state of optimum dispersion at which the permeability is lowest. The thickness of the films of water increases until the optimum dispersed state is reached, therefore the thickness is limited for any given soil grain at the dispersion state. Thus for a given mean grain size (dm), we have the same thickness of film of water at the most dispersed state resulting to the lowest permeability. When the compactive effort is changed, the resulting mean grain size (dm) will have a different permeability value unless the mean grain size remains the same.

The results, therefore indicate that the most dispersed state in these soils occurs slightly above the optimum water content after which the permeability depends on the workability of the soil. The results presented in Figure 4, also show the effect of increased compactive effort on the structure of the soil. This is indicated in the abrupt decrease in permeability at a lower water content which implies that the degree of dispersion increased with increase in compactive effort.

INFLUENCE OF CELL PRESSURE AND BACK PRESSURE

Figure 5, shows the variation in permeability with cell pressure. lt indicates that permeability values vary in a consistent and reasonable manner with increase in cell pressure. The figure, also shows that at lower

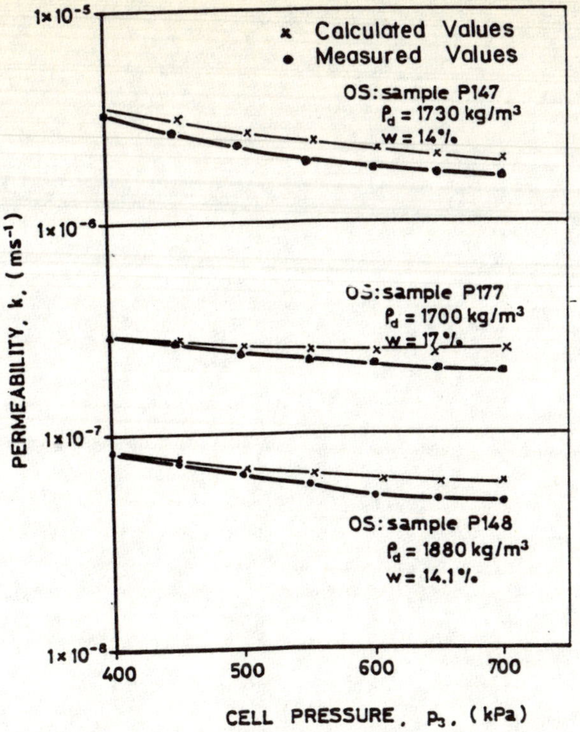

Figure 5. Permeability as a function of cell pressures

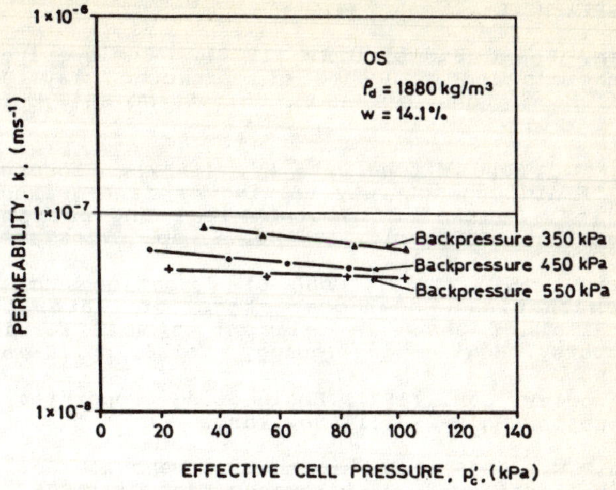

Figure 6. Influence of back pressures on permeability

cell pressure, the calculated values are in good agreement with the measured values. Both methods indicate that permeability values of these soils decrease with a decrease in void ratio, which is in agreement with results presented by Lambe and Whitman (1969).

Similiar observation is indicated in Figure 6, which is typical of most of the specimens tested. The results shows that at a high back pressure of about 550kpa, the influence of the effective cell pressure ,σ_3, is very minimal. On the other hand, at a lower back pressure, the effective cell pressure affects the permeability values obtained, indicating that high back pressure is required to actually saturate the soils from partial saturated state.

5.PRACTICAL IMPLICATIONS

The abrupt decrease and rise in permeability when the structure of the compacted soils is within the optimum dispersion limit can be of practical importance especially in the analyses of seepage, pollution migration and drainage problems. The results presented indicate that the permeability of these soils may vary over several orders of magnitude depending on the compaction conditions.

In the insitu and undisturbed state, these soils, due to cementation of the particles, have high permeability and are collapsible upon saturation. In the compacted state, the results show that these soils have low permeability and collapse under high presure is not very likely. However, the permeability is likely to increase with time after placement.

It must be remembered that the permeability values obtained in the laboratory represent generally the matrix hydraulic characteristices of the compacted soil. However, in the field, structural defects such as fissures, dessication crackings, or variations in the soil type are not represented in the laboratory specimens. The difference in results from laboratory and in-situ measurements could be reduced by improving construction monitoring and control.

6. CONCLUSION

Engineered fills are compacted by specified proceedures and/or to specified end results. The main objective of the specification is to produce compacted properties in the fill which are necessary to its safe and efficient performance. Among these properties, hydraulic conductivity is the most variable. The results obtained from this study have shown that the variation in permeability is influenced mostly by the orientation of the particles of the soil.

At low molding water content, the adhesive force between the particles and the water film is strong, resulting in a flocculated particle arrangement which gives high permeability values. Within the optimum dispersion state, the adhesive force is weak and the resulting permeability values are low.

Variation in permeability of two or three orders of magnitude may develop within the narrow range of optimum disperesd state for a particular compactive effort depending on the degree of dispersion of the soil structure.

High back pressure is essential to ensure saturation of these soils and for an effective comparative study of the different soils.

7. REFERENCES

Barden, L and Sides, G.R. (1970)." Engineering behaviour and structure of compacted clay",J. Soil Mech. Fdns.Div. ASCE, Vol.96, No.SM4, pp. 1171-1201.

Dunn, R.J. and Mitchell, J.K., (1987). Closure on "Fluid conductuvuty testing of fine-grained soils", Journal of Geotechnical Engineering Division, ASCE, Vol. 113, No. 6, pp. 672-673.

Gidigasu, M.D.,(1972)."Mode of formation and geomechanical characteristics of laterite material of Ghana in relation to soil forming factors", Engineering Geology, Vol.6, pp.79-150.

Hogentogler, C.H.,(1937).Engineering properties of soils. Mcgraw-Hill, New York.

Lambe,T.W.,(1958)."The engineering behaviour of compacted clays", J.Soil Mech. Fdns.Div;ASCE, Vol.84, No. SM2. Proc. Paper 1655.

Lambe, T.W. and Whitman, R.V.,(1969). Soil Mechanics. John Wiley and Sons, Inc. New York.

Mitchell, J.K., et. al.(1965)."Permeability of compacted clay". J.Soil Mech. Fdns. Div.,ASCE, Vol. 91, No.SM4, pp.41-65.

Nnadi, G.N.,(1987)."Geotechnical properties of tropical residual soils",Lulea University of Technology, Sweden, Research Report, Tulea 1987:09

Nnadi, G.N.,(1987). "Geotechnical properties of tropical residual soils",Lulea University of Technology, Sweden, Research Report, Tulea 1987:09

Nnadi, G.N.,(1988)."Mechanical and hydraulic properties of compacted tropical residual soils", Lulea University of Technology, Sweden, Research Report,Tulea 1988:02.

Ola, S.A.,(1980)."Permeability of three compacted tropical soils", Q.J.Engineering Geology,London, Vol.13, pp.87-95.

Olsen, H.W.,(1962)."Hydraulic flow through saturated clays", Clay and Clay minerals, Vol.9,Pergamon Press, New York.

Seed, H.B., et.al.,(1960)."The strength of compacted cohesive soils".Proc., ASCE Research Conference on Shear Strength of Cohesive Soils,Boulder, Colorado. pp. 877-964.

Engineering properties of a saprolitic soil from granite

H.NOVAIS-FERREIRA, Senior Research Officer, LNEC, Portugal
A.VIANA DA FONSECA, Research Assistant, LNEC, Portugal

SYNOPSIS A study has been made to full characterize a residual soil from granite in order to use it as construction or foundation material. For the former purpose some aspects were analized and discussed: mineralogical identification, particle-size distribution, Atterberg limits, "in situ" densities, compaction and CBR characteristics. For the later purpose results from extensive oedometric, direct shear and triaxial tests are discussed. Compressibility and shear strength characteristics of undisturbed and remoulded samples are compared. Stress path and type of consolidation influence is also discussed. A reference is made of k_0 value and suction potencial.

1 - INTRODUCTION:

The present work is a second part included in a more general study to characterize residual soils resulting from weathering of granite from a region of the North of Portugal (Oporto City). The soil was studied as foundation and embankment material. "In situ" and laboratory tests results are presented.

2 - SOIL,IDENTIFICATION,COMPACTION AND CBR

2.1 - The soil studied is as residual soil from granite which is an alkaline and leucocratic rock, mainly moscovitic; the predominant feldspar is the albite; important magnetitic intrusion appear frequently. The weathering process transforms the granite in a discoloured soil (white-yellowish colour) with a high void ratio; a mineralogical analysis by x-ray diffraction ($k\alpha$ of Cobalt), shows the presence of crystaline quartz minerals, feldspar, mica and kaolinite. A 105 - 110° C oven-drying of the samples did not change the size distribution of particles, but it did change the potassic and sodic feldspars by iron action. Direct shear tests showed a slight decrease of the shear strength of the oven-dried samples. Consequently it was decided to dry the soil in the laboratory environment, for all the remoulded samples to be tested.

The particle size distribution is indicated in Table I. The Atterberg limits and specific gravity are indicated in Table II.

The fraction of soil passing n° 200 ASTM sieve (0,075 mm) lies under the line A in the plasticity chart. This may be explained (Horta da Silva, 1974) by a decisive influence of the high percentage of mica and feldspar which retain water in the internal planes of cleavage. This water is non effective as regards plasticity, however as it is expelled by oven-drying at 110° C, it increases the values

TABLE I

Particle size distribution

size distribution parameters		Type of soil: natural	Type of soil: after compacted
accumulated percentages in particular sieves	# 9,5 mm (3/8")	100	100
	# 4,75 mm (n° 4)	95,4	94,7
	# 2,000 mm (n° 10)	83,2	81,8
	# 0,425 mm (n° 40)	62,5	60,5
	# 0,075 mm (n° 200)	37,1	36,1
	0,002 mm (silt/clay)	6,0	7,0
$U = D_{60}/D_{10}$		92,5	114,3
$Cc = (D_{30})^2/(D_{10} \cdot D_{60})$		1,62	2,16
Specific gravity γ_s = 2,63 g cm^{3}			

TABLE II
Compaction energies characteristies

Designation	E_1	E_2	E_3	E_4
Weight of the rammer P (kg)	2,49			4,57
Falling height H (m)	0,305			0,450
Number of layers, C	3			5
Number of blows per layer, N	12	25	55	55
Energy of compaction P. H. C. N. (kgm)	27,3	57,0	125,3	570,6

of liquid and plasticity limits. The spongeous structure of the iron oxids has similar effect.

2.2 - Compaction tests were executed in CBR mould, according to ASTM D 698 - 78 and ASTM 1557 - 78) using four different compaction energies (see table III). The compaction results, are represented in fig. 1.

2.3 - With samples obtained from the compaction tests, CBR tests were executed (ASTM D 1883 - 73 (1978)). On soaking and penetration an overload of 2,774 kPa was used (2 weights of 2,53 kg), corresponding to a 30 cm thick pavement.

In fig. 2 log CBR (2.5 mm) is correlated to dry density, and the CBR is found to be fairly sensitive

TABLE III
Compaction curves coefficients

Compaction energies			E_1	E_2	E_3	E_4
Granite residual soil	dry branches	A_d	0.867	0.678	0.664	0.163
		B_d	- 0.0125	- 0.0045	- 0.0062	- 0.0030
	wet branches	A_h	0.450	0.468	0.457	0.439
		B_h	0.0086	0.0074	0.0078	0.0078
Mean soils	dry branches	$\overline{A}_d$	0.728	0.704	0.669	0.619
		$\overline{B}_d$	- 0.0054	- 0.0059	- 0.0066	- 0.0076
	wet branches	$\overline{A}_h$	0.38 - 0.41			
		$\overline{B}_h$	0.0104 - 0.0117			

The following straight lines are obtained (Novais-Ferreira, 1962).

a) coordinates of compaction curves peaks:
$(1/\gamma_{dmax}) = 0{,}398 + 0{,}0115\ w_{opt}$ (r = 0,9997)

b) relationship between specific gravity and compaction energy:

$(1/\gamma_{dmax}) = 0{,}732 - 0{,}033.\ \log E$ (r = - 0,997)
$\gamma_{dmax} = 1{,}272 + 0{,}09859 \log E$ (r = + 0,9999)

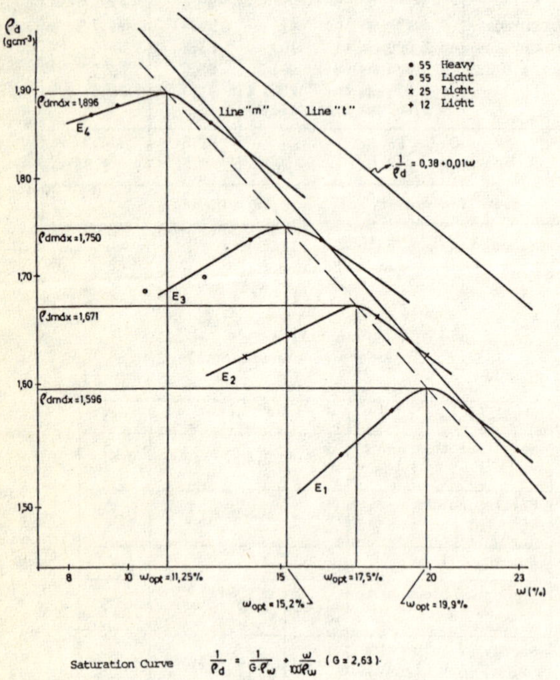

Fig. 1 - Compaction curves
(ordinate scale proporcionate to 1/ d)

to the compaction moisture content for the lowest energy (E_1). This influence decreasesas energy increases.The log CBR versus γ_d curves (fig. 2), corresponding to the compaction curves, lie in a zone limited by two straight lines forming a small angle and meeting at a common point: γ_d = 2,30 g/cm^3; CBR = 543% (Novais-Ferreira, 1962).
Fig. 3 represents the expansion variation z with d for the several compaction energies, and with w. The expansion varies directly with the dry density and inversly with the moisture content, with a stronger sensitivity in the dry branch. The expression that relates z to w is of the hyperbolical type.

$$\frac{w - w_o}{z - z_o} = -2{,}512 - 0{,}162\ (w - w_o)$$

with : w_o = 80%
z_o = 2.70%

3 - UNDISTURBED SAMPLES

3.1 - The samples were picked in a recent excavation, 8 meters below the initial surface level. A 1,5 meter trench was opened. The escavated material was used as remoulded sample. With a hidraulic jack reacting on the opposite wall of the trench several intact samples were picked by horizontal nailing of samples of two sizes (ϕ = 10 cm and l = 70 cm; ϕ = 8,2 cm and l = 40 cm).The samples obtained show a large dispersion in moisture content, which is in accordance with the "in situ" tests results. However, the soil dry density in the samples takes values around 1.60 g/cm^3. Saturation degree is high (70 - 90%) which is in agreement with results from other authors (Silva Cardoso, 1986).

3.2 - Compression characteristics

3.2.1 - Oedoemetric tests executed with undisturbed samples, saturated from the beginning, gave the following results.

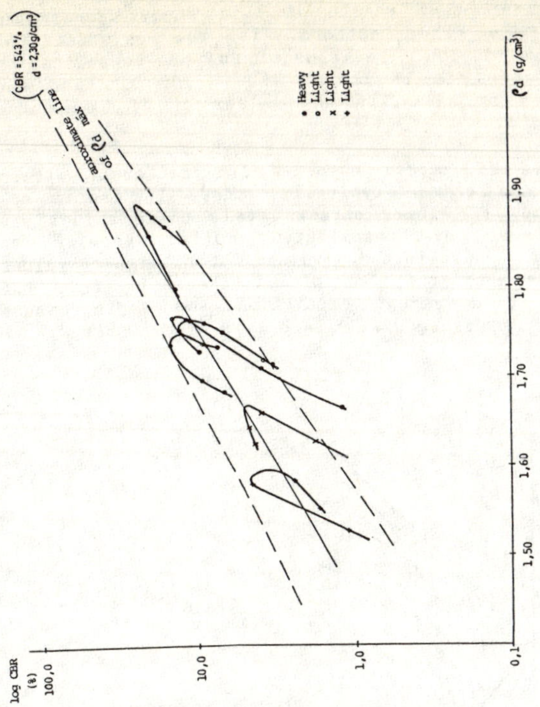

Fig. 2 - Correlation between log CBR and d (2,5 mm penetration)

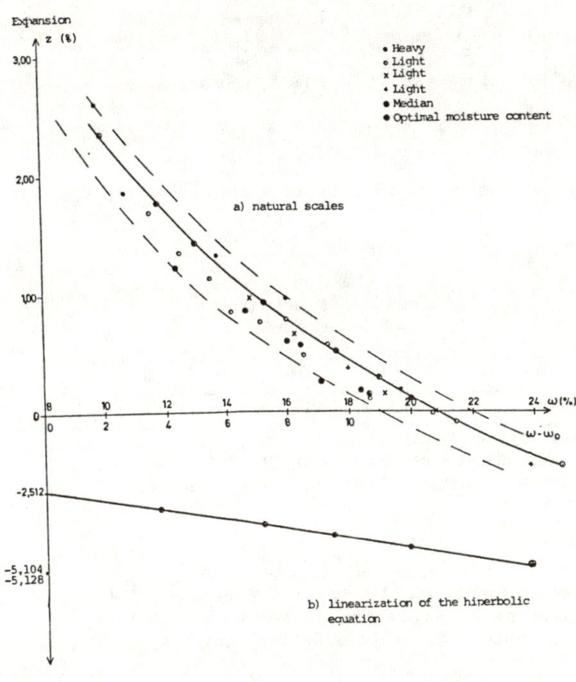

Fig. 3 - Expansion/moisture content correlation

a) <u>Coefficient of consolidation</u> (c_v). There's an initial zone with values around: c_v = 0,6 - 3,23 x 10^{-2} cm^2/seg growing with the initial dry density and with the applied stress, this being more obvious for lower values of initial dry densities. This behaviour is normal and characteristic of silty materials, with low percentage of clay.

b) <u>Compression index</u> (C_c). There seems to be two zones:

i) Stress values lower than 100kPa (overconsolidated zone: OCR > 1). The compression index takes values between: C_c = 0,024 - 0,037 it was verified that, for a particle size distribution with a maximum diametre of 9,51 mm (# 3/8", ASTM), it is very difficult to obtain an oedometric sample without disturbance. More disturbed oedometric samples assumed C_c values as great as 0,093.

ii) Stress values greater than 316,8 kPa (normally consolidated zone). Values of C_c between: C_c = 0,080 - 0,232

iii) The perconsolidation stress (σ_{pc}), determined from the oedometric tests, is around 100 - 150 kPa. Triaxial tests will give greater values which confirms the idea of partial sample disturbance in oedometric tests.

c) The secondary consolidation is non determinant (c < 0,01), as it would be expected.

3.2.2 - Oedometric tests, with undisturbed samples, were executed with the soil saturated from the beginning and with its natural moisture until the end. Comparing the values obtained, there was no evidence of significant differences, although the coeficient of consolidation (c_v) was slightly smaller. The equivalence between C_c values is justified by the high degrees of saturation of samples with natural moisture.

Vargas (1971) indicate a critical saturation degree (S_r = 70 - 80%), separating two different compression behaviour. For S_r lower than that value, the cemented structured between particles (coloidal cementation remainning from the weathering of original rocks) gives an aditional resistance to deformation. Reaching values of S_r greater than that, the cemented structure disappears, with the hydration of the colloidal material, and the soil becomes more compressible.

In natural state of moisture, with high saturation degrees, the possible suction contribution to the stabilization of the soil is neglegible because suction values are as low as 20 - 30 kPa, innefective in avoiding compression.

3.2.3 - Two series of oedometric tests, with undisturbed and saturated samples, were executed by loading each in one of two perpendicular directions (paralel and perpendicular to the ground surface). Similar values of the coeficient of consolidation, c_v, and compression index, C_c, for the two load directions, were obtained. This may mean that the soil has no ineherent anisotropy for compression characteristics, in accordance with the uniformity of rock weathering process and in opposition to layer deposition of sedimentary soils.

3.3 - Triaxial test

3.3.1 - Non-drained triaxial tests with measurement of pore pressure were executed after isotropic consolidation under σ_c near 50, 100, 200, 300, 400 and 600 kPa. Classical compression paths (ac-axial compression) were adopted - first series of tests. In order to study the stress path influence, four possible paths were carried out, after consolidation, at σ_c near 100kPa (ac-axial compression; l_e - lateral extension; ae-axial extension; and l_c - lateral compression) - second series of tests - and, to study the influence of the type of consolidation, an ac-axial compression test was made after anisotropic consolidation for σ_a near 200 kPa (axial) and σ_r near 100 kPa (radial) - third series of tests. The samples were saturated before consolidation by the application of a 300 kPa back - pressure; Skempton's B parameters of 92 - 95% were obtained. Two failure criteria were used: peak strength (max σ'_1/σ'_3) and strength at the inversion of pore pressure ($\partial u/\partial\sigma = 0$).

The Mohr envelope obtained (between σ_c = 50 KPa and σ_c = 600 kPa) presents a curvature in the earlier confinement stresses becoming a straight line when reaching values of σ_c around 100 kPa (fig. 4). This linear development of the curve gives

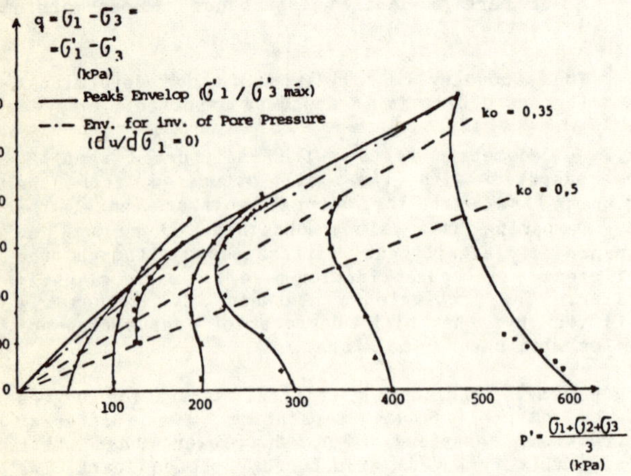

Fig. 4 - Envelop of the classical compression stress paths (σ a increasing; σ 1 = const)

the idea of existence of cohesion with values of: $\varphi' = 26^o$ and c' = 59 KPa), which is non traditional in soils with no plasticity and particle size distribution typical of granular soils (Silty - Sand, SM, from ASTM classification). This fact is in accordance with some authors (Vargas, 1971) that explain the existance of effective cohesion in undisturbed residual soils by the presence of cemented connections (coloidal type) between the larger particles giving the soil an aditional shear strength in lower stresses.

Tests made with "ae" and "lc," stress paths, seem to confirm this possibility (fig. 5), by showing a tendency for higher values of tensile strenghts than one would expected for a granular soil.

Some reserves must be made to these results of tensile strenght. Additional tests are necessary.

Another important result of this study, confirming ideas from other authors, is the existence of overconsolidation. The value of the pre-consolidation stress seems to be around 150 - 200 kPa and was controled by analysing the values of Af Skempton's parameter.

In table IV values obtained for different stress levels and common typically values for soils are presented. For lower stress levels these values are typical of overconsolidated soils, growing to normally consolidated values for higher values of σ_c. Note that we picked the soil from the bottom of a recent excavation, 8m depth, where geostatic pressure was σ_g = 8 x 18,5 = 150 kPa.

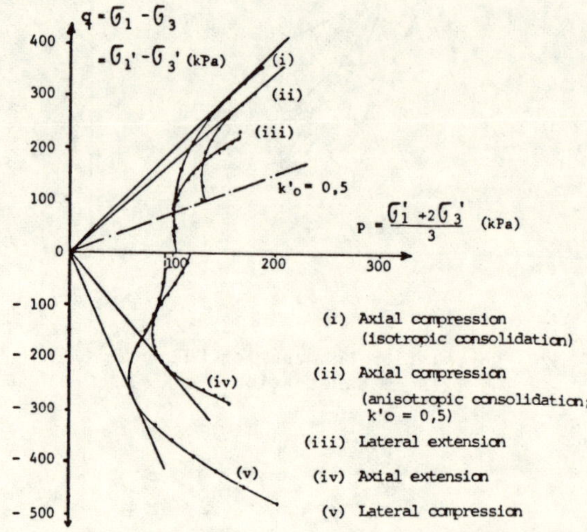

Fig. 5 - Stress paths possible for σc = 100 KPa

TABLE IV
Skempton's Af parameters
(undisturbed samples)

σ_c (kPa)	57	96	200	299	400	600
Af	- 0,005	0,060	0,160	0,391	0,479	0,536

normally consolidated clays: Af = 0,5 - 1,0
overconsolidated clays (OCR < 4): Af = 0,0 - 0,5
overconsolidated clays (OCR > 4): Af = - 0,5 - 0,0

In fig. 6 a diagram of σ'_1/σ'_3 versus (%) is presented. There are two types of curves: for $_c$ < 100 kPa a peak is presented; for σ_c > 200 kPa the peak is not noticeable. Notice that σ'_1/σ'_3 = $(q/\sigma'_3) + 1$

This probably is the result of overconsolidation with $\sigma_{pc} \simeq 150$ kPa. The origin of overconsolidation is different from the one existing in clayey soils. Its existence may be due to the referred "solid skeleton - coloidal structure" alied with a vertical discharge resulting from weathering process together with some erosion.

The fact reported by Lambe; Whitman (1969) that there might be an "interlocking decrease as the confinning stress increases, because the particles become flattement at contact points, sharp corners crushed and particles brooken "was studied by making a particle size analysis (with sedimentation) of the triaxial tests remaining material. It was then verified that there is no evidence of that particle size evolution.

The results obtained in the third series of tests, anisotropical consolidation ($k' = \sigma'_v / \sigma'_h = 0{,}5$) seems to demonstrate that the soil is non sensitive to this type of consolidation (at least for values smaller than the impulse at rest: $k' = 0{,}5$, $k'o$ - defined below).

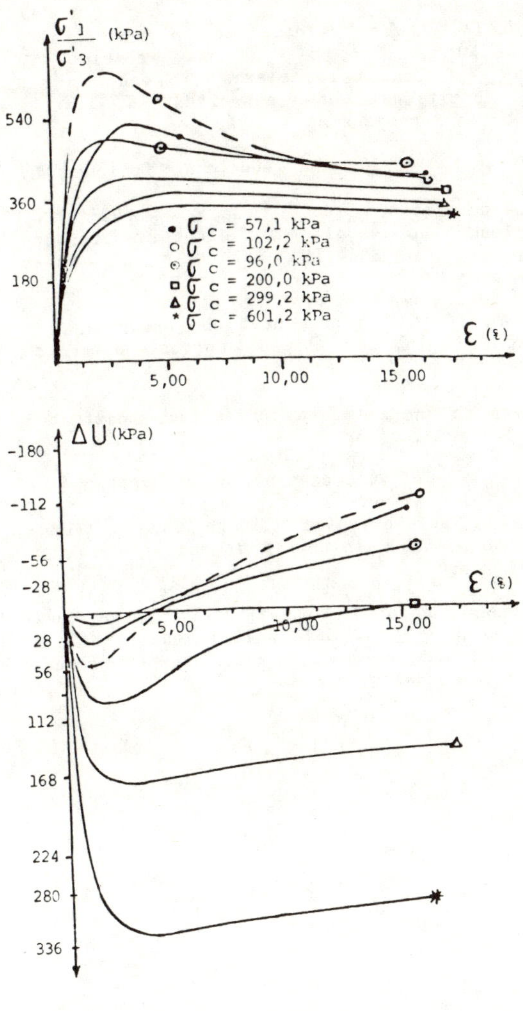

Fig. 6 - Stress/strain normalized curves and pore pressure/strain curves for increasing levels of consolidation (undisturbed samples)

There seems to be (fig. 6) a large influence of stress paths in shear strength of this soil (typical of granular soils). Effective shear strength for lateral extention ("le") stress paths is considerably smaller than the one obtained for axial compression ("ac") strength. This may be due to a greater trend to volume increase in shear phase with an increase in void ratio and consequent decrease in shear strength. The fact that $p' (= (\sigma'_1 + 2. \sigma'_3)/3)$; $\sigma_1 = \text{const})$, increases while σ_3 decreases, implies the following relation between stress variations:

$$\Delta u = 2/3 \, \Delta \sigma_3$$

Greater values of shear strength in tensile paths ("ae" and "lc") occur for little deformations assuming also lower values for higher deformations. This is expressed by the stress path inflection, in the p' - q diagram (fig.6), for residual strengths considerably lower that peak strengths.

There is a strong development of pore pressure values in the lc tensile paths which can greatly explain, the excessive strength values and difference between "lc" and "ae" values. Axissimetric load conditions contributes to this pore pressure development.

3.3.2 - - Direct shear tests were conduted with samples previously consolidated in the oedometric to 200 kPa. Some samples were contaminated with migmatitic soil. The results were:

a) Granite residual soil; $\varphi' = 34.0^o$ and $c' = 26.5$ kPa;

b) Migmatitic soil; $\varphi' = 32.8^o$ and $c' = 18.0$ kPa.

Taking into account the difference between those values, special attention should be given to the existence of this migmatic intrusions in field (see strength path for σ c = 400 kpa in fig. 4).

3.3.3 - A triaxial test was carried out with lateral restraining of deformations, using a Bishop's ring, on an undisturbed sample. The axial load was taken to $\sigma_1 = 1$ MPa. Until $\sigma_1 = 260$ kPa the sample had its natural moisture content; from $\sigma_1 = 260$ kPa until $\sigma_1 = 540$ kPa, a 300 kPa back-pressure for saturation was applied. On fig. 7 values of k'_o (relation between effective stresses) are indicated. k'_o values

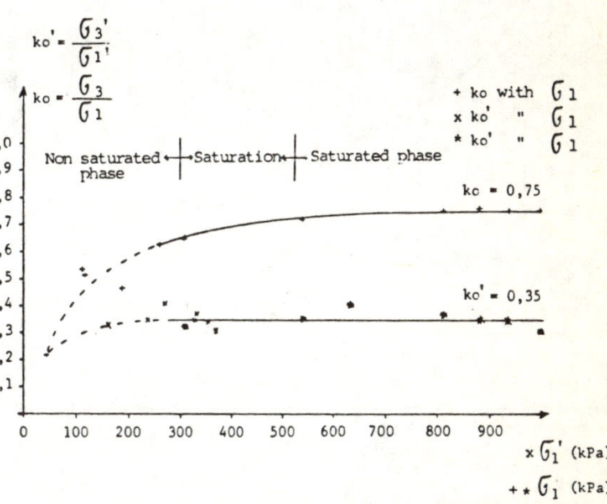

Fig. 7 - Variation of earth pressure impulses at rest with axial stress

stabilize at 0.35 and k_o values at 0.75. We must emphasize that the constancy of k_o and k_o simultaneously only can be achieved with u varying in accordance with: $u = a\ \sigma_1$; $a = 1.555$. This value was almost respected: $a = 1.511$; [1.3 - 1.7].Vaughan and Kwan (1984) assimilate a rock to a quasi-elastic structure whose stiffness is given by the isotropic elastic theory. The weathering is simulated by reducing values of the compression modulus E', and Poisson's ratio (ν'). Loss of strength due to weathering is simulated by reduction of cohesion c', with a constant effective angle of internal fiction. A small loss of stiffness (10% of the initial value) causes "in situ" horizontal effective stress to tend towards the limiting value given by the coefficient of earth pressure at rest in accordance with the elastic theory $k'_o = \nu' / (1 - \nu')$. Because the Poisson's ratio will tend to decrease with weathering, k_o will assume low values, lower than those usually assumed for normally consolidated soils (according to Jaky, $k_o = 1 - \sin\varphi'$). The value of 0,35 obtained by the test in laboratory, is in fact considerably lower than that given by the Jaky's formula (which would be about 0,5).

In opposition to higher values of k'_o for overconsolidated soils (in clayey soils, k_o may often take values as high as $k'_o = 3{,}0$), this small value for a soil with OCR greater than one, as we have proved above, confirms the different origin of this overconsolidation. In fact, while traditional overconsolidated soils have high dry density, this residual soil from granite presents values for γ_d as low as 1,60 g cm^{-3}. The cemented coloidal structure will be the justification for this behaviour.

3.3.4 - Suction tests showed that for the initial test conditions, the negative pressure is very low

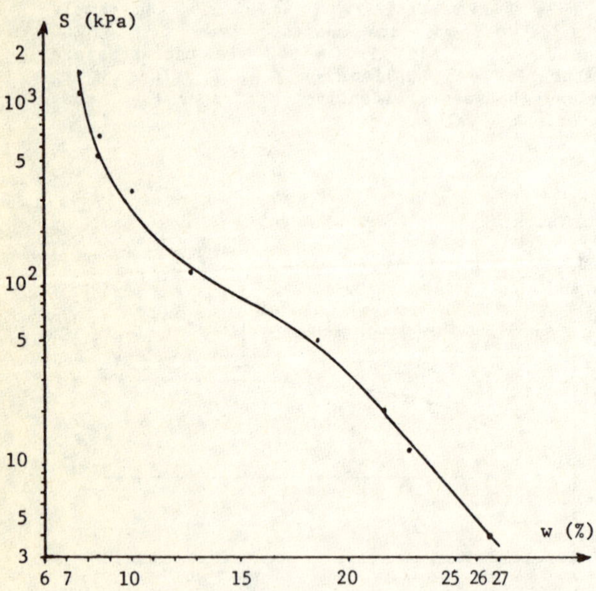

Fig. 8 - Suction curve (variation of negative pore pressure with soil moisture content)

(about 50 KPa), and so, non-saturated conditions (though with high degree of saturation), correspond to k'_o values that are not very different from those in saturated conditions.

The analysis of the suction curve shape (fig. 8), shows no existence of platform (as known, the platform is characteristic of the clayey materials, and the higher the plasticity index the larger the platform is). Non-existence of the platform reveals non-uniformity of the pore diametres which is a characteristic of spread particle size distribution. A steeper slope of the suction curve indicates higher sensitivity of soil to negative pore pressure variation (which induce stabilization) with the moisture content.

4 - REMOULDED SAMPLES - Shear tests:

4.1 - The soil was recompacted to dry density of $\gamma_d = 1{,}60$ g cm^{-3}, near to the existant "in situ" value, and tested in triaxial in each of the following conditions of consolidation, after saturation by back-pressuring:

i) σ_c = 108,202 and 400 kPa, followed by shear,
ii) Preconsolidated to σ_{pc} = 700 kPa, unloaded to consolidation effective pressure of σ_c = 304 kPa and taken to shear failure.

In figs 9 and 10, it is presented the shear envelop and stress strain normalized curves obtained, respectively.

In opposition to the obtained curves for undisturbed samples it is verified that:

a) There is no evidence of peak strength in the σ_1'/σ_3' versus - ε curves;
b) There is no inversion of the pore pressure variation, which initially increases and then stabilizes;

c) The shear envelop is approximately linear, with c' = o and $\varphi' = 35{,}6^o$. This means that the disturbed and recompacted soil doesn't present effective cohesion.

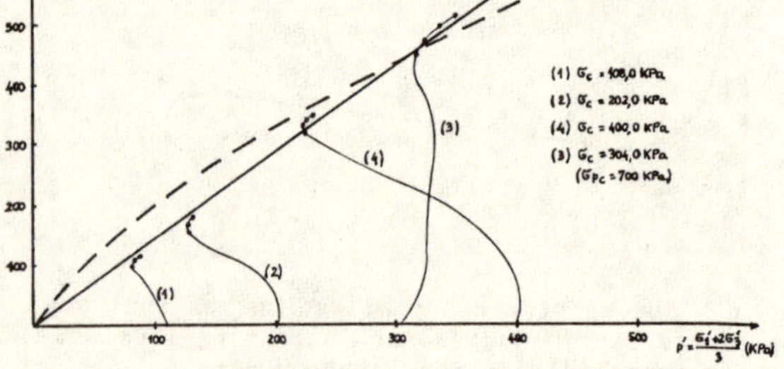

Fig. 9 - Mohr envelop for axial compression tests, on remoulded samples

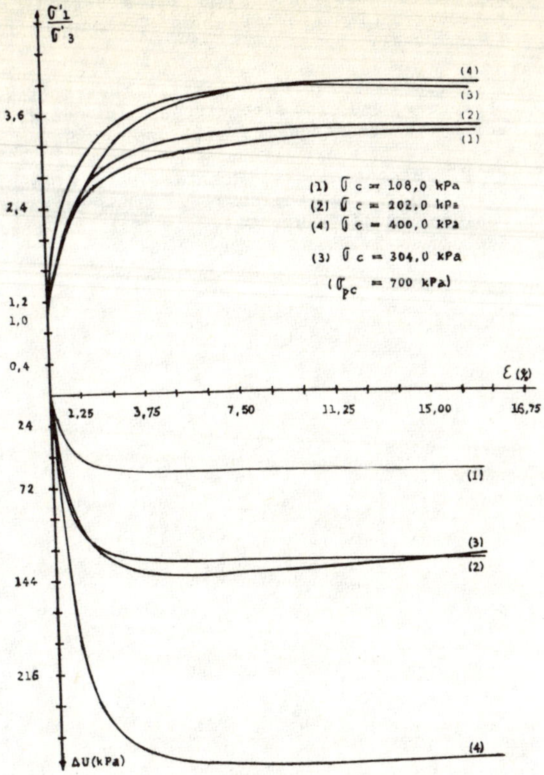

Fig. 10 - Stress/strain normalized curves and pore pressure/strain curves for increasing values of consolidation (remoulded samples)

This fact confirms the idea that the shear behaviour in an undisturbed residual soil differ very much from that of remoulded one. It also evidences the different origin of the effective cohesion in these soils with relation to the sedimentary clayey soils cohesion.

Values of Skempton's Af parameter are indicated in table V:

Table V
Skempton's Af parameters
(remoulded samples)

σ_c (kPa)	Af
108	0,550
202	0,757
400	0,849
304 (σ_{pc} = 700 kPa)	0,277

Comparing these values with those of table V it is observed that, for similar dry densities, undisturbed samples show an overconsolidated behaviour, while remoulded samples, excepting the one preconsolidated to σ_c = 700 kPa, behave as normally consolidated.

The overconsolidated struture of undisturbed samples is due to cemented connection, already referred, while the preconsolidated remoulded sample struture is perhaps due to interlocking between particles or grains, as a consequence of high pressures.

A particle size analysis of the triaxial tests remaining material was also made and there was no evidence of particle size evolution.

4.2 - Some direct shear tests were made with the soil passing n° 10 ASTM sieve and recompacted to dry densities of γ_d = 1,60 - 1,80 g cm^{-3} and the absence of effective cohesion was confirmed.

CONCLUSIONS

There is influence of drying process, for sample preparation, in strength characteristics of this soil.

The soil is a non-plastic silty sand but can present a ficticious plasticity due to retaining of water in internal planes of cleavage.

Compaction tests conducted to compaction curves with normal shape to medium soils. Therefore the compaction energy influence is expressed by semilogorithmics laws of the following type: $\gamma_{dmax} = a_1 + b_1 \log E$ and $w = a_2 + b_2 \log E$.

As a consequence the correlation between the dry density and moisture content for the various energies is expressed by a law of the type: $1/\gamma_d = a + b.w$.

Simultaneously, CBR tests show that exists a correlation of the following type: $\gamma = A + B. \log CBR$. This soil is very sensitive to the compaction energy. Values of expansion, when soaking CBR moulds, are related to w, following a hyperbolic law.

Oedometric tests show consolidation coefficients proper of silty soils, with an evidence of overconsolidation ($\sigma_{pc} \simeq$ 150 kPa).

The soil seems to have ineherent isotropy concerning to compressibility characteristics. The secondary consolidation is irrelevant.

Shear effective strength envelop, taken from saturated non-drained triaxial tests, with measurement of pore pressure, presents a curvature which, together with the Skempton's Af parameter variation with σ_c, confirms the existence of overconsolidation, mentioned above.

Tests with anisotropical consolidation reveal no induced shear anisotropy. The possibility of existence of cohesion due to cementation and tensile strength is also discussed.

A k'_0 test confirms hypotheses that k'_0 values (0,35) for this residual soil are smaller than the usually given by the common theories (Jaky's law). This is an indirect confirmation of the cemented structure between grains.

Suction tests reveal non-existence of platform, characteristic of spread particle size distribution, and a steep slope of the curve, indicating higher sensivity to variation with moisture content.

Triaxial tests with disturbed and recompacted samples conducted to a straight linear shear envelope.

This residual soil, as refered in literature for other soils, presents an effective cohesion "in situ", an overconsolidation ratio greater and k_o values smaller than expected. This behaviour, different from sedimentary soils, may be explained by particle cementation resulting from the weathering process of the original rocks, because it disappears with remoulding.

Soil parameters for earth works analysis, taken for "in situ" foundation or embankment materials, must be determined from distinct studies in well characterized conditions.

REFERENCES

Brand, E.W., Phillipson, H.B. (1982). "Sampling and testing of residual soils. A review of international practice". Technical Comittee on sampling and testing of residual soils. Spounsored by Southeast Asian Geotechnical Society.

Croney, D. (1952). "The movement and distribution of water in soils". Géotechnique, 3 (1), pp. 1-16.

Lambe, T. W., Whitman, R.V. (1979). "Soil Mechanics, SI version". J. Wiley & Sons.

Law, K.T., Hoetz, R.D. (1978). "A note on Skempton's A parameter with rotation of principal stresses". Géotechnique, 28 (1), pp. 57-64.

Novais - Ferreira, H. (1962). "A variação de baridade nos solos". Memória nº 173. LNEC, Lisboa.

Novais - Ferreira, H. (1985). "Characterization, identification and classification of tropical lateritic and saprolitic soils for geotechnical purpose". Tropical's 85. 1st Int. Conf. on Geomechanics in tropical lateritic and saprolitic soils. Brazilian Society for Soil Mechanics - ISSMFF.

Novais -Ferreira, H; Viana da Fonseca, A. (1988) "Engineering properties of a residual soil from granite". Proc. Int. Conf. on Engineering Problems of Regional Soils. Beijing, China, Aug. 11-15.

Silva, J.A.H. da (1974). "Residual soils from granite. Subsidy to their geologic - geotechnical knowledge". Publication nº 195. Lab. Nac. de Eng. de Angola. Luanda.

Townsend, F. C. (1985) "Geotechnical characteristics of residual soils" Journal of Geotechnical Engineering, vol 111, nº 1, Jan. 1985, pp. 77-94.

Vargas, M. (1971) "Geotecnica dos solos residuais" Revista Latino-Americana de Geotecnia, nº 1, pp 20-41.

Vaughan, P.R., Kwan, C.W. (1984). "Weathering, structure and in situ stress in residual soils". Géotechnique, 34 (1), pp. 43-54.

Vaughan, P. R.; Maccarini, M.; Mokhtar, S. M. (1988) " Indexing the engineering properties of residual soils" Quartely Journal of Eng. Geology, vol 21, nº 1, pp. 69-84.

Prediction of compressibility of tropical soils

N.S.PANDIAN, Indian Institute of Science, India
A.VATSALA, Indian Institute of Science, India
T.S.NAGARAJ, Indian Institute of Science, India

ABSTRACT : Residual tropical soils usually do not pose any stability problems. However, settlement plays a major role in the design and construction in these soils. The generalised compressibility equations have been developed within the framework of diffused double layer theory to predict the compressibility behaviour of saturated, uncemented fine grained soils in both normally as well as over consolidated ranges. It has been also shown that the equations derived for uncemented normally consolidated fine grained soils can be applied to natural soils containing coarser particles also, using floating matrix concept. These equations have been used to predict the compressibility behaviour of field samples which are in various stages of decementation/re-cementation due to geological processes and the predictions are found to be within the required level of accuracy. Their predictive capability with respect to the preconsolidation pressures which are due to cementation (as against the stress induced ones in the laboratory samples which formed the basis for the equations) are also shown to be reasonable.

1 INTRODUCTION

In a broad sense, soil formation is the result of the vast geological cycle of erosion, transportation, deposition and earth movement, occuring continuously and relentlessly throughout the geological time scale. Soils can broadly be divided into two groups on the basis of their origin, viz.,

i) residual - those derived in situ from weathering of rock material and,

ii) transported - those transported by wind, water and ice to new locations, redeposited and densified into various land forms.

Soils being particulate materials, the state in which they are encountered is influenced by past stress history, environment and time. The low shear strength and high compressibility of soils relative to intact rock material from which they originated are a consequence of high void volumes and weak bonding forces. Upon loading there would be appreciable relative movements between soil particles or their aggregated units resulting in pronounced volume reduction with the consequent increase in shear strength.

Design for stability in geotechnical engineering, when reduced to its simplest form, can be visualised as a comparison between loading and the ability of in situ soils to withstand the same within the permissible constraints of deformation, displacements and changes in volume. The ability of soils to withstand external loading is primarily reflected by their strength and compressibility characteristics determined under simulated field and environmental conditions. While each field situation is considered as independent and, extensive laboratory testing programme is contemplated, probing in detail tǒ the origin of the soil and other physical parameters of the soil would not be of any serious consequence. On the other hand, geotechnical engineers have always been striving to evolve simple methods to predict soil behaviour with a minimum input of the parameters that are commonly determined in practice. Empirical approaches that have been developed from time to time are satisfying this need. Any further attempt to interpret, generalize and predict soil behaviour within the framework of a scientific base will go a long way to

i) form an independent check on the laboratory investigations

ii) increase the level of confidence in handling the test data and,

iii) in specific cases to minimise, if not avoid time consuming laboratory tests.

This paper reports an attempt made to examine the possibility of applying the physical models already developed to predict the compressibility characteristics of normally consolidated uncemented fine grained soils from water contents corresponding to liquid limit, to predict the compressibility characteristics of undisturbed tropical residual soils. Before further discussion on the experimental programme, it is necessary to briefly discuss the differences between stress dependent normally and, over-consolidated transported soils and residual soils.

2 SEDIMENTARY SOIL DEPOSITS

As soils are carried by water and deposited elsewhere, the underlying layers get densified under their own weight. This causes the expulsion of interstitial fluids with a reduction in void ratio. The state of stress under which a sediment consolidates depends upon the overburden pressure. Consolidation of clay layer is governed by Terzaghi's principle of effective stress. The relationship between void ratio and effective pressure has been identified by Terzaghi (1941) as the sedimentation curve. Fig. 1 shows a sedimentation compression curve in an arithmetic scale. Terzaghi defined a clay as normally consolidated if it had never experienced a stress higher than the existing in situ effective stress. If clay samples are normally consolidated, the points plot on the line ABC: In the case of an over-consolidated clay, the maximum past pressure is larger than the present one and such a clay deposit would plot on the curve CD. The e-log p relation can be linearized in the zone of engineering interest.

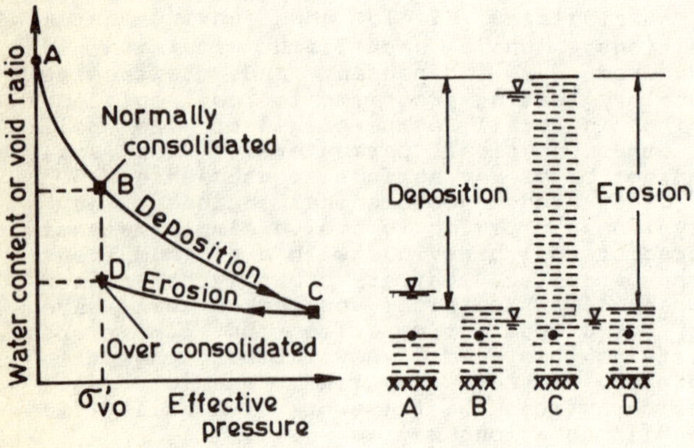

Fig. 1 Sedimentation Compression Curve

3 RESIDUAL TROPICAL SOILS

Exposed rocks are eroded and degraded by physical and chemical weathering. There would be gradual downward transition from soil to rock. The factors controlling the process of weathering are characteristic of parent rock, topography, climatic conditions, vegetation and time. If there is no relative displacement of soil from the locations where they are formed, such deposits are regarded as residual soils, mostly reflecting the compositional characteristics of parent rock material below. Soil formation in tropical environment adds one more dimension of complexity. The environment in tropics are hot and humid with evaporation in excess of precipitation. Due to wetting and drying cycles, the soil deposits are desiccated causing partial re-cementation. Thus residual tropical soils exhibit pseudo-overconsolidation compressibility behaviour. The e-log p relation would be bilinear with the stress at the intersection of two linear plots being the yield stress or pseudo-preconsolidation pressure. Due to partial weathering, soils can still retain some degree of cementation. This is also reflected in the above mode of pseudo-overconsolidated soil behaviour. The compressibility response cannot be related to existing overburden pressure of the soil. This is the main difference between the stress-dependent overconsolidation of transported, sedimented and consolidated soils and, environment dependent pseudo-overconsolidation effect of tropical residual soils.

4 PREDICTION OF COMPRESSIBILITY BEHAVIOUR

Nagaraj and Srinivasa Murthy (1983, 1985, 1986) based on diffuse double layer theory, have developed a generalised equation for the compressibility of fine grained normally consolidated uncemented soils, in the form.

$$e/e_L = 1.122-0.2343 \log p \qquad (1)$$

where e = void ratio, p = consolidation pressure in kPa and e_L = the void ratio at liquid limit. They have extended this further to generalise the compressibility behaviour of stress induced overconsolidated fine grained soils, of the form.

$$e/e_L = 1.122-0.188 \log p_c-0.0463 \log p \qquad (2)$$

where p_c is the preconsolidation pressure in kPa.

Fig. 2 schematically indicates the generalised normally consolidated line ABC representing equation (1) and the average rebound recompression line BD with a slope of 0.0463. The consolidation pressure at B represents the preconsolidation pressure.

Even though these equations have been developed for fine grained soils, it has been shown by Srinivasa Murthy, Nagaraj and Bindumadhava (1987) that they can be applied to soils containing coarser fraction also using the modified liquid limit based on the principles of floating matrix. The modified liquid limit of fine grained soils containing fractions coarser than 425μm can be determined using the equation.

$$W_{Lm} = W_L \times F/100 \qquad (3)$$

where W_L is the liquid limit of the soil on fraction finer than 425μm, F is the percent fine passing 425μm sieve and W_{Lm} is the modified liquid limit of the soil having a fraction coarser than 425μm equal to (100-F)%.

They have shown that the presence of coarser fraction linearly reduces the liquid limit and compressibility.

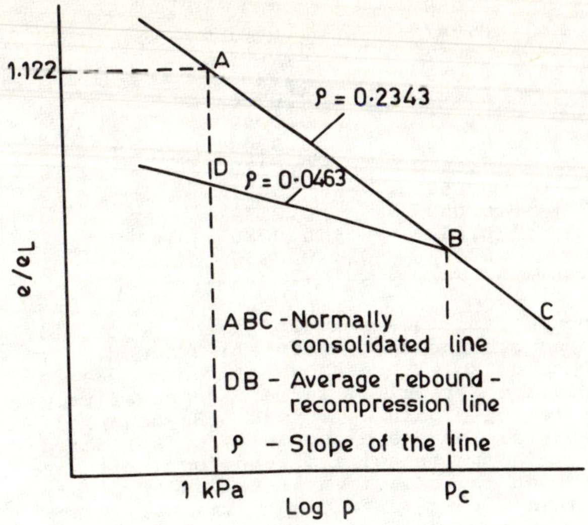

Fig. 2 Schematic Representation of Generalised Normally Consolidated Behaviour

In this paper an attempt is made to use these equations to predict the behaviour of tropical soils having varying degrees of decementation due to weathering and recemetation due to environmental factors.

5 INVESTIGATION DONE

Detailed soil investigations were carried out over an extensive area to house an industrial complex in the southern part of India. The area falls in the Archean age and is mostly occupied by granites and basalts along with kanker, bands of quartz, pegmatites and amphibolites. The soil is tropical in nature and its residual formation is due to in situ weathering of the rock, subsequently subjected to cycles of wetting and drying. This is typical of natural soil formations in tropical regions. The investigations consisted of laboratory as well as field tests. However, in this paper, the data relevant for discussion (i.e. physical properties and oedometer consolidation test results) only are presented.

For the consolidation tests, undisturbed samples were taken from different depths and the conventional oedometer tests done using a load increment ratio of one, the maximum load intensity being 800 kPa. Bulk density, dry density, natural moisture content, specific gravity and Atterberg Limits were also determined. Data obtained from 12 locations which are typical of the area investigated are presented in Table I.

Table I

Sl. No.	Depth (m)	γ_b (gm/cc)	G	e_o	LL	% passing ISSieve 425
1	0.35	1.83	2.65	0.770	60	75
2	1.70	1.72	2.70	1.000	61	88
3	0.71	1.79	2.70	0.730	51	69
4	0.68	1.88	2.65	0.710	48	85
5	0.63	1.88	2.65	0.670	60	58
6	0.60	1.89	2.65	0.710	64	78
7	0.85	1.88	2.70	0.695	65	57
8	0.48	1.78	2.70	0.890	66	79
9	0.73	1.76	2.65	0.860	67	85
10	0.55	1.93	2.65	0.780	62	76
11	0.60	1.93	2.65	0.760	59	75
12	0.75	1.93	2.65	0.920	70	86

Here LL refers to the liquid limit obtained using the fraction passing IS Sieve 425 and e_o is the in situ void ratio.

6 ANALYSIS OF THE RESULTS

The results are analysed to check the validity of equations (1) and (2). The liquid limit LL is modified using equation (3) and presented as modified LL. The void ratio corresponding to this modified liquid limit is given as e_L. The slopes of the compression and rebound lines of the e-log p plots as obtained from the experimental results together with those predicted are presented in Table II.

The overburden pressure is computed using the depth of sampling and bulk density. The preconsolidation pressure is obtained graphically using the e-log p plots and given as p_c experimental. Using equation (2), the preconsolidation pressure is predicted and reported as p_c predicted. These are reported in Table III.

7 DISCUSSION OF RESULTS

The predicted as well as experimental values of C_c and C_R are reported in Table II. Fig. 3 presents the values of C_c. The predicted and experimental C_c values of the first seven locations (serial numbers 1 to 7) agree reasonably well (shown by x in Fig. 3). In the remaining five locations, the experimental values deviate much from the predicted values and are always less than the predicted ones (shown by ⊗ in Fig. 3). Even among the seven locations having good agreement, excepting in one case, the experimental values are less than the predicted values. The reason for this can be discerned if the process by which soil was formed here is understood.

As stated earlier, the soil formation is because of decementation of rocks due to weathering. Subsequently, due to the process of wetting and drying over a geological time scale which is typical of tropical soils, certain amount of recementation might have also taken place. It is a well known fact that cementation resists compression. Fig. 4 schematically presents the e-log p plots for a stiff cemented soil. In the figure "ab"

Table II

Sl. No.	p_o	Modified LL	e_L	C_c Predicted	C_c Experimental	C_R Predicted	C_R Experimental
1	6.41	45	1.1925	0.2794	0.2750	0.0552	0.0048
2	29.24	54	1.4580	0.3416	0.3200	0.0675	0.0601
3	13.60	35	0.9450	0.2214	0.2500	0.0438	0.0204
4	12.69	41	1.0865	0.2546	0.2400	0.0503	0.0300
5	11.75	35	0.9275	0.2173	0.2000	0.0429	0.0200
6	11.34	50	1.3250	0.3104	0.2600	0.0613	0.0099
7	15.98	37	0.9990	0.2341	0.2200	0.0462	0.0100
8	8.54	52	1.4040	0.3290	0.2340	0.0650	0.0110
9	12.76	57	1.5105	0.3539	0.2000	0.0699	0.0094
10	10.60	47	1.2455	0.2918	0.2040	0.0577	0.0139
11	11.58	44	1.1660	0.2732	0.1499	0.0540	0.0080
12	14.48	60	1.5900	0.3725	0.2490	0.0736	0.0250

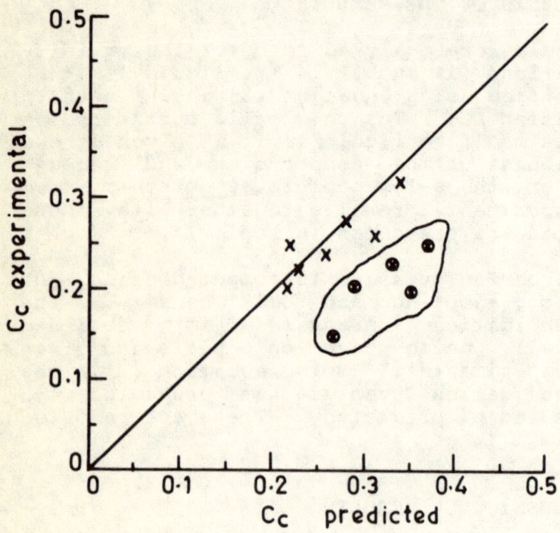

Fig. 3 Plot of Predicted Vs Experimental C_c

Table III

Sl. No.	p_c predicted (kPa)	p_c experimental (kPa)	p_c predicted / p_c experimental
1	113	100	1.13
2	88	100	0.88
3	34	70	0.49
4	132	160	0.83
5	59	70	0.84
6	355	275	1.29
7	70	100	0.70
8	133	160	0.83
9	246	185	1.33
10	148	160	0.93
11	108	105	1.03
12	253	240	1.05

corresponds to the e-log p path prior to pronounced breakdown of cementation bonds. At "b", considerable breakdown of fabric due to decementation results in a comparatively steeper e-log p path upto "c". Beyond "c", the behaviour is one of normally consolidated system. In the transition zone "bc", it is natural that the experimental values of C_c are less than the predicted values since cementation bonds are present to varying degrees resisting compression. The results reported in this paper are likely to be in the transition zone "bc" which explains why the experimental values of C_c are less than the predicted values. If the loading intensity is increased still further, breakdown can become still prominent approaching a completely particulate state. Subsequently, the experimental values of C_c will be closer to the predicted values.

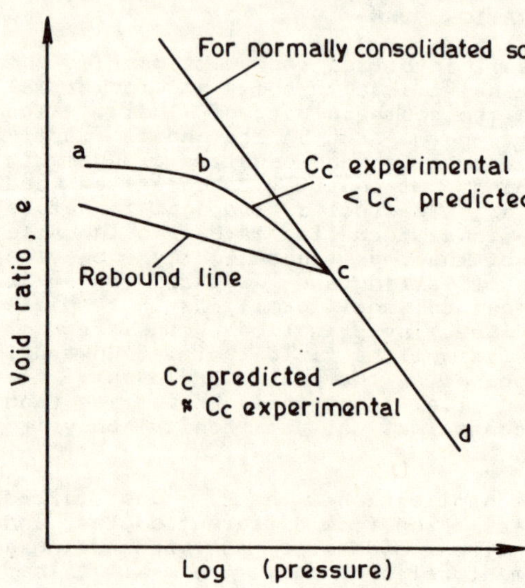

Fig. 4 Schematic Representation of the Compressibility Behaviour of Cemented Soils

Further, the equations were developed based on laboratory test results. Remoulded samples of fine grained, uncemented, saturated soils with initial water content equal to liquid limit water content were used. Hence, in addition to being devoid of any cementation bonds, they were also free from initial fabric effects. But, the field samples, in addition to some cementation and fabric effect, unlike laboratory samples, contained coarser fraction of soil also.

In the case of C_R, the rebound slope, all the experimental values are less than the predicted values. An important reason for this is the presence of cementation bonds in the soil system which do not allow the rebound to occur freely. Further, if it is remembered that the equations were developed for fine grained soils wherein surface forces dominate the gravitational forces, another reason for the measured low values of C_R can also be discerned. In the field samples, due to the presence of coarser particles, the gravitational forces have become dominant making the rebound slopes flatter. In the case of compression, the magnitude of gravitational forces of soil particles was negligible compared to the consolidation loads and hence their effect was not felt.

The values of preconsolidation pressures p_c predicted as well as experimentally obtained together with the ratio of predicted value to experimental value are reported in Table III. Excepting a couple of locations, the agreement is fairly good. It is to be mentioned that the preconsolidation pressures were stress induced in the laboratory samples, whereas, they were due to varying degrees of decementation/recementation for the field samples.

The e-log p curves experimentally obtained as well as predicted (pertaining to serial numbers 2 and 5) are presented in Fig. 5 and 6. For these locations, the modified liquid limit values are 54 and 35. These figures illustrate the capability of the equations to predict the compression and rebound paths fairly well.

The parameters used in the expressions do not require any sophisticated testing facility or time consuming experimentation. The in situ void ratio, the liquid limit, the fraction passing 425µm, the bulk density of the soil and the depth of sampling - are all the data required. These form part of the routine testing in geotechnical engineering practice and can be easily determined even in the field.

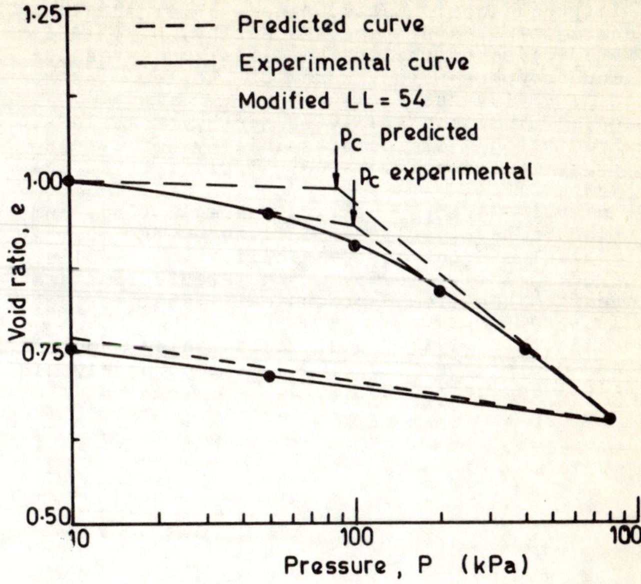

Fig. 5 Void Ratio Vs log Pressure, Predicted and Experimental

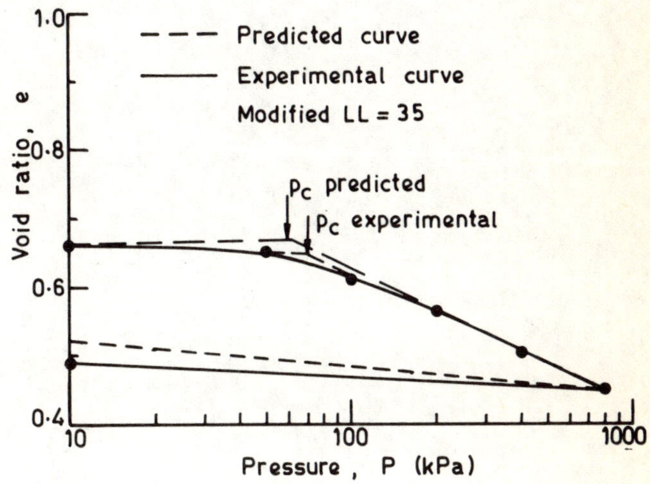

Fig. 6 Void Ratio Vs log Pressure, Predicted and Experimental

8 CONCLUSION

With very simple tests, it is possible to predict the "e-log p" curves and preconsolidation pressures of field soils within the accuracy usually needed in Civil Engineering practice. The predicted values of C_c are on the safer side.

The equations developed for uncemented, saturated, fine grained remoulded soil samples can be used to predict the behaviour of residual tropical soils having coarser fractions within the accuracy required in field operations.

9 REFERENCES

Nagaraj, T.S. and Srinivasa Murthy, B.R. 1983. Rationalisation of Skempton's compressibi-

lity equation. Geotechnique. 33.4. 433-443.
Nagaraj, T.S. and Srinivasa Murthy, B.R. 1985. Prediction of the preconsolidation pressure and recompression index of soils. ASTM, GTJ. 8. 199-203.
Nagaraj, T.S. and Srinivasa Murthy, B.R. 1986. Prediction of compressibility of overconsolidated soil. ASCE, GT Divn. 112. 484-488.
Srinivasa Murthy, B.R., Nagaraj, T.S. and Bindumadhava, 1987. Influence of coarse particles on compressibility of soils. Proc. Intl. Symp. on Prediction and Performance in Geotechnical Engineering. 196-200.
Terzaghi, K. 1941. Undisturbed clay samples and Undisturbed Clays. J., Boston Society of Civil Engineers. 28. No.3

Consolidation behaviour of a residual soil

I.PANNILA, Educational Assistant, The Open University, Nawala, Nugegoda, Sri Lanka
A.THURAIRAJAH, Professor, Civil Engineering Department, The Open University, Nawala, Nugegoda, Sri Lanka

ABSTRACT The results of one-dimensional consolidation tests carried out on undisturbed samples of a residual soil obtained from the central hill country region of Sri Lanka are discussed. Tests have been carried out under different consolidation pressures and pressure increment ratios. The settlement under each pressure increment was measured for a period of one week. Settlement time curves obtained consist of the following three phases ; i) an immediate settlement, ii) primary consolidation settlement and iii) secondary consolidation settlement.

Settlement generally varied linearly with logarithm of time for large values of time before decreasing rapidly to zero. Results show that the consolidation behaviour of the residual soil can be defined satisfactorily by the Gibson and Lo model. Values of soil parameters required for this model have been determined graphically from the tests.

INTRODUCTION

About 90% of the land mass of Sri Lanka is made up of highly crystalline rocks of granites, charnockites and gneisses belonging to one of the most ancient and stable part of the earth crust, the South Indian Shield. In-situ weathering of these rocks under intense tropical conditions i.e. high temperature and rainfall, has resulted in the formation of residual soils which cover most parts of the land area. Since the degree of weathering depends on the depth from ground surface, residual soils consist of materials with different grades of weathering, varying from fresh rock to completely weathered material.

In the hill slopes of the central region of the country, transported residual materials, colluvium, consisting of gravels, cobbles and boulders in a matrix of sand, silt and clay are found deposited over the weathered in-situ material. This material is also included under residual soils since it has similar characteristics to the in-situ weathered material.

Residual soils are considered problematic soils due to their heterogeneous and anisotropic nature. Research studies associated with the engineering behaviour of residual soils are very limited.

In this paper, the results of one-dimensional consolidation tests carried out on undisturbed samples of a residual soil obtained from the central hill country region of Sri Lanka are discussed.

EXPERIMENTAL DETAILS

One-dimensional consolidation tests were carried out on undisturbed samples of a residual soil taken from Hunasgiriya area in the central highlands of Sri Lanka. Block samples were obtained from a depth of 1 m below the ground level. Classification tests carried out on air-dried samples gave the following results ; liquid limit = 62.7%, plastic limit = 25.3% ; clay fraction = 27.1% ; silt fraction = 16.9 %; sand fraction = 38.5% ; gravel fraction = 17.5%. The soil has a specific gravity of 2.71.

Standard fixed ring type consolidometers having 75 mm internal diameter and 20 mm high brass rings were used for the tests. Soil samples were obtained by pushing in the brass rings into the block samples. Porous stones were kept at the top and bottom ends of the sample in the consolidometer, so that drainage can take place from both ends of the sample during consolidation. In order to prevent fine particles being forced into the pores of the stones, filter papers were placed between the porous stones and the soil sample.

The inside of the consolidation ring was lubricated with silicone grease to reduce the side friction. As constant temperature facilities were not available in the laboratory the tests were carried out under normal laboratory conditions. The average degree of saturation of the soil samples was about 93.1 %.

The soil samples were consolidated under initial pressures of 24.4, 48.8, 97.7, 195.4, 390.8, 781.6, 1563.3 and 3126.5 k Pa, and settlement observations for each pressure increment were taken for a period of seven days. Pressure increment ratios of 1/8, 1/4, 1/2, 1, 2, 4 and 8 were used.

SETTLEMENT - TIME CURVES

The settlement-time curves obtained from the consolidation tests consisted of the following three phases ;

i) an immediate settlement, probably due to the collapse of the structure of the residual soil ;
ii) primary consolidation settlement, due to gradual dissipation of excess pore water pressure ; and
iii) secondary consolidation settlement due to readjustment of soil particles after excess pore water pressures have been dissipated.

The curve in Fig. 1 shows the variation of the degree of consolidation U with the time factor T_v, predicted from Terzaghi's one-dimensional consolidation theory. The points in the figure shows the variation obtained from experiments on four typical samples initially consolidated under a pressure of 195.4 k Pa and subjected to pressure increment ratios of 1/4, 1/2, 1 and 2. Taylor's square root of time fitting method was used to determine the coefficient of consolidation and 100% primary consolidation for these tests. The experimental points lie close to the theoretical curve up to about 90% primary consolidation. Beyond this, the observed settlement is greater than the settlement predicted by Terzaghi's theory.

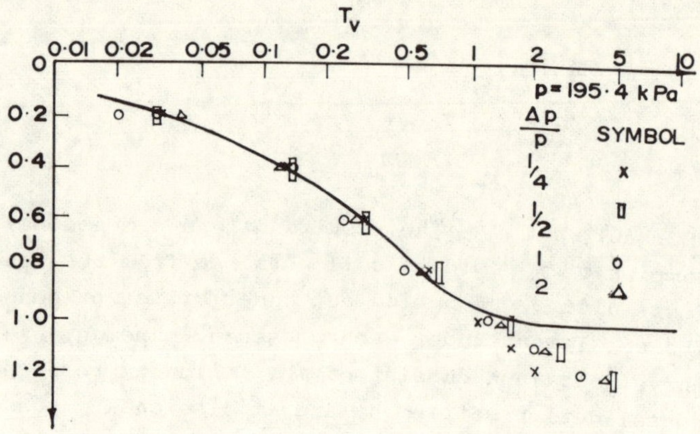

Fig. 1 - Variation of U with T_v during consolidation tests

A typical settlement - square root of time curve for a consolidation test where a pressure increment ratio of 1 has been applied on a soil sample that has been consolidated under a pressure of 781.6 k Pa is shown in Fig. 2. Generally, it was found difficult to separate the primary consolidation settlement from the total settlement by this method.

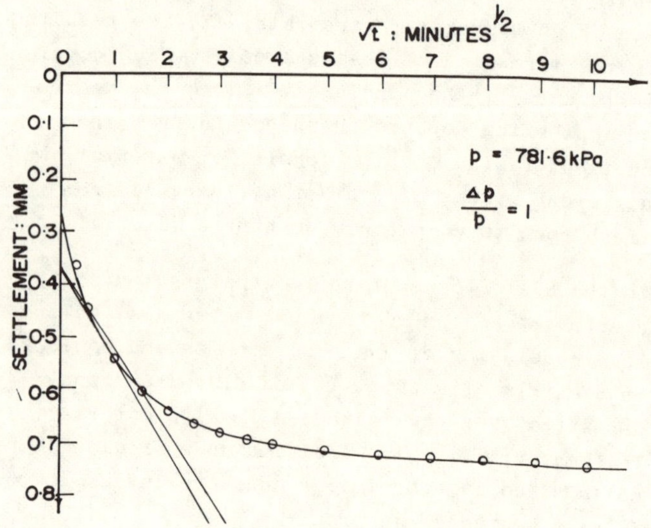

Fig. 2 - Settlement - Square Root of Time curve

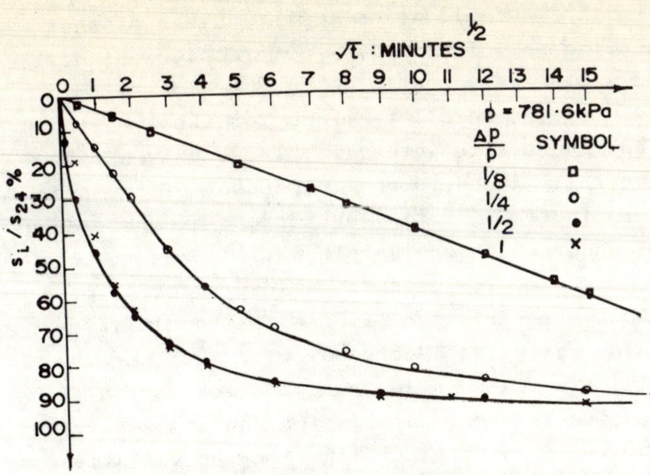

Fig. 3 - Variation of s_i/s_{24} with Square Root of Time

Fig. 3 gives some of the results of the consolidation tests carried out on a sample initially consolidated under a pressure of 781.6 k Pa and subjected to consolidation under different stres-s increment ratios. Here s_i is the settlement at any time minus the immediate settlement and s_{24} is the settlement in 24 hours minus the immediate settlement. It can be seen from the figure that the portion of the settlement time curves obtained with the dissipation of the excess pore water pressure is more representativ-e of Terzaghi's theory for low stress increment ratios.

One-dimensional consolidation tests on a number of clays show that the settlement time curves can be classified into three types (Lo, 1961) ; in type I, the rate of secondary compression gradually decreases with the logarithm of time until ultimate settlement is reached; in type II, the rate of secondary compression stays proportional to the logarithm of time over an appreciable range, then decreases rapidly to zero ; in type III, the rate of secondary compression increases with the logarithm of time and then decreases rapidly.

Fig. 4 shows some typical settlement-time curves obtained for four pressure increment ratios applied on soil samples consolidated initially

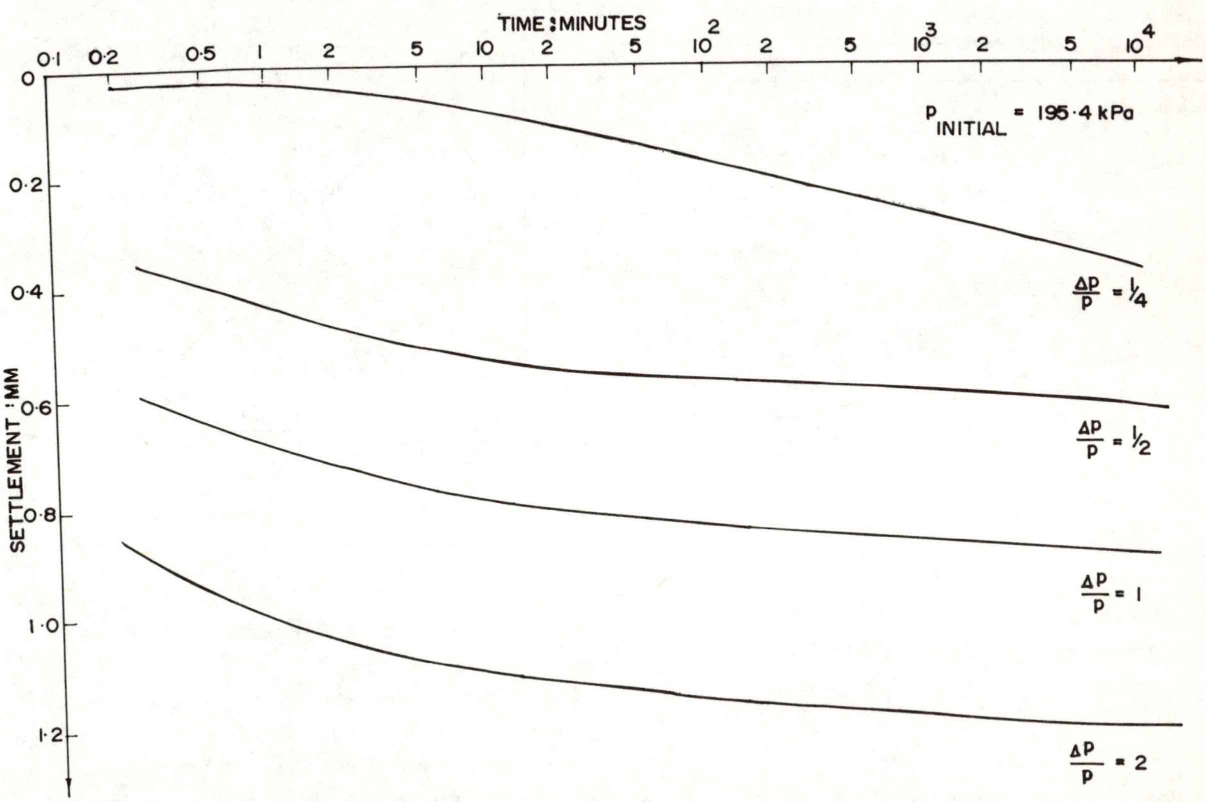

Fig. 4 - Settlement-Time curves for different pressure increment ratios.

under a pressure of 195.4 k Pa. The shape of the settlement curve is seen to depend on the pressure increment ratio applied. Most of the curves obtained are found to be of type II while type III curves are obtained for small pressure increment ratios.

The voids ratio change during the secondary compression stage is plotted against logarithm of time in Fig. 5 for a typical pressure increment ratio of 1/2 applied on soil samples initially consolidated under different pressures. The slope of the straight lines drawn through the experimental points seems to increase with the increase in consolidation pressure. One-dimensional consolidation tests carried out on laterite showed similar behaviour, Thurairajah and Wijeyakulasuriya (1979). Thurairajah (1967) found that the slope of the secondary tail decreased with increasing consolidation pressure for Thunukkai clay. On the other hand, Newland and Allely (1960) found that the slope of the secondary tail was independent of the consolidation pressure for Whangamarino clay.

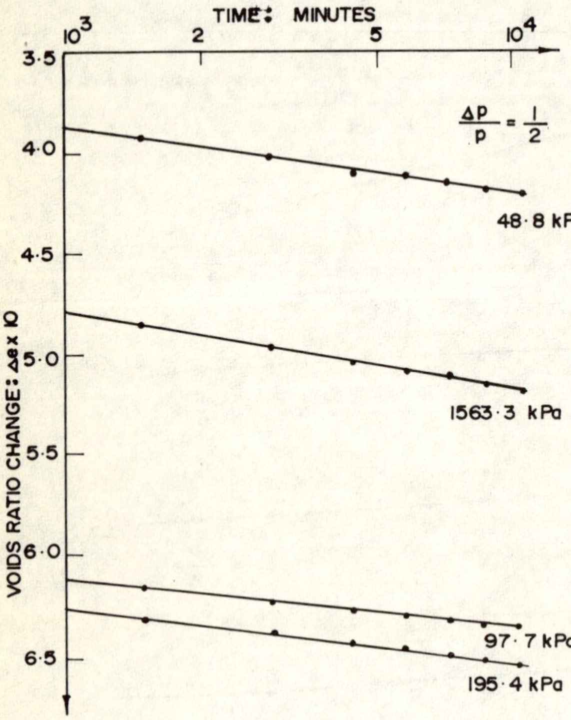

Fig.5 - Voids Ratio change vs. Time during secondary compression stage.

From the consolidation tests, values of the coefficient of secondary compression c_s, defined as the voids ratio change per logarithmic time cycle, were determined for the 1000 to 10,000 minute time interval. These values were generally found to be independent of the pressure increment ratio, but depend on the pressure under which the sample was allowed to consolidate i.e. the ultimate voids ratio. Fig. 6 shows the variation of c_s with the ultimate voids ratio. c_s decreases with increase in ultimate voids ratio, but wide scatter in the results is seen. Statistical analysis carried out to fit a straight line for the variation of c_s with the ultimate voids ratio e gave the following equation for the line :-

$$c_s = 7.29 \times 10^{-3} - 1.06 \times 10^{-2} \; e.$$

The regression line has a sample correlation coefficient of - 0.81.

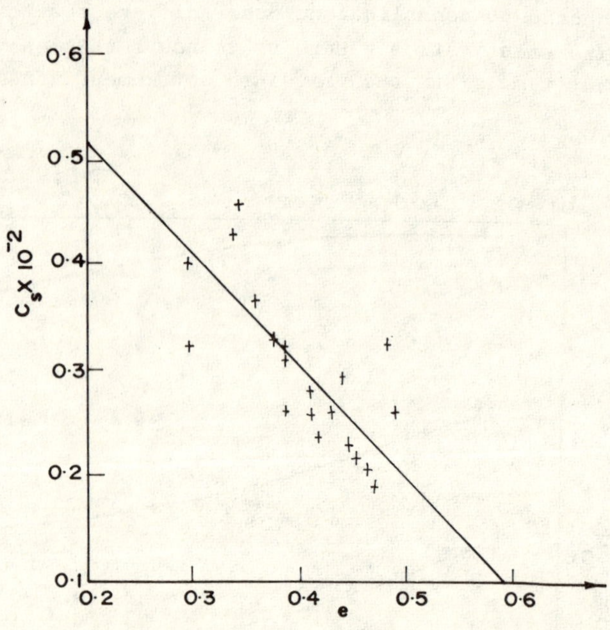

Fig. 6 - Variation of c_s with the ultimate voids ratio.

In Fig. 7, the ratio of immediate compression D_i to the ultimate compression D_u is plotted against the pressure increment ratio. The points plotted are the average values for each pressure increment ratio. The ratio of immediate compression to ultimate compression D_i/D_u increased from 7% for a pressure increment ratio of 1/8 to about 77% for a pressure

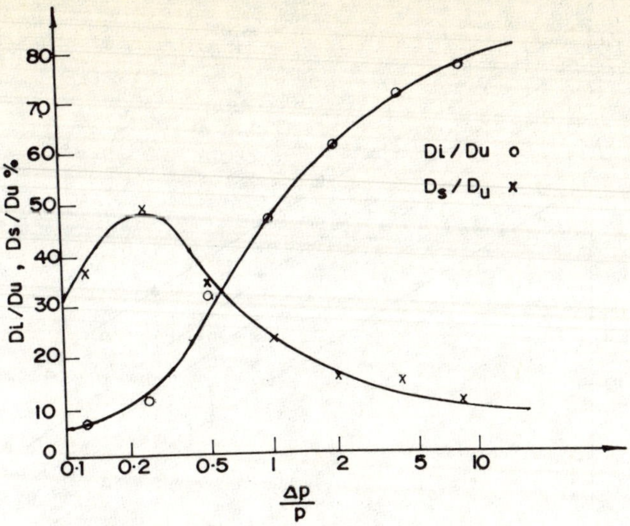

Fig. 7 - Variation of immediate compression and secondary compression with pressure increment ratio

increment ratio of 8. This behaviour is presumably due to the collapse of the structure of the residual soils at high pressure increment ratios which leads to larger immediate settlements.

Fig. 7 also shows the variation of the average value of the ratio of secondary compression per logarithmic time cycle D_s (time interval between 1000 and 10,000 minutes) to the ultimate compression D_u, with the pressure increment ratio. The ratio D_s/D_u decreased from about 50% for a pressure increment ratio of 1/4 to about 10% for a pressure increment ratio of 8. However, D_s/D_u was found to increase within the range of pressure increment ratio of 1/8 to 1/4. This behaviour is different from that observed for a remoulded clay (Thurairajah 1967) where D_s/D_u was found to decrease continuously with the increase in stress increment ratio.

The above results should be taken into consideration when estimating settlements in practical problems, as the pressure increment ratio would vary with the depth in the soil strata.

SOIL PARAMETERS

According to Gibson and Lo's (1961) theory for consolidation, the settlement s_t at any time t after the application of a pressure increment Δp is given by ;

$$s_t = H \quad \Delta p \left\{ a + b\,(1 - e^{-\lambda t/b}) \right\} \quad \ldots (1)$$

for large values of t, where H is the thickness of the soil sample, and a, b and λ are soil parameters.

If s_T is the settlement at any time T after the application of Δp, where $T = t + t_1$, then

$$s_T - s_t = H \;\Delta p\; b\, e^{-\lambda t/b}(1 - e^{-\lambda t_1/b}) \quad .. (2)$$

For large values of t_1, and neglecting $e^{-\lambda t_1/b}$,

$$\log (s_T - s_t) = \log (H \;\Delta p\; b) - .434\left(\frac{\lambda t}{b}\right) \quad ..(3)$$

A plot of log $(s_T - s_t)$ against t gives straight line graph with gradient $- 0.434\ \lambda/b$ and intercept log $(H \;\Delta pb)$ for large values of t.

Differentiating equation (1) gives,

$$\frac{ds_t}{dt} = H \quad \Delta p \quad \lambda \; e^{-\lambda t/b} \quad \ldots (4)$$

Therefore,

$$\log \left(\frac{ds_t}{dt}\right) = \log (H \quad \Delta p\, \lambda) - .434 \quad \left(\frac{\lambda t}{b}\right) \quad ..(5)$$

A plot of log $\left(\frac{ds_t}{dt}\right)$ against t gives a straight line graph with gradient $-0.434\ \lambda/b$ and intercept log $(H \quad \Delta p\, \lambda)$.

Settlements were measured for a duration of 10,000 minutes during the consolidation tests. s_T for T = 15,000 minutes was estimated assuming that the secondary compression curve varies linearly with logarithm of time in this time range.

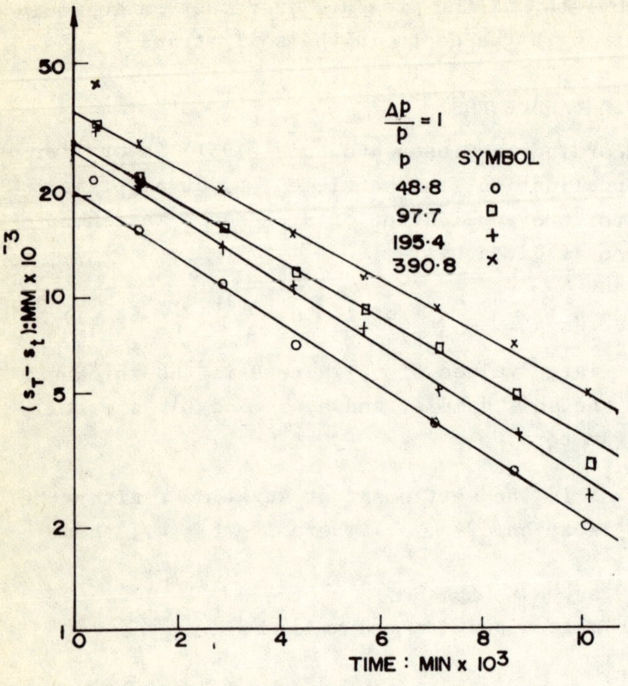

Fig. 8 - $(s_T - s_t)$ vs. Time

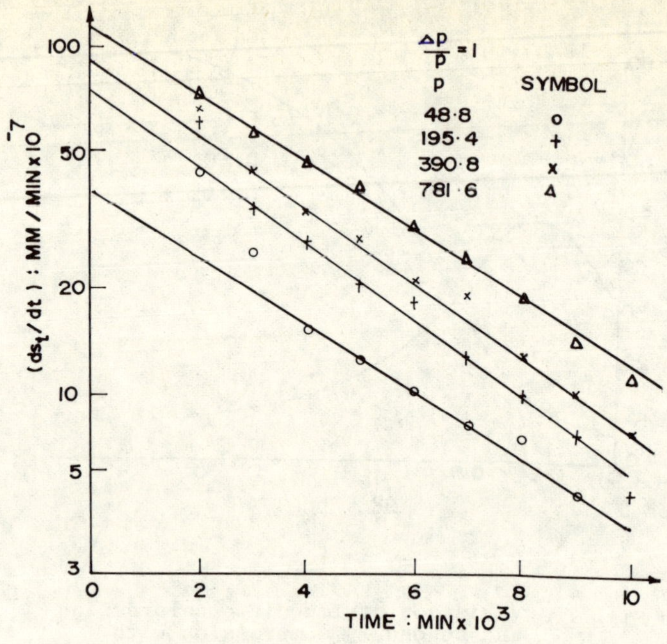

Fig. 9 - (ds_t/dt) vs. Time

In Fig. 8, log $(s_T - s_t)$ is plotted against t for four typical consolidation tests carried out with a pressure increment ratio of one. The points lie reasonably well on a straight line for large values of time for each test, confirming that Gibson and Lo's theory can be applied to residual soils.

Values of soil parameters estimated for some typical tests are presented in Table I.

In Fig. 9, log (ds_t/dt) is plotted against for four typical consolidation tests carried out under pressure increment ratio of 1. The points lie reasonably well on a straight line. Values of soil parameters estimated for some typical tests are presented in Table II.

Table I - Values of Parameters for Gibson and Lo's Theory

p kPa	$\Delta p/p$	a m^2/kN x 10^{-6}	b m^2/kN x10^{-6}	λ/b 1/min x 10^{-4}	λ $m^2/kN/min$ x 10^{-10}	c_v m^2/min x 10^{-5}
48.8	1/2	757	92	2.26	208	2.59
195.4	1/2	363	20	2.16	44	3.58
781.4	1/2	76	6	2.31	14	4.05
1562.9	1/2	40	3	1.93	6	5.32
3125.8	1/2	21	2	2.02	5	1.76
48.8	1	479	23	2.27	52	9.23
97.7	1	422	17	2.05	34	6.92
195.4	1	259	9	2.27	19	3.72
390.7	1	132	6	1.94	11	4.69
781.4	1	64	5	2.12	10	3.27

Table II - Values of Parameters for Gibson and Lo's Theory

p kPa	$\Delta p/p$	a m^2/kN x10^{-6}	b m^2/kN x10^{-6}	λ/b 1/min x10^{-4}	λ $m^2/kN/min$ x10^{-10}
48.8	1/2	806	81	2.48	201
97.7	1/2	671	32	2.47	79
195.4	1/2	362	21	1.79	38
781.4	1/2	77	5	1.79	9
1562.9	1/2	50	4	2.43	10
48.8	1	483	19	2.20	42
195.4	1	260	9	2.55	23
390.7	1	132	6	2.51	15
781.4	1	64	4	2.15	9

Based on the series of one-dimensional consolidation tests carried out, the following observations can be made about these soil parameters.

i) λ/b is approximately constant for the soil, being independent of the consolidation pressure and the pressure cincrement ratio. The average value of λ/b is 2.11.

ii) The viscosity of the soil skeleton 1/λ increases with increase in consolidation pressure for any pressure increment ratio.

iii) The parameters a and b decrease with increase in consolidation pressure for any pressure increment ratio.

iv) The ratio of secondary compressibility to primary compressibility b/a decreases with increase in pressure increment ratio for any consolidation pressure.

Similar behaviour was observed for soil parameters determined for laterite. (Thurairajah and Wijeyakulasuriya, 1979).

CONCLUSIONS

The settlement time curves obtained from one-dimensional consolidation tests on the residual soil consist of three phases:immediate settlement, primary consolidation settlement and secondary consolidation settlement. The ratio of the immediate settlement to the total settlement increases with the increase in pressure increment ratio while the ratio of secondary consolidation settlement to the total ultimate settlement generally decreases with increase in pressure increment ratio. During the secondary consolidation stage, the change in voids ratio varies linearly with logarithm of time. The slopes of these lines generally increased with increase in consolidation pressure.

Tests show that the consolidation behaviour of the residual soil can be defined satisfactorily by the theory proposed by Gibson and Lo(1961). Soil parameters for this theory have been determined graphically for each test.

ACKNOWLEDGEMENTS

The authors express their thanks to the laboratory staff Mr. A. Rajaguru, Ms. Pradeepa Kudahetty & Ms. H. Liyanagamage for their assistance in the experimental work, and Ms. Mahendra Wijesinghe and Ms. S.S. Wijayalath for the assistance in preparing the paper.

REFERENCES

Gibson, R.E. and Lo, K.Y. (1961) - A theory of consolidation for soils exhibiting secondary compression. Acta Polytechnica Scandinavica, 296/1961

Lo, K.Y. 1961. Secondary compression of clays. Jour. Soil Mech. & Found Div. ; Proc. ASCE, Vol. 87, SM 4, pp. 61 - 87.

Newland, P.L. and Allely, B.H. 1960. A study of the consolidation characteristics of a clay. Geotechnique, Vol. 10, p. 62-74.

Thurairajah, A. 1967. One dimensional consolidation characteristics of Thunukkai clay. Proc. 3rd Asian Reg. Conf. Soil Mech. Found Eng., Vol. 1 pp. 318 - 322.

Thurairajah, A. and Wijeyakulasuriya, C.V. 1979 Consolidation characteristics of laterite. Proc. 6th Asian Reg. Conf. on Soil Mech. Found. Eng., Singapore, Vol. 1, pp. 91-94.

Properties of granitic soil in Aswan, Egypt

A.M.RADWAN, Civil Engineering Department, Helwan University, Cairo, Egypt

SYNOPSIS Site investigation was carried out in Aswan, Egypt, in an area within the outcrop of weathered pink to gray fine grained granite. The granitic soil in Aswan are derived from in-situ weathering of several types of granite. Mineralogical and chemical properties of soil in the investigated site are presented. These properties, particularly at depth, were found to preserve original features of parent rock. Index properteis of successive soil horizons down to parent rock are also outlined. Variation of these properties as related to their degree of decomposition is presented. Shear strength parameters of the granitic soil were measured. Variation of these parameters for different soil layers was presented. It was found that the obtained values depended on the void ratio and degree of decomposition of granite. Use of field and laboratory test results in determining the allowable bearing pressure for foundations in the studied granitic soil is discussed.

INTRODUCTION

Construction of residential buildings at a site in Aswan required special consideration in determining the geotechnical properties of soil. Early surficial examination of old excavations indicated the occurrance of an heterogeneous soil formed of a mixure of particles of different sizes and varying characteristics. The soil formation was noticed to be overwhelmed by the presence of granitic rock in the area.

Granitic rock of varying characteristics outcrop in several regions in Egypt. Geologists classify these rocks in different catergories based on their geological setting (e.g. Samuel et al,1983; El Shazly, 1964; Ghobrial, 1982; Ghobrial and Girigs, 1982; El Shazly et al, 1977). At present it is generally agreed that the granitic rocks of Aswan are present in four phases. In chronological sequence, begining with the oldest, they are termed : granodiorites, coarse grained granites, fine grained granits and leucogranites (Taher, 1976). The crystalline basement complex covers most of the area and outcrops along the banks of the Nile. The nubia sandstone covers unconformably some parts of the crystalline basement.

Several research works are found in the literature dealing with the distribution, structure, petrography and geological history of Aswan granites (e.g. El Shazly, 1954; Gindy, 1954; Higazy and Wasef, 1956; Zalata, 1966; El sokkary, 1970). A geological map of Aswan presented by Attia (1955) is shown in Fig.1.

The study site in located within an area where the outcrop of pink to gray fine grained granite is abundantly noticed. The outcropping rocks frequently shows primary foliation exhibited by the parallism of biotite flakes. Mural jointing and bouldery weathering are also common features.

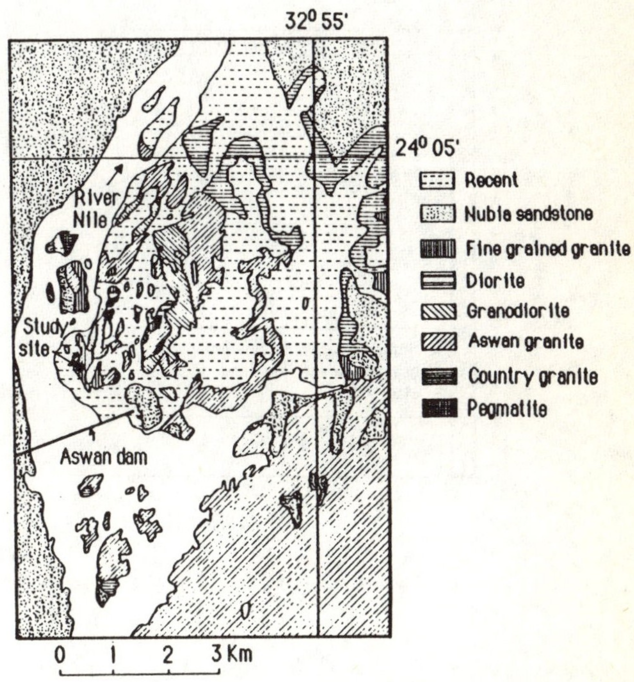

Fig.1 Geological Map of Aswan Area (After Attia, 1955)

The fine grained granite is in average about 1 mm in grain size, however, in few places medium grained and coarse grained boulders are found. It is doubtful whether these boulders are products of the in-stu parent rocks, or transported from neighbouring location.

FIELD WORK AND SOIL DESCRIPTION

Site investigation included execution of eight boreholes and twelve test pits. Boring extended to depth of 10 to 15 m, depending on the depth of "hard" granite. Standard penetration tests were carried out in the boreholes where granular soil was encountered. Disturbed spoon samples were collected from such material. A small number of undisturbed samples could also be obtained. Undisturbed core samples were obtained from the rock material. Recovery of cores depended on the degree of decomposition of rock.

Excavation of test pits was carried out manually and was terminated at about 3 m depth due to presence of groundwater. Disturbed samples and undisturbed block samples were collected. The fabric of top soil layers was clearly identified by observation of the freshly cut sides of test pits.

Typical borehole log is given in Fig.2. The soil formation can be described as follows :

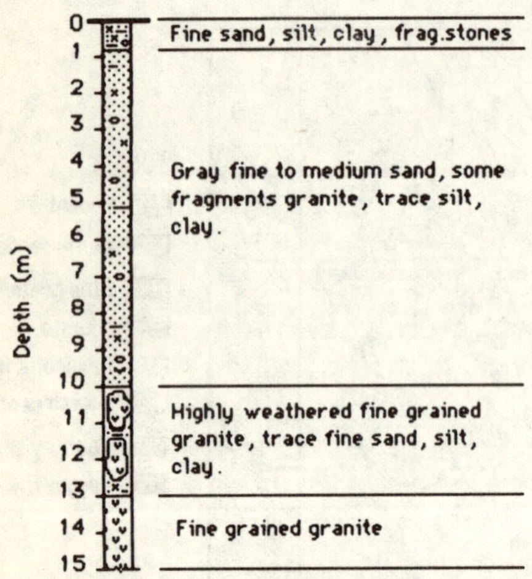

Fig.2 Typical borehole log

Layer (a) : Top soil : This soil to depth 0.3 to 0.8 m is a mixture of fine sand, silt, clay and fragments of stones. Its colour is generally grayish.

Layer (b) : The layer extends to depth varying between about 6 to 10 m. It is formed of gray fine to medium sand which includes some fragments of granitic stones and trace of silt and clay.

Layer (c) : Underlies layer (b) and extends to depth varying between about 8 and 13 m. It is formed of highly weathered fine grained granite including trace of fine sand, silt and clay.

Layer (d) : Parent rock formed of hard intact pink to gray fine granite.

Soil layer (a) is believed to be a mixture of materials transported from different locations. The soil texture in layer (b) tends to reflect the original granite rock. Disturbance of samples extracted from the boreholes, and remolding of intact blocks cut from test pits, yields such material which obviously has multiple mineralogical and chemical composition.

The present soil formation resembles the description quoted by Ruenkrairergsa and Watcharnon (1985) for granitic residual soil. Also, description of the present soil profile is qualitatively similar to that described by Deere and Patton (1971). Accordingly the soil material present in the shown typical borehole log between depths 0.8 and 13 m can be termed "residual soil".

MINERALOGICAL AND CHEMICAL COMPOSITION OF GRANITIC SOIL

The granite outcropping at the site is typically composed of quartz, alkali feldspars, plagioclase, biotite and hornblende. Taher (1976) pointed out that zircon and iron oxides form the common accessory minerals in these rocks. Fig.3 shows the average results of modal analysis of the fine grained granite.

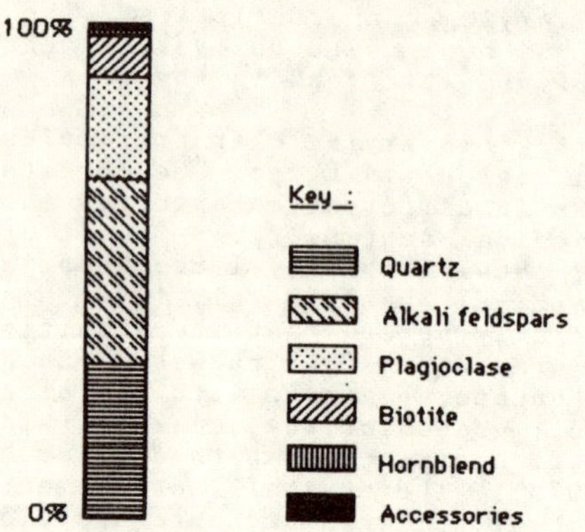

Fig.3 Average Modal Composition of Fine Grained Aswan Granitic Rock

Chemical analysis was carried out on three sets of samples obtained from each soil, or rock, layer. Average results are presented in Table I. The results show that the following phenomena occur as the soil layer becomes nearer to ground surface :

(a) Rapid decrease of Na, Mg and Ca, and slower decrease of K and Si.

TABLE I

Chemical Analysis of Granitic Soil from Aswan

Soil layer	Percentage by weight													
	SiO_2	Al_2O_3	Fe_2O_3	FeO	MgO	CaO	Na_2O	K_2O	TiO_2	MnO	P_2O_5	H_2O	CO_2	Sum
a	67.54	18.73	2.65	0.40	0.18	0.06	0.45	2.02	0.22	0.01	0.01	6.45	1.45	100.17
b	68.02	16.67	2.21	0.75	0.31	1.11	1.21	2.78	0.24	0.00+	0.05	5.53	0.98	99.86
c	68.95	15.10	1.93	1.12	0.46	1.19	3.43	3.83	0.27	0.02	0.08	2.92	0.59	99.89
d*	70.01	13.55	1.24	1.87	0.73	1.59	4.56	4.99	0.35	0.01	0.12	0.76	0.00+	99.78

+Concentration is under detection limit
*Bed rock : fine grained granite.

(b) Apparent increase of Al, and actual increase of H_2O and CO_2.
(c) Increase of oxidation of Fe^{2+} to Fe^{3+}.

According to the relation given by Parker(1970) the weathering index can be calculated as :

$$W.I. = [(Na)_a/0.35 + (mg)_a/0.9 + (K)_a/0.25 + (Ca)_a/0.7] \quad .... (1)$$

where : W.I. = weathering index (%),
$(X)_a$ = atomic percentage of element.

For each soil layer the percentage of each element was determined and the atomic percentage of each of Na, Mg, K and Ca was found. The average weathering for each soil layer is given in Table II.

TABLE II

Weathering Index of Granitic Soil Layers

Soil layer	a	b	c	d
Average W.I.%	16.23	26.86	42.23	53.90

It is clearly noticed that the weathering index decreases as the soil layer becomes nearer to ground surface. Parker (1970) mentioned that the weathering index for fresh granite is about 80% and reduces to 30% when completely weathered.

INDEX PROPERTIES OF GRANITIC SOIL

Index properties were obtained by testing both disturbed and undisturbed samples. Variation of these properties for the different soil layers is presented.

Particle Size Distribution

Grading curves of samples obtained from different depths are shown in Fig.4. It is noticed that the soil above bed rock is mainly sand. However, samples from ground surface to 0.8 m depth (layer a) includes 12 to 22% fines (silt and clay). For the following depths down to 6 to 10 m (layer b) the soil is generally described as fine to medium sand with some

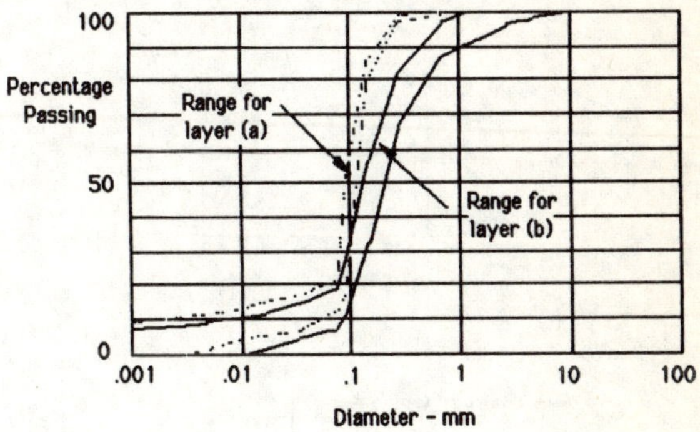

Fig.4 Grading of Granitic Soil

gravel and trace to little amount of fines (6 to 18%). Fig.5 shows that for layer (b) there is a tendency for the percentage passing sieve size 0.075 mm to decrease with depth. Sueoka et al (1985) stated that the change of particle size distribution with depth reflects the degree of weathering of the granitic soil. For layer (a)

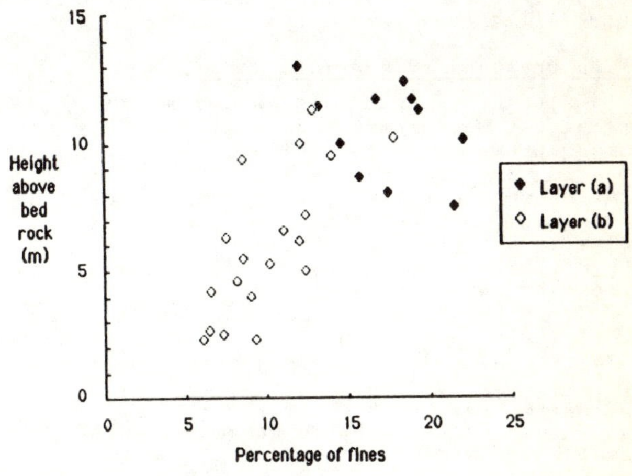

Fig.5 Variation of Percentage of Fines With Height Above Bed Rock

the percentage passing sieve size 0.075 mm is independent on the height above bed rock, which gives more evidence that this layer is not purely a product of the in-situ bed rock.

The texture of the granitic soil of Aswan is compared with that reported for other regions, Fig.6. It is shown that the area corresponding to residual soil on the plot lies within the area corresponding to residual soils of Hong Kong, however, Aswan granitic soil is coarser, which confirms that the variability of texture is much greater in the tropical region, as stated by Sueoka et al (1985).

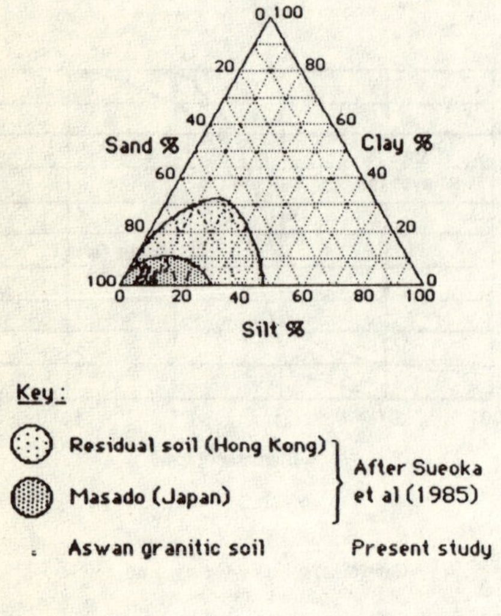

Fig.6 Texture of Aswan Granitic Soil

Specific Gravity

Results of determination of specific gravity values for samples obtained from different depths are plotted in Fig.7. It is noticed that the specific gravity is independent on depth. All values vary between 2.67 and 2.73.

In-situ Dry Density and Void Ratio

The in-situ dry density was determined by carrying out measurements on intact specimens obtained from undisturbed samples. Variations of in-situ dry density and void ratio with height above bed rock are presented in Figs. 8 and 9. It is noticed that the in-situ dry density is highest for the bed rock (layer d). The overlying layer (c) has in average lesser dry density, which is about 90% of that of layer (d). The decrease of in-situ dry density from the original value of parent rock is noticeable for layers (a) and (b). The average in-situ dry densities of layers (a) and (b) are about 60 and 63% of layer (a), respectively. Fig.9 shows clearly that the void ratio decreases with depth, becomming minimum for parent rock.

Lumb (1962) pointed out that variation of grading and natural void ratio are related to the degree of decomposition of the soil. De Sola (1985) stated that the most striking consequence of weathering is loss of unit weight. These two statements are clearly applicable to the investigated granitic soil.

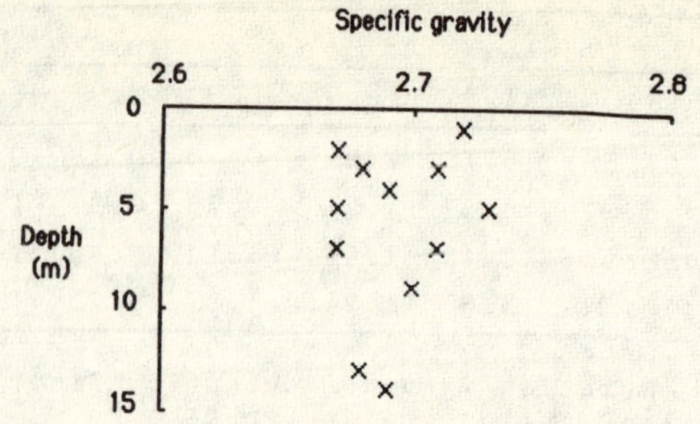

Fig.7 Specific Gravity of Granitic Soil

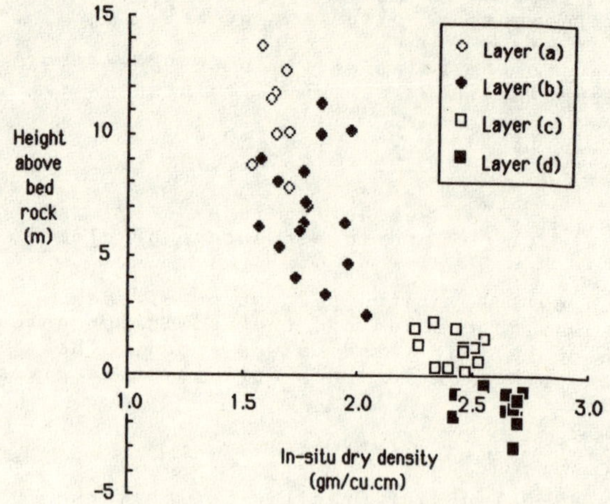

Fig.8 Variation of In-situ Dry Density With Height Above Bed Rock

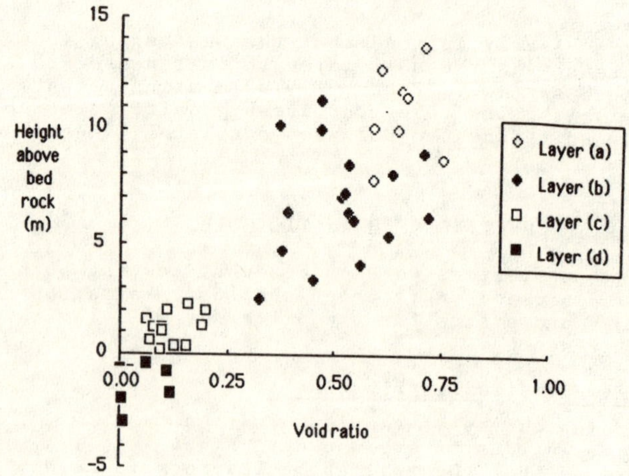

Fig.9 Variation of Void Ratio With Height Above Bed Rock

Relative Density

After determining the in-situ dry density and void ratio, the maximum and minimum void ratios were determined for typical specimens from layer (b). The average of these values were found to be 0.975 and 0.435, respectively. Hence, the relative density values were obtained. Variation of these values with depth is presented in Fig. 10. It is noticed that below groundwater table there is a tendency for relative density to increase with depth. The boundary of minimum relative density is constructed.

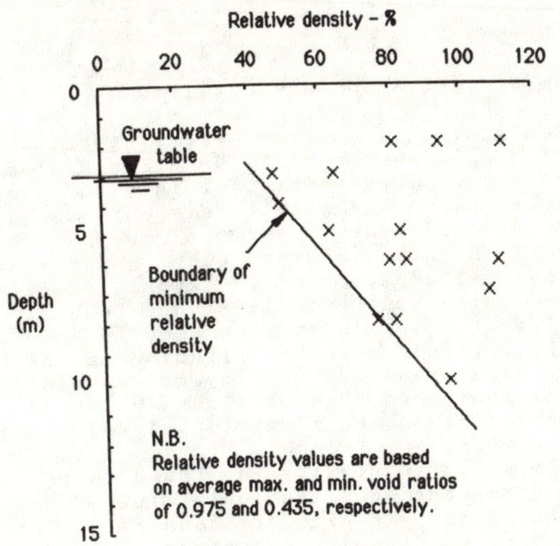

Fig.10 Variation of Relative Density of Fine to Medium Granitic Sand With Depth

The variation of relative density with blow count obtained from standard penetration tests, after being corrected for the effect of overburden (Peck et al, 1974) is presented in Fig.11.

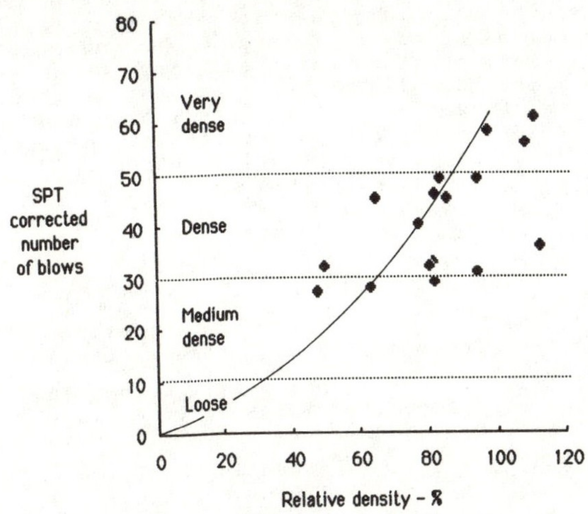

Fig.11 Relative Density Versus Corrected Number of Blows for SPT

For comparison, the conventional relation(Bowels, 1977) is also plotted. The plot generally implies that the obtained results of standard penetration tests can be used in analysing allowable foundation stress applying conventional procedures. Furthermore, the variation of corrected number of blows obtained from standard penetration tests with in-situ dry density is given in Fig.12. The plotted values are compared

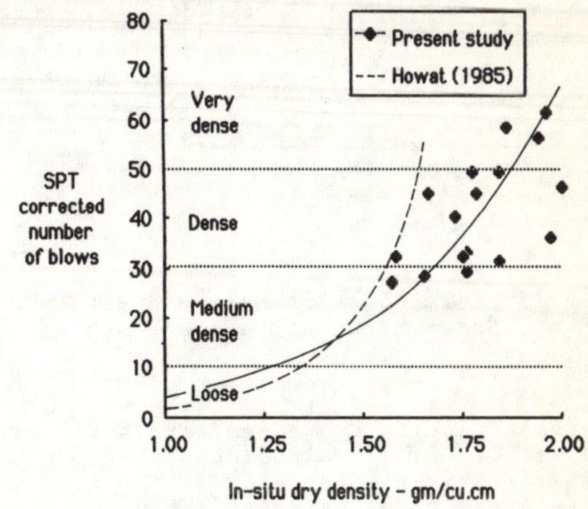

Fig.12 In-situ Dry Density Versus Corrected Number of Blows for SPT

with the relation found by Howat (1985) given by:

$$\text{In-situ dry density} = 0.93 + 0.46\times\log_{10}N^{*} \quad (2)$$

where : N^{*} = corrected number of blows obtained from standard penetration test.

For comparison, the results obtained in the present study are tentatively represented by :

$$\text{In-situ dry density} = 0.35 + 0.7\times\log_{10}N^{*} \quad (3)$$

SHEAR STRENGTH OF GRANITIC SOIL

The shear strength parameters of granitic undisturbed samples extracted from layers (a), (b), (c) were determined using a standard shear box 60x60 mm internal dimensions, following standard procedure for quick shear described by Akroyd (1957). Specimens above groundwater table were trimmed from undisturbed block samples. These specimens were tested at both natural water content and after soaking with water.

Due to the difficulty in trimming tube samples to fit into shear box, specimens obtained from layers under groundwater table were cut to dimensions slightly less than the dimensions of the shear box. The specimens were placed inside the shear box surrounded by gypsum plaster in the manner described by ISRM (1974).

Variation of shear strength parameters obtained from shear box tests with void ratio for all specimens from layers (a), (b) and (c) are presented in Fig.13. It is shown that the angle

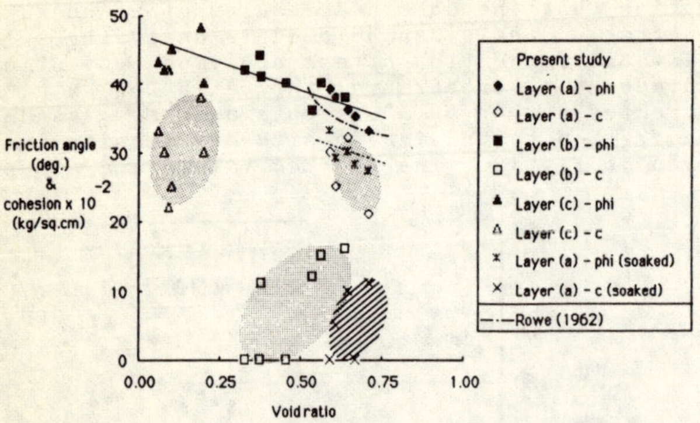

Fig.13 Relations Between Void Ratio and Shear Strength Parameters

of internal friction decreases with the increase of void ratio. Fig.13 also shows that the obtained results are generally in the same trend of the relation given by Rowe (1962) for medium to fine sand.

The cohesion parameter is found to be minimum for layer (b) and increases for layers (a) and (c). The increase of cohesion for layer (a) may be attributed to presence of higher fines content. In layer (c) the cohesion is believed to be an inherent property of the parent rock.

The effect of soaking specimens with water on their shear strength is also presented. It is found that the angle of friction reduces noticeably, and the cohesion is almost lost. Such phenomena were also found by Fkuda (1983) who mentioned that shear strength of decomposed granitic soil would decrease due to submergence.

ALLOWABLE STRESS FOR SHALLOW FOUNDATION

The allowable stress for foundations for ordinary reinforced concrete buildings is known to be dependent on soil properties, depth of groundwater, and size of foundation. Considering the fact that the mechanical properties of soil layers above groundwater will change when exposed to water, it was decided to take foundation level coinciding with groundwater level.

From Fig.10 the average relative density for soil at depth 2 to 3 m, which is equal to anticipated average footing width, is 50%. According to Vesic (1973) the expected mode of failure is local shear. Therefore, settlement rather than shearing strength is generally the criterion for estimating allowable bearing stress for foundation. From Fig.11 the number of blows in standard penetration test corresponding to relative density of 50% is 20. Using the charts for estimating allowable bearing pressure for 25 mm settlement for footings in sand (Peck et al 1974) and taking into consideration effect of groundwater depth, the allowable pressure was found to be 1.1 kg/cm^2. It is to be noted that according to the relation given by Peck et al (1974) the angle of internal friction corresponding to blow count of 20 is estimated by approximately 33°. From Fig.9 the maximum void ratio of layer (b) is about 0.5. Then, from the results of shear box tests, Fig.13, the corresponding angle of internal friction is approximately 38°. The difference between estimated and measured values, which is about 15% may be due to presence of fragments of granite in the tested soil.

CONCLUSIONS

The main conclusions are outlined as follows :

(1) The studied granitic soil, except for top material, was found to be a residual soil formed from weathering of in-situ parent fine grained granite. This soil was divided into four layers : top soil, sandy granitic soil, highly weathered rock, and parent rock
(2) Chemical analyses of the granitic soil layers indicated that each layer has been affected by weathering in varying amount.
(3) The particle size distribution of the studied granitic soil layers was discussed and the percentage of fines was found to be dependent on the elevation of soil above bed rock.
(4) The specific gravity was found to be independent on elevation of tested material. Contrarily, in-situ dry density of all soil layers was found to increase, and void ratio decease, with such elevation.
(5) Standard penetration test results were analysed and found to be compatible with conventional relations for sandy soil. The results can be used to estimate the allowable bearing stress for footings on sand.
(6) From the results of shear box tests, the angle of internal friction of the studied granitic soil was found to decrease with the increase of void ratio. Cohesion was found to be preserved in the highly weathered granite material, lost in sandy granite, but regained at top soil layer. The angle of internal friction measured using the shear box is about 15% higher than that estimated from standard penetration test blow count.
(7) For soil above groundwater table, soaking reduced angle of internal friction and almost eliminated cohesion.

REFERENCES

Akroyd, T.N.W. (1957). Laboratory testing in soil engineering. Soil Mechanics Ltd.

Attia, M.I. (1955). Topography, geology and iron core deposits of the district east of Aswan. Geol. Surv. Egypt, Cairo, 262.

Bowles, J.E. (1977). Foundation analysis and design. McGraw-Hill Book Co., New York.

Deere, D.U. and Patton, F.D. (1971). Slope stability in residual soil. Proc. 4th. Pan-American Conf. on SMFE, 1, 87-170.

De Sola, O.C. (1985). Weathering in the Guri-area - Venezuela. Proc. 1st. Int. Conf. Geomechanics in Tropical Lateritic and Saprolitic Soils. Brasilia, Brazil, Feb. 11-14, 2, 155-164.

El Shazly, E.M. (1954). Rocks of Aswan area. Geol. Surv. Egypt. Mines and Quarries Dept. Cairo, 23.

El Shazly, E.M. (1964). On the classification of the precambrian and other rocks of magmatic affiliation in Egypt. Int. Geol. Congr., Report of the 22 Session, India, Section 10, 88-101.

El Shazly, E.M.; El Sokkary, A.A. and Khalil,S. (1977). Petrography and geochemistry of some altered granite pebbles from Atud area, eastern desert. Egypt. J. Geol., 18, No.2, 105-114.

El Sokkary, A.A. (1970). Geochemical studies of some granites in Egypt. Ph.D. Thesis, Dept. Geol., Fac. Sci., Alexandria Univ., Egypt.

Fkuda, M. (1983). Decrease of shear strength and of stability of earth-structure of decomposed granite soil due to submergence. Soils and Foundations, 31, No.1, 95-100.

Ghobrial, G.A. (1982). Petrochemistry of the granitic rocks of Wadi Umm Had area, eastern desert, Egypt. J. Geol., 26, No.2, 121-140.

Ghobrial, G.A. and Girgis, M.H. (1982). Granitoid rocks of Wadi El Bolla area, eastern desert, Egypt. J. Geol., 26, No.2, 107-120.

Gindy, A.R. (1954). The plutonic history of the Aswan area. Egypt. Geol. Mag., 91, 484-497.

Higazy, R.A. and Wasfy, H. (1965). Petrogenesis of granitic rocks in the neighbourhood of Aswan, Egypt. Bull. Inst. Desert, 6, 209-256.

Howat, M.D. (1985). Completely weathered granite - soil or rock? Qua. J. Eng. Geol. London, 18, No.3, 199-206.

International Society for Rock Mechanics - ISRM (1974). Suggested methods for determining shear strength. Commision on Standardization of laboratory and field tests, 5-17.

Lumb, P. (1962). The properties of decomposed granite. Geotechnique 12, No.3, 226-243.

Parker, A. (1970). An index of weathering for silicate rocks. Geol. Mag., 107, 501-504.

Peck, R.B., Hanson, W,E, and Thornburn, T.H. (1974). Foundation engineering. 2nd ed., Wiley, New York.

Rowe, P.W. (1962). The stress-dilatancy relation for static equilibrium of an assembly of particles in contact. Proc. Royal Soc., A269, 500-527.

Ruenkrairergsa, T. and Watcharnon, W. (1985). Stress-strain and strength characteristics of undisturbed granitic soil. Proc. 1st. Int. Conf. Geomechanics in Tropical Lateritic and Saprolitic Soils. Brasilia, Brazil, Feb. 11-14, 1, 371-382.

Samuel, M.D.; Saleeb, G.S.; Hilmy, M.E. and Moussa, H.E. (1983). Study on cretaceous foreland granite and associated rocks at Bakriya, eastern desert. Egypt. J. Geol., 27, No. 1-2, 11-20.

Sueoka, T.; Lee, I.K.; Muramatsu, M. and Imamura, S. (1985). Geomechnical properties and engineering classification for decomposed granite soils in Kaduna district, Nigeria. Proc. 1st. Int. Cof. Geomechanics in Tropical Lateritic and Saprolitic Soils. Brasilia, Brazil, Feb. 11-14, 1, 175-186.

Taher, R.M. (1976). Petrological studies on some granitic rocks of Aswan area, Egypt. M.Sc. Thesis, Dept. Geol., Fac. Sci., Ain Shams Univ., Egypt.

Vesic, A.S. (1973). Analysis of ultimate loads of shallow foundations. J. Soil Mech. and Found. Div., ASCE, 99, No. SM1, 45-73.

Zalata, A.A. (1966). Geology of pegmatite occurrances at Aswan and Rod El Biram areas, U.A.R. M.Sc. Thesis, Dept. Geol., Fac. Sci. Ain Shams Univ., Egypt.

Atterberg limits for Nairobi soils

B.K.SAHU, Dr, University of Zimbabwe, Zimbabwe
F.J.GICHAGA, Prof., University of Nairobi, Kenya
T.G.VISWESWARAIYA, Dr, Kalpataru Institute of Technology, India

ABSTRACT: Atterberg Limits have been found to be very useful as index properties which may be correlated with more fundamental properties such as swelling and compressive characteristics of soils. Seed et al (1964) had suggested the following relationship between liquid limit (W_L) and and plasticity index (PI)

$$PI = N (W_L - n)$$

Where N and n are constants, which depend upon the mineral composition of soil. In the present study constants N and n have been determined for Nairobi soils. It has been found that while constant n has unique value, the constant N has two distinct values for two soils having different mineral composition.

1 INTRODUCTION

Engineers have found it convenient to express the plastic properties of soil in terms of plastic limit (W_P), liquid limit (W_L) and plasticity index (PI) as first proposed by Atterberg in 1911.

The potential value of Atterberg Limits for the use in soil mechanics was first indicated by Terzaghi in 1926 when he noted that "The results of the simplified test (Atterberg Limits) depend precisely on the same physical factors which determine the resistance and permeability of the soils (shape of the particles, effective size, uniformity)". The limit values although based on simple empirical tests, have been found to be very useful as index properties which may be correlated with more fundamental properties such as shear strength and compressibility.

The comprehensive studies on Atterberg Limits by various investigators have shown the dependence of these limits on mineral composition of soils. Seed et al (1964) suggested the following relationship between liquid limit and plastic index:

$$PI = N (W_L - n) \quad \text{.................} \quad (1)$$

where N and n are the constants which will depend upon the mineral composition.

In the present study, effort has been made to quantify these constants N and n for Nairobi soils.

2 GEOLOGY AND SOIL FORMATION

The City of Nairobi is in an area of successive volcanic activity. The type of geologic formations in this region is called the Basement System because of the volcanic rock formation on the original sedimentary rock which forms the base. These sedimentary rocks are hard, fine grained and black in colour. The volcanic rocks formed above the sedimentary rock base are in the following order:-

i) phonolite
ii) trachyte
iii) tuff
iv) clay stone

The phonolites at the bottom are the hardest while the clay stones at the top being the softest.

The parent material, volcanic ashes, are converted into soils with different minerals in the following sequence, (Gonzalez De Valego 1981):

Volcanic ashes
↓
Montmorillonite
(Black Cotton Soils)
↓
Halloysite
(Red Coffee Soils)
↓
Kaolinites

Investigations reveal that, in Nairobi, the weathering of volcanic ash, in general is upto halloysites. Kaolinite is found only in traces mixed with either montmorillonite and/or halloysite. Predominantly two types of soils are found in Nairobi. One is red in colour and is locally known as Red Coffee Soil. It contains mineral halloysite, and is formed in the regions where drainage conditions are generally good. The other soil is black in colour and is locally known as Black Cotton Soil. It contains mineral montmorillonite and is formed in relatively flatter areas where drainage is poor. It is highly expansive in nature. Red Coffee Soils are generally highly compressible but sometimes exhibit slight swelling characteristics due to the presence of mineral montmorillonite. The Atterberg Limits of these soils vary to a great extent due to their mineralogical composition. Red Coffee Soils, when dry, look like murram due to aggregation of clay particles and are often considered as non plastic soils, but when soaked in water they develop plasticity. The Atterberg Limits of these soils vary depending upon mineral composition and clay content whereas constants N and n will depend only on mineral composition.

3 EXPERIMENTAL WORK

Using similar approach to that of Seed et al (1964), an attempt is made in this paper, to arrive at suitable values of N and n (Equation 1) for Nairobi Soils.

Liquid Limits of soil samples were determined as per B.S. 1377 (1975) using Casagrande Apparatus, while plastic limits were determined by measuring water contents of the soils which developed cracks when rolled in 3mm threads. Soil samples were collected from 14 different sites of Nairobi. From each soil sample, 6 soil specimens of different clay contents were prepared. This was achieved by adding 0%, 5%, 10%, 15%, 20% and 25% (by weight) of sand to the natural soil samples. The liquid limits and plastic limits of each soil sample were determined. Liquid limits were plotted against the plasticity indices on an ordinary graph paper. It was found that there was a linear relationship between the liquid limit and the plasticity index. A typical plot is shown in Figure 1.

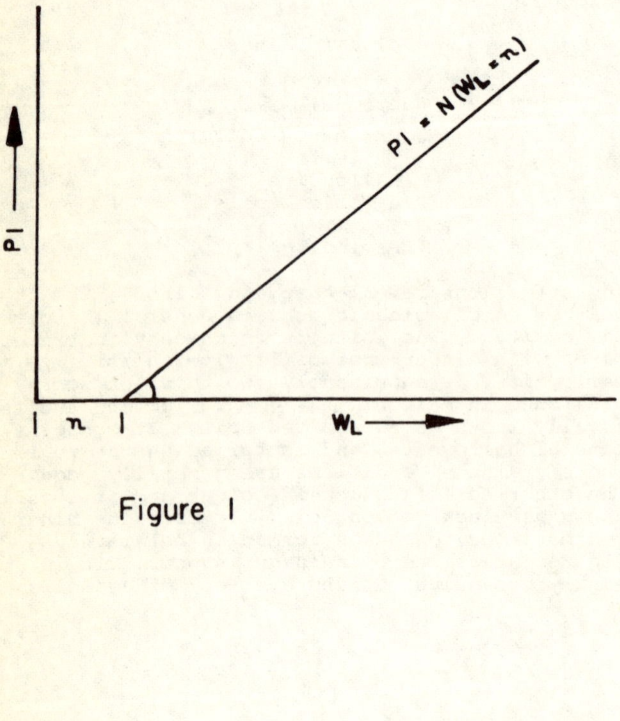

Figure 1

The slope of the straight line was taken as N while the intersection between the line and the abscissa (Figure 1) was taken as n. The values of N and n so obtained are shown in Table I.

The qualitative analysis of clay mineral composition was done by x-ray diffractometer at National Agricultural Laboratory, Nairobi. The results of the laboratory investigations are presented in Table I.

4. DISCUSSION

It was interesting to note that for all the samples the value of n was found to be 10. However, the values of N varied from soil to soil . Based on the values of N, the soils could be classified into two major groups. One group of four soils had the value of N between 0.81 and 0.82 with an average of 0.815. These soils containted predominantly montmorillonite clay mineral. The other group of ten soils had the range of values of N from 0.60 to 0.63 with an average of 0.617. Out of these ten soils, six soils Sr No.9 to 14 were red in colour (Red Coffee Soils) and contained mineral halloysite. Mineral montmorillonite was completely absent showing that the parent material has been weathered upto the stage of halloysite. Although Four soil samples (Sr No.5 to 9), had N between 0.60 and 0.62, they contained traces of montmorillonite along with kaolinite, halloysite and mica.

From the above observations it may be concluded that soils having N close to 0.82 contain mineral montmorillonite and will exhibit swelling characteristics. Soils with constant N close to 0.61 contain halloysite as the predominant clay mineral and montmorillonite if present will be in traces. These soils will either have no swelling characteristics or will have low degree of expansion.

Substituting constants N and n in equation (1), the relationships obtained were as follows:

$PI = 0.817 (W_L - 10)$ .... (2) for Black Cotton Soil

$PI = 0.617 (W_L - 10)$ .... (3) for Red Coffee Soil

If equations (2) and (3) are compared with the following equations:-

$Cc = 0.009 (W_L - 10)$ .... (4) Skempton (1944)

$Cc = 0.007 (W_L - 10)$ .... (5) Skempton (1944)

relationships between compression index Cc and plasticity index are obtained as follows:

$PI = 0.817 A$ .... (6)

$PI = 0.617 A$ .... (7)

$Cc = 0.009 A$ .... (8)

$Cc = 0.007 A$ .... (9)

where $A = (W_L - 10)$

TABLE I   n = 10 for all the samples

Sr No	LOCATION	N	MINERAL COMPOSITION AS SHOWN BY X-RAY ANALYSIS
1	MUTHURWA	0.81	Predominantly well crystallized montmorillonite & fair amount of mica & traces of kaolinite & metahalloysite.
2	INDUSTRIAL AREA	0.82	Predominantly well crystallized montmorillonite & traces of metahalloysite and sepiolite.
3	NAIROBI HOSPITAL	0.82	Predominantly well crystallized montmorillonite & traces of mica & halloysite.
4	TECHNICAL TEACHERS COLLEGE GIGIRI	0.81	Moderate amount of well crystallized montmorillonite & fair amount of mica & traces of metahalloysite and kaolinite.
5	UNIVERSITY OF NAIROBI	0.62	Traces of well crystallized montmorillonite & metahalloysite and mica.
6	DAGORETTI	0.60	Traces of montmorillonite & mica, metahalloysite Kaolinite.
7	MATHARE VALLEY	0.62	Traces of montmorillonite & kaolinite & mica.
8	NAIROBI RIVER	0.62	Traces of montmorillonite & halloysite and mica
9	STATE HOUSE	0.63	Amorphous in nature with mineral halloysite
10	LANGATA/KAREN	0.63	- do -
11	WESTLANDS	0.63	- do -
12	HIGHRIDGE	0.62	- do -
13	KABETE	0.60	- do -
14	KASARANI	0.62	- do -

From equations 6, 7, 8, and 9, we obtain:-

$Cc = \frac{0.009}{0.817} \times PI$ from equation 6 & 8

$= 0.011\ PI$ .... (10)

$Cc = \frac{0.007}{0.817}\ PI$ from equation 6 & 9

$= 0.009\ PI$ .... (11)

$Cc = \frac{0.009}{0.617}\ PI$ from equation 7 & 8

$= 0.015\ PI$ .... (12)

$Cc = \frac{0.007}{0.617}\ PI$ from equations 7 & 9

$= 0.011\ PI$ .... (13)

The limitations of equations 10, 11, 12 and 13 can be derived from the limitations of equations 2, 3, 4 and 5. A more general equation holding good for all soils and all samples (remoulded/undisturbed) can be given by an average equation of equations 10, 11, 12 and 13 i.e.

$Cc = 0.0115\ PI$ .... (14)

It is interesting to note that Sahu et al (1978) and (1984) had suggested a similar relationship as follows:-

$$Cc = 0.011\ PI \quad (15)$$

His correlation was based on the studies of clay mineral mixtures of kaolinite, Illite and montmorillonite, which had also shown good correlation for natural soils.

5 CONCLUSIONS

The equation $PI = N (W_L - n)$ holds good for Nairobi soils. Constant n has unique value of 10 for all the soils whereas N varies depending upon the mineral composition. An average value of constant N was found to be about 0.815 for soils containing montmorillonite as the predominant clay minerals. These soils are Black Cotton Soils. For Red Coffee Soils which contain halloysite as the predominant clay mineral, the average value of constant N was found to be about 0.617.

Since the method of determining plastic limit is rather crude, one can determine plasticity index on the basis of liquid limit alone using equation (2) and (3). The plasticity index so determined can also be used to determine compression index of a soil using equation 14 or 15

REFERENCES

British Ssandard 1377 (1973) Methods of Testing Soils for Civil Engineering purposes. British Standards Institution.

Gonzalez De Valego, L.I. Kiminez Salis, J.A. and Leguey Jiminz (1981), "Engineering Geology of Tropical Soils of La Laguni Tenerife", Engineering Geology, vol. 17.

Sahu, B.K. (1978) "Basic studies on Mineral Mixtures in Relations to Engineering Behaviour of Natural Soils" Ph.D Thesis at I.I.T. Kanpur.

Sahu, B.K. Yudhbir & Raymahashay, B.C., (1984), "Consolidation Characteristics of Clay Mineral Mixtures". Proc. 1st National Conference on Science and Technology of Buildings, Khartoum, Sudan, December, 1984.

Seed, H.B. Richard J. Woodward Jr. and Raymond Lundgsen (1964). "Clay Mineralogical Aspects of the Atterberg Limits" Proc. ASCE Jr. of S.M. and F.E. Volume 90 No. S.M. July 1964 Part I.

Skempton, A.W. (1944), "Notes on the compressibility of Clays". Quart Jr. Geological Society London.

Visweswaraiya, T.G., Sahu, B.K. and Gichaga, F.J. (1987). "Geotechnical Properties of Nairobi Soils". Technical Report submitted to National Council of Science and Technology, Kenya.

Stress-strain and strength properties of a compacted, residual volcanic soil

H.K.TAM, Geotechnical Engineer, Trafalgar House Engineering Services, London, UK
R.P.BRENNER, Principal Geotechnical Engineer, Electrowatt Engineering Services Ltd, Zürich, Switzerland

ABSTRACT: Three series of strain-controlled, drained triaxial compression tests were conducted with the aim to investigate the stress-strain and strength properties of compacted residual soil materials in the low stress range, i.e. with consolidation pressures from 10 to 200 kPa. The specimens in the three series had different initial dry densities, but the same initial moisture content. The test results are presented in various forms of stress-strain plots. They reveal that the material behaves like an overconsolidated clayey soil. The overconsolidation ratio increases with higher initial dry density and lower pre-shear consolidation pressure. It can be observed that the initial dry density affects mainly the effective cohesion intercept, while the effective angle of internal friction remains essentially unchanged. The results are also compared with those of consolidated undrained tests, and the difference in the strength envelopes is found to be insignificant.

1 INTRODUCTION

Shear strength is an important component of slope stability, however its determination on residual soils, or on colluvium (transported residual material) is difficult, because of the great variability in composition these soils exhibit. Laboratory tests can only be carried out on the matrix material of residual soils or of colluvium, and the effect of large-sized particles cannot be determined. It is, therefore, possible that shear strength parameters obtained in the laboratory underestimate the correct in situ values. But inspite of these shortcomings of laboratory tests, they are still useful to establish a likely range of shear strength. While for undisturbed specimens multi-stage tests have been introduced to eliminate the effect of soil heterogeneity on the test results (Lumb, 1964), single-stage tests are amply suitable for compacted soil, because the reproducibility of specimens is usually quite satisfactory (Wong, 1978).

Slope failures in residual soils, frequently triggered by high-intensity rainfall, occur usually on relatively shallow slip surfaces, which means that the effective stresses along the failure surface are low, i.e. typically in the range from 30 to 200 kPa. Cell pressures used for triaxial tests must, however, relate to the correct in situ ranges, if the measured strength values are to be meaningful. However, routine testing at low confining pressures, say less than 100 kPa, is not convenient, because the triaxial tests are then difficult to control satisfactorily.

Strength envelopes of certain undisturbed residual soils may be curved at low effective stresses, which may then cause an underestimation of the strength in the low stress range if the projection of the straight-line relationship obtained from the high stress range is used (Brand, 1985). With compacted soils, however, this may less likely be the case.

In this paper, the results of three series of high-quality, drained triaxial compression tests, each with five different consolidation pressures in the range from 10 to 200 kPa, are presented. The test specimens were pre-formed from a remoulded residual soil.

The test results were analyzed with respect to the influence of the initial soil density and of the pre-shear consolidation pressure on the stress-strain and strength characteristics. In addition, the strength parameters are also compared with those of consolidated undrained tests on specimens prepared from the same material and with the same initial conditions.

The tests reported herein form part of a comprehensive testing program on the stress-strain and strength properties of residual soil materials, which was initiated in order to gain a better insight into the stability of fill slopes subjected to heavy rainfall. Complementary results obtained with other types of test have been reported elsewhere (Brenner et al, 1985). It must be pointed out here, however, that rain-induced slope failures take place under conditions of more or less constant total stress (Brand, 1981, 1982) and the stress paths occurring actually in the field are thought to be quite different from those in conventional triaxial tests.

2 EXPERIMENTAL PROCEDURE

2.1 Material tested

The soil used in this study was taken from a borrow area in Tai Po, Hong Kong. The material can be described as a matrix of reddish brown silty clay with a high percentage of voids, containing occasional small subangular fragments of volcanic rock. There are, however, also inclusions of completely weathered volcanic rock of up to 100 mm in size. The material can be classified as a colluvium derived from a residual volcanic soil.

For laboratory testing, the soil was passed through a 20 mm BS (British Standard) sieve. The relevant physical and chemical properties of the material are summarized in Table I. All classification tests were conducted in accordance with the procedures given in BS 1377:1975 and were performed on air-dried soil.

Figure 1 shows the grain size distribution of the material employed in this investigation. The granulometry of weathering grade VI Tai Po volcanic soil (TPV) falls within a relatively narrow band, which lies at the upper boundary of the grading limits given by Lumb (1965). The latter were obtained from a large number of samples from various locations in Hong Kong.

Table I Material properties

Property	Unit	Typical range
Sand (2.0-0.06 mm)	% by weight	18 ± 1
Silt (0.06-0.002 mm)	% by weight	47 ± 2
Clay (< 0.002 mm)	% by weight	35 ± 2
Specific gravity	-	2.71 ± 0.02
Liquid limit	%	58 ± 2
Plastic limit	%	29 ± 2
Linear shrinkage	%	11
Organic matter	%	0.28 ± 0.03
pH value	-	4.5 ± 0.3

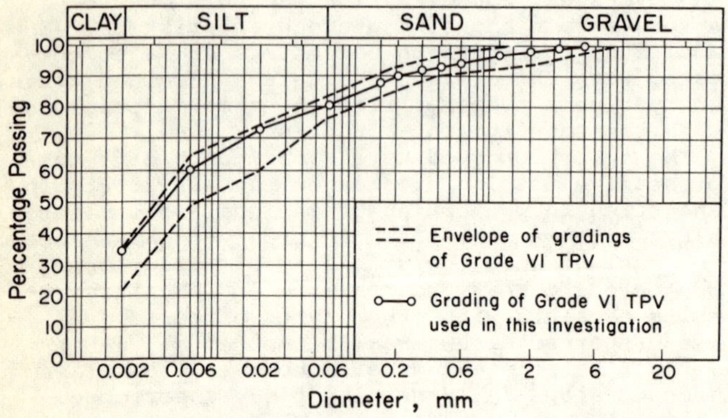

Figure 1. Grain size distribution of Tai Po decomposed volcanic soil, weathering grade VI

2.2 Testing program

Three series of isotropically consolidated, drained, strain-controlled triaxial compression tests, each with five different consolidation pressures, were carried out. The three series had different specified initial dry densities but the same initial moisture content. Table II summarizes the testing program.

Table II Testing program

Series no.	Isotropic consolidation pressure, σ_c' (kPa)	Initial conditions		
		%MDD	ρ_d (Mg/m^3)	w (%)
CDS-1 CDS-2 CDS-3	10, 30, 60 100, 200	95 100 105	1.47 1.55 1.63	24.5 24.5 24.5

ρ_d = initial dry density, w = water content

2.3 Specimen preparation

A reference moisture - dry density relationship of the material was established from twelve standard Proctor tests performed according to BS 1377:1975, Test 12. The maximum dry density (MDD) and the optimum water content (OMC) were found to be 1.55 Mg/m^3 and 24.5 % respectively. These values served as the basis for the initial conditions given in Table II.

Prior to compaction, the air-dried soil was mixed with water in a pan in order to achieve a moisture content of 24.5 % for compaction. The wetted soil was then stored in plastic bags for equalization and checked regularly for achieving the correct water content.

Specimens for the triaxial test were then prepared by statically compacting, in three layers, known weights of material into a mould of known dimensions. Typical specimen dimensions after extrusion from the mould were 49.6 mm in diameter and 99 mm in height.

2.4 Testing procedure

The test specimens were set up between a top and bottom low-air entry porous stone and fitted with two 0.3 mm thick rubber membranes which were separated by a coating of silicone grease. Both, cell pressure and back pressure, were increased simultaneously in steps of 10 kPa during saturation. The cell pressure was always kept two kPa above the back pressure to prevent excess water from getting into the sample. To achieve full saturation, a back pressure of 350 to 400 kPa applied for 48 hrs was sufficient.

Isotropic consolidation was performed in stages to the levels specified in Table II. The next stage was applied when 95 % pore pressure dissipation had been achieved from the previous stage. The duration required for full consolidation varied between 1 to 3 days, depending on the initial density of the specimen and the consolidation pressure.

For shearing, the values of c_v and t_{100} calculated from the triaxial consolidation phase served as a guideline for choosing the initial strain rate. Pore pressures were monitored continuously and the shearing rate was reduced in case appreciable build-up of measured pore pressure occurred. Each specimen was sheared to failure at an appropriate strain rate with top cap open to the back pressure line. The volume change of the specimen during shearing was recorded. All test results were corrected for effects of rubber membrane stiffness, as suggested by Bishop & Henkel (1962). The compression modulus of the rubber membrane per unit width was determined to be 0.35.

3 TEST RESULTS

3.1 Definition of parameters

In the following presentation of the test results the stress and strain parameters used are defined as follows:

$p' = (\sigma_1' + 2\sigma_3')/3$ = mean normal effective stress (σ_1', σ_3' are principal effective compressive stresses

$q' = \sigma_1' - \sigma_3'$ = deviator stress

$p_e' = \exp[(N-\upsilon)/\lambda]$ = equivalent pressure (λ = neg. slope of normal consolidation line, υ = specific volume (= 1+e), N = specific volume at p' = 1.0 kPa, see e.g. Atkinson & Bransby, 1978)

$\varepsilon_a = \Delta L/L = \varepsilon_1$ = axial strain

$\varepsilon_v = \Delta V/V = (\varepsilon_1 + 2\varepsilon_3)$ = volumetric strain

$\varepsilon = \varepsilon_a - 1/3\ \varepsilon_v = 2/3\ (\varepsilon_1 - \varepsilon_3)$ = shear strain

$\eta = q'/p'$ = stress ratio

σ_c' = pre-shear consolidation pressure (confining pressure)

3.2 Stress-strain characteristics

Figures 2, 3 and 4 show the variation of deviator stress, obliquity and volumetric strain with axial strain for the three test series. A summary of the relevant test data is given in Table III.

From Figs. 2a, 3a and 4a it can be seen that the q' vs. ε_a relationship depends on the preshear consolidation pressure. Also, with increasing initial density, ρ_d, the peaks become more pronounced and the strength increases.

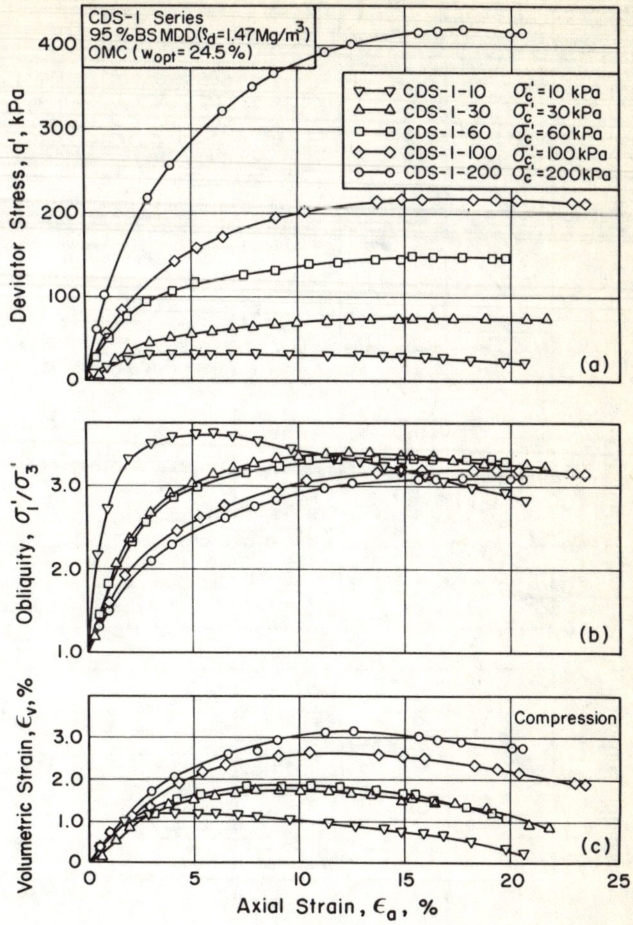

Figure 2. CDS-1 series: Variation of (a) deviator stress, (b) obliquity, and (c) volumetric strain with axial strain

Table III Summary of results of isotropically consolidated drained (CDS) triaxial tests

Test series	σ_c'	Initial conditions				At end of consolidation				At $(\sigma_1'/\sigma_3')_{max}$ or q'_{max}				
		ρ_d	S	e	w	ρ_d	S	e	w	ε_a	ε_v	t'	s'	E_{50}
	(kPa)	(Mg/m³)	(%)		(%)	(Mg/m³)	(%)		(%)	(%)	(%)	(kPa)	(kPa)	(MPa)
CDS-1	10	1.47	79.0	0.84	24.5	1.37	100.0	0.98	36.0	5.9	1.15	13.1	23.1	1.63
	30	1.47	80.6	0.84	25.0	1.42	100.0	0.91	33.5	13.3	1.65	36.0	66.0	1.99
	60	1.47	78.6	0.84	24.5	1.45	99.9	0.87	32.2	16.0	1.55	70.5	130.5	3.95
	100	1.47	78.3	0.84	24.3	1.45	100.0	0.87	32.2	18.0	2.35	110.0	210.0	4.00
	200	1.47	80.4	0.85	25.0	1.48	100.0	0.83	30.5	18.5	2.85	210.0	410.0	7.59
CDS-2	10	1.56	87.9	0.74	24.1	1.43	99.9	0.89	32.9	4.0	0.45	21.5	31.5	1.79
	30	1.56	89.0	0.74	24.3	1.49	100.0	0.82	30.4	6.0	0.66	48.0	78.0	3.09
	60	1.56	89.6	0.74	24.5	1.49	100.0	0.82	30.3	8.0	0.84	84.0	144.0	4.67
	100	1.56	87.6	0.74	23.9	1.51	100.0	0.80	29.4	9.6	1.15	120.5	220.5	5.83
	200	1.55	89.8	0.75	24.8	1.50	100.0	0.81	29.9	17.8	2.20	218.0	418.0	9.22
CDS-3	10	1.62	97.7	0.67	24.2	1.50	100.0	0.81	29.9	3.7	0.20	29.8	39.8	1.92
	30	1.63	98.6	0.67	24.3	1.53	100.0	0.78	28.7	4.6	0.85	58.5	88.5	2.88
	60	1.64	95.3	0.65	23.0	1.56	100.0	0.74	27.4	5.3	0.60	101.1	161.1	4.33
	100	1.63	99.8	0.66	24.5	1.55	100.0	0.75	27.7	6.0	0.45	141.0	241.0	7.92
	200	1.62	99.8	0.67	24.7	1.57	99.9	0.73	26.9	7.6	1.05	238.0	438.0	12.63

Note: S = degree of saturation, e = void ratio, w = water content, $t' = (\sigma_1'-\sigma_3')/2$, $s' = (\sigma_1'+\sigma_3')/2$

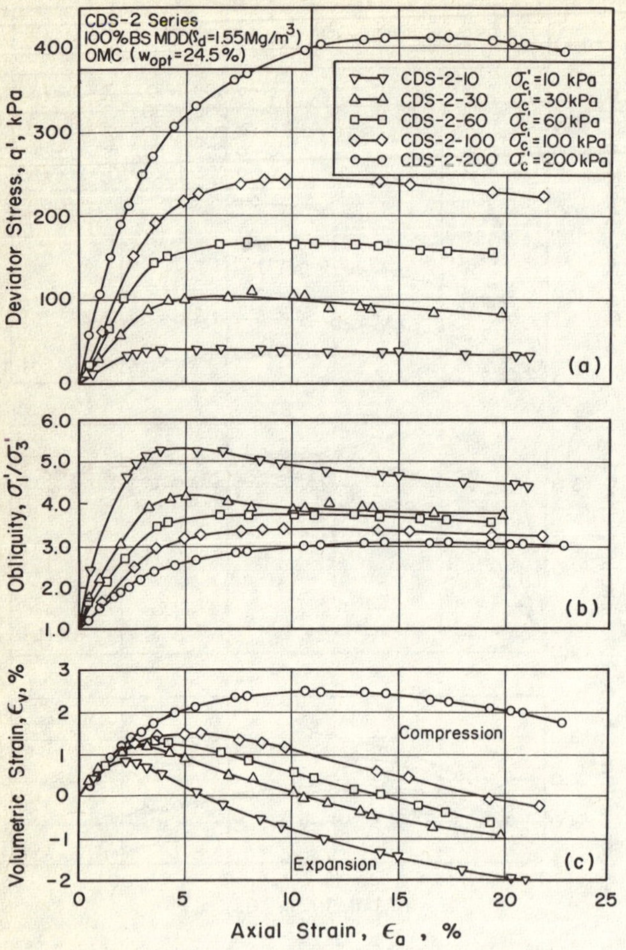

Figure 3. CDS-2 series: Variation of (a) deviator stress, (b) obliquity, and (c) volumetric strain with axial strain

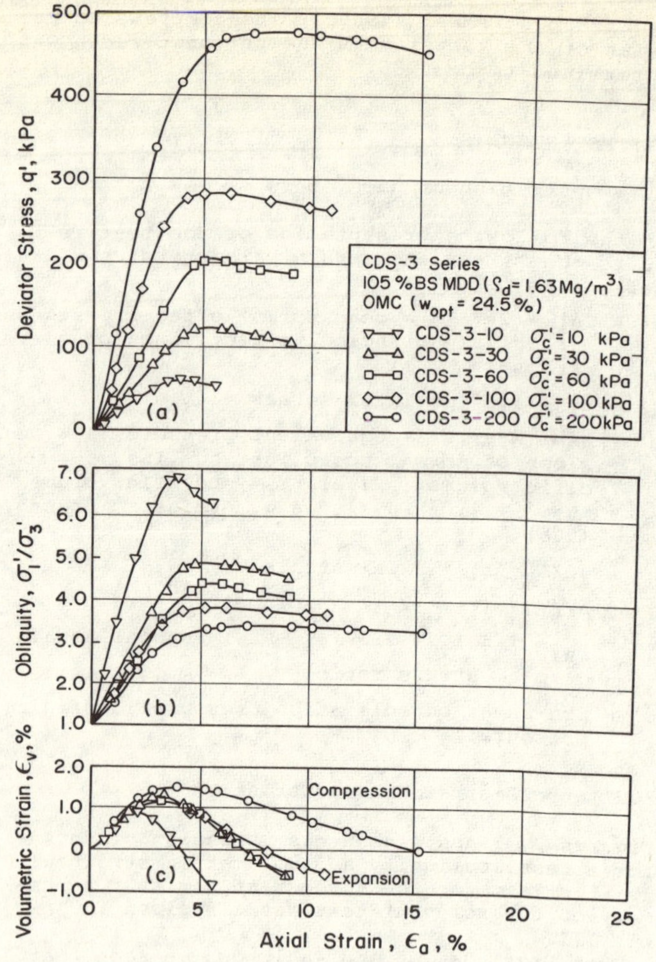

Figure 4. CDS-3 series: Variation of (a) deviator stress, (b) obliquity, and (c) volumetric strain with axial strain

The obliquity vs. axial strain relationships (Figs. 2b, 3b and 4b) depend also on the pre-shear consolidation pressure and the peak values decrease as σ_c' increases. Higher densities also increase the maximum obliquity. This behaviour indicates that the compacted specimens have the characteristics of an overconsolidated clay with variable overconsolidation ratio (OCR) depending on the consolidation pressure and the initial density. The effect of overconsolidation on the strength decreases with increasing consolidation pressure.

Figures 2c, 3c and 4c show the variation of volumetric strain with axial strain. It can be seen that all samples first contract, but those with higher densities (i.e. CDS-2 & CDS-3 series) become dilatant at higher strains. For a given density the samples with the lower consolidation pressure are more dilatant, and for samples at the same consolidation pressure, those with the higher initial density are more dilatant.

Figure 5a gives the normalized deviator stress q'/σ_c', vs. axial strain relationship for the CDS-2 series only. It is observed that a unique behaviour does not exist and that the values of q'/σ_c' decrease with increasing σ_c'. Similar graphs were obtained with series CDS-1 and CDS-3, but the spacing of the curves increases with increasing density. This is again an indication of the overconsolidated nature of the compacted specimens of this soil.

Similarly, Fig. 5b presents the stress ratio vs. axial strain relationship for the CDS-2 series. Again, there is no uniqueness in these curves and the difference increases the lower σ_c' and the higher the initial dry density.

Figures 6a and 6b give, respectively, the variation of deviator stress and stress ratio with shear strain for series CDS-2. The trend of these curves is very similar to those found when these quantities are plotted vs. the axial strain (Figs. 3a and 5b).

Finally, Fig. 7a presents the deviator stress and Fig. 7b the stress ratio vs. the volumetric strain relationship for series CDS-2. These plots again show that at the initial stage of shearing, the specimens will contract with increasing deviator stress and stress ratio, respectively. Subsequently, they start to dilate, whereby the stress ratio remains more or less constant during shearing.

Figure 8 compares the maximum effective principal stress ratio with the consolidation (confining) pressure, σ'_c, for the three test series. The denser specimens produce more pronounced changes in $(\sigma'_1/\sigma'_3)_{max}$, particularly at low consolidation pressure.

The influence of density and confining pressure on the secant modulus, E_{50}, of the test specimens is shown in Fig. 9. It appears that E_{50} in-

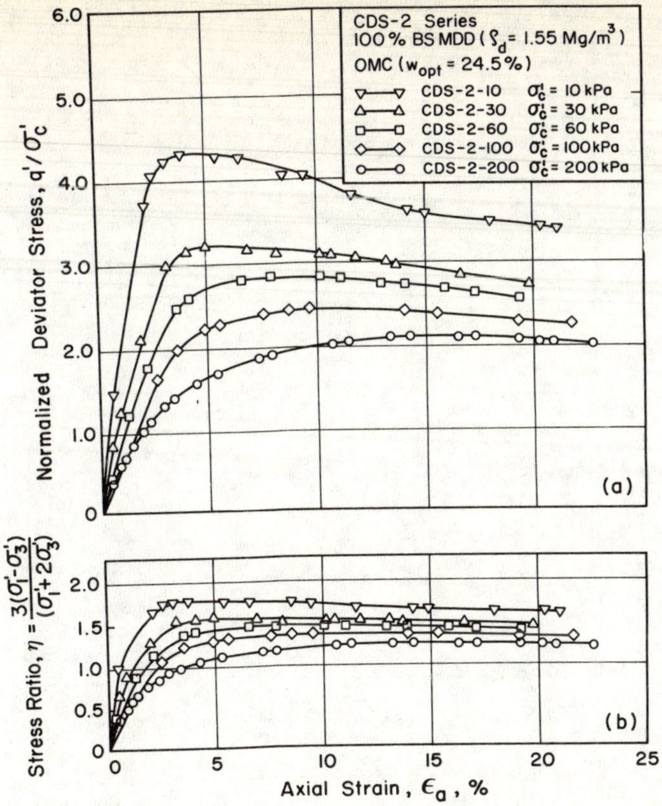

Figure 5. CDS-2 series: Variation of (a) normalized deviator stress and (b) stress ratio with axial strain

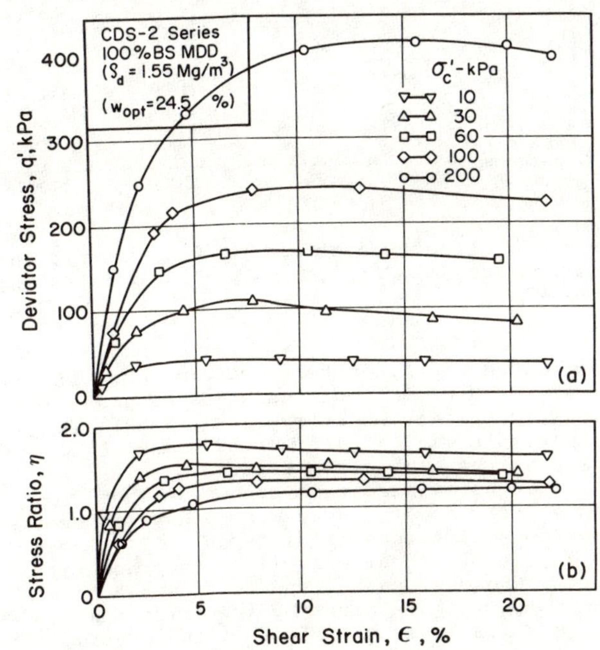

Figure 6. CDS-2 series: Variation of (a) deviator stress and (b) stress ratio with shear strain

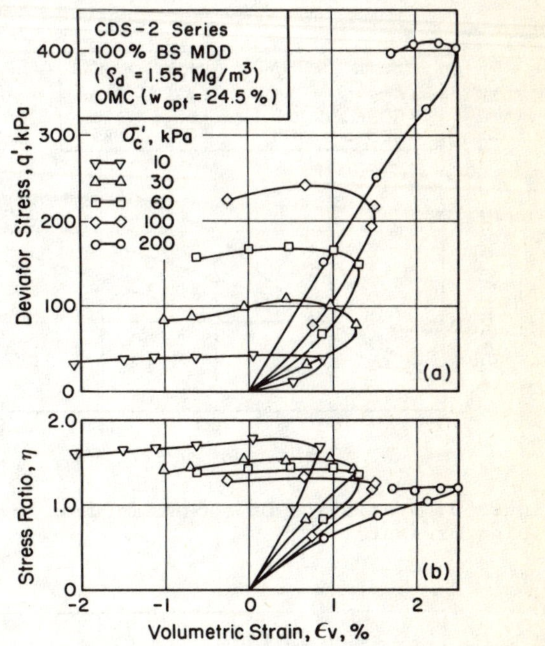

Figure 7. CDS-2 series: Variation of (a) deviator stress and (b) stress ratio with volumetric strain

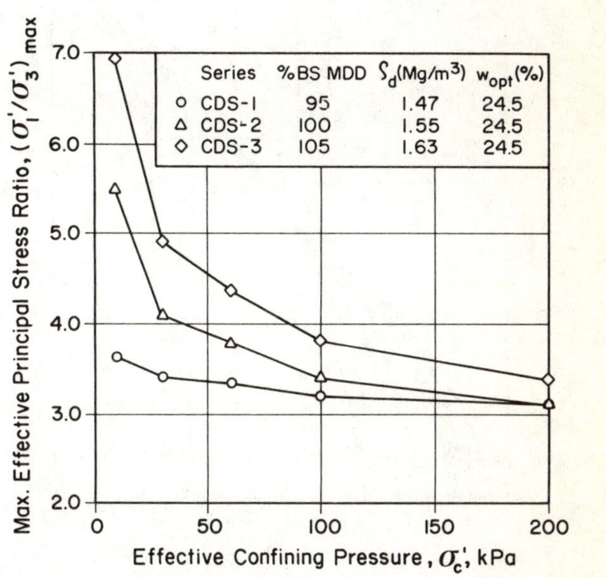

Figure 8. Effect of confining pressure on maximum effective principal stress ratio

creases more or less linearly with the confining pressure, and also that higher densities produce higher moduli (at least in the higher confining stress range).

3.3 Stress paths and strength parameters

In Fig. 10, the normalized stress paths of all fifteen tests have been plotted in a q'/p'_e vs. p'/p'_e space using the current value of υ of the specimen. The values used for λ and N were 0.1 and 2.37 respectively. They were computed from

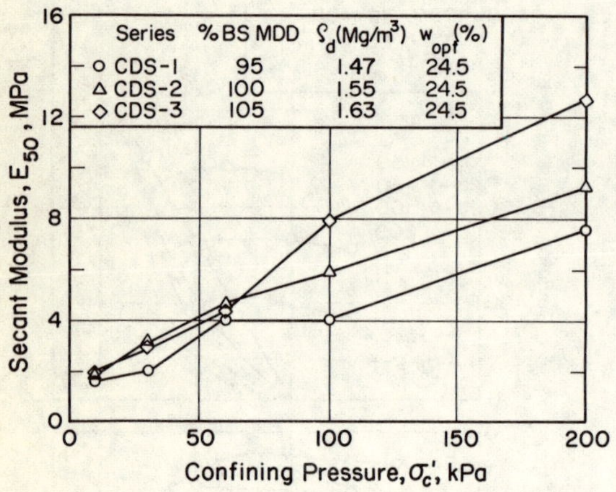

Figure 9. Variation of secant modulus with confining pressure

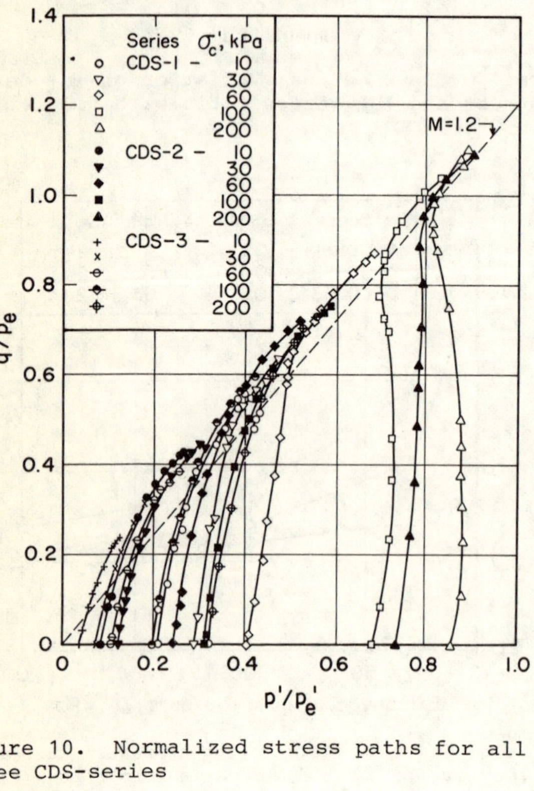

Figure 10. Normalized stress paths for all three CDS-series

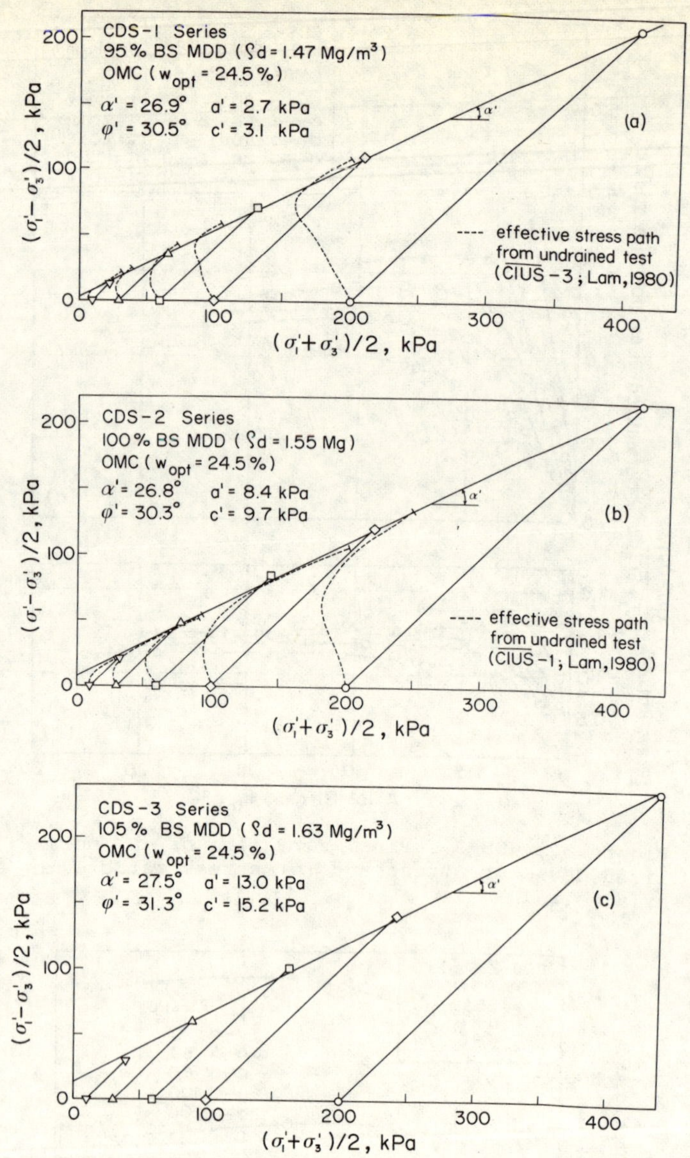

Figure 11. Effective stress paths and strength envelopes for (a) CDS-1, (b) CDS-2, and (c) CDS-3 series

consolidated undrained tests carried out with the same soil and with specimens prepared under more or less identical conditions (Lam, 1980). The test paths in Fig. 10 are curved, because p'_e increases as the sample compresses.

A unique normalized path cannot be obtained, because the compacted specimens behave as heavily overconsolidated with different OCR's depending on the pre-shear consolidation pressure. However, the test paths tend to converge on a line whose slope is given by the frictional constant M = 1.2 (M is the slope of the critical state line and the above value corresponds to that found by Lam (1980) from undrained tests). [Note that for triaxial compression tests $M = 6\sin\phi/(3-\sin\phi)$].

The stress conditions at failure can be taken from Table III and are plotted in Fig. 11 in $(\sigma_1'-\sigma_3')/2$ vs. $(\sigma_1'+\sigma_3')/2$ coordinates. With these coordinates the stress paths of drained tests are straight lines inclined at 45°. The failure envelope has been drawn as a best fit straight line. The resulting strength parameters are recorded in Fig. 11. Also shown in Fig. 11 are the stress paths from consolidated undrained tests (CIUS series) obtained by Lam (1980). As usual, the undrained strengths are considerably smaller than the drained values.

The strength parameters obtained from the undrained tests (there was also an undrained test series carried out with 90 % BS max. dry density ρ_d = 1.395 Mg/m³) are plotted in Fig. 12 vs. the initial dry density, together with those found from the drained tests (CDS-series). It can be seen that for the drained tests the cohesion intercept increases with increasing density (Fig. 12a), whereas for the undrained tests the density does not seem to influence significantly the effective cohesion. The effective angle of internal friction, ϕ', on the other hand, appears to be almost independent of density for both, drained and undrained shearing. From these limited data, it can be tentatively concluded that within the range of densities considered, a higher degree of relative compaction influences mainly the effective cohesion with this particular type of material, while ϕ' remains more or less unchanged.

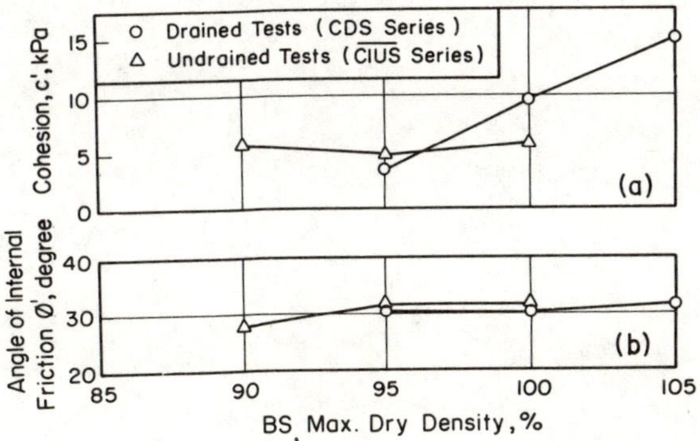

Figure 12. Variation of (a) effective cohesion intercept and (b) effective angle of internal friction with density for drained and undrained tests

4 CONCLUSIONS

The strength tests carried out in this investigation attempted to establish, from conventional triaxial tests, reliable strength parameters for a certain compacted residual soil material in the low stress range. Based on the data available from this study, the principal observations may be formulated as follows:

(1) The compacted residual soil material tested behaves as an overconsolidated silty clay with different overconsolidation ratios depending on the initial dry density and the pre-shear consolidation pressure. The OCR increases with higher initial densities and lower consolidation pressures. Samples with higher OCR tend to dilate at higher shear strains.

(2) The degree of compaction (initial density) influences mainly the effective cohesion intercept, while the angle of internal friction remains essentially unchanged. But even at the highest density used in this investigation, the cohesion intercept remained quite small. Also the strength envelopes for drained and undrained tests did not show significant differences.

ACKNOWLEDGEMENTS

The tests reported herein were conducted at the Material Testing Laboratory, Public Works Department, Hong Kong. The authors gratefully acknowledge the advice to the testing program given by Dr. E.W. Brand, and the cooperation in the laboratory provided by Messrs. M.D. Howat, J.M. Shen, K.L. Lai and C.H. Hui.

REFERENCES

Atkinson, J.H. & Bransby, P.L., 1978. The mechanics of soils - an introduction to critical state soil mechanics. McGraw-Hill, London, 375p.

Bishop, A.W. & Henkel, D.J., 1962. The measurement of soil properties in the triaxial test. 2nd ed., Edward Arnold, London, 227p.

Brand, E.W., 1981. Some thoughts on rain-induced slope failures. Proc. 10th ICSMFE, Stockholm, 3:373-376.

Brand, E.W., 1982. Analysis and design in residual soils. Proc. ASCE Specialty Conf. Engineering and Construction in Tropical and Residual Soils, Honolulu, pp. 89-143.

Brand, E.W., 1985. Predicting the performance of residual soil slopes. Proc. 11th ICSMFE, San Francisco, 5:2541-2578.

Brenner, R.P., Tam, H.K. & Brand, E.W., 1985. Field stress path simulation of rain-induced slope failure. Proc. 11th ICSMFE, San Francisco, 2:991-996.

Lam, K.S., 1980. Strength and suction-moisture retention of a compacted residual volcanic soil. M.Eng. Thesis no. GT-79-13, Asian Inst. of Technology, Bangkok, 184p.

Lumb, P., 1964. Multi-stage triaxial test on undisturbed soil. Civil Engineering and Public Works Review, 59(694):591-595.

Lumb, P., 1965. The residual soils of Hong Kong. Géotechnique, 15:180-194.

Wong, H.Y., 1978. Soil strength parameter determination. Hong Kong Engineer, 6(3):33-39.

Geotechnical properties at 'Kunigami Maaji' soils on Okinawa Island, Japan

H.UEHARA, University of the Ryukyus, Japan

INTRODUCTION

"Kunigami Maaji" Soils distributing on the Okinawa Islands have been used of old for the construction materials of roads ,fills and shallow foundation bearing layers in site. We have had also much troubles in the case of artificial slopes for road costructions etc. because of weatherings and erosions due to climatic conditions. Geneses, locations and some fundamental characteristics of the soils were introduced in the paper presented at the First International Conference on Tropical Soils in Brasilia(1985). "Kunigami Maaji" soils are divided into two groups by their geneses and geological characteristics in the engineering field even though those have been treated as same soils in the field of agriculture. One is residual soils of old rocks and the other is gravelly soils of dilluvial deposits.

The present paper deals with a part of engineering properties of Kunigami gravelly soils for the construction materials. Those soils are the formation containing the fragments of the older rocks such as cherts,slates,phyllites, and some limestones. Those gravelly soils develope on the terraces and terrains near the seashore as shown in Fig.1(b) with general maps of "Kunigami Maaji" soils in Fig.1(a).

Picture-1 Cut Slope in Kunigami Gravelly Soil Layer

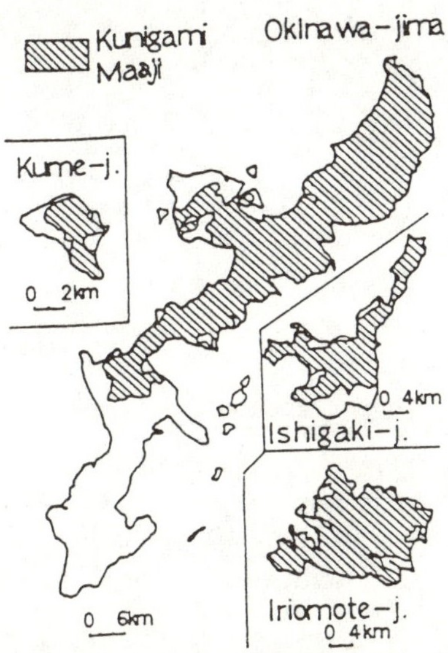

Fig.1a Location-Kunigami Maaji

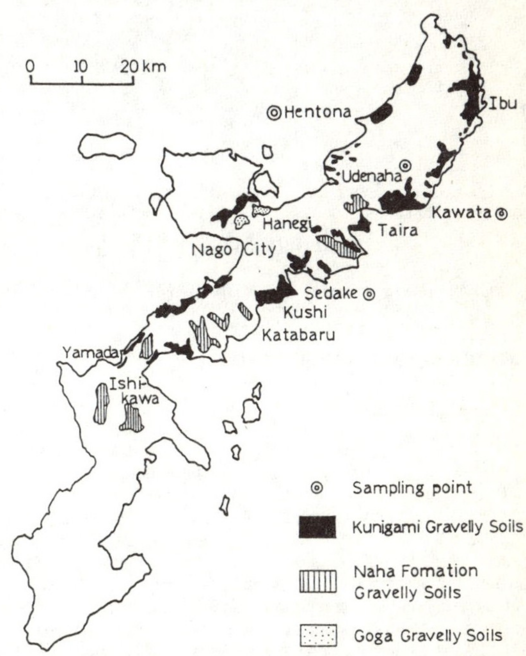

Fig. 1b. Distribution of Gravelly Soils on Kunigami area in Okinawa Island.

PHYSICAL PROPERTIES

Test samples were taken at the area of Udenaha, Kawata, Sedake, Hentona and the results of physical tests are presented in Table 1 and Fig.2. The specific gravity values of test samples finer or coarser than 2mm size are around 2.70. Fine grains of Kunigami gravelly soils are mostly classified as CL(Japan Unified System) on the plasticity chart.

Table - 1. Results of Physical Tests

Samples	UDENAHA	KAWATA	SEDAKE
W_n (%)	14.2	11.1	10.5
G_s	2.696	2.709	2.695
Grain<2.0mm	2.694	2.684	2.702
Grain>2.0mm	2.692	2.719	2.699
W_L (%)	38.7	49.8	51.5
W_P (%)	18.7	24.6	23.5
I_P (%)	20.0	25.2	28.0
W_S (%)	30.2	24.2	21.5
Gravel (%)	23.6	54.3	30.2
Sand (%)	42.4	32.0	40.9
Silt (%)	22.5	4.0	8.0
Clay (%)	11.5	9.7	20.9
60%Grain(mm)	0.62	6.9	0.80
30%Grain(mm)	0.029	0.42	0.091
10%Grain(mm)	0.0041	0.0051	---
maxGrain(mm)	19.1	50.8	38.1
U_c	150	1350	---
U_c	0.3	5	---

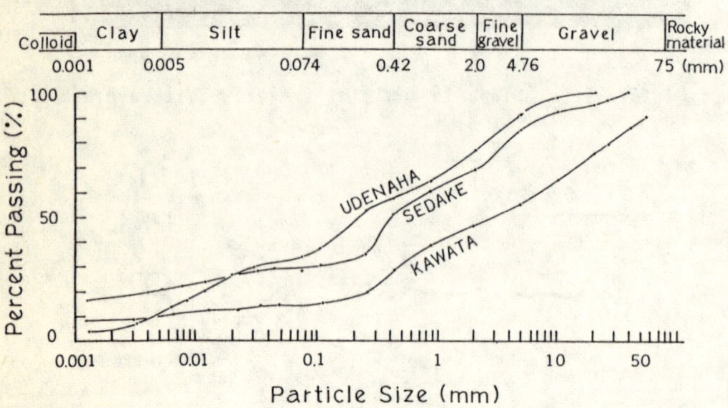

Fig.-2 Grain size curves for KUNIGAMI Gravelly Soils.

COMPACTION CHARACTERISTICS

1 Compaction Test

Compaction characteristics of most soils are changeable due to the test procedures such as compacting method, maximum diameter of soils, compacting energy, and repeated or unrepeated use of sample so on. Especially those of Kunigami Maaji (residuals and gravels) soils which are commonly in weathering and erosible conditions are quite influenced by test procedures. Consequently we have to found such variabilities for evaluation of compaction characteristics and applicability as materials of Kunugami gravelly soils(W.M.Lyle and E.T. Snerdon,1965).

Procedures;Japan Industrial Standard A1210(1970).

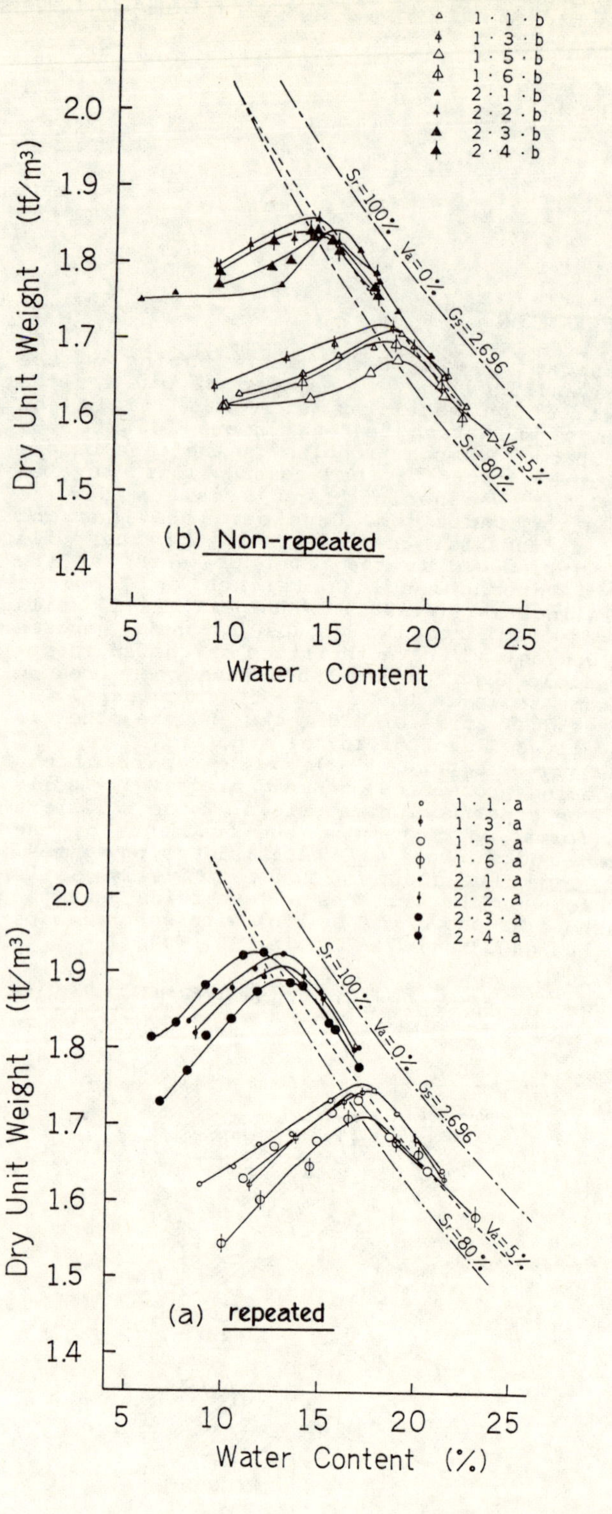

Fig.-3 Compaction Curves by Various Compacting Method on UDENAHA Gravelly Soil.

Factors of compaction effects selected in this study are as follows,

a) Maximum Diameter of the soil grains; D(mm)
b) Mold Diameter; M(cm)
c) Compaction Energy; E(kgf*m/m3)
d) Repeated or Non repeated use of the sample; ;[A],[B]

Tests were carried out according to Table 2, and those facters were adopted as

Dmax=4.76mm,19.1mm
M=10cm,15cm
$E=5.63*10^4$, $25.3*10^4$ $kgf*m/m^3$

then 8 types of test made 16 test results.

Table-2. JIS A 1210 Test Procedures

Test Type	Max Grain M.G (mm)	Mold Size M.S (cm)	Compaction Energy (m*kgf/m3)
1. 1	4.76	10	$5.63*10^4$
1. 3	19.1	10	$5.63*10^4$
1. 5	4.76	15	$5.63*10^4$
1. 6	19.1	15	$5.63*10^4$
2. 1	4.76	10	$25.3*10^4$
2. 2	19.1	10	$25.3*10^4$
2. 3	4.76	15	$25.3*10^4$
2. 4	19.1	15	$25.3*10^4$

2 Test Results and Evaluations

Test results are presented in Table 3 and Figs.3 (a),(b) show respectivly typical compaction curve examples for Udenaha soils by repeated use and non-repeated use of the samples.

In order to investigate the relationships between D, M, E and [A],[B], the experimental formulae were evaluated as follows,

Case1; Equations for max. dry density-optimum moisture content relationships with repeated use [A]

Case2; Equations for max. dry density-optimum moisture content relationships with non-repeated use [B]

1 $\gamma_{dmax}[B] = a_0 + a_1 D + a_2 M + a_3 E$ (tf/m3) eq. 1

Here coefficient a_i is dependent on each soil type as shown in Table 4. Applicability of eq.1 is examined on Fig.4 for Udenaha soil, and it was found to be coincidental for each soil.

2 Relationship between Wopt. and (γ_{dmax}) is presented by eq.2.

$$\frac{1}{\gamma_{dmax}} = \alpha + \beta w_{opt} \qquad \text{eq. 2}$$

Table - 3. Test Results

Test · Type	UDENAHA		KAWATA		SEDAKE	
	γ_{dmax} (tf/m3)	W_{opt} (%)	γ_{dmax} (tf/m3)	W_{opt} (%)	γ_{dmax} (tf/m3)	W_{opt} (%)
1. 1a	1.760	17.2	1.745	16.9	1.800	15.0
1. 3a	1.749	17.8	1.785	15.8	1.850	14.0
1. 5a	1.738	16.7	1.740	17.6	----	---
1. 6a	1.715	17.4	1.795	15.5	----	---
2. 1a	1.923	12.9	1.915	13.4	1.990	11.5
2. 2a	1.908	13.1	1.930	13.1	2.015	10.9
2. 3a	1.926	11.8	1.911	12.6	----	---
2. 4a	1.895	12.9	1.935	12.5	----	---
1. 1b	1.710	18.2	1.690	17.6	1.780	15.9
1. 3b	1.720	17.6	1.736	15.7	1.826	14.4
1. 5b	1.676	18.6	1.695	18.3	----	---
1. 6b	1.700	17.9	1.721	17.1	----	---
2. 1b	1.810	15.1	1.878	13.9	1.960	12.0
2. 2b	1.860	14.0	1.876	13.2	1.972	11.5
2. 3b	1.840	14.6	1.870	13.8	----	---
2. 4b	1.842	13.9	1.878	13.3	1.937	11.7
2. 5b	1.849	11.4	1.855	12.3	1.919	12.0

γ_d max: (tf/m3)
W_{opt} : (%)

Table - 4. a_i coefficients of each soil

	a_0	a_1	a_2	a_3
UDENAHA	1.670	$1.499*10^{-3}$	$-2.010*10^{-3}$	$6.950*10^{-9}$
KAWATA	1.656	$1.360*10^{-3}$	$-6.908*10^{-4}$	$8.403*10^{-9}$
SEDAKE	1.818	$2.022*10^{-3}$	$-8.534*10^{-3}$	$8.287*10^{-9}$

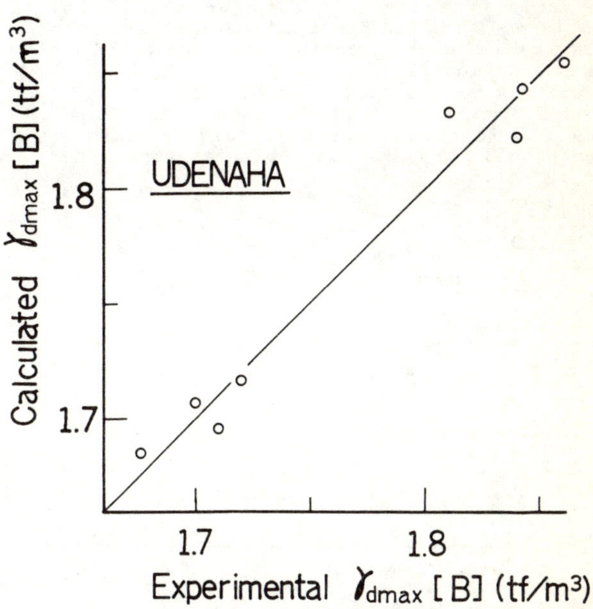

Fig.-4. Comparison between Experimental and Calculated $\gamma_{d.max}$ [B].

Applicability of eq.2 is examined on Fig.5. From Fig.5 writers can see that there is no difference between repeated use and non repeated use of samples, and no difference for each soil.

3 Relationship between (γ_{dmax}) of repeated use [A] and (γ_{dmax}) of non-repeated use [B] of the samples is presented by eq.3.

$$\gamma_{dmax}[A] = \alpha_1 + \beta_1\, \gamma_{dmax}[B] \qquad \text{eq. 3}$$

There is no difference for each soil, and applicability of eq.3. is examined on Fig.6.

4 Relationship between wopt. of repeated use [A] and Wopt. of non repeated use [B] of the sample is introduced by eq.2 and eq.3 as follows.

$$W_{opt}[A] = \frac{1}{\beta}\left[\frac{\alpha + \beta W_{opt}[B]}{\alpha_1(\alpha + \beta W_{opt}[B]) + \beta_1}\right] - \alpha \qquad \text{eq. 4}$$

Here, if the coefficients $\alpha\, \beta\, \alpha_1\, \beta_1$ are substituted into eq.4, then eq.4 is approximated by eq.5.

$$w_{opt}[A] \fallingdotseq \alpha_2 + \beta_2\, w_{opt}[B] \qquad \text{eq. 5}$$

Applicability of eq.5 is examined on Fig.7.

3. Summary

Now writes can summarize test results from the above discussions as follows,

1 Using a_i value in Table 4, (γ_{dmax}[B] can be calculated by eq.1
2 (γ_{dmax})[A] can be calculated by eq.3
3 Wopt.[B] can be calculated by eq.2
4 Wopt.[A] can be calculated by eq.4 or eq.5

Those relationships are shown in Fig.8 and Fig.9, and then each coefficients and the relativity of each equation are explained in Table 5 and Fig.10.

Now writers can get rather good correlation of the experimental values with the calculated values for Kunigami Gravelly soils on Okinawa island, and it may be applicable for some other gravelly soils.

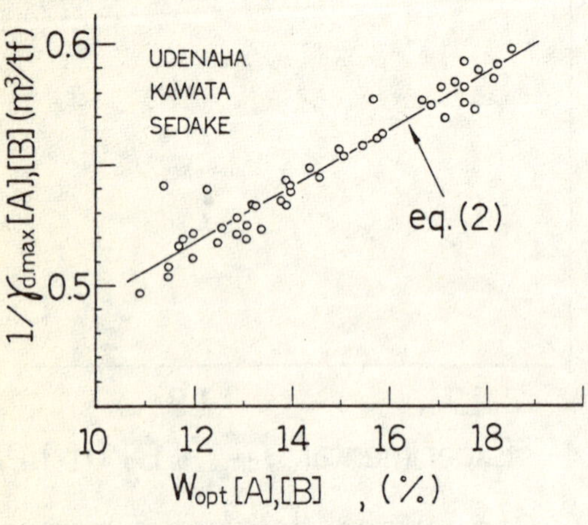

Fig-5 Relationship between W_{opt} and $1/\gamma_{d.max}$

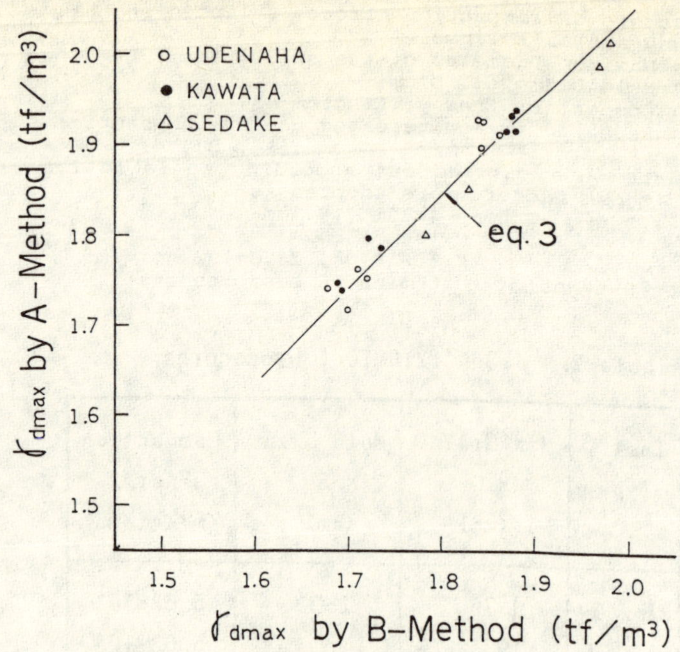

Fig.-6. Comparison between A-Method and B-Method for $\gamma_{d.max}$ of Kunigami Gravelly Soils.

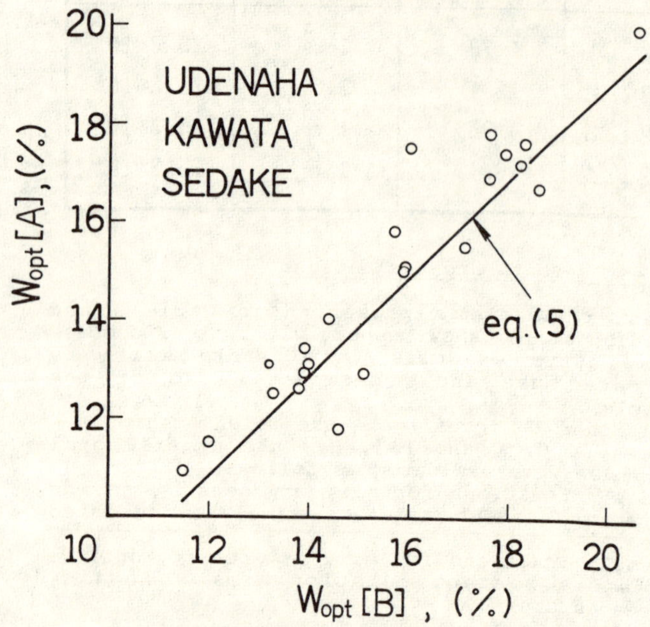

Fig-7 Comparison between W_{opt}[B] and W_{opt}[A] of Kunigami Gravelly Soils.

Table 5 Coefficient $\alpha, \beta, \alpha_1, \beta_1, \alpha_2, \beta_2$.

α	0.379	α_1	-0.0838	α_2	-1.446
β	0.0149	β_1	1.071	β_2	1.020

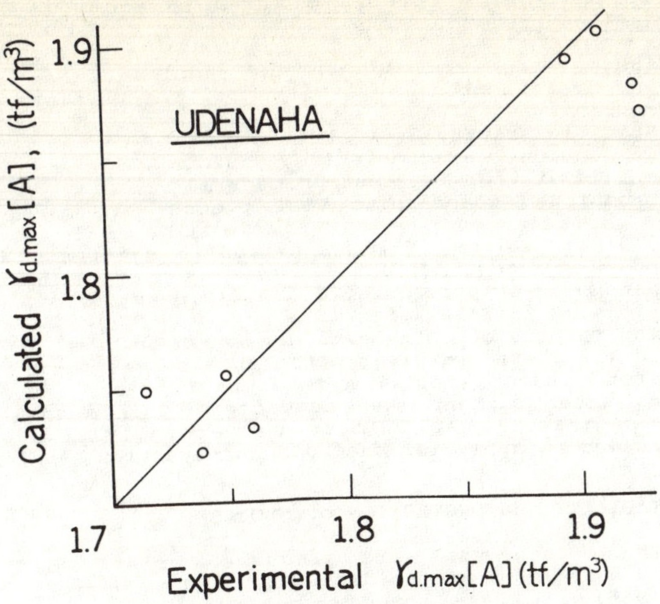

Fig-8 Comparison between experimental and calculated

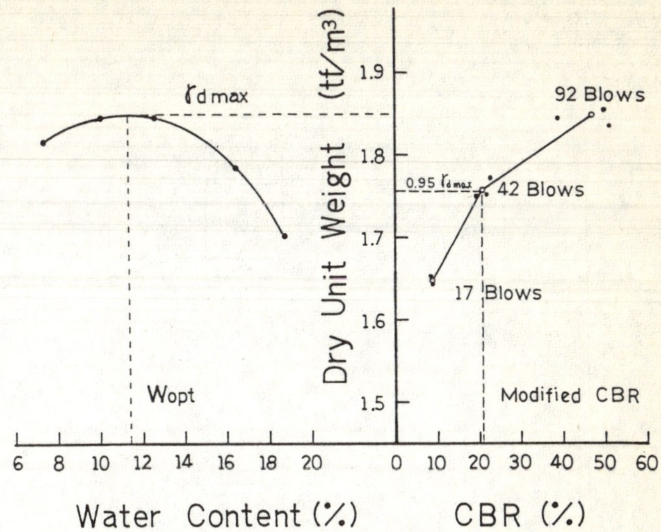

Fig.-11 Relationship between γ_d, CBR and UDENAHA Gravelly Soil.

Table-6. Results of CBR Test

		17	42	92
	N_b	17	42	92
UDENAHA	γ_d (tf/m3)	1.648	1.757	1.847
	$CBR_{2.5}$ (%)	8.8	17.3	38.7
	CBR_5 (%)	8.7	21.2	47.4
	Mod. CBR (%)	21		
KAWATA	γ_d (tf/m3)	1.628	1.762	1.838
	$CBR_{2.5}$ (%)	9.3	21.7	40.7
	CBR_5 (%)	9.1	26.3	47.9
	Mod. CBR (%)	24		
SEDAKE	γ_d (tf/m3)	1.712	1.870	1.939
	$CBR_{2.5}$ (%)	4.4	17.1	18.6
	CBR_5 (%)	4.5	20.0	26.4
	Mod. CBR (%)	17		

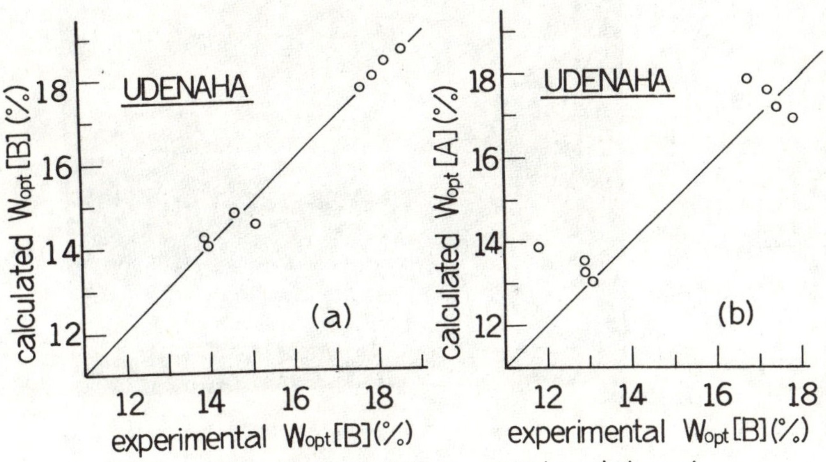

Fig-9 Comparison between experimental and calculated W_{opt} .

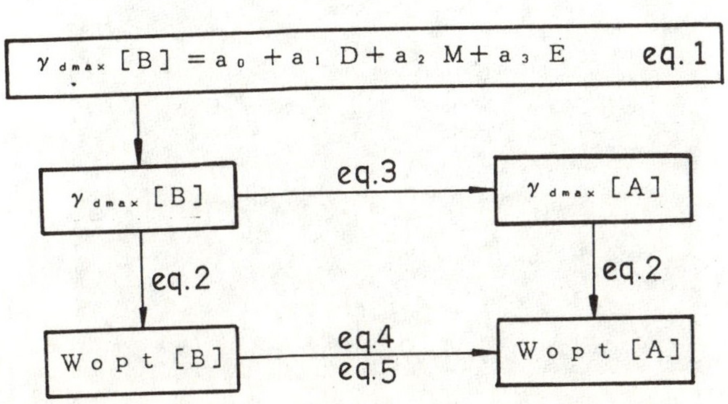

Fig.10 Relativity of each equation.

STRENGTH CHARACTERISTICS

1 Modified CBR

Representative values of modified CBR for each gravelly soil are presented in Table 6, and the example case of Udenaha soil is shown in Fig.11.

Here writers can get the value 21 for Udenaha, 24 for Kawata, and 17 for Sedake soil. These values are applied commonly for the base course of roads (20-30) in the case of Japanese standard. However, those are not so good for high quality roads, and from Fig.2(soil grain distribution curves) and also from Table 3 (max dry density of each soil), writers can not find any special relativity btween modified CBR and physical properties, compacted densities of those soils. This problem must be solved somehow relating with soil properties effects.

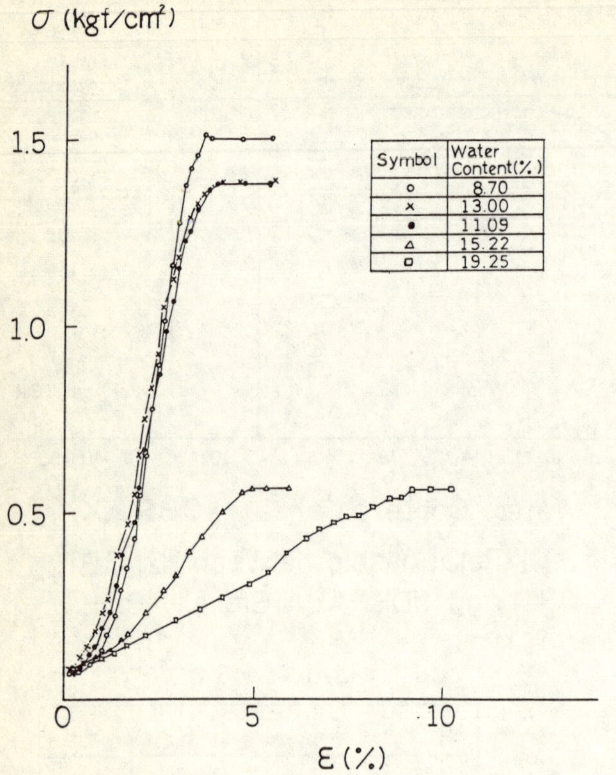

Fig. 12 Unconfined Compression Test by various Water Contents.

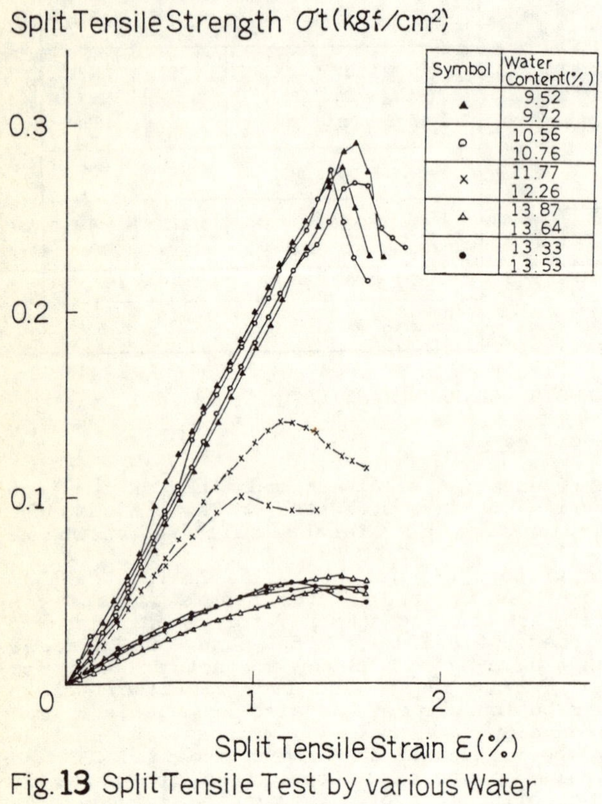

Fig. 13 Split Tensile Test by various Water Contents.

2 Unconfined Compressive Strength and Split Tensile Strength

These tests were made on the samples of Hentona area. Hentona gravelly soil has physical propretis as follows
Specific gravity Gs=2.71
Compaction characteristics; γ_{dmax} =1.750tf/m^3, Wopt=14.9% for 1-1 test procedure.
Unconfined compressive test pieces were made by means of 1-1b procedure(JIS A 1210;Dmax=4.76mm, M=10cm, E=5.63*10 kgf*m/m^3), and split tensile test pieces were made by means of 2-4b procedure (JIS A1210;Dmax=19.1mm, M=15cm, E=25.3*10 kgf*m/m^3).
The results of both tests are shown in Figs.12 and 13. There are clear difference between high water content (>optimum moisture content) and low water content when compacted. This means that gravelly soils must be carefully controlled to apply for the road construction.

STABILITY and PROTECTION of SLOPES

These phenomena were discussed briefly by H.Uehara(1985) and in detail by W.U.Lyle et al (1965), so here some protection performances will be shown by pictures (Pictures 2-6).

Picture-2

Picture-3

Picture-4

Picture-5

Picture-6

REFERENCES

UEHARA,H(1985); Geotechnical Characteristics of "Kunigami Maaji" Soils on Okinawa Island. Proc. 1st. Int. Conf. on Tropical Soils, Brasilia pp.187-198.

UEHARA,H & HARA,H(1984); Engineering Properties of "Kunigami Gravwlly Soils", 19th Annual Meeting, Japanese Society of Soil Mechanics and Foundation Engineering pp.165-166 (in Japanese)

W.M.Lyle & E.T.Smerdom(1965); Relation of Compaction and Other Soil Properties to Erosion Resistance of SOils, Trans. of the ASAE, pp.419-422

Session 3
Stability of slopes and excavations in tropical soils

Refinement of hydrological factors for the design of cut slopes in tropical soils

M.G.ANDERSON, Bristol University, UK
M.J.KEMP, Bristol University, UK
D.M.LLOYD, Bristol University, UK

ABSTRACT

The importance of combined instrumentation and hydrological modelling is reviewed in the context of tropical slope stability. A specific instrumentation design is outlined that meets the hydrological modelling requirement, in particular the use of automatic monitoring positive and negative pore water pressures. A 2-dimensional hydrology-slope stability model is described, and subsequently applied to simulate two design storms on three slope profiles for typical residual soil conditions. The results of the simulations are displayed in two design charts.

1 INTRODUCTION

There is little doubt that the overall knowledge relating to the hydrology of tropical soils is relatively poor compared to that of other elements of geomechanics (see for example, Brand, 1985). Recent years have seen the development of instrumentation for the determination of both positive and negative soil water pressures (Anderson and Kneale 1987 review such progress). Such development and research has, however, been usually focussed on individual site investigation requirements. Whilst this in many respects is appropriate research, it is necessary to broaden and refine hydrological research within tropical soils if a more general methodology is to be developed. This paper seeks to present research which aims to coordinate instrumentation development for pore pressures with a hydrological modelling approach that facilitates the establishment of hydrological design charts for slope stability.

The basis of generating design charts of this form has to be an appropriate hydrological model capable of representing the physical processes of soil water movement both on and within a hillslope. Anderson and Pope (1984), Anderson and Howes (1985) and Anderson et al (in press) have provided the outline and prototype of such a scheme which is further refined in this paper. Such models of course require parameterising, and hence necessitate the development of both field instrumentation and laboratory strength testing. The model structure capability and design chart requirements therefore dictate the style, scope and research design of the instrumentation required. This research scheme then requires validation and subsequent wide-area application. The modelling - instrumentation - hydrological design chart package is currently in the development phase and is centred on Malaysia with the cooperation and collaboration of JKR, Kuala Lumpur. Validation and wide-area application will be undertaken within the next twelve months, and it is therefore the current development of the package that is reported here.

The research presented here has the following three objectives:

(i) to stress the need to develop a hydrological modelling - instrumentation - design chart package for wide-area application in the tropics

(ii) to outline specific instrumentation designed, developed and installed by the authors to meet the modelling requirement, and

(iii) to outline the modelling development that has been completed and to present resultant hydrological design charts.

2 INSTRUMENTATION DESIGN

A combined slope hydrology - slope stability modelling scheme has been developed and as such requires field validation. Such a scheme demands accurate and frequent field monitoring of both soil water pressures and soil suctions at high temporal resolutions. A review by the Hong Kong Geotechnical Control Office (1984) of equipment for monitoring soil water conditions revealed that there was really no satisfactory automatic equipment, at that time, capable of measuring both positive and negative pressures, with the one possible exception of a tensiometer system. Described below is a combined automatic tensiometer-piezometer system developed by the authors which successfully fulfills these criteria.

The combined tensiometer-piezometer system is shown in figure 1. To simplify the description of the system three major components can be identified:

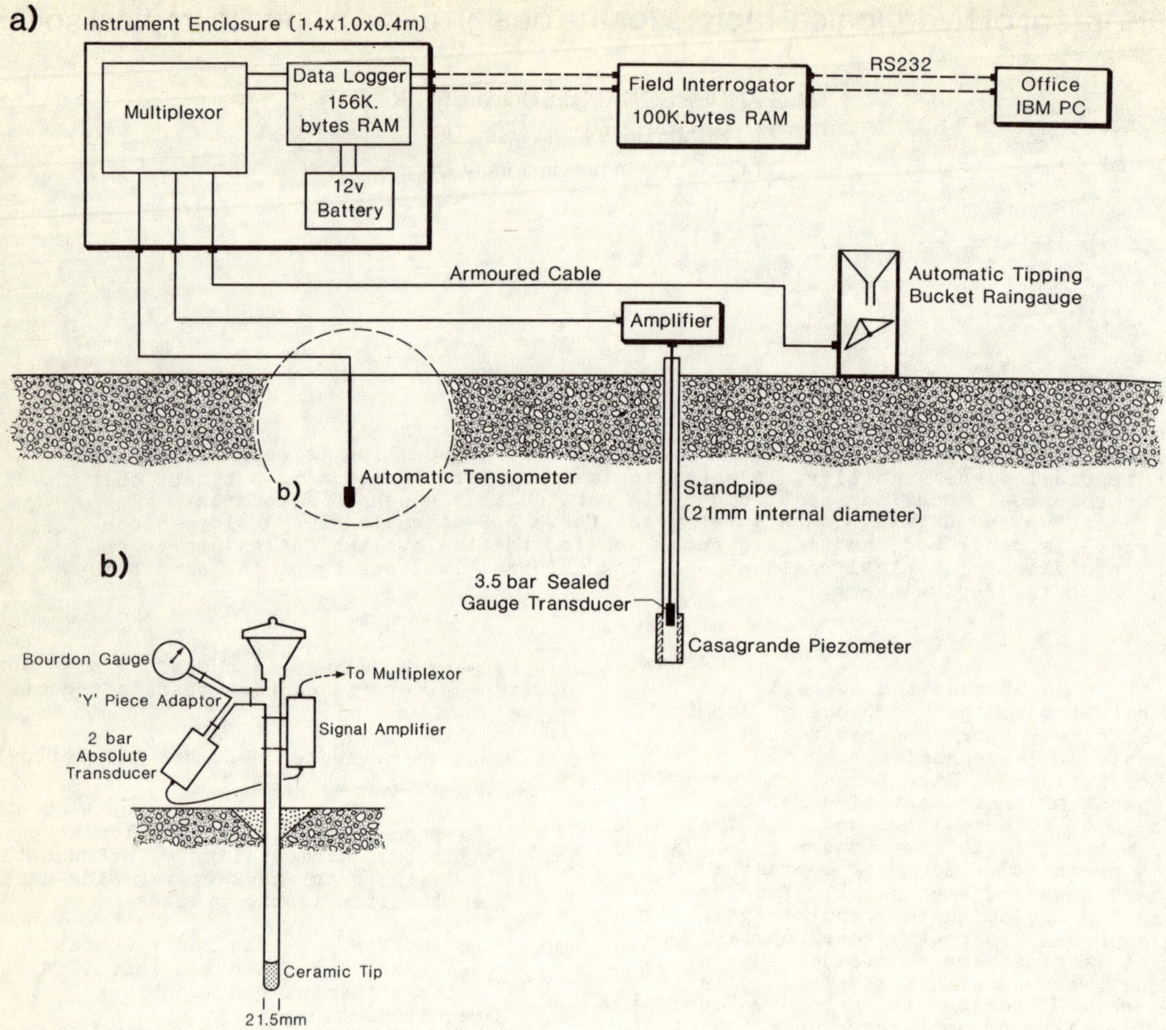

Figure 1 Combined automatic tensiometer-piezometer system illustrating: (a) system configuration and (b) tensiometer details

(i) the pore pressure sensors and related components
(ii) the multiplexor and data logger
(iii) the field interrogator and data communication

The instrument sensors adopted for both the tensiometers and piezometers were electronic transducers. In order to enable instrument sensor separations of up to 500 metres, the millivolt signal is amplified by a factor of ten to remove the problem of a drop in potential resulting from the cable specification and length. This also allows sensors to be interchanged or relocated without the need for recalibration.

Jetfill tensiometers (Soil Moisture Corporation, USA) were used for the measurement of soil suction. These were adapted in that a threaded 'y'-piece allows the internal water column to be linked to both the standard bourdon gauge and a 2 bar absolute pressure transducer. The signal amplifer is mounted in a box on the side of the tensiometer tube (see figure 1b). The 2 bar absolute transducers used are unaffected by moderate (but not large) temperature changes so each is insulated as protection against direct insolation making it necessary to protect the transducer from direct sunlight. Additionally, atmospheric pressure is monitored to allow atmospheric pressure compensation of transducer readings.

Piezometer water level monitoring is undertaken using a sealed gauge pressure transducer. The transducer is designed so that it can be lowered down the standpipe (minimum I.D. 21mm) and located at a position below minimum water level. Using a 3.5 bar sealed gauge transducer a piezometric head range of 35 metres can be monitored to an accuracy of 4 mm with a response commensurate with the piezometer and material specification. Sealed gauge pressure transducers are sensitive to large temperature changes but within the piezometers this does not occur. The signal amplifier is located at the top of each piezometer standpipe.

On site, the transducers are connected to the multiplexor using four core armoured cable, the multiplexor being directly connected to the data logger. The data logger used in this study (supplied by Geotechnical Instruments Ltd) can accommodate 32 input channels, two of which are dedicated to the monitoring of atmospheric pressure and input from the tipping bucket raingauge. It has a memory of 156 Kbytes and can be programmed to record at any reading interval between one minute and 24 hours. At full capacity the data logger can accommodate 1,250 sets of 32 readings before down-loading becomes necessary. The system is designed to operate on a very low power supply and can be effectively run off a standard 12 volt car battery.

In order to receive stored data and to program the data logger with reading sequence instructions, the system is connected to a hand held field interrogater (a small microprocesser). This interogater has a 100 Kbyte RAM for the temporary storage of data before downloading to an office micro computer. When connected to the data logger the interrogator can be used to monitor each instrument channel providing an instant equipment check against manual readings (bourdon gauge, dip meter). It also enables in situ calibration of additional sensors.

This automatic combined tensiometer-piezometer system has been installed on a cut slope some 40 km east of Kuala Lumpur on the Karak highway, Malaysia. Many of the slope features including the material (residual porphoritic granite), are typical of other slopes along the highway, of which a number have experienced instability during periods of heavy rainfall (see figure 2). There is much evidence to suggest that these failures are infiltration induced (Brand, 1984) and therefore require

Figure 2a Field site photograph showing instrument and granite locations

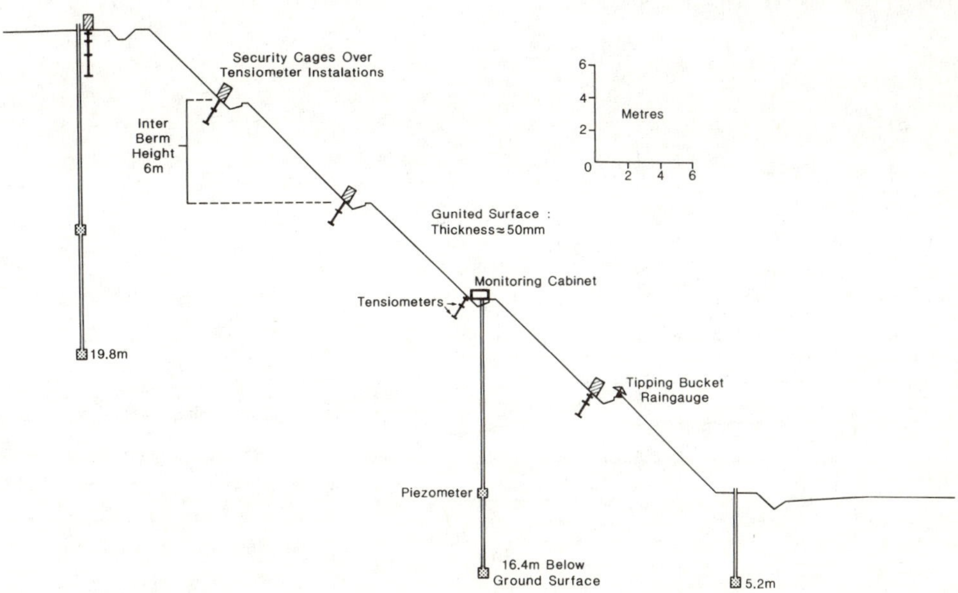

Figure 2b Slope cross section with instrument positions

detailed near-surface pore pressure monitoring in addition to groundwater monitoring. Whilst general instability processes may be known (i.e. infiltration), the actual soil water conditions at failure are not, thus posing questions such as;

i) Is suction maintained in the near-surface during intense storm events and, if so, what are the typical values?

ii) To what extent can the soil water conditions be improved and suctions maintained by slope cover modification (e.g. hydroseeding or concrete cover).

It is apparent that an automatic tensiometer-based system with additional piezometer monitoring, as in the system described above, is essential if such questions are to be satisfactorily answered.

For a comparative study, a 15 metre strip of gunite (concrete cover) was constructed down the centre of the slope, with adjacent slope material being hydroseeded (see figure 2a). Twenty one tensiometers (of which 14 were automatic) were located in groups down two transects on the slope, one of these being down the centre of the gunite strip. An additional five piezometers were installed on site of which three were automatic (see figure 2b).

The flexibility offered by this system allows instrument configuration to be optimised in respect of on site conditions. For example the resolution of the data can be enhanced during the monitoring period by instrument relocation and the provision of additional equipment, as indicated from analysis of results. Similarly, the whole system can be deployed at another site without the loss of any expensive components. Auxillary instruments can be added to the system provided the output is in mV analogue form. In the example installation, an on site tipping bucket raingauge was linked into the system. By integrating all the equipment to the single data logger correct time resolution of the data is maintained throughout.

The design and application of the automatic combined tensiometer-piezometer system has provided hydrological data of suitable temporal and spatial resolution necessary to meet the demands set by current modelling requirements.

3 MODEL DEVELOPMENT AND HYDROLOGICAL BASIS OF DESIGN CHARTS

In order to evaluate the effect of return period storms on slope stability, a real time 2 dimensional simulation model has been developed that uses predicted changes in the soil water conditions (which in turn are capable of field validation using the equipment described in 2 above). By linking a finite difference hydrology simulation model to a slope stability analysis (using Spencer's method of slices), a combined model has been produced that is capable of calculating the worst stability conditions (given by the minimum factor of safety), for particular storm and slope conditions. A definition diagram for this combined model is illustrated in figure 3.

It is necessary to take this approach as a large number of instability problems result from a rapid change in the soil water conditions in response to storm events. These problems are not consistent with the classical engineering approach of a static groundwater level with zero pore pressures above the phreatic surface. The combined 2-dimensional

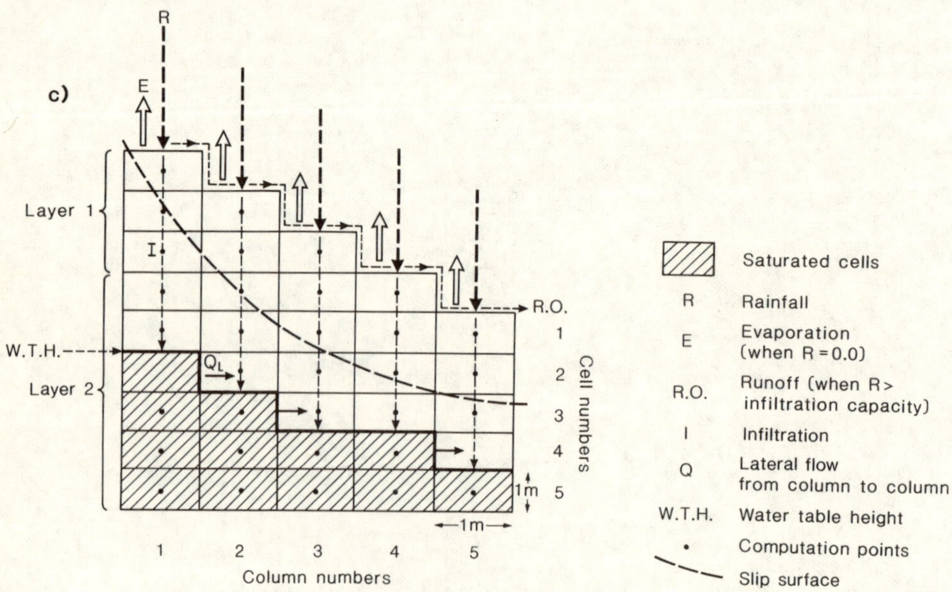

Figure 3 Schematic representation of the two dimensional hydrology model

hydrology-stability model has the capability of dynamically transferring water within both the saturated and unsaturated zones. This enables the minimum failure surface to occur within the unsaturated zone (where appropriate) in response to an infiltration wetting front (as is often the case on steep tropical slopes), rather than a change in the basal groundwater level.

The configuration of the hydrology component of the model is based on a grid system of a series of columns of unit width (figure 3). Each column is subsequently divided up into a number of similar cells of defined width and depth, with the centre of each cell representing a computation point. To represent the slope profile the size and the number of cells in each column have to be specified such that there is a maximum of a one cell difference between consecutive columns. This cell system is further divided up to allow for two hydrologically distinct layers. The depth of the first layer is represented by a specified number of cells; the second layer is given by the remaining cells in each column.

To simulate the slope soil water conditions within both layers it is necessary to specify the hydrological parameters of saturated hydraulic conductivity (K_s), soil water content at saturation (θ_s) and the suction/moisture curve. The surface boundary conditions are defined by the precipitation, maximum evaporation rate and the surface detention capacity.

The dynamic soil water model is driven by changes in the surface boundary conditions. Precipitation supplies water into each column at a rate up to the infiltration rate. However, in the case of the infiltration rate being exceeded, water is stored on the surface until the detention capacity is reached, after which overland flow is initiated. After rainfall ceases soil water removal through evaporation can be simulated.

The hydrology model redistributes soil water vertically within each column and laterally within the saturated zone depending on the cell water contents (shown in figure 4). Within the unsaturated zone the flow is vertical, obeying Darcy's Law, using the Millington-Quirk formulation to caluclate the unsaturated hydraulic conductivities. The height of the water table is obtained from the number of continuous saturated cells above the base. The flow between saturated cells is controlled by the hydraulic head and the specified saturated hydraulic conductivity. In the event that the column becomes totally saturated, a return overland flow component is determined using the slope angle for the hydraulic gradient.

At the end of each simulated hour, the pore water pressure/suction for each cell is recalculated from the cell water contents using the suction/moisture curve. These pore water conditions are automatically transferred into the stability model as piezometric heads located at points by the centre of each cell. Given the soil strength and density for each layer the model searches for the minimum factor of safety from a grid of slip surface centres and a number of slope circle radii. The overall minimum factor of safety for the simulated event is obtained from the hourly minimums. It is important to note that the factor of safety and the slip circle are dependent on the soil water conditions, therefore making it necessary to undertake the search procedure.

The combined model was run for a number of type conditions to enable hydrological design charts to be produced for two storm events. The storms selected were of 1 and 8 hour duration with 10 and 100 year return periods respectively. In addition, three slope angles and antecedent groundwater levels were chosen (shown in figures 5 & 6) each with three soil permeabilities (table 1). The minimum factor of safety was calculated for each combination of slope conditions given a 30 hour model

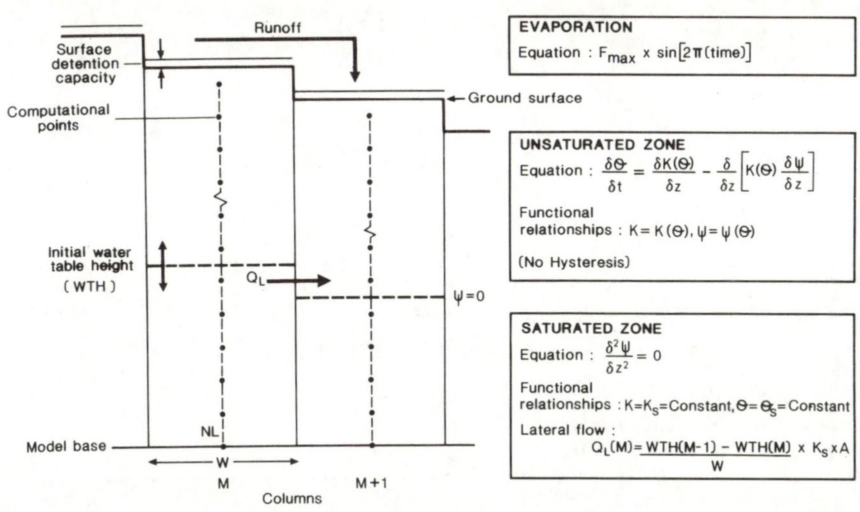

Figure 4 Two dimensional soil water model: computation points and governing equations

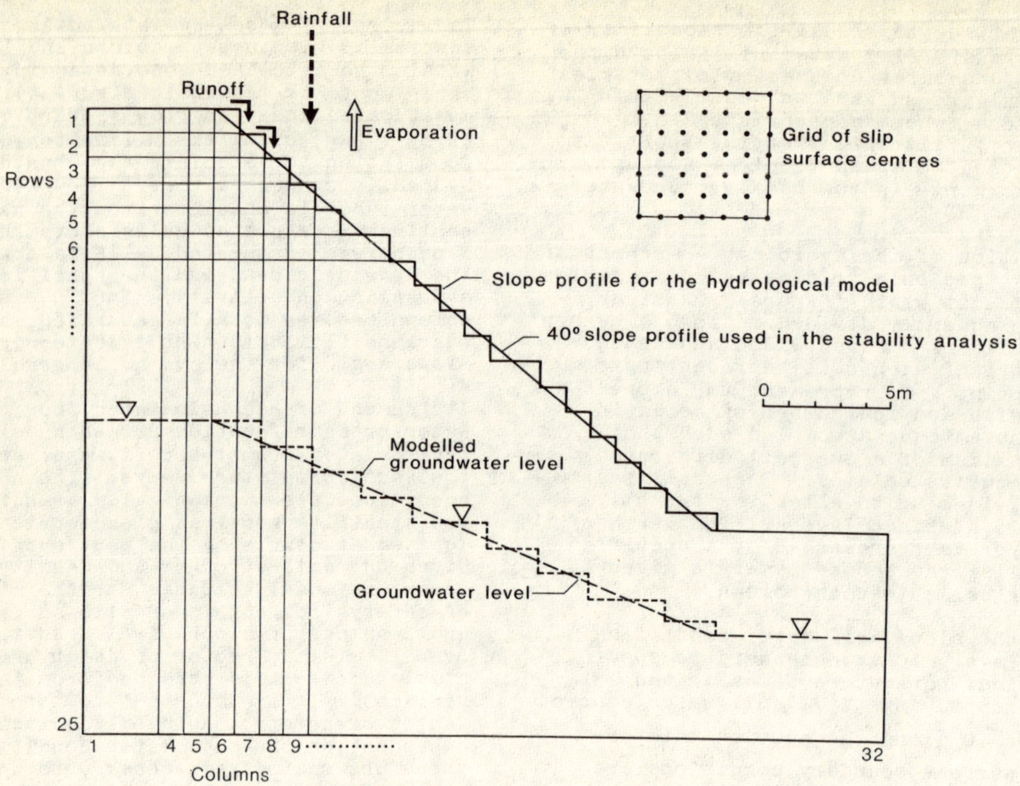

Figure 5 Slope, groundwater and modelling profiles used for the 40° slope simulation

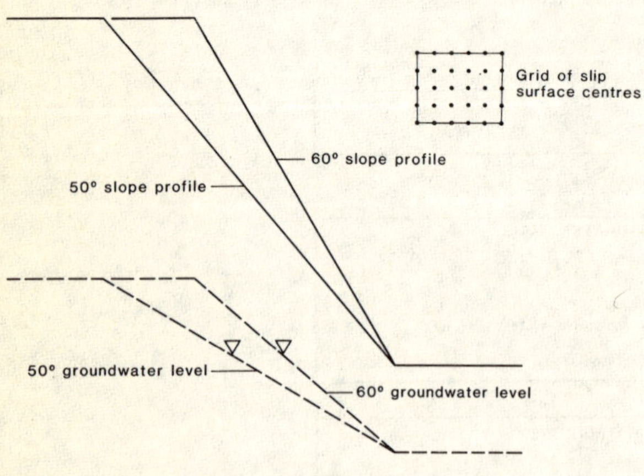

Figure 6 Slope, groundwater and modelling profiles used for the 40° slope simulation

Table 1 Model parameter list

Rainfall rate	a	1 hr 10 yr -100mm hr^{-1}
	b	8 hr 100 yr - 50mm hr^{-1}
Saturated hydraulic conductivity (ms^{-1})	a	1×10^{-5}
	b	1×10^{-6}
	c	1×10^{-7}
Saturated moisture content (m^3/m^{-3})		0.342
Detention capacity (m)		0.04
Shear strength values C(KPa)/ø(degrees)		0/36
		2/34
		4/32
		6/30
		10/26
Slope Angle (degrees)		40
		50
		60

simulation time. The results of these simulations are displayed in two charts (figures 7a & 7b) in which F/tano is plotted agianst c/tanø.

Since the charts (figures 7a & 7b) show a high degree of similar linearity, the prediction of F* may be determined by considering a multiple regression function to describe the relationship between the dependent and independent variables. The multiple regression equations for the two storm simulations are given in table 2, and are shown to be statistically significant ($r^2 > 99.5\%$). It is important to note that these equations are only of relevance to correlations that fall within the limits of the descrete data set.

The importance of individual independent variables cannot be taken from the regression equation itself, since comparison of the regression coeffiencents is meaningless. However comparisons can be made using the standardized partial regression coefficients, represented as B_j values. The B_j value is defined in terms of the regression coefficient and the standard deviation, of the dependent and independent variables such that:

$$B_j = b_j(S_j/S_o)$$

where b is the regression coefficient, and S_j and S_o are the standard deviations of the independent and dependent variables. The magnitude of the B_j value is, therefore, a direct measure of the importance of the various variables.

The partial regression coefficients for the independent variables of each multiple regression equation are given in table 2. From the results it is apparent that slope angle has the strongest effect on F* (albeit a negative effect). Shear strength is the next most significant variable, being more important than $\log K_s$. However, in comparing the two storm simulations the partial regression coefficients for the slope angle and strength change little, this contrasts the increase in the importance of $\log K_s$ which changes some 460%. This finding highlights the increase in importance of the hydrological properties during long duration large storm events.

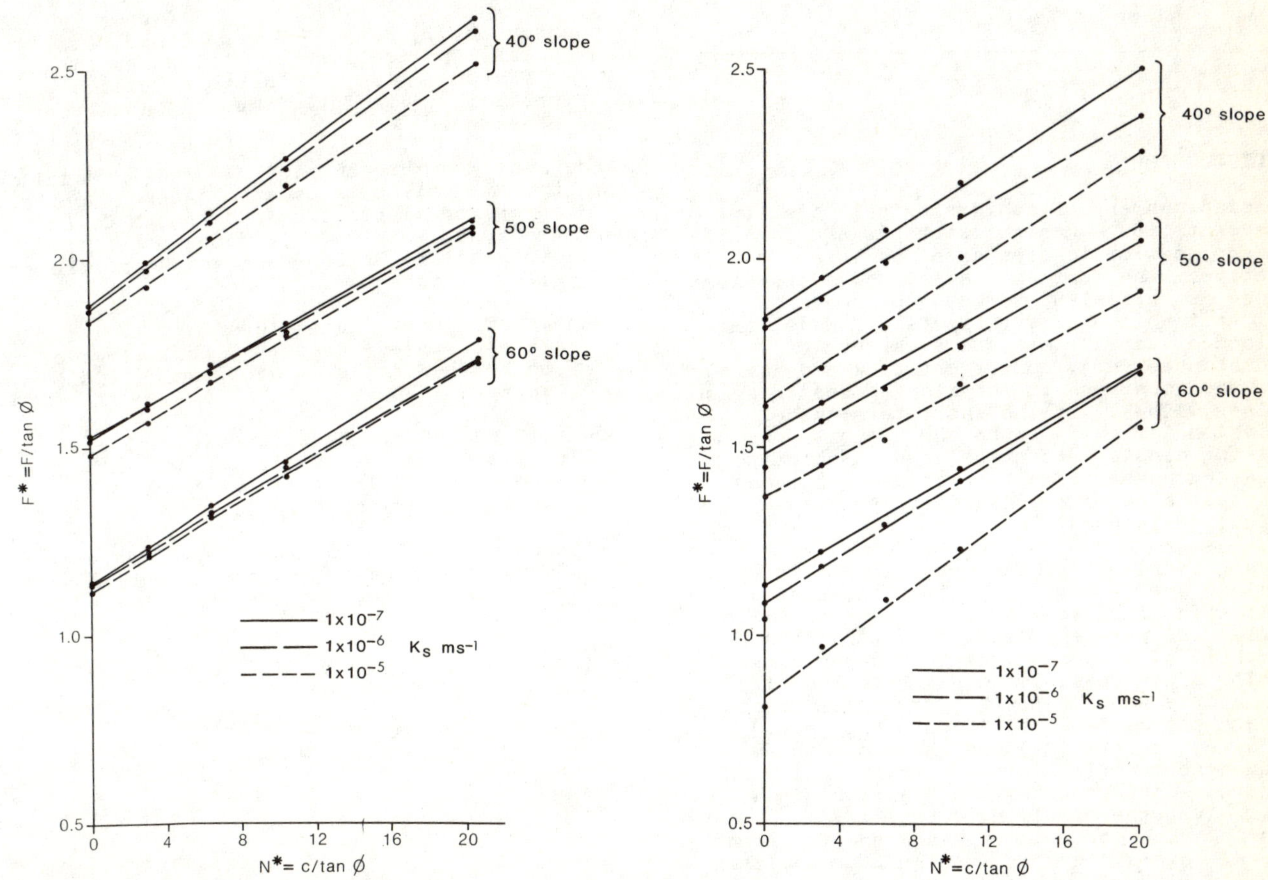

Figure 7 Slope design charts for (a) 1 hour 10 year storm and (b) 8 hour 100 year storm

Table 2 Regression equations and partial coefficients for predications of F*

Regression Equations

1 hr 10 yr $F^* = 3.29 - 0.0237 \log K_s + 0.0312N^* - 0.0386A$

$r^2 = 99.5\%$

8 hr 100 yr $F^* = 2.60 - 0.109 \log K_s + 0.0303N^* - 0.0368A$

$r^2 = 98.6\%$

Standardised Partial Regression Coefficients B_j

	$\log K_s$	N^*	A
1 hr 10 yr	-4.33×10^{-3}	6.53×10^{-4}	-7.06×10^{-4}
8 hr 100 yr	-2.00×10^{-2}	6.35×10^{-4}	6.72×10^{-4}

where

$F^* = F/\tan\phi$
$N^* = c/\tan\phi$
K_s = saturated hydraulic conductivity
A = slope angle

4 DISCUSSION

The establishment of a hydrological modelling - instrumentation - design chart package for wide area application is seen as a feasible objective. The instrumentation now available is seen as sufficient to meet the needs of model validation and verification. Modelling development can accommodate a variety of circumstances (e.g. convergent flow and surface cover modifications to the slope - see Anderson and Shen, 1987). In the short term, model validation is likely to be the element requiring closest attention, and to which the increasing number of pore pressure instrument based case studies being undertaken should be carefully related.

However, hydrological factors relating to detailed soil water movement in the context of infiltration induced landslides cannot all be modelled and parameterised (e.g. pipeflow). Nevertheless, evidence from this paper suggests that it is now possible to begin to identify those slopes where certain hydrological factors dominate the slope stability process (e.g. hydraulic conductivity) and therefore the more standard geotechnical properties of soil strength, slope geometry etc play a dominant role. Given the modelling capability of exploring such controls for various recurrence interval rainfall events (design storms), then the validation of such a design package, and its subsequent modification to include other hydrological aspects, is seen as an important research priority in enabling generalised wide-area hydrological design charts to be successfully applied.

ACKNOWLEDGEMENT

The work reported in this paper was funded by Overseas Development Administration, on scheme R4125 award to Dr M.G. Anderson. The assistance of JKR and LLM in Kuala Lumpur, Malaysia is gratefully acknowledged.

REFERENCES

Anderson, M.G. and Pope, R.G. 1984. The incorporation of soil water physics models into geotechnical studies of landslide behaviour. Int. Soc. Soil Mech. Fndtn. Eng: 349-353.

Anderson, M.G. and Howes, S. 1985. Development and application of a combined soil water - slope stability model. Q. J. Eng. Geol: 18: 225-236.

Anderson, M.G. and Kneale, P.E. 1987. Instrumentation of pore pressure and soil water suction, in Anderson, M.G. and Richards, K.S. (eds): Slope Stability - Geotechincal Engineering and Geomorphology, Wiley, Chichester : 77-112.

Anderson, M.G., Kemp, M.J. and Lloyd, D.M. (in press) Applications of soil water finite difference models to slope stability problems. Vth Int. Symp. on Landslides, Lausanne.

Anderson, M.G. and Shen, J.M. 1987. Modelling the effectiveness of a soil-cement protective cover for slopes, in Anderson, M.G. and Richards, K.S. (eds.): Slope Stability - Geotechnical Engineering and Geomorphology, Wiley, Chichester: 231-264.

Brand, E.W. 1984. 'Landslides in S.E. Asia: A state-of-the-art report', Procedings of the 8th South East Asian Geotechnical Conference, Bangkok, 17-59.

Brand, E.W. 1985. Geotechnical engineering in tropical residual soils. Proc. Int. Conf. Geomechan. in Tropical Lateritic and Saprolitic Soils, Brasilia.

Geotechnical Control Office (1984) Geotechnical Manual for Slopes (2nd edn). Public Works Department, Hong Kong 245 pp.

Parameter selection for stability analysis

L.E.P.DE CAMPOS, Associate Professor, University of Bahia, Brazil
M.S.DE SOUSA MENEZES, Consulting Engineer, Brazil
E.C.FONSECA, Assistant Professor, University of Bahia, Brazil

ABSTRACT: Slope stability analysis for residual soils requires the knowledge of the behavior of non saturated soils and of the contribution of suction to the increase of the shear resistance of such materials. Since there is not enough reliable data on the relations between moisture content and suction, and moisture content and hydraulic conductivity, a procedure that allows the indirect estimate of the contribution of suction to the stability of slopes has been used in Salvador, Brasil, and is presented in this paper.

1. FOREWORD

The estimate of the stability of natural slopes and cuts has led constructors and designers to a conflict, since many design methods are considered to be conservative, by constructors, due to the fact that steep and very high cuts have been stable for many years, without any kind of retaining structures. This fact, in many cases, is due to the utilization, by designers, of results of shear tests run on "saturated" or "flooded" samples, while, on the field, suction represents a significant contribution to the stability of slopes or cuts in wich soils are partly saturated.

It is then necessary to determine the geological-geotechnical properties of the soil mass and, mainly, to find out how the process that leads to rupture takes place on slopes and cuts in partly saturated soils.

In this paper, the term "flooded" is used to indicate that samples in the shearing ring are immersed in water during a 24 hour period, before running direct shear tests. It is known that, in most cases, the degree of saturation does not reach values above 90% or 95%, after immersion, when non disturbed samples of residual soils are used. In traxial tests, the same occurs, even when back pressure is applied for the same period of time.

2. GEOLOGICAL AND GEOTECHNICAL ASPECTS OF SOILS IN SALVADOR

Salvador is situated in the "Reconcavo Basin", which is defined as a graben, limited by the Maragogipe fault at west and the Salvador fault at east, both with a SW-NE direction. Most part of the city is built on the oriental horst of the tectonic depression.

Metamorfic and igneous rocks of different ages occur and are very fractured; gneisses are predominant.

Most soils are residual of such rocks; these soils are, nearly always, clayey silts or silty clays of mediun to hard consistency. Other soil types do occur, but are not of interest to this paper.

Direct shear tests, run on non disturbed samples of such soils, have shown that the cohesion intercept for samples tested at the natural moisture content and, simultaneously, after immersion, varies from 30 kPa to 60 kPa and from 0 to 30 kPa, respectively, while the friction angle seems not to be sensitive to flooding and presents values that vary from 26 to 34 degrees.

3. FAILURE MECHANISM

The occurrence of landslides on natural slopes and cuts in Salvador is related to rainfall and most of them take place at the end of the so called rainy season. Due to such facts, the hypothesis that the failure mechanism is related to the build up of neutral pressures derived from flow nets has been the starting point for slope stability analysis.

However, in most of the studied cases, the water table is located much below the rupture surface and the fore mentioned mechanism does not explain the occurrence of landslides.

Morgenstern and Matos (1975), Lumb (1962 and 1975), Wolle (1980), Campos et alii (1983), Campos (1984) and others have identified the possibility of failure due to the reduction of suction, which represents a significant part of the shearing resistance and is associated to the saturation of the slope, in tropical soils. In this case, the flow of water, determined by infiltration, is taken into account and it is admitted that rupture occurs along the surface limited by the moisture front.

These authors seem to agree that the following sequence takes place:

- before the rainy season, the slopes and cuts in residual soils are stable and partly saturated; the saturation degree, in Salvador, varies from 50% to 70% as an average. Since these soils are partly saturated, negative pore pressures (suction) build up and act as apparent cohesion which, in many slopes, is the predominant stability factor;

- as rainfall begins, absorption of water by the surface layers of the slope or cut takes place and the saturation degree increases, with the subsequent increase of hydraulic conductivity;

- the saturation zone advances to larger depths, generally a few meters thick, and is called saturation or moisture front. It is evident that the advance of this saturation front depends on the initial degree of saturation of the soil, on its void ratio, grain size distribution and plasticity, besides the intensity and duration of the rainfall and on the nature of the vegetation cover;

- after a certain depth is attained by the saturation front, a water flow takes place, parallel to the slope surface and in the upper zone of the front, where saturation occurs; the hydraulic gradient is equal to the sinus of the slope angle. Due to capillary and gravitational forces, however, the hydraulic gradient increases and the flow of water in the vertical direction is predominant; the flow parallel to the slope can then be neglected;

- once the rainfall is over, a redistribution of the water that infiltrated takes place and the average moisture content is larger than it was before the rainfall;

- the advance of the moisture front, when another rainfall of the same intensity takes place, will be faster, since the degree of saturation of the soil near the surface is higher than it was before the first rain.

In other words: during the rainy season, there is a continuous and gradual increase of the degree of saturation, that is responsible for the reduction of suction and the consequent reduction of the shear resistence of the soil, which can lead to the failure of a stable slope.

4. STABILITY ANALYSIS METHOD

The failure mechanism just described must be taken into account when the stability factor is determined; this implies in considering the advance of the saturation front, which depends on the type of soil, on the initial moisture content, on the geometry of the slope and on the vegetation cover. Many of the authors mentioned before have presented a sequence of calculations to be followed and demonstrated that this mechanism exists, through back analysis of failures in Brazil and other countries.

According to Campos et alii (1983), the simulation of such mechanism corresponds to the association of a method of estimating infiltration to a method of slope stability analysis, that would allow for the determination of the variation of the safety factor with the advance of the moisture front. Since there is no available data on the relations between hydraulic conductivity and moisture content, for brazilian residual soils, such routine cannot be used yet.

Due to such impossibility, one of the possible ways of taking into account the effect of suction in a slope stability analysis consists of running shear tests on samples with their natural moisture content and on "flooded" samples. The difference between the cohesion intercepts can be considered to be the portion of shear resistance due to suction. This procedure has been used and leads to safety factor values that correspond to near minimum and near maximum safety factors. The use the maximum safety factor value is evidently dangerous, while the minimum value is definitly conservative. It seems prudent to choose some intermediate value, between these limits.

5. AVAILABLE RESULTS

Slope stability analysis, using Bishop´s simplified method and the parameters determined through the described procedure, were made during the last three years, for more than a dozen of cut slopes, 8m to 12m high and with average slope angles of sixty degrees. Safety factors were always smaller then 1.5 and, sometimes, below 1.0, when results of tests run on "flooded" samples were used. For samples tested at their natural moisture contents, safety factors were always above 1.5, indicating that the slopes were stable and confirming field observations; many cut slopes, in similar soils, have been stable since 1970.

These results indicate that the slope failures occur due to the reduction of suction and the consequent reduction of shear resistance, when saturation takes place.

The use of test results obtained from "flooded" samples leads to a reduction of the slope angle or to the construction of retaining structures. The use of test results obtained from samples at their natural moisture content are subject to restriction, due to the limited amount of available laboratory and field data; they have led to solutions where the slope angle is maintained, but a careful drainage system is required, in association to the protection of the surface of the slope with some kind of "impermeable" coverage, such as cement mortar 1.5cm to 2.5cm thick, for sandy or clayey silts, or vegetation protection for silty clays, as a rule of thumb.

6. CONCLUSIONS

Slope stability analysis, for non saturated residual soils, can be developed using traditional methods, such as Bishop´s simplified method, and adopting safety factors estimated through the use of shear resistance parameters whose values are situated between the limits corresponding to tests run on samples at their natural moisture content and on "flooded" samples. Through such procedure, a portion of the soil cohesion, due to suction, is indirectly taken into account.

The use of shear parameters obtained through tests run on "flooded" or saturated samples leads to design of retaining structures or the reduction of the slope angle when, in many cases, the cheapest and acceptable solution would be a good surface drainage associated to a slope protection that would minimize infiltration, through vegetation coverage or the use of some "impermeable" cover on the slope.

The proposed procedure requires that the slope should be homogeneous and isotropic, as do most current methods. It must not be used before the identification of the type and orientation of all discontinuities which are, frequently, the determining cause of slope failures in residual and non saturated soils.

This procedure requires more theoretical studies and field observation, since it has been in use for approximately and only three years, in specific soil conditions and when the water table is far below the possible rupture surfaces.

None of the slopes to which this procedure was applied - and that already outnumber ten - has shown any sign of instability. Permanent maintenance of the slope protection and of the drainage system is of fundamental importance for long time stability.

7. REFERENCES

CAMPOS, L. E. P., 1984, "Influência da sucção na estabilidade de taludes naturais em solos residuais", Tese de mestrado, Pontifícia Universidade Católica do Rio de Janeiro (MSCE thesis)

CAMPOS, L. E. P.; VARGAS Jr., E. A. and ROMANEL, C., 1983, "Um estudo da variação do coeficiente de segurança de um talude em solo residual em função da intensidade da chuva", IV Congresso Latino-Americano de Métodos Computacionais para Enegenharia, Santiago - Chile (IV Latin American Congress on Computational Methods for Engineering).

CAMPOS, L. E. P.; GESTEIRA, C. S. and SOBRAL, G. S. F., 1984, "Aplicação da teoria do fluxo em meios não saturados no estudo de estabilidade de taludes", XII Encontro sobre Escoamento em Meios Porosos, Maringá, vol. II, pp. 176-185.

LUMB, P., 1962, "Effect of rain storms on slope stability", Symposium on Hong Kong Soils, pp.73-87.

LUMB, P., 1975, "Slope failures in Hong Kong", Engineering Geotechnical Journal, vol. III, pp. 31-65.

MORGENSTERN, N. R. and MATOS, M. M., 1975, "Satability of slopes in residual soils", V Panamerican Conference on Soil Mechanics and Foundation Engineering, Buenos Aires, vol. III, pp. 367-383.

WOLLE, C.M., 1980, "Taludes naturais - Mecanismos de instabilização e critérios de segurança", Tese de Mestrado, Universidade de São Paulo (MSCE thesis).

A study on some engineering properties of lateritic soil

R.H.CHEN, National Taiwan University, Taipei, Taiwan

ABSTRACT: The Linkou terrace in Northwestern Taiwan is composed of a very thick layer of gravels underneath a layer of lateritic soil of several meters. Recently, this area has developed rapidly for residential and industrial purposes. However, erosion problem on natural slopes is serious especially after heavy rainfall. Hence, a series of testing program is performed in order to know engineering properties and the erosive potential of lateritic soil. The main topics concerned are: undrained shear strength, swelling potential, and permeability of compacted lateritic soil affected by compaction condition and change in degree of saturation due to water permeance. Moreover, both physical and chemical methods are used to investigate the erosive characteristics of lateritic soil and lateritic gravels. A case of failure in a compacted lateritic soil slope was also examined. The result shows that the failure was triggered by surface erosion of lateritic gravel layer.

1 INTRODUCTION

Lateritic soil in Taiwan can be found in many regions on the island: the northwest, the west, and the southwest (Fig.1). The Linkou terrace situated in Northwestern

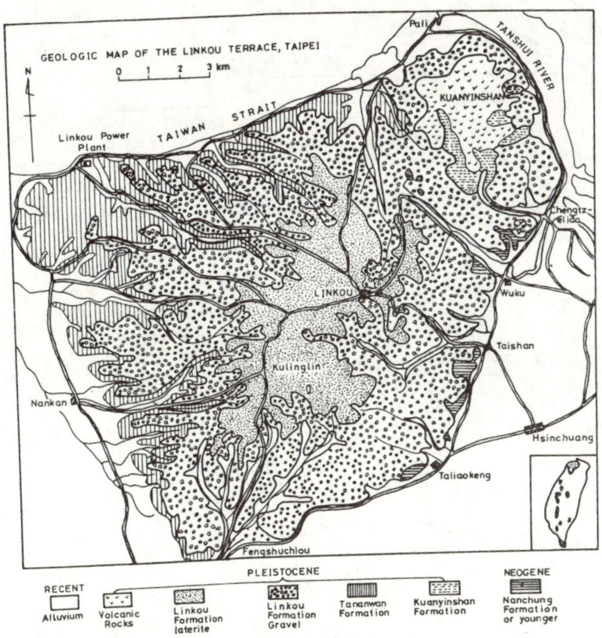

Fig.1 Geologic map of the Linkou terrace (Ho 1969)

Taiwan is the largest among them. Recently, this terrace has developed rapidly for residential and industrial purposes. Consequently, plants and natural slopes have been cut and more land are bare. Hence erosive problem in this area becomes more and more serious. The rainfalls on August 9th-10th, 1982, with 365 mm of rain in two days, brought devestation in the form of landslides, debris flow, and flood at Linkou city and its vicinities. Serious surface erosion was observed everywhere on natural slopes. There were also a few slumps occurred in compacted lateritic soil. It is believed that the above phenomenon was induced by the rain water.

In order to control better the erosive problem and slope instability in this area, the engineering properties of the soil and its erosive potential must be studied. Both physical and chemical methods are used to investigate the effects of rainwater and its permeation into the soil. The shear strength, permeability, and volume change characteristics of compacted lateritic soil under various degree of saturation and compacted conditions are studied. These results together with the understanding of the erosive properties of the soil and the collected field data will be used to look into a case of slope failure.

2 GEOGRAPHICAL AND GEOLOGICAL BACKGROUNDS

The Linkou terrace is a high coastle terrace, 220 m - 250 m above sea level, developed between Taipei and Hsinchu in Northwestern Taiwan. The Linkou Formation comprises all the gravels and the lateritic mantle of a few meters to about ten meters exposed in the Linkou terrace. The terrace is underlain extensively by Pleistocene clastic sediments (Ho 1969). In some clear profile, from top to bottom, it can be seen: lateritic soil, weathered gravel, and unweathered gravel. This indicates that lateritic soil was likely weathered and leached from gravel and its fillings. The sediments are mostly loosely cohered, little consolidated. The surface of the gravel was generally oxidized and stained by a layer of red lateritic clay. The gravel is intimately mixed with variable portions of sand-silt-clay (Woo, et al. 1982, Lee-Wei 1982).

3 METHODS OF STUDY

The mothods of study comprises a series of testing programs. As to the shear strength, the main topic concerned is that the effect of initial water content (when sample is compacted) and the degree of saturation (due to subsequent water permeation) on the undrained shear strength. Besides, the swelling potential and the permeability in function of initial water content are also studied. For the erosive potential, the three main topics concerned are: erosion of lateritic soil by permeant water, surface erosion of lateritic soil by rainfall, and erosive potential of lateritic gravel. Each topic is briefly deseribed below:

(a) Undrained shear strength- This test investigates the characteristics of undrained shear strength of lateritic soil compacted at various initial water contents and at subsequent final water contents. The compaction tcst, UUU (unsaturated-unconsolidated-undrained) test, and SUU (saturated-unconsolidated-undrained) test are performed. The UUU tests are for both remolded and undisturbed samples, other tests are for remolded samples only. The compaction curve of AASHTO compaction energy is composed of six points at different water contents. To study the effect of the degree of saturation, various amount of water was allowed to permeate into the compacted sample (for samples compacted dry of optimum only) to reach to different degree of saturation before running the triaxial test.

(b) Volume change- The compacted soil samples are saturated first. Then the sample is put in the oven to dry. The change in volume between two conditions is thus determined.

(c) Permeability- Permeability test is run using the triaxial cell so that the coefficient of permeability of the sample under various confining pressures can be studied.

(d) Erosion of lateritic soil by permenant water- The experiments include SCS dispersion test (Soil Conservation Service dispersion test), and pinhole test. The purpose is to study whether lateritic soil is dispersive.

(e) Surface erosion of rainfall- A water-falling equipment is used to erode artificial slopes. Three inclination of slopes considered are 20°, 30°, and 40°. The intensities of rainfall are controlled at 250, 300, and 350 mm h^{-1}. Samples are prepared by compacting lateritic soil in a wooden box to the desired density at optimum water content. After that, the sample is surface dried by a heater for 1 h, left for cooling down, and then subjected to rainfall for 1 h. This dry-wet cycle is repeated for four times.

(f) Evaluation of the erosive potential of lateritic gravel- The samples obtained are from various depth beneath the ground surface. Physical tests include determination of Atterberg limits, soil classification, and grain size distribution. Chemical tests include digestion process, estimation of the concentrations of metallic ions in the liquid of soil after digestion process, and determination of SAR values.

4 LABORATORY RESULTS

The results of laboratory study on each topic are summerized hereafter. Firstly, the engineering properties of compacted lateritic soil will be presented.

The compaction curves of two air-dried water contents are shown in Fig.2. The curve of lower air-dried water content (ω_a = 7.7%) has higher maximum dry density and lower OMC. (optimum moisture content or optimum water content) than the one at ω_a = 14.5%.

(a) Strength vs. degree of saturation

(i) The stress-strain curves of compacted samples with various degree of saturation (not shown) can be distinguished into two kinds: the one with a peak point (strain-softening) and the other without a peak point

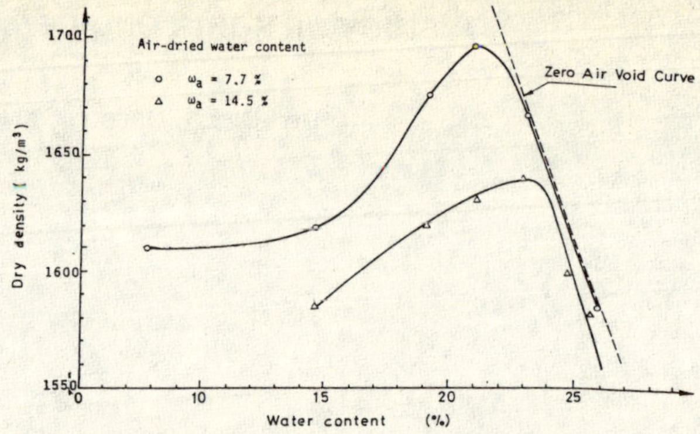

Fig.2 The compaction curve of lateritic soil of Linkou

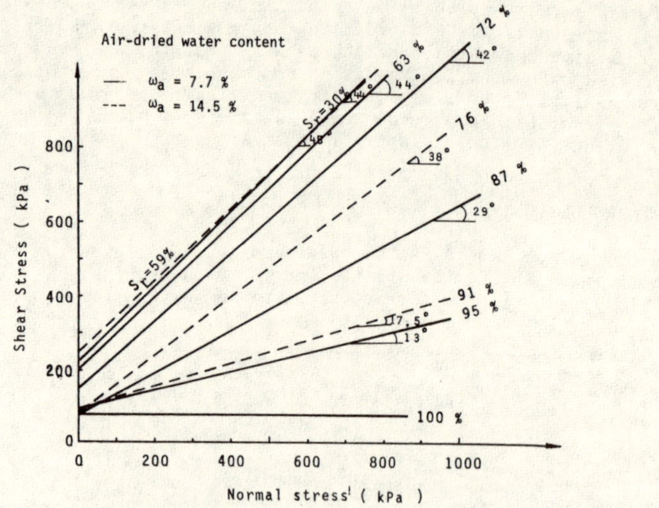

Fig,3 The Mohr envelopes at various degree of saturation

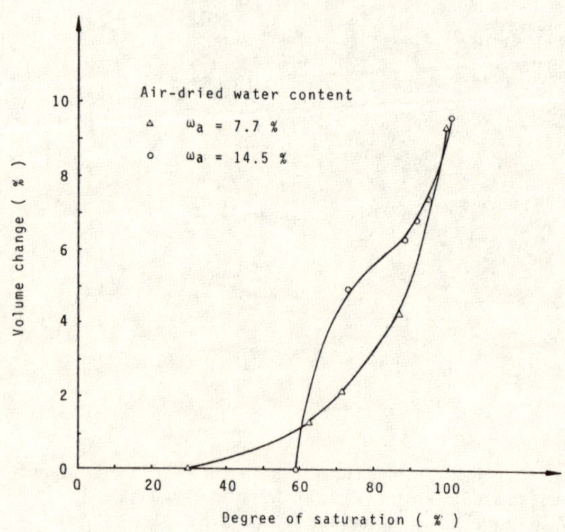

Fig.4 The relationship between volume change and degree of saturation

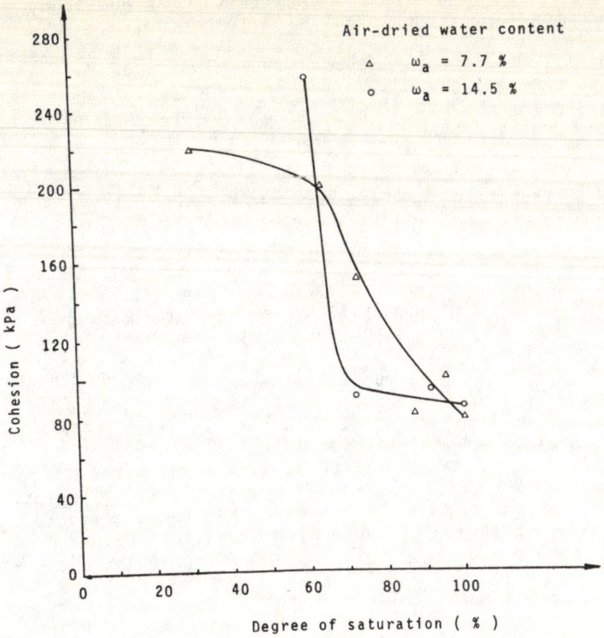

Fig.5a The undrained shear strength parameter (cohesion) for samples at various degree of saturation

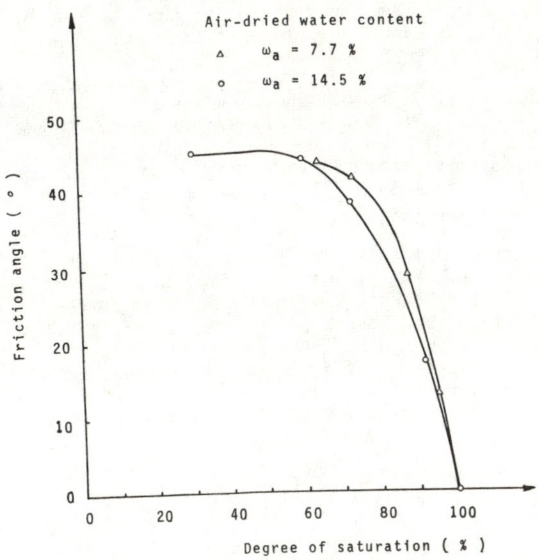

Fig.5b The undrained shear strength parameter (friction angle) for samples at various degree of saturation

(strain-hardening). The curves with a peak point are for samples compacted dry of optimum ($S_r \leq 72\%$ and $S_r \leq 76\%$ for $\omega_a = 7.7\%$ and 14.5%, respectively). The curves without a peak point are for samples compacted wet of optimum ($S_r \geq 87\%$ and $S_r \geq 91\%$ for $\omega_a = 7.7\%$ and 14.5%, respectively). These two typical curves are obtained no matter what confining pressure is (σ_3 = 50, 100, 200 kPa). The Mohr envelopes from undrained shear tests at various degree of saturation are shown in Fig.3. Fig.4 shows that the rate of volume change increases rapidly

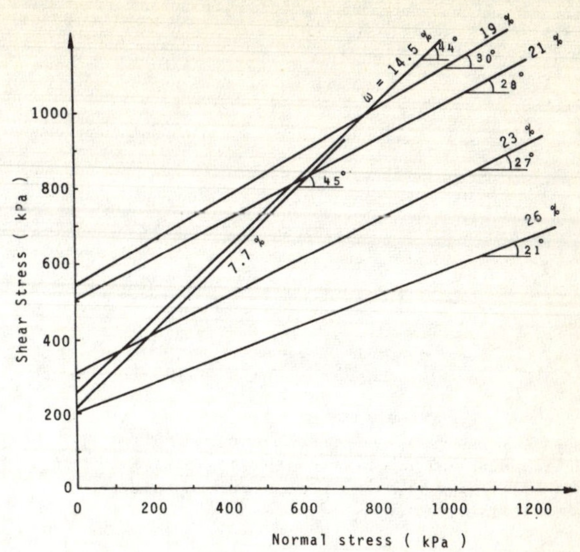

Fig.6 The Mohr envelopes for samples compacted at different water content (with the same compactive energy)

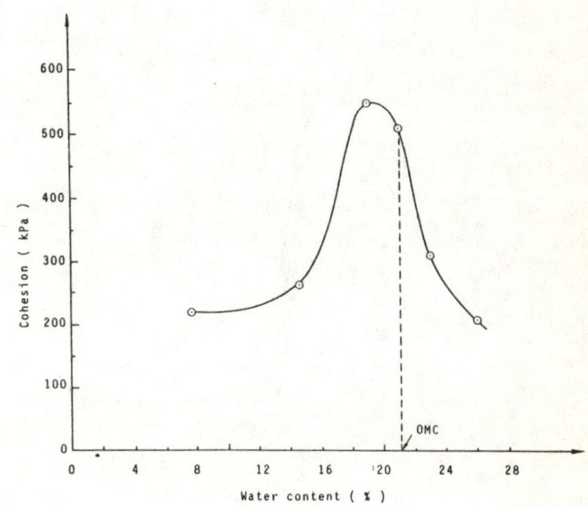

Fig.7a The undrained shear strength parameter (cohesion) for samples compacted with constant compactive energy

when $S_r > 63\%$. This behavior also causes more significant decrease in strength (Fig.3).

The influence of the degree of saturation on strength parameters c and ϕ can be seen from Fig.5: both c and ϕ values decrease as degree of saturation increases, especially when $S_r > 63\%$. This result is the same as that concluded by Lumb (Lumb 1962) in his study on lateritic soil in Hongkong: cohesion is influenced by both the degree of saturation and the void ratio, however, the friction angle is influenced only by the void ratio.

(ii) Strength vs. various water content for the samples on the same compaction curve— This test considers the compaction curve of $\omega_a = 7.7\%$ only. The Mohr envelopes obtained from UUU tests are shown in Fig.6. The c and ϕ

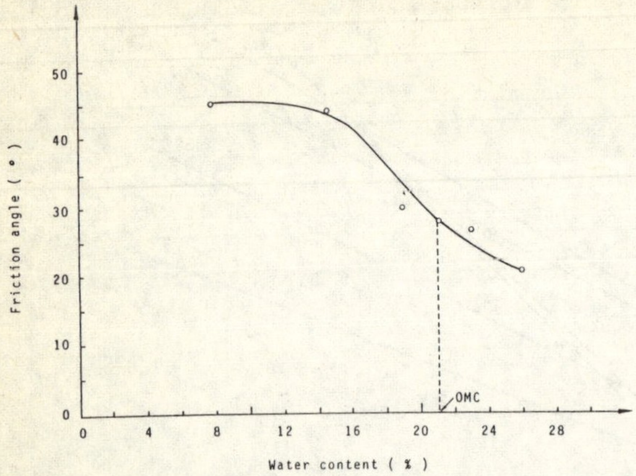

Fig.7b The undrained shear strength parameter (friction angle) for samples compacted with constant compactive energy

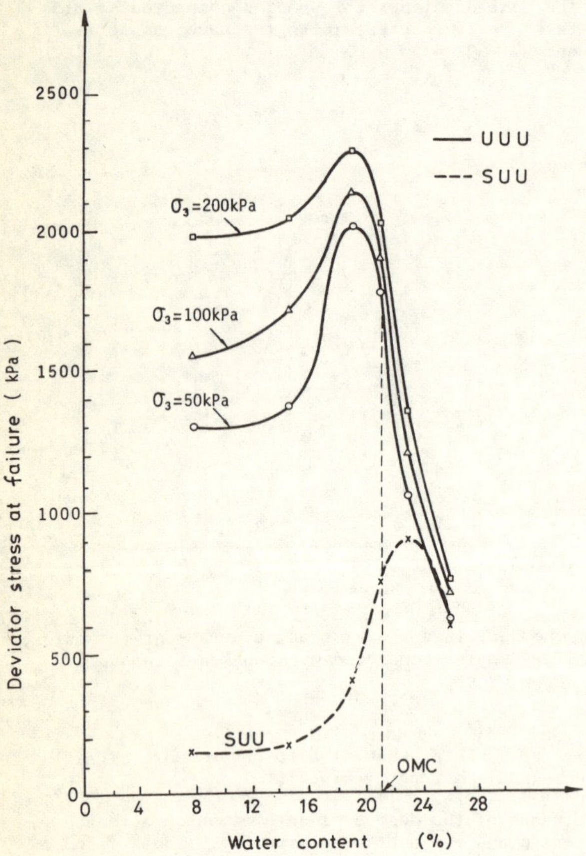

Fig.8 The UUU and SUU shear strengths of samples compacted at different water content

values are also replotted on Fig.7. The sample with the lowest water content, ω = 7.7% has the highest frictional angle (ϕ = 45°) and the lowest cohesion (c = 220 kPa) which shows the strength characteristics of a granular material. This is because when the soil is dry, its structure comprises mainly aggregates. The cohesion comes from the cementation of aggregates. In addition, the frictional angle decreases with water content increases. This might be due to that the aggregates is broken up. Consequently, the characteristics of a cohesive material gradually reveals. The cohesion reaches to its maximum at ω = 19%, then decreases with water content increases.

As for UUU test on natural soil (ω_n = 23%, ρ_d = 1520 kg/m^3), the strength parameters are c = 130 kPa and ϕ = 24°. This result will be used in the case study below.

The influence of saturation on the shear strength is significant as can be seen from Fig.8. The strength drops off remarkably when soil becomes saturated, especially for samples compacted dry of optimum.

(b) Volume change- The amount of volume change after saturation depends on the compaction condition. It is known that swelling potential of a soil is affected by the orientation of its clay particles: flocculated structure is more potential to swell than disperse structure. Hence, for the same compaction curve, samples compacted dry of optimum swells more than those compacted wet of optimum (Fig.9). Fig.9 also shows that volume change of more air-dried soil (ω_a = 7.7%) is much less.

This is because soil air-dried more has stronger cementation. However, it is in more water deficiency. Therefore, when this soil is compacted wet of optimum, it swells somewhat more because the cementation is destroyed and more water is absorbed due to more water deficiency.

(c) Permeability- The result of permeability test is shown in Fig.10. The coefficient of permeability decreases remarkably from the dry side to the wet side of OMC. Apparently, the soil structure is the main cause.

(d) Erosion potential of lateritic soil- The grain size distribution of lateritic soil shows that its clay content is only 1%. The percent dispersion obtained from SCS dispersion test is 1/60. Hence, the dispersive potential of lateritic soil is very small. On the other hand, the pinhole test shows that latertic soil belongs to ND1 soil (nondispersive).

(e) Experiments of rainfall- The following results show that the dry-wet process affects the resistance of lateritic soil to erosion: (i) For soils subjected to more dry-wet cycles and longer duration of rainfall, the percent of coarser particles increase in the grain size distribution of particles eroded. (ii) Drier soil has more amount of coarser particles produced and eroded. (iii) Steeper slopes has more amount of coarser particles eroded.

(f) Erosion potential of lateritic gravel- This test shows the following results: (i) Among 16 samples tested, only 4 samples have high SAR values (63 - 74) and high percentage sodium (84% - 87%). Note that these four samples were all obtained from shallow depth. (ii) The Atteberg limits of lateritic gravels are mostly in the regions of CL soil and their clay mineral content is mainly illite. However, lateritic soil is classifred as MH soil and its clay mineral content is kaolinite (Fig.11). Among the four samples of high SAR value, represented by solid cicles on Fig.11., there are three samples classified as CL soil and its clay mineral content is illite. (iii) The digestion process of lateritic gravel is slow (Fig.11) which indicates that most amorphous Fe_2O_3 existing in lateritic gravel hide in the lattices of SiO_2. This is quite different from lateritic soil which has the fastest digestion process. This phenomenon indicates that most amorphous materials coat on the surface of clay minerals of lateritic soil. (iv) The grain size distributions of lateritic gravels fall in two

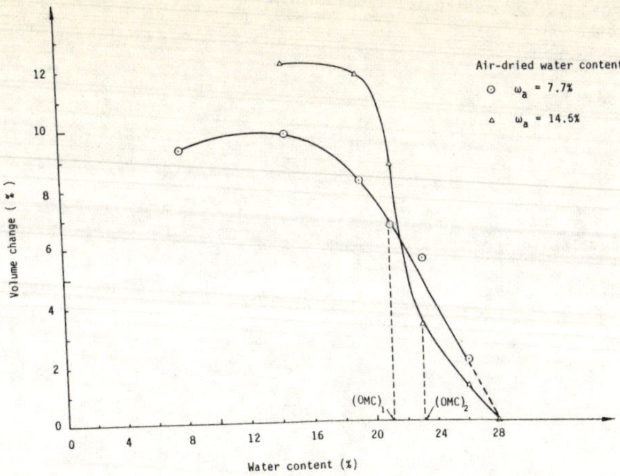

Fig.9 The influence of the water content of compacted sample on its volume change

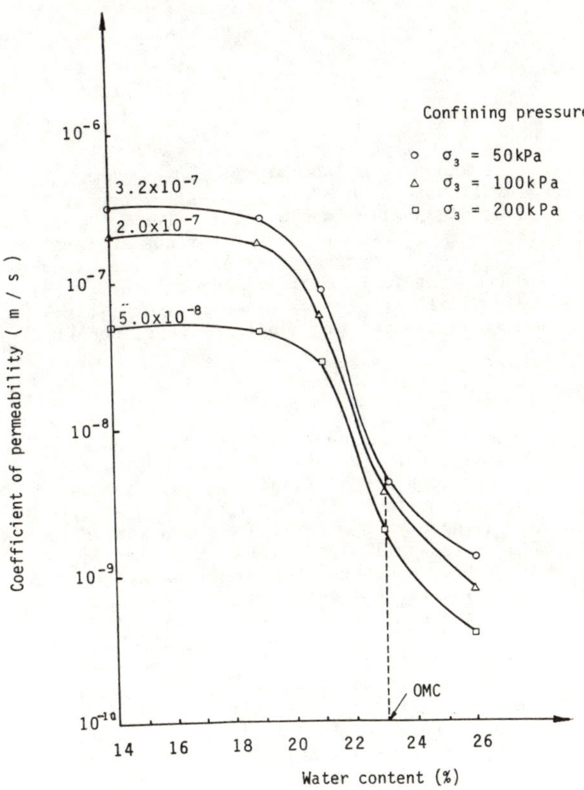

Fig.10 The coefficient of permeability of samples compacted at various water content

zones in Fig.12. In zone I, soils behave like that of clayey soils with high SAR values. In zone II, soils behave like that of sandy soils with high coefficient of permeability. Both types of weathering patterns have high erosion potential.

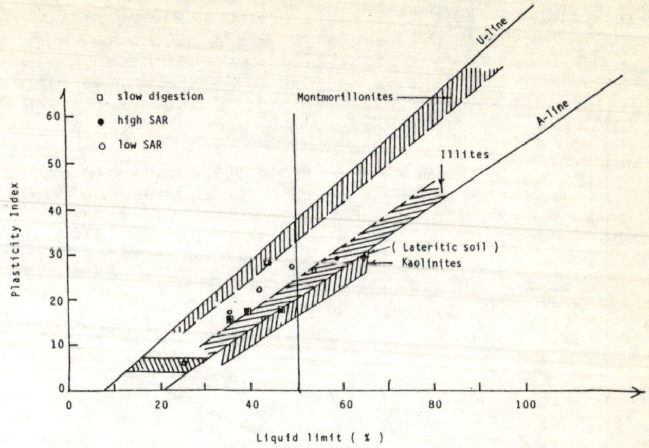

Fig.11 Plasticity chart for samples tested

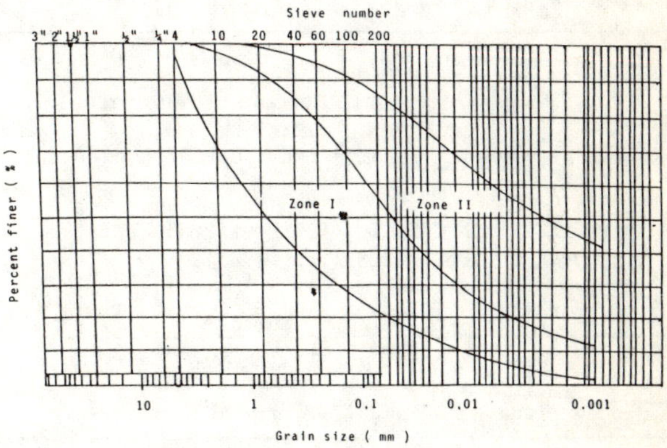

Fig.12 The grain size distributions of lateritic gravels

5 CASE STUDY

The following case record is a slope failure in compacted lateritic soil at Linkou industrial area. The profile of the slope is shown in Fig.13. The compacted soil overlies natural lateritic soil which is underlain by very thick layer of gravel deposits. The thickness of compacted soil and natural lateritic soil are 4 m and 6 m, respectively. The slopes of compacted soil, natural lateritic soil, and gravel deposits are 1/1.5, 1/1.5, and 1/2 (vertical/horizontal), respectively. The water content of compacted soil is 19% (OMC-2%). The depth of infiltration, being lack of data, is reasonably assumed to be 6m. The saturated compacted soil (within the depth of infiltration) has SUU shear strength of 200 kPa obtained from Fig.8. The saturated and unsaturated natural lateritic soil have SUU and UUU shear strength of c equal to 30 kPa (Hung 1967); and c equal to 130 kPa and ϕ equal to 24°, respectively. The SUU shear strength of gravel deposits has c equal to 15 kPa and ϕ equal to 24° (Woo 1982). The unsaturated gravel deposits is assumed to be a stiff layer in stability analysis. The tension crack is estimated to be 2.5 m from the field investigation. There is no shear strength in tension crack. Pore water pressure is assumed exist in tension zone. Since the ground water table is very deep, it is not considered in the study.

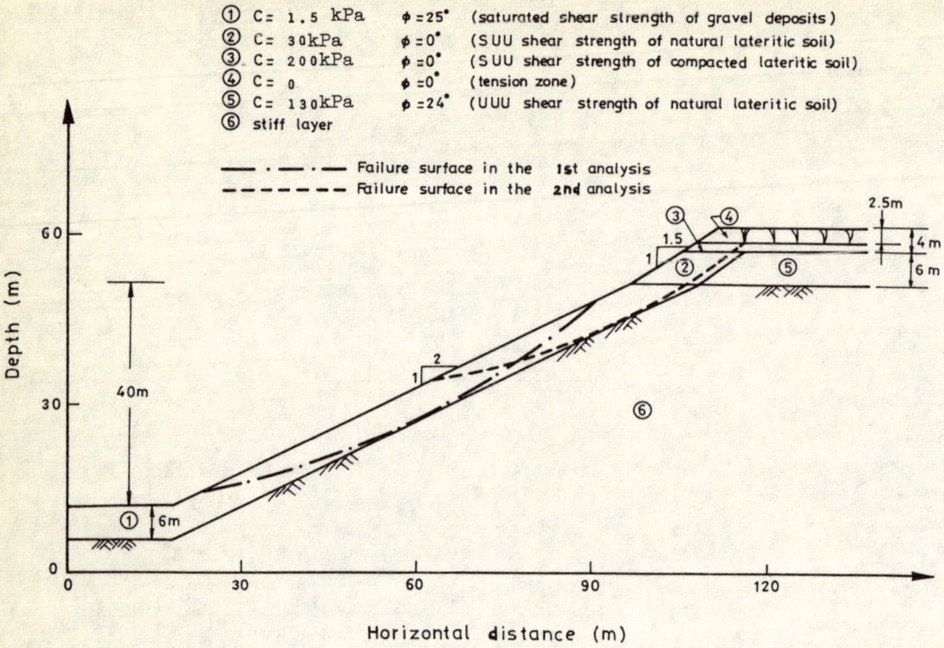

Fig.13 Failure profile of compacted slope at Liniou industrial area

TABLE I INFILTRATION CAPACITY FOR DIFFERENT SURFACE COVER OF VEGETATION (Yang 1982)

	Forest	Grass	Bared Surface	Trail
Average (mm/hr)	257	191	99	11
Range (mm/hr)	87 - 395	24 - 281	22 - 193	2 - 29

The stability analysis was made by computer program STABL II (Boutrup 1977). In the first analysis, the failure mass was within the layer of gravel deposits only. The factor of safety was 0.995. The second analysis was made on the profile after the first failure occurred. The failure mass included both gravel deposits and lateritic soil. The factor of safety was 0.953. This indicates that failure of compacted lateritic soil was very likely due to failure of lateritic gravel which removed the lateral support of lateritic soil.

In order ot find out the possibility of failure in gravel deposits, it has to go back to review the collected data. Since the coefficient of permeability of gravel deposits is high, it means that the infiltration capacity is high. Table I shows the infiltration capacity of soils in different extent of surface cover. For bared cover of slopes, the average infiltration capacity is 99 mm/hr. Hence, the duration of rainfall must be long enough and the rain intensity should be high, otherwise the depth of infiltration will not be deep enough and only surface erosion occurs.

On August 9th and 10th, 1982 the amount of rain were 143.8 mm/day and 221.3 mm/day, respectively. The rainfall concentrated at 1 AM to 3 AM on August 10th, with the intensity as high as 100 mm/hr. It is thus quite possible to have a deep depth of infiltration for two-day raining. Therefore, for those slopes steeper than 25°, it is very likely to fail if the rainfall sustained long enough.

Moreover, it is known from laboratory studies that lateritic gravels have high erosion potential especially for the soil within shallow depth. This finding is consistent with the field condition that serious surface erosion was observed on many natural slopes after heavy rainfall. Hence, the failure of compacted lateritic soil slope was very likely triggered by the loss of lateral support induced by erosion of underlain gravel deposits.

CONCLUSION

This paper studies the engineering properties and erosive charateristics of lateritic soil and gravel deposits at Linkou terrace by both physical and chemical methods. The engineering properties of lateritic soil are quite different depending on the water content and the degree of saturation of the soil. For soil more air-dried, the concretion is harder and hence the soil behaves like granular material with some cohesion. As the water content increases, the soil behavior is more close to that of cohesive material. Besides, the cohesion is more sensitive to water content of the soil than frictional angle. Consequently, it is probably better to compact lateritic soil with water content less than OMC in order that the compacted soil has less swelling potential, better resistance to erosion, less permeability and hence less influenced by rain water.

The study on erosive potential of lateritic soil shows that its erosive potential is very small. The erosion test using a rainfall simulation equipment also shows that the erosive potential of laterites is considered low. As for lateritic gravels, they are classified as CL soil with clay mineral content mainly of illite. Their erosive potential is higher than that of laterites. Another observation from digestion process indicates that most amorphous ferric ion oxide existing in lateritic gravels hide in the lattice of SiO_2. This is quite different from laterites in which most amorphous materials coat on the surface of clay minerals. In addition, from grain size distribution of lateritic gravels, two types of weathering patterns are observed: one behaves like sandy soil and another like clayey soils. However, both soil types have high erosive potential.

Finally, by using the laloratory data for slope stability analysis on a case record, the result of analysis is consitent with the field condition. It is thus concluded that the failure of compacted lateritic soil slope was due to erosion of underlain gravel deposits.

ACKNOWLEDGEMENTS

The author gratefully acknowledges the research assistants Messrs. C.L. Wang and J.S. Yang for their complete laboratory work. The research was initiated by Prof. J.J. Hung of National Taiwan University and sponsored by National Science Council, Republic of China.

Thanks are also given to Miss T.M. Cheung for typing the manuscript and to Miss M.L. Lee for helping in drawing figures.

REFERENCES

Boutrup, E., (1977), Computerized Slope Stability Analysis for Indiana Highways, JHRP Report No.77-35 and 77-26, Purdue University, U.S.A.

Hung, J.J., (1967), Physical Properties of Lateritic Soil in Northern Taiwan, Final Report, National Science Council, R.O.C. (in Chinese).

Ho, C.S., (1969), Some Stratigraphic-Structural Problems of the Linkou Terrace in Northern Taiwan, Proc. of the Geological Society of China, No.12, pp.65-80.

Lee-Wei Boring and Testing Engineering Company, (1982), Boring Report of Linkou Residential Construction (in Chinese).

Lumb, P., (1962), Effect of Rainstorm on Slope Stability, Porc. Symp. Hong Kong Soils, Hong Kong, pp.33-87.

Woo, S.M., et al., (1982), Engineering Problems of Gravel Deposits in Taiwan, Proc. ASCE Specialty Conference on Engineering and Construction in Tropical and Residual Soils, Hawaii, pp.500-518.

Yang, W.C., (1982), Effect of Hydrological Environment on the Construction on Slopes, World Construction and Engineering, pp.37-44. (in Chinese).

Three cut slope failures on relict discontinuities in saprolitic soils

J.W.COWLAND, Geotechnical Control Office, Hong Kong
A.M.CARBRAY, Geotechnical Control Office, Hong Kong

ABSTRACT: This paper presents engineering geological observations made at the time of failure of three cut slopes in saprolitic soils. The form of all three failures were observed to be strongly controlled by relict discontinuities orientated to the regional pattern. In addition, observations on the influence of structural control over the groundwater flow were made at two of the failures.

INTRODUCTION

Three temporary slopes, in Hong Kong, two for the construction of water service reservoirs and one for the construction of a water treatment works, have recently failed (one in 1986 and two in 1987). All three were formed in completely and highly weathered insitu tuffs, and all failed on distinct relict joints. Although the tuffs had weathered to an extent that they would usually be regarded as soils, the mode of failure was in each case more akin to that expected for a rock slope.

Weathering processes and the classification of the resulting weathering profile are reviewed by another paper to this conference (Irfan & Woods, 1988), to which the reader should refer for a more complete introduction. These three failures occurred in weathering zones composed of completely weathered (Grade V) material with some highly weathered (Grade IV) material, also known as saprolite (Sowers, 1963). An excellent description of saprolite is contained in Sowers (1985). The major part that relict joints play in slope instability in saprolites has previously been highlighted by Brand (1985), Hencher et al (1984) and Flintoff & Cowland (1982).

This paper presents observations made at the time of these three failures, including, for the second and third failures, observations of structural control over groundwater flow.

CASE HISTORY 1 - TSUEN WAN ULTRA HIGH LEVEL FRESH WATER SERVICE RESERVOIR

On the 12th August 1986, the fourth in a series of landslips occurred in the temporary cut slopes forming part of the construction works for the Tsuen Wan Ultra High Level Fresh Water Service Reservoir. Shortly afterwards the Geotechnical Control Office was called in to evaluate the probable cause of the failure, and to advise on remedial work.

The location of the service reservoir is shown in Figure 1. The design included a fourteen metre high permanent slope cut in soil with an overall slope angle of thirty degrees. In addition, at the base of this permanent slope, a nine metre high temporary sixty degree soil slope was cut to permit construction of the reservoir, as shown in Figure 2.

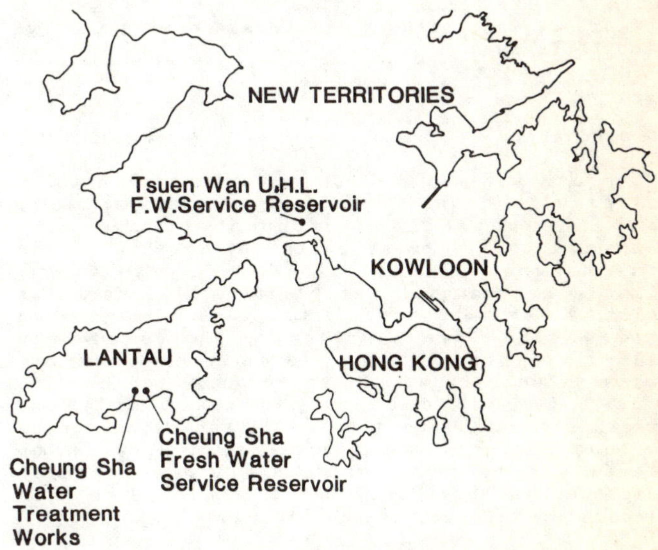

Figure 1. Case history locations.

Geology

The hillside in the immediate vicinity of the service reservoir has been mapped as fine to coarse ash tuffs, tuff breccia and tuffite. Three drillholes carried out prior to construction showed that these rocks had been weathered to a depth of about 25 m, and were overlain by a mantle of about 4 m of colluvium.

Cutting for the permanent slope at the top involved mainly the removal of colluvium (see Figure 2). The lower temporary slope was cut entirely in completely weathered tuff. After failure this weathered tuff was observed to be a reddish brown, with yellowish brown mottling, clayey silt. It had weathered uniformly to a grade V material throughout the depth of the cut. No zonal variation could be seen in the back scar of the failure.

The regional jointing trends have been superimposed over the topographic survey of the fourth failure in Figure 3. There was close alignment of joint sets J1 and J3/J4 to the orientation of the side and back face failure scars.

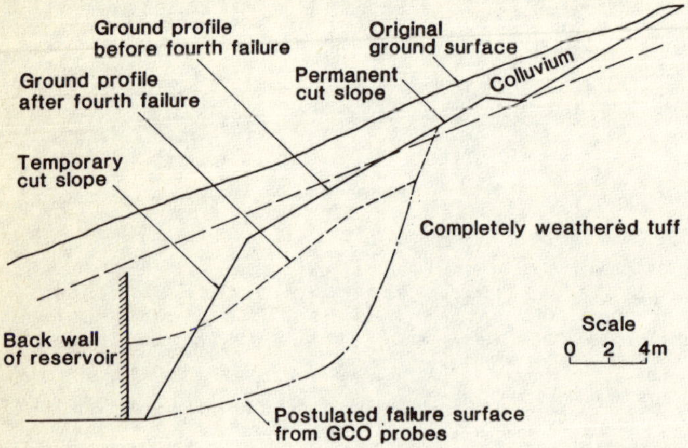

Figure 2. Case history 1 - Cross-section through fourth failure.

Groundwater

The site was located on an extensive natural slope with a well developed vegetation cover, and the potential catchment to the site extended to the crest of the hill behind, some one hundred metres higher. Unfortunately, little groundwater data is available for this site. Although piezometers were installed in three site investigation boreholes in July 1984, only one reading was taken in each; one week later. From this the groundwater surface was inferred to be within the completely weathered tuff some ten to fifteen metres below ground surface. Fluctuations in groundwater level due to seasonal rainfall are normally expected in Hong Kong, but there is no data to indicate their extent on this site.

Time of failure related to rainfall

The rainfall at the two nearest raingauges on the day of failure, and the three days prior to each failure, is shown in Table 1. For three of the failures, the maximum rainfall occurred one to two days prior to the failure.
Observations made after fourth failure

The fourth failure involved about 2000 m^3 of material, and it was considerably larger than the first three. The shape of the fourth failure surface was readily apparent, and a topographic survey was undertaken to fix its position, Figure 3. In order to try to determine the depth of the base of the failure, four GCO probes (Geotechnical Manual for Slopes, 1984) were put down on a section through the failure.

From the field inspection it was readily apparent that the shape of the fourth failure surface was strongly influenced by relict joints. A relict joint acted as a release surface on one side, and another exerted a controlling influence on the position of the back scar. The failed mass remained largely intact, and from the failure surface inferred from the GCO probe results (Figure 2), and these observations, it is considered that the fourth failure was akin to a planar failure.

Table 1. Case history 1 - Rainfall on day of failure, and the three days before.

Date		Rainfall in mm	
		Gauge No.	
		N3	N14
May 1986 (1st failure)	10	1.0	3.5
	11	155.5	262.0
	12	52.5	52.5
	13	0	2.5
June 1986 (2nd failure)	23	0	3.5
	24	4.0	5.0
	25	88.0	35.0
	26	75.0	108.0
July 1986 (3rd failure)	3	66.5	58.0
	4	86.5	111.5
	5	0	2.5
	6	0	3.5
August 1986 (4th failure)	9	5.0	3.5
	10	–	21.5
	11	46.5	102.5
	12	7.0	10.0

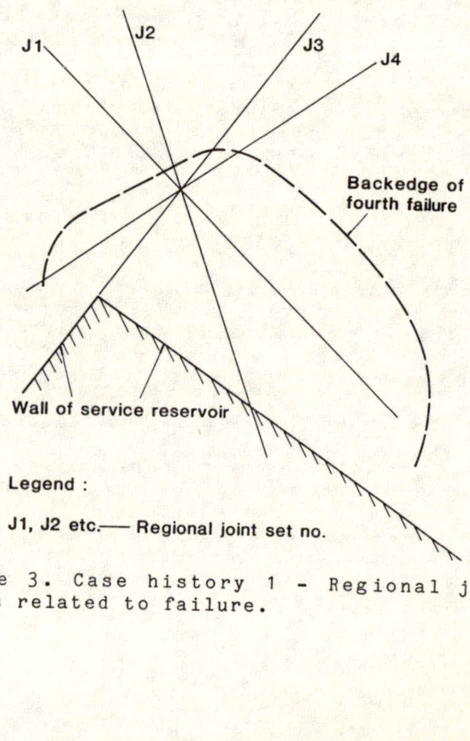

Figure 3. Case history 1 - Regional jointing trends related to failure.

CASE HISTORY 2 - CHEUNG SHA WATER TREATMENT WORKS

On 23rd September 1987, whilst making a routine site visit, it was observed that substantial cracks were developing in the slope that was being excavated for Cheung Sha Water Treatment Works on Lantau Island.

The location of the water treatment works is shown in Figure 1. To form a platform for the water treatment works, a twelve metre high permanent slope was being cut in soil with an overall slope angle of thirty degrees. In order to control the groundwater, in what was known to be a wet slope, cut-off drains were installed at the top and middle of the slope during excavation. Whilst cutting a trench for a third cut-off drain at the toe of the slope, see Figure 4, minor slippage of the sides of the trench is reported to have occurred. On the day of the site visit the upslope face of the trench was 5 m high, and stood at an angle of about 75°, see Figure 5. Substantial linear cracks, and groundwater springs, were noted in the slope above this trench. After rapid backfilling of the trench to prevent the impending failure, the slope was inspected.

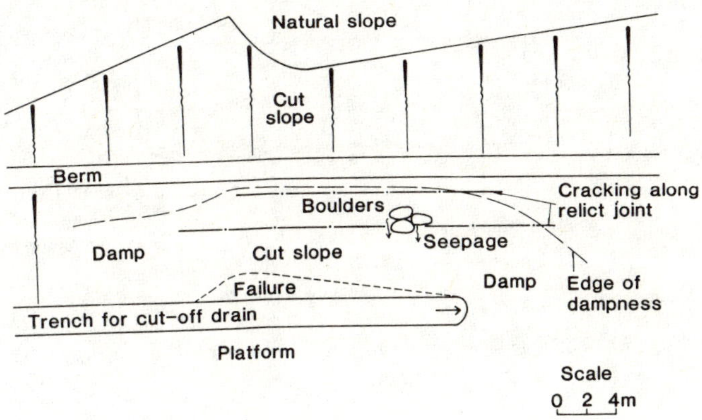

Figure 4. Case history 2 - Plan of cut slope and failure.

Geology

Geological mapping by Allen & Stephens (1971) assigned the bedrock in this area to a grouping they called undifferentiated volcanic rock (RB). In their memoir, it is stated that this grouping was introduced on Lantau Island as the rocks could not be mapped in detail due to the thick vegetation and debris deposits which mantled the slopes, masking the underlying geology. Only one of the eight drillholes carried out prior to construction proved bedrock, which was described as a greenish grey speckled white medium grained volcanic. The other drillholes found a thickness of completely and highly weathered material of at least 10-15m.

Examination of the cut slope revealed that the weathered material was a completely weathered (grade V) tuff. Whilst some areas of highly weathered (grade IV) tuff were present near the base of the slope, all of the weathered material could be described as saprolite (Sowers, 1985). A few corestones occurred randomly within the slope.

The drillholes, trial pits, and the cut slope itself showed the completely weathered tuff to be covered with a layer of colluvium, see Figure 5. Examination of the twelve trial pits revealed that the colluvium varied in thickness from 1.7 to 3.1 m, and that it was comprised of boulders set within a greyish brown sandy silt.

Groundwater

The site was located on an extensive natural slope with a well developed vegetation cover. The potential catchment to the site extended to the crest of the hill behind, some three hundred metres higher. A large catchwater follows the 110 m contour a short distance upslope of the site, but this was dry on 23rd September 1987.

Piezometers were installed in four of the drillholes in November 1983, and groundwater was monitored once a month for the next six months (the dry season). From this monitoring, a groundwater surface was found 5 m below the ground surface. In addition, the nine trial pits excavated in January 1987 (also the dry season) encountered a perched body of groundwater at the base of the colluvium.

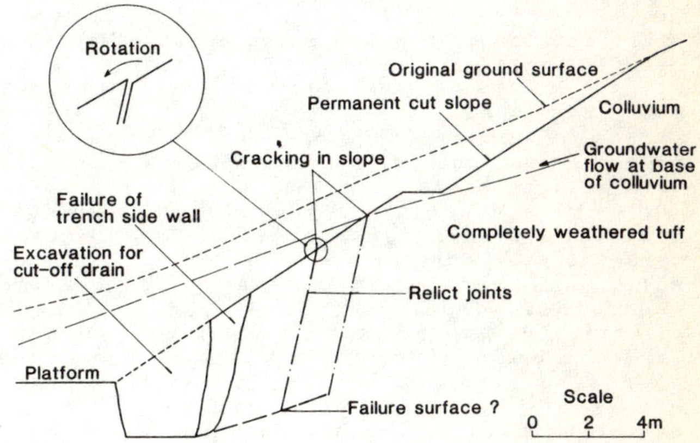

Figure 5. Case history 2 - Cross-section through failure.

On 23rd September 1987, the face of the cut slope was damp and significant seepages of groundwater were issuing from around some of the corestones in the face. These seepages dried up in the following dry season, but the face of the cut slope remained damp. Measurements were made of the amount of water being collected by the cut-off drains in September and November 1987 and it was found to be relatively constant at around 225 cm^3/sec (20 m^3/day).

Time of failure related to rainfall

There are only three raingauges on Lantau, and they are all some distance from this site, so it is not possible to know the specific rainfall on this site on the day of failure. Nevertheless, from the rainfall recorded at these, and other, gauges it can be seen that the 23rd September 87 was a relatively dry day; although a sprinkling of rain occurred on the

22nd, and the 21st was moderately wetter. Indeed, at the time the authors visited the site on the 23rd it was clear and sunny.

So the failure was not directly related to rainfall. However, the slope was very wet, and the failure did occur towards the end of the wet season. Our observations suggest that the slope was cut into an outflow from a reservoir of groundwater further up the slope.

Observations made on the day of failure

The first thing that attracted the authors' attention on arrival at the site was the wetness of the slope surface. The face of the cut slope was damp, and significant seepages of groundwater were issuing from around some of the corestones in the face. Whilst inspecting these springs of groundwater, a linear relict joint was observed, running along the strike of the slope. Closer inspection of this relict joint revealed a small crack, immediately upslope of it, running along its entire exposed length of 22 m. Examination of the rest of the slope revealed another parallel linear crack, of a similar length, further upslope. A plan of the slope is shown in Figure 4, and a cross-section in Figure 5. Returning to the lower relict joint, it was noted that its accompanying crack was becoming larger, and the authors quickly retreated from the slope. Temporary stabilization of the slope was achieved by rapid backfilling of the trench. It is estimated that about 1000 m^3 of material had started to move down the slope.

Examination of the slope after stabilization revealed that both of the cracks were associated with relict joints, and that the lower of these two joints contained a 10 mm thick completely weathered mineralised infill. The strike of these joints corresponded with one of the three main regional jointing trends, and they were found to be dipping at about 75° in the dip direction of the slope. In addition, a lip had formed along the lower relict joint (see inset to Figure 5) indicating that rotation of the blocks of saprolite had started to occur in a manner akin to a toppling failure, even though the dip direction of the relict joints was out of the slope.

CASE HISTORY 3 - CHEUNG SHA FRESH WATER SERVICE RESERVOIR

On 7th November 1987, a failure occurred in the temporary cut slope being formed to facilitate the construction of Cheung Sha Fresh Water Service Reservoir on Lantau Island. Shortly afterwards the Geotechnical Control Office was called in to evaluate the probable cause of the failure, and to advise on remedial work.

The location of the service reservoir is shown in Figure 1. The seventy degree temporary cut slope was 5 m high at the time of failure, and had been cut at the base of a ten metre high permanent thirty degree slope, as shown in Figures 6 & 7.

Geology

This site is adjacent to the treatment works discussed in Case History 2, and the general geological comments made for that site apply equally to this one.

Examination of the cut slope revealed that the weathered material was a completely weathered (grade V) tuff, with some areas of highly weathered (grade IV) tuff near the base of the slope. In comparison with the treatment works cutslope, less corestones were present, and a 600 mm wide relict dyke was exposed in the north-east corner of this slope, see Figure 6. The dyke had a uniform dark brown colour, was fine grained, and was probably a basalt. Associated with it were a number of thin off-shoots. It was in a completely weathered state, dipped into the slope at 70°, and had an east-northeast/west-southwest strike. The main regional trend in Hong Kong is northeast-southwest.

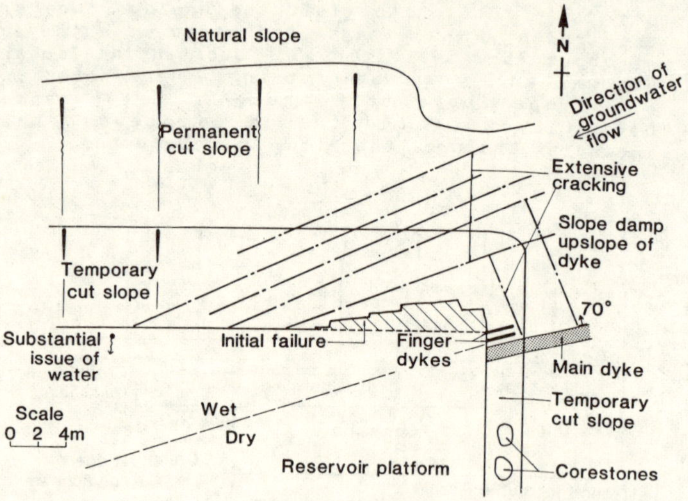

Figure 6. Case history 3 - Plan of cut slope and failure.

Groundwater

Again the general comments made for the treatment works slope (case history 2) apply equally to this cutslope. However, there were some very interesting observations associated with the relict dyke.

The site is located on slopes with a southerly facing aspect, and a number of poorly defined ridges and valleys run obliquely across these slopes in a northeast-southwest direction. The water courses in these valleys are ephemeral. One of the best defined of these valleys passes just to the north-east of the site, where it turns south.

On inspection of the cutslope, on 9th November 1987, it was observed that the surface of the cut slope (which had an east-west strike) was damp to the west, or upslope, of the dyke, and dry to the east, or downslope, of the dyke. In addition, a substantial spring of water was issuing from around an area of highly weathered (grade IV) tuff in the north-west corner of the platform. This spring was measured, on 12th November 1987, and was found to be producing about 400 cm^3/sec (35 m^3/day).

Most interestingly, the north-west corner of this newly formed platform at the base of the excavation was wet and soft underfoot, whilst the rest of the platform was dry and firm. The boundary between the wet and dry areas formed a remarkable 30 m long straight line, also striking east-northeast/west-southwest, see Figure 6. The presence of very soft light brown silty clay deposits within the relict joints striking in the same direction, described later, indicates that groundwater flow occurred through these joints. Those observations led us to conclude that there was a very marked uni-directional flow of groundwater which was not directly downslope. The flow appeared to be in a west-southwest direction, parallel to the relict joints and the relict dyke (see Figure 6).

Time of failure related to rainfall

Apart from a light sprinkling on 1st November 87, the first half of November was very dry in Hong Kong. So the failure was not related to rainfall.

Our observations suggest that the slope was cut into an outflow from a reservoir of groundwater further up the slope, which was probably linked to the valley which changed direction just to the north-east of the site.

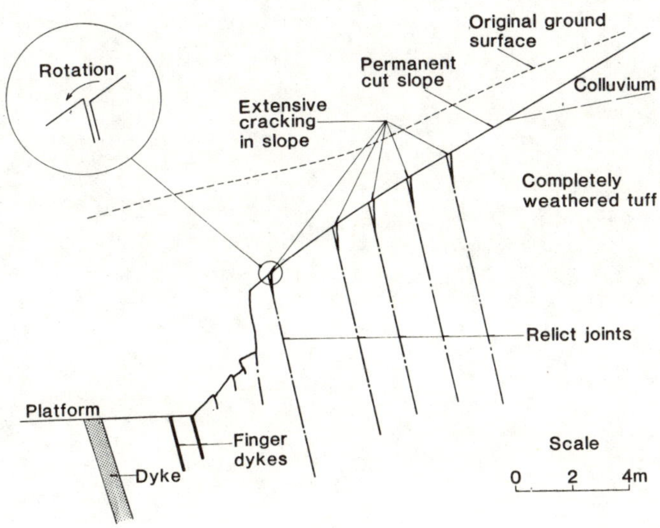

Figure 7. Case history 3 - Cross-section through failure.

Observations made shortly after the failure

On 9th November 1987, it was observed that some blocks of saprolite had fallen from the north-east corner of the temporary slope. The volume of material that had fallen was about 20 m^3, and it was soft enough to be able to easily push in a geological pick by hand. The rectangular shape of these blocks indicated that the failure had been influenced by relict joints.

In addition, on the face of the slope above, a number of large open cracks were visible, see Figures 6 & 7; indicating that a substantial slope failure had been initiated. It is estimated that approximately 2000 m^3 of material had started to move. These cracks were parallel to each other, and had an east-northeast/west-southwest strike, except for two in the northeast corner that had a northwest-southeast strike (release surfaces). The cracks were probed, and they were found to have an open depth of 1.0 to 1.4 m.

Closer examination of these cracks found them to be relict joints with slickensided surfaces and a mineralised infilling of manganese oxide. Some of these joints, particularly near the highly weathered (grade IV) tuff in the north-west corner, contained a thicker infilling which was very soft and had a waterbourne appearance.

The five most prominent relict joints were all striking parallel to each other, and to the relict dyke, and they were dipping at about 70° - 75° towards the northwest (i.e. back into the slope) see Figure 7. In addition, a lip had formed along the open crack at the top of thse relict joints (see inset to Figure 7) indicating that rotation of the intervening blocks of saprolite had started to occur. The mode of failure was toppling.

Some more minor relict joints were visible in the blocks of saprolite that had fallen from the north-east corner, and in the exposed failure surface. These minor relict joints had more variable dip angles and dip directions.

CONCLUSIONS

All three of the failures presented in this paper occurred on relict discontinuities in completely and highly weathered tuffs. The mode of failure for one was planar, and for the other two was toppling.

Observations made immediately after two of the failures suggested that these two slopes had cut into a specific outflow of water from a reservoir, or body, of groundwater further up the slope. On one of these slopes, a clearly defined boundary was observed between wet and dry areas, which formed a remarkable 30 m long straight line running at an angle to the dip direction of the slope. This boundary clearly suggested that the flow of groundwater in this slope was strongly uni-directional, and not directly downslope.

These three case histories demonstrate that it would be prudent for a designer of a cutslope in saprolitic soils to consider the possible effect of any relict discontinuities that may be present. Some indication of the likely relict discontinuities may be gained from the regional jointing system.

ACKNOWLEDGEMENT

This paper is published with the permission of the Director of Civil Engineering Services of the Hong Kong Government.

REFERENCES

Allen P.M. & Stephens E.A. (1971). Report on the Geological Survey of Hong Kong. Hong Kong Government Press, 116 p, plus 2 maps.

Brand E.W. (1985). Predicting the performance of residual soil slopes (Theme Lecture). Proceedings of the Eleventh International Conference on Soil Mechanics and Foundation Engineering, San Francisco, vol. 5, pp 2541-2578.

Flintoff W.T. & Cowland J.W. (1982). Excavation design in residual soil slopes. Proceedings of ASCE Speciality Conference on Engineering and Construction in Tropical and Residual Soils, Honolulu, pp 339-556.

Geotechnical Control Office (1984). Geotechnical Manual for Slopes, Geotechnical Control Office, Hong Kong, 295 p.

Hencher S.R., Massey J.B. & Brand E.W. (1984). Application of back analysis to some Hong Kong landslides, Proceedings of 4th International Symposium on Landslides, Toronto, pp 631-638.

Irfan T.Y. & Woods N.W. (1988). The influence of relict discontinuities on slope stability in saprolitic soils. Proceeding of the 2nd International Conference on Geomechanics in Tropical Soils, Singapore, in press.

Sowers, G.F. (1963). Engineering properties of residual soils derived from igneous and metamorphic rocks. Proceedings of the 2nd Pan-American Conference on Soil Mechanics and Foundation Engineering, Brazil, vol. 1, pp 39-61.

Sowers, G.F. (1985). Residual soils in the United States. Sampling and Testing of Residual Soils : A Review of International Practice, ed. E.W. Brand and H.B. Phillipson, pp 183-191, Scorpion Press, Hong Kong.

Natural slope problems related to roads in Java, Indonesia

W.HEATH, Transport & Road Research Laboratory, Crowthorne, UK
B.S.SAROSO, Transport & Road Research Laboratory, Crowthorne, UK

ABSTRACT

In relation to the high number of landslides which cause serious problems to Indonesia's highway network study was carried out in order to determine common patterns of failure and to establish methods of identifying potential high risk areas of slope instability in Java.

It consisted of a regional appraisal and a detailed study of soil and groundwater conditions at a few selected sites in Central and Western Java. The shallow soil slopes on which such failures often occur generally consist of colluvium resting on a relatively impermeable shale. The high weathered soils have a high clay content which, from mineralogy tests on samples recovered from test sites, proved to have an expansive 2:1 lattice structure and clays were generally montmorillonites and vermiculites. Other soil tests showed these soils to have a high activity value of up to 2.48.

Four modes of failure and associated groundwater conditions, consisting of mass creep, lateral slides, mud flows and a rotational slump of the road section, were identified as being characteristic of most failure sites. The report describes the instrumentation results and observations used in determining these features of slope failure.

1. INTRODUCTION

During 1981 a joint study commenced between the Indonesian Institute of Road Engineering and the Transport and Road Research Laboratories Overseas Unit into the problems of slope instability in Indonesia. The general aim of the study was to seek overall patterns of failure on soil slopes, which were connected with highway problems, and to determine the mechanisms of such failure in order to improve the methods of dealing with landslides. In addition it was anticipated that such a study would provide a means of identifying areas of potential risk and that this would assist in the preparation of hazard maps for highway alignment and maintainence purposes.

The study concentrated on landslide problems in the most developed area of Indonesia, that of Java, and besides examining widespread occurrences of slope failure a number of specific sites were selected for detailed examination.

As is often common with large landslide studies there were difficulties initially in determining which were the most appropriate investigation and testing procedures for use on the sites. In this respect the limited background information relating to slope stability problems in Indonesia proved to be a severe handicap. Consequently the number of slides being investigated grew and a restricted range of investigation procedures appeared to provide advantages in terms of the resources available. These included; mapping, sub-surface soil sampling and testing, the use of instrumentation to determine the rate and depth of failure and an investigation of groundwater conditions including the detailed study of pore water pressures at one test site.

Subsequently a range of groundwater and soil information about different slopes was collected and related to more general observations of slope failure throughout Java. In this respect the empirical records dealing with large scale slope movements, in those other parts of the world where similar slope problems exist, were of value.

2. BACKGROUND TO THE STUDY

The first phase of the study consisted of the selection of four representative sites for investigation, and included the preliminary mapping of these sites and limited sub-surface exploration. Reference to this stage has been made by Saroso B.S. et al (1983). It was planned that the second stage include a more detailed investigation of the sites including the determination of the depth and rate of failure and the groundwater conditions on the slopes. It was also proposed that the shear strength of slope materials should be determined to some degree of precision. Prior to the start of the second stage in 1983 fifteen piezometers were installed at one site in four groups along the slope. The selection of piezometers and their distribution on the slope were determined as part of a joint study involving Bristol University.

During the second stage it was recognised

that instrumentation results and observations were more relevant from sites that were failing rather than sites with a past history of failure. This meant the rejection of one previously selected site. Additional sites, where continuous slope movement caused problems to roads, were selected including one at the Puncak, km 88, on the main Jakarta to Bandung road. Table I shows the slides which were instrumented and observed in detail during the second stage of the investigation.

Table I. Slopes included in a detailed study

SITE	ROAD LOCATION	MORPHOLOGY OF SLOPE ABOVE ROAD	UNSTABLE PART OF SLOPE
1) (Km61)	BANDUNG-CIREBON	STEEP SLOPES OF	COLLUVIUM AND SHALE
2) (Km64)	" "	VOLCANIC BRECCIA	
3) (Km21)	BANDUNG-JAKARTA	STEEP ANTICLINE OF	COLLUVIUM AND SHALE
4) (Km24)	" "	REEF LIMESTONE	
5) (Km24)	WELLERI-SUKAREJO	STEEP SLOPES OF VOLCANIC BRECCIA AND SEDIMENTS	COLLUVIUM AND SHALE
6) (Km88)	JAKARTA-BANDUNG	VOLCANIC BRECCIA AND ASH	OLD SLIDE DEBRIS
7) (Km54)	BANDUNG-CIAMIS	ANDESITE BRECCIA	VOLCANIC COLLUVIUM
8) (Km46)	BANJAR-WANGON	VOLCANIC BRECCIA	VOLCANIC COLLUVIUM
9) SEMARANG	KRAPYAK-SEMARANG TOLL RD.	VOLCANIC ASH/SHALE OLD SLIDE MASS	MAINLY SLIDE DEBRIS

Much of the instrumentation and investigation took place at Site 1. Tomo, km 61, and Site 4. Citatah, km 24, two sites close to the Indonesian road research laboratories in Bandung. At Tomo highly porous deposits, composed of volcanic breccia which is generally old lahar, feed groundwater into colluvium which overlies folded beds of shale belonging to the Subang formation. Citatah is different only in that it is a section of a reef limestone anticline which feeds groundwater into a similar sequence of colluvium and shale.

At Tomo the road failed at four locations, within the sector from km 61.7 to 67, during the course of the study. Failures at the instrumented part of the site, km 61, consisted of a 100 m long translational slide, mud slides and continuous periods of slow earch deformation and creep. Within the old lahar deposits above the road failure only occurs when the lower slope becomes oversteepened. No such failures occurred during the period of the study. Figure 1 provides information about this site with details of the main failures and locations of instrumentation.

2.1 Geology and geomorphology:

The island of Java forms part of an active volcanic mountain chain, the Sunda Arc, where diastrophism has resulted in uplifted and folded sequences of Tertiary and more recent sediments to produce a physically rugged and complex landscape. This is modified by an extensive alluvial coastal plain in the north and a series of old alluvial lake basins in the central region. The main geotectonic activity has been in the southern region of Java where the terrain consists of extremely complex morphostructural units within an uplifted sedimentary and frequently active volcanic mountain system.

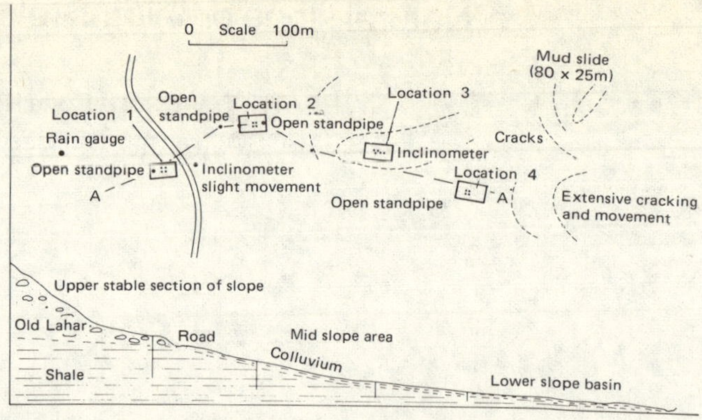

Fig. 1: Details of the slide at site 1; Tomo

The geomorphological features of many slopes in Java are similar in that they consist of folded marls or weathered volcanic deposits overlain by sequences of colluvium derived from volcanic breccia and ash. Massive deposits of these porous breccia and ash materials occur on the steeper sections of slopes to form huge collection areas of groundwater. The contribution of extensive groundwater retention zones, reservoirs, to slope failures has been described by Denness B (1973) in relation to failures in Colombia. In Java the lower areas of such slopes are invariably covered with extensive bodies of weak colluvium, an extremely variable material, consisting of boulders, cobbles and gravel in a sand/silt and clay matrix. Such conditions account for a considerable proportion of the total road network area which includes 3,200 km of main and 9,000 km of secondary roads servicing a population of 90 million. Figure 2 shows the percentage of the road network for West Java within areas of landslide risk.

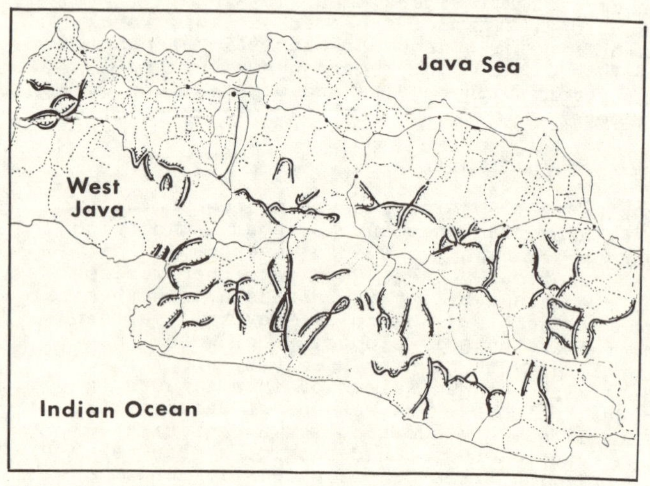

Fig. 2. Road-sector map of slope hazards in West Java

2.2 Characteristics of slope failure:

Within the lower slope regions failure generally follows a predictable pattern and is related to both high rainfall and the excess groundwater released from highly porous deposits on the upper slopes.

In terms of published references, Brand E.W. (1984), has outlined the extent of information for the whole of South-east Asia. The only published reference dealing extensively with slope failures in Indonesia is that by Wesley L.D (1977) and relates to problems in homogeneous soils on steep volcanic residual-soil slopes. Within the typical soils of Indonesia, derived from pyroclastic materials and marl sediments, Mohr E.C.J (1944) has commented on the high shrinkage, to approximately half the wet volume of these materials, and relates this to montmorillonite clays. Such clays, within the phreatic zone of groundwater, tend to form in alkali soil conditions which possibly indicates the influences the high pH calcium carbonate marls have on the weathering of volcanic materials.

Rouse W.C et.al (1986) have published results of a soil study in Dominica, West Indies, in environmental conditions similar to those in Indonesia. In particular the reported distribution of amorphous clay minerals at the higher elevations and 2:1 lattice clays, including smectite, on the lower footslopes, where high levels of groundwater accelerate the weathering process, can also be identified in Indonesia but in much less distinct patterns.

2.3 Rainfall related to slope failure:

The climate of much of Indonesia is quite seasonal being influenced by the equatorial tropical convergence and characterised by a prolonged wet season from September to May. Rainfall records have been collected over a period of sixty years, Berlage Jr.H.P (1949), and indicate that average yearly rainfall varies between 1,500 and 5,000 millimetre. Table II shows the general conditions of rainfall in Java.

Table II. Features of rainfall in Java (Berlage)

GENERAL	ANNUAL RAINFALL	INFIL-TRATION	EVAPO-TRANSPIRATION	PERIOD OF WET=SEASON
WEST JAVA (NORTH)	1,798mm	690mm	980mm	8 MONTHS
WEST JAVA (CENTRAL)	1,966mm	1,360mm	710mm	8 MONTHS
WEST JAVA (SOUTH)	3,454mm	2,820mm	410mm	10 MONTHS
EAST JAVA (NORTH)	1,740mm	830mm	1,340mm	7 MONTHS

In Java the principal influence on the amount of rainfall is the ground elevation and the period of the tropical convergence. Within the lower slope area it is between 1500 mm and 3,000 mm per annum. From the records of Berlage, peak rainfall generally occurs in January and then again between the months of March and May. This corresponds to the periods when the majority of landslides are reported. In terms of the effects of rainfall infiltration, and longer term groundwater conditions, such failures appear to have two distinct components.

General observations, Newspaper reports and inclinometer results support the view that most slides occur during or immediately after a period of exceptionally heavy rain. Both Brand E.W (1984), in terms of Hong Kong slides, and van Bemmelen R.W. (1949), in terms of lahar slides in Indonesia, have reported upon precipitation threshold levels of 70mm within an hour or less to initiate serious landsliding. Levels of rainfall exceeding these values are not uncommon in Java.

In contrast vast quantities of groundwater are retained within the highly porous upper slope deposits of breccia and ash throughout the wet season. This is dissipated onto the lower slope deposits maintaining a saturated condition and promoting creep failure. It is this, the least spectacular and difficult to observe aspect of slope failure, which is the fundamental factor of all slope instability affecting roads in Java.

3. GROUNDWATER AND INSTRUMENTATION

A difficult aspect of the study was the measurement of groundwater pore pressure and permeabilities within the restricted horizon of unstable colluvium. At the main instrumented slope, site one, groundwater levels measured from open standpipes at five positions on the slope were collected for a period of two years, see Figure 3. These indicated a high build up in the level of groundwater at the start of each wet season with the exception of an area of slope above the road. The most notable variation in groundwater level occurs at the mid-slope positions where an increase of more than 5 metres in the phreatic level was recorded. Similar patterns of groundwater build-up were also observed at sites three and five where open standpipe levels were also monitored.

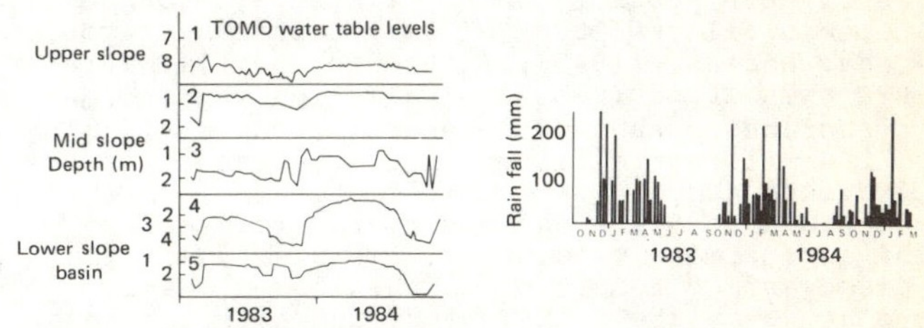

Fig. 3: Standpipe groundwater levels

3.1 Groundwater flow and pore pressure:

Groundwater flow conditions and the build up of pore pressure in the colluvium are relevant to slope failure investigations particularly in the design of drainage as a means of controlling movement. In this respect the model developed by Whipkey J.F and Kirkby M.J (1978), of the build up of a saturation front appears particularly relevant to slopes in Java. it presupposes that subsurface flow conditions generally require an impeding layer or a progressive decrease in permeability with depth before any appreciable flow occurs. A horizon of weathered shale meets such conditions on the majority of unstable slopes in Java. Figure 4 shows what are essentially the conditions during a period of steady long-duration rainfall. As the precipitation continues a zone of saturation within the impeding layer gradually increases and consequently reduces the diffusing gradient or rate of percolation into the layer. This causes a saturated layer to back-up within the upper more permeable colluvial zone and progressively extend upslope. However it does not extend uniformly because of differences in hydraulic conductivity.

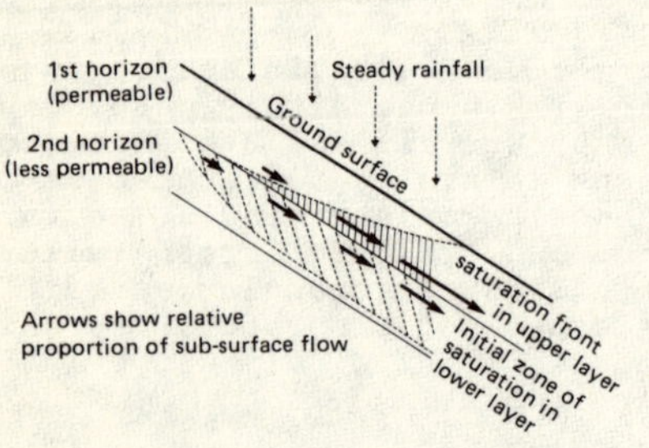

Fig 4: Slope saturation model. (Whipkey et.al)

The characteristic variations in patterns of instability on a given slope in Java may perhaps be explained in terms of the spatial distribution of this pattern of saturation across a slope, perpendicular to the forward movement of the saturation advance. This is likely to vary considerably, over periods of years, and may be influenced significantly by disturbances to the slopes groundwater such as deforestation and land development. Such models, in terms of Java's slope failure problems, appear to be extremely relevant but are particualrly complex to develop and prove.

3.2 Piezometers:

The piezometers used to determine groundwater pore pressure were the standard coarse ceramic filter (250mm x 50mm) Casagrande type, Casagrande A. (1949), fitted to 19 millimetre PVC open standpipes. Installation was in the prescribed manner with bentonite end seals and a 200 millimetre filter of coarse sand. It provided a hydrostatic piezometric response, to reach an 90% equalised hydraulic pressure, of approximately 10 hours. This was a serious disadvantage resulting in the loss of data relating to the transient groundwater response to storm events.

3.2.1 Recording piezometric levels: In general the instrumentation of unstable slopes is made difficult by the inaccessibility of the sites and therefore methods of recording the data from instruments are necessary. In Java a recently developed acoustic method was used to measure and record the level of piezometric water in the standpipes. For depths up to 10 m it was claimed to measure water levels to an accuracy of 5 mm under constant conditions. However assumptions made in the design of this technique, and referred to by Anderson M.G and Kneale P.E (1987), regarding the type of standpipe and influence of temperature led to practical difficulties and serious errors. Fortunately additional pore pressure data was obtained by taking manual readings at three day intervals and this proved to be more reliable.

3.3 Dye tracer:

Investigations were made using the dye fluoresceine LT, a yellow dye with the colour index; Acid yellow 73, constitution reference 45350 and a chemical base of the sodium salt of hydro-o-carboxy phenyl fluorine. The purpose was to determine permeabilities and groundwater flow paths across the slides.

4. SOIL PROPERTIES

4.1 Clay mineralogy:

X-ray diffraction examination was used to determine the clay mineralogy of soils on a number of failing slopes in Java. For each test three samples were prepared, one randomly orientated, and the second and third with the minerals orientated and glycolated and orientated. These allow the principal clay phases and the presence of swelling sheet silicates to be identified. The tests indicated relatively low levels of amorphous minerals and little Halloysite as being a characteristic of these lower slope materials. Table 3 shows the mineralogy of samples which have been examined. It also includes details of the shale at two sites.

Table III. Mineralogy of soil samples from colluvium within the slope failure zone.

SITE:	COLLUVIUM SOIL TYPE	MAXIMUM AMOUNT CLAY MINERAL		SECONDARY CLAY		QUARTZ %	ASSOCIATED CATION
1&2	BRECCIA/SHALE	SMECTITE	35%	KAOLINITE	15%	30%	Ca
3&4	L/STONE/SHALE	VERMICULITE	40%	MUSCOVITE	5%	20%-40%	Na
6	VOLCANIC ASH	GIBBSITE	20%	ATTAPULGITE	10%	40%	Mg
				SMECTITE	5%		
7	ASH/SHALE	SMECTITE	60%+	KAOLINITE	10%	1%	
1TO4	SHALE	MUSCOVITE	40%	MIXED	10%	50%	
7	SHALE	KAOLINITE	54%	SMECTITE	34%	6%	

Similar clay mineralogies have been reported in East Java, Subardja and Buurman P (1980). Gibbsite was related to mid slope weathering profiles and the smectites to basin positions. The landslide slopes on which smectite clays are prevalent can be readily identified by the high rate of shrinkage and cracking of the soil that occurs during the dry season.

Vermiculite was also found on shallow lower slope positions, in East Java, and defined as a pre-smectite stage of mineralogy. Unlike smectite clays which are widely associated with low values of shear strength and consequently slope failures, the vermiculites have received considerably less attention. However the clay structure, whilst not as active, is similar to montmorillonite in having a weak expanding 2:1 lattice and a high cation exchange capacity.

Gibbsite was found to be associated with old volcanic ash soils at one landslide site. The presence of gibbsite in such soils has been attributed to the severe tropical weathering of igneous basic rocks with the rapid removal of silicates, Harrison J.B (1934). Collapse, by saturation, and a weakening of cementation has been reported by Brink A.B.A and Kantley B.A (1961) and Vargus M (1973) for lateritic soils and Foss I (1973) for andosols. However there is still no direct test evidence of soil collapse to support field observations in Java.

4.2 Soil tests:

Soil samples were collected from a range of depths, up to 20 metres, including those positively correlated with slickenslides or planes of failure determined from inclinometers. Table 4 shows a range of soil test results.

Table IV. Soil test data from slopes

SOIL TYPE	DEPTH FEATURE	CLAY %	MINERALOGY OF MAIN CLAY		w%	LL	PI	c′ kN/m	φ Angle
** JAVA: (Main Test Site. 'Tomo' Site 1)									
Colluvium	2m Gravelly	22	Smectoid	30%	45	81	46	57	8
"	3m Clayey	47	"	35%	40	83	49	84	4
"	5m W/shale	43			30	81	46	45	14
Shale	11m	20	Muscovite	40%	28	110	75	58	23
**JAVA: (Other Test Site. 'Citatah' Site 3)									
Colluvium	2m Gravelly	48	Vermic-	30%	30	50	26	22	14
"	5m Gravelly	56	ulite	40%	18	46	27		
			Site 4						
Colluvium	1m	31	Vermic-			71	40	24	10
"	4.5m	31	ulite			70	27	10.5	13
"	8.5m	32	"			51	29	28	4.
	10m	25	"			49	27	33	4.5
"	13.5m	23	"			44	27	29	7.5

It is noticeable from these results that the smectites are not always associated with the highest values for either the liquid limit or plasticity Index. However in terms of activity, Figure 5 such soils have a high average value of 2.48, which is within the range of 1-7 for a calcium-cation associated montmorillonite. In this respect the soil test and x-ray diffraction results support the conclusion of a soil with a high smectite clay content. The less active clays including the vermiculite samples have an average activity of 0.87.

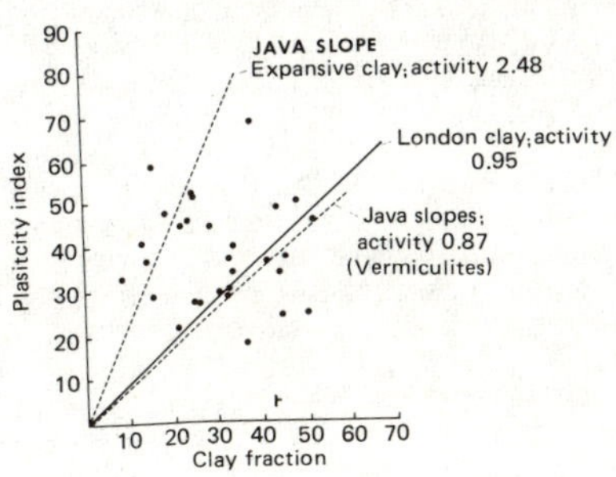

Fig 5: Activity value of soils

Laboratory soil grading tests were considered to be unrepresentative in terms of the clay-size proportion of the colluvium from failing slopes. From tests performed on a much greater sample than that shown in Table 4 an average clay-size content of 26% was determined. This had a standard deviation of 12% implying that few samples had clay contents greater than 40%. Other evidence including x-ray mineralogy examination, the high activity values of the soil and its physical properties all indicate a considerably higher amount of clay.

In this respect the fines in the soils appear to form clumped particles, larger than 0.002 mm, which resist breakdown during normal grading tests. Terzaghi K (1958) noted that Javanese volcanic residual soils had characteristics very similar to Kenya soils in terms of the clays forming aggregated particles that distorted the value of the soils liquid limit and were difficult to break down. Whilst it is difficult to draw parallels between the colluviums and the residual soils this ability of such soils to behave similar to soils with a large granular fraction, and consequently high cohesion, but readily break down under specific conditions should be considered in any future research into slope stability problems in Java.

The values obtained from pre-consolidated drained triaxial tests included low angles of shear resistance within the range 5° to 15°, and relatively high values of soil cohesion. They may reflect an orientation of the soil structure from the effects of the slow creep. The angle of internal friction is clearly related to the soils cohesion, see Figure 6. This shows the intercept between shear angle and cohesion for what are considered to be samples of colluvial near planes of failure. Figure 6 shows the increased range of plasticity for samples recovered from shallow depths and this is reflected in the range of soil moisture values. A correlation also exists between the soils liquidity index and horizons of failure within the soil profile.

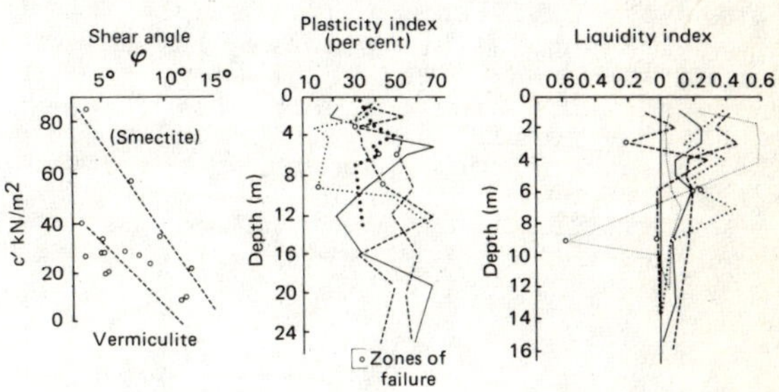

Fig 6: Range of soil properties

5. GROUNDWATER INFLUENCE ON SLOPE FAILURE

Rainfall conditions at a number of slopes are shown in Table 5. The information suggests that the total amount of annual rainfall had some influence on the number of slides. Also it was established that at least 50% of reported slides occurred during or after periods of heavy rain when daily rates exceeded 100mm. The rate of infiltration of such rain into the slopes is also relevant to slope stability. This together with groundwater flow conditions were determined from dye tracer tests. Such tests at four sites, within the upper slope area, provided samples of the dye, at dilution levels of between 0.3 and 2.3 parts-per-million recovered in bore holes placed up to 200 metres down slope. Intrinsic permeability, estimated from the dye tracer recovery, was as high as 5,000 m/hour in some tests indicating that fissure and natural pipe flow was a significant factor in groundwater movement. The distribution of dye concentration on an unstable slope is shown in Figure 7a and provides a basis for estimating groundwater flow with depth. The relative position of the spring line agrees with observations made at the surface. In hydro-morphological terms it is also significant that, as previously identified by Deere D.U and Patton F.D (1971), failures are generally within such groundwater discharge zones.

Table V. Features of rainfall at specific sites

SPECIFIC LANDSLIDE SITES	ANNUAL RAINFALL	SEASONAL RAINFALL	NUMBER OF LANDSLIDES
SITE ONE	1983. 2,329mm	1982/83. 2,644mm	NONE
	1984. 2,753mm	1983/84. 1,815mm	TWO
		1984/85. 3,222mm	NONE
SITE THREE	1983. 2,390mm	1982/83. 2,353mm	NONE
	1984. 2,340mm	1983/84. 2,280mm	THREE
		1984/85. 2,017mm	NONE
SITE FIVE	1982. 1,643mm	1982/83. 1,388mm	NONE
	1983. 1,862mm	1983/84. 2,513mm	SIX
DAILY RAINFALL RATES			
	DAYS EXCEEDING 50mm/24 HOURS	DAYS EXCEEDING 100mm/24 HOURS	PEAK RAINFALL PERIODS
SITE ONE	THIRTEEN	SEVEN	246mm/2 DAYS
SITE THREE	TWENTY FOUR	FIVE	168mm/1 DAY

The recovery of dye from an apparent slip plane was also achieved at one site, Figure 7b, indicating the preferential movement of groundwater along failure planes. The soil samples, of what was a coarse gravelly material, showed a high concentration of the tracer. Subsequently an inclinometer in the same borehole confirmed movement at the depth the tracer had been recovered.

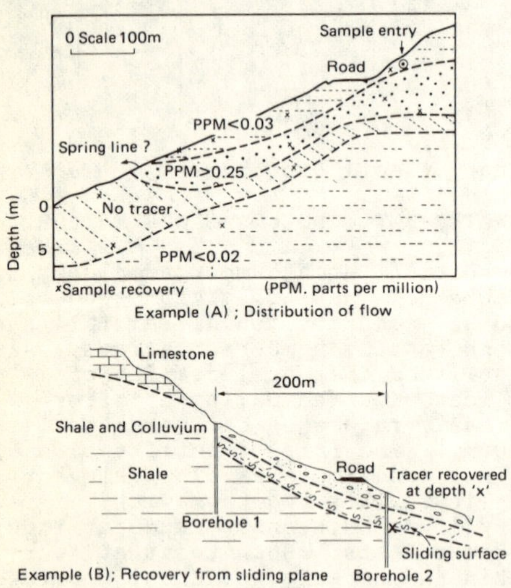

Fig 7: Results from dye tracer tests

5.1 Pore pressure and groundwater at Tomo site:

The hydrological pattern affecting slope behaviour has been determined from three instrumented sites on a failing slope at Tomo, see Figure 1. Fluctuations in pore pressure over a period of one year are shown in Figure 8. Within the mid-slope zone rapid transients in piezometric levels, at the beginning of the wet season, (about November, see Figure 8) are noticed and attributed to the priming of permeability in the soils. An explanation by Bouwer H (1978), is that entrapped air in the soil mass blocks conductivity paths and causes permeability to be less than when the soil is fully saturated. Consequently an initial build up of pressure or head of water is needed to overcome this.

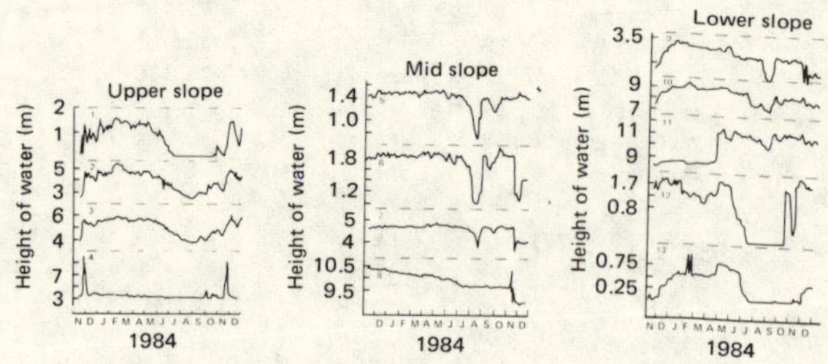

Fig 8: Piezometer levels at one site

5.1.1 Upper stable slope area: This zone is above the road on an area of steep volcanic breccia which overlies shale at a depth of 8.75m. Within this area there were considerable fluctuations in pore water pressure at depths up to seven metres.

Infiltration peaks were recorded on at least three occasions during storms, when open standpipe piezometer water reached to almost the maximum levels indicating pore water pressures in excess of 15 kN/m^2 at a depth of 1.7 metres, 50 kN/m^2 at 5.5 metres, and 60 kN/m^2 at 7.15 metres. In the shale there is little response and its not normally associated with the infiltration wetting front.

5.1.2 Mid-slope area: Within this area of the slope the phreatic level generally remained high for most of the year and piezometric levels rise above the tops of the standpipes frequently during the wet season. Figure 9 the left hand side, shows the variation in piezometric pressure and rainfall together with transient increases in groundwater pore pressure for three piezometers. The right hand side of Figure 9 illustrates increases in pore pressure and a five day pattern of rainfall. Piezometers in the shale showed little response in what is obviously a relatively impervious horizon. The uppermost 300 mm of colluvium, within the A horizon of the soil mass is also relatively impervious as is evident from the flooded paddy conditions on this and similar slopes. Hydraulic conductivity is therefore mainly confined to a narrow horizon which extends to a depth of 5 metres. In this respect the limited depth of the soil profile is probably more relevant than the basin morphometry in determining groundwater movement on this part of the slope.

Soil void ratios are in the range of 0.8 and 1.2 and therefore with natural moisture contents of up to 40% the soil is saturated. In this condition slow deformation and creep occurs. Whether it is further increases in pore pressure which causes the translational slides to develop was never determined because of the poor response of the piezometers.

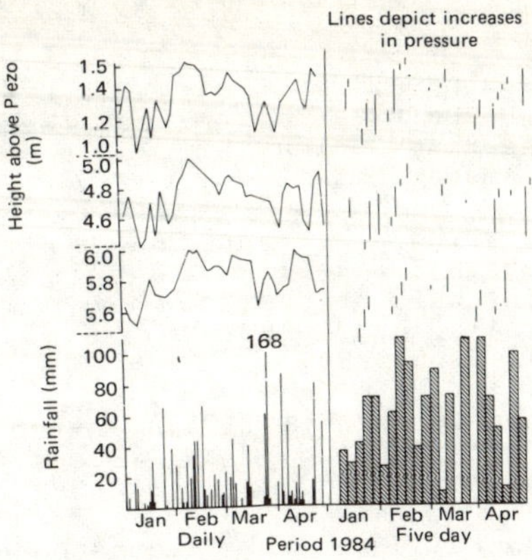

Fig 9: Response of pore pressure to rainfall

However such slides can be modelled in terms of shear stress/strain forces acting on the slope. The piezometer data indicates that the pore pressure at a depth of two metres could be equivalent to or exceed a 2.5 metre head of water, equal to a gauge pressure of 25 kN/m², during heavy rain. Based on an infinite slope analysis (Skempton and DeLory 1957) a calculation of the slope safety factor for a 2 metre deep translational slide shows a factor of safety well below unity for such piezometric pressures. A factor of safety of one occurs when the depth of water is at 0.2 metres. This is based on a 4.5:1 slope gradient and soil parameters of c' equal to 9.8 kN/m² and φ equal to 13°. These soil shear strength values were obtained from drained triaxial tests of samples recovered from just below the failure plane.

Mudslides occur further down such slopes and moisture levels of 59% have been determined from saturated samples. The liquid limit of these soils, determined from laboratory tests, was 78%. The difference possibly relates to the clumping of relatively dry soil particles within a very saturated matrix. The soils recovered from mudslides have a liquidity index of about 0.58 per cent which corresponds with that of a larger sample reported by Brunsden D (1984), in which the majority had a liquidity index of 0.4 per cent or more.

5.1.3 Lower slope basin: Less reliability could be placed on long term pore pressure data from this location because of a translational slide which occurred in the area. In 1985 it was found that the piezometer pipes had sheared at a depth of two metres. From the data it appeared that failure occurred during April 1984. New ceramic piezometer tips, connected to metal pipes, were installed in March 1985. In March 1986 the pipes attached to piezometers at depths greater than two metres had disappeared and were eventually located, folded and horizontal, at a depth of two metres. The sliding mass had pushed these pipes over. Piezometers above the two metre level were not damaged and indicated no abnormal groundwater conditions. Slope movement was calculated to be 1.8 metres over a period of twelve months of which approximately half a metre was attributed to the translational slide and the rest to rapid creep. Surprisingly there were no other signs on the slope to indicate that such a large amount of movement had occurred.

5.2 Hydrological patterns of slope failure

The basin morphometry and colluvium/shale soil conditions, as described, are widespread and relate to the described mechanisms of failure on all similar slopes. The hydrological aspects of what appears to be a common pattern of groundwater movement in permeable soils is relevant to slope stability conditions in Java. This pattern can be summarised as follows:

1) All failing slopes are in areas where groundwater is retained in highly porous deposits and supplied to the slopes continuously.

2) The groundwater on such slopes is confined to narrow layers of colluvium by impervious sediments and clays. Measurements show that a high pore water pressure and soil saturation occurs within these layers.

3) Any disruption to flow caused by the collapse of fissures or natural soil pipes, etc leads to rapid translational sliding.

4) At the toe of such slopes groundwater flow is disrupted by collapse and the resultant ponding causes mud-flows to develop.

5) On the road section of the slopes there is a loss of support and the road-fill fails in a circular slump. The moment of rotation is generally through the centre section of the road pavement.

6. CONCLUSION

The present study has provided results which indicate that the mechanism of slope failure affecting much of the road network in Java is related to a saturation of the soil mass promoting slope creep. This is induced by upslope deposits which have a rechargeable high storage capacity and provide a considerable hydraulic yield to the unstable slopes. Slopes in general are sensitive to such saturated conditions because of the high percentage of active clays that the soils contain.

The instrumentation results generally support the view of a high infiltration and discharge capacity in the breccia and a confined high yield aquifer developing in the colluvium. The test data has therefore helped to conform the mechanism of failure although the slow piezometric response resulted in a loss of some data relating to sudden transient changes in storm induced groundwater surges. These conditions appear to result in translation failures and mudslides. On the road section of such slopes, where landslides are most easily observed, the failure mechanism is totally different and occurs as a rotational slump.

ACKNOWLEDGEMENTS

This report describes research that is part of a joint cooperation programme between the Transport and Road Research Laboratory (TRRL) and the Institute of Road Engineering, Indonesia (IRE). The paper is published by permission of the Directors of TRRL and IRE.

The authors wish to express their gratitude to the former Director of IRE, Ir HAB Hasibuan, without whose generous cooperation and assistance the work would not have been possible. Also to colleagues in the geotechnical unit of IRE. Soil mineralogy testing was carried out at Queen Mary's College, London.

REFERENCES

Anderson M.G and Kneale P.E (1987). Instrumentation of Pore Pressure and Soil Water Suction. Slope Stability. (Ed.) Anderson M.G and Richards K.S John Wiley and Sons Ltd. London. P 77-112.

Bemmelen R.W. van (1949). The Geology of Indonesia. The Gov. Printing Office, The Hague, Netherlands. 3 Vols. 1027pp. plus 16 plates and 47 maps.

Berlage Jr. H.P (1949). Rainfall in Indonesia. Verhandelingen No 37. Department van Verkeer, Energie en Mijnwezen Meteorologische en Geophysische Dienst Koninklijk Magnetisch en Meteorologisch Observatorium. Netherlands.

Bouwer H. (1978). Groundwater Hydrology. Pub. McGraw-Hill, London. p43-45.

Brand E.W. (1984). Landslides in Southeast Asia; A State of the Art Report. Fourth Int. Symp. on Landslides. Toronto.

Brink A.B.A and Kantley A.B (1961). Collapsible Grain Structure in Residual Granite Soils in South Africa. Proc. 5th Intern. Conf. Soil Mechanics and Foundation Engineering, Paris, 1961. p611/614.

Brunsden D. (1984). Mudslides. Slope Instability; Ed. Brunsden D and Prior D.B Pub. John Wiley and Sons, London. p.363-418.

Deere D.U and Patton F.D (1971) Slope Stability in Residual Soils. Fourth Panamerican Conference on Tropical Soils, San Juan, Puerto Rico. p87-170.

Denness B (1973). Landslip Problems in Relation to Roads and other Route Planning. Institute of Geological Science, London. IGS.547 100 11/72. pp.70.

Hanna T.H (1985). Field Instrumentation for Geotechnical Engineering. Series on Soils and Rock Mechanics, Vol. 10. pp843. Trans Tech Publications. Germany.

Harrison J.B (1934). The Katamorphism of Igneous Rocks under Tropical Conditions. Imperial Bureau of Sciences, Harpenden. England. 1934.

Mohr E.C.J (1944). The Soils of Equatorial Regions Pub. J.W. Edwards, Ann Arbor, Michigan, USA.

Rouse W.C., Reading A.J and Walsh R.P.D (1986). Volcanic Soil Properties in Dominica, West Indies. Engineering Geology, 23. Elsevier Science Pubs. Amsterdam. p1-28.

Saroso B.S., Dowling J.W.F. and Heath W. (1983). Terrain Evaluation for Road Engineering in Indonesia. 4th Conf. of the Road Eng. Ass. of Asia and Australia. Aug. 1983. Jakarta, Indonesia. Vol. 5 p575-599.

Skempton A.W and DeLory F.A (1957). Stability of Natural Slopes in London Clay. Proc. 4th Int. Conf. Soil Mech. London, 2, 378-381.

Subardja and Buurman P. Soil Catenas on the West and North-east Slopes of the Lawu Volcano in East Java. Buurman P. (Ed.) (1980) Red Soils in Indonesia. Agricultural Research Reports 889. Bulletin No 5. Soil Research Institute, Bogor, Indonesia. p49-67.

Tavenas F. and Leroueil S. (1981). Creep and Failure of Slopes in Clay. Canadian Geotechnical Journal. 1981-02 v18 (1) p106-120.

Terzarghi K (1958). Design and Performance of the Sasumua Dam. Proc. Inst. Civ. Eng. London. No. 9 p369-393.

Vargas M (1974). Structurally Unstable Soils in Southern Brazil. Proc. 8th Intern. Conf. on Soil and Foundation Engineering, Moscow, Vol. 2 p239-246.

Wesley L.D (1977). Shear Strength Properties of Halloysite and Allophane clays in Java, Indonesia. Geotechnique 27 No 2. p125-136.

Whipkey R.Z and Kirkby M.J (1978). Flow Within Soil. Hillslope Hydrology. Ed. Kirkby M.J. Pub. John Wiley and Sons. Chapter 4. p121-145.

The influence of relict discontinuities on slope stability in saprolitic soils

T.Y.IRFAN, Geotechnical Control Office, Hong Kong
N.W.WOODS, Geotechnical Control Office, Hong Kong

ABSTRACT: This paper reviews the subject of relict discontinuities and their influence on slope stability. Particular attention is paid to those characteristics of relict discontinuities, e.g. infilling, coating and slickensides, which can result in low strengths along these features. Methods of detecting relict discontinuities during a site investigation are discussed and the importance of examining large scale exposures in order to obtain a fuller understanding of the pattern of relict features is emphasised. Methods of taking account of relict discontinuities in slope design are also examined. Case histories from Hong Kong and Malaysia are presented to add to the literature on this subject.

1 INTRODUCTION

In tropical and subtropical regions of the world rocks are deeply weathered to soil-like materials as a result of warm, wet climatic conditions. Despite the intense state of decomposition of the constituents of saprolitic soils, relict discontinuities inherited from the fresh parent rock are commonly well preserved. These structural weaknesses may exert considerable influence on the stability of slopes cut in these materials. This is evidenced by the large number of slope failures where relict discontinuities have been observed to form a part or the whole of the failure surface. In such cases, the relict discontinuities are invariably oriented unfavourably with respect to the slope face. They generally also contain soft infilling material of significantly lower strength than the surrounding intact materials. It may also be argued that relict discontinuities act as preferential seepage paths within which high porewater pressure may develop at points of constriction or where inflow exceeds the transmissivity of the discontinuity.

In site investigations carried out for the design of new slopes or for the stability assessment of existing slopes, relict discontinuities are frequently not detected. This may be due to the use of inappropriate drilling/sampling techniques or to difficulties in recovering such thin weak zones. Even when detected during the site investigation, the existence and orientation of relict discontinuities is rarely taken into account in stability analyses. Instead, the geological model adopted for the analysis normally assumes materials of uniform strength discounting these planes of weakness and the influence that they may have in dictating the position and shape of potential failure surfaces.

2 WEATHERING PROCESSES AND CLASSIFICATION OF THE WEATHERING PROFILE

2.1 Weathering Processes

Rock can weather by processes of chemical weathering (decomposition), physical weathering (disintegration) or, to a lesser extent, by biological activity. The type of weathering which predominates in a region is largely dependent upon climate, particularly temperature and rainfall. These factors generally also determine the type of weathering product.

In tropical and subtropical climates chemical weathering is the dominant process. Physical effects, such as the opening of discontinuities, the formation of new discontinuities by rock fracture, the opening of grain boundaries and the fracturing of individual grains, generally accompany decomposition.

Continued weathering activity, prevailing over long periods of time, results in conversion of rock to a soil which finally loses its original rock fabric and becomes a true residual soil. In tropical regions, intensive weathering may form a hardened, rock-like surface layer (laterite) as a result of removal of silica and bases, and cementation of the residual soil by iron and aluminium oxides.

Soils formed from insitu weathering of rocks occur to some degree in most countries of the world, but they are particularly common, and of great thickness, in humid subtropical and tropical countries, such as for example Brazil, Hong Kong, India, Malaysia and Nigeria, where soil forming processes are still very active. Extensive deposits of such soils are also present in regions where hot humid conditions once existed; for example, during the Tertiary in northern Europe and Britain (Irfan and Dearman 1978, Dearman and Turk, 1985, Kany and Herrmann 1985).

2.2 Hydrothermal Alteration

Like chemical weathering, hydrothermal alteration is a process which affects rocks of all types. However, the results of hydrothermal alteration are observed most commonly in areas rich in ore deposits and areas of geothermal activity. Hydrothermal alteration is the result of hot aqueous solutions penetrating the rock mass, especially along discontinuities, and producing mineralogical changes. The solutions are generated either within or at a greater depth (e.g. a cooling magma source) than the affected mass. Hydrothermal alteration commonly extends considerably deeper than chemical weathering.

The effects produced by hydrothermal alteration and chemical weathering are similar and the end-products are often difficult to distinguish except where hydrothermal activity leads to the development of a special mineral assemblage. The effects of the two processes can however be distinguished by the type of structural control and the material variations recognised in the field and in the laboratory (Dearman et al 1978).

2.3 Weathering Profile

In engineering terms a weathered rock profile contains a whole range of geological materials grading from rock to soil. Weathering profiles themselves can vary considerably from place to place because of variations in rock type, structure, topography, rate of erosion, groundwater conditions, and climate.

Although variable, a typical weathering profile generally consists of four major zones (Table I):

(a) a mantle of structureless soil (residual soil) the upper part of which may be transformed into laterite. This is underlain by;

(b) a zone of soil-like material which retains features of the original rock structure and fabric and which may contain less weathered corestones of various sizes. This zone is commonly known as saprolite and is generally underlain by;

(c) a transitional or partially weathered rock zone consisting mostly of rock material separated by friable soil-like material along discontinuities or a continuous network of rock separated by seams of soil of varying thickness. This is underlain by;

(d) a zone consisting wholly of rock which may be stained along discontinuities or throughout but in which the strength of the intact rock is similar to that of fresh rock.

IDEALIZED PROFILE	RUXTON & BERRY, 1957 GRANITE	SOWERS, 1963 IGNEOUS & METAMORHIC	DEERE & PATTON 1971 ALL ROCKS		DEARMAN 1976* ALL ROCKS	FAILURE CHARACTERISTICS MODE	FAILURE CHARACTERISTICS TYPE
	SOIL	SOIL	I RESIDUAL SOIL	IA A-HORIZONS	VI SOIL OR TRUE RESIDUAL SOIL	Soil Fabric Controlled	Erosion Circular Non-Circular
	I RESIDUAL DEBRIS			IB B-HORIZONS			
	IIA RESIDUAL DEBRIS WITH CORESTONES (< 10% Corestones)	SAPROLITE		IC C-HORIZONS (SAPROLITE)	V COMPLETELY WEATHERED		
	IIB (10 - 50% Corestones)	PARTIALLY WEATHERED ROCK	II WEATHERED ROCK	IIA TRANSITION FROM SAPROLITE TO WEATHERED ROCK	IV HIGHLY WEATHERED	Relict Discontinuity Controlled	Planar Wedge Toppling Boulder falls
	III CORESTONES WITH RESIDUAL DEBRIS (50 - 90% rock)			IIB PARTLY WEATHERED ROCK	III MODERATELY WEATHERED	Discontinuity Controlled	Rockfalls Planar Wedge Toppling and Combinations Circular (Closely Jointed)
	IV PARTIALLY WEATHERED ROCK	SOLID ROCK	III UNWEATHERED ROCK		II SLIGHTLY WEATHERED		
	FRESH JOINTED ROCK				I FRESH ROCK		

* Rock Mass Weathering Grading Scheme. Grades do not necessarily occur in the sequence indicated. All other schemes are zonal (layer) schemes.

Table I. Comparison of engineering classification of weathering profiles

2.4 Classification of the Weathering Profile for Engineering Purposes

Numerous classification schemes have been used to describe and characterize the weathering profile and to grade the weathered rock for various engineering uses. In tropical countries, the emphasis has been towards the classification of the (a) and (b) soil zones, i.e. residual soil and saprolite (e.g. Gidigasu 1975, Gordon 1984) as it is these materials that are most commonly encountered in engineering works involving slopes, foundations, etc. In subtropical countries, with distinct wet and dry seasons (e.g. Hong Kong, Australia), or countries which once had subtropical climates (e.g. Europe), the weathered soil zones are generally thinner and the fresher rock zones are more often encountered in excavations (particularly in mountainous areas). In such regions, the behaviour of the weathered and jointed rock, as well as that of the soil, needs to be known, and attention has been given to the description of the full weathering profile.

A sixfold weathering grade classification scheme, first proposed by Little (1969), and later refined by Dearman (1976), to characterize the various components of the weathering profile has found favour with various international engineering bodies (e.g. ISRM 1978, IAEG 1981, BSI 1981) and has been recommended for various engineering purposes. A critique of the scheme was recently made by Dearman (1984).

A comparison of this mass weathering grade scheme with other common zonal classification schemes is shown in Table I. The behaviour of each grade of weathered rock with respect to typical failure modes is also briefly summarized in the table. This paper concentrates on the soil zones of Grade IV and Grade V, otherwise known as saprolite (Sowers 1963). The strength along relict discontinuities or the macro- and micro-structures formed as a result of wethering, as well as the soil matrix strength, dominate the behaviour of these soils in slopes.

3 RELICT DISCONTINUITIES

3.1 Origin

Rocks contain structural discontinuities such as joints, faults, fractures, cleavage and foliation, and lithological structures such as bedding planes, unconformities, laminations, dykes, mineral alignment, etc. Weathering action is usually concentrated along these weaknesses, since they allow the weathering agents to penetrate easily throughout the rock mass. In granite, for example, weathering starts with penetration along the joints, usually following a regular pattern in which joint-bounded blocks are reduced by chemical decomposition to soil. In other hard rocks, weathering of rock material may follow a similar sequence although the pattern of weathering may be greatly influenced by the structural features and lithology of the rock mass. Even within one rock type, the pattern of weathering may be quite different and events early in their history may provide a legacy that influences later behaviour (Dearman et al 1978). The bedding plane spacing in sedimentary rocks affects the rate of weathering and weathering generally extends deeper along shear/fault zones. These structural features of rocks are generally preserved in the saprolite as relict structures.

3.2 Infillings and Coatings

Many types of infillings and minerals may be present along relict discontinuities. These may be inherited from the parent rock, washed in from higher horizons or formed in place by weathering or alteration. Inherited infillings may be hard or soft depending on their origin, e.g. hard quartz veins or soft clay fault gouge. Infillings washed in from above are generally clayey materials that either coat or completely infill open relict discontinuities. The thickness of infillings can vary from a very thin veneer to up to tens of millimetres. Chemical analysis of black seams in saprolitic soils by Masssad and Teixeira (1985) showed that they generally contain fine grained weathering products coated with iron-manganese oxides and a humic substance leached from the upper horizons.

3.3 Slickensides

Slickensided or polished relict discontinuities in

saprolites are commonly reported in the literature (St. John et al 1969, Koo 1982a),and the authors' observations in granitic and volcanic saprolites in Hong Kong and Malaysia confirm this.

Slickensides may be very old features formed as a result of tectonic movements in the parent rock and preserved in the weathered rock. They may also have been caused by differential movements occurring within the saprolitic soil as a result of the weathering process, i.e. stress relief resulting from gradual erosion of the overlying weathered materials. The rapid removal of large thicknesses of weathered soil by mechanical excavation may also result in the development of slickensiding and polishing. It is sometimes difficult to distinguish between an old, preserved slickenside and a recent one; often, more than one direction of polishing (striation) is found on the same relict discontinuity surface.

4. EFFECT OF RELICT DISCONTINUITIES ON SLOPE STABILITY

4.1 General

One of the most significant geologic factors affecting the stability of slopes in saprolites is the presence of relict discontinuities. A literature search has shown that a large proportion of landslips in saprolites are associated with relict discontinuities (St. John et al 1969, Koo 1982a, Massad and Teixeira 1985). Relict discontinuities generally have lower shear strengths than the surrounding weathered intact soil material. A number of landslips, some very large (Wolle 1985), have occurred along slickensided relict discontinuities with shear strengths near residual values.

Sowers (1985) pointed out that many cut slopes, which would otherwise be considered stable, fail as a result of water ingress into relict joints opened up by stress relief and that failure occurs along the lower strength relict discontinuities. Invariably, rainfall acts as the final triggering agent and many failures occur during or shortly after heavy rainfall. However, groundwater behaviour of these soils is complex due to permeability variations in the soil matrix and the influence of geological structure.

Relict discontinuities affect the stability of slopes when they are positioned unfavourably in relation to the slope or when they contain filling material that can influence movement of groundwater. The major adverse effects of relict discontinuities on slopes are :

(a) lower shear strengths than the intact soil either through the clayey or more intensely weathered infill or along infill-weathered rock contacts,

(b) particularly low shear strength values along slickensided discontinuities,

(c) reduction of the mass strength, and

(d) locally high water pressures due to damming effects by clayey infills or pipe development resulting from groundwater flow.

4.2 Shear Strength of Relict Discontinuities

Weathering affects the shear strength of discontinuities in a rock in two main ways :

(a) it leads to a reduction in the discontinuity wall strength, i.e. the strength of the discontinuity asperities, and

(b) it causes a reduction in the basic friction angle of rock surfaces through changes in mineralogy.

As weathering continues, the strength on a discontinuity surface approaches a residual value. This was illustrated by Deere and Patton (1971) on a hypothetical joint weathered to different degrees (Figure 1). In highly and completely weathered rocks, the strength of irregularities along discontinuities is reduced so that even at relatively low stress levels these irregularities are very easily sheared off, and the peak strength approaches the residual value. The residual friction angle, ϕ_r, may also vary because of mineralogical changes in most rocks with weathering. In addition, discontinuities in the soil zones generally contain thin clayey coatings or infillings which may be polished, resulting in a further reduction in shear strength. As a result, the strength available along relict discontinuities is invariably lower than that of the intact soil and their cohesion intercept is generally zero.

St. John et al (1969) were the first to study the shear strength along relict discontinuities. They determined, by consolidated undrained triaxial testing, that a relict joint with a black seam in a gneissic soil gave $\phi' = 14.5^\circ$, whereas on a slickensided surface the value of ϕ' was reduced to 10.5°. Wolle (1985) reported a ϕ_r' value of 16° from direct shear testing of a slickensided surface with smectite clay infill along which a large failure had occurred in a saprolitic soil formed from migmatite in Brazil.

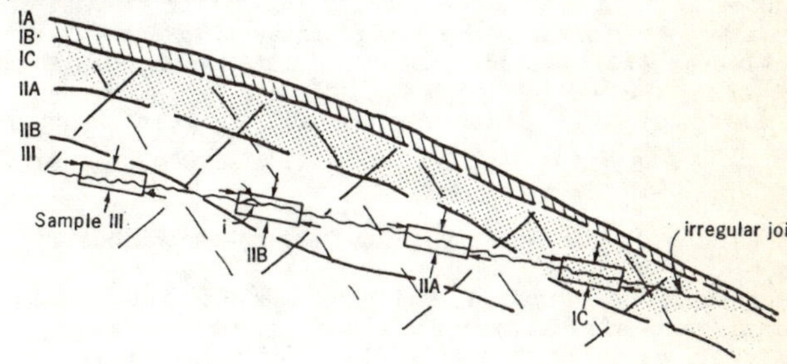

a) Joint with different degrees of weathering (see Table 1 for zone numbers)

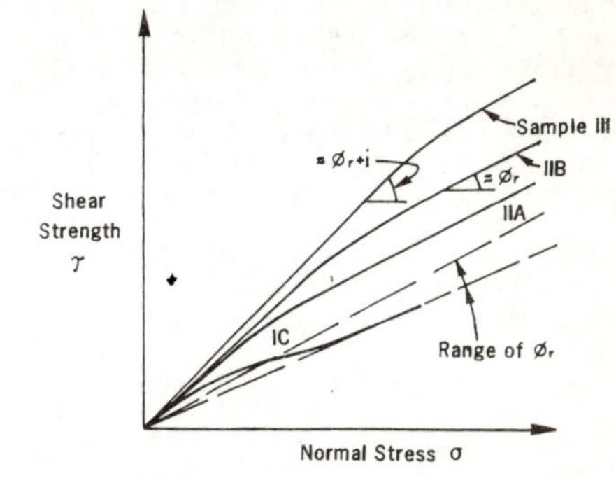

b) Shear Strength Envelopes for a Weathered Joint

Figure 1. Effect of weathering on shear strength of discontinuity (from Deere and Patton 1971).

Koo (1982b) obtained design peak strength parameters of $c' = 0$, $\phi' = 20^\circ$ from direct shear testing of smooth relict joint surfaces in weathered volcanic rocks in Hong

Kong, in contrast to intact soil strength of c' = 6 kPa and ϕ' = 32°. In highly weathered rocks, where the irregularities on discontinuity surfaces may still be strong enough to offer some resistance to shearing, higher strength values can be expected. For example, Richards (1976) reported values of ϕ' = 38° and c' = 30 kPa along clean, highly weathered joints in granitic saprolite from Hong Kong, and Serafim and Lopez (1961) determined 0' values of 45° - 57° (peak?) on highly weathered, rough joint surfaces in granite.

In some rocks, the shear strengths along relict discontinuities have been found to vary not only with depth within a profile but also with time. In a phyllitic soil in Ghana, Ayetev (1985) found that the shear strength increased by a few degrees over a period of 12 years, as a result of slight laterization and limonite cementation along the discontinuities, whereas in the highly weathered zone the shear strength decreased appreciably due to leaching effects. De Fries (in Deere and Patton 1971) and Sowers (1985) observed field evidence of strain softening properties of soils derived from phyllites and micaceous schists. This phenomenon has a bearing on the long term stability of slopes in saprolitic soils in areas where laterization is not active. Further chemical changes through weathering may occur within short periods of time along discontinuity planes and also within the soil matrix itself.

Some large scale failures have occurred along kaolinite vein bearing joints, in igneous and volcanic rocks in Hong Kong (Brand 1985, Irfan et al 1987) Kaolinite veins or other mineralized veins in these rocks are mainly of hydrothermal origin. Kaolinite is a layer lattice mineral which, when saturated, has a basic friction angle which is reduced greatly due to the lubricating action of water; it has a tan ϕ' sat/tan ϕ'dry ratio between 0.6 and 0.4 (Horn and Deere 1962). Experimental work by Kanji (1976) gave friction angles of between 12° and 22° along joints infilled with kaolinite, the lower values being for polished samples.

4.3 Failure Mechanisms Involving Relict Discontinuities

Discontinuity controlled failures, characteristic of rock slopes, are also common in slopes formed in saprolitic soils. In addition to simple planar or wedge sliding type failures, involving one or two planes, more complex failures, conditioned by a number of unfavourably oriented relict discontinuities, may occur. Failures may also be initiated by the opening up of discontinuities as a result of stress relief. The damming up of ground water along relict discontinuities may directly cause landslips even though the discontinuities themselves do not act as release planes. Unlike discontinuity controlled failures in rock, it is common to find that the failure surface passes from one relict discontinuity to another, through the intervening intact weathered rock material. Generally, failures are rapid and commonly happen during or shortly following heavy rainfall. They may however also occur in prolonged dry weather during excavation, due to removal of support, if, for example, relict discontinuities dip very steeply towards the cutting (Greenway et al 1987). Slow-moving failures have also been reported (Wolle 1985, Irfan et al 1987).

Wedge or planar sliding failures generally occur in soils which are not closely fractured. Koo (1982a) described the movement of wedges or blocks of soil along persistent relict joints in acid volcanic rocks containing slickensided black seams. Wedges of saprolite, sliding along relict discontinuities in phyllites, granites and gneisses in Brazil have also been reported by St John et al (1969) and Wolle (1980). Sheeting joints, formed as a result of stress release, generally with orientations subparallel to the slope surface, i.e. dipping unfavourably towards the cutting, can also cause planar slides, particularly in areas of uniform geologic environment where such discontinuities are of common occurrence.

In complex folded metamorphic rocks, relict joints, as well as relict cleavage planes, foliation planes and faults, can act as release surfaces for both simple and complex failures. Large slides controlled by relict stuctures are generally rare in these environments due to the highly variable nature of foliation planes and other structures. A number of small-scale wedge and planar sliding, toppling and complex failures, conditioned by schistosity, faults, bedding planes and joints, occurred along a 100 m long strip of a steep temporary slope cut for a large water tank in complex folded and weathered metasedimentary rocks in Hong Kong (Greenway et al 1987). In metamorphic rock, formed from interbedded competent and incompetent rocks, bedding slips are common and these are generally retained as slickensided structures.

The strain softening effects of weathering can play an important role in the initiation of some failures in slopes that may be stable when first formed. As explained previously, the shear strength along a discontinuity may decrease with time, particularly in areas where laterization is not active. Signs of instability may develop only after many years due to a gradual reduction in shear strength, with movements slowly accelerating until a major landslip finally occurs. Costa Nunes and Fernandes (1985) described a slowly developing landslip which occurred along a relict discontinuity in a metamorphic saprolite, years after the excavation was made.

In jointed saprolitic soils containing vertical or near vertical relict discontinuities, the toppling of single blocks or progressive mass toppling are other possible modes of failure. Figure 2 illustrates how such forms of toppling failure, initiated by the opening up of discontinuities by stress relief, can occur. Massad and Teixeira (1985) described the case of a mass toppling failure in weathered gneiss in Brazil in which movement was initiated by the opening up of near vertical schistosity planes.

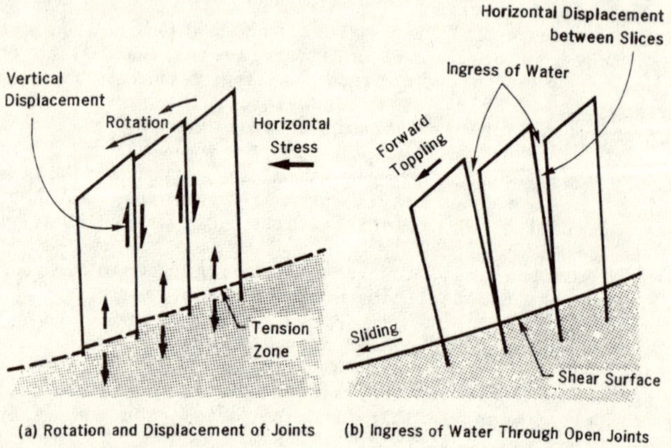

Figure 2. Mass toppling phenomenon in saprolitic soils (from Brand 1985).

5. DETECTION OF RELICT DISCONTINUITIES

In the planning of a ground investigation it is essential at the outset that the person responsible is aware that relict discontinuities exist in saprolitic soils and understands their significance with respect to slope stability. When dealing with saprolitic soils, it would seem a sensible philosophy to assume that relict discontinuities exist and that they will be detrimental to stability until proved otherwise.

Sandroni (1985) and others have stated that relict discontinuites are not easy to find and, if found, may not be easy to sample and test. The authors contend that, provided sufficient sample area is available for examination, identification of these features does not present a problem. Existing exposures in cuttings provide a large surface area for inspection and, at the same time, the performance of that slope is a useful guide to the likely performance of proposed cut slopes. The relative heights of the existing and proposed cuttings must however be taken into account, as well as other factors such as material type, slope orientation, upslope catchment, etc.. Existing cut slopes are particularly valuable as they allow individual discontinuities, as well as the overall pattern to be examined. Discontinuity line surveys can be carried out which can then be related to the proposed orientation and gradient of the future cut slope. Detailed observations can also be made with respect to persistence, joint surface regularity, nature of any infilling, etc.. Where existing cut slopes are covered by some artificial surfacing, to prevent erosion or infiltration of rainfall, the surfacing can be removed to expose the underlying slope forming materials. In Hong Kong, where a high proportion of cut slopes are covered by 'chunam', a lime-soil plaster, removal of chunam strips is a commonly used technique in the stability assessment of existing slopes. Even where not artificially surfaced, cut slopes often develop a natural surfacing produced by the deposition of iron oxides or the growth of organic matter (Bulman 1967). The deposition of such natural materials can completely obscure the underlying soils very rapidly, especially in tropical conditions. The existence of slip scars on existing cut slopes can be especially valuable if the failure mechanism can be deduced. Unfortunately, the planar shape of a failure scar can be quickly masked by erosion, or indeed, may not survive the actual failure (Flintoff and Cowland 1982).

On project sites, exposures of saprolitic soils may not be available. However, outcrops of the less weathered parent rock may occur nearby, e.g. in stream courses. The pattern of joints in the parent rock can provide a useful guide to the pattern of relict discontinuities in the overlying saprolitic soils (Cowland and Carbray 1988). In many cases, exposures of either the saprolitic soil or the less weathered parent rock will not be available for examination or, if available, may be located at such a distance from the site of the proposed cutting that their relevance may be questionable. In such circumstances, the use of trial excavations offers the best means of examining the soil mantle and the nature of relict discontinuities. These excavations may form an integral part of the proposed cutting (Paige-Green, 1984). The use of large trial excavations for major projects in the USA has been mentioned by Sowers (1985). He reported on the excavation of 20 m deep trenches in one project to establish the pattern of relict discontinuities. The term 'test pit' or 'trial pit' is generally used to mean excavations of small size dug by hand or by small excavation plant. Although they provide a smaller surface area for examination, trial pits can be excavated at less cost and, providing they can be taken deep enough, will still provide valuable information. In Brazil, according to Sandroni (1985), the excavation of hand-dug trial pits up to 20 m deep is commonplace, providing that the water table is not encountered. Harris (1985) discussed the use of trial pits for the collection of data on relict discontinuities in saprolitic soils in Hong Kong with emphasis on the ability to locate the trial pits at different orientations to avoid directional bias.

The drilling of boreholes is the most common ground investigation technique in use. However, the methods available for drilling boreholes vary widely in their ability to allow detection of relict discontinuities. The practice of alternate wash drilling (or augering) and sampling still persists in many countries as a common technique for investigating proposed cuttings. The only satisfactory method for drilling boreholes to detect relict discontinuities is one which allows 100% core recovery. The use of a retractable triple-tube core barrel, in combination with an air-foam flushing medium, to obtain high quality samples of saprolitic soils has been described by Phillipson and Chipp (1982). The identification of individual relict discontinuities in core samples is of limited value for design purposes; such important factors as persistence, waviness and orientation cannot be determined. As far as the authors are aware, the impression packer has not been successfully applied to saprolitic soils. Also, whilst it may be possible to identify discontinuities of similar character and dip direction in adjacent boreholes, it is rarely possible to follow individual discontinuities by interpolating between boreholes.

The mapping of relict discontinuities as a slope is cut is highly recommended and may indicate a need for revision of the original design or some form of preventive work to avoid subsequent failures (Powell and Irfan 1985). A systematic inspection, e.g. a discontinuity line survey, should be undertaken by an engineering geologist.

6. METHODS OF ANALYSIS

Numerous relict discontinuity failures have been reported in the literature since St. John et al (1969). However, very few authors have described the methods applied to analyse these failures. A brief review of methods of analysis available for both back analysis of slope failures and for calculating factors of safety of new slopes containing relict discontinuities will be given here. A detailed discussion of the methods of analysis can be found in the references quoted below.

6.1 Methods of Analysis For Failures Along Distinct Planes.

Stability along distinct discontinuities (e.g. simple planar and wedge failure modes) on those slopes where adversely oriented discontinuities can be recognised, may be analysed by limit equilibrium methods commonly applied to rock slopes, as outlined by Hoek and Bray (1981) using appropriate shear strength properties. A combination of numerical and graphical methods and stereographic projections can be applied to carry out simple wedge and planar failure analyses. For more complicated two dimensional and three dimensional discontinuity controlled failure modes of stepped planar, polygonal plane and stepped wedge form, the methods of analysis summarized by Kovari and Fritz (1984) can be used. To apply such methods however, it must be assumed that the weathered soil mass above the failure plane(s) acts as a rigid block and that no internal deformation of the soil mass takes place both before and during the failure.

Back analysis of a discontinuity-controlled failure in saprolitic soils, as with other failure types (Hencher et al 1984), may give grossly misleading results if the conditions at failure, including the failure surface characteristics, are not determined accurately. The planar shape of the failure surface can be quickly masked by erosion, or indeed, may not survive the actual failure. Usually, the groundwater conditions at the time of the failure are not known. Realistic groundwater assumptions are required for the calculated parameters to be meaningful.

6.2 Methods of Analysis for Failures in Closely Jointed Saprolites

In closely jointed saprolites, in which failure is not controlled by one or two dominant discontinuities, but where the joints contribute to a general weakening effect on the soil mass, slope failures can occur along an irregular or roughly arcuate surface. In these circumstances, conventional methods of stability analysis, developed for soil slopes (e.g. Bishop 1955, Janbu 1973), can be used for slope design and back analysis, provided that the non-linear characteristics of the failure envelope are taken into account (Hoek 1980).

Failure in a closely jointed saprolite, and particularly in the highly weathered material, is controlled by sliding, rotation, crushing and splitting of the individual pieces within the mass. A realistic determination of the strength of such a jointed mass for stability analysis purposes is not possible. Hoek and Brown (1980) have developed an empirical strength criterion for approximating the non-linear characteristics of the rock failure envelope, based on experience and limited testing. This empirical strength criterion can be used to determine the shear strength of the weathered rock mass.

For circular failure modes, Bishop's method of analysis can be applied (see Hoek 1980) and non-circular surfaces can be dealt with by using one of the commonly adopted methods such as that given by Janbu (1973) or Morgenstern and Price (1965). The influence of a tension crack, which may develop as a result of the opening up of steeply dipping relict discontinuities, may be taken into account in these methods of analysis. The use of probabilistic methods to account for the adversity of the relict jointing pattern on the mass strength has been discussed by Lumb (1971), Koo (1982b) and Harris (1985). It should be noted however that the approach adopted by these authors does not allow for a stepped failure mode.

The design charts given by Hoek and Bray (1981) for circular failure types can be used for preliminary design purposes.

Figure 4. Case history No. 1 - Junk Bay Road. A photograph of failed cut slope.

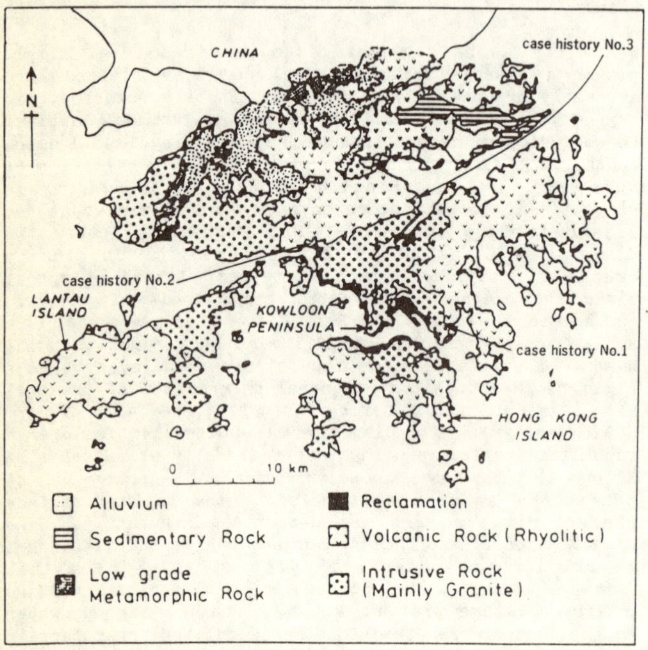

Figure 3. A simplified geological map of Hong Kong giving location of case history sites.

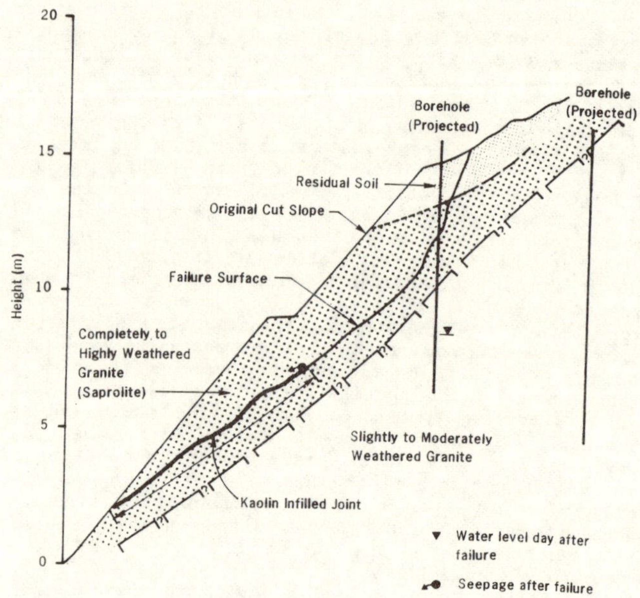

Figure 5. Case history No. 1 - Junk Bay Road. A sketch of failed slope (after Hencher et al 1984).

7. CASE HISTORIES

7.1 Case Histories from Hong Kong (Figure 3)

Case History No. 1 - Junk Bay Road. In 1982, following heavy rainfall, a large failure involving some 1300 m^3 of debris, occurred in a 30 m high cutting along Junk Bay Road in Kowloon (Figure 4). The slope was composed of highly to completely weathered granite (Hencher et al 1984). The slope had been previously investigated in early 1982 and piezometers installed. A slope stability analysis carried out at that time, using Janbu's Method (Janbu 1973), gave a factor of safety of above 1.0.

Failure occurred along a persistent, moderately steeply dipping relict discontinuity containing a kaolin infill up to 20 mm thick (Figure 5). The existence of this potential plane of weakness had not been recognised during the original ground investigation. High water pressures are thought to have developed along the discontinuity directly prior to failure,although this had not been detected by a piezometer located near the head of the failure. This case history highlights the difficulty of correctly locating instruments for recording groundwater pressures in weathered rocks.

Case History No. 2 - Tuen Mun Highway.A 70 m high cut slope along a major highway failed for the second time in 1983 (Brand 1985).The failure involved an area of approximately 10,000 m^2 consisting of completely weathered volcanic tuffs. The previous failure had occurred in 1975 during construction of the road and, on that occasion, involved weathered granodiorite overlying the volcanic tuffs. Following the first failure, the slope was cut back from the original slope angle of 60°, to a gentler gradient of 32°. A stability analysis carried out in 1978, using ground investigation data collected for the design of remedial works for the original failure, gave a minimum factor of safety of 1.3 (Figure 6).

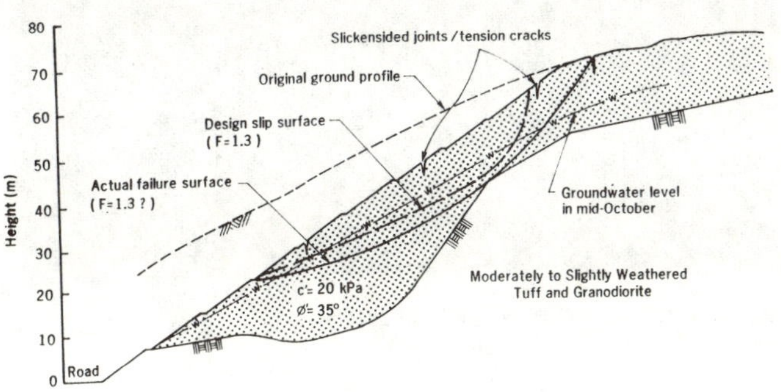

Figure 6. Case history No. 2 - Tuen Mun Highway. A sketch of failed slope (after Brand 1985).

The head scarp of the 1983 failure coincided with sub-vertical relict joints, indicating the influence of such features as release surfaces/tension joints on the formation of the landslip. The failure was attributed to stress release, dilation of relict joints, and consequent ingress of water. The mode of failure was considered to have involved the rotation and displacement of individual slabs, and, finally, the shearing of weak weathered intact material between joints (i.e. a mass toppling process as depicted in Figure 2). A similar mechanism was also postulated for the earlier failure in 1975. The presence of the vertical relict joints was not considered in the stability analysis carried out in 1978.

This case history highlights the importance of geological structure in controlling failure mechanisms in saprolitic soils, and also the need to take these features into account when postulating the shape of potential failure surfaces in stability analyses.

Case History No. 3 - Tai Po Road. On the 17th July 1973, during heavy rainfall brought about by Typhoon Dot, a large landslip occurred on an 18 m high cut slope on the Tai Po Road, the major traffic route between the urban areas of Hong Kong and the border with Mainland China at that time. The landslip debris completely blocked the road for about 24 hours.

Prior to the failure, the slope was up to 18 m high and consisted of a series of 7.5 m high faces each inclined at 60°. A study revealed the presence of clay-coated relict joints on the failure surface, dipping out of the slope at between 35° and 50° to the horizontal. It was concluded that the failure was principally due to the presence of these adversely orientated planes of weakness, although movement did not occur along one single joint, but 'stepped over' from one to another, thereby involving shearing of the intact material between. It was considered that high water pressure may have developed along the relict joints directly prior to failure.

Figure 7. Case history No. 3 - Tai Po Road. A photograph of slope showing outline of potential wedge failure.

A back analysis of the failure was carried out, resulting in the derivation of cohesian values of 26 kPa to 57 kPa for corresponding ϕ' values of 40° to 15°. However, the validity of the results of the analysis is questionable, as extreme groundwater conditions at the time of failure were assumed (e.g. a 6 m deep water-filled tension crack at the head of the slip) and the analysis did not take into account the waviness of the failure surface.

In mid-1987, signs of distress, including cracking and bulging of the chunam surfacing and severing of concrete U-channels, appeared on the steep cut slope on the southern flank of the old landslip scar. These signs of distress clearly indicated that some slope movement had occurred. An examination of the materials beneath the chunam revealed the presence of well-defined relict discontinuities in the completely weathered granite. The area of movement was bounded on one side by a slickensided, clay coated relict joint, dipping out of the face at 45° to 50°. The other side coincided with a well-defined discontinuity (possibly a shear zone) which extended at least 15 m across the face of the slope and dipped steeply into the area of movement. These features defined a wedge (Figure 7).

7.2 A Case History from Malaysia

A large number of cut slope failures occurred during the construction of a highway in the State of Kelantan, in the north-east of Peninsular Malaysia (Figure 8). A large number of the failures involved slopes formed of completely to highly weathered volcanic and metasedimentary rocks, and the majority of the failures were controlled by the presence of well-defined relict discontinuities.

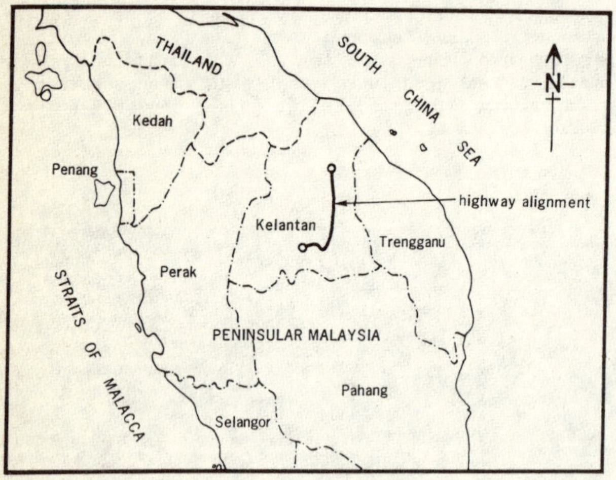

Figure 8. Case history No. 4 - Malaysia. Location map.

The ground investigation for the road alignment and for the design of cut slopes was limited to shallow auger drilling along the centreline, together with observations of the few existing cut slopes in the vicinity of the proposed road. The only deep drilling carried out was in areas where strong bedrock, which would require blasting during construction, appeared to be at a shallow depth. Slopes were initially cut at 2 (vertical) on 1 (horizontal) with benches at 10 m vertical intervals, in keeping with standard cut slope design practice at that time in Malaysia. Although geological factors played little part in cut slope design, it was recognised that the stability of cut slopes would be controlled more by relict discontinuities than by the intact strength of the material between them. It was also recognised that some failures could occur and that maintenance would probably be required.

Construction was carried out between 1979 and 1982. Initially, cut slope performance was quite satisfactory. However, it was noted during an appraisal of cut slope stability, carried out in late 1980, that few of the fully formed slopes had been subjected to a major rainstorm. During the 'monsoon season' of late 1980 to early 1981 the majority of the cut slopes in the northern section of the alignment failed, together with others elsewhere.

The slopes in the northern section of the highway were cut mainly in completely to highly weathered volcanic pyroclastic rocks of Permo-Triassic Age, ranging from fine-grained tuffs to coarse agglomerates. These rocks had been highly folded and also affected by low grade metamorphism. In their completely weathered state the rocks had a characteristic purple colour and were of very stiff to hard consistency, remoulding to a slightly clayey silt of low plasticity. Insitu, the highly to completely weathered rock was cut by well-defined, closely spaced, relict discontinuities which were planar, smooth and generally steeply dipping.

Figure 9. Case history No. 4 - Malaysia. A photograph of wedge failure.

Failures ranged in size from minor slips, involving only a few cubic metres to major failures in which whole cut slopes, more than 100 m long and up to 25 m high, were affected. Virtually all failure surfaces coincided with relict discontinuities which were so dominant that most failures involved relatively little shearing through the intact material between. The failures took the form of simple planar slides or wedge failures (Figure 9), and the persistence of individual joints was such that individual slides, up to 30 m long along a single joint were observed. Of particular significance, was the presence of a white 'soapy' clay coating on almost all of the joint surfaces examined.

The majority of cut slopes had a thin (usually up to 3 m) mantle of structureless residual soil. On most slopes affected by instability,the uppermost portion of the failure surface terminated abruptly at the completely weathered/residual soil interface indicating the absence of the relict joint structure in the residual soil mantle. In those cases where failure did extend up through the interface, the failure surface adopted more of an arcuate shape through the residual soil mantle.

Close attention was paid to newly formed cut slopes and it was observed that, within a short time following excavation, relict discontinuities on the recently exposed cut face would begin to open up, presumably as a result of stress relief, probably accentuated by near-surface desiccation. This allowed rainfall run-off to infiltrate more easily into these planes of weakness and help promote failure.

All failed slopes were subsequently reformed at a gentler slope angle of 1 on 1. Also, the design of those slopes which had not been cut at that stage was modified to the same angle. After that, the incidence of cut slope failures was minimal.

During the study of the failed slopes, no relationship was found between slope performance and the nature of the ground beyond the slope crest. Also, the height of individual slopes appeared not to be a controlling factor with respect to their stability, except of course with respect to the volume of slip debris. On those slopes which had not been affected by instability it was found that the pattern of relict discontinuities was favourable to stability.

ACKNOWLEDGEMENTS

This paper is published with the permission of the Director of Civil Engineering Services of the Hong Kong Government. The authors acknowledge the comments made by Messrs P.G.D. Whiteside and J.B. Massey in the preparation of this paper.

REFERENCES

Ayetev, J.K. (1985). Slope stability investigation and design parameters studies in phyllites in Ghana. Bulletin of the International Association of Engineering Geology, no. 31, p.19-23.

Bishop, A.W. (1955). The use of the slip circle in the stability analysis of slopes. Geotechnique, vol. 5, p.7-17.

Brand, E.W. (1985). Predicting the performance of residual soil slopes (Theme Lecture). Proceedings of the Eleventh International Conference on Soil Mechanics and Foundation Engineering, San Fransisco, vol. 5, (in press).

British Standards Institution (1981). Code of Practice for Site Investigations (BS5930). British Standards Institution, London, 147 pp.

Bulman, J.N. (1967). A survey of road cuttings in Western Malaysia. Proceedings of the First South-East Asian Conference on Soils Engineering, Bangkok, p 289-300.

Costa Nunes, A.J. & Fernandes, C.E.M. (1985). Slides in saprolites from gneissic rocks with relic slip surfaces, the B.N.D.E.S. case history. Proceedings of the First International Conference on Tropical Lateritic and Saprolitic Soils, Brasilia, vol. 2, p.49-55.

Cowland, J.W. & Carbray, A.M. (1988). Rock type failure of cut slopes in completely weathered tuffs. Proceedings of the Second International Conference on Geomechanics in Tropical Soils.

Dearman, W.R. (1976). Weathering classification in the characterisation of rock : a revision. Bulletin of the International Association of Engineering Geology, no. 13, p.123-127.

Dearman, W.R. (1984). State of weathering : the search for a rational approach. Proceedings of the Twentieth Regional Meeting, Engineering Group Geological Society on Site Investigation Practice : Assessing BS5930, p.132-142.

Dearman, W.R. & Turk, N. (1985). Sampling and testing of residual soils in the United Kingdom. In Sampling and Testing of Residual Soils : A Review of International Practice (ed. E.W. Brand and H.B. Phillipson), Scorpion Press, Hong Kong, p.175-181.

Dearman, W.R., Baynes, F.J. & Irfan, T.Y. (1978). Engineering grading of weathered granite. Engineering Geology, vol. 12, p.345-74.

Deere, D.U. & Patton, F.D. (1971). Slope stability in residual soils. Proceedings of the Fourth Pan-American Conference on Soil Mechanics and Foundation Engineering, Puerto Rico, vol. 1, p.87-170.

Flintoff, W.T. & Cowland, J.W. (1982). Excavation design in residual soil slopes. Proceedings of the ASCE Speciality Conference on Engineering and Construction in Tropical and Residual Soils, Honolulu, p.339-556.

Gidigasu, M.D. (1975). Laterite Soil Engineering. Elsevier, Amsterdam, 570 p.

Gordon, F.R. (1984). The laterite weathering profiles of Precambrian igneous rocks at the Worsley Alumina Refinery site, South West Division, Western Austrialia. Proceedings of the Fourth Australia-New Zealand Geomechanics Conference, Perth, vol. 1, p.261-266.

Greenway, D.R., Powell, G.E. & Irfan, T.Y. (1987). A case history of slope design in complex weathered low grade metamorphic rock. Geotechnical Engineering, vol. 18, p.145-166.

Harris, R. (1985). Mass or massive : a technique for jointed soils. Ground Engineering, vol. 18, no.5, p.24-31

Hencher, S.R., Massey, J.B. & Brand, E.W.(1984). Application of back analysis to Hong Kong landslides. Proceedings of the Fourth International Symposium on Landslides, Toronto, p.631-638.

Hoek, E. (1980). An empirical strength criterion and its use in designing slopes and tunnels in heavily jointed weathered rock. Guest Lecture presented at the Sixth Southeast Asian Conference on Soil Engineering, Taiwan.

Hoek, E. & Bray, J.W. (1981). Rock Slope Engineering (3rd edition). Institution of Mining and Metallurgy, London, 358 p.

Hoek, E. & Brown, E.T. (1980). Underground Excavations in Rock. Institution of Mining and Metallurgy, London, 527 p.

Horn, H.M. & Deere, D.U. (1962). Frictional characteristics of minerals. Geotechnique, vol. 12, p.319-335.

IAEG (1981). Rock and soil description and classification for engineering geological mapping. Bulletin of the International Association of Engineering Geology, no. 24, p.235-274.

ISRM (1978). Suggested methods for the quantitative description of discontinuities in rock masses. International Society for Rock Mechanics Committee on Laboratory Tests. International Journal of Rock Mechanics and Mining Sciences and Geomechanics Abstracts, vol. 15, p.319-368.

Irfan, T.Y. & Dearman, W.R. (1978). The engineering petrography of a weathered granite in Cornwall, England. Quarterly Journal of Engineering Geology, vol. 11, p.233-244.

Irfan, T.Y., Koirala, N.P. & Tang, K.Y. (1987). A complex slope failure in a highly weathered rock mass. Proceedings of the Sixth International Congress on Rock Mechanics, Montreal, p.397-402.

Janbu, N. (1973). Slope stability computations. In Embankment Dam Engineering : Casagrande Volume (ed. R.C. Hirschfield and S.J. Poulos), Wiley, Newyork, p.47-107.

Kanji, M.A. (1976). Shear Strength of Soil-Rock Interfaces. M.S. Thesis, Department of Geology, University of llinois, Urbana, USA.

Kany, M. & Herrman, R. (1985). Sampling and testing of residual soils in the Federal Republic of Germany. Sampling and Testing of Residual Soils : A Review of International Practice (ed. E.W. Brand and H.B.

Phillipson), Scorpion Press, Hong Kong, p.57-63.
Koo, Y.C. (1982a). Relict joints in completely decomposed volcanics in Hong Kong. Canadian Geotechnical Journal, vol. 19, p.117-123.
Koo, Y.C. (1982b). The mass strength of jointed residual soils. Canadian Geotechnical Journal, vol. 19, p. .
Kovari, K. & Fritz, P. (1984). Recent developments in the analysis and monitoring of rock slopes. Proceedings of the Fourth International Symposium on Landslides, Toronto, vol. 2, p.1-16.
Little, A.L. (1969). The engineering classification of residual tropical soils. Proceedings of the Seventh International Conference on Soil Mechanics and Foundation Engineering, Mexico, vol. 1, p.1-10.
Lumb, P. (1970). Safety factors and the probability distribution of soil strength. Canadian Geotechnical Journal, vol. 7, p.225-242.
Massad, F. & Teixeira, H.R. (1985). Deep cut on saprolitic soils conditioned by relict structures. Proceedings of the First International Conference on Tropical Lateritic and Saprolitic Soils, Brasilia, vol. 2, p.381-391.
Morgenstern, N.R. & Price, V.E. (1965). The analyjsis of the stability of general slip surfaces. Geotechnique, vol. 15, p.79-93.
Paige-Green (1984). Investigation for cut slopes for roads in South Africa. A review of techniques. Proceedings of the Eighth Regional Conference for Africa on Soil Mechanics and Foundation Engineering, Harare. p.475-479.
Phillipson, H.B. & Chipp, P.N. (1982). Air foam sampling of residual soils in Hong Kong. Proceedings of the ASCE Speciality Conference on Engineering and Construction in Tropical and Residual Soils, Honolulu, p.339-356.
Powell, G.E. & Irfan, T.Y. (1986). Slope remedial works in weathered rocks for differing risks. In Rock Engineering and Excavation in an Urban Environment, Institution of Mining and Metallurgy, London, p.347-355.
Ruxton, B.P. & Berry, L. (1957). Weathering of granite and associated erosional features in Hong Kong. Bulletin of the Geological Society of America, vol. 68, p.1263-1292.
Richards, L.R. (1976). The Shear Strength of Joints in Weathered Rock Ph. D. Thesis, University of London.
Sandroni (1985). Sampling and testing of residual soils in Brazil. In Sampling and Testing of Residual Soils : A Review of International Practice (ed. E.W. Brand & H.B. Phillipson), Scorpion Press, Hong Kong, p.31-48.
Serafim, J.L. & Lopez, J.J.B. (1961). In situ shear tests and triaxial tests of foundation rocks of concrete dams. Proceedings of the Fifth International Conference on Soil Mechanics and Foundation Engineering, Paris, vol. I, p.533-539.
Sowers, G.F. (1963). Engineering properties of residual soils derived from igneous and metamorphic rocks. Proceedings of the Second Pan-American Congress on Soil Mechanics and Foundation Engineering, Brazil, vol. I, p. 39-61.
Sowers, G.F. (1985). Residual soils in the Unites States. In Sampling and Testing of Residual Soils : A Review of International Practical (ed. E.W. Brand and H.B. Phillipson), Scorpion Press, Hong Kong, p.183-191.
St. John, B.J., Sowers, G.F. & Weavers, C.H.E. (1969). Slickensides in residual soils and their engineering significance. Proceedings of the Seventh International Conference of Soil Mechanics and Foundation Engineering, Mexico, vol. II, p.591-597.
Wolle, C.M. (1980). Taludes Naturais-Mecanismos de instabilizacao criterios de Sequranca. M.Sc. Thesis, EPUSP, University of Sao Paulo.
Wolle C.M. (1985). Slope stability. Peculiarities of geotechnical behaviour of tropical lateritic and saprolitic soils. Progress Report, Theme 3.2, Brazilian Society for Soil Mechanics, Sao Paulo, p. 164-214.

Stability of excavated slopes in the weathered Jurong formation of Singapore

K.W.LO, Department of Civil Engineering, National University of Singapore, Singapore
C.F.LEUNG, Department of Civil Engineering, National University of Singapore, Singapore
K.HAYATA, Taisei-Shimizu-Marubeni Consortium, Japan
S.L.LEE, Department of Civil Engineering, National University of Singapore, Singapore

ABSTRACT: Within the usual range of excavation depths in the Jurong formation of Singapore, sedimentary strata may be exposed in a variety of opposing attitudes and contain complex patterns of discontinuities as a result of past tectonic activity. Furthermore, due to a highly erosive climatic regime, severe weathering may have taken place to exacerbate degradation zones caused by hydrothermal alteration. In this mixed setting, a conventional slip circle approach may not provide meaningful results as indicated by analyses of four slopes in highly to completely weathered materials at an actual study site. By contrast, a vectorial approach based on Goodman's multiple free block model (1976) gave close comparisons of back-analysed material properties for these slopes.

INTRODUCTION

A proper evaluation of stability of excavated slopes in weathered rocks of the Jurong formation, and hence provisions for their security, is as yet essentially an unresolved issue. The common practice is to apply shear strength parameters based on some qualitative assessment of laboratory and/or field test results, often relying heavily on individual experience and judgement, in traditional slip circle analyses. However, in view of the inherent marked heterogeneity of these sedimentary rocks, such an approach is susceptible to under- or over-design of an indeterminate degree due to basic lack of rigour.

In these circumstances, a systematic study was conducted on a wide variety of slopes in this formation to develop better understanding of factors affecting, as well as more appropriate models for determining their stability. In this context, a detailed examination of slope behaviour was conducted on a substantial excavation site in the central business district of Singapore, typifying ground conditions and problems encountered in such construction.

GEOLOGICAL SETTING

Within the normal range of excavation depths of slopes in the Jurong formation (Fig. 1), sandstone, siltstone and shaley mudstone interbedding may be revealed in a multiplicity of orientations, each containing an extensive network of discontinuities brought about by faulting and folding in the geologic past. Due to exposure of the formation to a highly humid and otherwise erosive equatorial climate, correspondingly variable distributions of relatively intact to completely weathered materials may be encountered. For the purpose of construction of the recent mass rapid transit (MRT) system, the formation was subdivided on the basis of rock type and degree of weathering, as also indicated in Table 1.

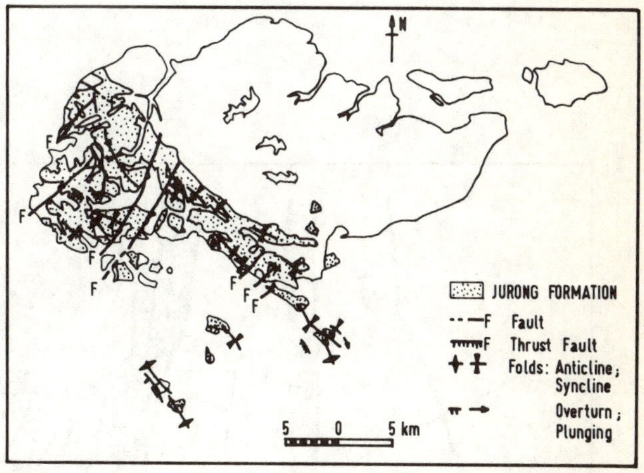

Fig. 1 Jurong formation and tectonic activity

Study site (Fig 2)

Some twelve facies of the formation were identified during MRT construction (Provisional Mass Rapid Transit Authority, 1983), of which the Rimau facies, consisting of sandstone in varying degrees of cementation with occasional shaley mudstone layers, was prevalent at the study site. At the western end, however, interbedded sandstone and siltstone typifying

Table 1 : Rock weathering classification (After Anon, 1970)

Geological Classification	Grade	Designation	Description
S1	IA	Fresh	No visible sign of weathering.
	IB	Faintly Weathered	Weathering limited to the surface of major discontinuities.
	II	Slightly Weathered	Penetrative weathering developed on open discontinuity surfaces but only slight weathering of rock material.
S2	III	Moderately Weathered	Weathering extends throughout the rock mass but the rock material is not friable.
	IV	Highly Weathered	Weathering extends throughout rock mass and the rock material is friable.
S4	V	Completely Weathered	Rock is wholly decomposed and in a friable condition but the rock texture and structure are preserved.
	VI	Residual Soil	A soil material with the original texture, structure and mineralogy of the rock completely destroyed.

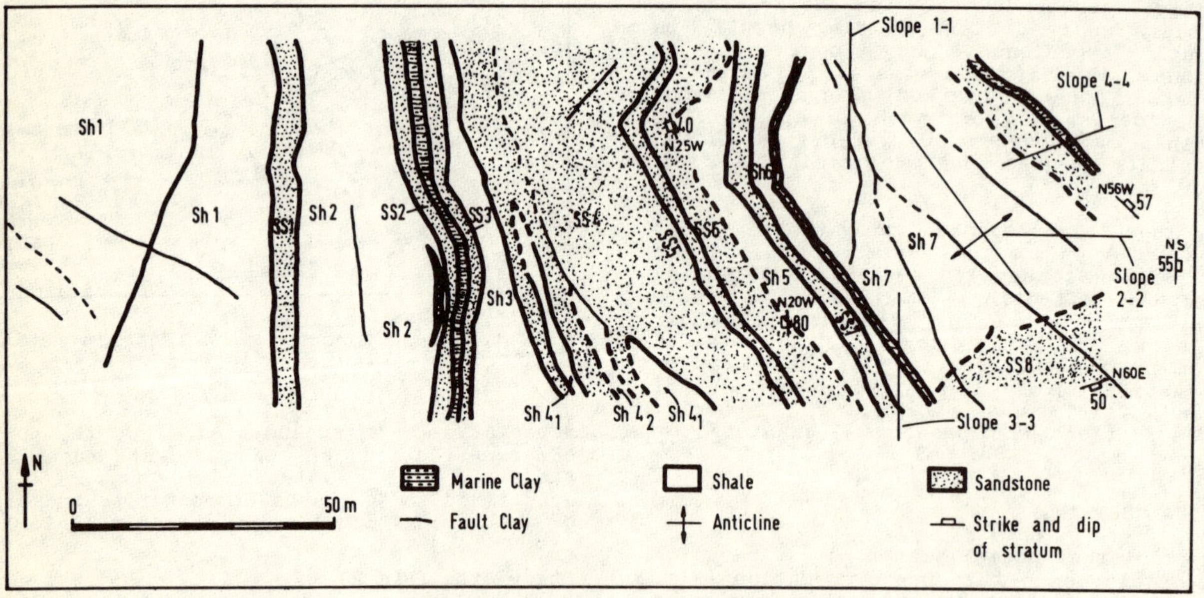

Fig. 2 Geology of study site

the Queenstown facies outcropped.

Subsurface conditions consisted of predominantly weathered rock with little soil cover. Major structural and facies changes over the length of site reflected the effects of faulting. The reddish-purple sandstone and siltstone at the western end graded from highly to slightly weathered materials over a depth of about 7m. At the eastern end, on the other hand, the whitish-grey to yellowish-brown sandstone and thin shaley mudstone interlayering occurred to depths exceeding 25m, grading from highly to slightly weathered condition within the top 10m. Generally, rock exposures were extensively fractured. At mid-section of the northern site area, very dense to lightly-cemented sand, that is completed weathered sandstone, occurred to a depth of about 24m. Along the western part of southern side, on the other hand, completely weathered gravel-sized shalcy mudstone and siltstone fragments embedded in very stiff to hard silty clay matrix overlay highly weathered and fragmented shaley mudstone and siltstone between 5 and 10m depths. The rock masses were fissile and although they possessed the reddish-purple hue typical of the Queenstown facies, did not possess the same high slake durability.

Wide differences in rock types of core samples logged at various borehole locations confirmed extreme transverse as well as longitudinal variability of subsurface materials. Away from the excavation area, a groundwater table fluctuated between 3 and 5m below ground surface due to tidal movements of the nearby sea frontage. A practically permanent water level existed at the invert of an adit towards which the groundwater table was drawn so that the phreatic surface was everywhere in excess of 5m depth from excavated slope surfaces.

The geological map of Singapore (PWD Singapore, 1976) indicates a strike of N10° W and dip of 80° E at a location near the site. However, rock cores revealed strata inclinations varying between 0 and 45 degrees reflecting past tectonic activity. In the event, an examination of exposed rock surfaces at mid-section along the excavation revealed sandstone (SS) stratum SS1 dipping at 90 degrees and striking at N10° W in the western end of the site. The corresponding orientations of SS2 were 40° W and N25° W, whereas for SS8 at the eastern end it was 50° SE and N60° E respectively. Beyond SS8, a reversal in trend of strata was implied by a dip measurement of 55° E and strike along NS implying an anticlinal axis running across shaley mudstone (Sh) stratum Sh7.

Many faults were found to occur at the site, the eastern half of stratum Sh2 and strata SS2, SS3 and Sh3 forming fault breccia inter alia. The nature of degradation of insitu materials suggested the outcome of hydrothermal alteration, as distinct from a purely weathering regime of progressively reducing degree with depth. For instance, strata SS4 and SS6 having lost their iron content, altered to a sandy consistency of quartz, kaolin and sericite minerals irrespective of depth from rock or ground surface. It was also observed that adjacent zone of shaley mudstone strata had altered to firm grey clay, as also the case at similar site locations. A reddish-brown clayey accumulation of iron compounds also formed as a discrete layer between the persisting hard shale and grey clay.

Strata SS1, SS2, SS3 and SS7 generally consisted of arkosic sandstone with open cracks to 10m depth and iron oxides deposited on crack surfaces. Weathering was relatively slight so that rock fragments were correspondingly hard. Strata SS4 and SS6 possessed a whitish colouration and lacked any significant cohesion. Stratum SS5 outcropped between SS4 and SS6 and was slightly harder than the latter sandstone strata. Stratum SS8 consisted of sandstone and occasionally alternating thin shaley mudstone layers. Discontinuity faces contained iron oxide deposits, although the rock constituents themselves were fairly hard. Shaley mudstone layer Sh1 was almost black in colour and relatively intact, parallel joints striking at N40° W to N70° W and dipping 70 to 80 degrees therein. The eastern and western halves of Sh2 were fairly distinct, rocks of the former being brownish-black and slightly weathered while those of the latter contained widespread fault breccia and had similar characteristics as stratum Sh3. The sub-divisions of stratum Sh4 were markedly layered although strongly weathered, whilst strata Sh5 and Sh6 had altered and weathered to a grey to reddish-brown colouration. In the case of Sh7, only cracked surfaces were reddish-brown, reflecting an advanced state of weathering, although below about 10 m weathering was only slight. Since this stratum was situated near the anti-clinal axis, closely-spaced joints of 2 to 5 cm intervals had developed. It was found that exposure by excavation and subsequent wetting and drying caused the stratum to soften significantly.

GROUND CHARACTERISTICS

Fig. 3 provides an indication of ground properties at a selected slope section, according to which wide variations reflected rapid changes in weathering. A conservative estimate of undrained shear strength of completely weathered insitu materials (Fig 4) indicated quite considerable excavation heights may be sustained in the short term. However, the presence of ground discontinuities indicated above implies some appropriate form of surface protection would be necessary for such heights to be attainable.

The influence of faulting on weathering of adjacent zones was deduced from standard penetration blow counts "N" measured in various boreholes on site. For instance, N<50 at several such zones where core recovery also revealed relatively large proportions of fairly permeable sandstone and siltstone altered to sand and silt. Furthermore, a friable zone of low blow count was registered just above fault breccia. From these findings, it was inferred that the severity of weathering was significantly increased in the vicinity of fault zones where local permeabilities were correspondingly elevated by a well-developed system of discontinuities providing additional drainage paths resulting in ground softening. Other factors considered to be major contributors of slope collapses on site were the generally advanced state of weathering arising independently of the above groundwater softening and the existence of degradation zones due to hydrothermal alteration. Following results of geochemical laboratory tests provide supporting evidence for the existence of these adverse conditions.

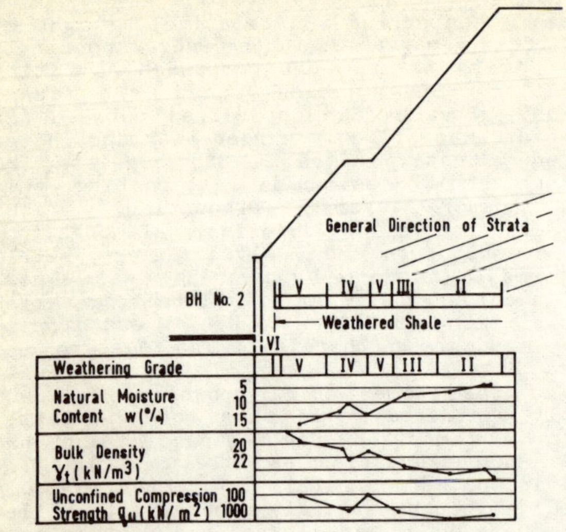

Fig. 3 Ground properties at borehole 2

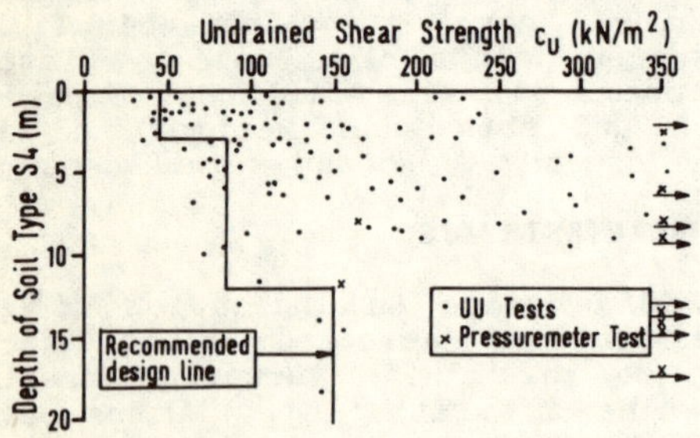

(after PMRTA, 1983)

Fig. 4 Undrained shear strength of completely weathered rock

Laboratory test results

The reddish-purple colour of shaley mudstone was due to the presence of aluminium and free iron oxides, a high content of which reflected severe weathering. These oxides tended, however, to lower cation exchange capacity (CEC) and liquid limit.

The mineral content of insitu rocks was established by X-ray diffraction, the powder method being adopted for identifying either coarse particle sizes or fine particles with well-developed crystalline structure, and the orientation method for the latter particles with poorly-developed crystalline structure, or clay minerals. All shaley mudstones contained α-quartz and hematite (Fe_2O_3), some including Gibbsite. Most sandstones contained α-quartz, some having minute quantities of Gibbsite. The hematite content of shale in the eastern area of the site was higher than in the west, reflecting higher degree of weathering of the former area. Sandstone strata, on the other hand, contained no significant amounts of free oxides, implying eluviation as a result of hydrothermal alteration and reduction processes. Quantitative analyses of free oxides also substantiated X-ray diffraction findings, namely that the amount of free oxides of shaley mudstone in the eastern area of the site was, as a whole, greater than in the west.

Moreover, the greatest quantities of mixed-layer sericite and montmorillonite were to be found in the eastern area of the site, decreasing towards the west until only sericite was detectable. Sandstones, on the other hand, consisted mainly of kaolinite with some sericite. From the distribution of minerals, a history of hydrothermal alteration could be inferred according to the proposition by Utada (1978). Thus, from the paragenetic relation of insitu minerals, it was apparent that at an earlier stage of the geologic past, intermediate hydrothermal water percolated through the ground from the western area of the site, followed more recently by similar inundation of acidic hydrothermal water, the cumulative effect being significant alteration of insitu material focused around the eastern area (Fig 5).

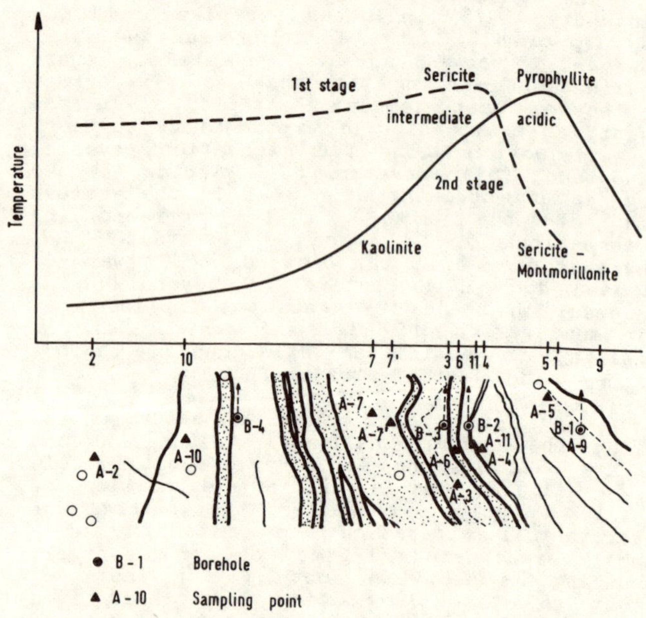

Fig. 5 Hydro thermal alteration on site

The CEC of shaley mudstone in the eastern area of the site was greater than in the west due to the presence of expansive mixed sericite and montmorillonite layer minerals of the former. However, at the western end, virtually no expansive minerals were detected, hence CEC was low. On the other hand, sandstone strata contained mainly kaolinite, thereby registering the lowest CEC levels. It was also found that CEC values after de-ironing were generally higher than before, larger increments of CEC being associated with higher initial values. Also, the trend was of decreasing free Fe_2O_3 content towards the central sandstone stratum except where layers of iron oxide accumulated due to de-ironing of the stratum by hydrothermal alteration, for instance within the stratum itself as well as in shale strata Sh2 and Sh3. The relative particle sizes of mixed-layer sericite and montmorillonite after de-ironing generally indicated a predominance of particles of less than 0.2 μ dimension, whilst sericite and kaolinite minerals contained more of the 0.2 to 1 μ and 1 to 2 μ size fractions. Thus, in reflecting mineral content of respective strata, the eastern end of the site contained the most expansive materials, becoming less so westwards towards the central sandstone stratum which was found to be relatively inactive. A medium to strong expansibility of the more active materials to the east was indicated.

Sandstone strata contained only very small quantities of exchangeable cations, as to be expected in view of their low CEC values. Generally 2-valence exchangeable cations such as Mg^{2+} and Ca^{2+}, which were least likely to be eluviated, were predominant in shaley mudstone, although the latter had apparently weathered to the core. It was also found that concentrations of Mg^{2+} and Ca^{2+} conformed with the CEC distribution. Swell tests were also conducted to complement the above geochemical findings. Accordingly, for shaley clay, volumetric swelling strain was found to correlate fairly positively with natural moisture content and increase most rapidly at low stress levels. Furthermore, conservative estimates of order of magnitude of swelling due to site excavations, and hence potential for softening due to subsequent wetting were high compared with the CBR limitation of 30% for suitability of fill material. Also, based on Cox's (1984) loosening mechanism following swelling of blocky ground, and Hoek and Bray's (1981) consequential failure model, it was deduced that there would be a tendency for expansive near-surface slope materials of insitu fractured rocks to shear at an overall friction angle given by angle of internal friction "ϕ" plus a significant roughness angle "i" corresponding to loosening. Furthermore, high angle sliding collapse would tend to be associated with those slopes having low cohesion and high overall angle of friction "ϕ+i".

Groundwater observations indicated wetting and drying cycles which would tend to aggravate any of the above conditions of potential slope instability. Thus, the repeated shrinkage and swelling of ground resulting from fluctuations of groundwater seepages measured would lead to significant reduction in shearing resistance of expansive materials in addition to the adverse effects of stress relief due to excavation mentioned above. Such an outcome is indicated by the ground properties of Fig 3 reflecting progressive weakening from ground surface.

SLOPE ANALYSES

Slope behaviour

The long-term stability of unsupported slopes in the Jurong sedimentary formation depends on relative proportion of reasonably intact rock to that in an advanced state of decomposition, disposition of local bedding planes, degree of development of discontinuities and ambient groundwater conditions. Thus, where highly-weathered sandstone strata are predominant, near-vertical slopes may stand unsupported. On the other hand, where completely decomposed mudstones prevail, unsupported slopes with surface vegetal cover may stabilise at only about 45 degrees to the horizontal.

In the case of more recent exposures, excessively steep excavations may become unstable, although the more extreme events tend to be associated with unsupported and unprotected cuts in completely weathered materials, adversely orientated to bedding and other discontinuity planes. More often, however, localised slippages take place in the form of mudslides of very wet material in tongue or lobe-shaped masses of limited volume oozing from between intact rock formations and exposing a striated shear surface after completion of ground movement. The appearance of the slipped mass resembles isolated fan-shaped debris such as typically encountered at the bases of ravines. Otherwise, the ground may merely creep, resulting in a small mid-slope bulge but with no distinct backscar at the crest; a slight bulge may also develop at the toe.

General slope failure, on the other hand, is often the result of long-term decrease in shear strength coupled with increase in shear stress of completely weathered materials. In these circumstances, exposures of rocks at backscars of initially steep cuttings often reveal friable materials, although essentially retaining the original rock structure. Also, many arenaceous and rudaceous units contain significant proportions of feldspar, both as clasts and in the matrix, which are susceptible to kaolinisation under humid equatorial conditions, which when moist become partially plastic. Cuttings in rock slopes may also generate fracturing and further opening of insitu discontinuities aligned sympathetically to the excavation face as a result of stress relief. Weathering via these discontinuities might then accelerate weakening along joint fractures with time. On discontinuity surfaces exposed by wedge failure, decomposition of sandstones and siltstones forming the rock mass may be evident, with extremely weak, friable and fine, decomposed material penetrating 5 to 10mm inwards from joint faces. Rotational slumps may also occur on steeply-cut rock slopes, the surfaces of blocky debris and exposed discontinuity faces of backscars revealing intense decomposition. In the near-surface zone, a clayey gouge may sometimes be washed into open joints by percolating rainwater.

Pitts (1983) reported on slides at terraced rock slopes of 50 to 70 degrees inclination formed from ridges of folded silty mudstones, sandstones and conglomerates of the Tengah and Queenstown facies at NTI campus during heavy rainstorms of November 1982. He concluded that since the water table was generally low at slopes, failure took place within the vadose layer, and landslides were thus directly attributable to effects of heavy rainfall in the partially saturated zone. Generally similar observations as above have been made on stability of excavations in sandstones, siltstones, shaley mudstones and conglomerates of the Ayer Chawan, Rimau, Tengah and Queenstown facies at NUS campus on Kent Ridge. It is noteworthy that two cuts of about 20m height each stood virtually unprotected in the weakest formations typically found there at inclines of 55 to 60 degrees to the horizontal over a period of several months.

Method of Analysis

Goodman (1976) presented an analysis of sliding rock mass consisting of multiple free blocks using stereographic projection based on friction cone concept. Figure 6 shows the

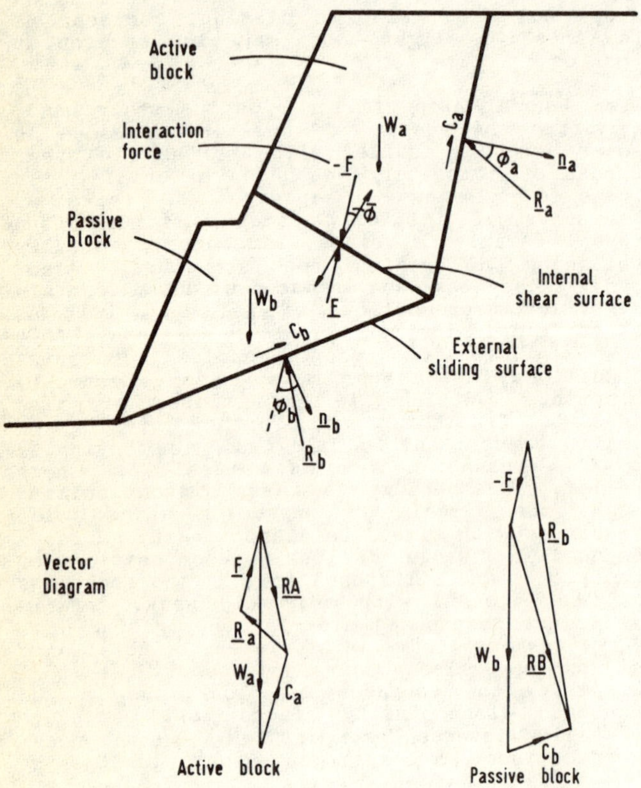

Fig. 6 Analysis of sliding of two blocks

basic case of two sliding blocks, one acting as a passive supporting block and the other acting as an active loading block. If the active block cannot be sustained by friction and cohesion on its base, there is a transfer of supporting interaction force from the passive to the active block along an internal shear surface, thus affecting the stability of the whole rock mass. The principle of two-block sliding can be extended to deal with problems of rock mass slide of multiple blocks.

Goodman's approach has been adopted in the present work. However, as his analysis was generally graphical in nature, it was rather time-consuming and prone to human error. In order to automate the analysis, an alternative approach involving the vectorial method was employed in the present work. A computer program was developed in BASIC language for this purpose and the entire analysis carried out on a micro-computer. The vectorial method is briefly summarised as follows.

The forces that act on each potential sliding block are self-weight, uplift water forces, external forces such as earthquake forces, rock-bolt and anchor forces etc, and cohesive forces. Suppose $\underline{RA}$ is the vector denoting the resultant of the forces acting on the active block, a transfer of supporting interaction force, $\underline{F}$, from the passive block is required in order that the active block be stable. The equations governing limiting equilibrium are

$$\underline{R}_a = \underline{RA} + \underline{F} \quad (1)$$

and

$$\hat{\underline{R}}_a \cdot \hat{\underline{n}}_a = \cos \phi_a \quad (2)$$

where

$\hat{\underline{R}}_a$ is the unit vector of the net resultant force $\underline{R}_a$,

$\hat{\underline{n}}_a$ is the unit vector of the normal of the external sliding surface of the active block,

and

ϕ_a is the angle of friction of the sliding surface.

Similarly, the equations governing limiting equilibrium of the passive block are given as

$$\underline{R}_b = \underline{RB} - \underline{F} \quad (3)$$

and

$$\hat{\underline{R}}_b \cdot \hat{\underline{n}}_b = \cos \phi_b \quad (4)$$

where

$\underline{RB}$ is the resultant of all the forces acting on the passive block.

The factor of safety for individual sliding blocks is defined as

$$F = \tan \phi / \tan \phi_r \qquad (5)$$

where ϕ_r is the angle of friction required for limiting equilibrium and ϕ is the available angle of friction. A solution is obtained by imposing the same value of safety factor for all the sliding blocks. A more detailed account of the analytical approach based on the vectorial method will be the subject of a separate publication.

Back-Analysis of Collapsed Slopes

A back-analysis was performed in order to establish the strength parameters of the rock mass. For present purposes, the external sliding surfaces were chosen such that they followed closely slip surfaces observed at the site. On the other hand, the weakness planes obtained from geological surveys were taken as internal shear surfaces. It was found that the four slope sections examined consisted of sliding involving one to three free blocks, as illustrated in Fig. 7. By correlating strength parameters with clay content and the plasticity index of weathered rocks the apparent cohesion and angle of friction of weathered sandstone were found to be 6 kNm^{-2} and 32° respectively. On the other hand, the apparent cohesion of shale was found to be within a relatively small range of 20 to 25 kNm^{-2} while the angle of friction varied between 20° and 26°. It was evident that the shale had softened during the heavy rainfall before the slope collapse. Back-analyses were then carried out on the basis of these parameters, as a result of which factors of safety varying between 0.97 and 1.05 were obtained for the four slope sections.

For the purpose of comparison, conventional slope stability analysis using Janbu's (1954) non-circular slip surface approach also was employed. Similar slip surfaces for the four slope sections were used for both methods of analysis. Using the same strength parameters as those for the vectorial method, the factors of safety obtained for the four sections using Janbu's method are given in Table 2. As the values of safety factor by the latter method varied significantly from 0.87 to 1.34 for the four collapsed sections, it was evident that the vectorial approach was more appropriate in handling the present case study involving slides of slopes in weathered rock formations. Had circular slip surfaces been adopted as in common practice, the departures from a factor of safety of one would have been even greater.

CONCLUDING REMARKS

As discussed above, a major cause for slope instability of the Jurong formation has been unrestricted ingress of groundwater into potential slip zones during heavy rainstorms. This condition may be further aggravated by the existence of complex geological structures, degradation by hydrothermal alteration and continual weakening by agents of weathering in an agressive climatic regime. Nevertheless, it has been found that a nominal amount of protection afforded to excavation faces can lead to substantial improvement of slope

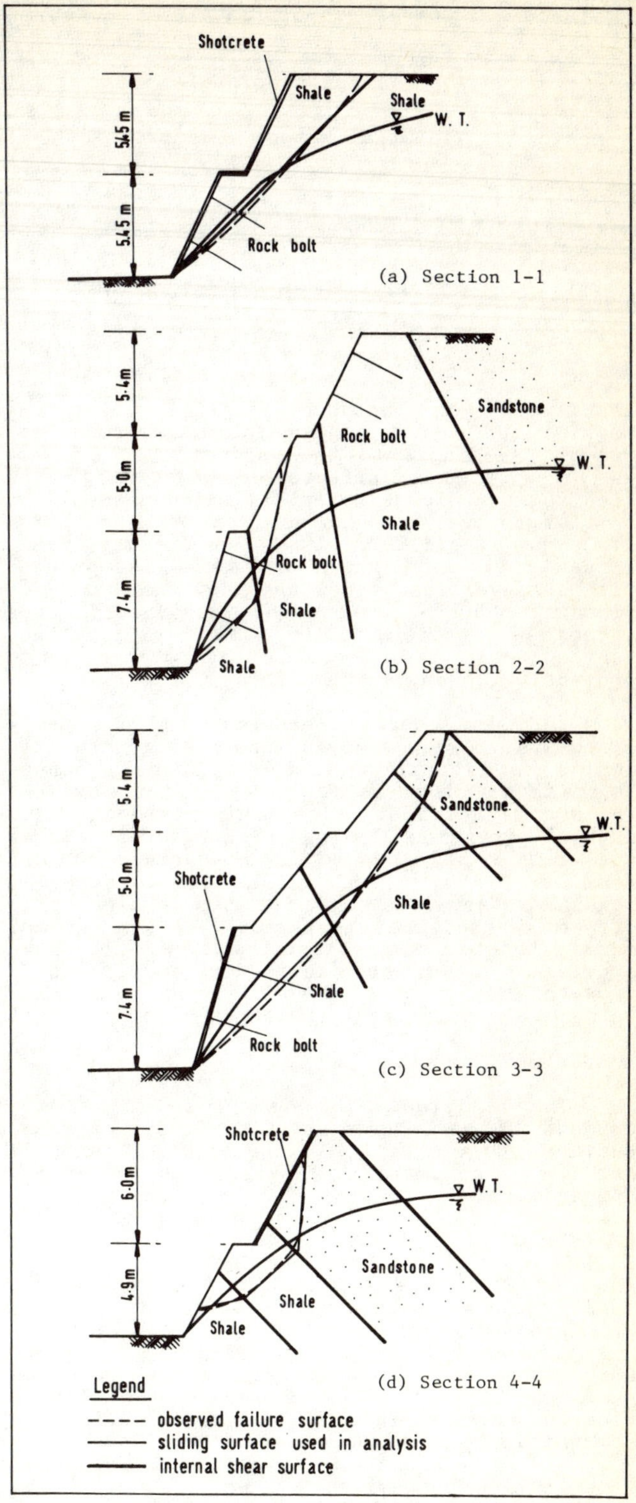

Fig. 7 Slope stability analysis

Table 2 : Comparison of Factors of Safety

Section	Factor of Safety	
	Vectorial Method	Janbu's Method
1-1	1.04	1.34
2-2	1.05	1.06
3-3	0.97	0.87
4-4	0.98	0.89

stability. For instance, vegetal cover may provide sufficient buffer to minimise surface run-off seeping into the vadose layer as well as other adverse effects of weathering, and bind together loose materials which present an initial source of weakness from the standpoint of slope stability. In these circumstances, slopes have been known to stand unsupported with near-vertical faces provided that no major flooding occurred. With the introduction of a layer of shotcrete in lieu of surface vegetation, considerable improvement in protective cover may be afforded.

For the above measures to be truly effective, however, any build-up of groundwater behind an excavation face should be avoided. Significant gains in factor of safety against failure may be achieved by fairly modest means such as provision of lined drainage channels at the crest of a slope to divert surface run-off from its vicinity, and drainage outlets set at regular intervals along a shotcrete blanket to off-set groundwater seepage. For deeper and more extensive excavations, however, the above provisions are likely to require supplementary rock-bolting or other form of ground anchoring to maintain stability of steeply-sloping faces.

ACKNOWLEDGEMENT

The assistance of Miss C Y Goh, former student of the Department of Civil Engineering, National University of Singapore, in carrying out the slope stability analysis is gratefully appreciated.

REFERENCES

Anon, 1970. The logging of rock cores for engineering purposes. Geological Society Engineering Group Working Party Report. Qtly Jour Engng Geol. 5: 297-367.

Cox, D.W. 1984. Discussion on 4th Geotechnique Symp. in print. Geotechnique. 34(2): 143-6.

Goodman, R.E. 1976. Methods of geological engineering. West Publising Co., St. Paul.

Hoek, E. & J. Bray 1981. Rock slope engineering (revised 2nd ed). London: The Institution of Mining and Metallurgy.

Janbu, N. 1954. Application of composite slip circles for stability analysis. Proc. European Conf. on Stability of Earth Slopes, Stockholm. 3: 43-49.

Pitts, J. 1983. The form and causes of slope failures in an area of West Singapore island. Singapore Jour Tropical Geography. 4(2): 162-168.

Provisional Mass Rapid Transit Authority, Singapore 1983. Detailed geotechnical study: interpretative report.

Public Works Department, Singapore 1976. Geology of the Republic of Singapore.

Utada, M. 1978. Hydrothermal alteration appearing in the Neogene sediments of Higashi-Aizu, minealisation area. North-east Japan Mining Geology. 28: 83-97. (In Japanese).

Soil nailing for slope stabilization in Singapore residual soils

S.B.TAN, Public Works Department, Singapore
S.L.TAN, Public Works Department, Singapore
Y.K.CHIN, Public Works Department, Singapore

SYNOPSIS: Over the past two decades, the rapid urbanisation of Singapore coupled with the intense utilization of scarce land has resulted in steep cutting and terracing of hillslopes. Under the continual influence of wet tropical climate, some of these cuttings have become gradually unstable. Although a number of methods can be used to stabilize residual soil slopes, the technique of soil nailing has gained increasing popularity over the last few years. This paper focusses on the application of soil nailing technique in Singapore. The installation and design aspects will be discussed and some practical applications of soil nailing will be highlighted. Some instrumentation results on the performance of the soil nails will also be presented.

INTRODUCTION

A number of methods have been used in the stabilization of slopes in Singapore. These include regrading of slopes, use of retaining walls, drainage improvement, anchored walls and the more recent soil nailing technique. In Singapore, the earliest known application of soil nailing was perhaps carried out in 1975 for the stabilization of an earth slope in a military camp at Choa Chu Kang. The soil nail system comprised a 50mm thick wire-mesh reinforced gunite facing held in place by 3 to 4m long 16mm diameter high tensile reinforcement bars grouted into holes spaced at 3.0m c/c. More recently, the technique was also used to stabilize a number of cut slopes (eg. slope behind caltex petrol station at Alexandra Road and slope above contiguous bored-pile wall at Bukit Teresa Close). At Gillman Heights, a high cut slope formed for the construction of the Ayer Rajah Expressway was also protected by soil nailing system. The height of the slope varies from 11.5 to 17.0m and the total length is about 215.0m. The details of the soil nailing projects have been reported by Tan et al (1985, 1987).

Several factors have contributed to the growing popularity of soil nailing as a slope stabilization technique, and these include:

(i) Ease of installation - the nails can be installed from stagings erected on the cut slope or can be phased in with the cutting of the slope.

(ii) Minimal disturbance to the existing slope and adjoining properties.

(iii) The layout and spacing of the nails can be varied easily to suit various shapes and heights of slopes.

(iv) The soil nails can be connected at the surface by a system of RC grid beams which apart from providing restraining action against deep and shallow failures, can also allow for turfing of the slope. The finished slope can therefore be made aesthetically pleasing.

(v) This technique is relatively attractive from the cost point of view.

Two soil nailing projects will be highlighted in this paper namely, the stabilization of a large slope failure at Lorong Terigu and the stabilization of the cut slope at the Singapore General Hospital (part of the Central Expressway project). Before proceeding with details of the project, the design methodology for soil nailing will be examined in some depth.

DESIGN METHODOLOGY

The design of a soil nail structure is a fairly complex problem as the nail is subjected to the action of shear, bending and pull-out. As the bending capacity of a bar is also related to shear and pull-out, the full solution to the problem would involve tedious mathematical derivations. A simplified approach is to consider only the action of pull-out and ignore any shear contribution eg. Shen (1981) and Gudehus (1981). A more general solution taking into account both the soil-reinforcement friction and the mobilization of the lateral earth thrust on the reinforcement, has been developed by Terrasol (1980). The solution satisfies global equilibrium but internal equilibrium is ignored. In PWD, a computer program basing on Bishop's simplified method has been developed to analyse the stability of nailed slopes using the multi-failure criterion proposed by Blondeau et al (1984). A brief review of the concept and approach is described below.

The equations are derived by using the Bishops's simplified method. The contribution from the soil nail is considered as a resultant force acting at the mid-point of the base of the slice. Though the soil nail intersect other slices, it is assumed that internal forces at the vertical slice boundaries are not duly affected by the nails. This assumption may appear simplistic but is necessary in view of the complexity of the problem. The forces acting on the slice is shown in fig 1 below.

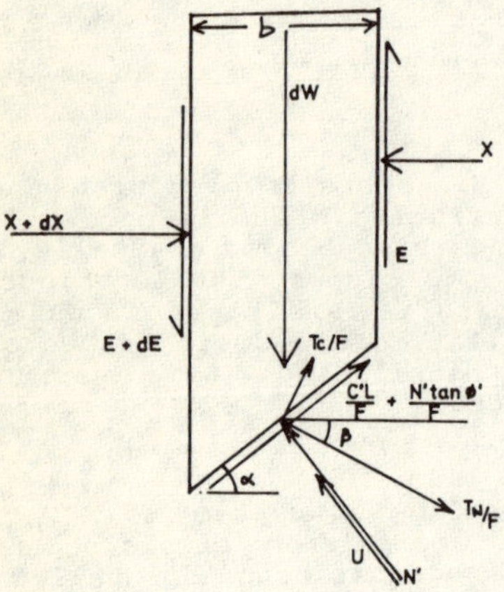

Fig 1. Forces acting on a slice.

In deriving the equation for the overall factor of safety of the slope, different partial safety factor can be incorporated for shear and pull-out failures of the soil nail. The partial safety factors can be predetermined (usually between 1.5 to 2) to control the allowable deformation of the soil nail system. Generally, a partial safety factor of 2 is assumed for passive failure of soil due to lateral movement of the nail. For pull-out failure of the nail, the partial safety factor should be the same as the overall factor of safety. The reason being that the nature of pull-out failure for the soil nail is similar to that of shear failure along the slip surface (both involving soil-soil slippage). For simplicity, a common factor of safety can be assumed for shear and pull-out failure of soil nail and slip failure. If this is done the following equations are obtained.

Resolving vertically,

$$N' = \frac{1}{m\alpha}\left[dW + dX + \frac{T_N}{F}\sin\beta - \frac{T_C}{F}\cos\beta - ub - \frac{c'b}{F}\tan\alpha\right]$$

$$\text{where } m\alpha = \cos\alpha\left[1 + \frac{\tan\phi'}{F}\tan\alpha\right]$$

Equating moments about the centre of circle gives:

$$F = \frac{\sum[T_C\sin(\alpha+\beta) + T_N\cos(\alpha+\beta) + c'b\sec\alpha + N'\tan\phi']}{\sum dW\sin\alpha}$$

Substituting for N' and simplifying:

$$F = \frac{\sum c'b + (dW - ub)\frac{\tan\phi'}{m\alpha}}{\sum dW\sin\alpha} + \frac{\sum\left[\frac{T_N}{F}\sin\beta - \frac{T_C}{F}\cos\beta \quad \frac{\tan\phi'}{m\alpha}\right]}{\sum dW\sin\alpha}$$

$$+ \frac{\sum(T_C\sin(\alpha+\beta) + T_N\cos(\alpha+\beta))}{\sum dW\sin\alpha} \qquad (1)$$

It is seen that the first term is the basic Bishop's equation and the remaining terms are contributions from shear and pull-out.

If $\beta = 0$ and $T_C = 0$ (as in R.E. Structures)

$$F = \frac{\sum\left[c'b + (dW - ub)\right]\frac{\tan\phi'}{m\alpha} + \sum T_N\cos\alpha}{\sum dW\sin\alpha}$$

If the shear force is factored by a different factor of safety say 2, the derived equation is slightly different.

$$F = \frac{\sum c'b + (dW - ub)\frac{\tan\phi'}{m\alpha} + \sum\left[\frac{T_N}{F}\sin\beta - T_C^*\cos\beta\right]\frac{\tan\phi'}{m\alpha} + \sum T_N\cos(\alpha+\beta)}{\sum(dW\sin\alpha - T_C^*\sin(\alpha+\beta))} \qquad (2)$$

$$\text{where } T_C^* = \frac{T_C}{2}$$

So far, no mention has been made with regards to the values of T_C and T_N. To evaluate these values, the multi-failure criterion proposed by Blondeau et al is used.

(i) Strength of inclusions

Assuming that the soil nail material obeys the Tresca type criterion, the following equation is obtained:

$$\left(\frac{T_N}{R_N}\right)^2 + \left(\frac{T_C}{R_C}\right)^2 \leq 1 \qquad (3)$$

where R_N = ultimate pull-out capacity of the bar (assuming grout has no tensile capacity).

R_C = ultimate shear capacity of grout and bar. For simplicity, $R_C = \frac{R_N}{2}$.

T_N, T_C = mobilized pull-out and shear resistances.

The equation (3) defines an ellipse in (T_N, T_C) space.

(ii) Nail pull-out capacity

The maximum value of T_N can be obtained from a pull-out test on site or alternatively can be estimated from soil properties.

$$T_N \leq \pi B L_a f_s \qquad (4)$$

where B = diameter of nail

L_a = effective adhesion length beyond the failure surface.

f_s = maximum shaft adhesion obtained from pull-out test or equal to αC_u of the soil if the former is not available.

α = value varying between 0.4 to 1 depending on the consistency of the soil. In stiff residual soils in Singapore, a value of 0.7 is reasonable.

The equation (4) defines a vertical line in (T_N, T_C) space.

(iii) <u>Maximum shear resistance of nail</u>

In determining the maximum shear resistance that can be mobilized by the soil nail, two cases are considered, namely the shear mobilized at the point of yielding of the soil or at the point of yielding of the nail. If one considers the yielding of the soil to occur first before yielding of the bar, then the maximum shear resistance that can be mobilized is:

$$T_C = \tfrac{1}{2} B\, l_o\, P_L$$

where l_o = transfer length = $\sqrt[4]{\dfrac{4EI}{K_S B}}$

and P_L = limit pressure at which the soil yields. This can be taken to be the P_L value obtained from the pressuremeter test or approximated to $9C_u$ (Broms 1964)

K_S = coefficient of subgrade resistance of the soil in kN/m^3.

The above equation assumes that the bar is infinitely long in relation to the transfer length l_o. For a typical soil nail of size 100mm with a 28mm diameter steel bar installed in medium dense silt with $K_S B$ = 30,000 kN/m^2, the value of l_o works out to be about 0.3m. This confirms that the long nail assumption is valid.

If the nail yields first before yielding of the soil, it is assumed that a plastic hinge forms at that point and the bar rotates further about that point until the yielding of the soil takes place.

The relevant equation derived by Blondeau et al (1984) is given as

$$T_C = 0.239\, B\, l_o\, P_L + 1.62\, \frac{M_{max}}{l_o} \quad ..(5)$$

Here, the value of M_{max} is assumed to be affected by pull-out and can be shown to be approximately equal to $[1 - (T_N/R_N)^2]\, M_p$ where M_p is the plastic bending moment of the nail. The above equation represents a parabola in (T_N, T_C) space.

All the above criteria can be summarised in the fig 2 for a typical soil nail of size 100mm, with a 28mm diameter bar and a medium dense silt with C_u = 80 kN/m^2 and $K_S B$ = 34,000 kN/m^2.

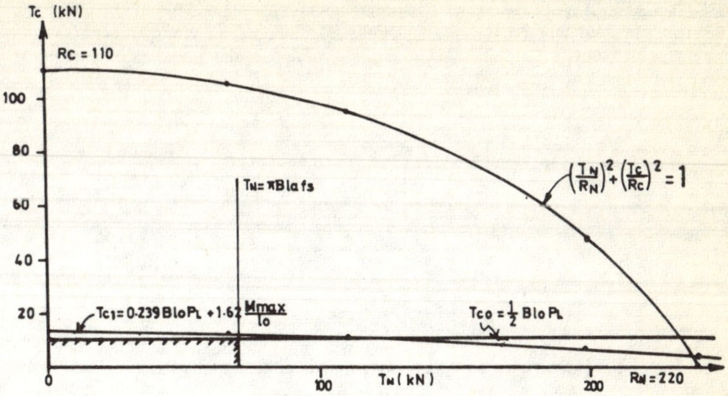

Fig 2. Soil nail failure criteria.

The hatched surface represents the failure envelope taking into account the various criteria. To define the maximum values of shear and pull-out, Blondeau uses the principle of maximum work and obtains the value of T_N, T_C at the point of tangency as shown in the fig 3.

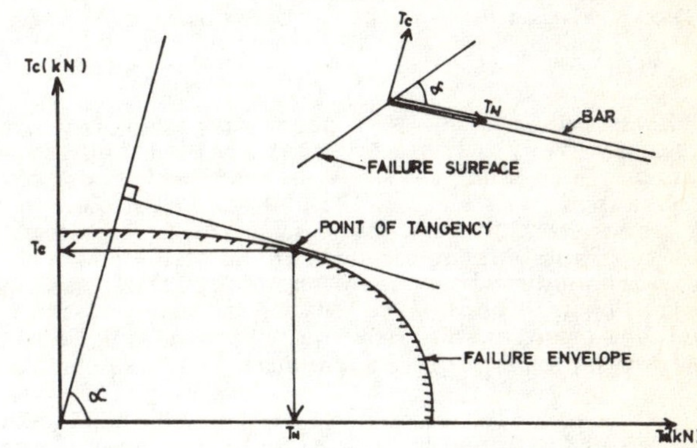

Fig 3. Determination of T_C and T_N.

These values of T_C and T_N are then used to calculate the overall factor of safety [equation (2)]. From the computer programming point of view, a simple routine can be written to generate the various failure criteria and to search for the critical failure envelope. The point of tangency is located by comparing the gradients at discrete points on the failure envelope with the angle $(90° - \alpha)$. The basic computer program for slope stability analysis using Bishop's simplified method can be easily modified to incorporate the soil nail failure criteria.

PROGRAM VERIFICATION

The computer program developed at the PWD can be verified by analysing a simple example of a vertical cut with two soil nails as shown in fig 4. Two cases are checked ie. with and without contribution from shearing of the nail. Mathematically, the FOS for the above cases is given in equations (6) and (7).

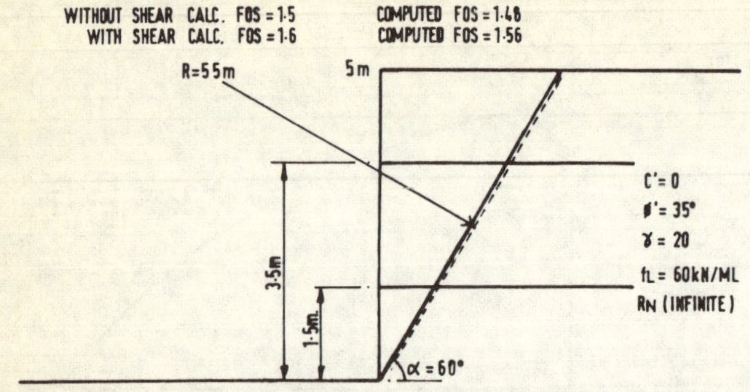

Fig 4. Test example for the PWD program.

Without shear

$$FOS = \frac{W\cos\alpha\tan\phi' + \Sigma T_N \sin\alpha\tan\phi'}{W\sin\alpha - \Sigma T_N \cos\alpha} \qquad (6)$$

With shear

$$FOS = \frac{W\cos\alpha\tan\phi' + (\Sigma T_N \sin\alpha - \Sigma T_c \cos\alpha)\tan\phi'}{W\sin\alpha - \Sigma T_N \cos\alpha - \Sigma T_c \sin\alpha} \qquad (7)$$

For simplicity, R_N is assumed to be infinite and soil yielding will occur first.

ie. $T_C = \frac{1}{2} B l_O P_L$.

For both cases, the PWD program gives factors of safety very close to that calculated from equations (6) and (7). A better solution can be obtained if the radius R can be further increased so that the failure surface approximates a plane surface. It can be seen that the shear component increases the factor of safety marginally by about 7 to 10%. It seems that an acceptable solution can be obtained by only considering the pull-out capacity of the nail.

The next example examined is the EPARIS wall analysed by Blondeau et al (1984) using the TALREN program. The wall details are summarised in the fig 5 below.

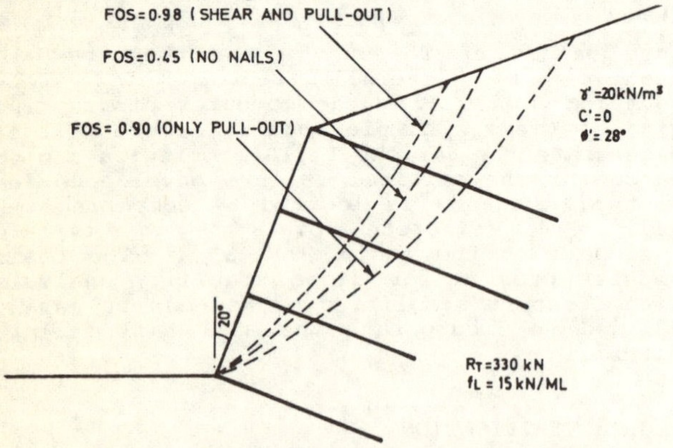

Fig 5. EPARIS Wall.

For the case of no nails, the PWD program gives a lower factor of safety of 0.45 as compared to 0.6 given in the paper. With only pull-out capacity, the PWD program gives almost the same value for FOS. As the paper does not provide information on the size and type of nail, it is assumed that the nail is 100mm in diameter with a 22mm high tensile steel bar. The factor of safety obtained by the PWD program considering both pull-out and shear is also very close of the corresponding value given in the paper.

SOIL NAILING IN THE CENTRAL EXPRESSWAY PROJECT

In the construction of Central Expressway a steep slope was formed near the Singapore General Hospital in residual soil of Jurong Formation. Although the slope will be stable in the short term, delayed failure in the long term due to wetting and softening effect is expected. It is therefore necessary to protect the slope against long term rain-induced failure. Soil nailing technique was chosen because of ease of construction, aesthetic consideration and economy.

Fig 6 shows the site plan of the 72m slope which was stabilized by 8m long soil nails spaced at 2m c/c. Surface grid beams of size 300x300m were used to tie up the soil nails to form an integral system. Fig 7 shows the cross section of the site and the soil profile obtained from boreholes 1 and 2. The soil at the site consists essentially of medium dense to dense silt of the Jurong Formation. The results of laboratory and field pressuremeter tests carried out on the soil are summarised in fig 8. It is interesting to note that the laboratory determined undrained modulus is higher than the field pressuremeter values and gives a E_u/C_u ratio of between 200 to 300. As suggested earlier, the limit pressure (PL) under passive failure can be approximated to 9 C_u. For the average measured C_u value of 100 kN/m^2 for the slope materials, the limit pressure can be taken to be about 900 kN/m^2. This value is comparable to the P_L values obtained from the pressuremeter test. With a plasticity index of between 20 and 23, the effective friction angle of the soil generally lies between 30 and 32°. From our past experience on similar soils, the effective parameters applicable for such soils would be $C' = 10$ kN/m^2 and $\phi' = 32°$.

Three sections of the slopes (see fig 6) were instrumented for the purpose of determining the distribution of internal forces in the soil nail. The detail of the instrumented soil nail is shown in fig 9. Vibrating wire strain gauges are bonded on either side of the reinforcement bar. The bar orientation is such that any bending of the nail in a plane orthogonal to the slope would be picked up by the strain gauges. The nails were instrumented at three locations along their length. Altogether six nails were instrumented with 2 nails per slope section. A typical plot of the measured stress with time in the reinforcement is shown in fig 10. As can be seen from the figure, the mid level strain gauges register consistently higher strains in all the six nails.

To determine the maximum shear resistance f_s, a pull-out test was carried out. Two nails were tested, one sleeved with a circular PVC pipe throughout the length of the reinforcement (unbonded) and the other a normal soil nail.

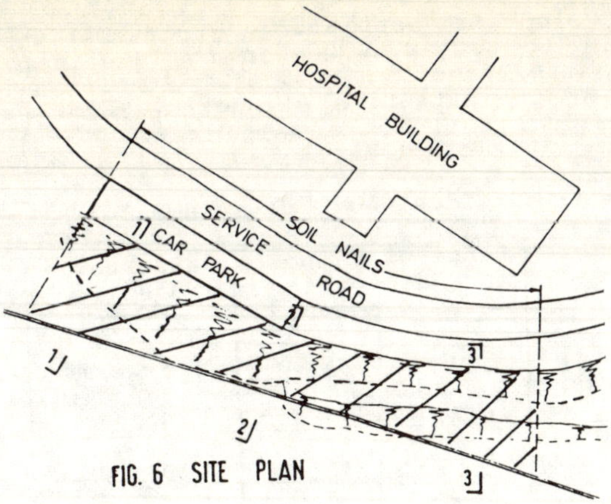

FIG. 6 SITE PLAN

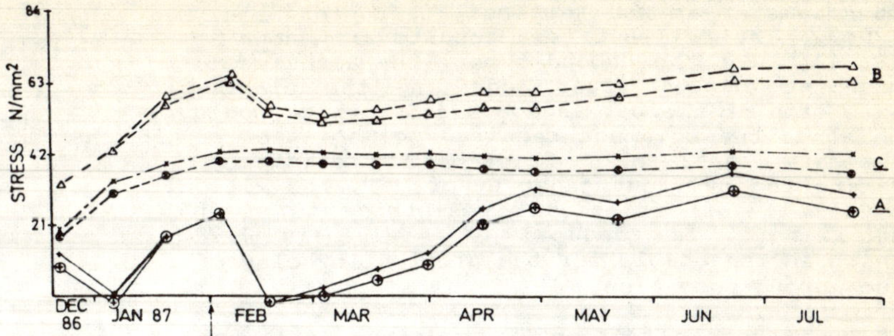

FIG. 10 VARIATION OF STRESS WITH TIME IN STEEL BAR FOR SOIL NAIL NO. 2

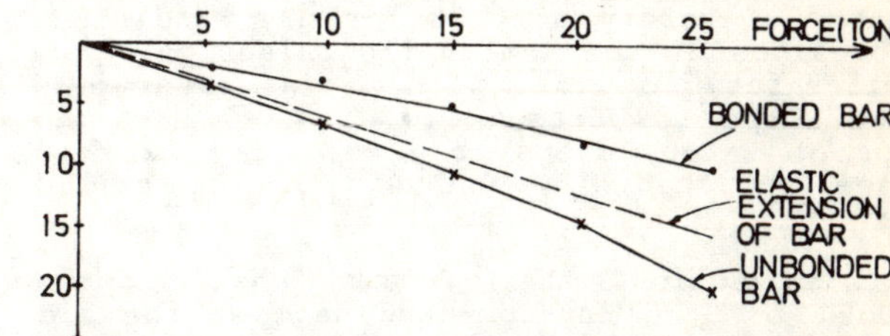

FIG. 11 SOIL NAIL PULL-OUT MOVEMENT CURVE

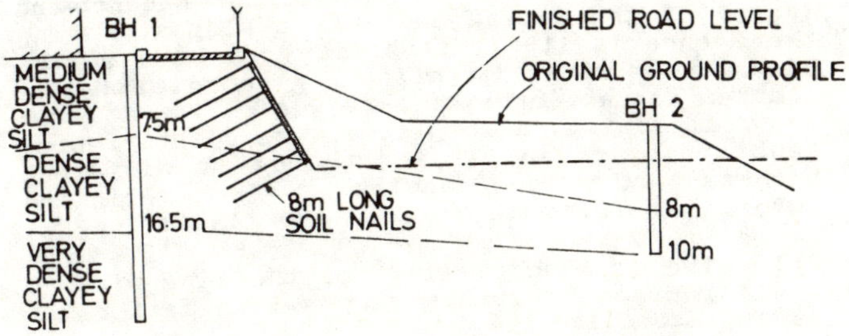

FIG. 7 CROSS SECTIONAL PROFILES

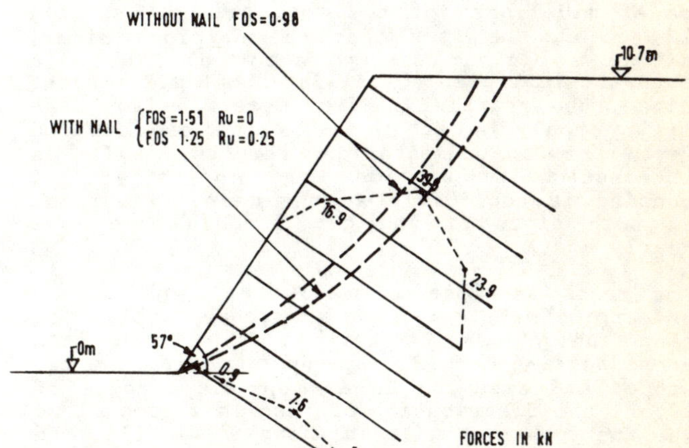

FIG. 12 CROSS-SECTION 2

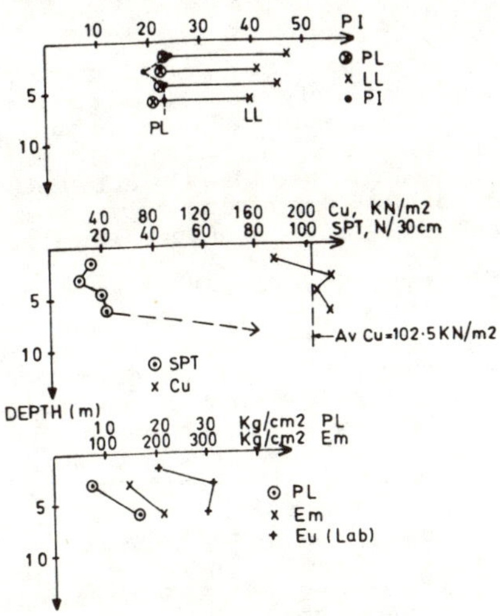

FIG. 8 SUMMARY OF SOIL PROPERTIES (SGH)

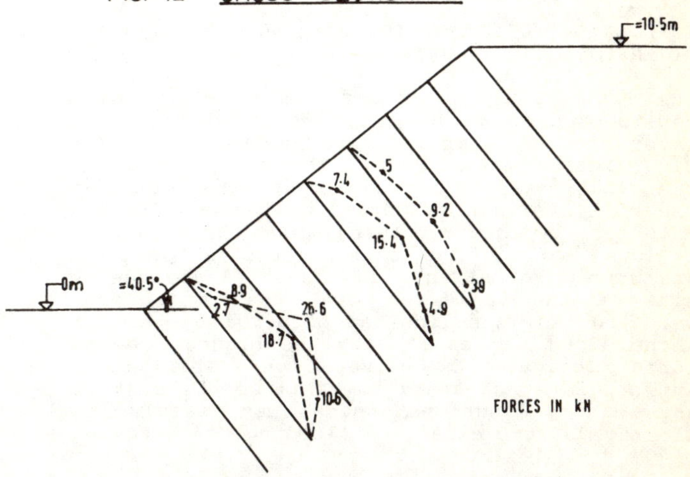

FIG. 13 COMBINED CROSS-SECTION 1 & 3

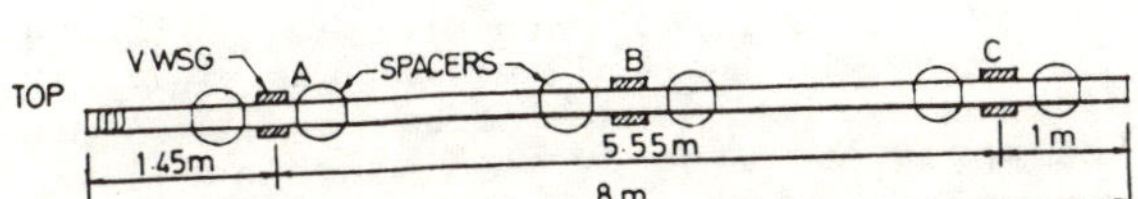

FIG. 9 TYPICAL DETAIL OF INSTRUMENTED SOIL NAIL

The diameter of the soil nail is approximately 125mm. The pull-out test results are shown in fig 11. For the unbonded bar, the actual soil movement at 25T after deducting the elastic extension of the bar is about 4.5mm which would be about the movement required to mobilize the ultimate nail-soil frictional resistance. Taking the pull-out capacity of 25T, the ultimate nail-soil frictional resistance is about 80 kN/m^2. The deduced α value is 0.78 which is reasonable for the stiff silty clay of the Jurong Formation.

At the steepest section 2, the slope angle and height are about 57° and 10.7m respectively. Without nails, the stability analysis assuming a critical circle passing through the toe gives a factor of safety of 0.98 for dry condition. The presence of the soil nails raises the factor of safety to 1.51. If partial saturation of the slope is assumed (ie. r_u = 0.25), the factor of safety drops to 1.25. Fig 12 summarises the stability analysis of the nailed slope.

For this section, the maximum tensile force recorded for the instrumented nail at the mid height of the slope is about 40 kN. From the distribution of tensile force along the nail and considering internal equilibrium of the bar, the potential sliding surface appears to be at mid-length of the soil nail. From fig 12, it can be seen that the critical circle giving a minimum safety factor of 1.51 also cuts the instrumented nail at about mid length. Hence, the critical surface determined by stability analysis is close to the one deduced from the instrumentation results. It is interesting to mention that the effect of bending is negligible as the measured strains at both sides of the bar at each level are nearly equal.

For cross sections 1 and 3, the slope angles are approximately 39° and 42° respectively. The slope height varies between 10 to 11m from cross section 1 to 3. By superimposing the two slopes and assuming an average slope angle of 40.5°, the distribution of tensile forces along the instrumented nails is shown in fig 13. The results suggest that the critical sliding surface is probably non-circular and roughly parallel to the slope.

SOIL NAIL PROJECT AT LOR. TERIGU - BUKIT TIMAH DIVERSION CANAL - PHASE I

The Bukit Timah Diversion Canal Phase I was constructed about 20 years ago to divert storm water from the Upper Bukit Timah Catchment to Ulu Pandan Canal. Due to the construction of the canal, high and steep slopes were formed in the hilly terrain traversed by the canal. The main soil types encountered along the canal are residual soils of Bukit Timah Granite Formation and Jurong Formation with recent fluvial deposits on the valleys. Over the years, many rain-induced landslips had occurred on the cut slope along the canal with the biggest one at Lorong Terigu. Extensive slope stabilization is therefore necessary to minimise the risk of any major landslips which may result in blockage of the canal and flooding. A detailed soil and slope investigation programme was carried out to gather data for the design of the slope stabilization measures. Statistical treatment of the effective strength parameters of the residual soil for a sample size N = 34 is shown in the table I below:

Table I: Residual soil data summary

	Range of values for 95% confidence interval	Value to be used in design
c': cohesion	9 to 14.8 kN/m^2	5 kN/m^2
Ø': friction angle	31 to 32.4°	30°
s : degree of saturation	91 to 92.5%	90%
n : porosity	0.46 to 0.52	0.5

Terrain mapping was also carried out to determine the characteristics (ie. the height and angle) of the failed and intact slopes. Using the Spencer's (1967) stability chart, a series of curves with different r_u conditions is plotted for a factor of safety of 1.2. This is shown in fig 14. Also plotted are points representing intact and failed slopes of different heights and slope angles. The following inferences can be drawn from the fig 14.

(i) The zone to the left of the line with ru = 0.5 represents stable slopes with factors of safety exceeding 1.2.

(ii) The zone between the line with ru = 0.5 and ru = 0.25 represents potentially unstable slopes if soil drainage condition is not improved. Generally, the height of existing slopes falling within this category vary from 4 to 10m.

(iii) The zone to the right of r_u = 0.25 represents potentially unstable slopes where 9 past failures had occurred including the large slide at Lorong Terigu. For slopes in this category, drainage measures will not be adequate and other measures for example soil nailing are necessary to ensure stability.

At Lorong Terigu, the failed slope is about 24m high and 100m in length. The slope at this location had suffered from a history of instability. A former deep seated slip had damaged the canal lining. The slope was subsequently stabilized by a row of sheetpile along the toe of the slope at the bank of the canal. Further slips occurred after the installation of the sheetpile wall. The most recent slip is a toe failure occurring above the sheetpile wall as shown in fig 15. The slip surface which was 5 to 6m at its deepest point was determined by Mackintosh probe. Back analysis using this slip surface gave values of c' = 12 kN/m^2 and Ø' = 32° for fully saturated slope. These values agree well with the laboratory test results given in table I.

The summary of the soil properties is shown in fig 16 from which it can be seen that the undrained shear strength increases with depth. For the upper 15m, the average undrained shear

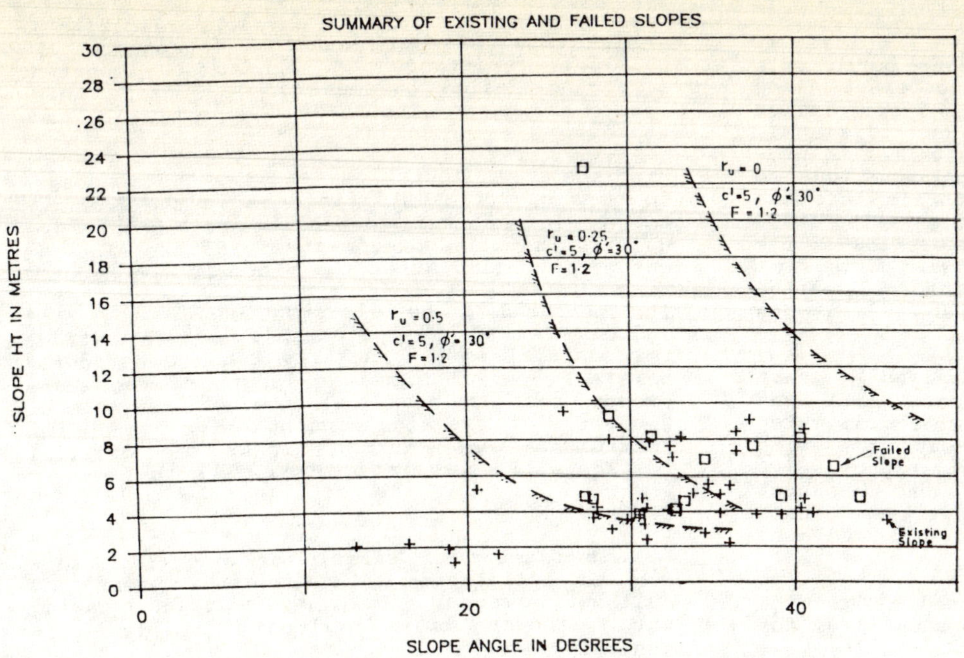

Fig. 14 PLOT OF SLOPE HT VS SLOPE ANGLE

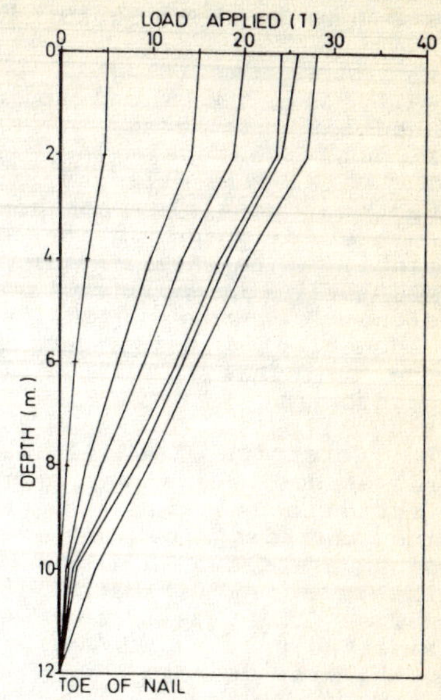

FIG. 18 LOAD DISTRIBUTION ALONG SOIL NAIL

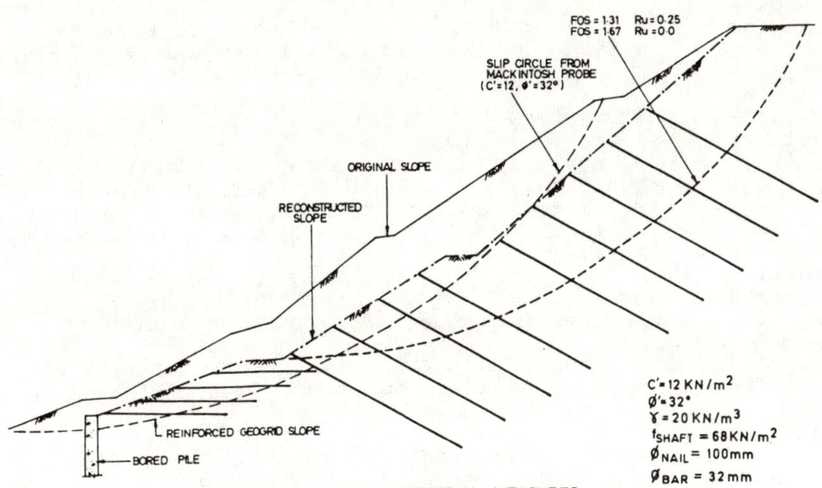

FIG. 15 SLOPE FAILURE AND REMEDIAL MEASURES AT LORONG TERIGU

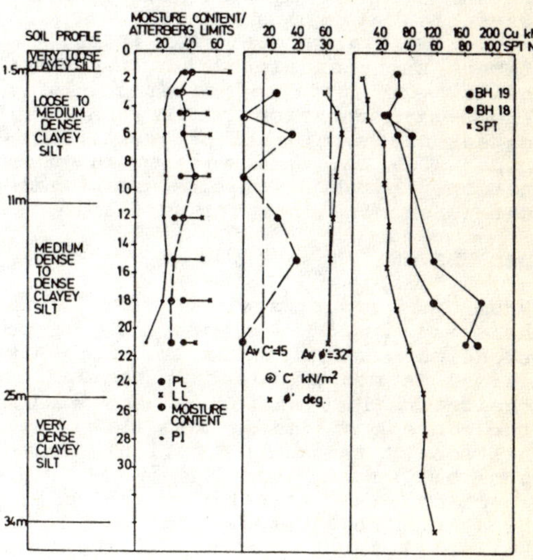

FIG. 16 SUMMARY OF SOIL PROPERTIES (LORONG TERIGU)

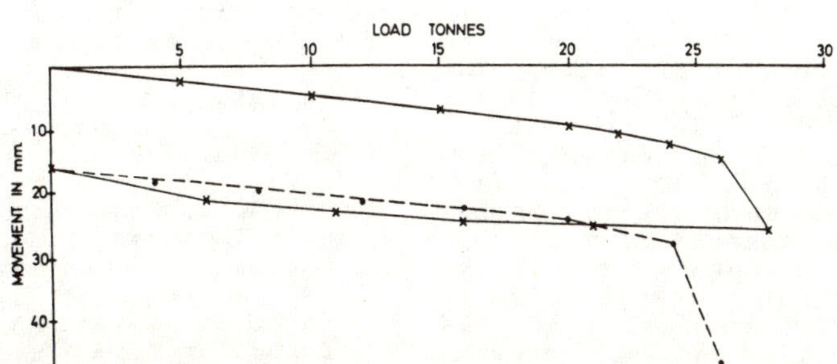

FIG. 17 INSTRUMENTED SOIL NAIL PULLOUT TEST AT LOR. TERIGU

strength is about 90 kN/m^2.

A system comprising of contiguous bored pile wall, soil nails and soil reinforcement using geogrids was used to stabilise the failed slope. The repair was carried out in stages. The bored pile wall was first installed to prevent any possible chokage of the canal due to further slippage of the slide debris into the canal. The soil nails were installed subsequently to stabilize the upper portion of the slope following which, the lower slope was reconstructed using geogrid reinforced earth. The soil nails were inclined at 30° to the horizontal.

To determine the ultimate nail-soil shaft resistance, an instrumented soil nail was tested to failure. The load deflection curve and the distribution of pull-out load with depth are shown in figs 17 and 18 respectively. The measured ultimate pull-out load is about 25T. This gives an average ultimate shaft resistance of 66 kN/m^2, from which an α value of 0.73 is deduced.

Using the PWD program, the computed FOS for the design slope with soil nails is about 1.31 for $r_u = 0.25$.

To measure the performance of the soil nails, a detailed instrumentation programme was implemented. This involved instrumenting soil nails installed at varying angles ie. 10°, 20°, and 30°. Strain gauges were installed at 5 different sections along the nail length. As the nails were installed after the slope was cut, the instrumentation results have so far indicated only marginal increases in strains in the bar. The detailed results of the instrumentation programme would be published at a later date when more data are available.

DISCUSSION

From our experience with the application of soil nails, it is clear that the soil nailing technique is effective for stabilising slopes formed in medium to stiff residual soils. The choice of inclination of the nail depends on the concept of design. As shown in the earlier sections, the contribution from shear is generally small and the nails would tend to function more effectively in pull-out. For greater effectiveness, the inclination of the soil nail to the horizontal should be as small as possible. However, there are practical problems in installing horizontal soil nails. So far, nails of up to 20° have been successfully installed.

In Blondeau's multi-failure criterion, the values of T_N and T_C are determined at the point of tangency to the failure envelop. It is obvious from the typical soil nail example that as long as the failure surface cuts the nail at an angle less than 90°, the mobilized value of T_N is generally high. Even for the case where the failure surface is perpendicular to the nail, the lateral deformation of the nail would result in a slight change in angle between the failure surface and the nail. This would imply that the point of tangency moves rapidly along the failure envelope resulting in a high value of mobilised T_N. In essence, the pull-out resistance is the most important governing design parameter. Solutions which ignore the shear effect would generally be adequate.

From the instrumentation results, there appears to be a significant difference in the measured forces in the nail depending on whether the nail is installed progressively as the slope is cut or installed after the slope has been cut.

The measured force distribution in the nail indicates that the potential failure surface for residual soils generally tend to be nearly parallel over the greater length of the slope except perhaps near the toe of the slope. However, it remains inconclusive from the instrumentation results whether the critical surface will emerge at or below the toe.

CONCLUSIONS

(1) The factor of safety for a nailed slope can be calculated by considering shear and pull-out effects. However, solutions taking into account only pull-out capacity are generally adequate.

(2) As pull-out capacity is the most important consideration in the design of nailed slopes, the angle of inclination of the nail should be as horizontal as possible.

(3) The method of soil nailing is found to be very effective for stabilizing medium to high slopes in residual soils.

REFERENCES

Blondeau F, Christiansen M, Guilloux A, Schlosser F, 1984. A Design Method for Reinforced Soil Structures, Proc of the Int. Conf. in In-situ Soil and Rock Reinforcement, Paris.

Broms, B.B., 1964. Lateral Resistance of Piles in Cohesive Soils, JSMFD, ASCE Vol 90 SM 2.

Gudehus, G&G Cartier 1981, Soil Nailing, some aspects of a new technique, Proc of the X conf of the ICSMFE, Stockholm.

Shen, C.K., Bang, S., Romstad, J.M., Kulchin, L. and Denatale, J.S. (1981). Field Measurement of an Earth Support System, JGED, ASCE, V.107, GT 12.

Spencer, E., 1967. A method of analysis of the stability of embankments assuming parallel inter-slice forces, Geotechnique XVII, No.1.

Terrasol, 1980. Dimensionnement des ouvrages en sol cloue, Programme Talren, Rapport Interne, Paris.

Tan S.B., Tan S.L., Yang K.S., Chin Y.K., 1985. Soil Improvement Methods in Singapore, Proc. of the 3rd Int. Geotechnical Seminar, S'pore 1985.

Tan S.B., Tan S.L., Lim T.L., Yang K.S., 1987. Landslide Problems and Their Control in S'pore, Proc. of the S.E.A. Geot. Conf., Bangkok.

Effect of root reinforcement on the stability of three fill slopes in Hong Kong

K.P.YIM, Geotechnical Control Office, Hong Kong Government, Hong Kong
L.K.HEUNG, Geotechnical Control Office, Hong Kong Government, Hong Kong
D.R.GREENWAY, Geotechnical Control Office, Hong Kong Government, Hong Kong

ABSTRACT: During the routine stability investigation of three existing vegetated fill slopes, the Geotechnical Control Office of the Hong Kong Government carried out studies to quantify the reinforcement effect of tree roots on the stability of the slopes. When the reinforcement effect was incorporated into the stability analyses, significant increases in factors of safety were obtained. Therefore, works required to upgrade the stability of the slopes to meet current safety standards were substantially reduced, and the existing pleasant cover of greenery was also preserved. This Paper presents the result of the root reinforcement studies undertaken for these slopes.

INTRODUCTION

Landslides have been a problem of great concern in Hong Kong for some time. Because of the scarcity of flat land , much development has taken place on the steeply sloping terrain. These factors in combination with the complex geology and the extremely high seasonal rainfall have caused serious landslide problems in the past. The geology has been well-described by Allen & Stephens (1971), and Lumb (1975) and Brand et al (1984) have described the landslide problems. Past practice in the design and analysis of slopes rarely considered the benefit to stability from vegetation. Usually, slope stabilization measures resulted in the removal of existing vegetation. For environmental reasons this practice is clearly undesirable, and now vegetation is being increasingly used as a surface protection for engineered slopes. Its cost is comparable with that of conventional surface protection measures, e.g. sprayed concrete.

In the past, it was difficult to quantify the effects of vegetation on slope stability, and the Geotechnical Control Office therefore initiated studies to quantify the effect of existing vegetation on some slopes. The three case studies described in this paper deal with the effect of root reinforcement on slope stability. Details of the slopes and tree species studied are summarized in Tables 1 and 2 respectively.

PRINCIPLE OF ROOT REINFORCEMENT

The role of vegetation in the stability and protection of slopes has been discussed by Gray & Leiser (1982), Greenway et al (1984), Lee (1985), Greenway (1987) and others. The various effects of vegetation on slope stability are listed in Table 3 and illustrated in Figure

Tree Species	Height (m)	Trunk Diameter (mm)	Root Tensile Strength Lower Bound Value (MPa)
(So Uk Estate)			
Acacia	10-12	250-540	10.6
Chinese Banyan	10-15	350-470	16.3
Candlenut	12-14	330-420	5.5
(Leighton Hill)			
Cinnamon Tree	13	130	30.0
Chinese Hackberry	7-12	180-560	20.0
Pond Spice	7-10	60-230	10.0
Camphora Tree	13.5	520	12.5
(La Salle Road)			
Acacia	8-12	75-300	17.1
Horsetail Tree	10-12	75-150	17.9

Table 2. Tree Characteristics

Slope Location	Slope Height (m)	Slope Angle (deg.)	Soil Type	Insitu γ_d (Mg/m^3)	Soil Shear Strength Parameters		Surface Protection Other than Vegetation
					C (kPa)	ø (deg.)	
So Uk Estate	10	34	Granitic fill	1.56	0	39	Chunam plaster
Leighton Hill	22	35	Granitic fill	1.40	0	35	Sprayed concrete
La Salle Road	25	35	Granitic fill	1.35	0	33	Chunam plaster

Table 1. Summary of slopes studied

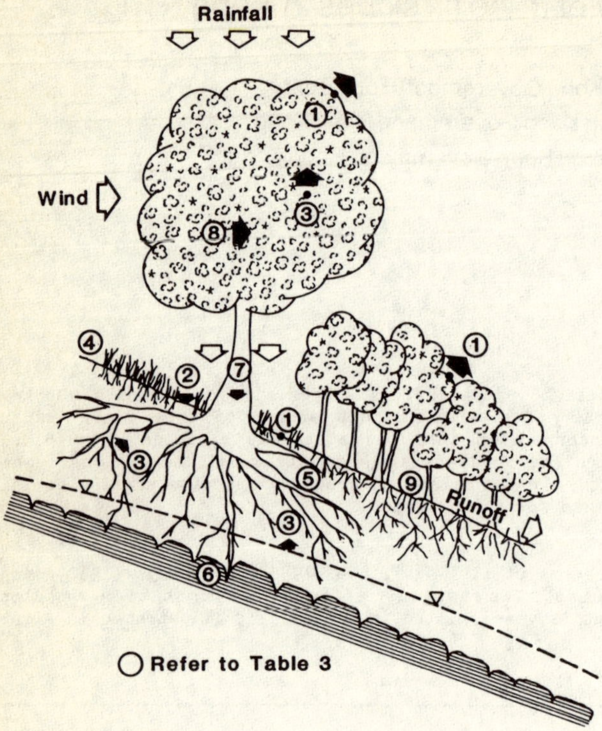

Figure 1. Slope-vegetation interactions influencing stability

Hydrological Mechanisms	Influence
1. Foliage intercepts rainfall, causing absorptive and evaporative losses that reduce rainfall valuable for infiltration.	Beneficial
2. Roots and stems increase the roughness of the ground surface and the permeability of the soil, leading to increased infiltration capacity.	Adverse
3. Roots extract moisture from the soil which is lost to the atmosphere via transpiration, leading to lower pore-water pressures.	Beneficial
4. Depletion of soil moisture may accentuate desiccation cracking in the soil, resulting in higher infiltration capacity.	Adverse
Mechanical Mechanisms	
5. Roots reinforce the soil, increasing soil shear strength.	Beneficial
6. Tree roots may anchor into firm strata, providing support to the soil mantle through buttressing and arching.	Beneficial
7. Weight of trees surcharges the slope, increasing normal and downhill force components.	Adverse/ Beneficial
8. Vegetation exposed to the wind transmits dynamic forces into the slope.	Adverse
9. Roots bind soil particles at the ground surface, reducing their susceptibility to erosion.	Beneficial

Table 3. Effects of vegetation on slope stability

1. One of the major mechanisms is root reinforcement.

The roots of plants tend to bind the soil together into a monolithic mass. On slopes, the root system penetrates through the surface soil mantle into firmer strata underneath, thus anchoring the surface soil to the slope and increasing its resistance to sliding. The mechanism can be explained by recent theories of reinforced fill.

A root-reinforced or fiber-reinforced soil behaves as a composite material. The reinforcements have relatively high tensile strength and the development of friction forces between the reinforcement and the surrounding matrix is the essential phenomenon that allows additional strength within the composite material to be mobilized. In other words, shear stresses in the soil mobilize tensile resistance in the reinforcement, which in turn imparts greater strength to the soil. The effect of the reinforcement on the stress-strain response of a soil is:

(a) to increase the shear strength of the soil, and
(b) to make the soil more ductile, i.e. able to resist continued deformation without significant loss of strength.

THEORETICAL MODELS OF ROOT-REINFORCED SOIL

A number of models have been proposed by various researchers for root reinforcement of soil (Gray & Leiser, 1982). The two models described by Gray & Ohashi (1983) are considered simple yet relatively comprehensive, and they were used in these studies. The two theoretical models, namely, the perpendicular root and inclined root models, are based on the following assumptions:

(a) A flexible, elastic root extends across a planar shear zone.
(b) The root may be initially oriented perpendicular to the shear plane, or at some arbitrary angle (i) measured between the root and the shear plane.
(c) The tensile force that develops in the root when the soil is sheared can be resolved into a tangential component, which directly resists shear, and a normal component, which directly increases the confining stress on the shear plane.
(d) The tensile strength of the root is fully mobilized, i.e. it fails in tension (breaks) rather than by pulling out.
(e) The soil friction angle ($\emptyset$) is unaffected by the reinforcement.

For a perpendicular root reinforcement model (see Figure 2), the shear strength increase in the soil as a result of the mobilization of tensile resistance in the roots is given mathematically as :

$$\delta SR = tR \ (\sin \theta + \cos \theta \tan \emptyset) \qquad (1)$$

where δSR = shear strength increase from roots, θ = angle of shear distortion, $\emptyset$ = angle of internal friction of soil, and tR = average tensile strength of roots per unit area of soil. A very useful relationship for determining tR is :

$$tR = TR \ (RAR) \qquad (2)$$

where TR = average tensile strength of roots, and RAR = root area ratio or fraction of soil cross-sectional area occupied by roots. This relationship provides a direct method for quantifying root reinforcement by simple measurement of the tensile strength of roots and the fraction of soil cross-sectional area occupied by roots.

Laboratory direct shear tests on various root-permeated soils conducted by Waldron (1977) indicated that the angle of shear distortion varied between 40 and 50 degrees. Wu (1976), in his observations of failures in root-permeated soils on slopes, discovered that the angle of shear distortion appeared to vary at most between 45 and 70 degrees. Sensitivity analysis of the perpendicular root reinforcement model (Equation 1) shows that the bracketed term is not particularly sensitive to the expected values of shear distortion angle (θ) and soil friction angle ($\emptyset$). The bracketed term in Equation 1 varies from 1.1 to 1.3 for $30° < \emptyset < 40°$ and $40° < \theta < 70°$ (see Figure 3).

In the field, roots may be inclined at many different angles to the shear plane. The inclined root reinforcement model (Figure 2) was therefore used to assess the effect of orientations other than perpendicular. The shear strength increase in this model is given by the equations :

$$\delta SR = tR \left[\sin(90 - \mu) + \cos(90 - \mu)\tan\emptyset\right] \quad (3)$$

$$\mu = \tan^{-1}\left[\frac{1}{k + (\tan i)^{-1}}\right] \quad (4)$$

where i = initial inclination of roots with respect to the shear plane, x = lateral displacement, z = thickness of shear zone, and k = shear distortion ratio (= x/z).

The result of a sensitivity analysis (Figure 4) shows that the maximum value of the bracketed term in Equation 3 occurs at a root inclination of around $45° + \emptyset/2$. This corresponds to an optimal inclination of about 60 degrees for a soil friction angle ($\emptyset$) of 30 degrees, and 65 degrees when $\emptyset$ = 40 degrees.

Gray et al (1982) and Greenway et al (1984) have indicated that the perpendicular root orientations assumed in Equation 1 may be equivalent to the random orientations of the roots that actually occur in the field. Both considered that the perpendicular reinforcement model would provide a satisfactory estimate of root reinforcement in soil. These studies adopted a similar approach, employing Equation 1 in the analysis.

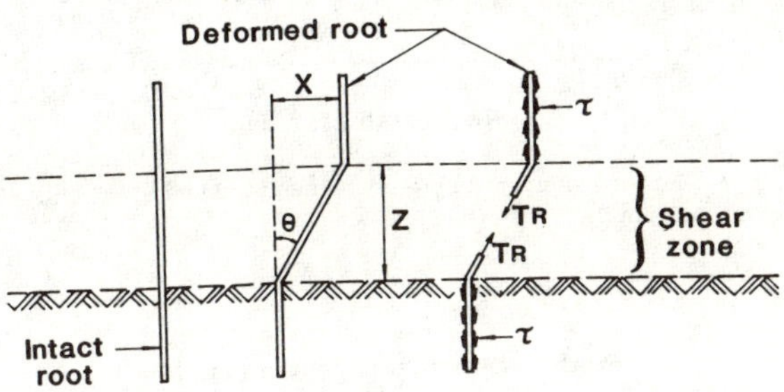

(a) Perpendicular orientation to shear surface

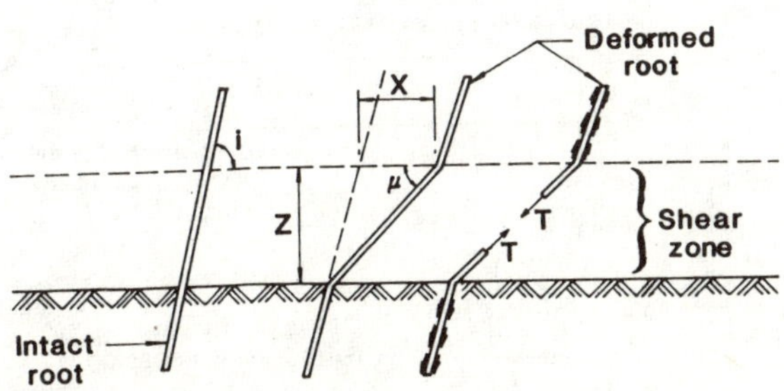

(b) Fiber orientated at angle (i) to shear surface

Figure 2. Root reinforcement models

CASE STUDIES

Case Study 1 - So Uk Fill Slope

The 10 m high, 34° slope was formed of poorly compacted granitic fill in 1958, as part of a public housing development. Trees of three different species grow on the slope, as shown in Table 2. Each tree species had been planted in a distinct area of the slope, so it was possible to isolate the roots of a single species in the subsequent investigation.

Nine hand dug trial pits, 1.4 m square in plan, were excavated to approximately 3 m deep midway between two or more trees of the same species. Each vertical face of the pit was examined over set depth intervals of 200 mm, and the number and diameter of roots were recorded. A root area ratio was computed for each depth interval to quantify the relative amount of roots present in the soil. The mean root area ratio from each trial pit is shown in Figure 5 as a function of tree spacing. In general, closer tree spacings yielded markedly higher root concentrations. The distribution of roots also varied with depth and tree species, as shown in Figure 6. The root area ratio was found to be greatest near the surface, owing to the many large roots (diameter > 10 mm) encountered there.

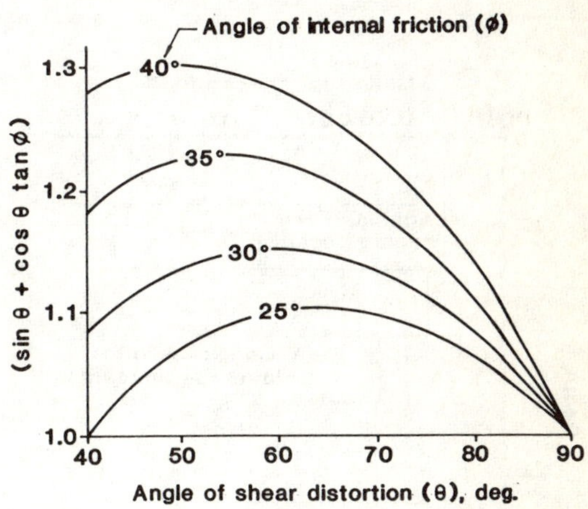

Figure 3. Range in value of bracketed term in equation (1)

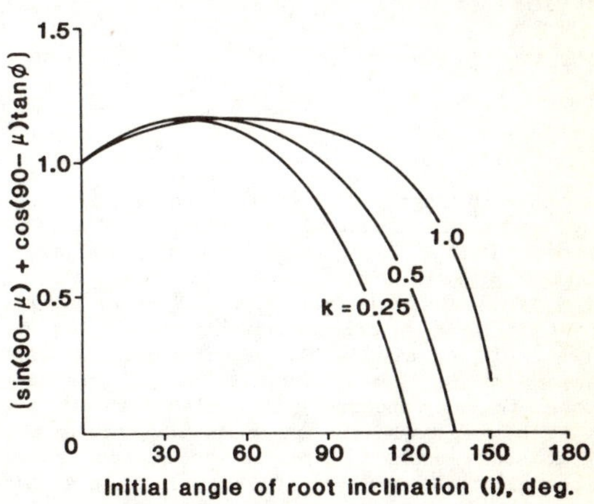

Figure 4. Range in value of bracketed term in equation (3)

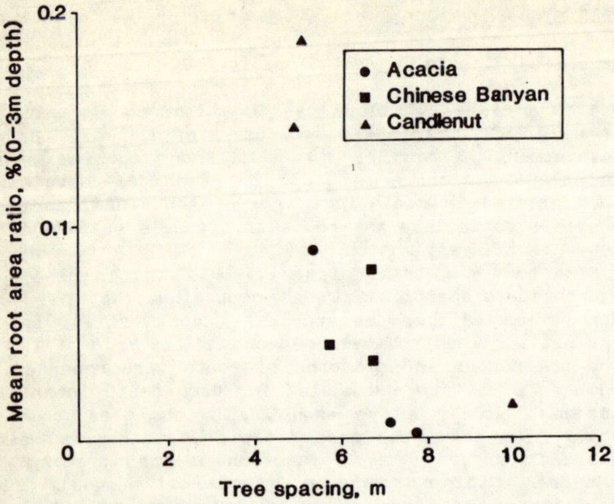

Figure 5. Relationship between root area ratio and tree spacing at So Uk

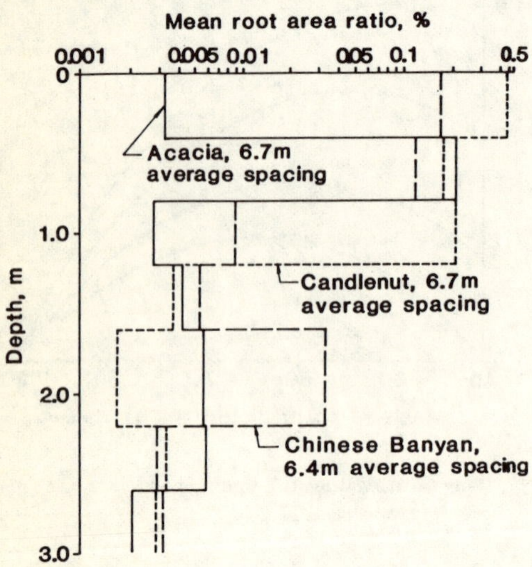

Figure 6. Average root ratio distributions at So Uk

The tensile strength of over 1000 roots encountered in the trial pits was measured in both the field and the laboratory. In the field tests, randomly selected roots of smaller diameters (generally less than 5 mm) protruding from the pit face were pulled by hand slowly and steadily with a calibrated spring similar to that described by Wu et al (1979). Root specimens of larger diameters (5 mm to 10 mm) approximately 200 mm long were transported to the laboratory in sealed plastic sacks and tested within 2 days, in order to keep them at their natural moisture content. During the tests, roots were pulled to failure in a Scott L6 Tester. This machine had a maximum capacity of 135 kg. The roots were gripped by roughened cams that increased the gripping pressure as tension was applied. Smaller roots were found to be strongest in terms of breaking stresses (Figure 7). The measured lower bound tensile strength of the tree species studied are given in Table 2.

The shear strength increase due to root reinforcement was predicted by employing Equations 1 and 2, taking the bracketed term in Equation 1 as 1.25 for a soil strength of $\emptyset = 39^\circ$. The predicted shear strength increase profile for each tree species is as shown in Figure 8; these increases are significant, being up to 35 kPa at the surface and decreasing to 1 kPa or less below a 2 m depth. These predicted increases are for trees spaced about 6.5 m apart, and would vary with other tree spacings.

Stability analysis, using Janbu's routine method, showed that the average depth of the critical slip was 1.3 m. The shear strength increases shown in Figure 8 at

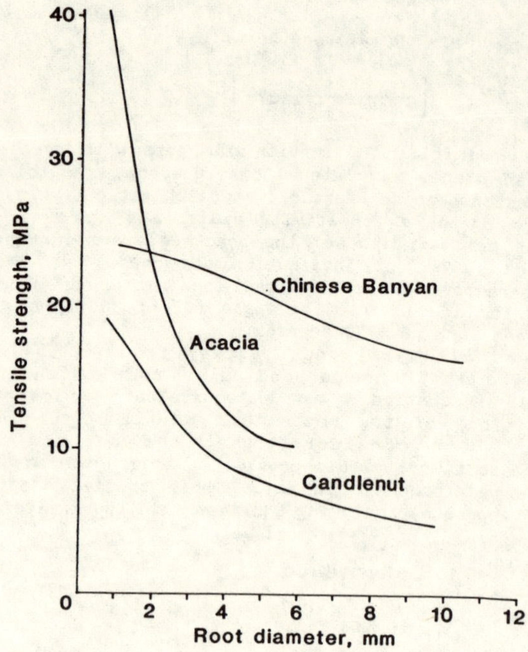

Figure 7. Relationship between tensile strength and root diameter at So Uk

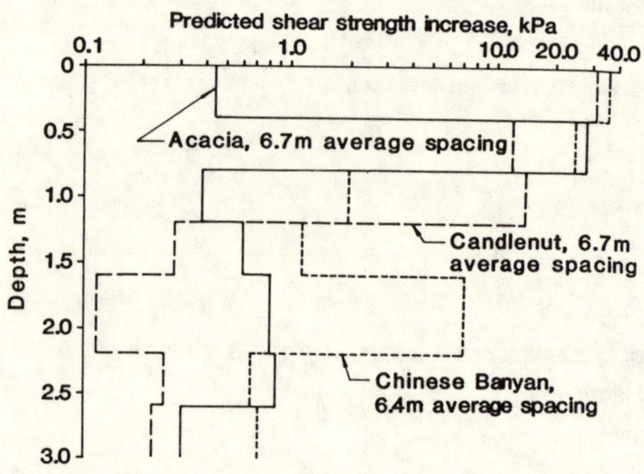

Figure 8. Predicted shear strength increases by tree species and depth at So Uk

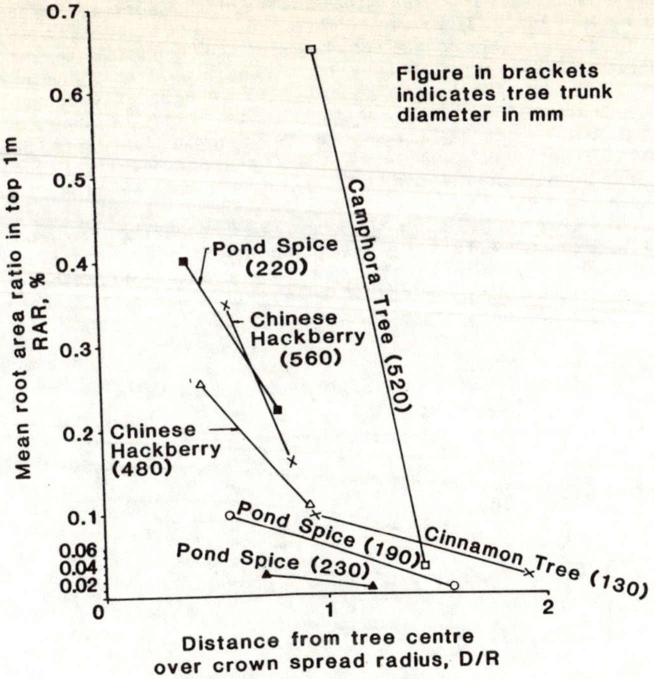

Figure 9. Mean RAR in top 1 m with radial distance from tree centre at Leighton Hill slope

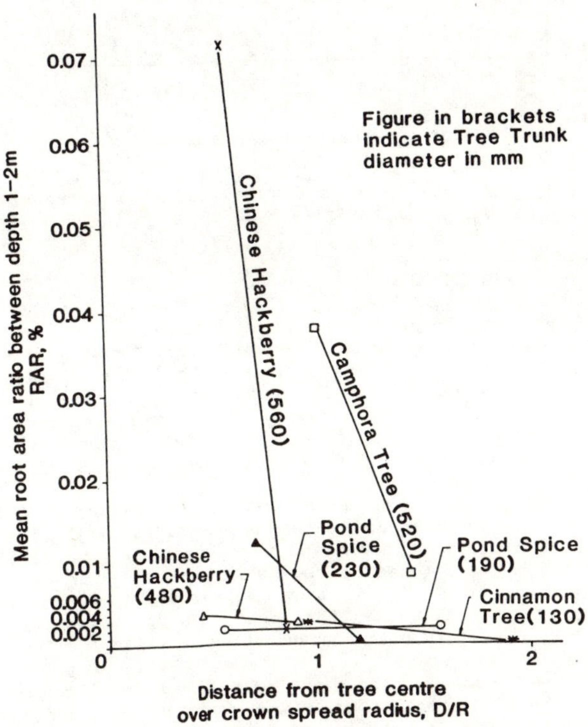

Figure 10. Mean RAR between depth 1 m to 2 m with radial distance from tree centre at Leighton Hill Slope

a 1.3 m depth are representative of points midway between trees and must therefore be adjusted for spatial variations along the slip surface. In the light of the observed spacing effect on root concentrations, much higher values probably exist closer to and beneath the trees. Based on these considerations, a shear strength increase in the range 1 to 5 kPa was adopted in the analysis of the critical slip. The calculated factor of safety increased by 8% for 1 kPa shear strength increase and by 38% for 5 kPa increase. This indicates that root reinforcement has a significant stabilizing influence on this slope.

Case Study 2 - Leighton Hill Fill Slope

The 22 m high, 35° slope was formed over 50 years ago. It supports a narrow access road to residential buildings further above the slope and overlooks two primary schools built close to the toe of the slope. Over 230 trees of 23 different species are growing on the slope. Average tree height is 9.7 m and average crown spread (the plan area projection of tree foliage)is 3.5 m. Over 90% of the trees are healthy, of good form and medium amenity value. Four major tree species (Table 2) were selected for the root reinforcement study.

Seven trial pits were excavated at selected positions near trees of major species to allow observation and measurement of the distribution and tensile strength of tree roots, using the same methods as used for the So Uk Case Study.

In general, the root area ratios of all tree species were found to be greatest in the surface 1 m, reducing substantially between a depth of 1 m to 2 m, and diminishing to practically zero below a 2 m depth. Most tree species (except Camphora) showed a marked decrease in root area ratio outside the tree crown spread area (Figures 9 & 10). Tensile strengths also generally decreased with increasing root diameter; the measured lower bound values are given in Table 2.

The bracketed term of Equation 1 was taken as 1.15 for a soil strength of Ø = 35° for the prediction of shear strength increases. The predicted mean shear strength increase for observed root area ratios (Figure 11) is significant, ranging from 4.2 to 102 kPa in the surface 1 m, and 1.4 to 2.2 kPa between a 1 m and 2 m depth.

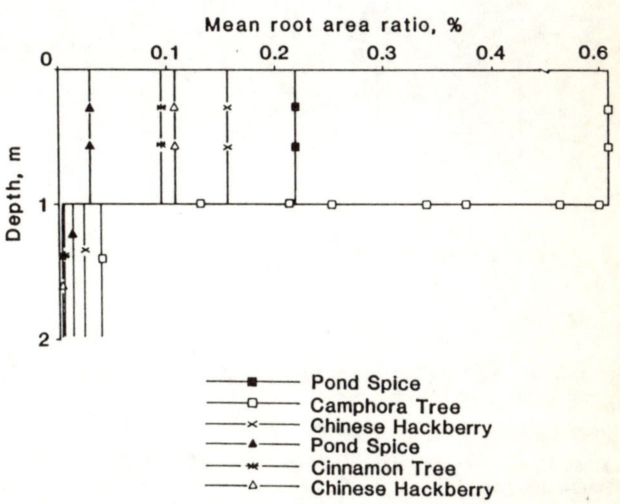

Figure 11. Generalised RAR with depth at tree crown spread edge at Leighton Hill

This investigation shows that a large portion of the root system of the tree species studied is concentrated within 2 m of the surface, and lateral roots extend outward as far as the margin of the tree crown spread. Camphora trees were found to possess the most extensive root system. Since the four species studied are the major species grown on the slope, it was postulated that the trees had a significant effect on stability to depths of at least 2 m, and in areas beneath the crown spread of the trees. The predicted shear strength increases were the minimum values expected at any point within the tree crown spread area at comparable depths. The total crown spread area on the slope is about 80% of the total area. It was therefore conservative to assume a shear strength increase of 2 kPa in the stability analysis of trial slips within the surface 2 m. Stability analyses of shallow slips for a typical slope section showed an improvement of 14% to 31% in the factor of safety.

Case Study 3 - La Salle Road Fill Slope

This 25 m high, 35° slope comprises poorly compacted granitic fill. It was formed in 1959 during the construction of an adjacent saltwater reservoir. The slope was considered in two parts. The southern part was covered mainly with Acacia spaced about 3 m apart, and the northern part with Horsetail Trees spaced about 12 m apart. All the trees were planted on the slope between 1967 and 1976 (Table 2).

Four and three trial pits were dug on the southern and northern parts of the slope respectively. With the exception of one trial pit, the pits were located where

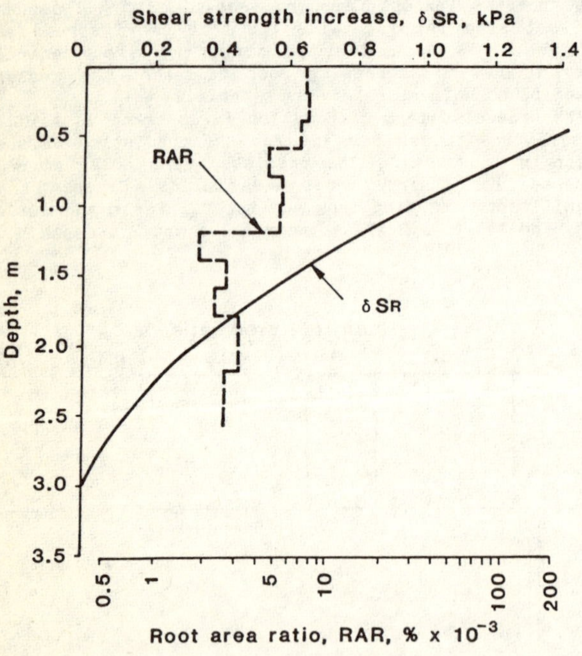

Figure 12. RAR distribution & predicted shear strength increase for horsetail tree at La Salle Road

the tree coverage was average or below average. The shear strength increases finally predicted were therefore considered to be conservative.

The average root area ratio and shear strength increase for the two types of tree were plotted against depth, as shown in Figures 12 & 13 respectively. It was found that the root area ratio was greatest near the surface due to the many large diameter roots encountered there . Also, the smaller the roots the higher the breaking tensile stress. The mean root tensile strength was 17.1 MPa and 17.9 MPa for Acacia and Horsetail Tree respectively, and the bracketed term of Equation 1 is 1.25. As shown in Figure 13, the predicted shear strength increases are significant, being up to 14 kPa for Acacia at the ground surface and decreasing to 1 kPa or less below a 2 m depth.

The slope was then analyzed incorporating the root reinforcement effect by using the additional strengths, as shown in Figure 13. The results of analyses indicated that the factor of safety of trial slips within the top 2 m of the southern slope increased by 6% .

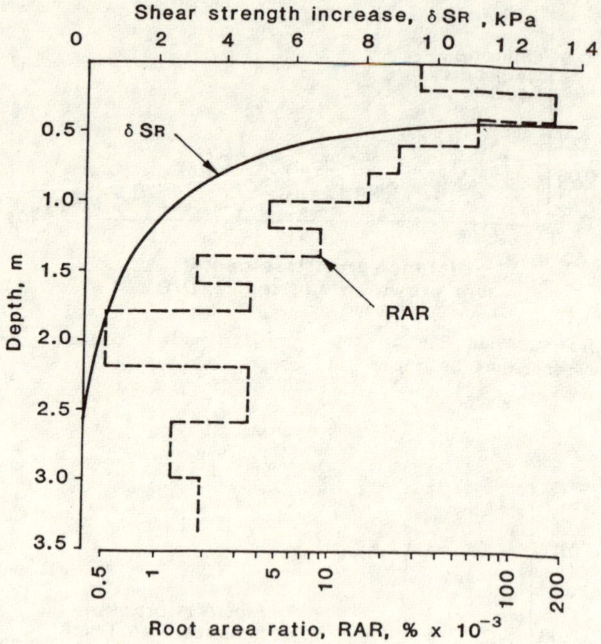

Figure 13. RAR distribution & predicted shear strength increase for acacia at La Salle Road

DISCUSSION AND CONCLUSIONS

This paper has briefly reviewed the principle of root reinforcement and described the mathematical models for root-reinforced soil. A simple method was used to measure tree root tensile strength and tree root distribution within a soil mass. The data collected, when coupled with the root-reinforcement model , provides a means of quantifying the effect of tree roots on slope stability.

The root studies have shown that closer tree spacings yield markedly higher root concentrations. Distribution of roots varies with depth and tree species. The root concentration was found to be, in general, greater near the surface, owing to the many large roots encountered there. It was also shown that most tree species showed marked decreases in root concentrations outside the tree crown spread area.

The tree root reinforcement effect was found to be species related. Smaller roots were found to be stronger in terms of tensile breaking strength. In addition, the many field tests did not encounter pull-out failure of the tree roots. The assumption in the root reinforcement model of full mobilization of root tensile strength therefore appears reasonable.

It has been found that tree roots have a significant stabilising effect on the three slopes investigated. They can provide effective resistance against relatively shallow failures. Quantification of the effect of tree root reinforcement has been particularly useful in these cases for increasing the calculated factors of safety of the slopes . Without inclusion of the effect of root reinforcement, more substantial and costly slope stabilization works would have been necessary to meet the current safety standards.

ACKNOWLEDGEMENTS

This paper is published with the permission of the Director of Civil Engineering Services Department of the Hong Kong Government. The authors also wish to express their grateful thanks to Mr S.K. Ng of the Geotechnical Control Office for this critical review of the manuscript and for providing valuable advice.

REFERENCES

Allen, P.M. & Stephens, E.A. (1971). Report on the Geological Survey of Hong Kong. Government Printer, Hong Kong, 116 p.

Brand, E.W., Premchitt, J. & Phillipson, H.B. (1984). Relationship between rainfall and landslides in Hong Kong. Proc. 4th Intl. Sym. Landslides, Toronto, Vol.1, pp. 377-384.

Gray, D.H. & Leiser, A.T. (1982). Biotechnical Slope Protection and Erosion Control. Van Nostrand Reinhold, New York, 271 p.

Gray, D.H. & Ohashi, H. (1983). Mechanics of fiber reinforcement in sand. J. Geotech. Eng., ASCE ,Vol.109, pp 335-372.

Greenway, D.R. (1987). Vegetation and slope stability. Slope Stability, edited by M.G. Anderson & K.S. Richards, pp 187-230. John Wiley & Sons, Chichester, UK.

Greenway, D.R. Anderson, M.G. and Brian-Boys, K.C. (1984). Influence of vegetation on slope stability in Hong Kong. Proc. 4th Int. Symp. Landslides, Toronto, Vol. 1, pp 399-404.

Lee, I.W.Y. (1985). A review of vegetative slope stabilisation. Hong Kong Engineer, Vol. 13, No. 7, pp 9-21. (Corrigenda, Vol. 13, No. 8, pp 49).

Lumb, P. (1975). Slope failures in Hong Kong. Quarterly J. Engr. Geol., Vol.8, pp. 31-65.

Waldron, L.J. (1977). Shear Resistance of root-permeated homogeneous and stratified soil. Soil Sci, Soc. Amer., Vol.J41, pp 843-849.

Wu, T.H. (1976). Investigation of Landslides on Prince of Wales Island, Alaska. Geotechnical Engr. Report No. 5, Dept. of Civil Eng., Ohio State University, Columbus, Ohio, 94 p.

Session 4
Foundations of buildings in tropical soils

Laterally loaded bored piles in residual soils and weathered rocks

M.F.CHANG, Nanyang Technological Institute, Singapore
A.T.C.GOH, Nanyang Technological Institute, Singapore

SYNOPSIS: In this paper the lateral load tests on six bored piles installed in the residual soils and weathered rocks of the sedimentary Jurong Formation of Singapore are described. These bored piles, approximately 680 mm in diameter and 6 or 7.7 m in length, were instrumented with strain-gauges and inclinometers and tested to failure. Various in-situ and laboratory tests were carried out at the test sites to obtain site-specific engineering parameters for the subsurface materials. The ultimate lateral load as well as the lateral load - deflection relationship were estimated using some common analytical design procedures, and comparisons were made with the field results. It appears that current analytical methods can give only fair estimates of pile behaviour under lateral loading conditions.

INTRODUCTION

Bored piles are commonly used in Singapore as vertical and lateral supports for tall buildings, bridges and elevated structures such as the Singapore Mass Rapid Transit System. The diameter of bored piles in Singapore varies from 0.4 m to 1.7 m. About 70% of these are installed in residual soils and weathered granite or sedimentary rocks, which are found in the western and northern parts of Singapore Island.

In recent years, instrumented large-scale load tests (e.g. Radhakrishnan et al 1985; Buttling & Robinson, 1987; Chang & Wong, 1987, and Chang & Goh, 1988) have been carried out to study the behaviour of axially loaded bored piles in the residual soils and weathered rocks of Singapore. However, litle is known about the actual performance of these piles under lateral loading. In an attempt to gain an insight into the behaviour of laterally loaded bored piles installed in the residual soils and weathered rocks of the Sedimentary Jurong Formation of Singapore, a number of lateral load tests were carried out on some instrumented bored piles. The bored piles were approximately 680 mm in diameter and 6 or 7.7 m in length. They were instrumented with inclinometers and strain-gauges. The test sites were relatively extensively investigated through ram sounding tests, standard penetration tests, pressuremeter tests and laboratory triaxial tests. In this Paper, the results of these lateral load tests are presented and comparisons made with the analytical procedures developed by Broms (1965), Suyama et al (1982) and Reese (1984).

DESCRIPTION OF TEST SITE

The test site is located within the campus of the Nanyang Technological Institute where the sedimentary Jurong Formation prevails. A total of six piles were installed at two locations approximately 80 m apart. Piles A-1, A-2 and A-3 were installed in Location A in linear alignment, as were Piles C-1, C-2 and C-3 at Location C.

The residual soils and weathered rocks present at both locations are very similar, although the stratifications are quite different due to the heterogeneous nature of the formation. The subsurface materials consist primarily of clayey silt and silty clay of low to medium plasticity (LL = 30 - 45, PI = 10 - 20). The fines content ranges from 48 to 90%. The materials are, in general, of high strength and low compressibility, behaving like "overconsolidated" soils. They are partially saturated within the depth of our current investigation.

Prior to the installation of the bored piles, a ram sounding test (Dahlberg and Bergdahl, 1974) was carried out at each proposed piling point. The ram sounding resistance profiles at the piling points are shown in Figure 1. The three piling points at each test location are only 1.8 m apart. It can be readily seen that the residual soils and weathered rocks of the Jurong Formation are highly variable and heterogeneous. Two boreholes were also drilled nearby the piling points at each test location and standard penetration tests (SPTs), "undisturbed" sampling (using U3 samplers), and pressuremeter tests (using OYO's Elastmeter) were carried out. The limited number of standard penetration resistance N-values, when compared with the nearby ram sounding resistance N_{20}-values (see Figure 1), seem to indicate an approximate correlation of $N = 2 N_{20}$, which is the same as that reported by Chang and Wong (1986) for both the residual soils of the sedimentary Jurong Formation and the residual granite soils of Singapore. Results of the pressuremeter tests, which include R_o (at-rest borehole radius), R_m (average borehole radius in the pseudo-elastic range), E_p (pressuremeter modulus), and P_l^* (net limit pressure), are summarised in Table 1. Uncon-

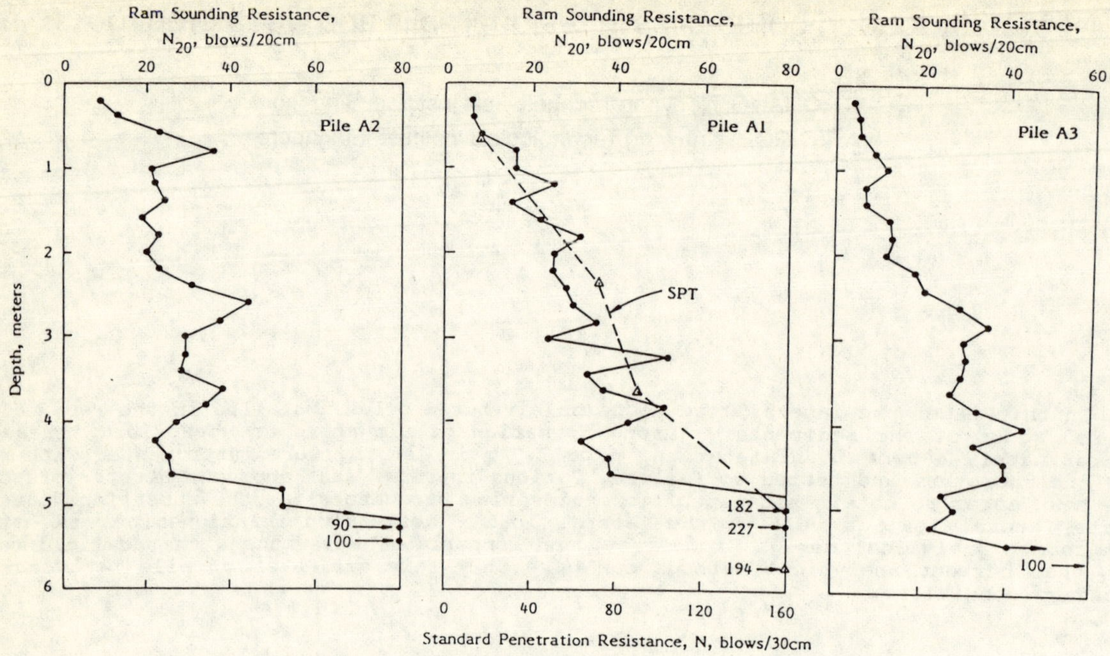

Ram Sounding Resistance Profiles at Location A

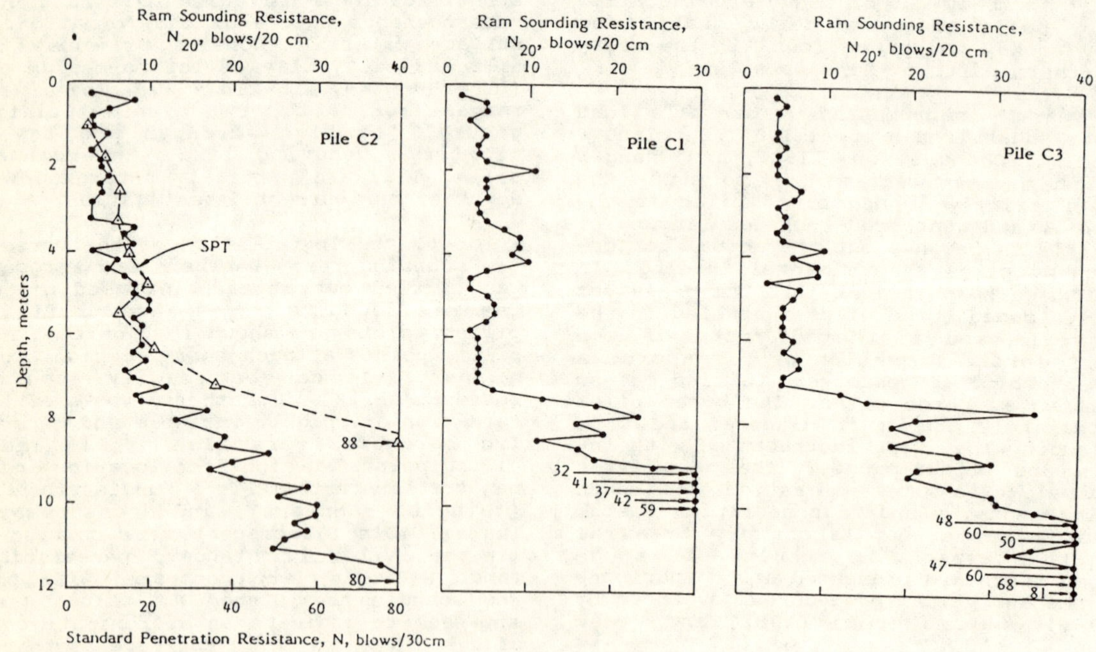

Ram Sounding Resistance Profiles at Location C

Figure 1 Ram sounding resistance profiles at piling locations solidated undrained triaxial tests were carried out in the laboratory on the "undisturbed" samples collected. From the stress-strain curves obtained, the strain corresponding to one-half of the maximum principal stress difference, ε_{50}, which is required in the analysis, was determined. The "equivalent" undrained shear strength measured in these tests did not seem to sufficiently reflect the variation of the materials with depth and among the piling points.

TEST PILE DETAILS

The reinforced concrete piles investigated were approximately 680 mm in diameter, with the piles within each group spaced at 1.8 m intervals. The lengths of the piles were 6 m and 7.7 m long at Locations A and C respectively. All the piles were cast in a dry condition immediately after the holes were prepared, the casting being carried out using a tremie pipe and high slump ready mix concrete of Grade 30. The compressive strengths of the concrete were determined from laboratory cube strength tests and the average values were 35.5 and 37.0 MPa at Locations A and C respectively. The elastic modulus of the concrete, as determined from laboratory modulus tests, was around 29 kN/mm^2 for piles at Location A and 33 kN/mm^2 for piles at Location C. The elastic modulus of the piles as back-calculated from axial load tests carried out earlier, ranged between 32 and 34 kN/mm^2, in general, at both test locations. For analysis purposes, the flexural rigidity, EI, of the piles is approximately taken as 320 $MN\text{-}m^2$ at Location A and 350 $MN\text{-}m^2$ at Location C.

All the piles were instrumented with 4 to 8 numbers of vibrating wire strain gauges. One Sinco 70 mm diameter inclinometer casing and another 50 mm diameter regular plastic casing were also installed in each pile to allow for the measurement of pile deflection during the lateral load test. All the piles which were originally designed for axial loading, were not heavily reinforced. For the middle piles, Piles A-1 and C-1, each was reinforced with 8 numbers of Y25 bars; whereas only 4 numbers of Y25 bars were used in the side piles, Piles A-2, A-3, C-2 and C-3. About one year prior to the implementation of the lateral load testing program, the middle piles were subjected to compression tests and the side piles to pull-out tests.

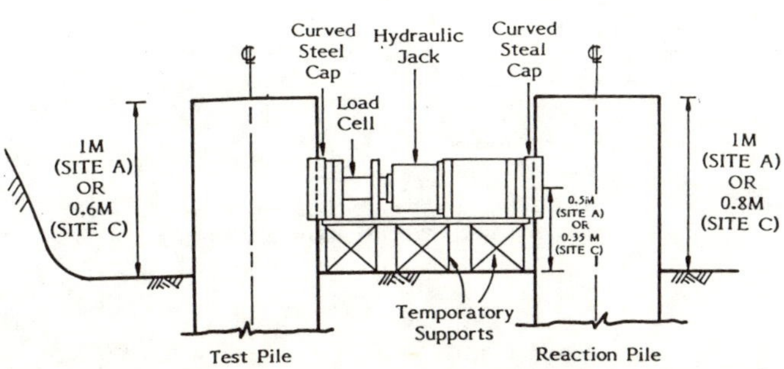

Figure 2 Schematic view of test setup

Table 1 Summary of pressuremeter test results

Location	Depth (m)	R_o (cm)	R_m (cm)	K_m (MN/m^3)	E_p (MPa)	P_l^* (kPa)	$P_l^*/12$ (kPa)
	0.7-1.3	3.825	3.883	324	16.7	1170	98
A	1.8-2.4	4.065	4.094	1048	56.9	2720	267
	3.7-4.3	3.905	3.945	1250	65.7	3660	305
	5.2-5.8	3.860	3.885	2707	139.8	3930	328
	1.4-2.0	4.060	4.175	64	3.6	410	34
C	2.9-3.5	4.020	4.140	170	9.3	1050	88
	4.5-5.1	4.020	4.130	156	8.6	860	72
	5.8-6.4	3.905	3.993	235	12.5	1225	102

TEST SETUP AND TEST RESULTS

Figure 2 gives a schematic representation of the test setup. Excavation was carried out to a depth of between 0.6 m and 1 m below the pile head at the two locations. Loading was applied simultaneously on the two outer piles with the center pile as reaction using two 200 ton capacity hydraulic jacks. In addition, load cells were used to monitor the loads. The applied load was located at 0.5 m above the cut ground at Location A and 0.35 m at Location C. Specially designed steel caps which fitted the curved pile surface were used to ensure proper transmission of the applied loads. On completion of the load tests on the outer piles, the center pile in each location was tested with one of the outer piles as reaction. The lateral resistance of the reaction piles were increased by backfilling the previously excavated areas behind the reaction piles.

Between nine and eleven dial gauges were used to monitor the lateral deflection of each pile as well as to check the eccentricity of the applied load. All the piles were subjected to maintained load tests. The piles were loaded to failure in four stages, each with a load increment corresponding to approximately 25% of the estimated failure load. Four equal load increments were used in each stage. Each virgin increment of load was maintained for at least 15 minutes until the rate of lateral movement was reasonably small. The test was stopped when the movement was found to increase continuously instead of slowing down without any further increase of load. In the case of the simultaneous testing of the two outer piles, one of the piles failed before the other. When this occurred, the piles were fully unloaded. In the subsequent loading cycle, the failed pile was utilised as reaction for the other pile up to a load slightly less than the measured failure load of the failed pile. This ensured that the center pile whcih was originally used as reaction would not be severely stressed when one of the outer piles failed, since it was to be tested subsequently.

Figures 3 through 8 show the lateral load versus groundline deflection curves for the six piles tested. It appears that for those piles at Location A, the measured ultimate load ranged from 260 kN to 410 kN and the corresponding maximum groundline deflection ranged from 18 mm to 28 mm. At Location C, the measured ultimate load was between 200 kN and 360 kN, only slightly smaller than those at Location A. The corresponding maximum groundline deflection of between 34 mm and 36 mm, however, are larger than

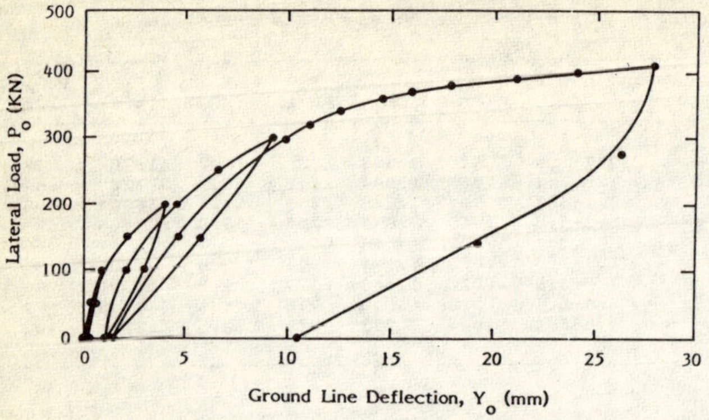

Figure 3 Measured lateral load-ground line deflection curve for Pile A-1

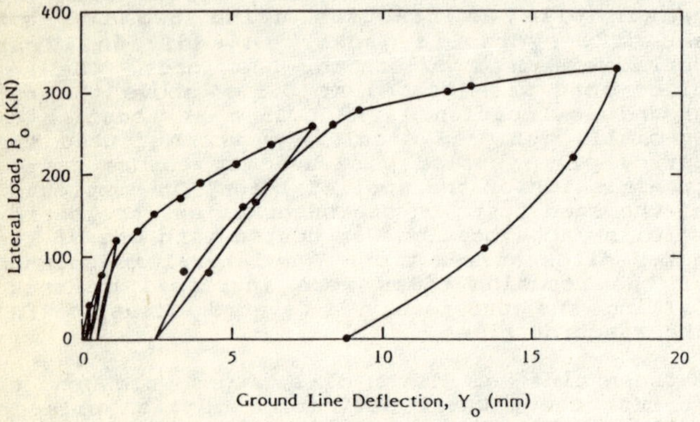

Figure 4 Measured lateral load-ground line deflection curve for Pile A-2

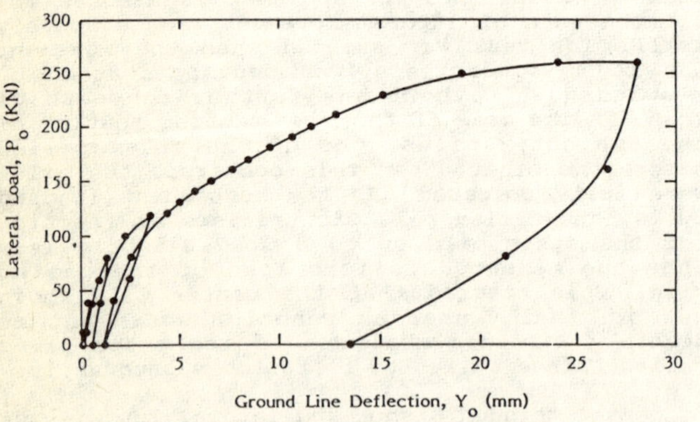

Figure 5 Measured lateral load-ground line deflection curve for Pile A-3

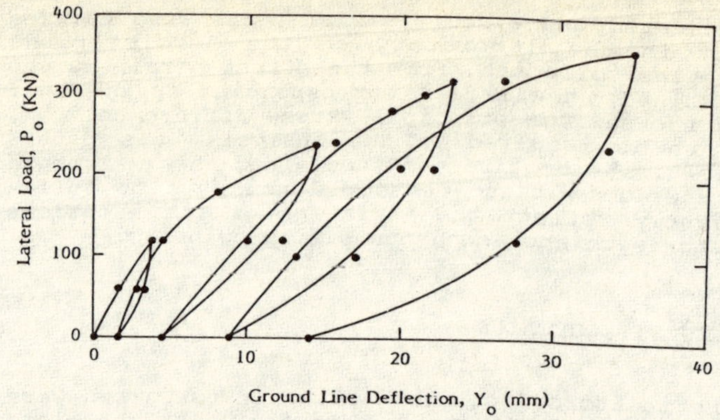

Figure 6 Measured lateral load-ground line deflection curve for Pile C-1

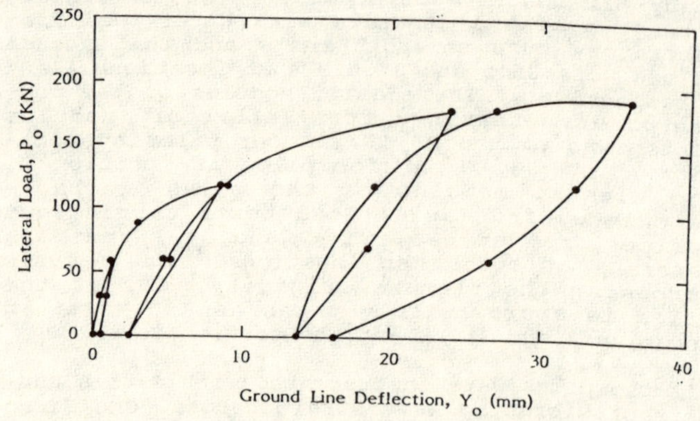

Figure 7 Measured lateral load-ground line deflection curve for Pile C-2

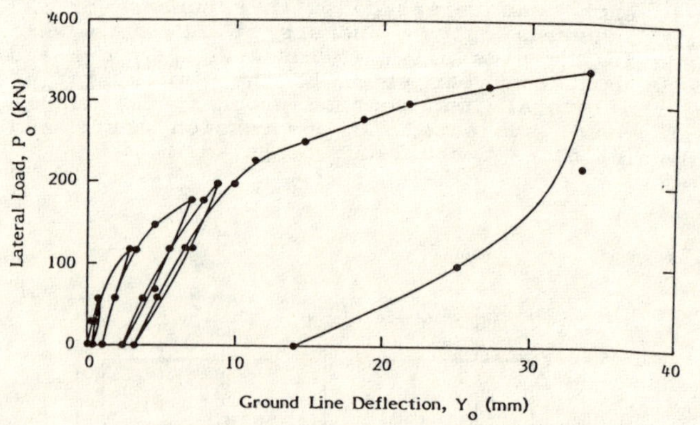

Figure 8 Measured lateral load-ground line deflection curve for Pile C-3

those at Location A. This seems reasonable as the materials at Location C were weaker than those at Location A.

ANALYTICAL METHODS AND RESULTS

Two commonly used methods for laterally loaded pile design are the procedures proposed by Broms (Broms, 1964a, 1964b, 1965) and Reese (1984). In Broms' method, the distribution of soil resistance with depth is assumed and statics is employed to compute the failure load. The approach developed by Reese models the soil surrounding a pile with a set of nonlinear independent springs. The response of the soil is represented by nonlinear functions commonly termed p-y (lateral resistance-pile deflection) curves. The set of differential equations of the pile bending is solved by the finite difference method. In this Paper, a finite difference program wich essentially followed the formulations by Reese (1984) was used. A semi-empirical procedure for the construction of p-y curves for stiff clays based on Reese and Welch(1975) was employed. As the yield moment of the pile can be calculated, the ultimate lateral load can be determined using plots of the lateral load against maximum pile bending moment obtained from the output of the program.

Prior to the load tests, the ultimate lateral capacity of each pile was estimated using Broms' method and the p-y curve method. The N-values were deduced from the ram sounding resistance profile using $N = 2 N_{20}$. The "equivalent" undrained strength, c_u, of the residual soils with $N \leq 75$ was estimated from the correlation : $c_u = 6N$ (kPa) as reported by Stroud(1974). It should be noted that the c_u values so obtained were very similar to those estimated from the net limit pressure using the correlation : $c_u = P_l^*/12$ normally considered for stiff clays (see Table 1). For materials with $N > 75$, the relationship between c_u and N as reported by Cole and Stroud(1977) was adopted. For the analysis using the p-y curve method, the strain at half ultimate load, ε_{50} was obtained from UU triaxial tests. The assumed soil properties are summarised in Table 2 and the computed results using the analytical methods are tabulated in Table 3. Also included in Table 3 are the interpreted limit loads which are defined as the lateral loads corresponding to the groundline deflections of 10 % of the pile diameter.

In many practical cases, the design of piles for lateral loading is governed by the limiting deflection requirement rather than the ultimate lateral capacity requirement. It is therefore important to be able to predict the lateral load versus groundline deflection relationship of a pile in the design. In addition to the p-y curve method just discussed, which is capable of making this kind of prediction, a simple hand calculation method proposed by Suyama et al (1982) was employed. The method is based on Chang's formulation (Chang, 1937) by treating the problem as a uniformly loaded beam resting on a perfectly elastic bed and it utilises the pressuremeter test (PMT) to obtain the coefficient of horizontal subgrade reaction, K, required in the analysis. A basic K-value, K_o, is first calculated from the K_m, R_o and R_m measured in the PMT for each of the soil layers

Table 2 Assumed pile and soil properties

Pile	EI ($MN\text{-}m^2$)	Depth (m)	N_{avg} (blows/ 0.3m)	c_u (kPa)	ε_{50}	Basic K K_o (MN/m^3)
A-1	320	0.0-0.8	16	96	0.0136	-
		0.8-1.7	40	240	0.0120	71
		1.7-3.0	56	336	0.0110	165
		3.0-4.8	84	540	0.0076	229
		4.8-6.0	240	2400	0.0062	390
A-2	320	0.0-0.6	28	168	0.0120	-
		0.6-2.4	42	252	0.0120	118
		2.4-5.0	60	360	0.0110	229
		5.0-6.0	240	2400	0.0062	390
A-3	320	0.0-0.8	10	60	0.0136	-
		0.8-2.3	20	120	0.0136	118
		2.3-3.8	60	360	0.0076	197
		3.8-4.7	76	456	0.0076	229
		4.7-5.4	52	312	0.0076	390
		5.4-6.0	240	2400	0.0062	-
C-1	350	0.0-3.2	9	54	0.0130	31.5
		3.2-5.8	12	72	0.0130	49.8
		5.8-7.4	8	48	0.0130	63.7
		7.4-7.7	34	204	0.0120	-
C-2	350	0.0-3.2	8	48	0.0130	37.2
		3.2-4.0	12	72	0.0130	55.7
		4.0-7.7	17	102	0.0130	-
C-3	350	0.0-3.6	9	54	0.0130	37.2
		3.6-7.4	12	72	0.0130	55.7
		7.4-7.7	45	270	0.0076	-

Table 3 Comparison of predicted and measured ultimate lateral loads

Pile	Ultimate Lateral Load (kN)			
	Measured (max)	Interpreted (at y/D = 10%)	Broms' method	p-y curve method
A-1	410	330	266	340
A-2	330	240	146	200
A-3	260	220	146	160
C-1	360	265	250	210
C-2	200	160	132	130
C-3	340	250	137	130

within the depth of influence, which is taken as approximately 1.5 times the depth at which the bending moment is zero. A weighted average or equivalent Ko- value is then calculated by assuming a triangular weight function with its base located at the groundline to take into account the fact that the pile develops most of its resistance from interaction with the soils close to the top. The design K-value, which is a function of the pile deflection, y, is then taken as $K_o/(y)^{0.5}$ with y expressed in cm.

In this investigation, both the p-y analysis and Suyama's analysis were carried out to predict the lateral load versus groundline deflection relationship for Piles A-3 and C-3. The p-y analysis was based on the material data listed in Table 2, as discussed earlier. In Suyama's analysis, a completely different source of soil data based on the pressuremeter test, as presented in Table 1, was used. Based on the field observations on these instrumented piles, the depth of zero bending moment averaged around 2.8 m at Location A and 3.2 m at Location C. The depth of influence was therefore taken as 4.25 m and 4.75 m below the cut ground surface for Location A and C respectively. The calculated equivalent basic K, or K_o values based on the K_o values shown in Table 2, were found to be 163 $MN\text{-}m^2$ and 40 $MN\text{-}m^2$ for Piles A-3 and C-3, respectively. The predicted horizontal load-deflection curves as compared to the measured curve are shown in Figure 9 and Figure 10 for Piles A-3 and C-3, respectively.

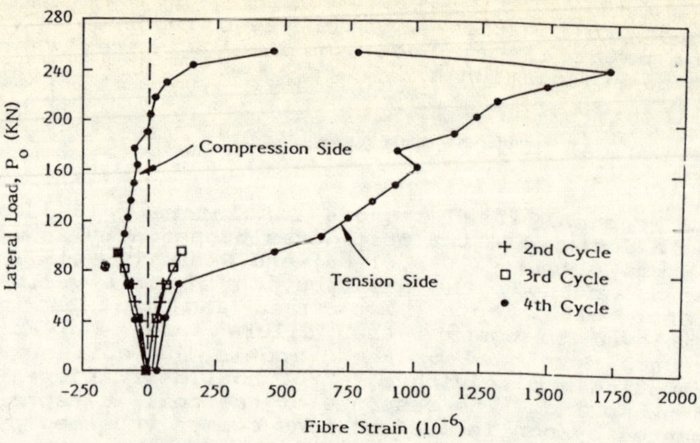

Figure 11 Lateral load-fibre strain measurements for Pile A-3 at 2.0 m depth

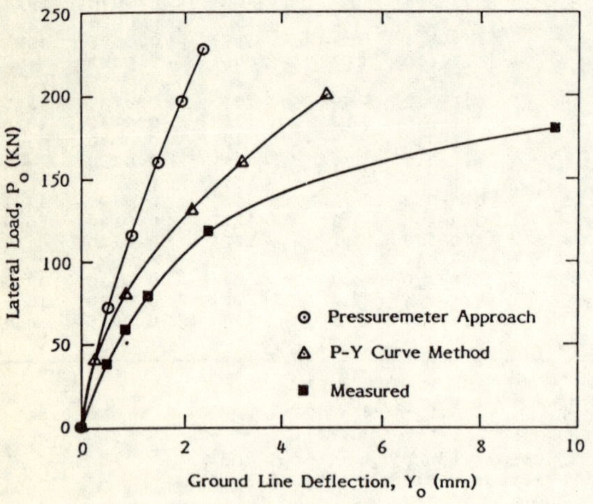

Figure 9 Variation of ground line deflection with lateral load for Pile A-3 - Measured versus predicted

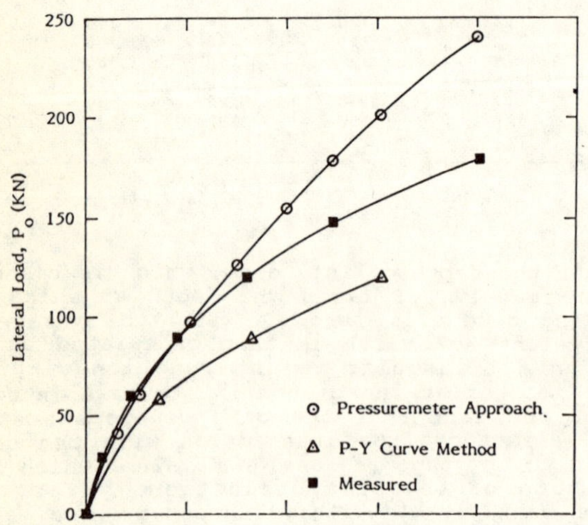

Figure 10 Variation of ground line deflection with lateral load for Pile C-3 - Measured versus predicted

DISCUSSION

The plots in Figures 3 through 8 show that the shape of the load-deflection curves is nonlinear and resembles that normally found in axial pile load tests. The pile behaviour is linear up to approximately one half of the limit load. The typical nonlinearity of the pile behaviour is illustrated in Figure 11 which shows the variation of the fibre strain with lateral load for Pile A-3 at a depth of 2 m below the pile head, which is close to the section of the pile with the maximum bending moment. A pair of strain readings were plotted for each load level. Positve strain and negative strain indicate tension and compression in the reinforcement bar of the pile, respectively. The strains indicate that the pile behaved linearly elastically at loads less than 70 to 90 kN or about one half of the limit load. Between 70 and 240 kN, the positive strain increased rapidly with load, indicating plastic deformation of the pile. A cracked section and elastic plastic stress condition is likely to have occurred. Above 240 kN, the positive strain decreased, possibly as a result of the breaking of the pile above this section. In the case of the negative strain, a reduction in the strain occurred between 100 and 190 kN. At loads higher than 190 kN, a reversal of the sign of the strain was observed. This could be attributed to the shifting of the neutral axis towards the compression zone. When the neutral axis shifted beyond the compression reinforcement, a reversal in strain occurred.

Figure 12 shows the deflected shape for Pile A-3 at various applied loads. All the other piles exhibitted similar behaviour. The results indicate that the mode of deflection resembles that of a long pile although all the piles are relatively short with length/diameter ratios between 9 and 11. Only minimal deformations occurred at depths greater than 3 m below the original ground surface or 2 to 2.4 m below the cut ground surface.

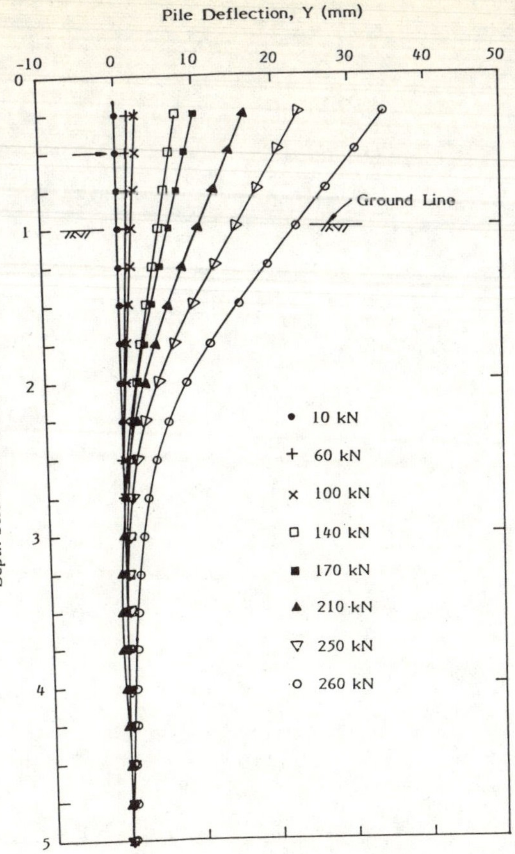

Figure 12 Variation of pile deflection with depth for Pile A-3

A comparison of the computed and interpreted failure loads in Table 3 shows that both Broms' method and the p-y curve method are able to predict fairly accurately the failure load. It appears that both these methods are likely to yield conservative estimates of the ultimate load of a pile. Although an attempt was made to correlate the net limit pressure, P_l^*, to the ultimate lateral load, no correlation seems obvious from this investigation.

The comparison of the predicted lateral load-deflection curves with the measured curves in Figure 9 and 10 seems to indicate the relative difficulty of predicting the behaviour of a laterally loaded pile. Both analytical methods give only fair predictions of the actual lateral load-deflection curves. Although the difficulty involved in the assessment of the design parameters for these relatively heterogeneous and stiff materials encountered could be partially responsible for the discrepancy, the main problem is possibly due to the fact that the behaviour of a pile subject to lateral loading has never been realistically modelled. According to Briaud et al (1983), the major components of soil resistance developed next to a laterally loaded pile are caused by the direct pressure in the direction of pile movement and the side shear drag on the side of the pile shaft. At working loads, the contribution resulting from the side shear accounts for the majority of the mobilised soil resistance. The direct soil pressures mobilised in front of the pile is predominantly in control only when the ultimate condition is reached. Yet, all the currently available analytical methods for laterally loaded piles idealise the soil reaction only as direct soil pressures acting across the pile width. It appears that a more realistic design approach should incorporate the effects of side shear in the formulation in order to produce a reasonable prediction of the lateral load-deflection characteristics.

CONCLUSIONS

In this Paper, the results of a series of lateral pile load tests in residual soils and weathered rocks of the sedimentary Jurong Formation have been presented. The results show that the behaviour of the bored piles under lateral loading are highly nonlinear. Comparison of the field data with the analytical results based on the p-y curve method and Broms' method indicates that both methods can be used to give reasonably conservative estimates of the ultimate lateral loads of bored piles using the site specific geotechnical data obtained from the standard penetration test and the UU triaxial test (for the p-y curve method). The p-y curve method and the method proposed by Suyama et al (1982) gives only fair predictions of the lateral load versus deflection relationship.

ACKNOWLEDGEMENT

The authors wish to thank the Nanyang Technological Institute for the financial support to study the behaviour of bored piles in Singapore. The technical assistance provided by Lam Cheow Shing, Lee Kwet Sang and several technical staff members of the NTI's Geotechnics Laboratories are appreciated.

REFERENCES

Briaud, J.L., Smith, T.D. and Meyer, B.J. (1983) Pressuremeter gives elementary model for laterally loaded piles. Proc. Int. Symp. on Insitu Testing of Soil and Rock, Int. Assoc. of Geology, Paris.

Broms, B.B. (1964a) Lateral resistance of piles in cohesive soils. J. Soil Mech. Found. Div., ASCE, Vol. 90, No. SM2, pp. 27-63.

Broms, B.B. (1964b) Lateral resistance of piles in cohesionless soils. J. Soil Mech. Found. Div., ASCE, Vol. 90, No. SM3, pp. 123-156.

Broms, B.B. (1965) Design of laterally loaded piles. J. Soil Mech. Found. Div., ASCE, Vol. 91, No. SM3, pp. 79-99.

Buttling, S. and Robinson, S.A. (1987) Bored pile design and testing. Proc. Singapore Mass Rapid Transit Conference, Singapore, pp. 155-175.

Chang, Y.L. (1937) Discussion on lateral pile

loading tests by L.B. Feagin, ASCE Transaction.

Chang, M.F. and Wong, I.H. (1986) Penetration testing in the residual soils of Singapore. Proc. Speciality Geomechanics Symposium on Interpretation of Field Testing for Design Parameters, Adelaide, Australia.

Chang, M.F. and Wong, I.H. (1987) Shaft friction of drilled piers in weathered rocks. Proc. 6th ISRM Congress, Canada, pp. 313-318.

Chang, M.F. and Goh, A.T.C. (1988) Performance of bored piles in residual soils and weathered rocks. Proc. 1st Int. Geotech. Seminar on Deep Foundations on Bored and Auger Piles, Ghent, Belgium, pp. 303-313.

Cole, K.W. and Stroud, M.A. (1977) Rock socket piles at Coventry Point, Market Way, Coventry. Piles in Weak Rock, ICE, U.K., pp. 47-62.

Dahlberg, R. and Bergdahl, U. (1974) Investigation on the Swedish ram sounding method. Proc. 1st Europ. Symp. on Penetration Testing, Stockholm, Vol. 2.2, pp.93-102.

Radhakrishnan, R., Leung, C.F. and Subrahmanyam, R.V. (1985) Load tests on instrumented large diameter bored piles in weak rock. Proc. 8th Southeast Asian Geotech. Conf., Kuala Lumpur, Malaysia, pp.2.50-2.53.

Reese, L.C. (1984) Handbook on the design of piles and drilled shafts under lateral loads. U.S. Dept. of Transportation, Federal Highway Administration, Virginia, U.S.A. 360pp.

Reese, L.C. and Welch, R.C. (1975) Lateral loading of deep foundations in stiff clay. Jnl. Geotech. Eng. Div., ASCE, Vol. 101, No. GT7, pp.633-649.

Stroud, M.A. (1974) The standard penetration test in insensitive clays and soft rocks. Proc. 1st European Symp. on Penetration Testing., Stockholm, Vol.2.2, pp. 367-375.

Sumaya, K., Imai, T. and Ohya, S. (1982) Development of LLT pressuremeter and its application in prediction of pile behaviour under horizontal load. Proc. Symp. on the Pressuremeter and its Marine Applications, Paris, pp. 61-76.

The insitu strength of saturated decomposed granite

M.D.HOWAT, Dames & Moore, Los Angeles, Calif., USA

ABSTRACT: Measurements of effective passive pressure were made in Gloetzl earth pressure cells cast on the inside face of diaphragm walls during excavations for deep basements in decomposed granite. Despite extensive pumping in deep wells installed inside the excavation, water pressures exceeded overburden pressures and conditions of zero vertical effective stress were reached. The measured "unconfined" passive pressure was up to an order of magnitude higher than that interpreted from triaxial tests. It appears that the preloaded temporary bracing for the excavations was in many cases overdesigned and calculated increases in strut loads were never found. A case is made for more economical temporary support using refined design and monitoring procedures.

1 INTRODUCTION

Eight of the shafts excavated for the construction of the Island Line of the Hong Kong Mass Transit Railway were entirely within granite decomposed by tropical weathering to the state of soft ground. These excavations were more than 30m deep below sea level within diaphragm walls up to 56m deep. Temporary bracing was designed on the assumption that the ground inside the diaphragm walls develops passive resistance to wall movement.

Howat and Cater (1985) found that water pressures inside the excavations of two advance contract shafts did not follow the design assumptions. Despite extensive pumping within the diaphragm walls, water pressures remained at high values, eventually exceeding overburden pressures as these reduced with excavation. When the effective vertical pressure reduced to zero, or even negative values, the decomposed granite could be said to be, for practical purposes, in an unconfined state.

Decomposed granite has traditionally been considered to lose its cohesion on saturation due to loss of capillary pore water tensions (Baynes and Dearman, 1978) and it is generally assumed that it has little or no unconfined compressive strength. Extrapolation of the results of backsaturated triaxial compression tests carried out prior to contract award indicated a nominal value of 20 kPa independent of dry density. Howat (1985) argued that this conclusion was partly the result of unrealistically severe backsaturation procedures, and partly the result of problems of interpretation of the test results.

Figure 1 shows the results of some of the triaxial tests carried out on samples of decomposed granite recovered from boreholes advanced in one of the shaft sites prior to the start of diaphragm walling. These tests were carried out without backsaturation, with

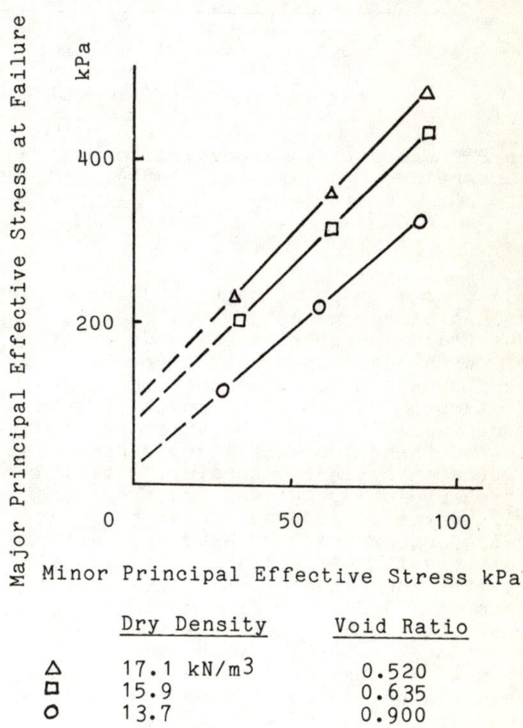

Figure 1. Triaxial tests on decomposed granite - From Howat & Cater (1985).

the top drainage line open to atmospheric pressure and with water pressures (both positive and negative) monitored through a high air entry ceramic at the sample base. These tests suggest a relationship between strength and density, which could be

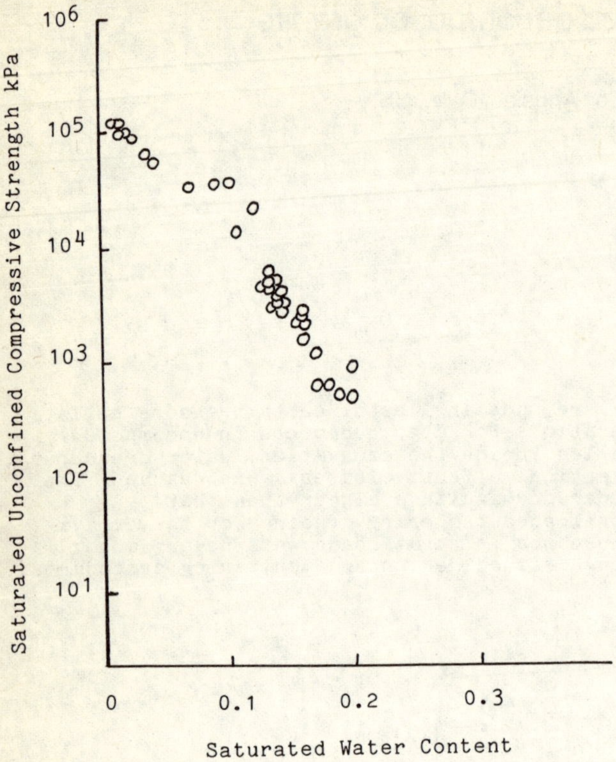

Figure 2. Laboratory strength of sandstone - From Dobereiner & de Freitas (1986).

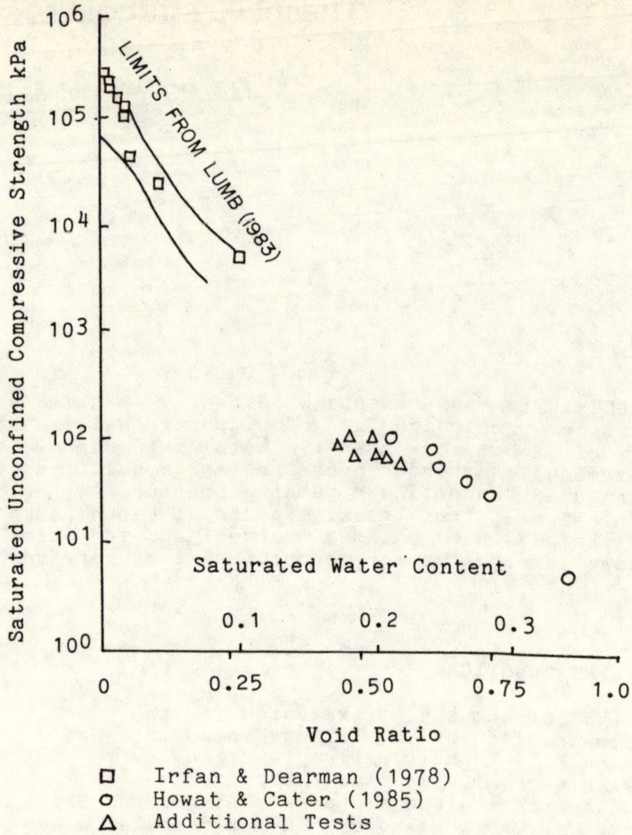

Figure 3. Laboratory strength of granite.

expressed as a graph of saturated unconfined strength plotted against saturated water content as found by Dobereiner and de Freitas (1986) for sandstones - see Figure 2.

Attempts were made to measure the unconfined strength directly but negative water pressures were found at the base of the specimens and these increased during compression making the analysis too complex. A minimum cell pressure of 30 kPa was required for the negative water pressures to disappear and unconfined strengths were extrapolated from the test results. It appears from Figure 3 that a limiting value of uniaxial compressive strength of 100 kPa is reached at a void ratio of about 0.6 and, unlike the sandstones, there is no steady increase in strength with decrease in void ratio towards the less weathered granites tested by Irfan and Dearman (1978) and Lumb (1983).

2 THE GLOETZL EARTH PRESSURE CELLS

The cast in situ diaphragm walls of eight of the shafts excavated for the construction of the railway were equipped with Gloetzl earth pressure cells twinned with pneumatic piezometers. These are shown diagramatically on Figure 4. The cells were calibrated by immersion in varying depths of sea water. Because the diaphragms of the Gloetzl cells are relatively stiff, the pneumatic pressure which triggers the return gas flow is consistently higher than that for the piezometers. The difference between the cell reading and the piezometer reading is taken to be an instrument constant called the Preliminary Value (PV).

The cell units were mounted on jacks fixed to the diaphragm wall panel reinforcement cage. Once the cage was lowered into position in the slurry trench, the cell was jacked out onto the earth face and kept in close contact with a positive jacking pressure. This was monitored during concreting and adjusted as required. After concreting, 18 of the 76 instrumentation units were found to be reading the same gross value in the cell and piezometer lines (i.e., the PV value had reduced to zero) indicating that the either the cell itself, or the connection to the pneumatic pressure line was ruptured and the Gloetzl cell was acting as a piezometer.

Three cells recorded no increase in reading after the cell was jacked out from the reinforcement cage, indicating that they had been installed in areas of overbreak and had reached the end of their travel without meeting the face of the trench. This problem did not occur on panels which had their profile checked with sonic surveying equipment. Occasionally, no return flow could be obtained from cells during or shortly after

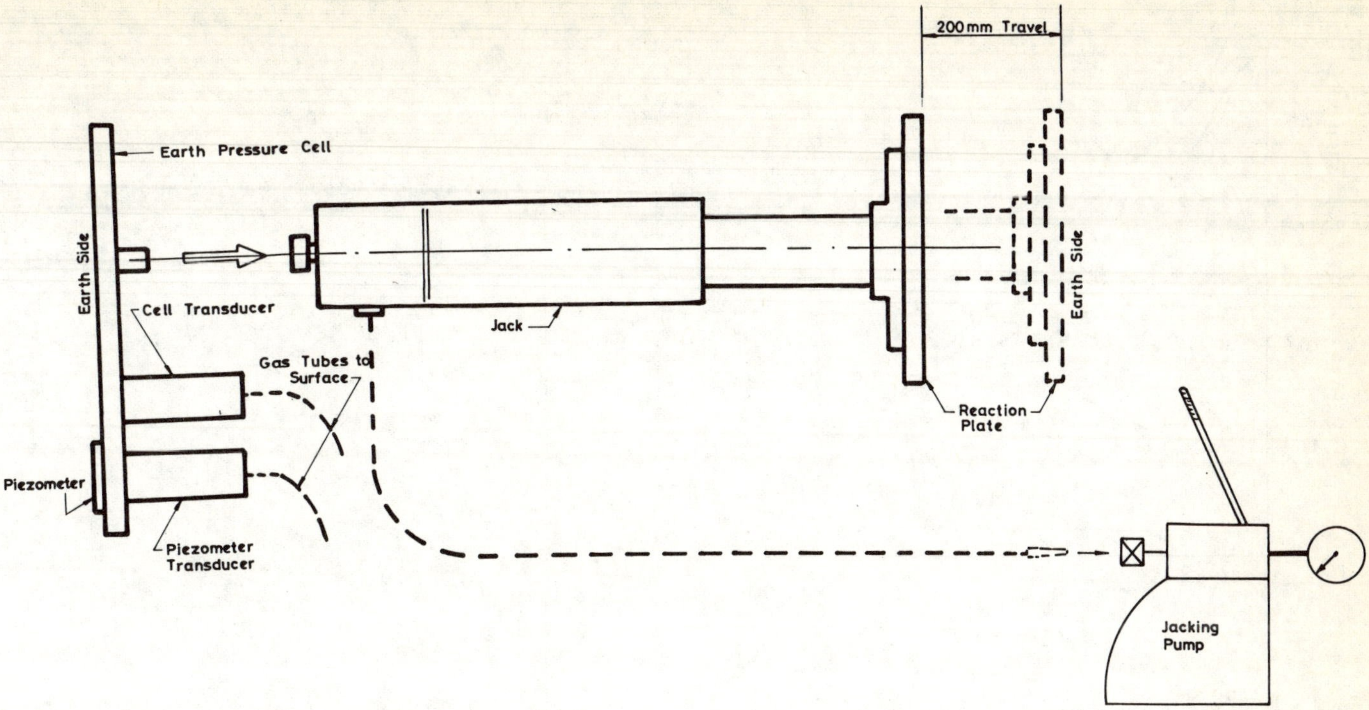

Figure 4. Diagramatic Arrangement of Earth Pressure Cell and Piezometric Unit

ground treatment injections, indicating they had been grouted up.

Some cells on the inside of the 30m deep excavations were uncovered - see Figure 5 - and the PV values were found to have increased by up to 25 kPa. This may have been a result of strain hardening or corrosion of the cells. The earth pressures quoted in this paper may therefore be underestimated by up to 25 kPa.

3 RESULTS

The Gloetzl cells and pneumatic piezometers were monitored as excavation progressed within the diaphragm walls. Howat and Cater (1985) noted that water pressures measured in the pneumatic piezometers exceeded overburden pressures inside the excavations long before final excavation level was reached and this was found to occur in all the other excavations instrumented with pneumatic piezometers. Vertical effective stresses therefore reduced to zero or even negative values, and yet horizontal effective earth pressures of between 15 and 500 kPa were still recorded. The very high water pressures recorded show that these unconfined strengths are not attributable to capillary pore water tensions and must therefore be a reflection of a true material cohesion.

The maximum values of these unconfined effective passive pressures are shown plotted against the void ratio of the decomposed gra-

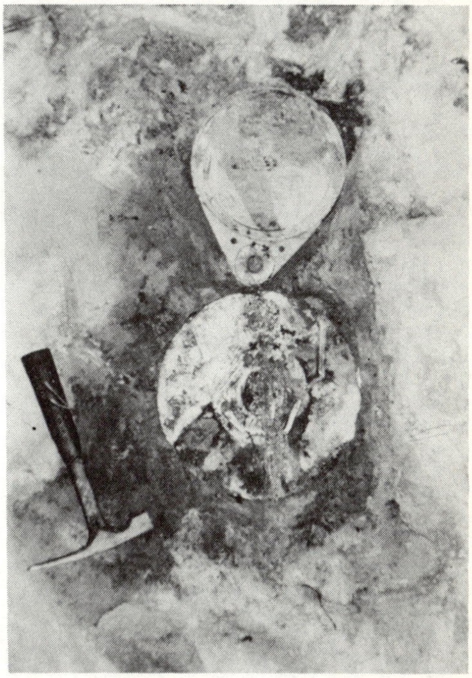

Figure 5. Exposed Earth Pressure Cell and Pneumatic Piezometer. Below is the Reaction Plate for a Second Unit cast in the Outer Face of the Wall.

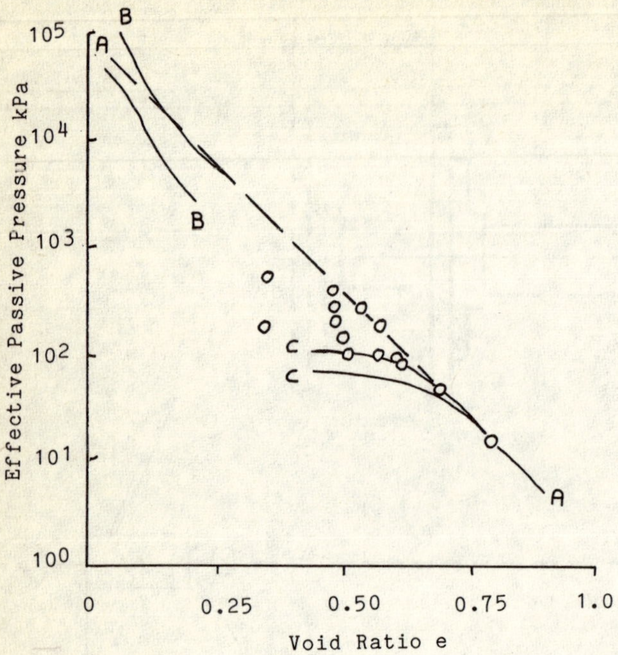

A-A $\log q_u = 5(1-e)$

B-B Laboratory unconfined compressive strength limits for granite (see Fig. 3)

C-C Laboratory unconfined compressive strength limits for decomposed granite (see Fig. 3)

Figure 6. Recorded effective pasive earth pressure under zero vertical effective stress - All case histories.

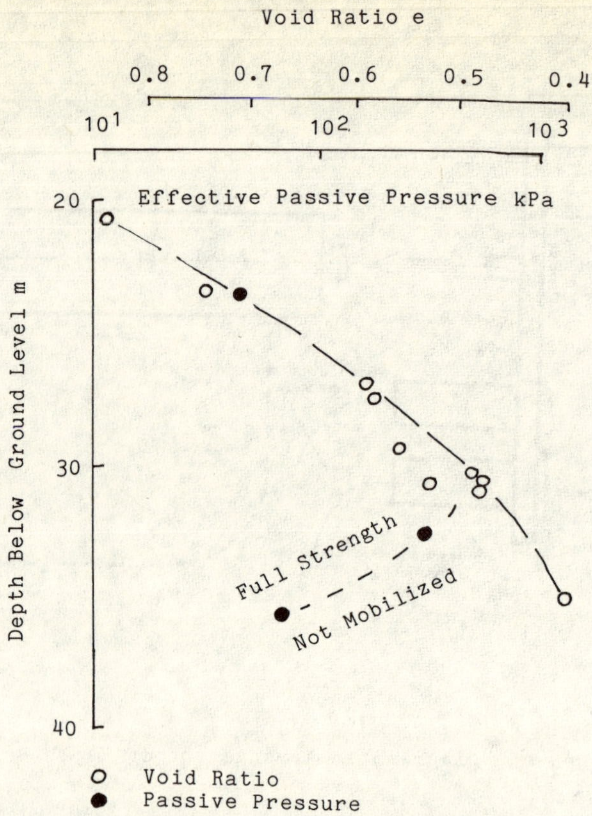

Figure 7. Recorded effective passive earth pressure under zero vertical effective stress - Case history 4.

nite found in adjacent boreholes on Figure 6. The greatest values fit an extension of the relationship found for laboratory tests on high porosity decomposed granite towards the values found by others for less weathered granite. This relationship can be expressed:

$$\log q_u = 5(1-e) \qquad (1)$$

where q_u is the unconfined compressive strength in kPa

and e is the void ratio of the decomposed granite

Many of the measured values of effective passive pressure at zero effective vertical stress are lower than the values given by Equation (1). Figure 7 shows a probable explanation for the lower values. It appears that the full unconfined strength was not mobilized at the lower levels of wall embedment as sufficient strength was available at higher levels to resist wall deflections.

4 PRACTICAL EFFECTS ON TEMPORARY SUPPORT

The temporary support design was based on nominal values of cohesion giving calculated unconfined compressive strengths of only 20 kPa. As a result extremely heavy temporary support was installed at the lower levels of excavation, generally 3 layers of struts preloaded to 1 MN/m run of wall. The struts were designed for working loads to increase to 2 MN/m run of wall as excavation progressed.

Many of these struts were monitored with vibrating wire strain gauges. In most cases the gauges indicated that a significant decrease in load was occurring as excavation progressed. In some cases negative values of strut load were deduced from the data and it was initially concluded that the strain gauges were defective.

In one case the strut load was affected by a ground treatment operation being carried out on the outer face of the wall. The strain gauge reaction to varying the number of injection pumps in operation at any one time is shown on Figure 8. This was the only case of recorded increase in strut load, and suggests that the strain gauges were in fact not defective.

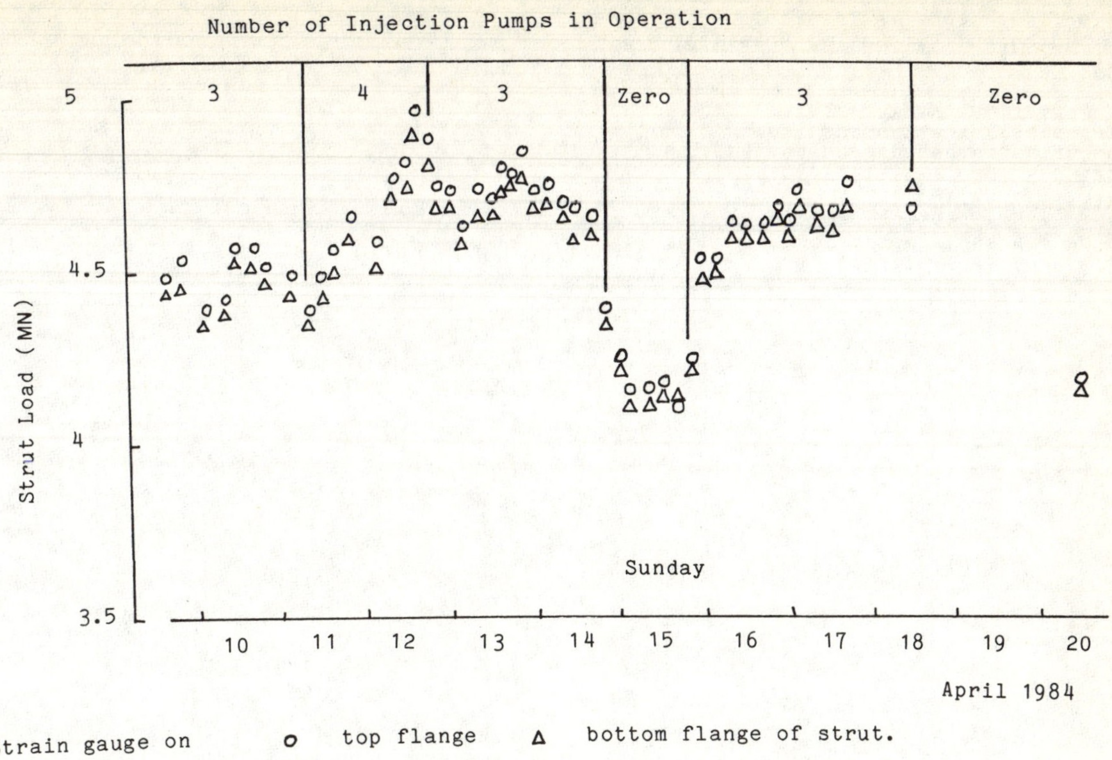

Figure 8. Influence of ground treatment injections on strut loads. Case history 4 - 16.5m below ground level.

5 CONCLUSIONS

Monitoring the passive earth pressures in Gloetzl cells cast into diaphragm walls shows that the strength of decomposed granite measured under zero vertical effective stress increases exponentially with decrease in void ratio, much in the same way as found by Dobereiner and de Freitas (1986) for sandstones.

It was not possible to measure the unconfined strength directly in the laboratory, because of the development of negative pore water pressures. Extrapolation of the results of triaxial tests carried out at cell pressures of 30 kPa or more gives unconfined strengths an order of magnitude lower than those found in situ for low void ratio decomposed granite.

It appears that the temporary bracing for the excavation was significantly overdesigned and that calculated increases in strut load never occurred, except where affected by high pressure grout injections on the outer wall face. It therefore seems likely that the strengths measured in situ are representative.

It is recommended that future temporary bracing for deep cuts in decomposed granite be designed using unconfined strength estimated from void ratio, not triaxial tests, and confirmed by monitoring with Gloetzl cells and pneumatic piezometers.

ACKNOWLEDGEMENTS

This study was the result of work carried out while the author was employed by the Hong Kong Mass Transit Railway Corporation. Sincere thanks are due to Mr. Frederick Wan and the staff of the Corporation's geotechnical instrumentation monitoring team who tirelessly collected all the data. The author is grateful to the Corporation for permission to publish. The views and conclusions expressed are not necessarily those of the Corporation.

REFERENCES

Baynes, F.J. & Dearman, W.R. (1978). The relationship between the microfabric and the engineering properties of weathered granite. Bull. Int. Assoc. Eng. Geol. 18: 191-7.

Dobereiner, L. & de Freitas, M.H. (1986). Geotechnical properties of weak sandstones. Geotechnique 36: 79-94.

Howat, M.D. (1985). Completely weathered granite - soil or rock? Q.J. Eng. Geol. 18: 199-206.

Howat, M.D. & Cater, R.W. (1985). Passive strength of completely weathered granite.

1st int. Conf. Geomech. in Tropical Lateritic and Saprolitic Soils, Brasilia, 2: 371-9.

Irfan, T.Y. & Dearman, W.R. (1978). Engineering classification and index properties of a weathered granite. Bull. Int. Assoc. eng. Geol. 17: 79-90.

Lumb, P. (1983). Engineering properties of fresh and decomposed igneous rocks from Hong Kong. Eng. Geol. 19: 81-94.

Skin friction on driven precast concrete piles founded in weathered granite

J.PREMCHITT, Geotechnical Control Office, Hong Kong
I.GRAY, Geotechnical Control Office, Hong Kong
J.B.MASSEY, Geotechnical Control Office, Hong Kong

ABSTRACT: The design of piled foundations in reclamation in Hong Kong is based on suitably conservative estimates of skin friction derived principally from studies elsewhere. There is sparse locally applicable data, particularly for the tropically weathered granites which frequently provide the bearing strata for the piles,but also for the overlying compressible deposits.Also, many of the overseas studies which provide the basis for current design practice were model studies or short term studies of working piles in different ground conditions. Long term field studies of working piles in Hong Kong are required to determine whether less conservative design assumptions can be used, particularly for older reclamation areas, leading to more economic foundation design. This paper describes the initial findings from such a study.

1 INTRODUCTION

Urban development in Hong Kong is increasingly spreading to land reclaimed from the sea. The main reasons are scarcity of flat land and severe problems associated with construction on steep slopes in the hilly areas (Lumb, 1975; Brand, 1985). Buildings on reclamation usually require deep foundations, consisting of piles penetrating to depths of up to 50 m. For these deep foundations, the skin friction on the shafts of the piles is a major determining factor in their bearing capacity and settlement characteristics.

Because of the sub-tropical climate, with high temperatures and heavy rainfalls in summer, the weathering processes in Hong Kong have been very active and have produced thick mantles of soil derived from insitu weathering of rocks in most parts of the territory. Driven piles in reclamations usually penetrate into, and bear on, the weathered rock. The weathered zones vary with depth, from residual soil having no remnants of the parent rock to lightly weathered rock containing less than 10% soil, but the bearing stratum of the driven piles normally consists of relatively strong extremely weathered granite having almost no corestones.

There exists extensive data on the performance of driven piles in alluvial and marine deposits (e.g. Broms, 1979), but there are very sparse data for piles in weathered rock, particularly on the development of skin friction. This is also true for negative skin friction in the compressible strata under reclamation in Hong Kong. In the absence of confirmatory data, a conservative design approach is usually adopted. Therefore, there is a real need for a study on actual working piles in reclamation to determine relevant parameters, in order to verify the conventional design assumptions.

A comprehensive long term study programme is being carried out by the Geotechnical Control Office, Hong Kong Government, in cooperation with the Architectural Services Department, to assess the development of skin friction on working piles in various subsurface strata, including weathered granite. These piles form part of the foundation of the new Public Works Central Laboratory, which is being constructed on the Kowloon Bay reclamation. The foundation consists of a total of 172 driven precast concrete "Daido" piles. The piles are prestressed, steam-cured spun concrete piles, tubular in section, with a 500 mm outside diameter. In this study programme, seven of the working piles were instrumented, and five of these piles were driven to depths greater than 40 m. Load tests were carried out on three of the instrumented piles, with loads up to 5400 kN.

The instrumented piles were driven, and will be incorporated into the building,in exactly the same manner as the other working piles. Study of the behaviour of these working piles will provide a great advantage over other investigations conducted elsewhere on test piles under experimental conditions. However, instrument installation at the construction site, where the contractors were working to a tight schedule, posed many constraints, including little control over timing and a high risk of damage to the instruments.

This paper briefly describes the instrumentation, and presents the results of the field measurement of the development of skin friction. Comparison is made between the skin friction developed in weathered granite and that built up in the other soils encountered at the site. The results of long term monitoring of the strain gauges are also presented.

2 WEATHERED GRANITE

Comprehensive discussion of Hong Kong weathered granites has been given by Ruxton & Berry (1957), Lumb (1965) and Hencher & Martin (1982). In summary, the sub-tropical climate enhances the decomposition of rock minerals in the granite by chemical action of water. The plagioclase, orthoclase and biotite are highly susceptible to decomposition, while the quartz remains unchanged. During the course of weathering, derived products such as colloidal silica, potash and soda are leached out in solution, together with some of the secondary clay minerals, resulting in an increase in porosity. The remaining soil consists largely of quartz and kaolinite, with final porosity generally greater than 0.50. In the late stages of weathering the soil consists mostly of sand and silt, with about 10-30% clay content.

The weathering processes are most active in the discontinuities of the rock mass in the zone above the permanent water table. Chemical action penetrating inward from the discontinuities results in 'corestones' remaining within the matrix of weathered materials. The thickness of the weathered granite profile is usually in the order of 30 m, and contains corestones having sizes greater than 1 m. The weathering profile is gradational. Examined from the ground surface downward, there is an upper zone consisting entirely of soil, gradually changing to lower zones with increasing content of corestones and more structure retained from the parent rock, until solid unweathered rock is found at depth. The exact configuration of zones varies from site to site.

3 PILE STUDIES

Typical subsurface strata encountered in Hong Kong reclamation are, from ground surface down, fill, marine deposits (clay and silt), alluvial deposits (sand and clay) and weathered granite, although the sequence of marine and alluvial deposits is more complex in places (GCO, 1986). Piles are usually driven through the compressible layers and founded at some depth into the weathered granite. Initial estimates of depth of pile penentration are normally made on the basis of data obtained from a ground investigation, particularly SPT N values. The actual founding depth is determined from the rate of pile penetration during pile driving according to a final set criterion (the Hiley formula). In practice, this usually corresponds to the zone consisting of a completely decomposed granite matrix with almost no corestones (extremely weathered), which generally has SPT N values well in excess of 30.

The load capacity of the pile is derived from the control of final set before termination of pile driving. This semi-empirical approach combines the end bearing resistance and positive shaft friction resistance, and the division of load between these is generally not known. The working load on the pile is estimated from building load (dead and live loads) plus allowance for future development of negative skin friction. The latter normally includes estimates for the fill and marine deposits. In newly reclaimed land, the adopted values of $\beta(K_o \tan \phi')$ for these two layers can be as high as 0.60 and 0.35 respectively, while even in 20 year old reclamation β values of 0.25 and 0.15 respectively have been used. These values were adapted from those commonly used overseas and from a few field tests in Hong Kong, e.g. Lee (1979) Lee & Lumb (1982). This practice could well result in conservative foundation design. Additional information on development of skin friction, both positive and negative, on working piles using strain gauge instrumentation is required to verify these parameters.

Cooke (1981), of the Building Research Establishment (BRE), UK stated two main reasons for pile instrumentation. The first is to improve general design methods by investigating individual test piles over relatively short periods of time. These piles are generally not part of buildings. The second is to check the validity of the design methods employed for a particular building. This usually requires long term observation of working piles. There are few reported cases of the latter, due to the difficulty of maintaining regular observations for a long time, and the high instrument failure rate (Cooke, 1981).

Studies of pile performance using strain gauge instrumentation have been carried out in many parts of the world. Most of these used test piles, for which long term performance under actual working loads was not monitored (Konrad & Roy, 1987; Ting & Toh, 1985; Reike & Crowser, 1987). Of these cases, only a few studied driven concrete piles (e.g. Mallard & Ballantyne 1976). Studies of actual working piles are also sparse (Cooke, 1981; Leung & Radhakhrisnan, 1985). The general conclusions from these studies are that, during initial stages, most of the pile resistance is derived from shaft friction, with a small contribution from end bearing. However, this distribution may change with time (Promboon et al, 1972). For test piles with immediate fill surcharge, development of negative skin friction is rapid (Fukuya et al, 1982), but there are no studies of working piles in an actual reclamation, where residual settlement could be much less than that of such a test site.

Two types of strain gauge have usually been used, namely electrical resistance (ER) and vibrating wire (VW) strain gauges. The operating principles of the two types are entirely different (Hanna, 1985). Cooke (1981) mentioned some advantages of VW gauges, but stated that they are not obviously more reliable than ER gauges. He found that a number of circuits of both types of gauge broke down after five years, but that some gauges of both types were still working. However, VW gauges have been used increasingly in recent studies.

4 INSTRUMENTATION

The piles used in this study were manufactured by Daido Concrete (HK) Ltd, and the instruments were installed in seven of these piles at the manufacturing plant. A brief description of the technique is provided here.

The piles are prestressed, steam-cured spun concrete piles. They are tubular in section, having 500 mm outside diameter. Two of the instrumented piles are 'thin' piles (300 mm inside diameter), and five of them are 'thick' piles (250 mm inside diameter). They are fabricated in lengths of 8 and 12 m, and these are welded together on site during driving to extend to the final depth. The piles are steam-cured at a temperature of about 200°C. This shortens the curing period, such that the entire process takes less than one week. The prestress in each finished section is 5.25 MPa, and the modulus of elasticity of the pile is 37.7 GPa. The allowable bearing capacities are 2300 kN and 2700 kN for the thin and thick piles respectively.

The strain gauges used were of the VW type, with additional ER type gauges as a low cost supplement and back-up. The two types also provided an independent check, one against another. The VW gauges were TES/5.5 (Gage Technique Ltd, UK), especially treated to resist temperatures of up to 240°C. Two types of ER gauges, referred to here as ER1 and ER2, were used. The ER1 gauges were OYO-ORH16B (Oyo Measurement Co. Ltd., Japan). The ER2 gauges were custom-made for this study using Kyowa KFH-5-C1-11 foil gauges (Kyowa Electronic Instrument Co. Ltd., Japan). In this latter type, there were two gauges incorporated in each unit.

Figure 1 shows an example of the pile instrumentation lay-out. The gauges were installed at five depths in each pile : one in each of three 8 m sections, and at two different depths in one 12 m section. They were distributed to three positions around the pile section approximately equally spaced 120° apart.

The number of instrumented piles and the number of gauges set in each pile were higher than the minimum required, to allow for possible losses of instruments in the short and long term. This also enabled better accuracy to be obtained in the calculation of stress and strain at each depth, and enabled comparison of results from similar piles.

A total of 115 gauges were installed in this study, consisting of 24 VW gauges, 44 ER1 gauges and 47 ER2 gauges.

5 PILING

The new Public Works Central Laboratory is situated in Kowloon Bay reclamation area, close to the Hong Kong International Airport. Reclamation was completed in 1967. Four boreholes were drilled at the site in 1983 as a part of the ground investigation for the foundation design. The borehole logs obtained are illustrated in Figure 2. The subsurface strata consist of fill, marine deposit, alluvial deposit and weathered granite. Each of the first three layers is approximately 10 m thick. The SPT N values are also given in the figure. The presence of some sand-size particles in the marine deposit may be indicative of disturbance and mixing with the fill during reclamation.

All piles were driven section by section, each 8 m or 12 m long. The first section was equipped with a cone-shaped steel shoe. The next section was set upright and welded to the top of the previously driven section. The instrumented piles were driven during the period from 30 March to 13 May 1987. Driving was terminated when the rate of pile penetration satisfied the final set criterion, which was calculated using the Hiley formula. Ground heave, in the order of 150 mm, was observed near each pile after driving. The hollow centres of all instrumented piles were subsequently filled with concrete grout.

The two thin-walled (300 mm inside diameter), instrumented piles were used for initial trials of the installation techniques. A large proportion of gauges in these piles was lost. As a result of further improvements after the initial trials, there were smaller losses for the remaining five piles (250 mm inside diameter). About 20% of all gauges in these piles were lost during manufacturing of the pile and during pile driving. Up to March 1988, there were further losses in the form of either complete breakdown or drifting of readings, mostly in the ER gauges, but it is expected that useful results will continue to be obtained from the remaining gauges. The types of gauge in descending order of durability were VW, ER1 and ER2. Therefore, the determination of strains after the load testing was based largely on VW readings, using ER gauge readings only as confirmatory data.

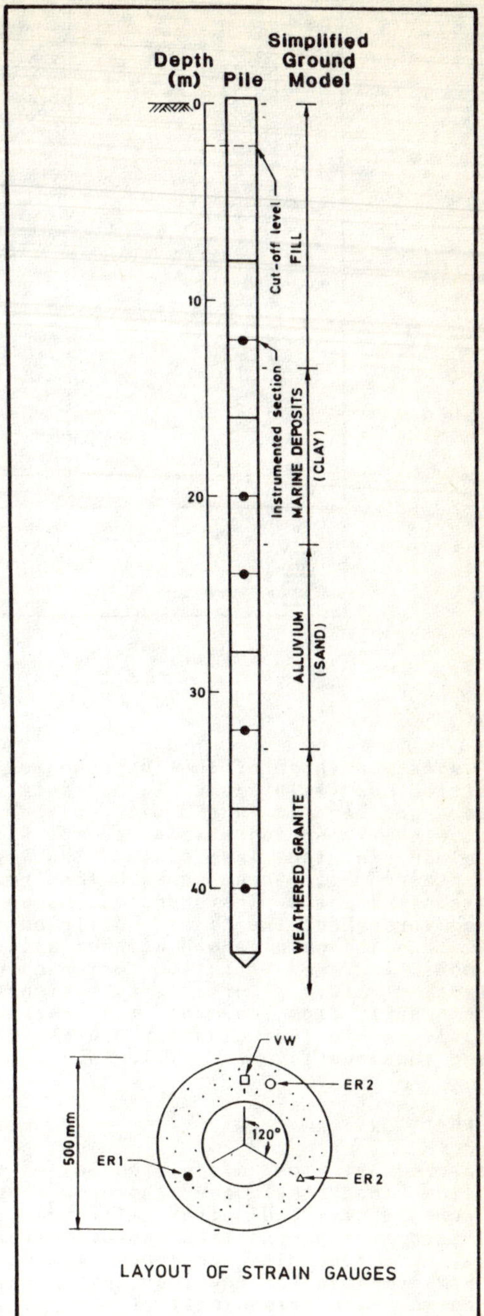

Figure 1. Details of Pile Instrumentation

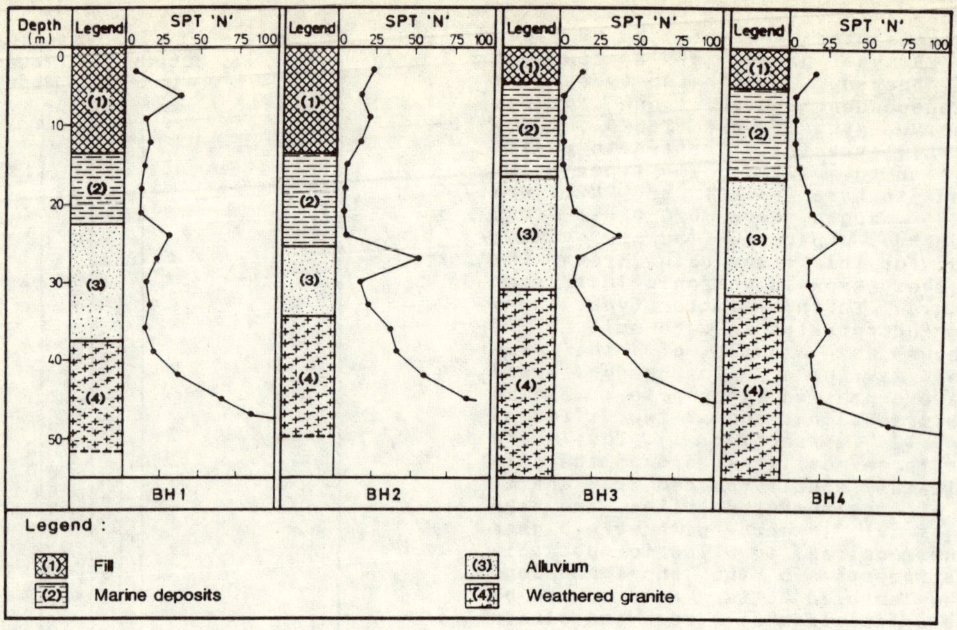

Figure 2. Borehole Logs and SPT N Values at the Site

Load tests on three of the instrumented piles were carried out (maintained load test). The load was applied in four cycles, with maximum load in each cycle increasing up to twice the design load in the last cycle. The load was then maintained for three days before removal. The strains in the instrumented piles were monitored throughout the tests. Distribution of strains along the pile length allowed estimation of the mobilized skin friction for each of the subsurface strata. For the sake of brevity, only the results from one of the instrumented piles (pile A, inside diameter 250 mm) will be discussed subsequently.

6 SKIN FRICTION

The measured strains along the pile length during the load test are shown in Figure 3. These were determined with reference to the initial zero strain condition immediately before the test. The four distributions were obtained at the maximum loads of the four loading cycles. Using the moduli of elasticity of the pile (37.7 GPa) and concrete grout in the central hole (11.5 GPa) the strains were converted to axial forces, and these are indicated on the scale at the bottom of Figure 3. The uppermost points were the applied loads, while all other points were derived from measured strains. At the maximum load, the axial forces at 11 m and 39 m depth were about 4000 kN and 450 kN respectively, showing a large decrease of axial force with depth. A small redistribution of axial force along the pile length occurred after the load had been maintained for three days. A larger redistribution would be expected to occur if the load had been maintained for a much longer period (Promboon et al, 1972; Leung & Radhakhrisnan, 1985).

It can be seen from Figure 3 that most of the

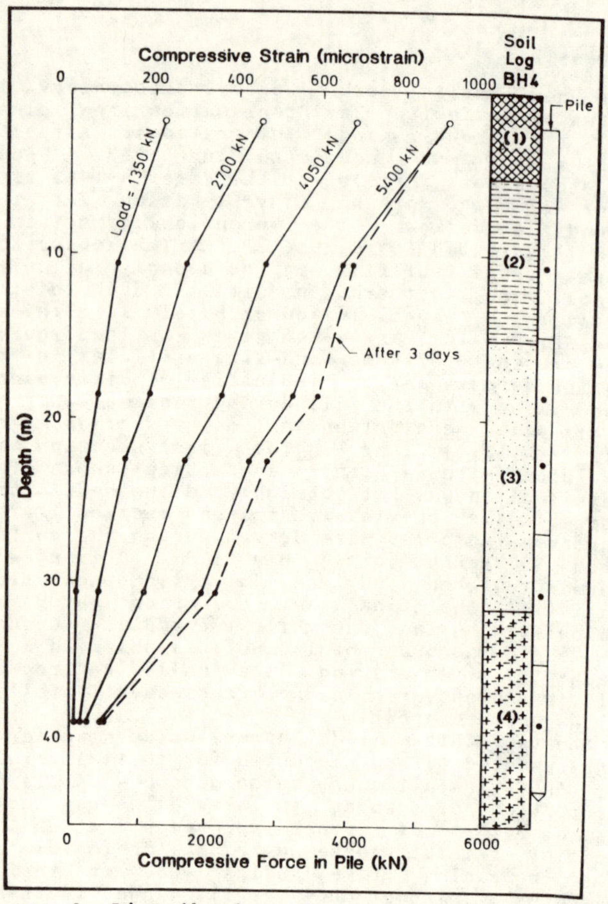

Figure 3. Distribution of Compressive Strain and Force in the Pile during Load Test

applied load was taken up by positive skin friction resistance, while end bearing resistance was small. It is estimated that more than 90% of the load was taken up by skin friction during the load test.

The unit frictional resistance was calculated from the difference in axial force at two gauge levels divided by appropriate pile surface area. This is plotted against the displacement between the pile and the ground for different depth ranges in Figure 4. The displacements were estimated from the cumulative elastic shortening of the pile from the pile tip, assuming that there were no significant pile tip penetrations into the ground during load test. A comparison between total calculated shortening and settlement measured from settlement gauges at the pile top showed that the error in estimated displacements was less than 10%.

From Figure 4, it is seen that the friction was approaching full mobilization at displacements greater than 5 mm. For weathered granite, the bottommost layer, the displacement was small and the friction curve did not show any flattening. Therefore, the developed friction was considered far from fully mobilized at the value of over 100 kPa, and the fully mobilized value could be much higher than this. In comparison, because of the large displacements, mobilized values for other layers could be estimated, and they are 80 kPa for fill, 65 to 95 kPa for the alluvial deposit and 45 kPa for the marine deposit.

There are few reported cases of determination of skin friction on piles in Hong Kong weathered granite. The friction derived from the results of a pile load test on a steel tube pile in Tuen Mun by Lee & Lumb (1982) indicated values greater than 60 kPa. On the basis of a load test on a bored pile at a site in the vicinity of the present study, Holt et al (1982) estimated the friction value to be more than 150 kPa. On the other hand, Fraser & Kwok (1986), using load test results from five instrumented bored piles at various locations, suggested that the maximum value of the skin friction could be in the order of 100 kPa. However, they emphasized the importance of the construction techniques for the bored piles on the skin friction obtainable in weathered granite. If the wall of the pre-bored hole was disturbed or there was poor contact between the cast in-situ concrete pile and the ground, the mobilized skin friction could be much lower than the maximum value.

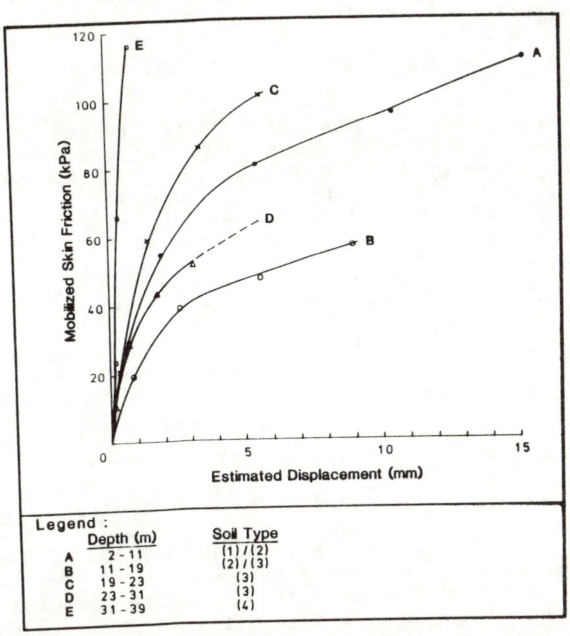

Figure 4. Relationship between Mobilized Skin Friction on Piles and the Estimated Soil/Pile Displacement

The comparison of the results from this study with the above cases confirms that the skin friction value in weathered granite for driven concrete piles should be higher than 100 kPa. In general, skin friction on bored piles is expected to be lower than that for driven piles because of disturbance and stress relief during preboring. Driven piles displace the soil by a volume equal to that of the pile and, therefore, the zone of soil adjacent to the pile will actually be more dense than the virgin ground. The skin friction value for a concrete pile is normally greater than for a steel pile in the same ground conditions (Tomlinson, 1977). This would explain the lower value derived from the study of Lee & Lumb (1982).

7 LONG TERM MONITORING

Strain gauge monitoring continued after pile driving and load test. The strains measured during the period from May 1987 to March 1988 are shown in Figure 5, for the pile discussed above. Large strain changes occurred during pile driving and during pile load test, but no significant changes in strain have occurred subsequently.

The piling work was completed in early July 1987, and the pile caps and superstructure were started in late November 1987. There were no construction activities in the interim period, and any changes in strain had to be due solely to negative skin friction effects. However, changes in strain were negligible during this five month period. The sensitivity of the instrumentation is demonstrated by the period from late November onwards, when work on the superstructure commenced. The measured strains responded to the excavation around the pile head and subsequent casting of the pile cap, even though the changes in load were relatively small. It can be concluded that development of negative skin friction at this site (a 20 year old reclamation) has been negligible during the period up to commencement of construction of the superstructure.

8 CONCLUSIONS

(a) Load tests on instrumented piles have enabled estimation of positive skin friction between weathered granite and driven concrete piles. The value determined is greater than 100 kPa, in comparison with maximum values of 80 kPa for fill, 65 to 95 kPa for alluvium and 45 kPa for marine deposit.

(b) The estimated value for weathered granite agrees well with those derived from past studies by Lee & Lumb (1982), Holt et al

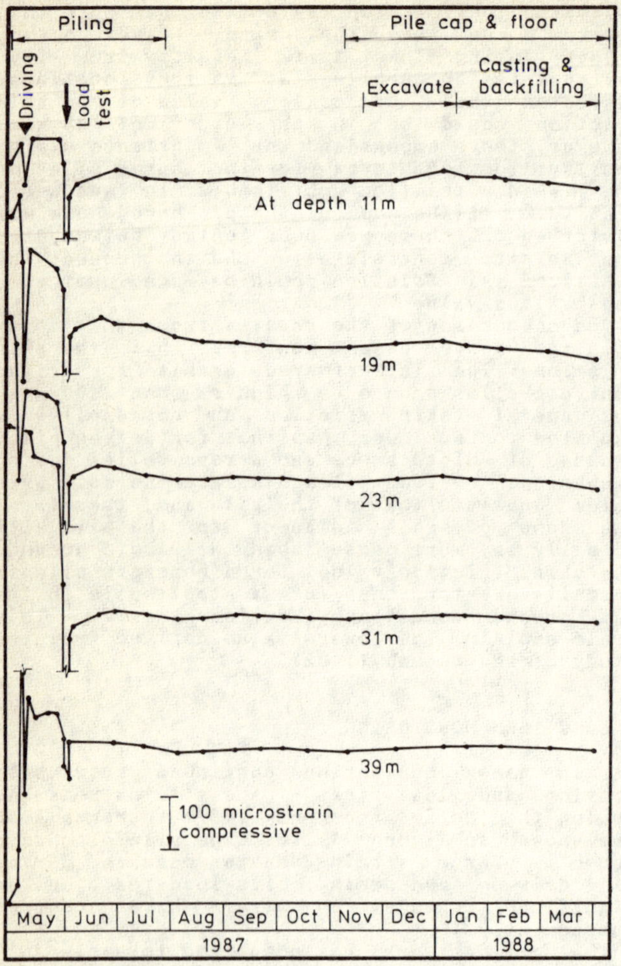

Figure 5. Strain Changes in the Pile during the Period May 1987 to March 1988

(1982) and Fraser & Kwok (1986).

(c) Long term monitoring has indicated negligible development of negative skin friction in the 20 year old reclamation during the five month period prior to construction of the superstructure.

ACKNOWLEDGEMENT

This paper is published with the permission of the Director of Civil Engineering Services of the Hong Kong Government. The authors also wish to thank the Architectural Services Department, Daido Concrete (HK) Ltd and Franki Contractors Ltd for their assistance in the installation of the instruments. Valuable contributions were made by a large number of individuals, particularly H. Kozuma (Daido Concrete (HK) Ltd), Michael Chan (Franki Contactors Ltd), W.K. Fung (Architectural Services Department), and Y.W. Ng and C.F. Chow (Geotechnical Control Office). The untiring efforts of Ms Kitty Ng in typing this paper is gratefully acknowledged. This project was initiated by the Pile Study Group of the Geotechnical Control Office.

REFERENCES

Brand, E.W., (1985). Landslides in Hong Kong. Proceedings of the Eighth Southeast Asian Geotechnical Conference, Kuala Lumpur, vol. 2, pp 107-122.

Broms, B. (1979). Negative skin friction. Proceedings of the Sixth Asian Regional Conference on Soil Mechanics and Foundation Engineering, Singapore, vol. 2, pp 41-76.

Cooke, R.W. (1981). Instrumented piles and pile groups. Piles and Foundations, edited by F.E. Young, pp 166-172, Thomas Telford Ltd, London.

Fraser, R.A. & Kwok, D. (1986). Mobilization of skin friction in bored piles in Hong Kong. Proceedings of the International Conference on Deep Foundations, Beijing, vol. 2, pp 2.65-2.74.

Fukuya, T., Todoroki, T. & Kasuga, M. (1982). Reduction of negative skin friction with steel tube NF pile. Proceedings of the Seventh Southeast Asian Geotechnical Conference, Hong Kong, vol. 1, pp 333-347.

Geotechnical Control Office (1986). Geology of Hong Kong Island and Kowloon. Hong Kong Geological Survey Memoir No. 2. Geotechnical Control Office, Hong Kong, 134 p.

Hanna, T.H. (1985). Field Instrumentation in Geotechnical Engineering. Trans Tech Publications, Clausthal-Zellerfeld, 843 p.

Hencher, S.R. & Martin, R.P. (1982). The description and classification of weathered rocks in Hong Kong for engineering purposes. Proceedings of the Seventh Southeast Asian Geotechnical Conference, Hong Kong, vol. 1, pp 125-142.

Holt, D.N., Lumb, P. & Wong, P.K.K. (1982). Site control and testing of bored piles at Telford Gardens an elevated township at Kowloon Bay, Hong Kong. Proceedings of the Seventh Southeast Asian Geotechnical Conference, Hong Kong, vol. 1, pp 349-361.

Konrad, J.M. & Roy, M. (1987). Bearing capacity of friction piles in marine clay. Geotechnique, vol. 37, pp 163-175.

Lee, P.K.K. (1979). Negative skin friction on piles. Asian Architect and Builder, vol. 8, no. 1, pp 31-35.

Lee, P.K.K. & Lumb, P. (1982). Field measurements of negative skin friction on steel tube piles in Hong Kong. Proceedings of the Seventh Southeast Asian Geotechnical Conference, Hong Kong, vol. 1, pp 363-374.

Leung, C.F. & Radhakrishnan, R. (1985). The behaviour of a pile-raft foundation in weak rock. Proceedings of the Eleventh International Conference on Soil Mechanics and Foundation Engineering, San Francisco, vol. 3, pp 1429-1432.

Lumb, P. (1965). The residual soils of Hong Kong. Géotechnique, vol. 15, pp 180-194.

Lumb, P. (1975). Slope failures in Hong Kong. Quarterly Journal of Engineering Geology, vol. 8, pp 31-65.

Mallard, D.J. & Ballantyne, J.L. (1976). The behaviour of piles in Upper Chalk at Littlebrook D Power Station. Géotechnique, vol. 26, pp 115-132.

Promboon, S., Brand, E.W. & Buchli, C. (1972). Observations on the deep bored piles for the Tha Chang Bridge, Bangkok. Proceedings of the Third Southeast Asian Conference on Soil Engineering, Hong Kong, pp 135-140.

Rieke, R.D. & Crowser, J.C. (1987). Interpretation of pile load test considering residual stresses. Journal of Geotechnical Engineering, American Society of Civil Engineers, vol. 113, no. 4, pp 320-349.

Ruxton, B.P. & Berry, L. (1957). Weathering of granite and associated erosional features in Hong Kong. Bulletin of the Geological Society of America, vol. 68, pp 1263-1291.

Ting, W.H. & Toh, C.T. (1985). An instrumented bored pile with soil improvement for increased shaft resistance. Proceedings of the Eleventh International Conference on Soil Mechanics and Foundation Engineering, San Francisco, vol. 3, pp 1483-1488.

Tomlinson, M.J. (1977). Pile Design and Construction Practice. Cement and Concrete Association, London, 413 p.

Some construction projects in the decomposed rock soils of Hong Kong

E.RAWCLIFFE, Wang Consultants, Singapore

SYNOPSIS

The decomposed soils of Hong Kong have good engineering properties but which some years ago appeared difficult to quantify accurately. Construction works were frequently designed on data based on Standard Penetration Tests and reasonably accurate descriptive data from boreholes. Guidance was provided in government regulations to assist in design of foundations. Two such cases and a third project carried out since formation of the Geotechnical Control Office are described in this paper.

ROCK AND SOIL TYPES

The rock types present in Hong Kong are listed by Berry (Ref. 4) as

Granitic Rock
Volcanic Group of Rocks
Red Bed Sedimentary Rocks
Older Sedimentary Rocks

Much of the literature on Hong Kong soils and rocks appears to concentrate on the granitic and volcanic types, possibly because sedimentary rocks are met with less frequently in construction works. However, it was reported in the magazine "Construction Today" Feb. 1988, that caverns up to 75m deep had been unexpectedly encountered in limestone rock at Tuen Mun in the north west of the territory.

The major part of the soil overlying rock is derived from weathering of the granite and volcanic rocks, but there is also a substantial area of marine deposits in the New Territories, where many new towns are being developed.

Chemical weathering is very active in Hong Kong and both granitic and volcanic rock can become completely decomposed to form a mantle tens of metres in thickness. The depth of weathering varies widely, depending on many factors such as mineral constituents of the rock, degree of fissuring and the supply of percolating water. The weathering extends outwards from water bearing fissures, gradually producing a matrix of fully decomposed rock surrounding isolated boulders of sound rock. In some cases these boulders, which may be small or very large, do not decompose and cause various problems in construction.

Grades of decomposition from I (fresh rock) to VI (soil) have been established for classification purposes. (Ref. 2 Table 2.4 p.16). The degree of decomposition is not a gradual downward change with depth and zones of decomposition from A to D have been established depending on the proportions of the various grades within a zone. (Ref. 2 fig. 2.4 p.15). These grades and zones largely depend on visual examination but nevertheless provide a very useful descriptive system. A feature of these weathered in place soils is that they retain the former fissure structure of the original rock.

The decomposition process and the effects of heavy rainfall result in masses of soil slipping down from higher steeper slopes to form less steep slopes in the foothills. The properties of this slipped soil, or colluvium, are similar in many ways to those existing before the slip, except that the material is generally in a looser condition and the fissure structure has been lost.

Marine Sediments. These are typically soft grey organic clays with shell fragments, sensitive to disturbance, found in coastal areas and river estuaries. In reclamation areas, such as those in the New Territories, care is required to ensure that un-necessary problems are not created by unsuitable filling methods over these clays. Close inshore or in estuaries, the sediments grade into sandy deposits of marine or alluvial origin.

Fill. Much of the coastline and large low-lying areas in the New Territories have been reclaimed using the natural decomposed soils, particularly decomposed granite, which forms a good quality fill. The properties of such fills are similar to those of the colluvium.

In general, the decomposed rocks have good engineering properties and are not difficult with which to work when adequate consideration is given to drainage.

TEXTILE FACTORY KOWLOON (1976)

This level site was 70m x 50m with a stone sea-wall forming one side. The column working loads were 20,000kN to 23,000kN.

Eleven boreholes were drilled and provided an adequate descriptive record of the soil formation. No undisturbed samples of soil were taken, although frequent Standard Penetration Tests were made. Bedrock was proved in each borehole to a depth of 5m and the core recovery and R.Q.D. values recorded. A number of specimens were prepared from the NX sized cores and their crushing strengths determined.

The soil stratification may be summarised as:

LAYER	THICKNESS	DESCRIPTION
1	2m - 7m	loose clayey silty sand with boulders - FILL. N=4
2	0m - 2m	soft marine CLAY N2 to N5. Cu 12 to 30kPa
3	0m - 5m	Alluvium. Firm to stiff silty sandy CLAY N6 to N35. Cu 35 to 200kPa
4	0m - 11m	Completely Decomposed Granite Ne 40 to Ne 140
5	5m+	grey and pink fine-grained granite ROCK R.Q.D. 19% av. Core Recovery 90% to 100%

G.W.L. was approximately 2m b.g.l.

A typical section is shown in Fig. 1, from which it can be seen that rock level was 3m to 20m below ground. The surface dipped 1 in 3.5 north to south and 1 in 10 west to east.

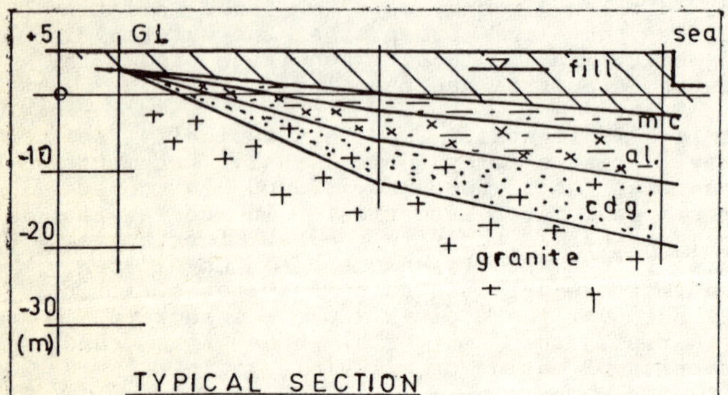

TYPICAL SECTION
TEXTILE FACTORY SITE FIG 1

A decision was made to use caisson foundations resting on fresh unweathered rock. Core samples at a depth of 1m into rock had an average crushing strength of 42,900kPa and a factor of safety = 8 yielded an allowable bearing pressure of 5,400kPa. However, to allow for the dip of the rock and the high degree of fissuring, this value was arbitrarily reduced using a factor of 0.75, giving a design allowable bearing pressure of 4,000kPa.

Ref. 5 Table IX recommended a safe bearing capacity of 5,000kPa for granite in hard sound condition with core recovery greater than 85%.

Hand-dug caisson foundations were adopted, generally about 2m in diameter and excavated 1m into fresh rock. Inflow of water was not a significant problem. Ref. 5 clause 38 recommended that plain or reinforced concrete for caissons should not be inferior to concrete having a works crushing resistance of 20MPa at 28 days and clause 34 recommended that the maximum permissible pressure on foundation concrete of this grade should be 5,300kPa. This grade of concrete was adopted for the caissons. Plain concrete was used for the shafts but reinforcement was used to tie the caisson shafts into pile caps and columns. The foundations have proved to be satisfactory.

This project demonstrates that, where there is adequate descriptive site investigation data, certain shortcomings - in this case lack of undisturbed sampling and testing of soils - can be overcome by experience of similar sites and use of design recommendations from government regulations to provide a safe, although probably conservative, solution to a foundation problem.

AIR FREIGHT GODOWN, KOWLOON (1976)

This was a structure with piled foundations and a basement requiring excavation to a depth of 5m. Working load per pile was 1,200kN.

Nine boreholes were drilled and provided an adequate descriptive record of the soil formation. No undisturbed samples were taken, although frequent Standard Penetration Tests were made.

The soil stratification may be summarised as:

LAYER	THICKNESS	DESCRIPTION
1	5m - 6m	firm sandy clay FILL. N4 to N15
2	2m - 4m	soft marine CLAY N2 to N6. Cu 12 to 37kPa
3	0m - 4m	firm silty CLAY (Alluvium) N11 to N24. Cu 70 to 150kPa
4	2m - 14m	medium dense silty SAND (Alluvium) N15 to N32. Ø 31° to 36°
5	14m - 26m	Completely Decomposed Granite Ne 36 to 59. Ø' 38° to 42°

Boreholes were drilled to a depth of 46m but did not encounter bedrock.
G.W.L. was approximately 1.5m b.g.l.

The problems to be solved on this site were:

(a) Design of a suitable piling system.
(b) Dewatering the excavation for basement construction.
(c) Support of the sides of the excavation.

Pile Design.

546mm diameter (21.5 inches) Colcrete piles were used. Their length of 35m ensured that the toes penetrated into dense CDG and the piles were formed by driving a steel tube with a detachable shoe and filling with a colloidal grout and crushed stone aggregate as the tube was withdrawn.

Since there was to be a basement, it was not necessary to consider the effects of downdrag, but for calculation purposes the soils down to a depth of 12m below ground were not considered in estimating shaft friction. The soils providing support were basically frictional i.e. the silty sand and the CDG. The Angle of Internal Friction Ø was determined for these soils from the S.P.T. results modified for depth by the Gibbs and Holtz method. The values from each borehole were checked and a typical range is shown in Fig. 2 with the appropriate assessment of Ø, Ka and Ko.

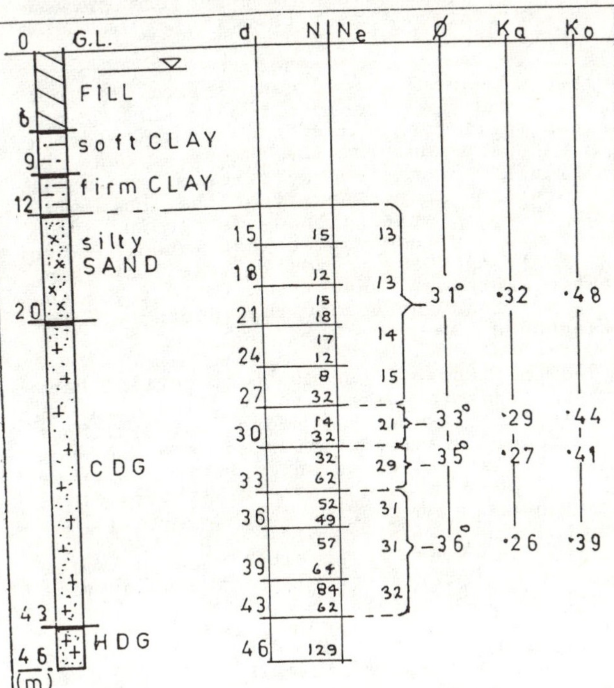

AIR FREIGHT GODOWN KOWLOON

Typical B.H. data and SPT results FIG. 2

From the values of Ø so determined, the shaft friction was estimated by the method proposed by Meyerhof:

Unit Shaft Friction (SF) = γ xKsxδxdxtan

γ = effective density
d = depth below ground surface
δ = angle of wall friction (=0.6xØ)
Ø = angle of internal friction
Ks = factor between 0.5 and 1 depending on relative density of the soil

The submerged density of the soil was taken as 9.4 kN/m3 and the value of Ks selected based on the relationship between S.P.T. and relative density given by Terzaghi.

Shaft Length (test pile) 35m		(works) 23m
Ultimate S.F.	0 - 12m	142kN
Ultimate S.F.	12m - 35m	1674kN
		1816kN

End Bearing.

This was assessed using Terzaghi bearing capacity factors. Since Nc=0 and N_γ was very small, the end bearing was based only on the Nq term.

Thus qu = γ x d (Nq-1)

where qu = ultimate end bearing capactiy
γ = effective density = 9.4kN/m3
d = depth to toe of pile = 35m or 23m
Nq = 44 (for Ø = 36°)

qu = 14,147kN/m2 and Qu = qu x A

The safe working load on the piles with a factor of safety of 3 was thus calculated as:

For Test Pile (3312 + 1816) 1/3 = 1709kN

For Project (2365 + 1674) 1/3 = 1346kN

A preliminary pile test and two works pile tests gave similar results and a typical result is shown in Fig. 3.

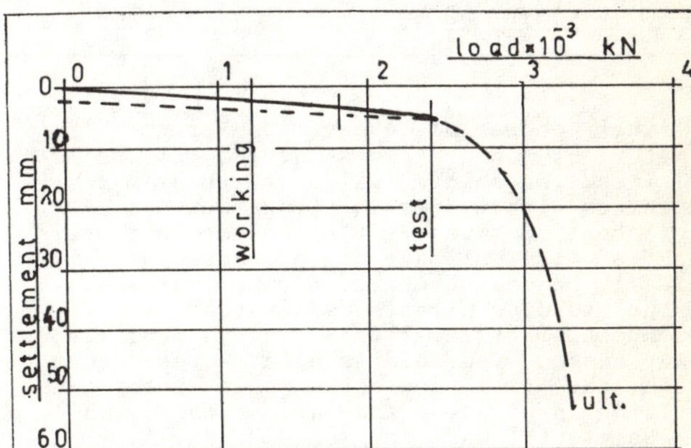

PILE LOAD TEST FIG.3

After installing the 551 piles, 18 wells were installed to dewater the site. 685mm diameter steel tubes with loose shoes were driven to 12m below ground and 250mm diameter slotted pipes surrounded by 10mm aggregate placed inside before withdrawing the tubes. Based on a coefficient of permeability k of 1x10 cm/sec it was calculated that sufficient drawdown would be obtained to dewater the site to 5m below ground level and the discharge per well would be a maximum of 14m3/hr.

The excavation was carried out leaving an embankment sloping at 30° around the site, this being calculated as sufficient to maintain stability. The pile caps and basement slab were constructed up to the toe of the embankment. Fig. 4.

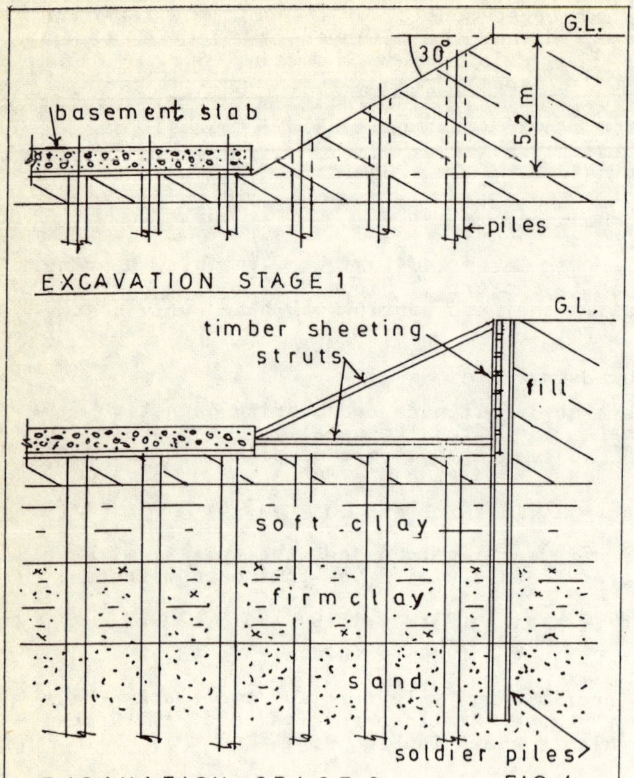

FIG. 4

The final phase of excavation around the perimeter of the site was carried out using steel I section soldier piles at approximately 2m centres with timber sheeting between. A gap of about 50mm was left between each timber layer to allow drainage of the retained soil should it become saturated during rainstorms and the soldier piles were strutted at the head and basement level from the basement slab already cast. The soldier piles were driven through the fill and soft clays into the underlying sand layer and later removed when the basement was completed. The soil properties used for design of the retaining wall were:

c'	=	0
Ø'	=	30°
bulk density	=	20kN/m3
submerged density	=	10kN/m3

This construction work was completed without difficulty and indicates that sheet piling is not always the cheapest solution, especially where the piles cannot be extracted, or when a boom in the construction industry drives up the price of steel.

SLOPE STABILITY AND ROADWORKS PROJECT HONG KONG ISLAND (1980)

The project involved the examination of the stability of existing slopes and structures and of new cut slopes over approximately 800m length of hillside, the design of appropriate improvement works and a flyover structure over the stabilised slopes.

The P.W.D. Geotechnical Control Office had been in operation for a number of years and published "Geotechnical Manual For Slopes" 1979 (Ref. 2) in draft form, a manual which is now widely accepted as setting out the guidelines for assessing slope stability and designing slope drainage.

Factors of Safety and Groundwater.

Factors of safety for slopes are based on a risk to life category and groundwater conditions associated with a 10-year return period rainfall. The safety factor for a 1000-year return period rainfall should not be less than 80% of that for the 10-year period.

Where the risk category is high as on this project, The minimum factor of safety for the 10-year return period is 1.4, but other factors, such as limitation of deformation to prevent structural damage, may require the use of a higher safety factor than the minimum.

It is a well-established fact that groundwater conditions have a significant effect on slope stability and the rise in phreatic surface due to water infiltration during heavy rainstorms, which are common in Hong Kong, can be significant. From the properties of the soil, the duration and intensity of the rainfall, the depth of a "wetting band" descending from the ground surface can be estimated. When this wetting band reaches the existing phreatic surface, the effect is to raise the level of that surface by the depth of the wetting band.

For design purposes, the depth of wetting band corresponding to 10-year to 1000-year return period rainfall should be added to the highest water level observed after an average wet season. Where there is an impervious surface to a slope such as chunam or ashlar, this condition does not arise since the soil is protected from entry of water from the surface. However, a possible rise in water level due to water entering the ground outside the site must be considered.

The wetting band principle does not apply to slopes in rock. Water pressure in rock joints must be investigated and for shallow depth where open joints in rock are present it can be assumed for design that such joints may fill with water up to ground level.

Ordnance Survey Maps to a scale of 1:600 were available for the area and also site investigation data from existing developments bounding the site, earlier road improvement works and landslide study reports. The ready access to this information, which has been a feature of Hong Kong administration for many years, was of great assistance in the geotechnical studies for this project.

Following site inspections to identify salient features of the area, a large number of sections were plotted to discover the most critical slopes. The 1:600 survey maps, larger scale plans of structures and site measurements where necessary, provided the information for plotting. Details of some typical sections are shown in Table 1.

Sec	L (m)	H (m)	L/H (deg)	Max Slope (deg)	Soil Mantle (m)
1	101	76	37	S 39 R 60	3 to 5
2	110	97	41	S 40 R 60	1 to 7
3	105	99	43	S 47 R 78	1 to 20
4	128	100	38	S 41 R -	5 to 21
5	118	74	32	S 55 R -	5 to 22
6	110	60	29	S 34 R -	3 to 11
7	32	23	36	S 65 R -	18
8	24	23	44	S 55 R -	32

S max. localised soil slope
R rock slope where exposed at toe

TYPICAL HILLSIDE SLOPES Table 1

Some of the features considered likely to present stability problems were:

(a) The condition of existing mansonry walls
(b) Existing localised steep cuttings
(c) Sheeting joints in the rock
(d) Poor existing slope drainage

A site investigation was carried out which, supplemented by the existing data, was adequate for the stability assessment of overall slopes and the many localised areas where stability appeared to be suspect. The ground conditions varied from fresh granite rock to completely decomposed rock and the following sampling methods were adopted in the rotary drillholes:

(a) In the less dense upper soils, driven U38 tube sampler.
(b) In the denser soils (where disturbance was probable using the normal driven sampler), Mazier triple tube core barrel sampler.
(c) In rock, a triple core barrel diamond drill to recover NX-sized cores.

Trial pits were dug to investigate the footings at the front of existing retaining walls and the properties of fill material behind. The thickness of masonry walls was determined by horizontal drillholes.

Field permeability tests were carried out in the soils at depths of 5m to 14m and pressurised water tests on selected rock joints. Stand-pipes and piezometers were installed in drillholes to monitor long term ground water level fluctuation.

A geological survey was carried out initially on the many exposed areas of rock on the hillsides and later extended to include the rock cores as they were recovered from the drillholes.

Rock core orientation tests were carried out in the drillholes to investigate the dip of the rock strata, the direction of dip and joint width. The Triefus Impression Device was used for this purpose, where an impression of the joints was obtained on a thermoplastic film pressed on the sides of the drill hole by an inflation pressure of 500 to 700 kpa, the direction being obtained by compass.

Soil Properties.

For a large part of this site it was found that the soils could be grouped conveniently into three types a, b and c, but at the east end of the site were some steeply cut slopes, visibly stable, but where these properties would be inadequate for stability. Further studies in this area yielded higher parameters d and e.

TYPE	DEPTH (m)	BULK DENSITY (kN/m3)	C' (kPa)	Ø' (deg)
a. Zone A structureless sand silt & clay	0-4	18.5	5	38
b. Zone B completely decomposed granite	4-10	19.0	15	40
c. Zone C highly decomposed granite	below 10	20.0	20	42
d. Zone B completely decomposed granite	0-8	19.2	15	42
e. Zone B highly decomposed granite	below 8	20.4	55	44

The shear properties were determined by consolidated drained triaxial compression tests on samples saturated before test by the application of a back pressure with a pressure response not less than 0.97.

Stability Analysis of Soil Slopes.

The average overall slope angle was 25° to 40° but localised cutting and fill slopes associated with existing roadways had angles of 55° to 75°. The roots of trees or bushes had shallow penetrations into the soil and the trees in particular cause damage to slopes when uprooted in typhoons. The surface soil was readily eroded by heavy rainfall and this was most prominent where road drainage gullies discharged downslope directly onto the soil surface. Where the surface was protected by an impervious layer, generally chunam, slopes remained stable under adverse weather conditions.

A live load surcharge of 10kPa was adopted for roadways in cut or on fill, but supports for elevated roads were taken down through the soils to found in granite rock. To prevent lateral loads from the columns being transmitted to the soils, the elements above rock level were provided with polystyrene sleeves.

The highest recorded water level in the wet season was determined from the piezometers. This was a relatively wet season, with numerous tropical storms, so the level was comparatively high. The depth of wetting band was calculated for 10-year and 1000-year return period rainfalls and added to the plotted water level to obtain the phreatic surfaces for design: the values were 2.5m and 6m respectively.

As a check on the sensitivity of the cohesion factor, analyses were repeated with different values of c'. The method of analysis was by the Janbu routine method for non-circular slip surfaces, using a programme adapted for an Olivetti P6060 computer.

In certain cases checks were made by circular slip analysis where this seemed appropriate and by a simple slide failure analysis at the soil-rock interface where the soil cover was thin and the rock surface relatively steep. It was possible to determine those soils slopes which were stable, those which had less than the required factor of safety and, in the latter case, to propose methods of improving the stability up to the desired standard. This was achieved by improved surface water drainage and impervious surfacing to reduce water level, and by flattening the slopes to about 40°, obtaining the space for this by realigning roadways on structures founded on rock, where necessary.

One of the effects of controlling water level was to maintain a degree of soil suction, which effectively maintained the value of c', which plays no small part in the stability of steep slopes in decomposed granite soils.

Some typical details of analysed slopes are given in Fig. 5.

Stability Analysis of Rock Slopes.

The geological survey established 11 joint sets J1 to J11 within the study area and three zones were established where particular joint patterns predominated. Stereoplots were made for each zone and the joint data recorded as follows:-

Strike direction	eg	N18° E
Dip direction	eg	NW
Dip angle	eg	<76°

Joint J2 was a sheeting joint and found to be the most significant with regard to slide failure, following the general direction of the rock slope. Joint J1 dipping at 70° to 80°, was assumed to form a rear boundary to a failure wedge. Most of the joints were open and moderate weathering along the joints was common. Weathering was more intense along sheeting joints and these were frequently infilled with completely weathered granite. A typical stereoplot is shown in Fig. 6.

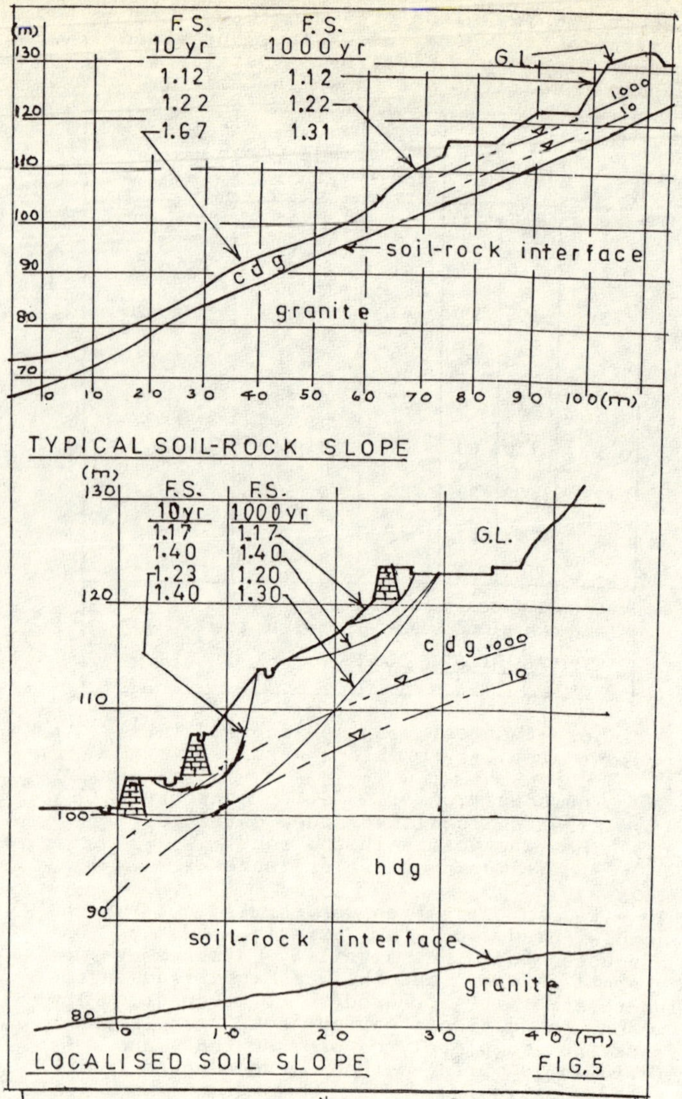

TYPICAL SOIL-ROCK SLOPE

LOCALISED SOIL SLOPE

FIG.5

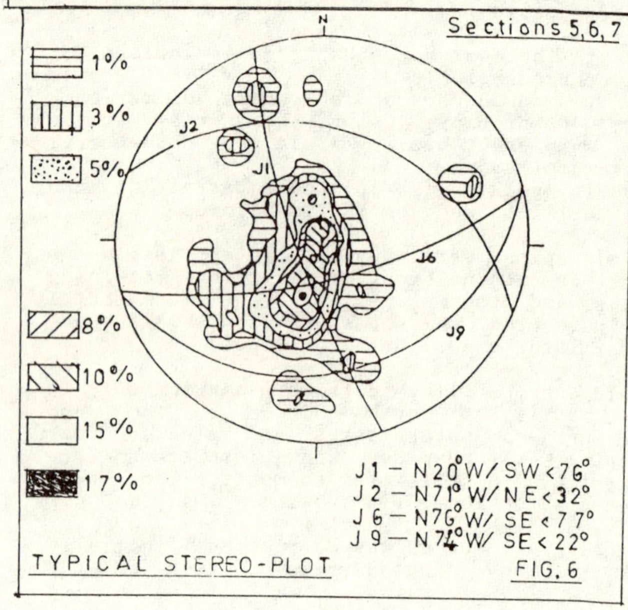

TYPICAL STEREO-PLOT

FIG.6

Laboratory tests on sheet joints were made to determine the shear strength along the joints using a standard soils laboratory shear box modified to perform tests on rock cores. Peak and residual properties were measured.

The rock properties assumed for design purposes were as follows:-

Density: Bulk 24.4kN/m3 submerged 14.6kN/m3
Compressive strength of weathered granite at joint 9800kPa.

	Peak	Residual
value of c'	26 kPa	0
value of ϕ'	33°	30°

For purposes of comparison, the shear properties for a slip surface passing through a rock mass with a discontinuous fissure system may be taken as:

$c' = 30$ kPa $\qquad \phi' = 39°$

A wedge failure on sheeting joints was taken to be the most critical, assuming that the joint must 'daylight' above the toe of the overall slope. Analysis was made by the method proposed by Barton, (Ref. 1) Bartons equation is

$T = \sigma \tan(\phi_R + i)$

where

i = JRC x $\log_{10}(\sigma_J/\sigma)$
T = design shear strength of joint
σ = effective normal stress
ϕ_R = residual angle of friction from tests
σ_J = compressive strength of rock at joint (measured or estimated)
JRC = Joint Roughness Coefficient
i = angle of first order roughness

The first order roughness of a joint corresponds to the major undulations on the bedding surfaces and this defines the value of JRC. Barton listed three categories as follows:

JRC = 20 Rough, undulating - tension joints, rough sheeting, rough bedding.
JRC = 10 Smooth undulating - smooth sheeting, undulating bedding.
JRC = 5 Smooth nearly planar - planar sheet joints, planar bedding.

The small very localised surface ridges and depressions along a rock joint have much greater values of i, which Barton termed second order projections and their presence is deemed not to have an influence on the shear strength of a joint at the low effective pressures normally acting.

When JRC = 10 $\qquad (\phi_R + i) = 50°$

The forces taken into account in assessing stability of rock wedges were generally as described in Ref. 3 and Fig. 7.

The dip of the sheeting joints varied 18° to 32° and for the major part of the site was about 30°. Cohesion along the rock wedge base was taken as zero. Calculations were carried out using a programmable HP65 calculator.

A typical section is shown in Fig. 8 and Factors of Safety in Table 2.

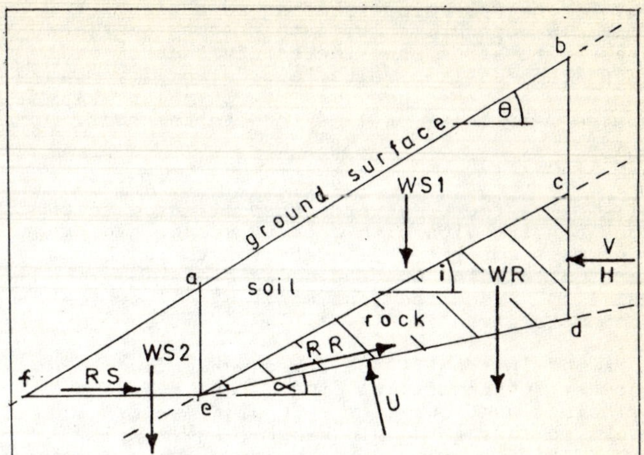

H	horizontal load - caissons
WR	wt rock wedge ecd
WS1	wt soil cover abce
WS2	wt soil wedge at toe fae
U	water uplift on ed
V	hydrostatic thrust
RS	friction and cohesion on fe
RR	friction on ed

FORCES- ROCK WEDGE ANALYSIS FIG. 7

Depth to Joint	Deg.			F. S.			
	α	i	θ	ϕ = 40°		ϕ = 50°	
				10 yr	1000 yr	10 yr	1000 yr
3	18	24	29	1.5	1.14	2.1	1.6
8	18	25	35	1.4	1.12	1.9	1.6
8	30	42	32	1.2	1.0	1.6	1.3
8	32	42	32	1.1	0.9	1.4	1.14
5	30	42	32	1.6	1.3	1.9	1.5
5	32	42	32	1.2	0.9	1.5	1.1
8	30	42	38	1.0	0.9	1.3	1.1

Factors Of Safety For Rock Slopes Table 2

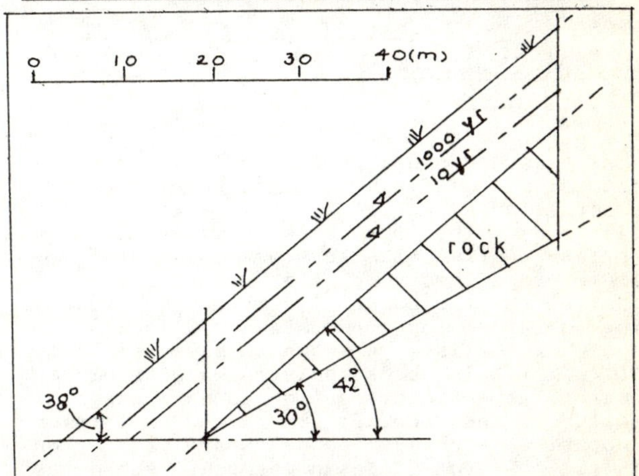

TYPICAL ROCK WEDGE FIG.8

Most of the rock slopes had an adequate factor of safety when JRC = 10, but in some localised areas with structures the safety factor fell below the desired value and stability was improved by the use of steel shear keys grouted into holes drilled through the rock.

Structures.

Existing structures were generally of granite block construction with dressed stone facing, or of mass concrete. A number of these showed signs of sliding or bearing problems and the stability required checking. A typical section is given in Fig. 9.

Signs of instability observed were downslope movement (basically sliding), cracks in the surfacing and subsidence behind walls and cracks following the blockwork jointing.

The material properties used for calculations were:

bulk density of soil	19kN/m3
for soil in front of wall	c' = 0 Ø = 35°
for soil beneath wall	c' = 0 Ø = 38°
bulk density of masonry	23kN/m3
for masonry	c' = 0 Ø = 42°
wall friction angle	2/3 Ø'
masonry sliding friction angle	2/3 Ø'
soil to base friction angle	Ø'
Passive Resistance of soil	½ Kp

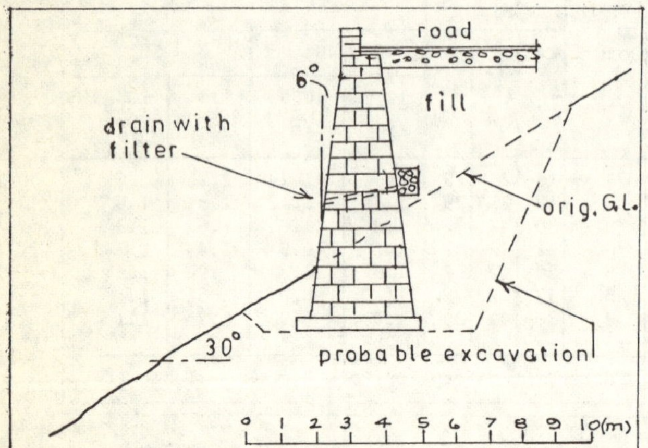

TYPICAL EX. RETAINING WALL FIG. 9

Checks were made on stability against overturning, sliding at the base, bearing pressure and shear across the walls at mid-height and at the base using an Olivetti P6060 computer. Active pressure was calculated by Coulomb wedge theory for a range of values of c' and Ø' for the backfill, the best agreement being obtained for c' = 5kPa and Ø' = 35° in cases where walls were known to have stability problems. This back analysis was useful in checking the condition of other existing structures and for designing new retaining structures. Checks were made with the backfill saturated, assuming that weepholes and the rudimentary drainage filters could become blocked.

On such steep slopes it is very difficult to provide strengthening buttresses, but in all cases it was found possible to achieve the required degree of stability by cutting down the height of each wall when incorporated into the overall slope improvement works. The factors of safety were, for sliding 1.5 and overturning 2.0. Stability against a localised slip failure was checked as part of the soil stability assessment programme.

New retaining structures were designed to the same factors of safety, using the soil parameters obtained from the site investigation. However, graded filters were provided for the full height of these walls and it was assumed that c' = 0 for the specified CDG backfill with Ø' = 35°.

Over 20 overall sections and many localised sections were analysed to assess existing stability and stability of the new cut slopes. Each separate flyover piled foundation was drilled to check for sheeting joints in the rock and strengthening measures adopted where calculations showed that such joints might reduce stability below the accepted minimum.

The improved site investigation methods and techniques of analysis adopted for this site were due very largely to the influence of the P.W.D. Geotechnical Control Office.

CONCLUSIONS

In the past, slopes in weathered granite existed naturally, or were cut, at such steep angles that there was always a risk of failure under the most adverse conditions which could occur. Compliance with the rules introduced by G.C.O. reduces this risk to acceptable limits. There was not the same problem with foundations, which tended to be designed conservatively.

REFERENCES

1. Barton N.R., 1973, Review of a new shear strength criterion for rock joints, Engineering Geology, Elsevier, Vol. 7.

2. Geotechnical Control Office P.W.D., 1979, Geotechnical Manual for Slopes, Hong Kong.

3. Hoek E. and Bray J., 1977, Rock Slope Engineering, Institution of Mining and Metallurgy, London.

4. Hong Kong Joint Group of the Institutions of Civil Mechanical and Electrical Engineering, 1962, Symposium on Hong Kong Soils.

5. P.W.D. Hong Kong, 1976, Building (Construction) Regulations.

6. Hong Kong Institution of Engineers, 1981, Guidance Notes on Hand-dug Caissons.

Geotechnical engineering in volcanic ash soils

L.D.WESLEY, Senior Lecturer in Civil Engineering, University of Auckland, New Zealand
T.MATUSCHKA, Senior Geotechnical Engineer, Tonkin and Taylor, Consulting Engineers, Auckland, New Zealand

ABSTRACT: Soils derived from the weathering of volcanic ash have special engineering properties due to the presence of the clay mineral allophane. A brief account is given of the main engineering characteristics of these soils and a description is given of several specific engineering projects carried out in volcanic ash soils. These projects include a goethermal power station in Indonesia, an oil tank farm and two earth dams in New Zealand.

INTRODUCTION

Soils derived from the weathering of volcanic ash and belonging to the pedological group andosol are found in many parts of the world, including Hawaii (Willis 1946, Lum 1982), New Zealand (Fieldes 1966), Indonesia (Wesley 1973), New Guinea (Wallace 1973), Japan (Uehara 1982) and the West Indies (Rouse et al 1986).

Population pressures and associated development programmes in many countries have meant that geotechnical engineers are encountering these soils with increasing frequency and a better understanding of their properties is developing. In this paper a brief review is given of the formation and the more distinct properties of volcanic ash soils, and several case records are presented of civil engineering projects in these soils. The terms andosol and volcanic ash soil will be used synonymously (interchangeably) in this paper.

FORMATION AND DISTINCTIVE PROPERTIES

The volcanic ash soils with which this paper is concerned are those in which the predominant clay mineral is allophane, or a combination of allophane and halloysite. As the name implies, these soils are normally derived from the weathering of volcanic ash. Although they appear to be most prevelant in the wet tropics, they are also found in countries of temperate climate such as New Zealand. In the tropics they are generally found at high altitudes, where the climate is wet and temperatures are moderate. These conditions inhibit the weathering process which at lower altitudes converts these soils into red clays. Volcanic ash soils are normally yellowish brown in colour, although in some situations the upper layers may be dark in colour due to the presence of some organic material. The name andosol originates from Japan and literally means "dark soil" (Thorp and Smith 1949).

The general characteristics of andosols have been described elsewhere (Wesley 1974, 1977, Morin and Todor 1975, Lohnes and Tuncer 1977) and only a brief summary will be given here.

Their distinctive characteristics are:

(a) Very high natural water contents and high values of L.L. and P.L.
(b) Irreversible decrease in plasticity and increase in particle size when air or oven dried.
(c) They are likely to have flat compaction curves without distinctive values of optimum water content or maximum dry density.
(d) Empirical relationships applicable to clays of sedimentary origin are likely to be misleading when applied to these soils.
(e) Engineering properties are generally good; they are normally of high strength and moderate to low compressibility.
(f) Properties are not greatly influenced by natural water content or Atterberg limits. Andosols with very high water contents may have the same strength and compressibility properties as those of low water contents.

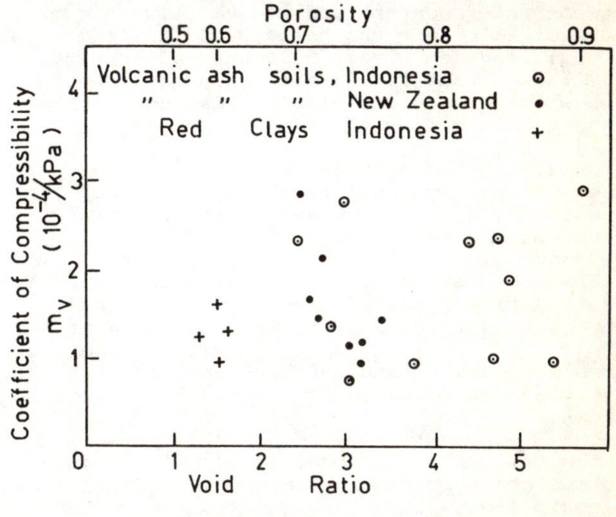

Fig. 1 : Compressibility versus void ratio

Point (f) above is illustrated in Fig. 1, where the coefficient of compressibility from a series of tests on undisturbed soils is plotted against void ratio. The values are for the stress range 0 to 200 kPa. It is seen that there is no clear evidence of increasing compressibility with void ratio. Limited data suggests that the same is true of the ϕ' value (Wesley 1974, 1977).

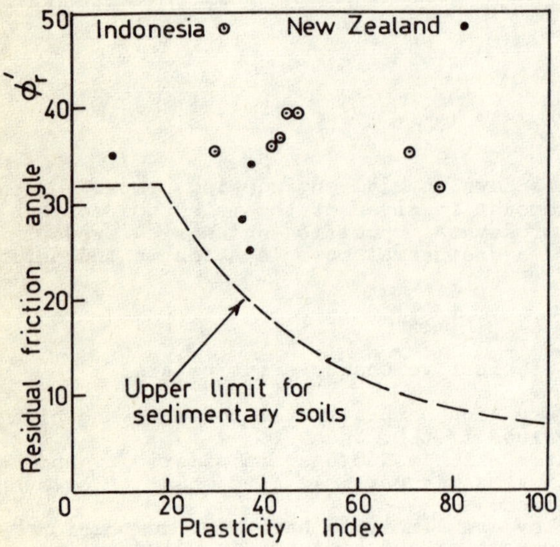

Fig. 2 : Residual friction angle for volcanic ash soils

Point (d) above is illustrated in Fig. 2, which shows residual strength (ϕ_r') value plotted against plasticity index. It is seen that the ϕ_r' values from volcanic ash samples in no way conform to the empirical relationship for sedimentary clays.

The high water content of volcanic ash soils appears to be directly related to the allophane content. Figure 3 shows the relationship between water content and the relative proportions of allophane and halloysite for Indonesian andosols and red clays. Lohnes and Tuncer (1977) show that for Hawaian andosols, the highest water content also corresponds to the highest allophane content.

It should not be supposed that all andosols are necessarily very similar, although it does appear that where the predominant clay mineral is allophane then strong similarities in behaviour are to be found. Particle size will also influence behaviour. Indonesian andosols are very fined grained with clay fractions above 80% and virtually no sand sized material; New Zealand volcanic ash soils on the other hand commonly contain considerable sand and silt and have clay fractions in the range of 25 to 50 percent.

In the following sections, an account is given of some specific engineering projects in volcanic ash soils.

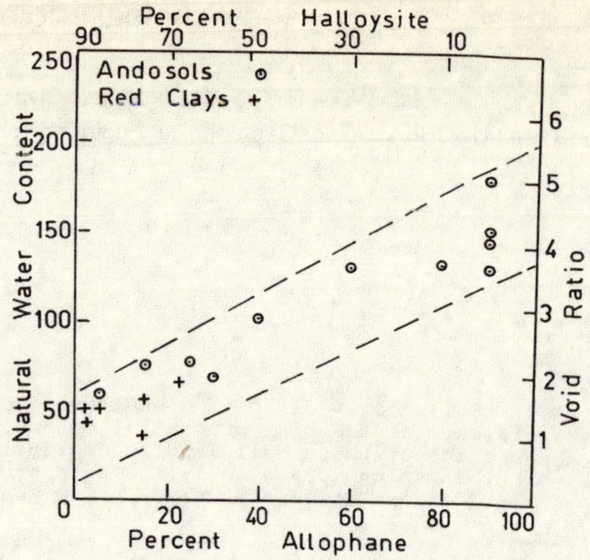

Fig. 3 : Water content versus allophane and halloysite content for Indonesian soils

KAMOJANG GEOTHERMAL POWER STATION, WEST JAVA, INDONESIA

General

The Kamojang power station has a current generating capacity of 150 MW. The first stage of 40 MW was completed in 1982, and the second stage of 110 MW in 1987. The site of the project is an old crater area some 1500 m above sea level in mountainous terrain about 40 km southeast of the city of Bandung.

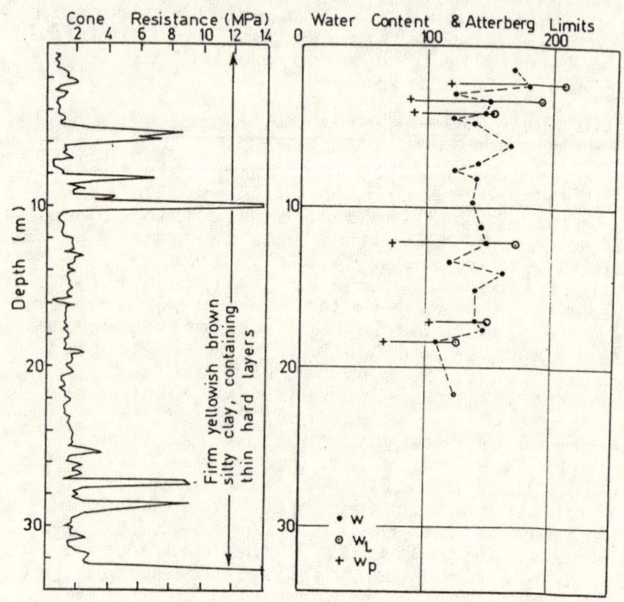

Fig. 4 : Kamojang project: soil conditions

The soil at the site, and for a large area of surrounding country, belongs to the Indonesian andosol group. In Java, andosol is found on nearly all of the older volcanic areas above an elevation of approximately 1000 m. In the area of the geothermal project, the soil extends down to depths of between about 10 m and 50 m, although in other parts of Java considerably greater depths may be found. Figure 4 summarises the soil conditions at the power station site. The andosol, best described as a silty clay (or clayey silt) extends to a depth of 32 m, and is underlain by sandy or gravelly silt, believed to be of weathered lahar origin. The Dutch cone penetrometer test shows the soil to be reasonably homogeneous, apart from some thin relatively hard layers found at depths of about 8 m and 10 m. These hard layers were encountered in all the penetrometer tests and boreholes, and consisted of weakly cemented silty sand layers. A limited number of natural water content and Atterberg limit tests were carried out, and the results are shown in Fig. 4. It is seen that the natural water content is generally between 120% and 150%, with considerable variation in Atterberg limit values.

Foundation Selection

The choice of foundation type for the power station was between a raft and piles. The considerable depth to hard material, and the remoteness of the site, made piles an unattractive proposition, and a raft foundation was adopted. The raft dimensions were 30 m by 40 m for the first stage, and 30 m by 65 m for the second stage. The rafts apply an average bearing stress of about 55 kPa and are founded 2.5 m below ground level; the net applied bearing pressure is thus very low, only about 20 kPa.

Soil Bearing Capacity

Dutch cone penetrometer tests, laboratory vane and undrained triaxial tests were used to measure the undrained shear strength of the soil; the values so obtained ranged from about 50 kPa to 120 kPa, with an average of 75 kPa.

For the purposes of bearing capacity estimation the undrained strength was taken as 75 kPa, giving an ultimate bearing capacity and an allowable bearing pressure of 450 kPa and 150 kPa respectively. The allowable pressure was thus well in excess of the applied foundation pressure.

Soil Compressibility and Settlement Estimation

Figure 5 shows the results of consolidation tests carried out on three undisturbed samples of volcanic ash soil. Two of the samples are from the Kamojang and Omata sites described in this paper; the third, included for comparison purposes is from the Tjipanundjang dam site in Indonesia (Wesley 1974). The curves in Fig. 5 suggest that all three samples have distinct "pre-consoldation" pressures, ranging from about 200 kPa (Kamojang) to about 500 kPa (Omata).

Based on the consolidation tests, the estimated settlement of the raft foundation was in the range of about 30 mm to 100 mm, depending upon whether the assumption was made that some swelling of the soil would occur between excavation and loading.

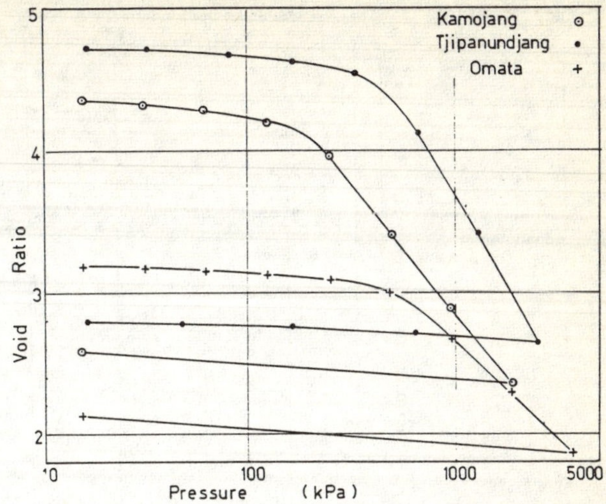

Fig. 5 : Consolidation characteristics

An attempt was made to measure the raft settlement during and after construction, but the actual settlement appeared to be negligible, and reliable measurements were not possible. When the second stage of construction was carried out, the raft was connected integrally with the existing raft. No sign of any differential settlement has been detected to date.

The performance of the raft foundations of the two stages of construction has been entirely satisfactory.

OMATA STORAGE TANK FARM, NEW PLYMOUTH, NEW ZEALAND

The Omata storage tank farm is located close to the port of New Plymouth on the west coast of the North Island. The tanks are used to store a range of petroleum products. This paper describes two of the tanks, which will be designated the SBPT (Shell, BP and Todd) tank and the Petrocorp tank. The SBPT tank is 60 m in diameter and 16 m high, while the Petrocorp tank is 43 m in diameter and 20 m high.

New Plymouth is located at the edge of a coastal plain formed by deposits of volcanic material originating from Mt Taranaki (Egmont) and from earlier volcanoes near Mt Taranaki. Deposits in the area consist of weathered volcanic ash overlying, or interbedded with lahar flow deposits.

Figure 6 summarises soil conditions at the site. At the surface over part of the site there is a layer of fine sand up to 5 m thick; below this extending to about 25 m is a thick layer of volcanic ash soil, which can be described as a slightly sandy, clayey silt. Below 25 m, a dense gravelly sand is found, considered to be a lahar flow.

The Dutch penetrometer tests show the ash layer to be of fairly uniform consistency. The results of a limited number of natural water content and Atterberg limit tests are given in Fig. 6. The average water content is between 90% and 100%.

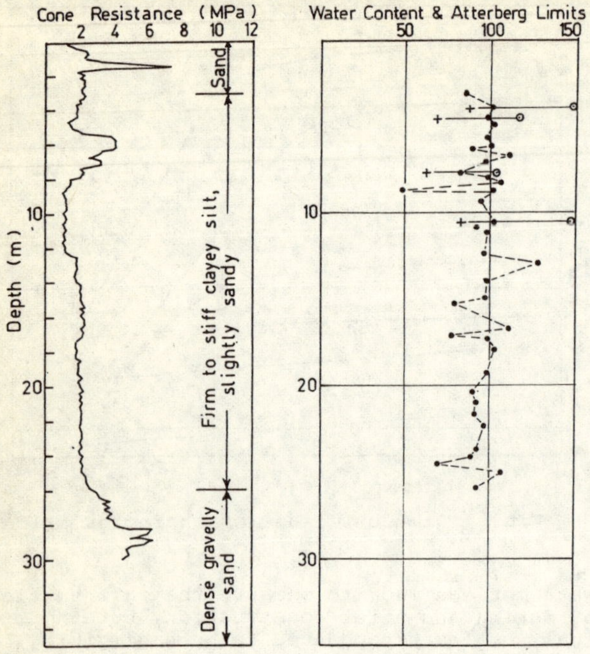

Fig. 6 : Omata tank site: soil conditions

Soil conditions at this site are therefore similar to those at Kamojang in Indonesia. However the Omata soil has a small proportion of fine sand in it while the Kamojang soil is virtually free of any sand sized soil particles. The lower natural water content of the Omata soil (90 to 100%) in comparison to the Kamojang soil reflects the coarser nature of the former.

Foundation Selection

The choice of foundations with steel storage tanks is somewhat limited, and normal practice of founding the tanks directly on the soil was adopted here. In the case of the SBPT tank, some excavation and recompaction of the surface sand layer was carried out, partly to remove old buried timber, and partly to densify the sand to improve its reistance to liquefaction under earthquake loading.

Soil Bearing Capacity

Dutch cone penetrometer tests, laboratory vane, and pressuremeter tests were used to determine the undrained shear strength; the values ranged from 75 kPa to 170 kPa with an average of about 100 kPa.

For design purposes, an undrained shear strength of 100 kPa was adopted; the ultimate bearing capacity was therefore in the vicinity of 600 kPa, assuming undrained loading. The applied pressures from the two tanks were expected to be as follows:

	Normal Operation (kPa)	Hydrotest (kPa)
SBPT tank	105	160
Petrocorp tank	160	190

During earthquake loading these pressures rise somewhat near the tank edges; however in all loading conditions the safety factor with respect to bearing capacity does not fall below 2.5 and this was considered to be adequate.

Settlement

A number of consolidation tests were carried out on undisturbed samples from both tank sites. A typical result is shown in Fig. 5. All the tests produce e - log p curves indicating apparent preconsolidation pressures in the range of about 400 kPa to 800 kPa. The average m_v value from these tests for the appropriate pressure range was about 10^{-4}/kPa, with slightly higher values close to the surface. Pressuremeter tests gave m_v values closely in agreement with the consolidation test values.

Based on these m_v values, estimates were made of expected settlement of the tanks; these are summarised in Table 1.

Table 1 : Omata Tanks Settlement Data

Tank	Position	Estimated Settlement (mm)		Actual Settlement in Hydrotest Loading	
		Normal Operation	Hydrotest	Maximum Pressure (kPa)	Settlement (mm)
SBPT	Centre Edge	220 80	315 115	155	160 51 (38 to 70)
Petrocorp	Centre Edge	160 75	215 100	200	200 36 (21 to 57)

The values listed for normal operational loading are of relevance only if the hydrotest loading does not produce full consolidation of the soil, that is if the time is not long enough to allow full pore pressure dissipation. The rate at which settlement will occur is difficult to predict with volcanic ash soils on the basis of conventional consolidation tests, because compression takes place too rapidly for reliable C_v values to be obtained from time versus settlement curves. In the present case the theoretical estimate for 90% consolidation was of the order of 1 to 2 years, although it was considered likely that the actual value could be much smaller, in the range of 1 to 2 months.

The actual settlements which occurred during the hydrotest loading (i.e. slowly filling the tanks with water) are shown in Table 1. It is seen that the actual settlement at the centre of the tank is substantially less than the predicted value for the SBPT tank and marginally less for the Petrocorp tank. The edge settlements are surprisingly small in comparison both to the centre values and the predicted values. More detailed information on the settlement behaviour of the SBPT tank is given in Fig. 7.

Figure 7(a) shows settlement, pore pressure, and tank loading plotted against time. The pore pressure is from one of a number of piezometers placed at a depth of 6.6 m beneath the edge of

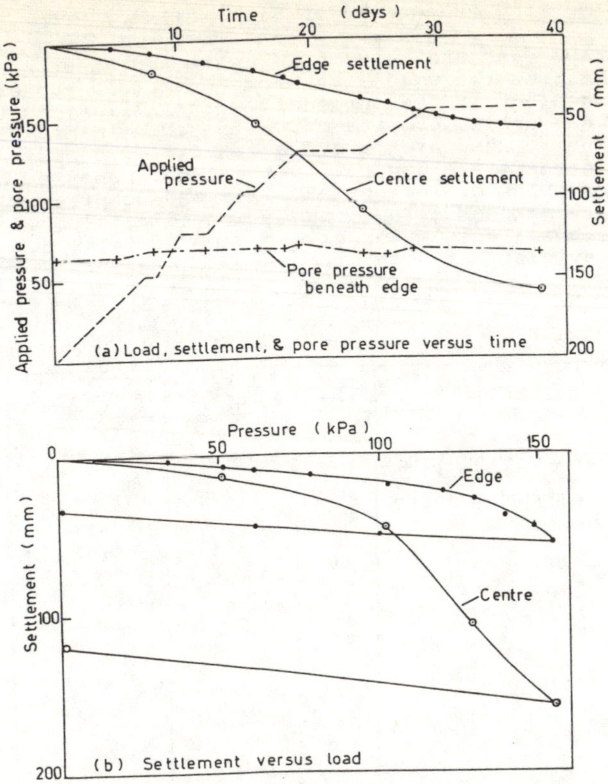

Fig. 7 : Settlement of Omata SBPT tank

the tank. In Fig. 7(b) settlement at the tank edge and centre is plotted against the tank pressure. These records indicate the following:

(a) Settlement appears to occur very quickly with no pore pressure rise. None of the piezometers placed around the perimeter of the SBPT tank showed significant pore pressure rise.

(b) There appears to be a critical pressure (or possibly a critical deformation) above which the soil compressibility increases significantly. This pressure is about 90 kPa and 130 kPa for the tank centre and edge respectively, corresponding to a settlement of about 30 mm.

(c) The critical pressure evident from the field records is much lower than the apparent preconsolidation pressure in the laboratory tests.

The settlement data from the Petrocorp tank is not as complete as that from the SBPT tank, but the limited data available is in agreement with the behaviour described above. The performance of the two tanks lends some support to the view that settlement estimates based on conventional tests will tend to overestimate the true settlement with residual soils, although the divergence is probably no greater than it often is with sedimentary soils. The performance confirms that rates of pore pressure dissipation are likely to be very rapid with undisturbed volcanic ash soils.

The performance of the two tanks has been satisfactory and illustrates that volcanic ash soils do not present any special problems when used as foundation materials for relatively heavily loaded structures such as oil tanks.

COMPACTION OF VOLCANIC ASH SOILS

Before proceeding to the next case record it is appropriate to make some comments on the compaction behaviour of volcanic ash soils. Two aspects of their behaviour are of special significance.

Firstly, normal compaction tests are likely to give compaction curves which are quite flat and do not show a peak dry density corresponding to an optimum water content. Secondly, some volcanic ash soils are sensitive and for a given water content will become progressively softer with increased compactive effort. These effects are illustrated in Figs. 8 and 9.

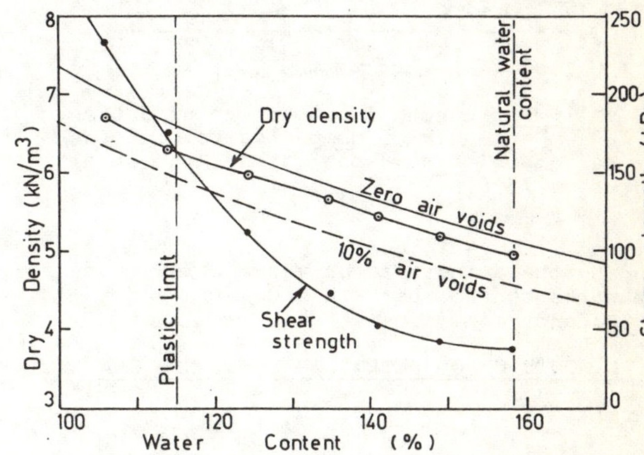

Fig. 8 : Compaction behaviour: Indonesian volcanic ash sample

Figure 8 shows a standard compaction test curve for an Indonesian andosol, as the soil is dried progressively from its natural water content of 160% to a value of 107%. At this latter water content the soil had become too dry and brittle to be considered acceptable for compaction. At each compaction point the shear strength of the compacted soil was measured using a hand shear vane; the values obtained are shown on the same figure. A shear strength of from about 70 kPa to say 200 kPa is normally acceptable for compaction so that the permissible water content range for this soil would be from about 110% to 140%.

Figure 9 shows the effect of repeated compaction on the strength of the compacted soil for a series of samples from Japan. The strength is indicated by the cone penetration measurements. It is seen that for most samples the strength decreases with compactive effort. The important conclusion to be drawn from Figs. 8 and 9 is that conventional methods of compaction control using maximum dry density and optimum water content are not appropriate for volcanic ash soils. A control method based on shear strength and air voids, along the lines proposed by Pickens (1980) appears admirably suited to these soils and has been

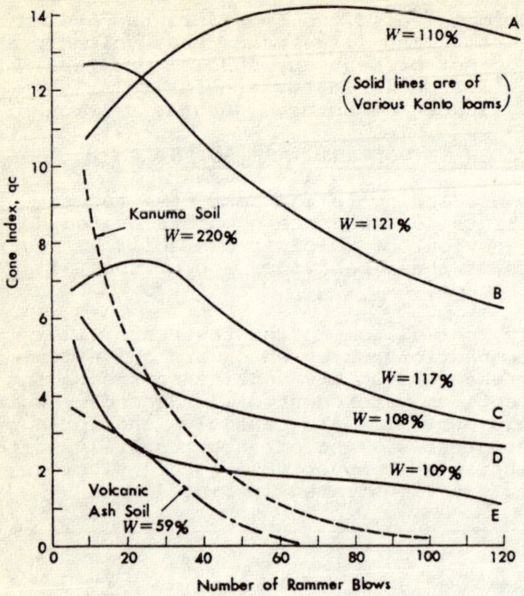

Fig. 9 : Influence of compactive effort on strength of compacted volcanic ash soil from Japan (after Kuno et al 1978)

found by the authors to be very satisfactory for earthworks control both in Indonesia and in New Zealand.

NEW PLYMOUTH FLOOD DETENTION DAMS, NEW ZEALAND

General

The city of New Plymouth, located on the west coast of the North Island of New Zealand, has had a flooding problem for many years. Two earthfill flood detention dams, the Huatoki and the Mangaotuka, were constructed in 1985/1986 as part of flood protection works for the City. They are located on the outskirts of the city across major streams which run through the city centre. The Huatoki dam is 25 m high with a crest length of 120 m and the Mangaotuka dam is 15 m high with a crest length of 80 m. Both are constructed from on-site volcanic ash soils and typical cross-sections are shown on Fig. 10.

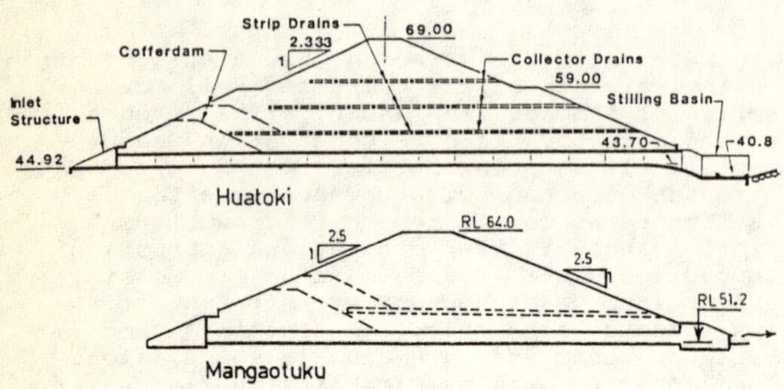

Fig. 10 : Cross-sections of New Plymouth dams

Geology and Properties of Site Soils

The two dams are about 5 km apart and about 10 km from the Omata tank farm described above; the geology of the three sites is similar. At the dam sites deposits of weathered volcanic ash up to 20m thick overlie lahar flow deposits. The streams are downcut to the lahar deposits, which being of low permeability and compressibility make excellent foundation materials. Some unsuitable material was removed from the valley floor at both sites.

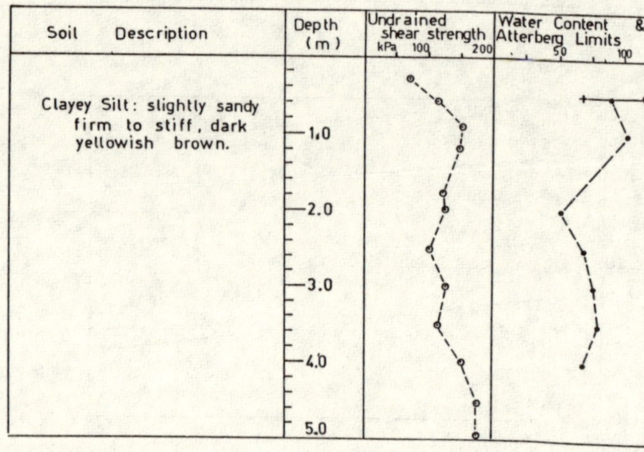

Fig. 11 : Typical soil profile in a borrow area

Site investigations for both dams concentrated on checking foundation conditions and determining the suitability of on-site ash soils for fill. Investigation pits were dug to investigate potential fill; a typical log is shown in Fig. 11. Water contents generally ranged from 50 to 105% and decreased with depth. Samples from ridge tops had lower water contents than those from near the base of slopes. Although the top 2 m of soil, approximately, had higher water contents, this material was more suitable for earthfill. Deeper material was more sandy and of a granular texture but greasy appearance, and released water and softened markedly on compaction.

Laboratory tests were undertaken to determine the properties of the compacted ash necessary for design. Results are summarised in Table 2. In addition, numerous compaction tests were undertaken to determine the water content limits corresponding to acceptable shear strength and air voids values.

Design

The design dam sections shown in Fig. 10 were determined from consideration of static long-term, rapid draw-down, seismic, and construction stability. Apart from infrequent short intervals (up to 5 days) when major floods occur, the dams are empty and a steady state seepage condition does not develop. For this reason, static long-term and rapid drawdown

Table 2 : Properties of Compacted Ash

Sample Depth (m)	Water Content (%)	Air Voids (%)	Test Type	Test Results
—	—	—	Specific Gravity	2.86 (average of three tests)
0–5	50–105	<10	Bulk Density	1.3–1.5 t/m^3
1.0	103	—	Clay Fraction	< 5 %
2.4	95	—	" "	< 2 %
1.0	102	3	Permeability	1.5×10^{-8} m/s
3.0	51	7	"	8.5×10^{-9} m/s
1.0	95	7	$\overline{B}$	0.67– 0.7
3.0	55	3	"	0.59
1.0	95	7	Pinhole Dispersion	Erosion Resistant
1.0	102	8	CU Triaxial	$c' = 32$ kPa $\phi' = 33.5°$
3.0	51	7	" "	$c' = 27$ kPa $\phi' = 37°$

stability is not a concern. The seismic response of the dams was evaluated using the simplified method of Makdisi and Seed (2978). Analysis of the Huatoki dam indicated that for 100 year and 1000 year return period earthquakes (peak ground accelerations of 0.25 g and 0.5 g) undrained shear strengths of 50 kPa and 95 kPa respectively were required to keep the safety factor at unity (i.e. no permanent deformation). Since the specified average shear strength for the fill was 100 kPa seismic stability was not critical. Dynamic strength tests on similar ash from a nearby hydro dam indicated no loss of strength when subjected to severe levels of shaking.

Stability during construction was found to be the most critical due to high pore pressures arising from compaction and loading ($\overline{B} = 0.7$). Historically there have been instances of slope instability in road embankments in Taranaki where construction has been rapid. To accelerate pore pressure dissipation internal sand filled strip drains were used as shown in Fig. 10. Spacing was selected to ensure stability. For stability estimates, a construction rate was assumed, and soil parameters taken as $c' = 10$ kPa, $\phi' = 35^o$, $C_v = 1.2$ mm^2/sec, $\overline{B} = 0.7$. Twelve piezometers were installed in Huatoki dam to monitor construction pore pressures. Settlement estimates indicated that about 65% of expected settlement would occur during construction.

Because of the large variation in water content of the fill, compaction control was by means of undrained strength and air voids (Pickens 1980). Average undrained strength was to be not less than 100 kPa with minimum single values of 85 kPa, and average air voids ratio was to be not greater than 8% with maximum single values of 10%. To prevent softening of the soil from overcompaction, maximum weave was specified as 40 mm, and maximum velocity of plant was 15 km/hr.

Construction

Compaction trials, undertaken to determine the most appropriate plant indicated that sheepsfoot, rubber tyred, or vibrating steel wheel rollers would be satisfactory; however, sheepsfoot rollers were used to ensure good bonding between successive lifts. It was found that placing the fill in thick layers helped prevent softening from overcompaction which tended to occur with thin layers. It was noted also that tractor pulled scoops were less likely to cause softening and excessive weave than motorized scrapers.

Careful control of borrow areas was maintained; large areas were stripped in advance of use to allow drying. While this approach worked well at these sites it may be undesirable at other locations where rainfall is more frequent or construction is not limited to dry summer months. Attempts were made to mix the wetter soil with drier by excavating on a sloping face which cut across the bedding rather than along it.

Pore pressures were monitored throughout construction and indicated that the strip drains functioned as intended, with no apparent instability. A typical result is shown in Fig. 12. A steady trickle of water was evident from the drains once they were covered by about 2 m of fill. Post construction settlements were measured and found to be less than predicted (less than 25 mm at Huatoki), probably due to the effectiveness of the strip drains.

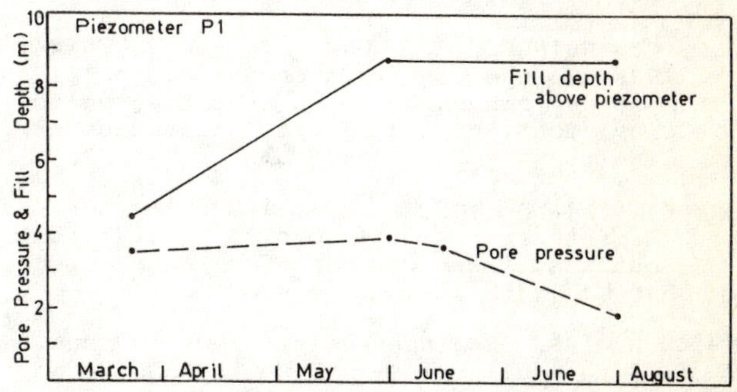

Fig. 12 : Construction pore pressures at Huatoki

No major probles were encountered in constructing either of the dams. The strength, permeability and consolidation characteristics of the weathered ash made it ideal earthfill for dams. Special attention is required during construction to ensure that softening from "overcompaction" does not occur, and special provisions such as strip drains may be desirable to control construction pore pressures.

DISCUSSION AND CONCLUSION

Volcanic ash soils are frequently regarded with suspicion by geotechnical engineers because of their unusual properties. Gidigasu and Bani (1974) for example identify these soils as "troublesome soils" while Morin and Todor (1975) describe them as "problem soils" and include them with other so-called problem soils of the tropics. However, their identification as problem materials arises more from the fact that they do not conform to certain empirical behaviour patterns associated with conventional clays of sedimentary origin than from any evidence that they perform poorly as engineering materials. The performance of the Tjipanundjang dam built in Java (Indonesia) in 1927 of volcanic ash soil (Terzaghi 1958, Wesley 1974) is a graphic demonstration of the good engineering properties of volcanic ash soils. It has been the purpose of this paper to describe several recent case records which help to confirm the very good performance of these soils as engineering materials.

ACKNOWLEDGEMENT

The authors wish to thank Tonkin and Taylor Ltd for permission to publish this paper and SBPT and Petrocorp for making available the settlement data on their oil tanks. The authority of the New Plymouth City Engineer to describe aspects of the New Plymouth dams is also gratefully acknowledge.

REFERENCES

FIELDES, M. 1966 The nature of allophane in soils, part 1, Significance of structural randomness in pedogenesis, New Zealand Journal of Science 9, No. 3, 599-607

GIDIGASU, M.D. and BANI, S.K. 1973 Geotechnical characteristics of troublesome lateritic materials, Proc. 8th Int. Conf. on Soil Mechanics and Foundation Engineering, Moscow, Vol. 4, 89-96

KUNO, G., SHINOKI, R., KONDO, T. and TSUCHIYA, C. 1978 On the construction methods of a motorway embankment by a sensitive volcanic clay, Conf. on Clay Fills, Insitution of Civil Engineers, 149-156

LOHNES, R.A. and TUNCER, E.R. 1977 Engineering characteristics of andosols, Proc. Fifth Southeast Asian Conf. on Soil Engineering, Bangkok, 305-312

LUM, W.B. 1982 Engineering problems in tropical and residual soils in Hawaii, ASCE Geotechnical Engineering Specialty Conference on Engineering and Construction in Tropical and Residual Soils, Honolulu, Hawaii, 1982, 1-12

MAKDISI, F.I. and SEED,H.B. 1978 Simplified procedure for evaluating embankment response, J. Geot. Eng. Div. ASCE 105, GT2, 1427-34

MORIN, W.J. and TODOR, P.C. 1975 Laterite and lateritic soils and other problem soils of the tropics, USAID 3682, Lyon Associates, Baltimore, U.S.A.

PICKENS, G.A. 1980 Alternative compaction specifications for non-uniform fill materials, 3rd Aus-N.Z. Conf. on Geomechanics Perth, Vol. 1, 231-235

ROUSE, W.C., READING, A.J. and WALSH, R.P.D. 1986 Volcanic soil properties in Dominica, West Indies, Engineering Geology 23, 1-28

TERZAGHI, K. 1958 The design and performance of Sasamua dam, Proc. Institution of Civil Engineers, Vol. 9, 369-394

THORP, J. and SMITH, G.D. 1949 Higher orders of soil classification; order, suborder and great soil groups, Soil Science 67, 117-126

UEHARA, G. 1982 Soil science for the tropics, ASCE Geotechnical Engineering Specialty Conference on Engineering and Construction in Tropical and Residual Soils, Honolulu, Hawaii, 1982, 13-29

WALLACE, K.B. 1973 Structural behaviour of residual soils of the continually wet highlands of Papua New Guinea, Geotechnique Vol. 23, No. 4, 471-494

WESLEY, L.D. 1973 Some basic engineering properties of halloysite and allophane clays in Java, Indonesia, Geotechnique, Vol. 23, No. 4, 471-494

WESLEY, L.D. 1974 Tjipanundjang dam in West Java, Indonesia, ASCE Journal of Geotechnical Engineering, Vol. 100, No. 4, 503-522

WESLEY, L.D. 1977 Shear strength properties of halloysite and allophane clays in Java, Indonesia, Geotechnique, Vol. 27,No. 2, 125-136

WILLIS, E.A. 1946 Discussion on: A study of lateritic soils by Fruhauf, B., Proceedings Highway Research Board, Vol. 26, 589-593

Session 5

Construction of dams, roads, airfields, harbours, land reclamation in or on tropical soils

Compaction and strength characteristics of a residual clay from Bali, Indonesia

L.BELLONI, Project Director, ELC-Electroconsult, Milano, Italy
D.MORRIS, Geotechnical expert, ELC-Electroconsult, Milano, Italy
G.A.BELLINGERI, Resident Team Leader, ELC-Electroconsult, Milano, Italy
I.PURWOKO, Project Manager, Bali Irrigation Scheme, Ball, Indonesia

Abstract: The clay for the core of Palasari Dam, Bali, Indonesia, is a residual soil of volcanic origin containing halloysite/metahalloysite. During the investigations no consideration had been given to the effect of pretreatment methods for laboratory tests (ie., drying, wetting, heating) on the soil properties. The conclusion reached at that time was that the clay would be wet and soft, having a moisture content much higher than laboratory optimum. A trial embankment proved the clay behaved differently, was not soft, and could be compacted satisfactorily in its natural state. The investigations have shown the sensitivity of the material to the method of preparation for testing. By drying or wetting the relict structure of the parent rock is destroyed and the clay softened. By preserving this structure, the clay handles well. Remoulding the clay without consideration of this aspect leads to an unrealistic appreciation of the true engineering properties of such material.

1. INTRODUCTION

Palasari Dam, a 40m high clay cored rockfill dam, is being built on the river Sangian Gede in north west Bali to provide storage of water for irrigation in the area. The embankment is a classic fill structure with a central clay core, protected downstream by a fine sand filter and transition zone of processed sandy gravel, and upstream by a transition of pit run sandy gravel. The shoulders are of quarry run rockfill. The cross section is shown in Figure 1.

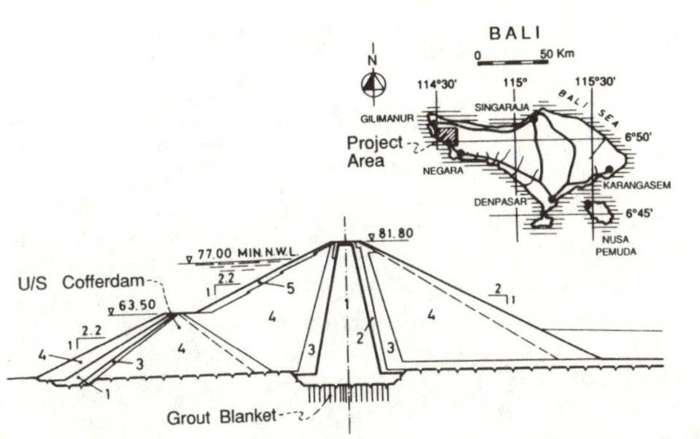

Figure 1. Typical Cross Section of Dam
1. Clay Core. 2. Fine Processed Sand Filter.
3. Pit run Sand and Gravel Transition.
4. Rockfill Shells.

The clay for the core is a red-brown lateritic material formed as the residue, from weathering, of the country rock; a fine to medium grained sandstone comprising clinopyroxene, volcanic glasses, carbonate shell debris and olivenes. This has weathered to produce a clay containing smectite, halloysite/metahalloysite, and some illite.

The early test results led to the conclusion that, in use, the material would be soft with a low shear strength, consistent with the measured high natural water content. Later tests, and experience of using the clay for a trial embankment showed this not to be the case. The test results from each of the phases of investigation are presented and conclusions drawn as to the reasons for this behaviour, and practical applications of the findings suggested.

2. DESIGN STAGE INVESTIGATIONS

A series of investigations were carried out over the period 1973-1985 by various bodies. These investigations included classification, compaction and shear strength tests carried out in accordance with normal practice. The results are summarised in Table I, and in Figures 2 and 3.

The material is essentially a clayey silt of medium to high plasticity and high activity.

Table I
MATERIAL CHARACTERISTICS FROM
PRE-CONSTRUCTION INVESTIGATIONS

Parameter		up to 1979	1983	1985
Clay Fraction	(%)	44	21	39
Specific Gravity	(-)	2.63	2.74	2.65
Liquid Limit	(%)	79	61	69
Plastic Limit	(%)	36	31	32
Activity	(-)	0.98	1.43	0.95
Natural Moisture Content	(%)	53	35	47
Optimum Moisture Content	(%)	43	30	40
Maximum Dry Density	(t/m^3)	1.17	1.36	1.17
Shear Strength				
Friction Angle	(°)	11.6	7.8	-
Cohesion	(t/m^2)	7.3	2.2	11.8(1)

(1) Unconfined Compression Value

During the investigations up to 1979, shear strength tests showed little difference between peak and residual values. This is in line with clays containing significant proportions of halloysite (Wesley 1977). However, few tests were carried out and the results could not be taken as meaningful for design purposes. The results indicated a material which was at approximately 10% wet of optimum in the natural state, and thus might be difficult to handle and compact. An early report of a field trial using material at about the optimum moisture content (43%) had indicated dry densities close to, but higher than, the laboratory values.

X-ray analysis of the clay in 1978 identified smectite and metahalloysite, with one sample containing a mixed layer illite-smectite-illite. The smectite is always in greater proportion than the halloysite mineral.

Halloysite and metahalloysite are normally considered as the same mineral, with the former taken to be the hydrated form (containing $.4H_2O$). Partial dehydration can occur at normal temperatures in nature and such material is referred to by some authorities as metahalloysite. Further dehydration continues at higher temperatures (notably 60°C and 110°C) and is known to continue up to 350°C.

Halloysite, a member of the Kandite group of minerals, has a single layer of water molecules between the structural sheets, which weakens their bond, and hence the stacking of these

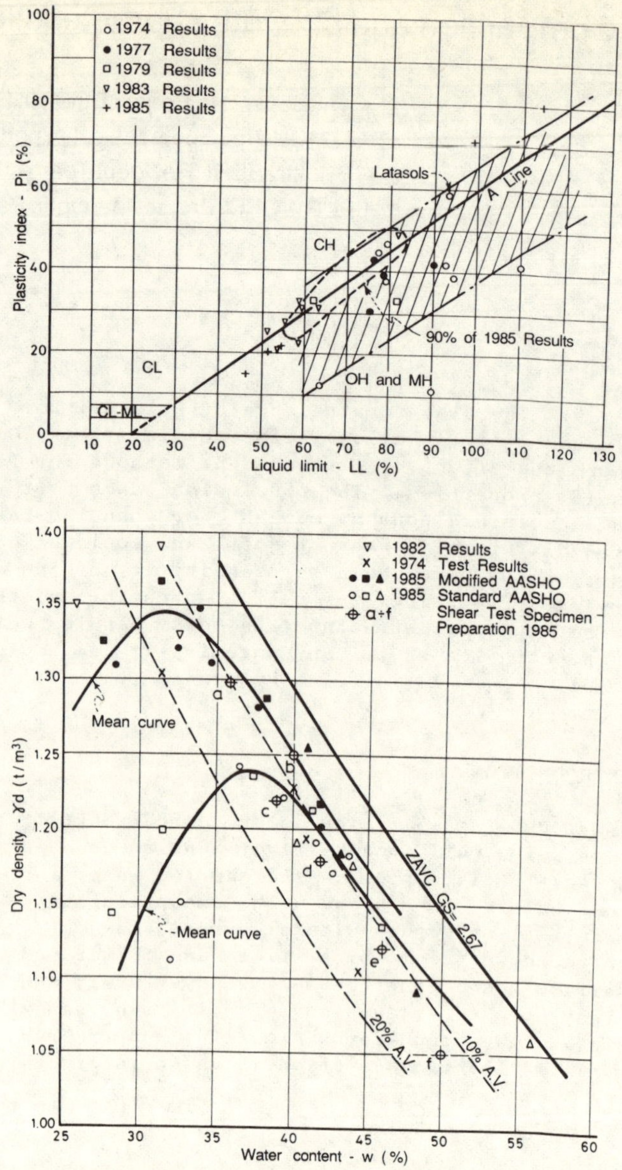

Figure 2. Preconstruction Test Results.
a) Plasticity Chart b) Compaction.

sheets is disordered and the interlayer distance is increased from the normal 7A° (Angstrom Units) to 10A° (Kirsch 1968). On heating to 110°C, some of the adsorbed water is lost and, at 200°C, the basal spacing is reduced by collapse to 7.4A°; close to that for the rest of the Kandite group. The rest of the water is lost gradually, and is not completely expelled until 400°C.

The minerals of the Kandite group may occur in compact massive blocks, in varmi-form or granular aggregates. Halloysite has a fibrous,

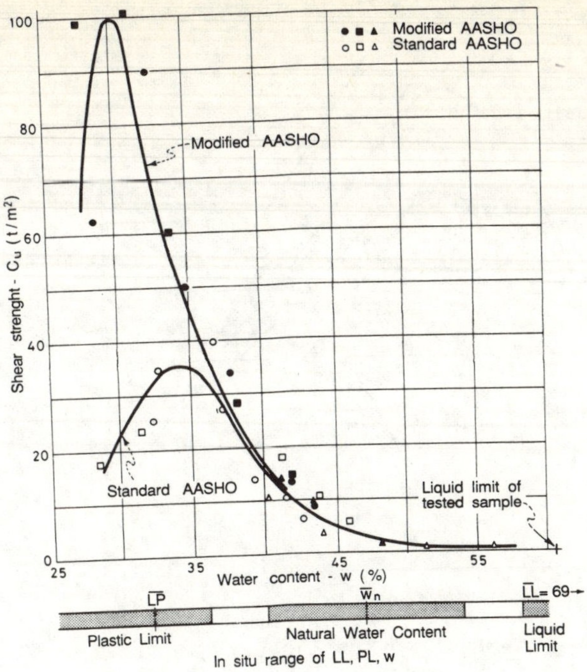

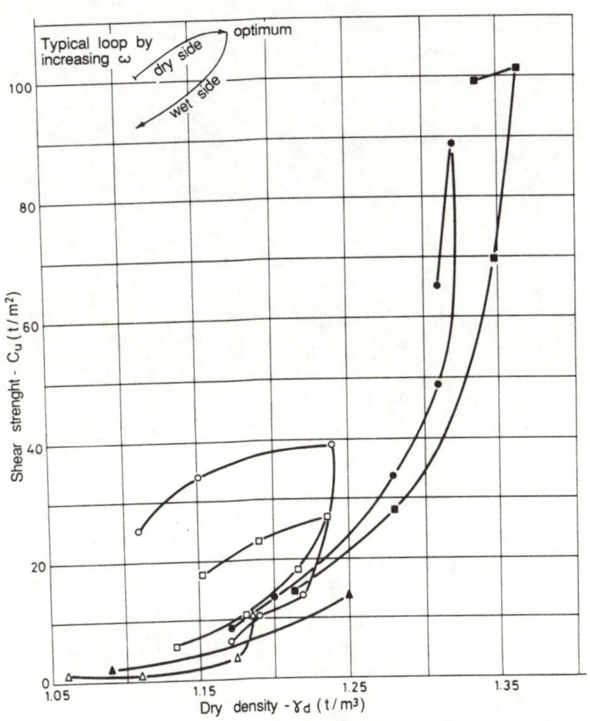

Figure 3. 1985 Test Results. Unconfined Compression Tests. a) C_u vs W. b) C_u vs γ_d.

rather than a platy morphology, with the fibres tubular or scroll like, and thus a skeleton like structure, containing adsorbed water can be visualised for the clay. This could be taken as analogous to the relict rock structure remaining intact after weathering, but with the various minerals being transformed into the weathered products. The resultant granular structure within the residual clayey soil, composed of resistant clumps, will have a significant effect on the shear strength.

The presence of halloysite/metahalloysite is normal in the residual clays of Indonesia and in soils formed from effusive volcanic products. Wesley (1973) describes some of the basic engineering properties of such clays from Java, Indonesia. In particular, the apparent high shear strength in the natural state, the variation in some of the Index Properties and reduction in Plasticity and Optimum Moisture Contents on drying. These change as the material transforms from halloysite to metahalloysite.

The earlier tests carried out for the dam had been preliminary, and more information was required for the final design. Thus in 1985 a further investigation was carried out and in 1987, before the start of dam construction, a trial embankment was built. The results of this latter are reported below, ELC-ADC (1988).

The 1985 investigation revealed some anomalous behaviour of the clay. In the borrow area, excavation of test pits by hand was difficult and they stood with vertical sides, indicating a high strength. In the laboratory, however, normal sample preparation and testing produced very low shear strengths of the order of 5 to 10 times less than in situ, even if remoulded to the same moisture content and density. The results are summarised in Figure 3.

A consideration of the relationship between C_u and σ'_v of the type:

$$C_u / \sigma'_v = (0.11+0.37I_p)$$

where C_u = undrained shear strength;
σ'_v = vertical effective pressure;
I_p = Plasticity Index

gives $C_u = 0.25\,\sigma'_v$ which would give a range of values for C_u of between 5 to 7.5t/m². This accords with the value of C_u for remoulded soil at the natural moisture content, as if it were normally consolidated in Figure 3. This could also be taken to represent a material with an SPT 'N' value of 8 to 10 blows per foot, a figure which is much lower than experienced in situ where the pits required mechanical breakers for excavation.

A high in situ shear strength is typical of Indonesian halloysitic clays (Wesley 1977) although the marked drop in remoulded strength is different to that found by Wesley; the activity of the clay reported by the latter was considerably less than at Palasari, and this could account for this different behaviour.

The compaction test results from the 1987 investigations are shown in Figure 4. These show a number of interesting features in that there is a marked difference between the optimum values for natural and remoulded soil curves A and B respectively. In this case, remoulded means air dried to a moisture content of 25% then water added for the test, while natural means progressively dried from the natural value in the borrow without the addition of water; in both cases the material was sieved to remove all particles greater than #40 sieve. The principal difference between the two curves is a significant drop in Optimum Water Content and rise in Dry Density on drying. This mirrors the results of Wesley (1973).

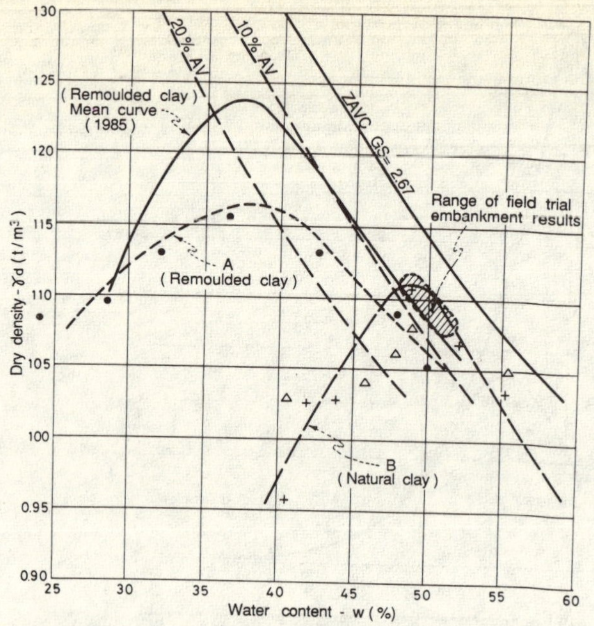

Figure 4. Compaction Test Results 1987. Standard AASHO Compaction Curves - Laboratory and Field Values.

3. 1987 TRIAL EMBANKMENT

In 1987, before the start of core placement, a trial embankment was built to verify the earlier test results from small scale laboratory samples and to optimise the Contractor's proposed equipment and working methods.

The material was used at its natural moisture content in the borrow and was compacted using the full scale equipment of, dozer for spreading, low ground pressure dozer, and pad foot roller. The moisture content varied between 49% and 56%, a figure which depended upon the rainfall. On average, 1.5% of the contained water was lost during loading, hauling and spreading and, thus, the actual average compaction value was about 48.5%. This is well within the range for field compaction shown in Figure 4.

The tests showed that compaction of the clay using normal equipment was possible and, in fact, produced a dense material having a significant shear strength as the equipment used did not sink very far into the soil. This confirmed the reported results of the earlier trial embankment and is consistent with the high values normally expected from halloysitic clays. The compaction values measured in situ are shown in Figure 4.

A number of Unconfined Compression Tests were carried out on the clay, both from the borrow and from the trial embankment. In an attempt to assess the effect of further wetting and remoulding of the soil, samples were prepared by

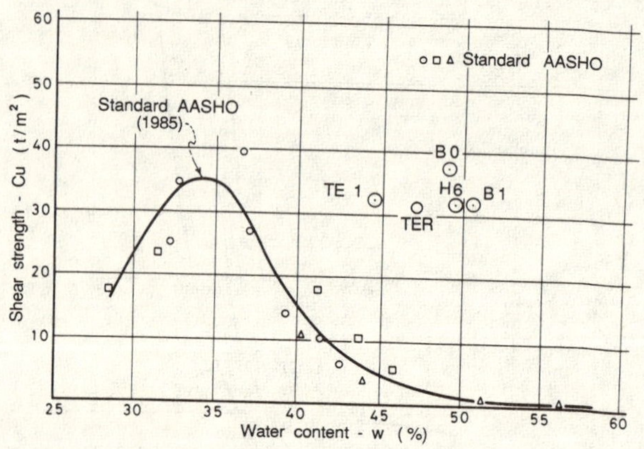

Figure 5. Unconfined Compression Test Results 1987.

C_u values of Samples Compacted with AASHO Standard Energy.

drying to a minimum moisture content value of 35% and breaking down all the 'clumps' in the soil then adding water to bring it back to the natural moisture content of the borrow; this did not subject the sample to a full remoulding cycle and the results show that there is no significant difference in strength for such a change in state, see Figure 5.

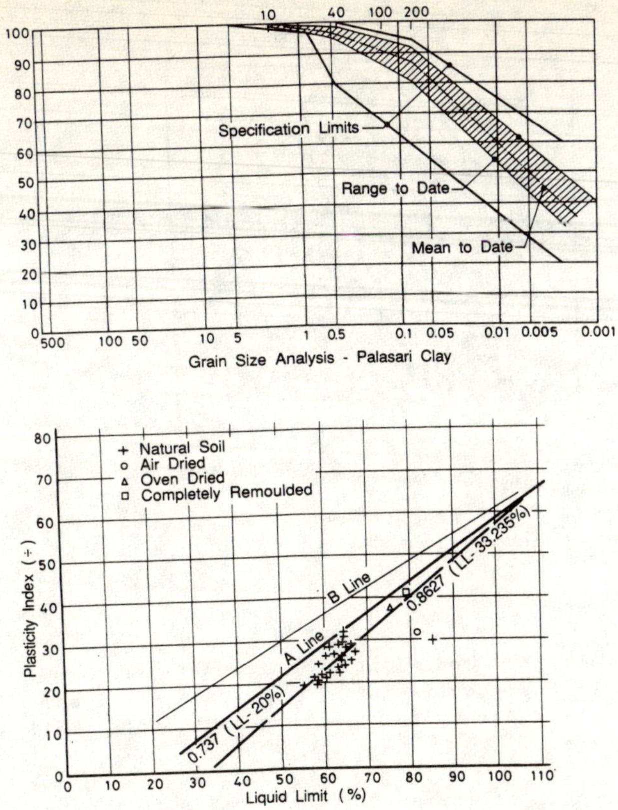

Figure 6. Construction Stage Test Results. a) Grading. b) Plasticity Chart.

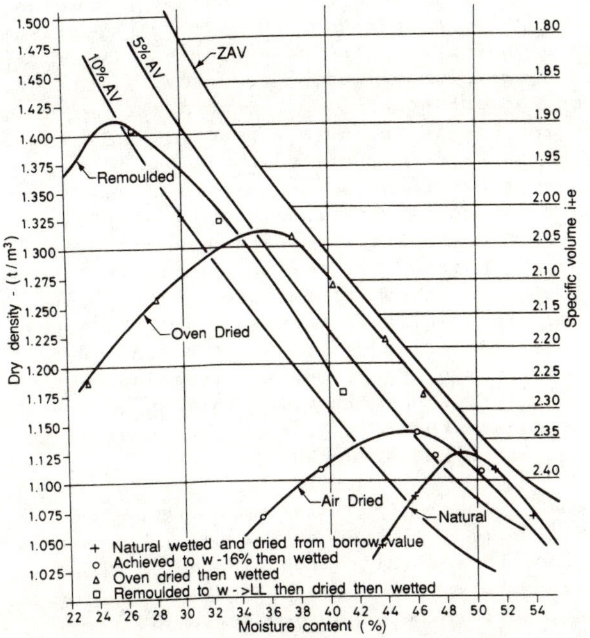

Figure 7. Construction Stage Compaction Curves.

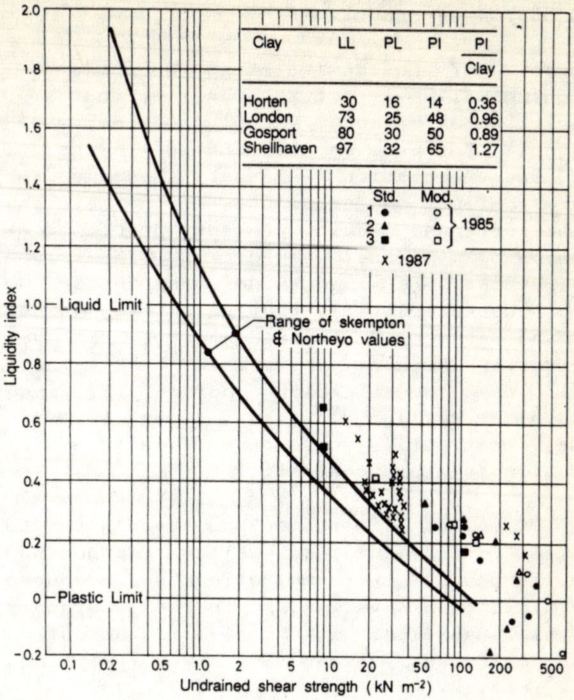

Figure 8. Relationship between Shear Strength and Liquidity Index. (After Skempton and Northey 1953).

4. EARLY CONSTRUCTION RESULTS

The experience gained during the early stages of construction confirmed the results of the trial embankment. The moisture content in the borrow remained reasonably static in the range 47-52%, which is very close to that obtained earlier. Even at the end of the dry season the value did not fall significantly below 47% except where the soil had been left exposed for a significant time. To allow for the moisture lost during transportation etc., a small quantity of water was added in the borrow by spraying over the clay during dozing and loading. This only served to replace that lost and did not increase the value beyond that normally in the borrow. If water was added at the dam it was noticed that the clay quickly became very sticky and unworkable. The quantity of water necessary to give 100% saturation is, on average, small ($30l/m^3$); and full saturation, coupled with the increase in saturation resulting from compaction, appears to make a significant difference.

A further series of Compaction and Unconfined Compression Tests were carried out this time on natural clay, and clay which had been dried by various means, or had been remoulded at moisture contents higher than the Liquid Limit. The results are shown in Figures 6 to 8.

5. DISCUSSION OF RESULTS

The results of the design stage investigation, based on normal soil testing practice, indicated a material which, though wet of optimum, would be suitable for use in the core of a dam 40m high. The expectation was that, based on the moisture content and shear tests, the material would be soft and would present problems for compaction in the field. The 1983 investigations were aimed at defining better the material characteristics, but gave lower values for Cu and Optimum Moisture Content. The 1985 investigations brought the values back to near those of the investigations before 1979, and confirmed the earlier view that, though suitable for use, construction would be difficult because of the high moisture content.

At that time, no importance had been given to the effect of remoulding, neither method nor degree, supposing the same effect was produced in the field by excavation as in the laboratory by normal handling, without specific attention to the different steps involved (wetting, heating, etc.).

In situ the material was stiff, and remained in this state after hauling. This did not accord with the behaviour expected from the results of the investigations. The structure of the clay can be described as blocky, and there is a tendency for the soil to break along predetermined planes or microfissures to leave exposed the matrix of the parent rock. Thus the soil can be described as a type of granular agglomeration rather than as a true clayey soil. This could be taken as agreeing with the 'cluster hypothesis' (Wesley (1973b), Wallace (1973), De and Furdas (1973[2])) in which the strength of the soil is due to a structure comprising an open skeleton of rock forming minerals, surrounded by clay particles and a viscous gel of highly hydrated clay minerals, Figure 9. The structural forms adopted by the Kandite minerals would seem to fit this pattern, especially the halloysite/metahalloysite members of the family. The loss of water from these minerals on heating, and the consequent collapse, could lead, on disturbing the soil by mechanical action, to the breakdown of the cementing bonds in the skeleton, the rearrangement of the individual particles and the reduction in volume of the viscous gel which does not reform. This could help to explain the reduction in Optimum Moisture Content of the material after heating, Figure 7. The clay fraction from the investigation results was of the order of 39%. This is in line with the values found by others from mechanical analysis of such a clay. Wesley (1973b) found that, though the normally measured clay fraction of Indonesian Clay's was of the order of 40%, the

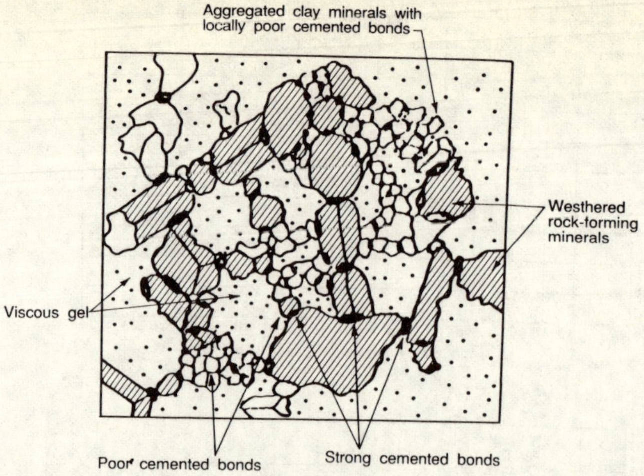

Figure 9. Idealised Residual Soil Structure. (After De and Furdas 1973).

proportion of individual particles finer than 1 could be 80%, confirming the presence of some sort of structural arrangement of soil particles.

The results of the early construction tests show that not only is drying an important agent in controlling the behaviour of the soil, but that the method of drying also has an important role. Figure 7 shows the Moisture/Density relationship established for various methods of pretreating the clay. The results for the natural and air dried (dried to 16%) samples are close, whereas those for the oven dried and the remoulded samples show a marked reduction in optimum moisture content. The oven dried value is, within normal experimental accuracy, similar to that of the 1983 tests; which would also be oven dried, in line with normal practice. It is to be noted that at higher moisture contents the line for these samples meets the others. The reason for this is not clear, but could indicate some sort of physical rearrangement of soil grains at moisture contents near to those pertaining in the natural state. The samples remoulded to a moisture content at the Liquid Limit then dried and wetted again, show by far the lowest value of optimum moisture content but a higher value of air voids throughout, which could indicate that complete re-ordering of the soil structure has occurred.

In the Plasticity diagram Figure 6, a progression of lower values for Liquid Limit is seen from the natural to the oven dried samples. This trend is reversed for the remoulded sample however, confirming the re-ordering of the soil structure to produce a more clay like material.

The results of the trial embankment, and from the early stages of construction, have confirmed that the clay, as taken from the borrow, is relatively easy to handle if kept at the natural moisture content and has a relatively high shear strength. The shear strength has been plotted on a diagram, taken from Skempton and Northey (1953), in terms of shear strength and Liquidity Index, and it is seen that the best fit line through the points lies above the envelope from the quoted paper; confirming the higher shear strength of halloysitic soils at normal moisture contents.

The addition of large amounts of water to the clay makes it difficult to handle and very sticky. A reason for this could be that the surface of the individual clumps of soil, which become dryer during handling than the interior, become very wet on the addition of water and the mechanical disturbance of handling and trafficking causes complete remoulding of this very thin surface layer with the attendant loss of shear strength and increase in Plasticity.

6. CONCLUSIONS

Based upon the observations of the actual material behaviour in the field the following comments can be made. In the natural state the clay retains some of the original structure of the parent rock, and when excavated results in clumps of individual soil particles. Handling of these clumps reduces the moisture content of the surface layer while retaining the interior at the natural value. The presence of these clumps imparts to the soil a significant shear strength allowing the easy circulation of construction traffic and of compaction. After compaction, the moisture content of the interior of the clumps prevails giving dry densities and moisture contents in line with standard compaction energies and with the values in the borrow. Complete remoulding of the soil results in a lower shear strength, probably as a result of the breakdown of the soil skeleton, and this leads to construction difficulties in handling and compaction. There is, therefore, a need to minimise the remoulding of the soil during handling to take advantage of its natural high strength.

From the above results and discussion it can be seen that the results of laboratory tests on residual soils of the type encountered at Palasari depend upon the method of pretreating the soil before test, especially the method used to dry the soil. This is a result of the presence of the halloysitic clay minerals. Care must, therefore, be taken in choosing the test procedure for soils of this type, as the use of high temperatures in pretreatment drying will give rise to values of compaction and shear strength parameters which will not pertain in the field and which may cause the rejection of soil which is perfectly usable. When soils of this type are encountered, therefore, tests should be carried out from the natural moisture content state, by wetting and air drying from the natural value as necessary to give values representative of those in situ.

Residual soils formed from effusive volcanic products and containing halloysitic minerals exist in large quantities over a wide area of the world. These soils have high natural moisture contents and have, as a result, tended, in the past, to be considered as unsuitable for use as fill, especially in embankment dams. The experience at Palasari has shown that such soils are usable and have a significant shear strength in spite of their high moisture contents. It is necessary to confirm in each case, however, by tests from the natural state, that the soil will be suitable for the intended purpose, and not rely on standard laboratory practice.

REFERENCES

De, P.K. and Furdas,B., 1973, Discussion- Structural Behaviour of Residual Soils of the Continually Wet Highlands of Papua New Guinea, Geotechnique 23, No.2, p601-603.

ELC-ADC Consulting Engineers, 1988, Palasari Dam - Trial Embankment Compaction Test Main Report, Bali Irrigation Sector Project - Palasri Sub-Project.

Kirsch, H., 1968, Applied Minerology, Chapman and Hall and Science Paperbacks.

Skempton, A.W. and Northey, R.D., 1953, The Sensitivity of Clays, Geotechnique 3, p30-35.

Wallace, K.B., 1973, Structural Behaviour of Residual Soils of the Continually Wet Highlands of Papua New Guinea, Geotechnique 23, No.1, p203-218.

Wesley, L.D., 1973, Some Basic Engineering Properties of Halloysite and Alloyphane Clays in Java, Indonesia, Geotechnique 23, No.4, p471-494.

Wesley, L.D., 1973b, Cluster Hypothesis and the Shear Strength of a Tropical Red Clay, Geotechnique 23, No.1, p109-113.

Wesley, L.D., 1977, Shear Strength Properties of Halloysite and Allophane Clays in Java, Indonesia, Geotechnique 27, No.2, p125

Interaction of geogrid reinforcement with gravel and decomposed granite backfill with applications

D.T.BERGADO, GTE Division, Asian Institute of Technology, Bangkok, Thailand
C.L.SAMPACO, GTE Division, Asian Institute of Technology, Bangkok, Thailand
N.MIURA, Department of Civil Engineering, Saga University, Saga, Japan
K.ONITSUKA, Department of Civil Engineering, Saga University, Saga, Japan

ABSTRACT: The problems associated with the use of decomposed granite as construction material and the interaction between Tensar geogrids with both gravel and decomposed granite were investigated in this study. The Tensar geogrids consisted of such types as GM1, SR2 and SS35. The laboratory tests included pull-out tests of geogrids using gravel and decomposed granite backfill. The pull-out tests were done using a new equipment designed and constructed for this study. The data obtained from the pull-out tests were used in the FEM model. The results of the load tests on unreinforced and reinforced gravel are also included in this study. The tests results were applied to an analysis of an actual box culvert construction with reinforced foundation using the FEM model. The reinforced foundation yielded less differential settlements than the unreinforced one. Moreover, the use of reinforced foundation resulted to a cheaper construction cost.

1. INTRODUCTION

Decomposed granite soil has been widely used as construction material for earth embankments and earthfills. Two of its most common engineering properties include: the loss of strength when wet and the stress-dependent failure envelope. This soil type can be improved using geogrid reinforcements. Reinforcements were also proposed for gravel backfill. To study the soil-geogrid interaction, pull-out tests were performed in the laboratory of Saga University using a new equipment designed and constructed for this study based from Bergado et al (1987). The results of the pull-out tests on gravel were applied to the analysis of a reinforced foundation of a box culvert construction on soft and subsiding ground. Additional input data were obtained from load tests of unreinforced and reinforced gravel in the laboratory.

2. THE DECOMPOSED GRANITE BACKFILL

The decomposed granite soil consists of breakable particles that tend to decrease in shear strength especially when wet. Numerous cases of slope failures have been documented in this soil during rainy season. This phenomenon are caused by the disappearance of the meniscus effect, the reduction of the cementation effect, and, particularly, the particle-crushing effect (Miura & O-hara, 1979; Miura et al, 1983; Onitsuka et al, 1988). Figures 1 and 2 show the typical results of the triaxial tests on Toyoura sand and decomposed granite, respectively. In the case of Toyoura sand, the axial and volumetric strains were not affected by the addition of water and there was no change in the stress-strain curves. In the case of the decomposed granite, the stress-strain curves changed upon the addition of water. In Fig.2, D_o and D_{40} were dry samples while $D_{40}WD$ stands for the specimen where water was supplied at 40% of the failure stress level for 120 hours. It was found that the deviator stress at failure of the wet samples was 20% less than the dry

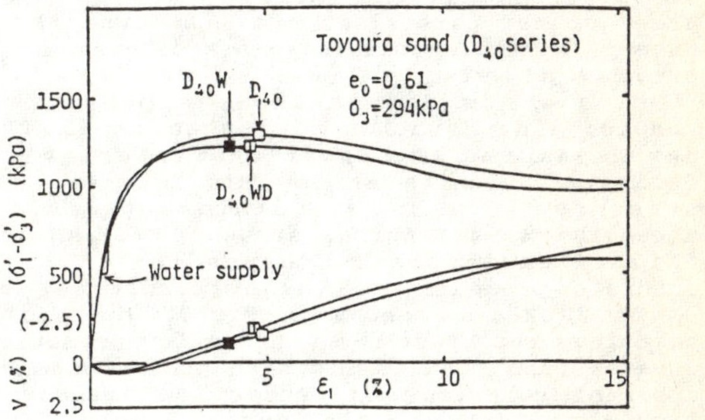

Figure 1. Stress-strain and volume change for Toyoura sand.

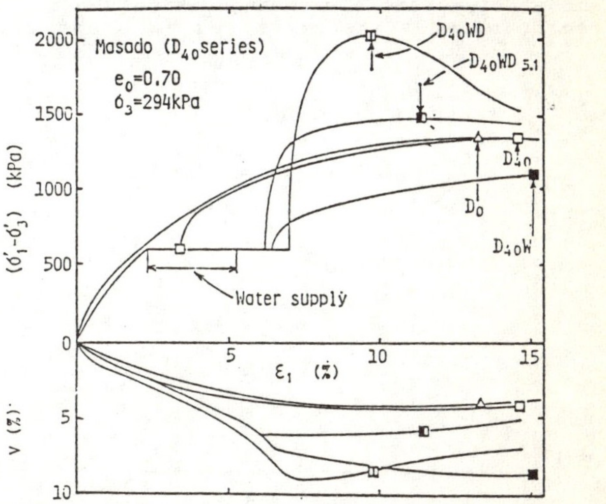

Figure 2. Stress-strain and volume change for decomposed granite soil.

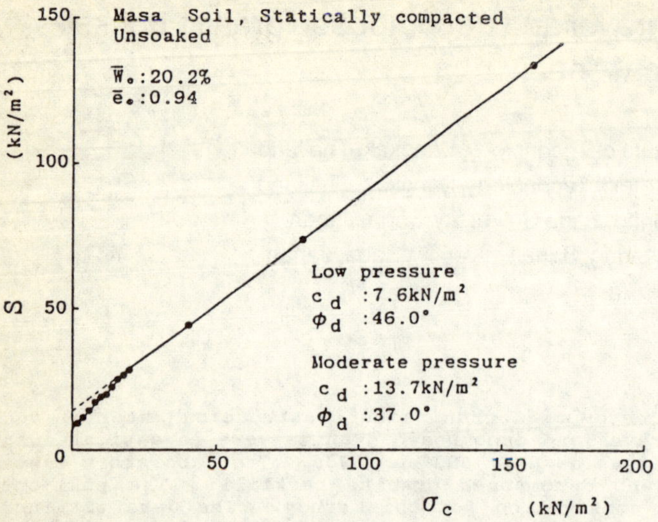

Figure 3. Failure envelope of decomposed granite soil.

sample. The failure stress of the dehydrated sample was 30% higher than that of the dry sample. The amount of particle crushing was measured through the increase of the surface areas of the particles. It was found that the rate of particle crushing was accelerated by the presence of water.

Using a new direct shear apparatus with reduced side friction, Onitsuka et al (1988) have measured the shear characteristics of decomposed granite at low and moderate normal pressures. The relationship between shear strength, S, and normal stress, σ_c , shows two failure envelopes corresponding to low and moderate pressures with the boundary at about σ_c of 20 kPa as shown in Fig.3. The angle of shearing resistance, ϕ_d, under low pressures is greater than that of moderate pressures and the apparent cohesion, C_d, under low pressures is less than that under moderate pressures. The change in the failure envelope is blamed on the amount of particle crushing. Figure 4 shows the relationship between the amount of particle crushing at different normal pressures. The amount of particle crushing during shear in the region of moderate pressures is greater than

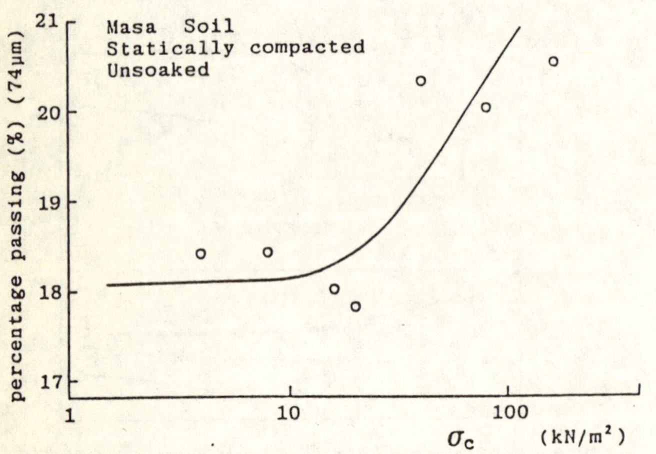

Figure 4. Percent passing #200 U.S. Sieve vs. normal stress

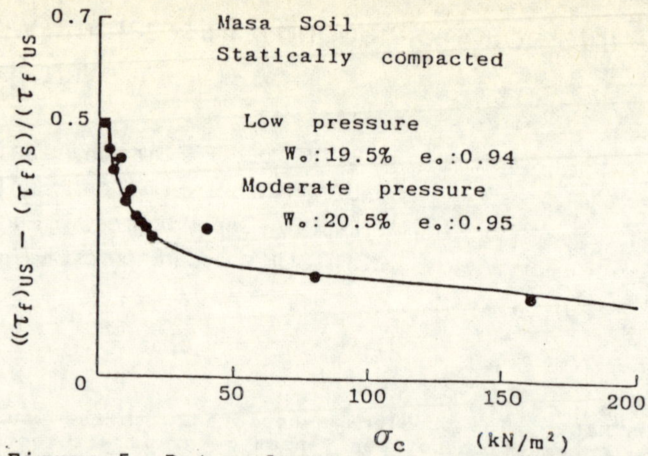

Figure 5. Rate of strength decrease of decomposed granite soil with normal stress.

that in the region of low presures. The increase of percent passing no. 200 US standard sieve (0.074 mm) was used to measure the amount of particle crushing.

The shear strength of decomposed granite also decreased upon soaking with water (Onitsuka et al, 1988). The shear strength, under soaked conditions is given as:

$$\tau_f = C' + \sigma \tan \phi' \qquad (1)$$

where C' and ϕ' are the shear strength parameters. During unsoaked conditions, the shear strength is given as:

$$\tau_f = C' + (\sigma + p^*) \tan \phi' \qquad (2)$$

where p^* is the suction pressure at zero air pressure. After soaking, the shear strength is smaller by an amount of $p^* \tan \phi'$. The degree of decrease in shear strength due to soaking is larger at low normal pressures than at moderate normal pressures. The reduction of shear strength due to soaking depends mainly on the decrease of apparent cohesion and not much on the angle of shearing resistance. The degree of decrease in shear strength due to soaking can be expressed as in terms of the peak strength of soaked soil (τ_{fs}) to that of unsoaked soil (τ_{fus}) as shown in Fig.5.

3. THE REINFORCED EARTH

Based on the results of past investigators (Ingold, 1982; Schlosser & Long, 1973; Hausmann, 1976; Yang, 1972), two basic concepts were proposed regarding the principle of reinforced earth, namely: the anisotropic cohesion concept and the enhanced confining pressure concept. In the former case, the additional tensile strength imparted by the reinforcement could be represented by an apparent cohesion. In the latter case, the horizontal plane of the reinforcement could not be assumed as principal plane because of induced stresses due to soil-reinforcement interaction resulting in an enhanced confining pressure producing additional strength. Geogrids made of steel or geotextiles consist of transverse and longitudinal members in which the transverse members run parallel to the face of the structure while the longitudinal members run perpendicular to the face. At first, this type of reinforcement has been

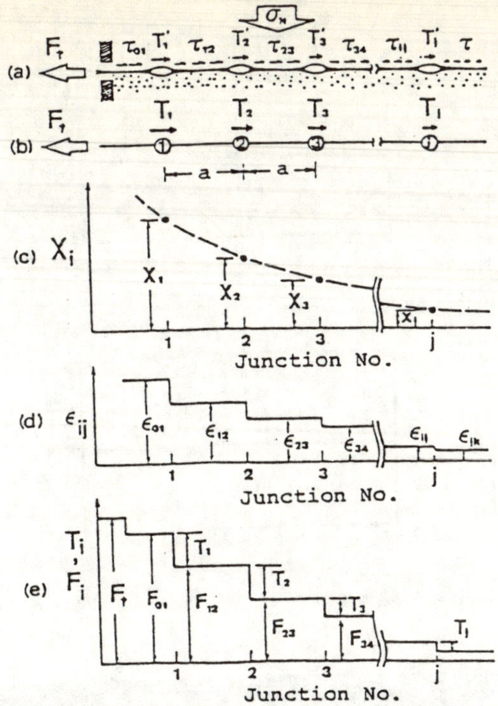

Figure 6. Analytical procedure of results obtained from pullout test.

developed using steel materials called welded-wire or welded-bar reinforcement (Chang et al, 1977; Forsyth, 1978; Peterson & Anderson, 1980). Later, polymer grids with high extensional stiffness and resistance to corrosion were utilized in the form of mesh reinforcement (Ingold, 1983; Jewel, 1985). Pull-out tests on decomposed granite and its applications has been presented recently by Bergado et al (1987) and Ochiai et al (1988).

4. INTERACTION OF SOIL AND GEOGRID REINFORCEMENT

When the geogrid is pulled out from the backfill soil, the pull-out resistance is mobilized in both grid junctions and ribs as shown in Fig.6a. The pull-out resistance on both sides of the geogrid are modeled to be concentrated at the grid junctions as shown in Fig.6b (Ochiai & Sakai, 1987). The pull-out force, F_t, exerted on the polymer grid produces displacement, X_i, at each grid junction. The displacements of the grid junctions can be measured in the pull-out test under constant vertical stress. The strain, ε_{ij} of the geogrid between junctions is

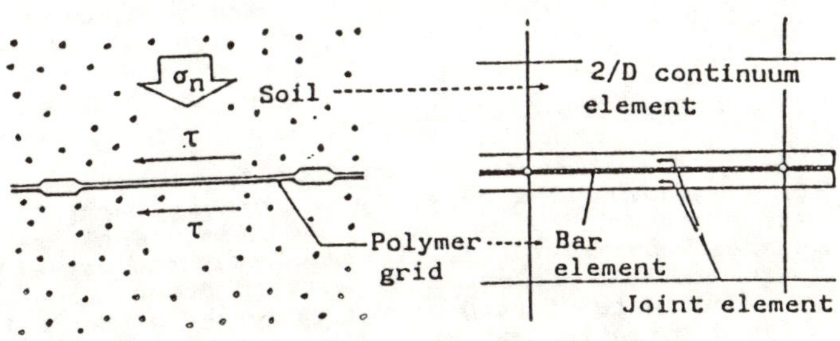

Figure 7. The FEM model of reinforced soil.

calculated by:

$$\varepsilon_{ij} = (X_i - X_j)/a \tag{3}$$

where a is the length between grid junction. Figure 6c shows the strain plots. The pulling force, F_t, between each grid junction that correspond to the strain, ε_{ij}, is determined using the standard stress-strain curves of the geogrids provided by Netlon (1984). The plot of the pulling force (F_i-F_{i+1}) represents the pull-out resistance, T_i, mobilized at each grid junction. The shear stiffness, k_s, can be obtained from the plot of T_i versus displacement, u, by the following equation:

$$2\,\tau = 2\,k_s\,u \tag{4}$$

where τ is the shear resistance.

5. THE FINITE ELEMENT MODEL

Figure 7 shows the finite element model of the reinforced soil (Ochiai et al, 1988). The geogrid is transformed into a joint element on both sides representing the interface with the soil and a truss element in the middle transmitting axial force only. The joint element expresses the transmission of the compressive force and sliding against shear displacement. The relationship between shear displacement $\{u\}_{s,n}$ of the joint element (length $= \ell$) and the incremental loading $\{F\}_{s,n}$ is given by the following expression:

$$\{F\}_{s,n} = [K]_{s,n}\,\{u\}_{s,n} \tag{5}$$

where the stiffness matrix $[K]_{s,n}$ is expressed by:

$$[K]_{s,n} = 1/4\ell \begin{bmatrix} k_s & 0 & k_s & 0 & -k_s & 0 & -k_s & 0 \\ & 2k_s & 0 & 0 & 0 & 0 & 0 & -2k_n \\ & & k_s & 0 & -k_s & 0 & -k_s & 0 \\ & & & 2k_n & 0 & -2k_n & 0 & 0 \\ & & & & k_s & 0 & k_s & 0 \\ & \text{sym.} & & & & 2k_n & 0 & 0 \\ & & & & & & k_s & 0 \\ & & & & & & & 2k_n \end{bmatrix} \tag{6}$$

The joint element has two unit stiffness, a normal stiffness, k_n, and a shear stiffness, k_s. The shear stiffness can be determined from the laboratory pull-out tests.

6. EXPERIMENTAL INVESTIGATIONS

Load tests on both unreinforced and reinforced gravel using double reinforcement and geocell are shown in Fig.8. Types 1, 2, and 3 refer to the unreinforced, geocell, and double reinforcement, respectively. The polymer geogrid was Tensar GM1. The gravel specimen has plan dimensions of 150 by 150 cm and 50 cm in height. The schematic diagrams for types 4 and 5 refer to the geocell and double reinforcement, respectively, overlying 40 cm thick reconstituted soft Ariake clay. A non-woven geotextile was laid at the interface between gravel and clay. The set-up for the load tests are shown in the photos given in Fig.9. The unit weight of the gravel was maintained at 1.8 t/m^3.

The schematic diagram for the pull-out test is shown in Fig.10. Based on this diagram and the previous research by Bergado et al (1987), a new equipment was constructed for this study (Fig.11). The size of the geogrid reinforcement was 80 cm

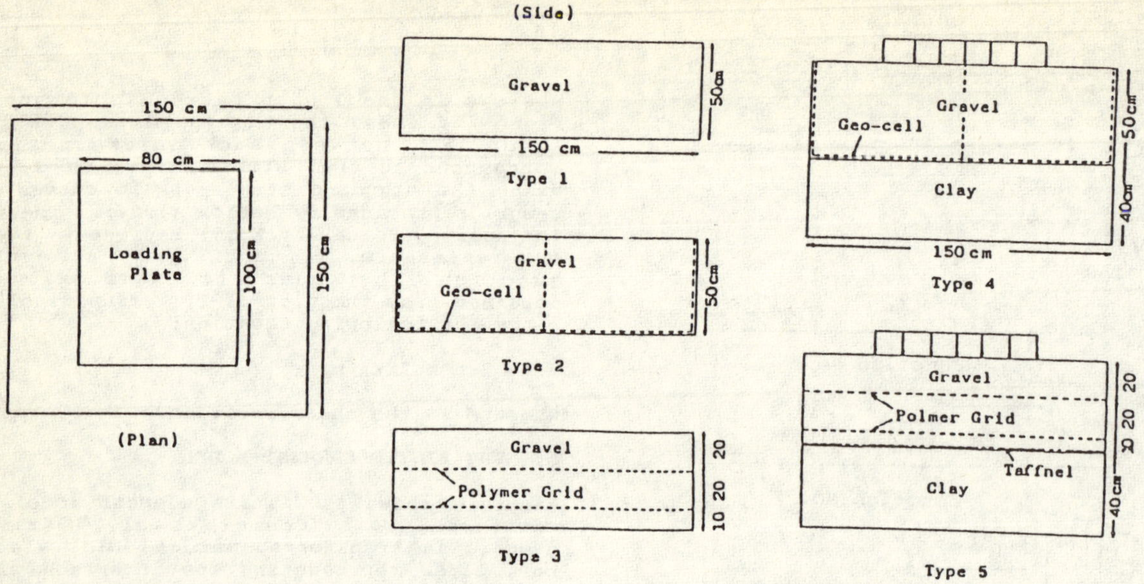

Figure 8. Load tests on unreinforced and reinforced gravel.

Figure 9. Set-up of load test in the laboratory.

Figure 11. Set-up of pullout test.

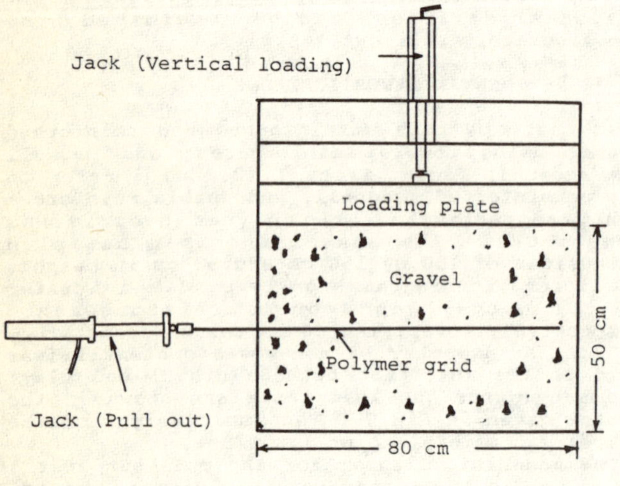

Figure 10. Diagram of pullout test machine.

by 75 cm. A total of 14 strain gages were attached to the geogrid in two lines. The strain gages consisted of Kyowa KFC-5-C1-11L300 and were attached to the geogrid by Kyowa strain gage cement CC-15A. The gages were further glued to the geogrid by using Araldite and covered by Hamatite. For the gravel backfill, a total of 4 tests were performed, namely: GM1 with 6 and 8 tsm normal pressure, SR2 with 6 tsm, and SS35 with 6 tsm. For the decomposed granite, on the other hand, SS35 was used under vertical stresses of 2,4 and 6 tsm for a total of 3 tests.

7. RESULTS AND DISCUSSIONS

The load-settlement curves from the load tests on unreinforced and reinforced gravel backfill are plotted in Figs.12a,b and the tests are summarized in Table 1. It seemed that the gravel with geocell produced the largest deformation. The double reinforced gravel performed better with less deformation and was subsequently recommended.

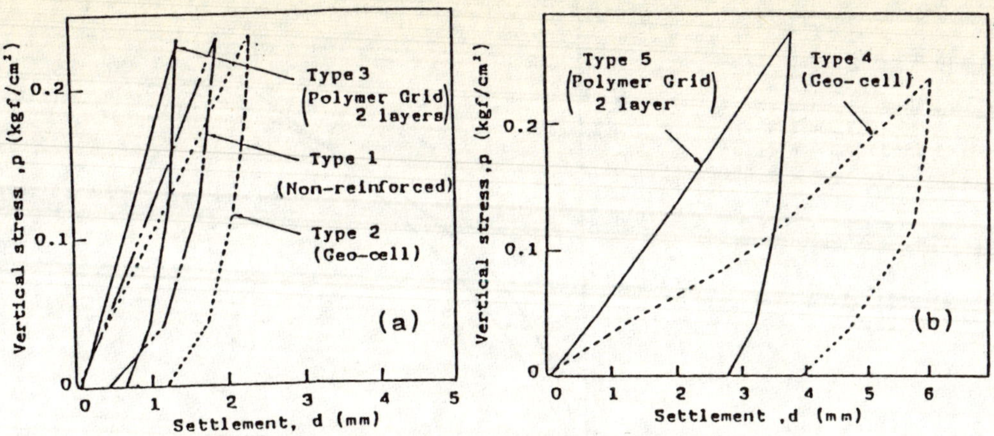

Figure 12 a,b. Load settlement curves from the load tests.

Table 1. Summary of load-settlement data for unreinforced and reinforced gravel.

	p/d (kgf/cm³)
Type 1	1.29
Type 2	1.00
Type 3	1.82
Type 4	0.48
Type 5	0.73

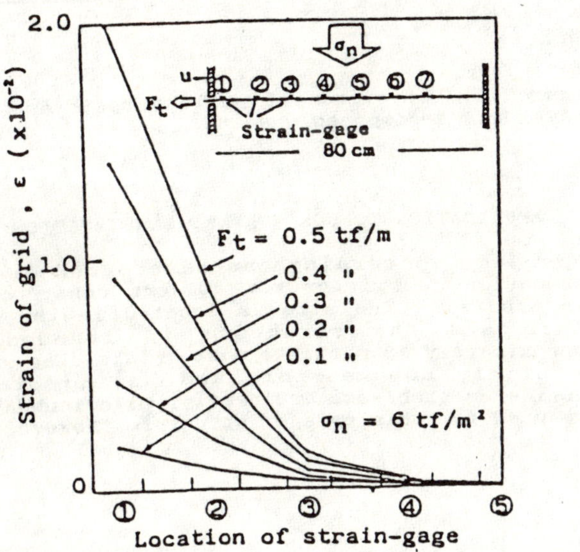

Figure 13. The strain plots from pullout test.

The pull-out force, F_t, for gravel backfill with Tensar GM1 reinforcement was plotted against strain. Figure 13 shows the strain plots under normal stress of 6 tsm at different pull-out force, F_t. The pulling force, F_i, between grid junction is plotted in Fig.14 and the corresponding pull-out resistance is given in Fig.15. The shear stiffness, k_s, calculated from the pull-out resistance is plotted in Fig.16 versus displacements for GM1, SR2, and SS35. The k_s obtained for SR2 was the highest followed by GM1 and SS35. The above process demonstrate the calculation of k_s. Similar set of data were obtained from the pull-out test using decomposed granite backfill. The corresponding data for k_s using Tensar SS35 is given in Fig.17.

Comparing Figs.16 and 17, it can be seen that the relationship between k_s and resistance force may not be linear for the case of decomposed granite backfill. Figure 18 shows the pull-out force for gravel and decomposed granite using the same SS35 Tensar geogrid. At low magnitudes of pull-out force, the gravel backfill yielded larger displacements, u, of the pull-out jack.

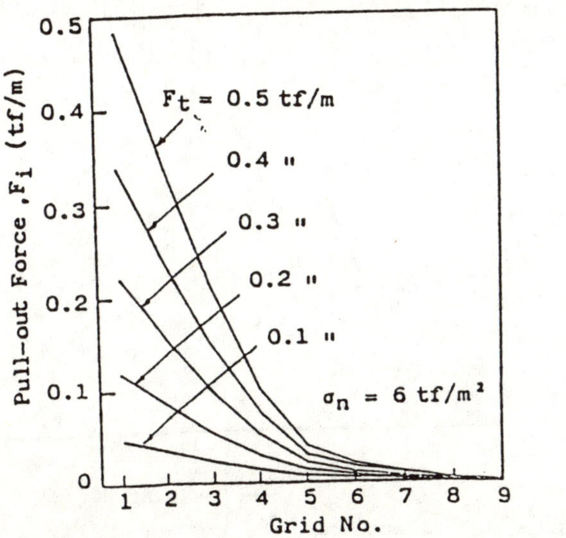

Figure 14. The pullout force between grid junctions.

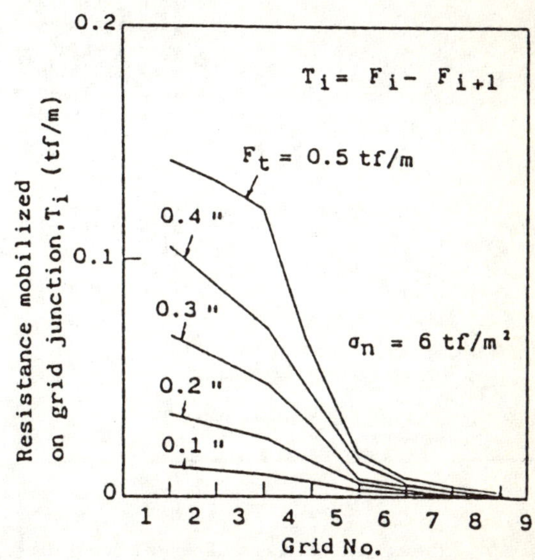

Figure 15. The pullout resistance on grid junctions.

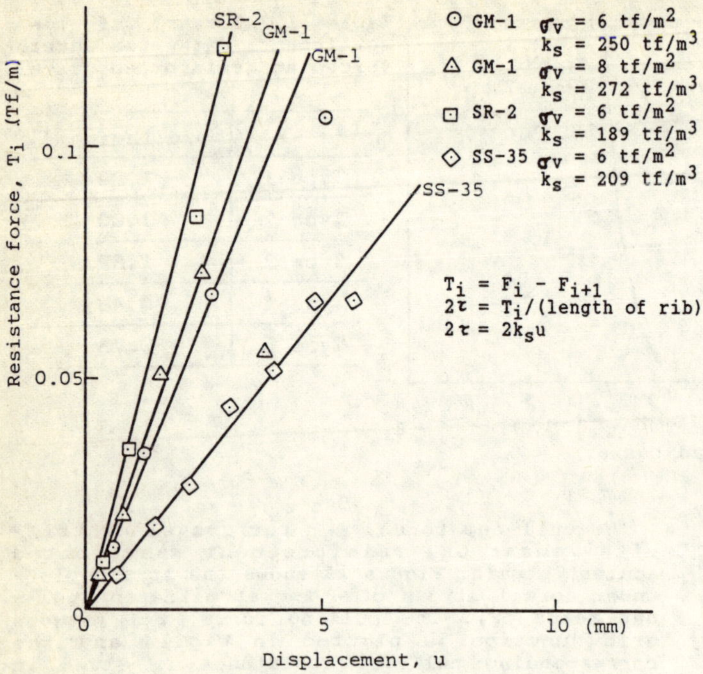

Figure 16. Shear stiffness for gravel backfill.

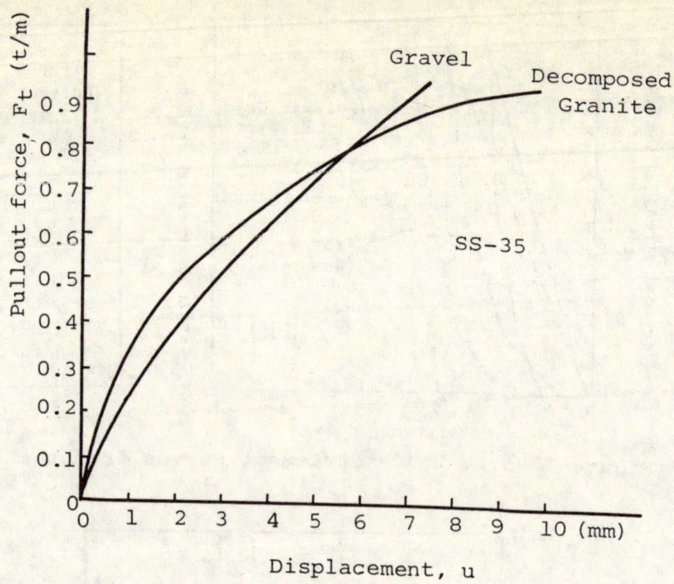

Figure 18. Comparison of pullout resistance for gravel and decomposed granite.

However, the gravel-reinforcement interaction was larger. Figures 19 and 20 show the shear resistance versus grid displacement relation for both backfill materials using SS35 Tensar geogrid. Larger grid displacements were observed in the decomposed granite backfill as the pull-out force was increased. This indicates that larger relative displacements between the reinforcement and the backfill occurred in the decomposed granite than in the gravel backfill.

8. APPLICATION TO BOX CULVERT CONSTRUCTION

A project concerning the reinforced gravel foundation design for box culvert construction was studied. The site is located in the Saga Plain and the structure is founded on approximately 15 m thick, soft Ariake clay which is highly compressible and having ground subsidence problems. Ordinarily, pile foundations are used for this type of structure. However, in

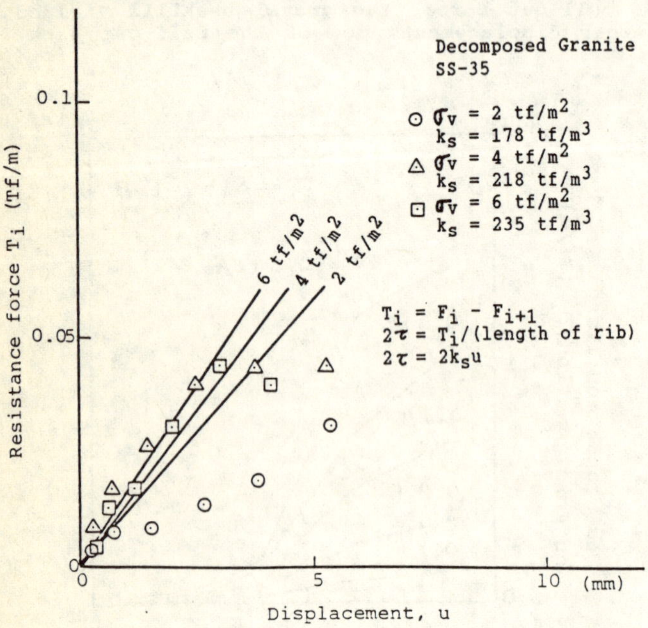

Figure 17. Shear stiffness for decomposed granite backfill.

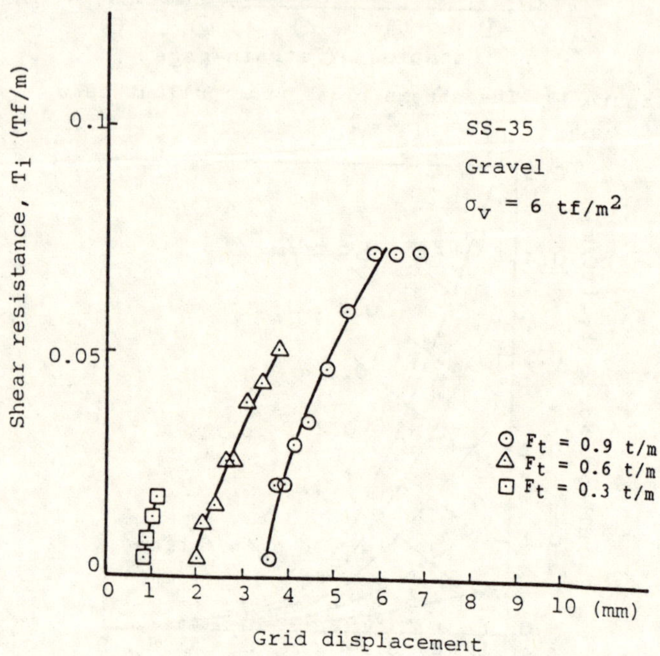

Figure 19. Shear resistance versus grid displacements for gravel backfill.

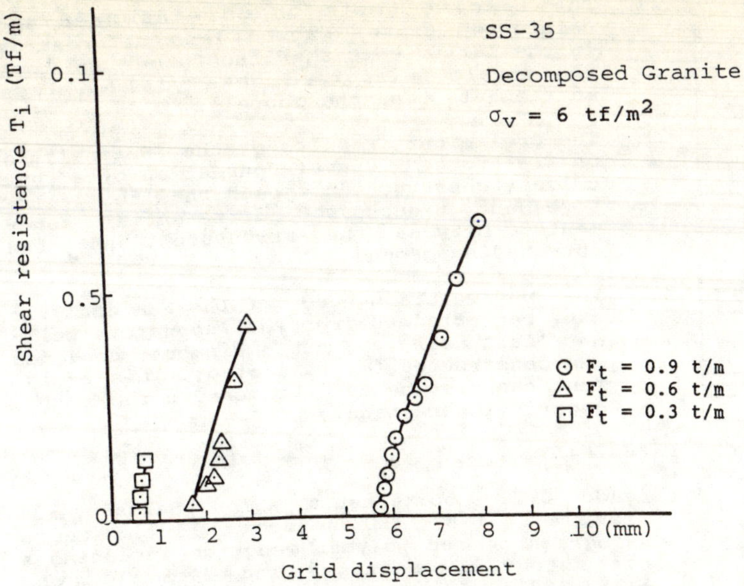

Figure 20. Shear resistance versus grid displacements for decomposed granite backfill.

this study, it was recommended to use a reinforced gravel foundation using two layers of Tensar GM1 geogrids (Fig.21). Not only that the differential settlements will be minimized but also this scheme is a lot cheaper. The FEM model described earlier was used for the analysis. The parameters used are tabulated in Table 2. As shown in Fig.21, the double reinforced gravel foundation has a thickness of 50 cm whose bottom is located 2.75 m below the ground level (line A-A in Fig.21).

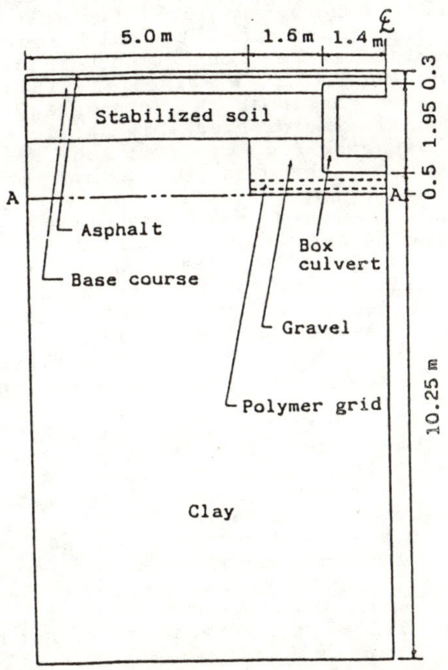

Figure 21. Box culvert construction with reinforced gravel foundation on soft Ariake clay.

Table 2. Parameters used in the FEM analysis.

	$E(t/m^2)$	ν	$\gamma(t/m^3)$
Asphalt	2.0×10^5	0.38	2.3
Base course	1.75×10^4	0.43	1.8
Stabilized soil	1.0×10^4	0.40	1.8
Gravel	1.5×10^4	0.43	0.8
Clay	3.0×10^2	0.45	0.6
Culvert	3.0×10^6	0.17	2.5
	$k_s(t/m^3)$	$k_n(t/m^3)$	
Joint element	2.50×10^2	1.0×10^4	
	$E(t/m^2)$	$A(m^2)$	
Bar element	1.8×10^5	1.3×10^{-3}	

The results of the FEM analysis are plotted in Figs.22, 23 and 24. The settlement magnitudes in Fig.22 refer to plane A-A in Fig.21. The settlements resulting from the body force plus external force are shown in Fig.22 for both unreinforced and reinforced cases at a uniform load of 6 tsm. The reinforced case yielded less differential settlements. A different loading scheme was used in Fig.23 wherein the culvert was not loaded. The settlements at line A-A (Fig.21) resulting from the external loads only are also shown with less differential settlements for the reinforced case.

Another analysis was done by prescribing a culvert displacement of 20 cm. The resulting settlement magnitudes for both cases are

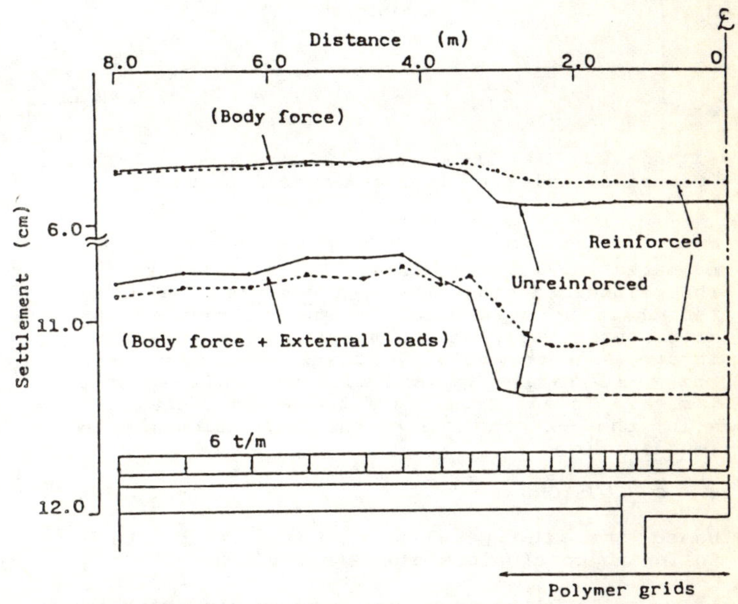

Figure 22. Settlements at line A-A under uniform loading.

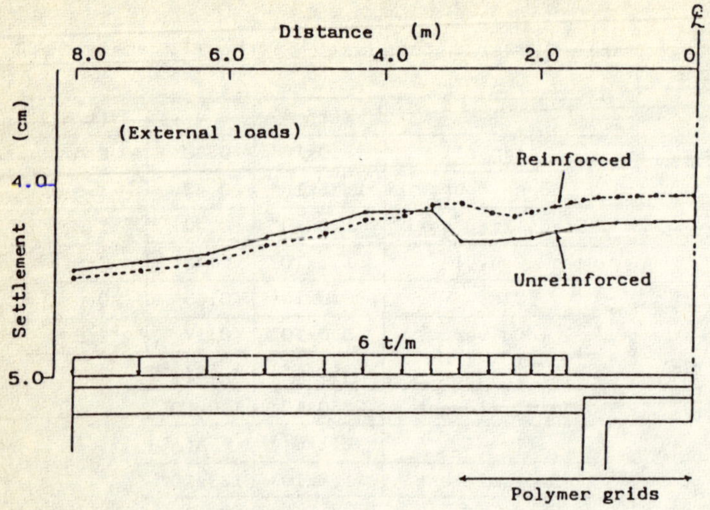

Figure 23. Settlements at line A-A without load in the box culvert.

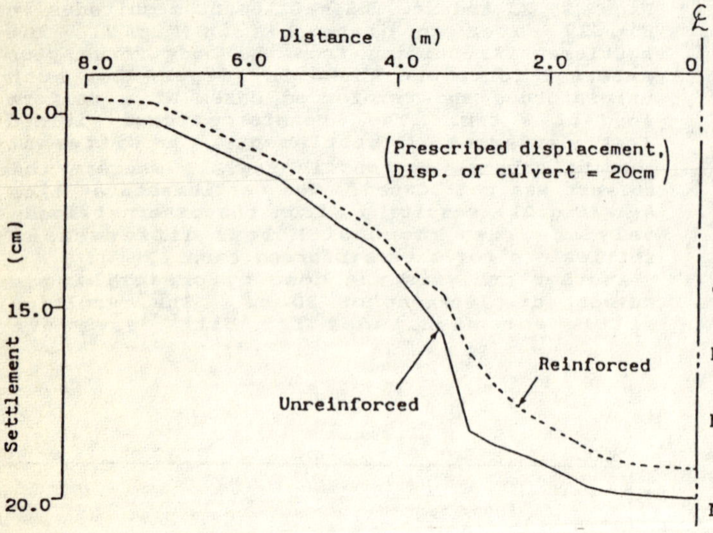

Figure 24. Settlements at line A-A under prescribed culvert displacement of 20 cm.

compared in Fig.24 yielding less settlements for the reinforced case. It is, therefore, demonstrated that the reinforced foundation reduced the differential settlements and that its effect is larger when the relative displacements between the surrounding ground and the culvert are larger. In addition, this foundation scheme is a lot cheaper compared to the conventional pile foundations.

9. CONCLUSIONS

Based from the results of this study, the following conclusions can be made:

1) Construction problems on decomposed granite backfill includes particle crushing, loss of apparent cohesion, and stress-dependent failure envelope especially when wet.

2) The loading tests on unreinforced and reinforced gravel showed less deformation in the latter case than the former and that the double reinforced case yielded less settlements than the geocell.

3) The preliminary results of the FEM analysis on reinforced gravel foundation of a box culvert construction showed that the reinforcement reduced the differential movements between the structure and the surrounding ground.

4) The parameters needed for the FEM analysis can be obtained from the laboratory pull-out test using the new equipment designed and constructed for this study. This equipment can also be used to measure the soil-geogrid interaction.

REFERENCES

Bergado, D.T., Bukkanasuta, A. & Balasubramaniam, A.S. 1987. Laboratory pull-out tests using bamboo and polymer geogrids including a case study. Geotextiles and Geomembranes. 5:153-189.

Chang, J.C., Hannon, J.B. & Forsyth, R.A. 1977. Pull resistance and interaction of earthwork reinforcement and soil. Trans. Res. Board 640, NAS, Washington, D.C.

Forsyth, R.A. 1978. Alternative earth reinforcement. Proc. ASCE Symp. on Earth Reinforcement, Pittsburg, U.S.A.

Hausmann, M.R. 1976. Strength of reinforced soil. Proc. 8th Australian Road Res. Conf., 8(13):1-8.

Ingold, T.S. 1983. Laboratory pull-out testing on grid reinforcement in sand. Geotechnical Testing J. 6(3):101-111.

Jewell, R.A. 1985. Material properties for the design of geotextile reinforced slopes. Geotextiles and Geomembranes. 2:83-109.

Miura, N. & O-hara, S. 1979. Particle-crushing of a decomposed granite soil under shear stresses. Soils and Foundations. 19(3):1-13.

Miura, N., Murata, H. & Harada, A. 1983. Effect of water on the shear characteristics of sandy soils consisting of breakable particles. Proc. of the Japanese Soc. of Civil Eng. 336:105-112.

Netlon 1984. Techniques and design philosophy for utilizing Tensar grids in the reinforcement of soil structures. Tech. Guideline of Netlon Ltd., England.

Ochiai, H. & Sakai, A. 1987. Analytical method for geogrid-reinforced soil structures. Proc. 8th Asian Regional Conf., Kyoto, Japan.

Ochiai, H., Hayashi, S., Ogisako, E. & Sakai, A. 1988. Analysis of polymer grid reinforced soil retaining wall. Proc. 6th ICONMIG, Innsbruck, Austria.

Onitsuka, K., Yoshitake, S. & Bergado, D.T. 1988. Shear characteristics of decomposed granite soil under low pressure and stability analysis of slope failures. Proc. Intl. Conf. on Problems of Regional Soils. Beijing, China.

Peterson, L.M. & Anderson, L.R. 1980. Pull-out resistance of welded-wire mesh embedded in the soil. Res. Report to Hilfiker Pipe Co. Dept. of Civil Eng., Utah State Univ., Utah, U.S.A.

Vidal, H. 1969. The principle of reinforced earth. Highway Res. Record No. 282. NAS, Washington, D.C.

Yang, Z. 1972. Strength and deformation characteristics of reinforced sand. Ph.D. Thesis. University of California, Los Angeles.

Residual soils as the core materials of a dam

Y.J.CHEN, Senior Engineer, Institute of Water Conservancy and Hydroelectric Power Research, Beijing, People's Republic of China

ABSTRACT: Residual soils in Guangdong Province of southern China were used to construct the sloping core placed on the upstream surface of Nanshui rockfill dam,having a height of 80.2 m. The climate,geomorphology as well as the properties of the residual soils near the dam site are presented. The design, construction and the behavior of the core during the first impoundment of the reservoir are reported.

INTRODUCTION

Large scale construction of hydraulic engineering generally started in the East and North China, where the transported and redeposited soils are predominant. Since 1958, the need for earth dam construction has extended on a large scale into much less familiar areas,including the Guangdong Province.During the preliminary design of Nanshui rockfill dam of Guangdong Province in 1959, it was found that some residual soils near the dam site possessed some peculiar properties having higher shear strength and lower compressibility than those of so called normal soils having the same density. Therefore, it was necessary to examine carefully the possibility of using this cheap material for constructing dam core at that time.

SITE CONDITIONS

Nanshui rockfill dam is located in Ruyuan County in the northern part of Guangdong Province, at approximately 25 degrees north latitude. It creats a reservoir having a capacity of 1050 cubic hm for flood control, irrigation and power generation. The recorded temperatures varied from -4°C in January to 40°C in August, had an average value of 19.8^{o}C. The average annual rainfall amounted to 1877 mm, nearly half of which occurred in April through June. There were about 160 raining days a year.

Nanshui is the name of one of the tributaries of Beijiang River. On this tributary the dam was built. It has a total length of 93 km with an average gradient of 5.2%. The catchment area above the dam site was 608 square kilometers. Almost no any soil deposited on the bottom of the V-shaped river channel at the dam site, where the elevation of the river bed ranged between 140 and 150 m. The tops of left and right abutments were 130 and 250 m higher than the river bed respectively. The slope angle of the river banks varied from 40° to 65°. Erosion gullies towards the river made the landsurface undulated abruptly.

The ground water levels during the dry season were 24-45 m and 30-70 m below the surface of the left and right banks respectively. The pH value and HCO_3 content of the ground water were 6-6.5 and less than 0.6 mg/l respectively, the former was higher than that of the water in the river. Permeability of rock bank within the upper 30 m was estimated to be approximately 10^{-6} cm/s based on the test results of injecting water into boreholes.

Rock out cropped on the banks below the elevation 180 m, whereas residual soils covered the landsurface at higher elevations. The rock formations near the dam site consisted of sandstone, siltstone, claystone and conglomerate, among which sandstone predominated over the others, and contained mainly quartz and a small amount of pyrite, silicious materials and sericite.

DAM

The rockfill dam had a height of 80.2 m, and a crest length of 215 m. The main part of rockfill was constructed by directional blasting, which created a pile having an average height of 62 m and a volume of 0.93 cubic hm, as shown by dashed lines in Fig.1a and 1b. Soon after blasting in 1960, a 18 m wide space downstream of cofferdam, as shown in Fig.1b, was cleaned; the upstream surface of the rockfill was sluiced by hose under high pressure, and was then smoothed by filling rock fragments in dents. The placement of filters and part of impervious core were followed up to elevation 195 m for retaining flood water temporarily. The thickness of this part of core was about one eighth of water head, but it was 1 m thick at elevation 195 m. After the additional 0.33 cubic hm of rockfill and a total of 0.32 cubic hm of impervious core, which was protected by concrete facing, were placed, the dam completed in 1969.

Four test pits were dug in the surface of blasted rockfill to measure the gradation, which was shown in Fig.2. The filters between the rockfill and the impervious core consisted

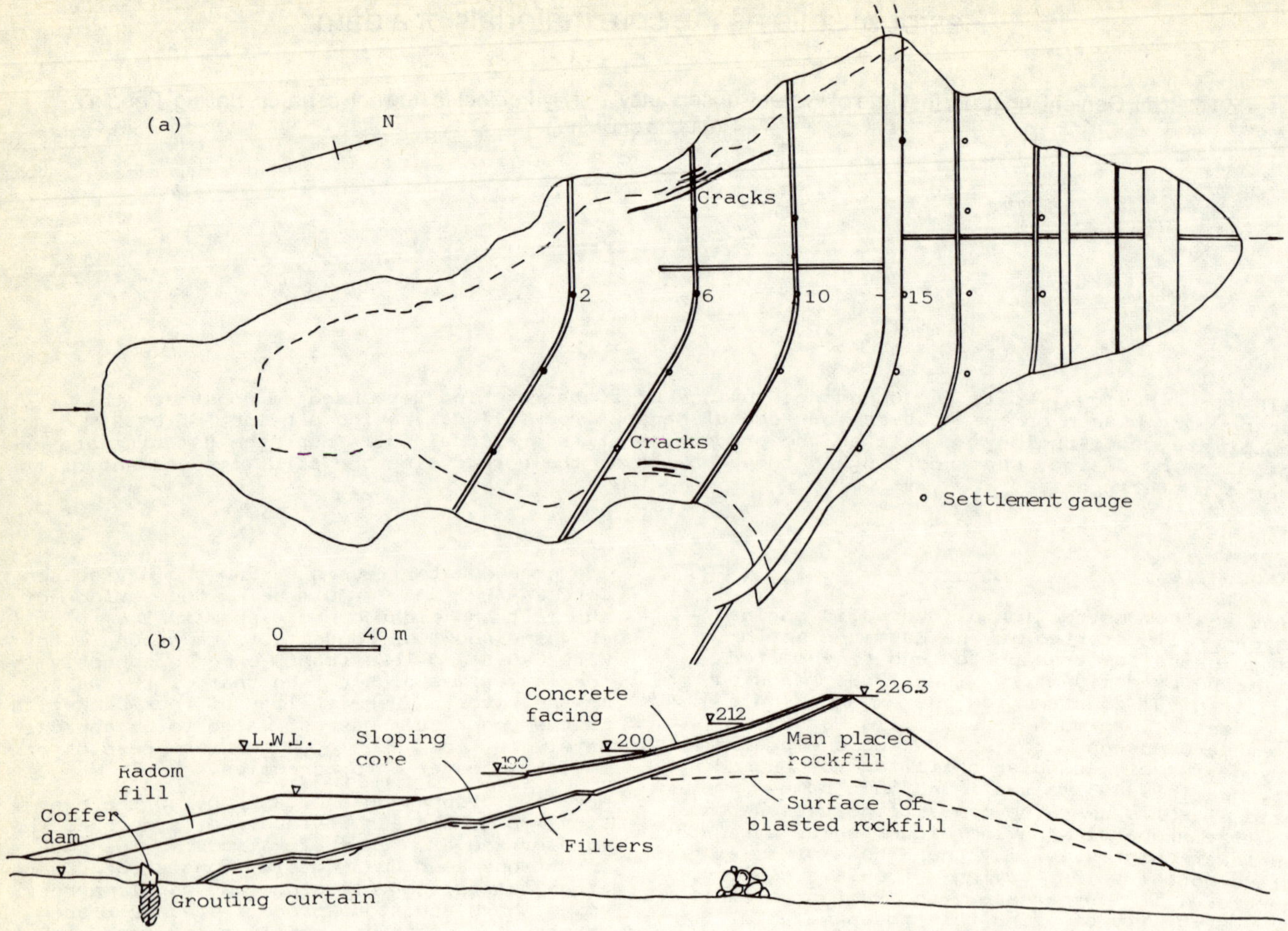

Fig.1 Plan and Cross Section of Nanshui Rockfill Dam

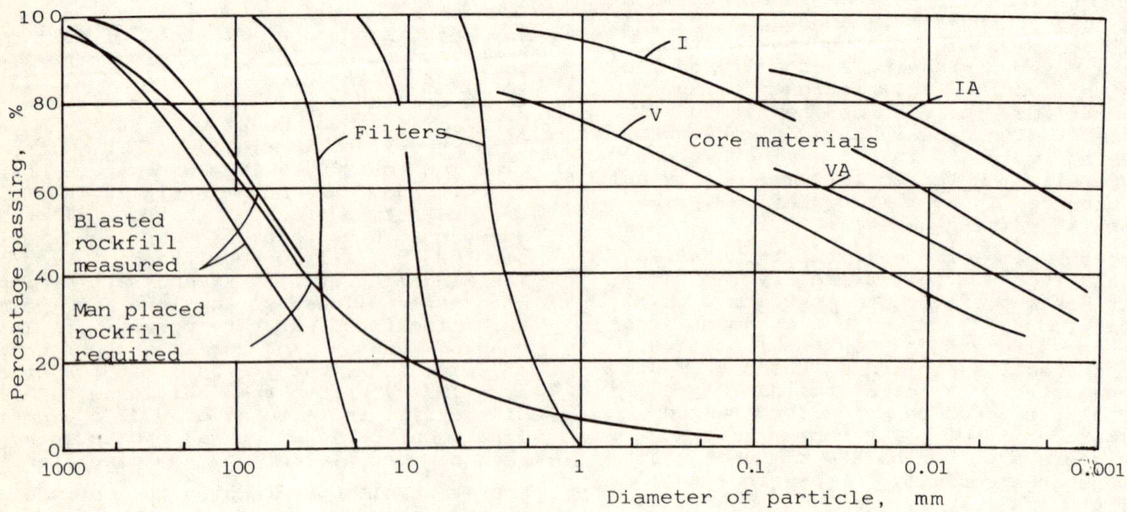

Fig.2 Gradations of Dam Construction Materials

TABLE I

Results of Identification Tests

Borrow area	Soil type	Liquid limit,% Disk	Liquid limit,% Cone	Plastic limit %	Specific gravity	Gradation,% > 2mm	0.05-2mm	0.005-0.05mm	< 0.005mm	< 0.002mm	Natural Water Content,%	Natural Dry unit weight,kN/m^3
I			53	31	2.78	2	22	25	51		31.6	17.2
II			56	33	2.70	4	26	24	46		30.0	16.5
III			46	28	2.72	41	5	16	38		20.0	15.6
IV			51	28	2.70	16	30	25	29			
V			46	26	2.74	20	30	20	30			
VI			51	30	2.76	1	12	35	52			
VII			39	22	2.67	24	41	19	19			
	IA	.65	49	28	2.78		17	15	68	57		
							(68	29	3	0)		
	VA		33	19	2.72		41	17	42	35		

of a 20 cm layer of less than 5 mm particles; a 20 cm layer of 5-20 mm particles and a 50 cm layer of 20-80 mm particles. Fig.2 shows the gradations of the filters as well as of the man placed rockfill.

RESIDUAL SOILS

Large amount of residual soils could be found near the dam site. Most of the soils distributed above elevation 180 m. The surface layer, having a thickness of less than 4 m, was sandy clay containing very small amount of gravel and was red brown in color. This layer was underlain by a 0.5-2 m layer of gravelly clay laying on yellowish highly weathered rock, some particles of which could be pulverized between fingers. Two typical cross sections through borrow areas are shown in Fig.3.

Field identification tests were conducted on numerous soil samples taken from seven borrow ares. The upper part of Table I shows the results obtained. Based on these results, two typical samples IA and VA were chosen and tested thoroughly in a well equipped laboratory. Fig.2 shows the gradations of samples IA,VA and the average gradations of the borrow areas from which these two samples were taken. The lower part of Table I shows the identification test results of samples IA and VA.

The specifications of compaction test were somewhat different from those of Proctor compaction test. The soil in a 1 liter mould was compacted by 25 blows of a 2.5 kg rammer dropping from the controlled height of 46 cm in 3 layers. Though the compaction efforts were approximately 37% larger than that of Proctor test, the maximum dry unit weights were found to be much smaller than corresponding values for redeposited soils of comparable liquid limits. Therefore it seemed necessary to increase the number of blow to 50 for achieving higher density. However, it was not very effective. The maximum dry unit weight was still as low as 15.0 kN/m^3 with the optimum water content of 27% for soil IA, and the corresponding values of soil VA were 15.9 kN/m^3 and 23%.

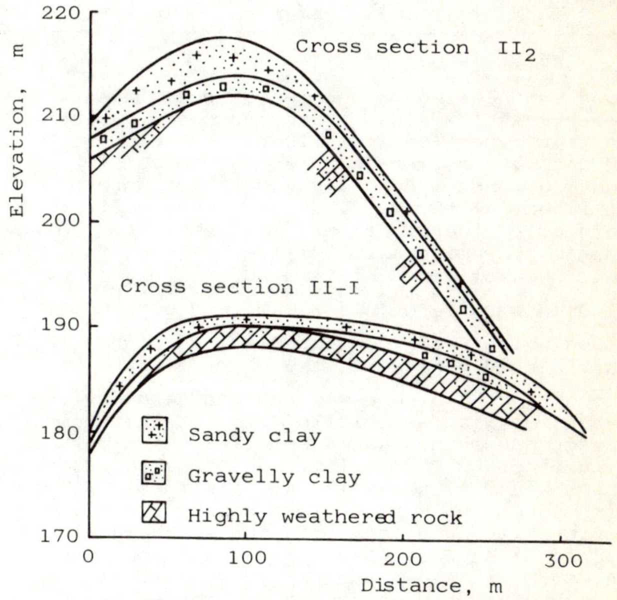

Fig.3 Cross Sections Through Borrow Areas

The required density of the fill was mainly determined by field compaction tests, which revealed that the dry unit weights of 14.5 and 15.7 kN/m^3 with optimum water contents of 30% and 23% could be achieved for soils IA and VA respectively by 19 passes of a 7t segmented roller on 25 cm layers. The coefficients of permeability and of compressibility of saturated specimens having above mentioned density were shown in Table II. Both soils were impervious enough to be the core materials, and had nearly the same compressibility. Undrained triaxial tests on as compacted specimens were conducted to determine the shear strength of the core during construction. Consolidated undrained triaxial tests as well as slow direct shear tests on saturated specimens were assigned to determine the effective and total shear strength parameters, as shown in Table II.

TABLE II

Indices of Engineering Properties

Sample	IA	VA
Dry unit weight, kN/m^3	14.5	15.7
Moulding water content, %	30.0	23.0
Coefficient of permeability, 10^{-6} cm/s	0.5	1.6
Coefficient of compressibility at 100-300 kPa, kPa^{-1}	0.00015	0.00016
Shear strength parameters:		
Undrained test, in respect of total stress		
Cohesion kPa	90-110	30-35
Friction angle, degree	10-11	9-12
Consolidated undrained test, in respect of total stress		
Cohesion, kPa	62-80	19-29
Friction angle, degree	20-23	23-24
Slow direct shear test		
Cohesion, kPa	75	25
Friction angle, degree	27.5	32.2

Good engineering properties of residual soils at relatively low density had just been explained by that time (Terzaghi, 1958), therefore similar investigations were carried out to verify the postulations. Both X-ray diffraction and differential thermal analyses revealed that the main clay mineral of soil IA was illite being a usual clay mineral. However, it was found that the soil IA contained 36.2% SiO_2; 15.6% Al_2O_3; 31.2% Fe_2O_3; 1.3% CaO; and 2.0% MgO, among which Fe_2O_3 content was much higher than that of redeposited soils.

Though the 76 g cone was the standard device to determine liquid limit in China at that time, Casagrande method was specially assigned to determine the liquid limit and thus the activity which was 0.67 for soil IA. Different dispersion agents were added into soil suspension during hydrometer analysis for investigating the degree of aggregation. The gradation of soil IA, shown in Fig.2 and the lower part of Table I, was dispersed by sodium carbonate, which increased the pH values of the suspension from 6-7 to 8-9, the former was the pH values of suspension with adding no any dispersion agent. The gradation of soil IA not being dispersed was shown in the parenthesis of Table I revealing high degree of aggregation, which might be directly related to Fe_2O_3 content, but to much less extent related to clay mineral.

Stability of the core during construction was analysed by sliding wedge method (Chen, 1964), in which the interslice forces were assumed to be parallel to the surface of their blocks, and the safety factor along the entire sliding plane was the same. Besides, the lower boundary of shear strength envelops of filters and core was used to determine the shear strength of the sliding plane passing the contact face between the filters and the core.

PERFORMANCE

After the first impoundment of reservoir in 1969, several transvers cracks appeared on the surface of the core adjacent to both abutments under the load of water, whose elevation rose up to 2.5 m higher than normal high water level. The cracks near the right abutment distributed between the elevations 200 and 215 m, as shown in Fig.1a. The largest cracks had the length of 28 m and a maximum width of 20 mm, which decreased to 2 mm at the depth of 3.9 m. The differential settlement between the opposite walls of the crack was as much as 40 mm. The cracks near the left abutment were even more and distributed between the elevations 190 and 212 m. The longest crack was 90 m, had a maximum width of 30 cm, which vanished at the depth of 4.2 m. The differential settlement between the opposite walls of the crack was as large as 140mm (DSIWREPG, 1977).

Settlement gauges were installed on the surface of concrete facing, as shown in Fig.1A. The measured settlements of gauge 2 at elevation 190 m, gauge 6 at elevation 200 m, gauge 11 at elevation 212 m and gauge 16 at elevation 226.3 m on the crest were 486, 374 317 and 262 mm respectively during the first impoundment. It was considered that the real settlements of the core were even larger than the measured values, because some concrete facings had suspended over the earth fill. The large settlement was caused by the high compressibility of man placed rockfill as well as the concaved upstream face of the dam. Though the above mentioned values of settlement were larger than those near the abutments, they revealed that the cracks appeared at the elevations where larger settlement occurred. The post settlement in the following eight years amounted to 20-30% of the total, so that no any cracks has appeared after the cracks had been grouted. Meanwhile the seepage measured at the downstream toe of the dam decreased from 46.3 l/s immediately after the first impoundment to 20.0 l/s soon after sealing the cracks and was getting smaller in the following years.

CONCLUSIONS

High Fe_2O_3 content probably caused the aggregation and thus good engineering properties at relatively low density of residual soils near Nanshui dam site at the boundary between tropic and temperate zones. The core made of these soils underwent large deflection and severely cracked, but has not been piped, because of high erosion resistance of the soils as well as of the protection of filters.

REFERENCES

Chen,Y.J.,1964. Improving the Stability Analysis of Earth Dam Having sloping Core, Journal of Water Resources and Hydropower, No.6,28-30, in Chinese.

Design and Survey Institute of Water Resources and Electric Power of Gaungdong Province,1977. Geological Review of Construction of Nanshui Dam by Directional Blasting, in Chinese.

Terzaghi,K., 1958. Design and Performance of the Sasuma Dam. Proc. I.C.E.,London, 9: 369-394.

Prevention of collapse in residual foundation soils of Guri embankment dams

T.CHOUDRY, Harza Engineering Co. International, Venezuela

Abstract: During the investigation stage of the Guri hydroelectric project, the geotechnical testing program established that some of the soils showed the collapse problem, i.e., they underwent pronounced settlements on wetting at a constant load. The project envisaged the construction of embankment dams with a maximum height of 90 m. over the soils which are generally characterized by a residual mantle of decomposed gneiss overlying rock.

In this article are described the different types of tests performed which confirmed the collapsible nature of these soils, and the measures adopted during construction to prevent the collapse.

The deformations during the construction of the embankment dams and the reservoir raising were recorded by means of different types of instruments installed in the foundation and the body of the dams. It was shown by these measurements that most of the settlements occurred during construction and no collapse took place while the reservoir was raised to its full height. This confirmed that the measures adopted to avoid collapse in the foundation soil achieved successful results.

In addition, a brief summary of deformations obtained by the different testing methods are compared with the actual instruments' measurements, showing that a Ko-type Triaxial test best simulates the real field settlements.

INTRODUCTION

Collapse phenomenon is generally defined as a sudden pronounced settlement due to an increase in the degree of saturation of a soil at a constant load.

During the investigation phase of the Guri hydroelectric project (Final Stage), it was found out that some of the foundation soils of the embankment dams were collapsible. The soils under this study are residual and saprolites formed by de decomposition of granitic gneiss. The weathering profile and geotechnical characteristics of these soils have been discussed by O. De Sola (1985), Prusza and Choudry (1979), and Prusza (1985). The collapsible nature of these soils was first depicted by one - dimensional consolidation tests, and later on, confirmed by field plate-load and laboratory triaxial tests.

The soil investigation included the following tests:

a) One-dimensional consolidation tests on block samples obtained from the walls of the cut-off trenches corresponding to the foundations of the right and left embankment dams. Some samples showed varying degrees of collapse due to inundation at different loads.
b) Plate-load tests, performed in the bottom of the cut-off trenches, confirmed the occurance of collapse phenomenon (Prusza and Choudry, 1979).
c) One-dimensional consolidation tests on samples obtained close to the plate-load test sites. Some samples showed collapse at different loads on saturation (K. De Fries, 1977).
d) Triaxial tests on specimens with diameters varying from 7,5 to 15 cms, obtained from block samples near the plate-load tests. The specimens were consolidated to a certain stress level by maintaining the effective stress ratio $K = \sigma_1'/\sigma_3' = 2$ or 2.5 and inundating the samples by gravity. Many samples showed collapse settlement varying from 1 to 5% of the height of the samples (K. De Fries, 1980).
e) To simulate the deformation, as it would occur in the field, Ko-triaxial tests were performed on saturated foundation samples. Increments of 10 kPa of the confining pressure were applied, with simultaneous increase in the deviator stess to avoid any lateral deformation (K. De Fries, 1982), and the corresponding settlements were recorded.

ANALYSIS OF THE TEST RESULTS

The load-settlement curves obtained by the above mentioned tests are summarized in figures No. 1 and No. 2. The settlements recorded in these tests at 500 kPa and 1000 kPa are shown in Table No. 1.

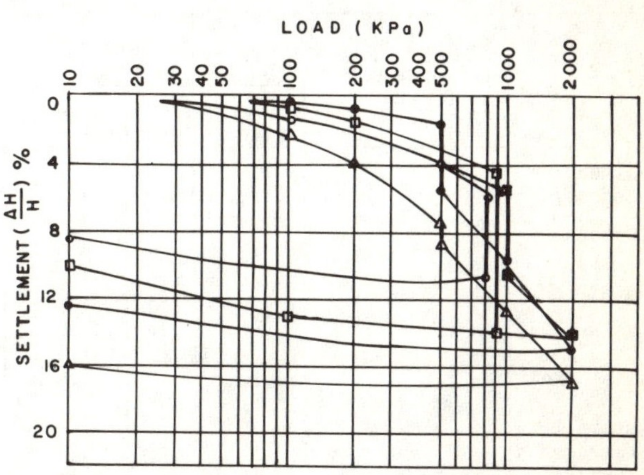

FIG.1 – SOME TYPICAL LOAD SETTLEMENT CURVES UNDER ONE-DIMENSIONAL CONSOLIDATION TESTS.

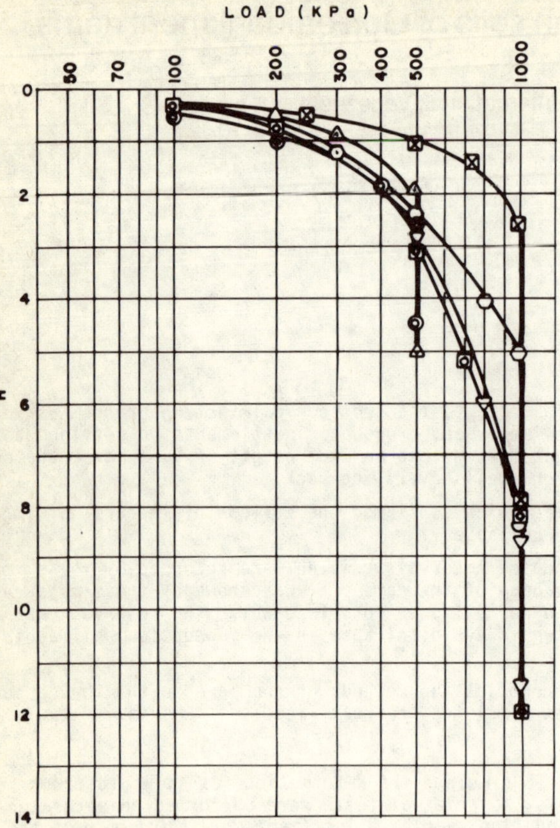

FIG.2– SOME TYPICAL LOAD SETTLEMENT CURVES UNDER TRIAXIAL TESTS.

Table No. 1: Comparison of collapse and deformations of foundation materials by different methods.

Type of test	Maximum Collapse at 1000 kPa	Deformation at 500 kPa		Deformation at 1000 kPa	
		Dry	Saturated	Dry	Saturated
One-Dimensional Consolid.	10	3	10	6.3	14.8
Plate-Load Test	9	1.2	6	3.3	7.9
Triaxial (σ_1'/σ_3' =2 or 2.5)	5	2	3.5	5.5	9.0
Triaxial (Ko)			2.2		4.6
Measurements by instruments (1987)	---		1.7		5

From these measurements it could be established that:

a) Some of the soils in the foundations of the embankment dams undergo pronounced settlement due to wetting, i.e., they show collapsible behavior.

b) The collapse is more evident in saprolites (decomposed gneiss) than in slope-wash materials.

c) The main indices of collapsible soils are the low degree of saturation, existence of a porous structure held together by capillary tension and low natural densities.

d) The Guri soils are not truly collapsible but rather fall in the category of conditionally collapsible, as they tend to collapse under a certain load.

MEASURES ADOPTED TO AVOID COLLAPSE

The existence of thick layers of saprolites under the foundations of the embankment dams, with low natural dry densities and low degree of saturation caused considerable concern as to the possibility of collapse phenomenon to occur during the raising of the reservoir. This could lead to differential settlements, causing cracks in the dams which could result in concentrated leaks. In addition to the removal of collapsible material in certain areas (Medina and Liu, 1982) and adopting changes in the alignments wherever it was economical and practical, it was decided to prewet the foundations where the soil was susceptible to collapse during the raising of the reservoir.

The prewetting was accomplished by flooding a series of canals, varying in depth from 0.5 to 2.0 m, excavated in the foundation of the embankment dams. In addition, drain holes of 0.5 m.in diameter were drilled spaced every 20 to 30 m. to the top of rock, along the bottom of the canals. These holes were filled with gravel and served as a means of accelerating the penetration of water in the soil. These canals were continuously flooded for a period of up to four months. The raising of water levels in the surrounding soil was measured by the observation wells drilled near the canals. Canals A, B, C, D, E, F, G, and J were excavated in the right dam foundation and canals K1, K2 were excavated in the left embankment dam foundation (Fig. 3).

EFFECT OF PREWETTING

The effect of prewetting the foundation soil of the embankment was studied by instruments in two stages: during the flooding of the canals and after the completion of the embankments and raising of the reservoir.

a) During the Flooding of Canals

By continuously flooding the canals it was observed that the phreatic level raised in the vicinity of the canals, indicated both by the levels in the observation wells and by the appearance of filtrations on the face of the cut-off trench which was being excavated near the canals. In some of these canals the flooding was abandoned because of the interference in construction caused by filtrations in the cut-off trench. In other canals the flooding continued until a stabilized phreatic level was achieved. (Fig. 4).

The settlement monuments located near the canals showed no settlement of the soil in the areas being prewetted. This confirmed our observation from the field and laboratory tests data that soils under study were not naturally collapsible and a certain load was required to cause the collapse to occur. The main benefit achieved

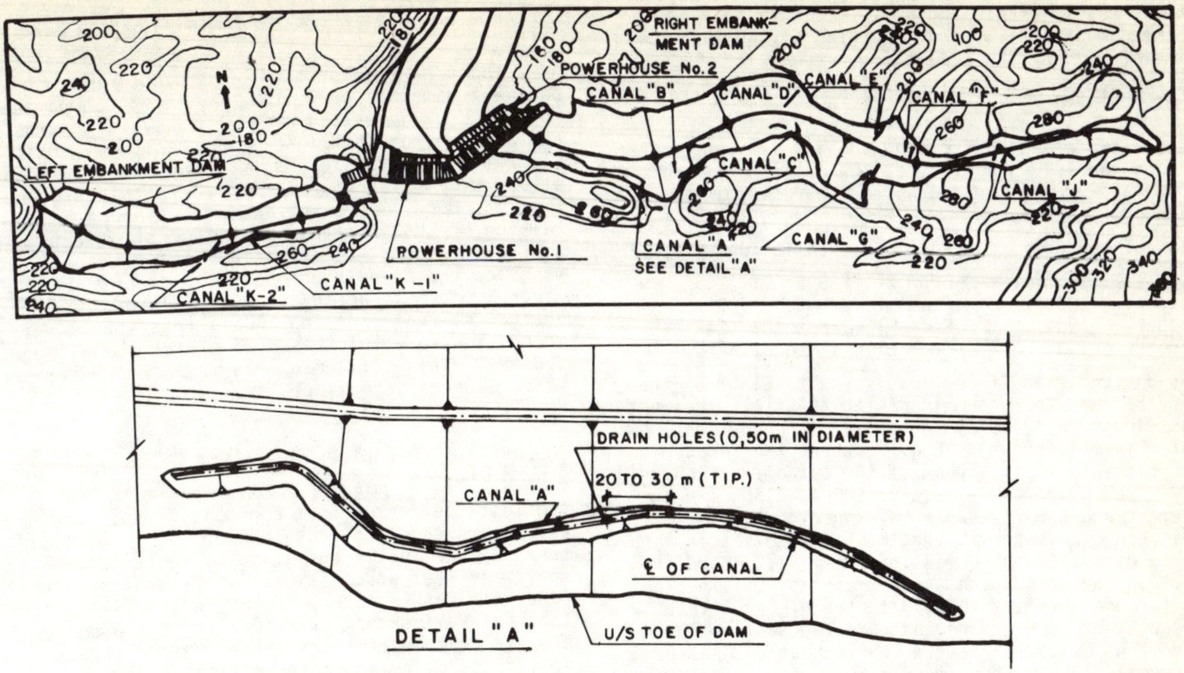

FIG. 3.–LEFT AND RIGHT EMBANKMENT DAM ALIGNMENTS WITH LOCATION OF PREWETTING CANALS AND A DETAILED PLAN VIEW OF ONE OF THE CANALS

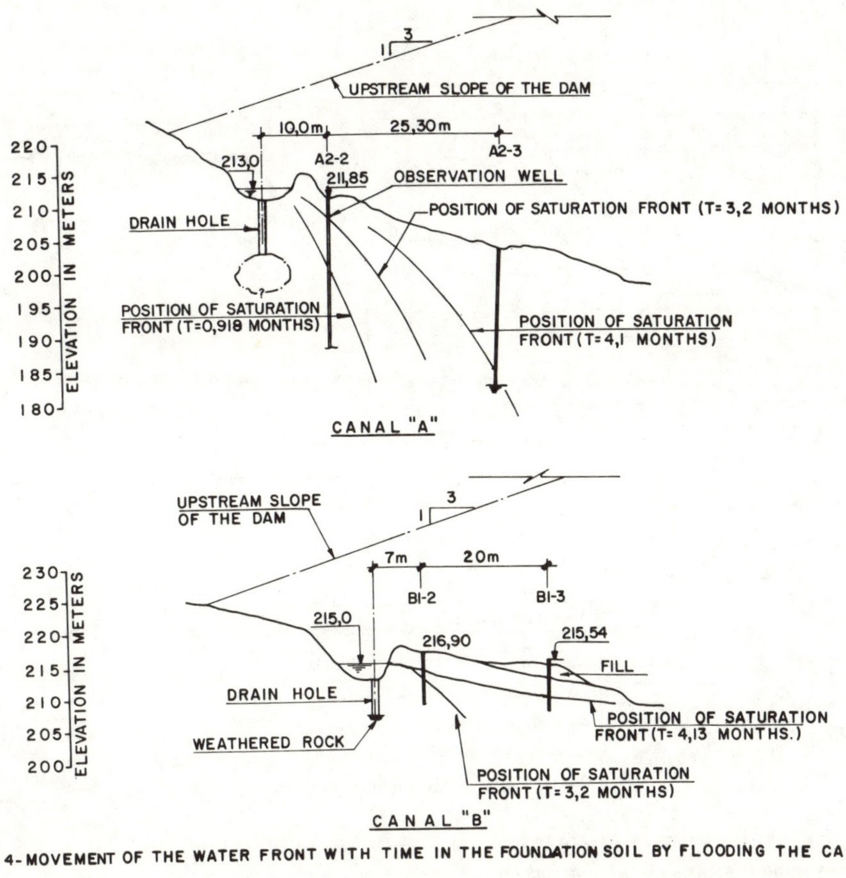

FIG. 4-MOVEMENT OF THE WATER FRONT WITH TIME IN THE FOUNDATION SOIL BY FLOODING THE CANALS

by prewetting was to enable the soil to settle in a wet state as the dam was built, thus avoiding any collapse at the time of raising the reservoir.

b) <u>After Embankment Dam Completion and Raising of the Reservoir</u>.

In the foundation areas upstream of the dam axes where prewetting by canals was performed, the settlements were recorded during construction and while the reservoir was being raised.

The raising of the Guri reservoir occurred in four stages:

May	1983	From El. 215 to El. 225
September	1983	From El. 225 to El. 235
October	1984	From El. 235 to El. 242
October	1986	From El. 242 to El. 270

In the foundation areas of low levels (up to El. 235) the raising of the reservoir, combined with the fill placement made it difficult to identify the cause of foundation settlement. This is shown in Fig. 5, where the foundation settlement close to the upstream of the cut-off trench is plotted for the instrumented section SMD-4, alongwith the fill placement and the reservoir raising with time. However, it could be observed that most of the settlement had occurred during the placement of the fill, before the water level in the foundation was effected by raising the reservoir (around April of 1983). In the higher levels of the foundation, such as the instrumented section SMD-6, the reservoir water level reached the foundation one year after the completion of the dam (Fig. 6). It can be observed that most of the settlement occurred during fill placement and no noticeable settlement took place due to reservoir raising.

COLLAPSE IN AREAS WHERE NO PREWETTING WAS PERFORMED

As a major portion of the foundation for the main embankment dams was prewetted, no direct comparison can be made with un-prewetted areas. However, towards the eastern abutment of the right embankment dam, an impervious blanket alongwith slope protection was placed on the upstream side of a natural ridge. In this area no prewetting was done. Some surface cracks appeared during the raising of the reservoir on the crest of this upstream blanket. A total differential settlement of 15.2 cms. has occurred to-date across these cracks. These settlements appear to be the result of foundation saturation during reservoir filling and is less than 1% of the foundation soil thickness (25 m) for a fill load about 400 kPa. This settlement has stabilized after the reservoir reached its maximum level.

COLLAPSE OF COMPACTED FILL

The tests performed on the compacted soil did not show any occurance of collapse phenomenon. The field measurements by the instruments located in the compacted fill confirmed that no additional settlement occurred when the reservoir water reached the compacted fill. It shows that the natural porous structure is destroyed during the process of compaction and the remolded structure of the soil does not collapse on wetting.

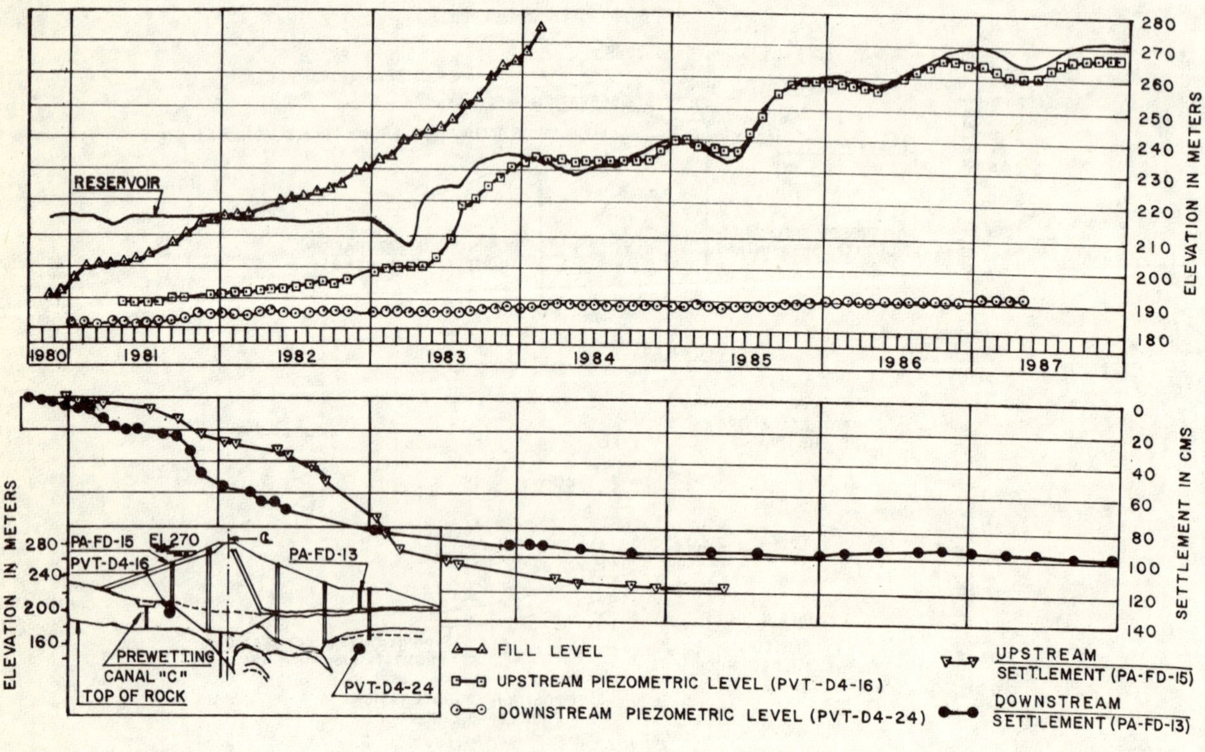

FIG. 5 — FOUNDATION SETTLEMENT AND PIEZOMETRIC LEVEL WITH RAISING OF THE RESERVOIR FOR THE RIGHT EMBANKMENT DAM INSTRUMENTED SECTION SMD-4

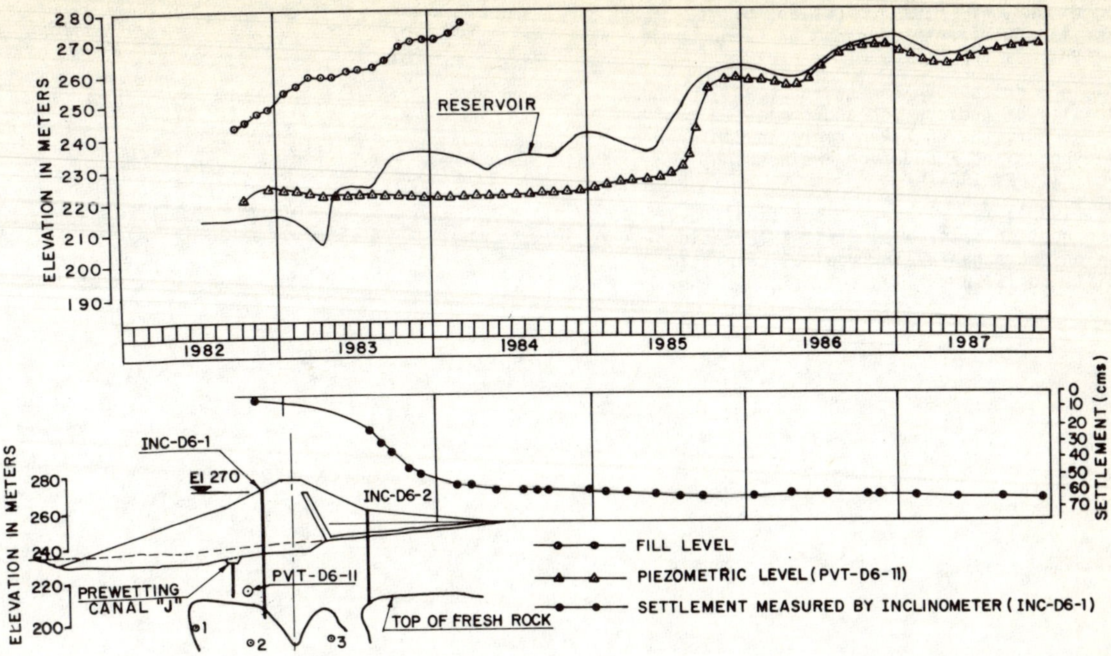

FIG.6.—FOUNDATION SETTLEMENT AND PIEZOMETRIC LEVELS WITH RAISING OF RESERVOIR FOR THE RIGHT EMBANKMENT DAM INSTRUMENTED SECTION Nº. SMD-6

TOTAL FOUNDATION SETTLEMENT

The total settlements at loads of 500 kPa and 1000 kPa, obtained by different laboratory tests conducted on the undisturbed samples are reported in Table 1. Also, the settlements measured in the plate-load tests and those recorded by the field instruments are shown in this table. It can be observed that one-dimensional consolidation test greatly overestimates the real settlement. However, the results obtained by Ko-triaxial tests show comparable results with the field measurements. These results were earlier presented by Prusza et. al., (1983) when the embankments were under construction, causing vertical stress in the order of 700 kPa. With the embankments completed and reservoir raised, it can be observed (Table No. 1) that Ko-triaxial tests very closely relate to the real settlement for vertical stresses of 1000 kPa, and can be used seccessfully to predict settlements in this type of soils.

CONCLUSIONS

On the basis of the geotechnical investigations carried out on the foundation soils at Guri, to study their collapsible behavior, the measures adopted to avoid collapse, and the settlement measurements obtained by the ins - truments placed in the dams, we can conclude that:

a) Some of the residual foundation soils show collapsible behavior when they are under loads exceeding 400 kPa.

b) The collapse seems to be due to a low degree of saturation and the existence of capillary tension in the porous structure of these soils, which is lost on wetting.

c) Prewetting the foundation materials by canals proved to be an effective method to avoid collapse. This caused most of the settlement to occur during fill placement thus avoiding any collapse during the raising of the reservoir.

d) The compacted soil does not show collapse as the original porous structure is completely destroyed in this process.

e) Ko-type triaxial tests indicate settlements very close to the real ones, whereas one-dimensional consolidation tests highly overestimate the real settlements.

REFERENCES

1. De Fries, K. (1977): Summary of some geotechnical properties of soils used in the construction of Guri dam, Oficina de Ingeniería De Fries, Caracas.
2. De Fries, K. (1980): Special Triaxial Tests-Ko. Parameters for predicting settlements, Oficina de Ingeniería De Fries, Caracas.
3. De Fries, K. (1982): Behavior of earth structures on the basis of instrument observations. Oficina de Ingeniería De Fries, Caracas.
4. De Sola, O. (1985): Weathering in the Guri area, First International Conference on Geomechanics in Tropical Lateritic and Saprolite Soils, Brasilia, Vol. II, PP. 155-164.
5. Medina, Jesus and Liu, Bernard S. (1982): The influence of collapsible foundation on the design of Guri embankment dams, Fourteenth Congress on Large Dams, Rio de Janeiro, Vol.II, PP. 241-259.

6. Prusza Z., Choudry, T. (1979): Collapsibility of residual soils, Thirteenth Congress on Large Dams, New Delhi, Vol. II, PP. 117-130.

7. Prusza, Z., De Fries K., Schmitt, R. (1983): The compressibility of undisturbed and compacted resi - dual soils at Guri, Seventh Pan American Conference on Soil Mechanics and Foundation Engineering, Vancouver, Vol. I, PP. 185-197.

8. Prusza, Z. (1985): Miscellaneous aspects of Guri soils as foundation materials, First International Conference on Geomechanics in Tropical Lateritic and Saprolite Soils, Brasilia, Vol. II, PP. 156-171.

Road construction using lateritic gravel in Western Australia

G.C.COCKS, Main Roads Department, Perth, Western Australia, Australia
G.HAMORY, Main Roads Department, Perth, Western Australia, Australia

ABSTRACT: Western Australia has an extensive network of bitumen surfaced roads carrying low and medium volumes of traffic. Lateritic gravels have been widely used in the pavement of these roads. The performance of these gravels has generally been better than might be predicted on the basis of particle size distribution and plasticity tests alone. In this paper the characteristics of some lateritic materials which have been used are presented. Evidence of the increase of the strength (self stabilisation) of lateritic gravels in road pavements is also presented.

1. INTRODUCTION

The State of Western Australia occupies the western third of the Australian continent. It stretches a distance of approximately 2400 kilometres in a north-south direction and approximately 1600 kilometres east-west. It has an area of a little over 2.5 million square kilometres and its population is approximately 1.4 million. Much of the state has a warm/hot arid to semi arid climate with areas of high rainfall restricted to the extreme north and the southwest corner. Rainfall varies from 1400 mm/year on the south coast to less than 200 mm/year in the interior. Annual evaporation is high (1200 mm - 4400 mm) and exceeds annual rainfall in all but a very narrow strip on the south coast.

The state's road network includes 15 728 km of highways and main roads and 150 139 km of secondary and local roads. The discussion in this paper is mainly limited to highways and main roads.

With such a vast road network and small population, a strong committment to low cost road construction has been necessary. This was achieved by the extensive use of natural gravels for pavement construction together with a thin (chip seal) bituminous surfacing.

This construction is well suited to the arid/semi arid regions of Western Australia. The term "natural gravel" used here means a gravelly material occurring in deposits which can be excavated with mobile construction equipment. Some processing may still be necessary during construction to remove or to break down oversize particles. However a distinction is made between these "natural gravels" and material produced by crushing hard rock and referred to as "crushed rock base". The use of crushed rock base as a structural layer in the pavement is largely restricted to the Perth metropolitan area. This material was used only occasionally in rural highways and only in areas where satisfactory natural materials were unavailable.

The natural gravels used in pavement construction are primarily those of pedogenic origin and include ferricretes (lateritic gravels), calcretes and silcretes. Of these, the ferricretes or lateritic gravels have had the greatest use in Western Australia.

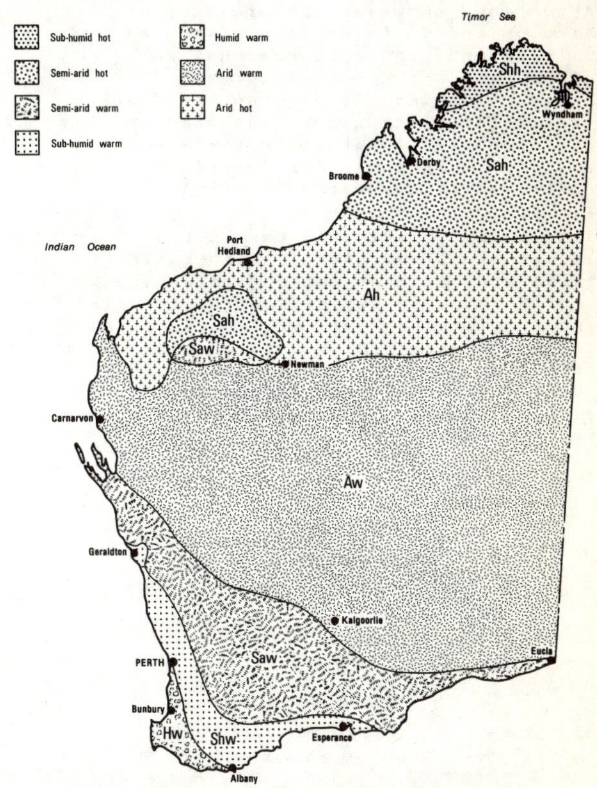

Fig 1. Climatic regions of Western Australia Thornthwaite's Method (After Gentilli 1972)

2. FORMATION AND OCCURRENCE OF LATERITIC GRAVELS IN WESTERN AUSTRALIA

The definition of the terms ferricrete and laterite have been the subject of considerable debate. Grant and Aitchison (1970) defined ferricrete as a "ferruginous natural material formed in a zone of iron oxide or hydroxide accumulation in the earth's crust. The accumulation of oxides or hydroxides of iron is by superficial physical - chemical processes and not by normal sedimentary, metamorphic, volcanic or plutonic processes". Wooltorton (1954) defined lateritic soils in terms of the molecular ratio of combined silica (SiO_2) to the sesquioxides (Fe_2O_3, Al_2O_3). The ratio is expressed in the form:

$$S/R = \frac{\frac{SiO_2}{60}}{\frac{Al_2O_3}{102} + \frac{Fe_2O_3}{160}}$$

The term "lateritic" is used loosely in Western Australia to include what are called "ferricretes" by Grant and Aitchison. Strictly speaking the term "laterite" as used by geologists in Western Australia encompasses the full lateritic profile from fresh rock to surface soil as shown in figure 2.

The veneer of pisolitic or nodular concretions overlying the duricrust is the material most commonly used in road construction. However the products of erosion of the duricrust are also used. In some locations, the duricrust has not developed and the pisolitic gravel sits directly on the completely weathered rock. The term "lateritic gravel" as used in this paper includes all natural gravels of lateritic origin including both in situ pisolites and transported material.

Indurated ferricretes and associated red-brown pisolitic gravels (lateritic gravels) are the most common surface rocks and soils found in Western Australia. They extend from the hot semi arid north of the state to the warm humid south west as shown in figure 3. In some areas the ferricrete has been removed by erosion and the lateritised zone is represented by isolated buttes and mesas.

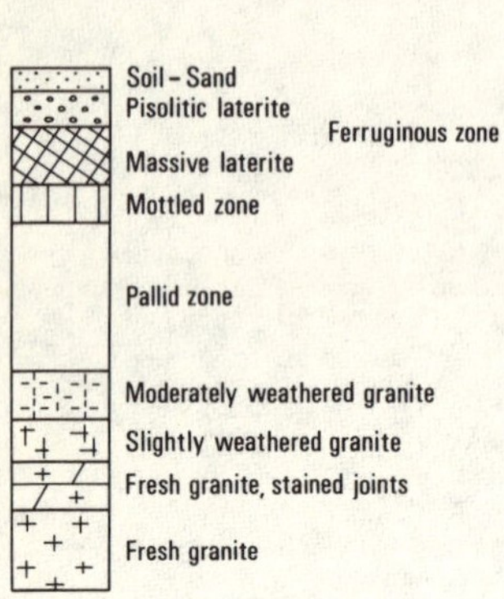

Fig 2. Typical lateritic profile

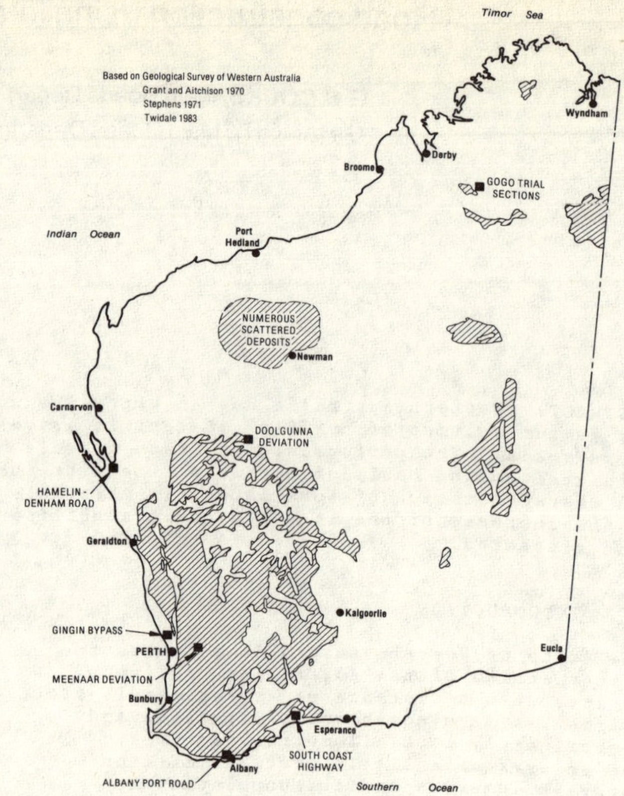

Fig 3. Occurrence of ferricretes and lateritic gravels in Western Australia

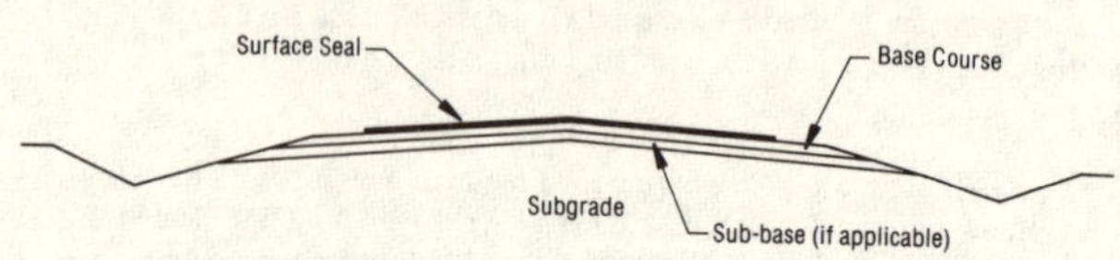

Fig 4. Typical rural road cross section

3. PROPERTIES AND PERFORMANCE OF LATERITIC GRAVEL AS BASECOURSE

Extensive use has been made of lateritic gravel as a basecourse layer in low and medium trafficked (less than 5000 vehicles per day) rural roads in Western Australia. A typical cross section is shown in figure 4.

A typical pavement consists of a basecourse of lateritic gravel 100-250 mm thick on a 8-12 m wide, typically 0.5 m high formation. The surfacing consists of a sprayed bituminous surface seal 6-8 m. The sprayed surfacing is normally constructed in two layers. The first coat (primerseal) consists of sprayed bitumen (cutback with kerosene and/or distillate) and covered with coarse sand or fine aggregate. After trafficking for 6 to 18 months a seal coat consisting of sprayed bitumen covered with single sized crushed aggregate (typically 10 mm or 14 mm nominal size) is applied.

The performance of pavements constructed with lateritic gravel basecourse has generally been satisfactory and quite frequently with lives in excess of 30 years. However some cases of unsatisfactory performance have also been observed. In Tables Ia and Ib, the basic properties of lateritic gravels used as basecourse materials for a number of projects, are summarised. The projects listed include examples of satisfactory and unsatisfactory performance.

3.1 Brand Highway (Gingin Bypass)

Construction of the Gingin Bypass was completed in April 1980 (the beginning of the wet season). Prior to opening to traffic it was surfaced with a fine aggregate primerseal. The primerseal treatment was left untrafficked for some weeks prior to opening to traffic. Although the weather had been dry during the construction period, heavy and consistent rainfall occurred during the period between application of the primerseal and opening to traffic.

Shear failures and wheel rutting in the lateritic gravel basecourse was evident within hours of the highway being opened to traffic. An investigation was initiated immediately and found that the average basecourse moisture content in failed areas was 106% of the optimum moisture content (OMC) (modified compaction). Strength tests on the basecourse indicated that failed areas had a class number (see section 4.1) of 3.0 or greater.

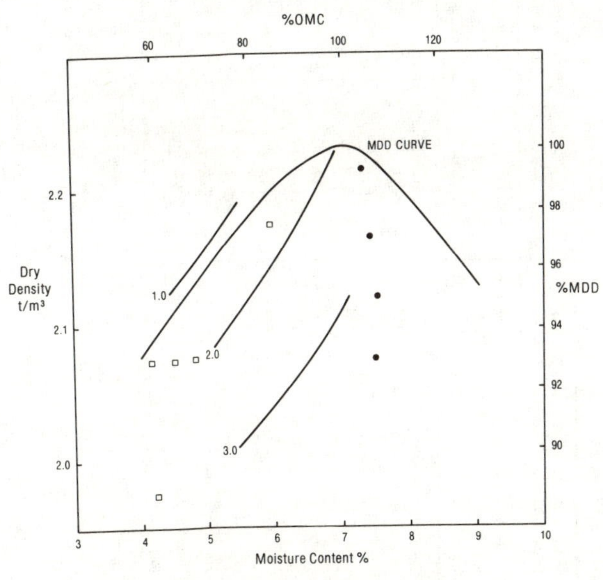

Fig 5. Brand Highway, Gingin Bypass Moisture/density relationship and WACCT class number contours of the lateritic gravel basecourse
□ Sound sites
• Failed sites

Such high moisture contents are unusual in Western Australian pavements. An examination of photographic records of the construction process showed that the compacted basecourse surface had been very dry and dusty at the time of spraying the bitumen for the primerseal. Surface tension effects in the primerseal due to the dust had led to holes in the bitumen layer. This is not uncommon when primersealing dry lateritic gravel basecourses. However the holes in the primerseal are usually kneaded closed by the action of traffic. In the case of the Gingin Bypass the road was not open to traffic for some weeks. The primerseal did not form a waterproof layer and in the weeks of heavy rain following the primersealing, the basecourse became almost saturated.

Within days of the failures occurring, the failed areas were ripped up, the basecourse layer dried out, recompacted and a new primerseal applied. To date the pavement has given satisfactory performance.

3.2 Great Eastern Highway (Meenaar Deviation)

The Meenaar deviation was constructed in 1984 using lateritic gravel in the basecourse. Material from two pits, Fullwood's and Bushell's was used. In the winter immediately following construction, runoff from the primerseal infiltrated into the unsealed portion of the road shoulder. This moisture then moved laterally under the shoulder seal resulting in a significant loss of basecourse strength. Shear failures and rutting developed in the road shoulders but did not extend into the trafficked lanes. No action other than to repair the failed areas was carried out.

Since 1984 the basecourse strength has been monitored on a regular basis using the Clegg Impact Testing Device (ITD).

The impact tests were carried out at various lateral positions on the sealed pavement following significant rain. At each test location a sample of basecourse material was taken for the determination of moisture content. As the basecourse moisture content varied, generally being greater near the seal edge following rain, it was possible to develop an impact value/moisture content relationship for each monitoring period. By interpolation, the impact value (IV) corresponding to moisture contents of 100% OMC and 70% OMC were determined. These interpolated IV's are plotted as a function of time in figures 6 and 7.

The results indicate a significant time related strength increase, ie self stabilisation. For a moisture content corresponding to 100% OMC the basecourse strength has increased from about 40 IV (low-medium) to over 60 IV (high).

Apart from failures in the basecourse under the sealed shoulder area in the first year after construction, the performance of the basecourse on this project has been satisfactory.

TABLE Ia

PROPERTIES OF LATERITIC GRAVELS AND OF CLAYEY SANDS UNSED IN THE PAVEMENT OF WESTERN AUSTRALIAN ROADS

Road Section Source Pit	Brand Highway Gingin Bypass Mooliabeenee Pit	Great Eastern Highway Meenaar Deviation Bushell's Pit	 Fullwood's Pit	Great Northern Highway Gogo Trial Section 7 Pit 2648	 Gogo Trial Section 11 Pit 2628	 Gogo Trial Section 12 Pit 2709
Construction	1980 April	1984 May		1982 November		
Traffic (veh/day) 1987	1716	1678		192		
% Heavy vehicles	13	12		26		
ESA to Dec 1987	5.1×10^5	1.6×10^5		7.4×10^4		
Annual Rainfall, mm	685	435		525		
Annual Evaporation, mm	2000	2200		3800		
Particle Size Distribution Sieve Size, mm						
% Passing 19.00	88-95	94-100	98-100	91	99	98
9.5	68-82	77-97	86-93	73	85	79
4.75	53-70	53-85	68-76	46	62	57
2.36	46-62	39-78	60-64	33	48	41
1.18	36-48	33-71	54-58	29	42	35
0.60	25-34	28-62	46-51	26	36	31
0.425	21-29	25-54	41-47	24	30	28
0.30	17-26	22-48	34-41	21	25	26
0.150	10-18	16-38	22-30	16	17	22
0.075	6-13	13-30	16-21	12	12	17
0.0135	3- 8	7-19	9-10	7	8	12
Liquid Limit	22-25	24.4-31.8	23.2-27.4	26.8	27.5	25.6
Plastic Limit	NP-15	12.5-14.9	14.4-17.7	13.3	13.8	13.8
Plasticity Index	NP-10	10.6-16.9	5.5-13.0	13.5	13.7	11.8
Linear Shrinkage	0.6-3.9	6.0- 7.5	2.4- 9.0	5.9	6.0	7.0
% Pass 0.425 x PI	NP-287	283-913	226-611	324	411	330
% Pass 0.425 x LS	13-113	150-405	108-423	142	180	196
Max Dry Density (Mod Comp) kg/m³	2146-2230	2086-2283	2027-2083	2150	2090	2230
Optimum Moisture Content, %	6.8-7.3	6.5-8.5	7.2-8.2	7.2	8.2	7.1
WACCT Class No at 95% MDD/110% OMC	3.3	3.0-3.1	≈3.2-≈3.5	≈3.2	≈3.3	3.0
95% MDD/100% OMC	2.9	2.5	2.5-3.1	2.6	2.5	2.3
95% MDD/ 80% OMC	2.0	1.3-1.8	1.1-2.0	1.4	1.1	1.0
Al_2O_3, % of fraction passing a 0.425 mm sieve XRF	13.3-17.9	20.6	28.4	16.3	18.0	17.0
ICP	12.8-17.7	20.4	29.5	16.2	9.5	8.8
Fe_2O_3, % " " " XRF	6.5-8.7	13.5	4.8	6.9	7.8	10.5
ICP	6.8-8.8	13.6	5.0	6.9	5.2	7.1
SiO_2, % " " " XRF-XRD	16-31	25	30	21	23	20
ICP	11.8-17.3	21.7	29.6	19.5	21.5	20.5
S/R XRF-XRD	1.2-2.6	1.5	1.6	1.7	1.7	1.4
ICP	1.1-1.3	1.3	1.5	1.6	2.9	2.6
Grading Modulus	2.0-2.3	1.4-2.2	1.7-1.8	2.31	2.10	2.14
Dust Ratio	0.29-0.45	0.52-0.56	0.39-0.45	0.50	0.40	0.61

TABLE Ib

PROPERTIES OF LATERITIC GRAVELS AND OF CLAYEY SANDS USED IN THE PAVEMENT OF WESTERN AUSTRALIAN ROADS

Road Section Source Pit	Hamelin-Denham Road Close to junction with NWCH Approx 5.0 km from junction with NWCH	Albany Port Road Causeway Napier's Pit	South Coast Highway SLK 289.74 Carlingup Road Pit	Great Northern Highway Doolgunna Deviation SLK 900.1
Construction	1980	1972	1979 June	1976
Traffic (veh/day) 1987	107	3837	537	154
% Heavy vehicles	4	5	14	24
ESA to Dec 1987	7.1×10^3	1.1×10^6	4.9×10^5	1.1×10^5
Annual Rainfall, mm	210	815	419	209
Annual Evaporation, mm	2700	1500	2100	4100
Particle Size Distribution Sieve Size, mm				
% Passing 19.00		92-95	96-98	92-99
9.5		73-77	79-80	87-97
4.75		54-55	48-50	79-97
2.36	100	46	29-34	62-77
1.18	99-100	41	25-28	49-63
0.60	94-95	36-37	23-25	45-58
0.425	82-84	31-32	22-24	42-55
0.30	68-70	25-26	21-23	38-50
0.150	31	17	18-19	25-36
0.075	15-16	10-11	11-13	14-25
0.0135	8-4	4-6	4-5	6-15
Liquid Limit	-	21.6	19.3	17-23
Plastic Limit	-	17.6	16.4	9-12
Plasticity Index	NP	NP-4.0	NP-2.9	6-11
Linear Shrinkage	0.5-1.0	0.4-1.8	0.3-1.2	2-5
% Pass 0.425 x PI	NP	NP-124	NP-64	252-624
% Pass 0.425 x LS	41-84	13-56	7-26	84-303
Max Dry Density (Mod Comp) kg/m³	2135-2196	2162	2370	2415-2483
Optimum Moisture Content, %	6.0-6.2	8.0	6.5	6.0-6.6
WACCT Class No at 95% MDD/110% OMC	3.7-3.9	NA	NA	≈2.9
95% MDD/100% OMC	3.5-3.8	NA	NA	≈2.2
95% MDD/ 80% OMC	2.9-3.4	NA	NA	1.0
Al_2O_3, % of fraction passing a 0.425 mm sieve XRF	6.34	13.7	6.50	11.2
ICP	6.15	6.62	2.65	5.25
Fe_2O_3, % " " " XRF	3.48	7.81	4.31	8.17
ICP	3.30	5.00	2.85	5.02
SiO_2, % " " " XRF-XRD	8	31	15	22
ICP	8.36	10.3	6.27	13.5
S/R XRF-XRD	1.6	2.8	2.8	2.3
ICP	1.7	1.8	2.4	2.7
Grading Modulus	1.0	2.12	2.38	1.48-1.79
Dust Ratio	0.18-0.21	0.35	0.50	0.33-0.45

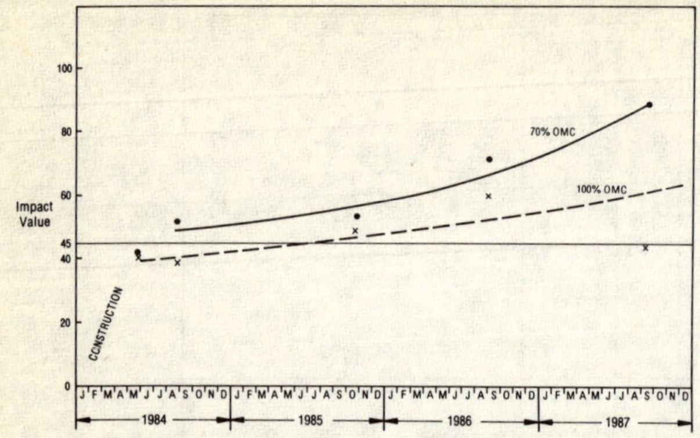

Fig 6. Great Eastern Highway, Meenaar Deviation
Change of strength of the lateritic gravel base course from Fullwood's Pit

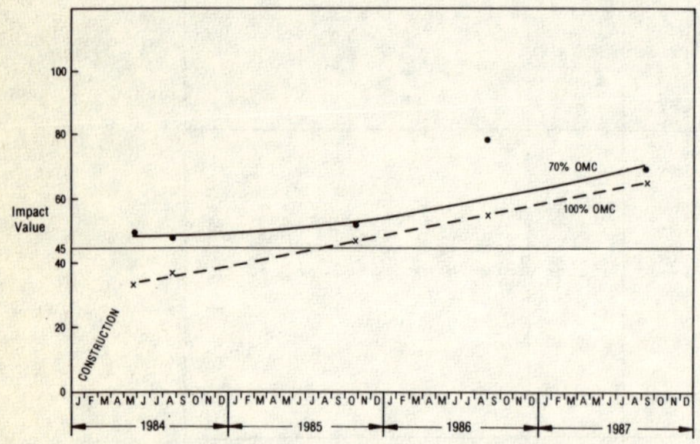

Fig 7. Great Eastern Highway, Meenaar Deviation
Change of strength of the lateritic gravel base course from Bushell's Pit

3.3 Great Northern Highway (Gogo Trial Sections)

In 1982 a series of thirteen trial sections, each 500 m long, were constructed at Gogo on the Great Northern Highway. Calcrete and lateritic gravels were used in the basecourse. A more detailed description of these trials and the early performance was described by Kilvington and Hamory (1986). Only the performance of three sections constructed of material described as lateritic gravel are presented in this paper.

The construction of the trials was carried out without significant incident and the performance of the basecourse has been satisfactory since opening of the highway to traffic. In April 1983 (end of the first wet season) an attempt was made to induce pavement distress by running a heavily laden vehicle along the seal edge. The accelerated loading was abandoned after 165 passes without visible damage to the pavement. The vehicle used had a 10.5 tonne load* on the dual tyred single rear axle.

* The legal maximum load for a single axle with dual tyres in Western Australia is 8.5 tonne.

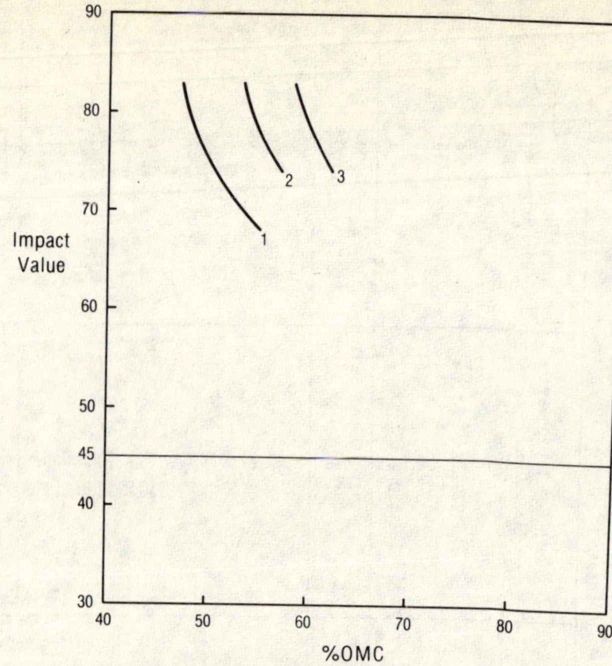

Fig 8. Great Northern Highway, Gogo Trial Section 7
Change of the impact values of the lateritic gravel basecourse from Pit 2648
Curve 1 – 1982 October/November
Curve 2 – 1983 April
Curve 3 – 1985 April

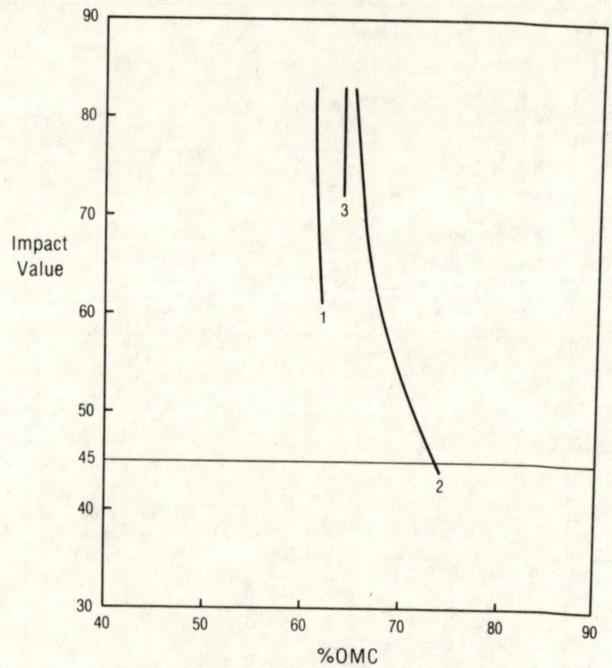

Fig 9. Great Northern Highway, Gogo Trial Section 11
Change of the impact values of the lateritic gravel basecourse from Pit 2628
Curve 1 – 1982 October/November
Curve 2 – 1983 April
Curve 3 – 1985 April

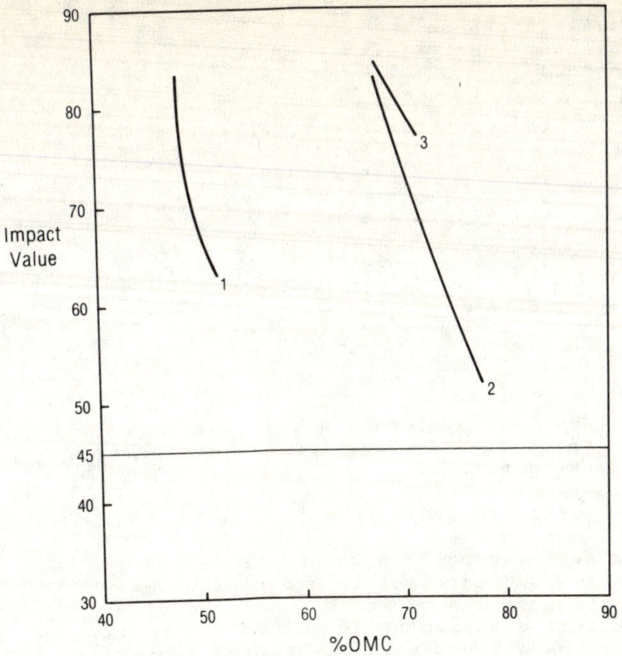

Fig 10. Great Northern Highway, Gogo Trial Section 12
Change of the impact values of the lateritic gravel basecourse from Pit 2709
Curve 1 – 1982 October/November
Curve 2 – 1983 April
Curve 3 – 1983 April

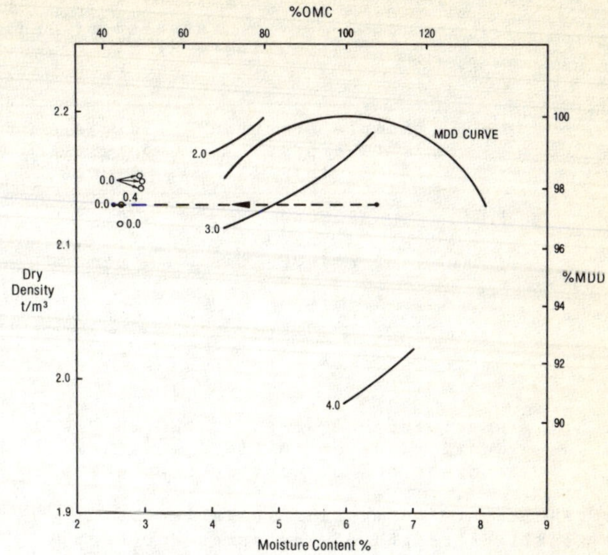

Fig 11. Hamelin–Denham Road. Moisture/density relationship and WACCT class numbers of the clayey sand basecourse
●----- Class number of specimens dried back in the laboratory
○ Class number of cores taken from the pavement

Since the completion of the trial sections, changes of strength with time have been monitored using the ITD and a similar technique to that described for the Meenaar deviation. The results are presented in figures 8, 9 and 10. The results show a very significant gain in strength with time, ie self stabilisation.

3.4 Hamelin-Denham Road SLK 6.0

The material used as a basecourse on the Hamelin-Denham Road is not a gravel. It is a red desert clayey sand. South of about latitude 21°S this material is known colloquially as "sand-clay", a misnomer as little clay may be present. Material of similar appearance and properties when found further north is known as "pindan sand".

This material is of aeolian origin and is not generally referred to as lateritic in Western Australia. However, the silica sesquioxide ratio of samples from the Hamelin Denham Road was less than 2 and their inclusion under the umbrella of the term lateritic may be justified. They are not included on the occurrence map (figure 3).

The section of the Hamelin-Denham Road referred to in this paper was constructed early in 1980. Despite the fine grading of the basecourse material, performance of the pavement has been good apart from edge fretting of the seal and some erosion of the unsealed shoulder by wind, water and traffic.

The decision to utilise this material was based on strength evaluation using accelerated triaxial testing. The method used was the Western Australian Confined Compression Test (WACCT) which is similar in its essential elements to the Texas Triaxial Test.

The class number of a material is a function of its moisture and density condition and contours of class numbers can be superimposed on a moisture/density graph as shown on figure 11.

The Hamelin-Denham Road is located in a particularly arid area. In this area the design moisture content under the seal of a well drained road is only about 50% of OMC. While this indicated a marginal performance (2.0 < Class No < 3.0), a decision to use the material was made bearing in mind the low traffic volume.

In November 1980 (approximately 6 months after construction) a number of cores were cut from the pavement and subjected to indirect tensile strength tests.

Core samples were taken at three sites and for each site tests were carried out on three sets of cores/specimens:-

Set 1: cores taken from the basecourse.

Set 2: specimens compacted at field moisture content to field density.

TABLE II
Results of Indirect Tensile Strength
Tests on Red Clayey Sand Basecourse Material,
Hamelin-Denham Road

Sample Location (Chainage) (m)	Field Density % MDD	Field Moisture % OMC	Indirect Tensile Strength (kPa) Set 1	Set 2	Set 3
206060	97	48	449-573	25	252
206900	95	44	292-346	30	228
207500	95	49	281-371	45	288

Set 3: specimens compacted at OMC to field density, dried to field moisture content and cured for three weeks.

The results are summarised in Table II. The tensile strength of the cores cut from the road and the laboratory "dried back" specimens is sufficiently high to infer a class number of 0.0 even with the most conservative estimate of angle of shearing resistance.

The high tensile strengths achieved in the field and obtained by drying back specimens compacted at high moisture content is further evidence of some form of self stabilisation mechanism.

3.5 Albany Port Road (Causeway)

The section of Albany Port Road referred to in this paper was constructed in 1972. It is located on a sand and limestone causeway constructed within the harbour. It provides the main access for heavy vehicles to the port loading facilities. Performance has been satisfactory. This is one of the more heavily trafficked roads constructed using lateritic gravel basecourse. Routine monitoring of strength change has not been carried out but impact values are available from tests in 1981 and 1987. The impact values demonstrate adequate and increasing strength of the basecourse.

3.6 South Coast Highway (SLK 289.74)

This section of South Coast Highway was constructed in 1979. Performance has been satisfactory.

Of interest with this material is the low Al_2O_3 and Fe_2O_3 contents determined by the wet chemistry/inductively coupled plasma (ICP) technique. By some chemically based definitions this material would not be considered a laterite. However its physical appearance is lateritic.

There has been no monitoring of changes in strength in the pavement.

3.7 Great Northern Highway (Doolgunna Deviation)

The Doolgunna Deviation was constructed in 1976. Construction proceeded without serious incident and performance has been satisfactory. The section is located in a very arid region.

4. STRENGTH TESTS

The performance of a base material in a sealed road pavement is principally a function of its strength. Methods of strength measurement in common use by the Main Roads Department of Western Australia (MRD WA) include:-

Accelerated Triaxial Test - laboratory
California Bearing Ratio Test - laboratory
Maximum Dry Compressive Strength - laboratory
Clegg Impact Test - field

4.1 Accelerated Triaxial Test [Western Australian Confined Compression Test (WACCT) Test Method WA 142.1]

This test is a slightly modified version of the Texas Triaxial Test. The WACCT may be used to assess the shear strength of a pavement material at the most adverse moisture and density conditions considered to be likely in the field.

A compacted cylindrical specimen (100 mm diameter x 150 mm high) of the material is placed in a triaxial cell in which the lateral stress is kept constant, and the vertical stress increased until failure occurs. The test is carried out on the material passing a 19 mm sieve. A number of tests must be performed at different lateral pressures, on specimens at the same density and moisture content, so that a complete failure envelope may be drawn. The failure envelope is used to assign a class number for a given moisture and density condition.

Other parameters such as cohesion, angle of shearing resistance and moisture sensitivity may also be determined.

The class number is a measure of the shear strength of a basecourse material. In terms of field performance, it relates to the probability of basecourse failure under heavy vehicles. If a material has a class number of 2.0 or less at the expected field moisture and density conditions, then satisfactory performance can be expected. A class number of 3.0 or more indicates a high risk of unsatisfactory performance.

Intermediate values of class number indicate marginal performance ie increasing risk with increasing class number.

4.2 California Bearing Ratio Test (Test Method WA 141.1)

The California Bearing Ratio (CBR) Test is one with which most readers would be familiar. The MRD WA test is identical with that in use in most parts of the world except that it has been metricated. The test is carried out on material passing a 19 mm sieve. Generally a soaked test is used for evaluation of basecourse materials. However unsoaked tests on materials compacted at 100% OMC have been used where highly favourable drainage conditions exist.

A minimum CBR of 80 is commonly specified for basecourse materials on sealed roads.

4.3 Maximum Dry Compressive Strength Test (Test Method WA 140.1)

The Maximum Dry Compressive Strength (MDCS) Test is commonly used on non plastic materials to ensure that they have adequate binding strength.

The test is carried out on compacted soil cubes approximately 70 x 70 x 70 mm. Specimens are compacted at a moisture content equal to OMC and a little above OMC and a little dry of OMC. After extrusion from the mould they are dried at room temperature for 6 hours then at 105°C for 12 hours before crushing. The test is carried out on material passing a 19 mm sieve.

An MDCS of at least 1700 kPa is considered indicative that a sound, tight, well bound surface can be achieved which is suitable to receive a bituminous seal. The MDCS is also regarded as indicative of the performance of the material as an unsealed shoulder. A material with a low value is likely to develop a loose surface.

4.4 Clegg Impact Testing Device

A major advantage of the Impact Testing Device (ITD) developed by Clegg (1976) is its portability. The device is similar in appearance to the compaction hammer used to carry out laboratory maximum dry density (MDD) tests. The hammer has a mass of 4.5 kg and is dropped from a height of 457 mm. As the hammer strikes the pavement, an accelerometer records the peak deceleration in g's. The deceleration is displayed on a scale of 10 g = 1 Impact Value (IV). MRD WA practice is to drop the hammer five times at each test site. The average of the IV's of the third, fourth and fifth drops is reported as the IV for the site.

The MRD WA has utilised the ITD for basecourse strength evaluation in the field since the late 1970's. Based on the observations of cases where pavement condition could be related to IV, the relationships detailed in Table III have been established.

TABLE III
Impact Values for Basecourse

Impact Value*	Basecourse Strength
> 75	Very High
60 - 75	High
45 - 59	Medium - High
30 - 44	Low - Medium
< 30	Low

* values listed refer to Impact Testing Devices fitted with a 0.001 μF capacitor.

The following relationships were established at sites with inadequate basecourse strength:-

Impact Value	Condition
6	70-80 mm deep wheel rut caused by one only loaded single axle truck.
20	5 mm deep wheel rut caused by one only loaded single axle truck.

5. BASECOURSE MOISTURE CONTENTS

An important factor influencing the strength of compacted basecourse material and as a result its performance, is the field moisture content. In 1969 the MRD WA established a network of 44 stations on existing sealed roads where moisture content in the various pavement layers (basecourse, subbase, subgrade) was monitored on a regular basis.

Each station typically consisted of 3 subsections providing a total of 132 locations where moisture content was monitored. Monitoring of moisture content continued until 1979 at most locations.

With the availability of more powerful statistical analysis computer packages, this data is now being re-analysed. To date only the data for well drained sections satisfying the following criteria have been analysed:-

- bituminous surface treatment
- good external drainage
- deep water table
- not subject to innundation
- subgrade permeability similar to or greater than basecourse permeability

With regard to the last point, direct measurements of permeability were not available. The assessment was based on classification tests.

Thirty one (31) sections satisfied these criteria. The resulting equation (Mackenzie 1988) was:-

$$MC = 0.70\ OMC - 0.02\ P_{2.36} + 0.29 \times 10^{-3}\ (AR-PE) + 0.58\ \frac{P_{0.425}}{P_{0.075}}$$

where MC = basecourse design moisture content %

OMC = basecourse optimum moisture content % (modified compaction)

$P_{2.36}$ = percentage by mass passing a 2.36 mm sieve

$P_{0.425}$ = percentage by mass passing a 0.425 mm sieve

$P_{0.075}$ = percentage by mass passing a 0.075 mm sieve

AR = annual rainfall (mm)

PE = potential evaporation (mm)

This equation had a correlation coefficient of correlation of 0.86 and a coefficient of variation of 0.15.

6. SELF STABILISATION

References to a phenomena variously referred to as self bonding, petrification, self cementation or self stabilisation, in relation to lateritic soils are common. Nascimento, Branco and De Castro (1963, 1965) described a "petrification degree" test to measure the development of water resistant bonds in lateritic soils during drying. Van Der Merwe and Bate (1971) carried out cyclic (alternative wetting and drying) CBR testing on lateritic gravels in Zimbabwe and concluded that laterites in that country did not possess self cementing qualities. LNEC et al (1959) found a positive correlation between swelling and silica sesquioxide ratio (S/R) ie lateritic soils with a low S/R were likely to swell less on wetting. Grant (1974) compacted samples of lateritic gravel and subjected them to wetting and drying tests. A number of samples showed evidence of self bonding. He concluded that initial bonding was due to solution and precipitation of the hydrated aluminium oxide. Netterberg (1975) reviewed the evidence for improvement in strength due to self stabilisation and concluded that it can be significant. Charman (1988) discusses the "self hardening" of lateritic materials used in road construction and presents a number of hypotheses to explain the phenomena.

While evidence of some form of time related strength gain has been obtained under laboratory conditions, there has been a lack of field evidence for roads. In the authors' view this is probably due to the expense and difficulty of measuring strength changes in the field by conventional techniques such as plate bearing tests, Benkelman beam deflection tests or field CBR tests. The Clegg ITD has a major advantage in this regard. It is relatively cheap and very simple to operate.

It is significant to note that at the three sites, with material from six different source pits, where changes in strength have been monitored, a significant increase of strength has been noted. The changes in strength reported for Meenaar Deviation (Bushell's and Fulwood's pits), the Gogo trial sections (pits 2628, 2648 and 2709) and the Hamelin-Denham Road were all of sufficient magnitude to influence the performance of the material as a pavement basecourse layer.

There are a number of hypotheses which can be proposed to explain the phenomena of self stabilisation in lateritic soils. The one favoured by the authors is as follows:

Field compaction of basecourse material is carried out at or near optimum moisture content (OMC) which for an arid climate, such as that in much of Western Australia, exceeds the long term equilibrium moisture content. As the wetted soil begins to dry by evaporation, capillary tension causes the remaining water to withdraw into the narrow spaces close to the soil grain contact points. The water carries with it aluminium oxide and iron oxide in solution and/or suspension. As the soil continues to dry these sesquioxides come out of solution and bond the larger soil particles together. This process is not rapidly reversible and on rewetting not all of the cementing material goes back into solution/suspension. The process continues with seasonal moisture fluctuations.

If this hypothesis is valid then the properties of the water used for compaction (particularly pH) may be a significant factor in determining the amount and rate of self stabilisation which occurs.

7. SILICA SESQUIOXIDE RATIO

The silica sesquioxide ratio (S/R) has been used by some authors (Wooltorton 1954) as a basis of definition for lateritic soils.

Lateritic Soils $S/R \leq 2$

A relationship between S/R and field performance was also suggested by Wooltorton (1954) who wrote "In general the higher the S/R, the more active the clay fraction and more troublesome the soil for road engineering purposes". An upper limit of 2 for S/R has also been included in some Brazilian (DNER) and Portuguese (LNEC) specifications for lateritic gravel basecourse materials.

A review of the literature found no consistency with regard to the fraction on which chemical analysis of the constituents of lateritic soils was carried out. In some cases the clay fraction only (< 2 μm) had been used, in others material passing a 2 mm sieve and in some cases the total material. The authors chose to avoid the clay fraction approach (< 2 μm) due to the difficulties in separating this material out in a manner which would not effect the composition of the material obtained. The results reported in this paper are for the fraction passing a 0.425 mm sieve. This was chosen as it is the

TABLE IV
SURVEY OF CHEMICAL COMPOSITION* OF SOME WESTERN AUSTRALIAN ROAD BUILDING MATERIALS

		LATERITIC GRAVEL			RED CLAYEY SAND			CRUSHED IGNEOUS ROCK		
		Al_2O_3	Fe_2O_3	SiO_2	Al_2O_3	Fe_2O_3	SiO_2	Al_2O_3	Fe_2O_3	SiO_2
XRF/XRD	$\bar{x}$	16.1	7.9	23.2	16.8	4.3	20.5	13.5	3.2	44.3
	s	5.2	2.3	5.2	6.3	1.1	17.7	0.1	0.6	3.1
	n	13	13	13	2	2	2	3	3	3
Wch/ICP	$\bar{x}$	12.8	6.8	16.8	6.4	3.3	12.3	2.0	1.6	6.8
	s	7.1	2.6	6.4	0.4	0.04	5.6	0.4	0.2	1.1
	n	13	13	13	2	2	2	3	3	3

* Expressed as percentage by mass of the fraction passing a 0.425 mm sieve.

same fraction as that on which liquid limit (W_L) and plasticity index (W_P) tests are performed.

The tests were carried out by the Chemistry Centre WA (CCWA) on samples of material representing the sections of road detailed in Table I. In addition tests were carried out on other lateritic materials, a red clayey sand and on the crushed granite rock base used in Perth for basecourse construction.

Two methods of analysis were used. The first used x ray fluorescence (XRF) to determine the Al_2O_3, Fe_2O_3 and total SiO_2 content. The quartz content was determined by x ray diffraction (XRD). The combined silica content used to determine the S/R was calculated as the difference by the XRF (total silica) and XRD (quartz) results.

The second method involved a wet chemistry dissolution with inductively coupled plasma spectrometry (ICP) to quantify the sesquioxide and silica proportions. The distribution parameters of the results of the two analysis are shown in Table IV.

A comparison of the S/R's obtained by the two methods is presented in figure 12. For all samples tested, the proportion of all oxides (Al_2O_3, Fe_2O_2, SiO_2) determined by the XRF/XRD method was equal to or greater than for the ICP method. In some cases the difference was similar for all oxides with the net effect that the S/R was similar by both methods. The reasons for the difference in oxide contents between the two methods are still under investigation. The extent to which the material will dissolve in the wet chemical attack is probably a significant factor.

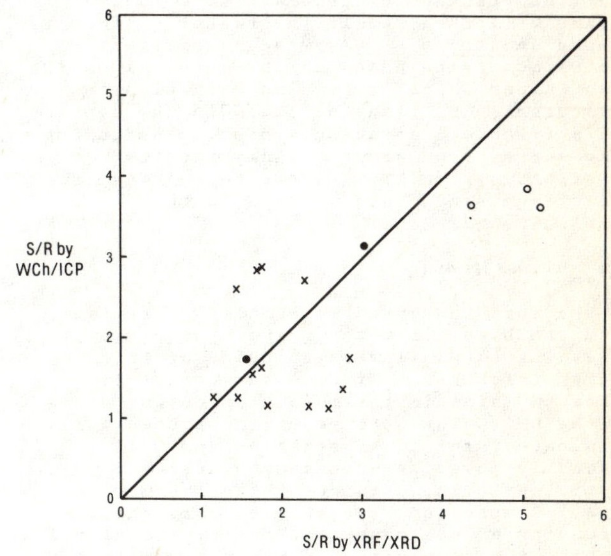

Fig 12. Comparison of S/R values obtained by x ray fluorescence/x ray diffraction and by wet chemistry/inductively coupled plasma spectrometry
x Lateritic gravel
• Red clayey sand
o Crushed igneous rock

8. CONSTRUCTION TECHNIQUE

The performance of lateritic material as a basecourse is not just a function of properties such as particle size distribution, Liquid Limit, Plasticity Index or CBR. The construction technique used to win, place and compact the material is also important. The following comments are based on observations by the authors:-

8.1 Variability

Lateritic gravels being natural materials are more variable than crushed rock products. In particular there is a change in properties with depth in the borrow pit. For this reason winning by scraper is rarely suitable. The material must be pushed up by dozer to ensure mixing then loaded and carted onto the road bed. Further mixing by windrowing with motor graders is then required to blend sandy and clayey portions of the material.

8.2 Oversize

Almost all lateritic gravels in Western

Australian contain some oversize material, ie particles retained on a 75 mm sieve. The presence of oversize particles or of high plasticity fines precludes the use of paving machines. Segregation of the material into coarse and fine fractions is also a problem with pavers. The MRD WA uses motor graders for spreading and to achieve the design level. Grid rollers are frequently required to breakdown the oversize particles.

8.3 Curing

Most specifications require the compaction of lateritic materials in the pavement to a density not less than 95% MDD (modified compaction). This density level is easier to achieve if the materials are cured at a moisture content close to the OMC prior to compaction. The curing will assist the establishment of a uniform moisture regime in the fine fraction of the material. Such regime will then result in greater uniformity of the density of the compacted pavement. The water may be added to the material on the road, after which it is placed in windrows for curing, or it may be added in the deposit via a trickle irrigation system. The curing time varies from about one day for low plasticity lateritic gravels to 4-5 days for red clayey sands such as used in the Hamelin-Denham Road.

8.4 Blade Mixing

On the road the cured materials are remixed with the blade of motor graders prior to spreading out in layers to build up the design thickness. If necessary water is added to maintain the moisture content close to the OMC. The wetting/mixing process will eliminate particle segregation and will gradually move the uncrushed oversize particles to the outside of the pavement edge. Compaction is continued throughout the spreading of the material. This process will ensure high densities and minimal vertical density gradient.

8.5 Compaction Moisture Content

If a tightly bound surface and the full strength potential is to be achieved, a compaction moisture content close to OMC must be used. While adequate density can be achieved using very heavy rollers compacting dry of OMC, the resulting surface is less likely to be satisfactory and the potential strength may not be achieved. Compacting dry of OMC also increases the risk of the unsealed portion of the shoulder being more permeable and admitting water. For some materials it is necessary to roll the finished compacted surface with light rollers at moisture contents greater than OMC (slurrying) to mobilise the fines so that a tight, impermeable, well bound surface can be achieved.

8.6 Drying Back

For some lateritic gravels and red clayey sands, the strength at a moisture content of 100% OMC is inadequate. This is not of serious concern provided adequate strength at the generally lower design moisture content is available. However the basecourse material is required to have sufficient strength from the first day of opening to traffic. It is therefore necessary to dry back the compacted basecourse prior to application of the primerseal. MRD WA specifications typically require dry back to 80% OMC.

8.7 Primersealing

The provision of a bituminous seal is essential for the satisfactory performance of a lateritic gravel basecourse. On some dry surfaces it is necessary to apply a very light spray of water immediately prior to the application of the primerseal. This will reduce surface tension effects which can cause holes to develop in the bitumen blend. Typical bitumen blends are 80% of bitumen, 15% of kerosene and 5% of distillate. Rolling by rubber tyred rollers or by traffic is necessary to knead any holes closed, so that the primerseal develops a continuous and stable mat. The application of a seal coat 6-18 months after primersealing is necessary to maintain a durable waterproof surface layer.

9. SELECTION CRITERIA

In a state with as much range in climate and traffic loading as Western Australia, there is no single specification or set of selection criteria which can cater for all situations. The suggested selection criteria presented in Tables V, VI and VII should exclude most unsatisfactory materials. The MRD WA specification for crushed igneous rock is included for comparison purposes.

10. CONCLUSIONS

The large area of Western Australia in combination with the low population density dictated the use of material not conforming to the standard selection criteria of basecourse material. The development of appropriate construction techniques was necessary to ensure adequate long term performance of such pavements. This low cost construction has largely been made possible by the arid/semi arid climate and by the widespread occurrence of lateritic gravels and other suitable natural materials.

The performance of lateritic gravels has been better than that which might be predicted on the basis of Plasticity Index and other classification tests.

Evidence of the increase of strength (self stabilisation) of lateritic materials in road pavements has been obtained. The change in strength is of sufficient magnitude to influence pavement performance.

The Clegg Impact Testing Device provides a suitable low cost technique for monitoring changes in strength with time.

Under well drained conditions in arid areas the traditional limits for Plasticity Index and particle size distribution may be relaxed.

TABLE V

SUGGESTED INTERIM SELECTION CRITERIA FOR LATERITIC GRAVELS BASED ON GRADING AND CLASSIFICATION TESTS[1]

Type of Material		Lateritic Gravel			Crushed Rock[2]
Design Traffic ESA		$<5 \times 10^6$	$<10^6$	$<5 \times 10^5$	NS
Rainfall Deficit[3]		>0	>1500	>3000	NS
Sieve Size mm					
% Passing	37.5	100[4]	100[4]	100[4]	
	26.5				100
	19.0	71-100	95-100	95-100	95-100
G[5]	13.2				70-90
R	9.5	50-81	50-100	50-100	60-80
A	4.75	36-66	36-81	36-81	40-60
D	2.36	25-53	25-66	25-66	30-45
I	1.18	18-43	18-53	18-53	20-35
N	0.60				13-27
G	0.425	11-32	11-39	11-39	
	0.30				8-20
	0.15				5-14
	0.075	4-19	4-23	4-23	4-10
	0.035	2-9	2-11	2-11	
Dust Ratio[6]		0.3-0.7	0.3-0.7	0.3-0.7	0.4-0.6
Liquid Limit[7] %		≯25	≯30	≯35	NS
Plasticity Index %		≯6	≯10	≯16	≯6
Linear Shrinkage %		≯3	≯5	≯8	≯3
$P_{0.425}$ x LS		≯150	≯200	≯250	NS
MDCS[9] kPa		≮1700	≮1700	≮1700	NS
$Al_2O_3 + Fe_2O_3$[10] %		>10?	>10?	>10?	NS
Los Angeles Abrasion		Note (11)	Note (11)	Note (11)	≯40

NOTES:
(1) Selection criteria apply to basecourse on well drained roads with a thin bituminous surfacing.
(2) Non lateritic, included for comparison purposes only. MRD WA Specification Contract 94/87.
(3) Rainfall deficit = potential evaporation - annual rainfall.
(4) Most deposits of lateritic gravel contain some oversize material which must be broken down by grid rolling.
(5) Dry Sieving and Decantation, Test Method WA 115.1.
(6) Dust Ratio = $\frac{P_{0.075}}{P_{0.425}}$ Material with a low dust ratio is likely to be harsh and the achievement of a satisfactory surface prove difficult.
(7) Liquid Limit, Plasticity Index and Linear Shrinkage tests on samples air dried at 50°C.
(8) For materials approaching the upper limit for plasticity index or $P_{0.425}$ x LS confirmation of suitability by strength testing is recommended.
(9) Maximum Dry Compressive Strength, Test Method WA 140.1.
(10) Al_2O_3 and Fe_2O_3 determined by ICP on the fraction passing a 0.425 mm sieve.
(11) No particular test for particle toughness is specified at this time. However the lateritic pebble must be hard and durable.
(12) NS = not specified.

TABLE VI

SUGGESTED INTERIM SELECTION CRITERIA FOR LATERITIC GRAVEL BASED ON STRENGTH AND CLASSIFICATION TESTS[1]

Design Traffic ESA		$<5 \times 10^6$	$<10^6$	$<5 \times 10^5$
Rainfall Deficit [2]		>0	>1500	>3000
WACCT[3]				
	Class No	≯2.0	≯2.0	≯2.3
	Cohesion[4] kPa	≮85	≮85	≮85
	Tensile Strength kPa	≮55	≮55	≮55
CBR[6]				
	Soaked	≮80	≮80	≮60
	Unsoaked	NS	NS	≮80
Maximum Size[7] mm		37.5	37.5	37.5
Grading Modulus[8]		≮ 1.5	≮1.5	≮1.5
Dust Ratio[9]		0.3-0.7	0.3-0.7	0.3-0.7
Plasticity Index %		≯6	≯10	≯16
Linear Shrinkage %		≯3	≯5	≯8
$P_{0.425}$ x LS		≯150	≯200	≯250
$Al_2O_3 + Fe_2O_3$ % [10]		>10?	>10?	>10?
Particle Toughness[11]		note 11	note 11	note 11

NOTES:

(1) Selection criteria apply to basecourse on well drained roads with a thin bituminous surfacing.

(2) Rainfall deficit = potential evaporation - annual rainfall.

(3) West Australian Confined Compressive Test, Test Method WA 142.1. Class number, cohesion and tensile strength assessed at 95% MDD (modified compaction) and at the design moisture content for the site.

(4) Cohesion and tensile strength may be reduced to 45 kPa and 30 kPa respectively provided the angle of shearing resistance is > 60°. These parameters are not as critical where the shoulders are sealed.

(5) The criteria for the assessment by WACCT may not be met when testing specimens immediately after compaction. In these cases the specimens should be compacted at 100% OMC, dried to design moisture content and cured for three weeks without further loss of moisture prior to testing.

(6) CBR specimens compacted to 95% (modified) at OMC. Test Method WA 141.1

(7) Most deposits of lateritic material include some oversize material which must be broken down by grid rolling.

(8) $$\text{Grading Modulus} = \frac{300 - (P_{2.36} + P_{0.425} + P_{0.075})}{100}$$

(9) $$\text{Dust Ratio} = \frac{P_{0.075}}{P_{0.425}}$$

(10) Al_2O_3 and Fe_2O_3 determined by ICP on the fraction passing a 0.425 mm sieve.

(11) No particular test for particle toughness is specified at this time. However the lateritic pebble must be hard and durable.

(12) NS = Not Specified.

TABLE VII

SUGGESTED INTERIM SELECTION CRITERIA FOR RED CLAYEY SAND FOR USE AS BASECOURSE ON WELL DRAINED SEALED ROADS

Design Traffic ESA		$<10^5$	$<3 \times 10^4$
Rainfall Deficit (1)		>2500	>2500
Sieve Size mm			
% Passing	4.75	100	100
G	2.36	70-100	70-100
R	1.18	50-79	50-100
A	0.60	36-63	36-100
D	0.425	30-56	30-84
I	0.300	25-50	25-71
N	0.150	18-40	18-50
G	0.075	13-31	13-35
	0.0135	5-15	5-15
Dust Ratio (2)		0.2-0.6	0.2-0.6
Liquid Limit %		$\ngtr$20	$\ngtr$20
Plasticity Index %		$\ngtr$8	$\ngtr$8
Linear Shrinkage		1-3	1-3
CBR (3)			
	Unsoaked	$\nless$80	$\nless$80
WACCT (4)			
	Class No	$\ngtr$2.0	$\ngtr$2.3
	Cohesion kPa	$\nless$85	$\nless$85
	Tensile Strength kPa	$\nless$55	$\nless$55
	Horizontal Separation of Class No Contours %	$\nless$1.3	$\nless$1.3
Maximum Dry Density (6) kg/m^3		$\nless$2100	$\nless$2100
Optimum Moisture Content		5-7	5-7
$Al_2O_3 + Fe_2O_3$ % (7)		>8?	>8?

NOTES:
(1) Rainfall deficit = potential evaporation - annual rainfall.
(2) Dust Ratio = $\frac{P_{0.075}}{P_{0.425}}$
(3) California Bearing Ratio carried out on specimens compacted at 90-100% OMC to 95% MDD (modified).
(4) Western Australian Confined Compression Test. Class number, cohesion and tensile strength assessed at a dry density ratio of 96% (modified compaction) and moisture content of 60% OMC.
(5) The criteria of the assessment by WACCT may not be met by testing specimens immediately after compaction. In these cases the specimens should be compacted at 100% OMC, dried to the design moisture content and cured for three weeks without further loss of moisture prior to testing.
(6) Maximum Dry Density, Test Method WA 133.1, similar to Modified AASHO.
(7) Al_2O_3 and Fe_2O_3 determined by ICP on the fraction passing a 0.425 mm sieve.

The Indirect Tensile Strength Test is a useful method to monitor the time related increase of strength of specimens compacted in the laboratory. The increases may then be related to similar changes of strength evidenced by testing cores cut from basecourse constructed of fine grained materials.

ACKNOWLEDGEMENT: The approval of the Commissioner of Main Roads, Western Australia, Mr A H Tognolini, to the publication of this paper is gratefully acknowledged. A number of people and organisations assisted with the provision of data and advice for this paper including Mr S Brice (Geological Survey of Western Australia), Mr M Costello (Chemistry Centre WA) and various officers of the Main Roads Department. The views expressed are those of the authors and do not necessarily reflect those of the Department.

REFERENCES:

CHARMAN J H (1988): "Laterite in Road Pavements". CIRIA Special Publication No 47, London.

CLEGG B (1976): "An Impact Testing Device for In Situ Basecourse Evaluation". Proc 8th ARRB Conf, Vol 8 Part 3.

DEPARTAMENTO NACIONAL DE ESTRADAS DE RODAGEM DNER (1974): "Especificacoes gerais para obras rodoviarias". 1 Reedicao, Servico de Publicacoes, Rio de Janeiro.

GENTILLI J (1972): "Australian Climate Patterns". Thomas Nelson, Melbourne.

GEOLOGICAL SURVEY OF WESTERN AUSTRALIA: 1:250 000 Geological Map Series.

GRANT K (1974): "The Composition of Some Australian Laterites and Lateritic Gravels". Proc 2nd International Congress of Int Assoc of Eng, Geol. Sao Paulo.

GRANT K & AITCHISON G D (1970): "The Engineering Significance of Silcretes and Ferricretes in Australia". Eng Geol 4, Elsevier, Amsterdam.

KILVINGTON D & HAMORY G (1986): "The Performance of Natural Gravel as Basecourse in a Sealed Pavement". Proc 13th ARRB/5th REAAA Conf.

LNEC, LEA, LEMMS - As laterites do Ultrarmar Portugues, LNEC, Memoria No 141, Lisboa 1959. (Laterites from the Portuguese Overseas Provinces).

LABORATORIA NACIONAL DE ENGENHARIA CIVIL, LABORATORIA DE ENGENHARIA DE ANGOLA, LABORATORIO DE ENGENHARIA DE MOCAMBIQUE, JUNTA AUTONOMA DE ESTRADAS DE ANGOLA (1969). Portuguese studies on engineering properties of lateritic soils, Proc 7th Int Conf Soil Mech Fndn, Engng, Spec Session Eng. Props. Lat. Soils, Mexico City, 1, pp 85-96, LNEC Memoria No 143, Lisboa, 1972.

MACKENZIE R (1988): Design Moisture Content Prediction for Basecourse Materials in Western Australia". Main Roads Department of Western Australia, Report 88/29 M (unpublished), Perth.

NASCIMENTO U, DE CASTRO E & RODRIGUES M (1963): "Swelling and Petrification of Lateritic Soils". Proc Reg Conf on Soil Mech and Fndn Engng, Salisbury.

NASCIMENTO U, BRANCO F & DE CASTRO E (1965): "Indentification of Petrification in Soils". Proc 6th ISSMFE, Montreal.

NETTERBERG F (1975): "Self Stabilisation of Road Bases : Fact or Fiction". Proc 6th Reg Conf Africa Soil Mech and Fndn Engng, Durban Vol 1 pp116-119.

STEPHENS C G (1971): "Laterite and Silcrete in Australia. A Study of the Genetic Relationships of Laterite and Silcrete and their Companion Materials, and their Collective Significance in the Formation of the Weathered Mantle, Soils, Relief and Drainage of the Australian Continent." Elsevier, Amsterdam.

THORNTHWAITE C W (1933): "The Climates of the Earth". Geogr. Rev Vol 23.

TWIDALE C R (1983): "Australian Laterites and Silcretes :Ages and Significance". Revue de Geologie Dynamique et de Geographe Physique Vol 24, F1, Paris.

VAN DER MERWE C R AND BATE I C (1971): "The Properties and Uses of Laterites in Rhodesia". Proc 5th Reg Conf Africa Soil Mech and Fndn Engng, Luanda, 1, Theme 2, 3-15.

WOOLTORTON F L D (1954): "The Scientific Basis of Road Design". Edward Arnold Publishers, London.

Grouting of canaliculae in residual soils and behaviour of the foundations of Balbina Dam

L.G.F.S.DE MELLO, Vector S.A., Brazil
J.M.M.FRANCO FILHO, Vector S.A., Brazil
C.R.ALVISE, Vector S.A., Brazil

ABSTRACT: The residual soils of the Brazilian Amazonic region show the presence of canaliculi in many sites. Treatment of these foundation conditions should take in consideration the local logistic and climatic conditions. The utilization of soil-cement grouting through manchette valves at pressures above the hydraulic fracturing pressures of the soil was conceived and used. Special in-situ and laboratory tests were planned in order to better evaluate the sensitivity of the variables involved. With the beginning of the raising of the water in the reservoir piezometric data is already available showing the improvements.

1 INTRODUCTION

The Balbina Hydroelectric Scheme will generate 250 megawatts in the Uatumã River, Amazon Basin.

The embankment dam is a homogeneous earthfill with average height of 30 meters. Its foundations consist of silty-clayey residual soils generated by the weathering of vulcanites. Among the geotechnical-geological features conditioning the design of the embankment dam it is emphasized the existence of localized points with permeability higher than 10^{-3} cm/sec.

Different phases of subsoil investigation associated the high permeabilities to the presence of tubular cavities, with diameters varying from millimetric to centimetric, intercommunicating and with an erratic geometric pattern. These cavities are generically called canaliculae and are thought to be generated mainly by termite activity. This activity is fossil and linked with the last glacial period (de Mello, L.G. et al, 1987).

2 FOUNDATION TREATMENT

Many different alternatives were studied prior to the definition of the foundation treatment to be used in the dam (Sathler, G. et al, 1985; Remy, J.P.P. et al, 1985).

Grouting of the residual soils using the tube a manchette technique was proved to be the best alternative costwise, when the analysis included the logistic and climatic peculiarities of the Amazon region.

The concept behind the treatment of the foundation through grouting aims at creating a zone under the embankment in which all localized permeability singularities are eliminated and the foundation can be treated as homogeneous, in terms of statistics of averages (de Mello, V.F.B., 1977).

Using the tube a manchette technique, grouting can be done localizedly, applying pressures high enough to generate hydraulic fracturing of the soil, with penetration and filling of the canaliculae and voids intercepted by the lenses of grout.

The advantages of this concept of treatment are: a) grouting can be intensified in regions where the intake is higher; b) treatment can be done all year round, despite of climatic changes; c) use of conventional equipment; d) the same boreholes can be used to treat the decomposed rock underneath the residual soil. The only disadvantage of this treatment is its pioneer characteristics, requiring a rigorous supervision of the work.

3 TREATMENT METHODOLOGY

The methodology used in the treatment resulted from the experience of the professional team involved, and was optimized many times as a function of the results and data collected.

The systematic used during the major part of work can be summarized as:

a) Water loss tests were done in the residual soils with pressure stages of 0.2; 0.4; 0.7 ... kg/cm^2 untill hydraulic fracture occurred. Through this testing the determination of the coefficient of permeability prior to treatment and relevant indications of the pressures in which treatment should be pursued were gathered.

   Special techniques to assure perfect contact of the expansible packers with the walls of the borehole were developed in order to avoid preferential vertical seepage which would invalidate all the testing.

b) Three lines of grouting holes were planned, with grouting starting from the external ones in order to confine and improve treatment in the central holes.

c) Grouting was done using double packers specially developed in the site, which would seal against a PVC tube left in the borehole and grouted in its external side to develop contact with the residual soil (as shown in Fig. 1).

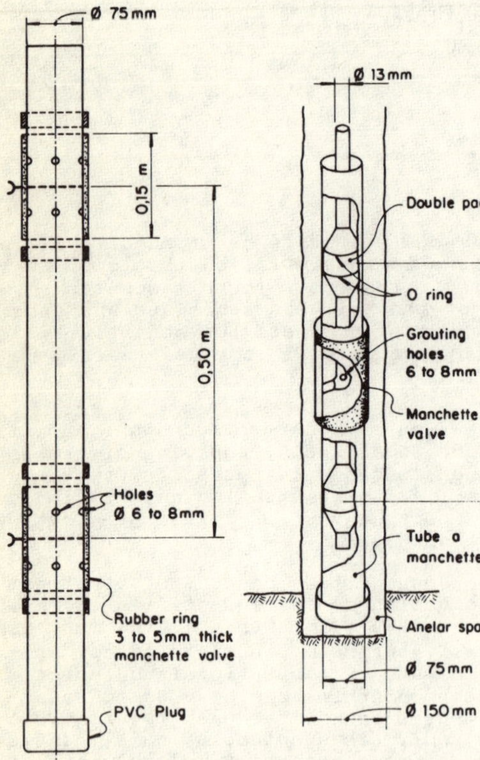

FIG. 1 - TUBE WITH MANCHETTE VALVE AND GROUTING PACKERS

d) The grouts studied to treat the soil were different than those used to fix the PVC grouting tube. Both used as much as possible the local clayey residual soils in order to reduce costs of treatment.

The specification of the soil-cement grout asked for:

1. uniaxial compressive strength at 28 days of 1.5 kg/cm^2;
2. viscosity ≤ 50 cP;
3. sedimentation factor ≤ 5%;
4. fluidity in Marsh cone of 40 to 45 seconds;
5. yield limits ≤ 0.1 g/cm^2.

e) Grouting was undertaken from the lowest manchette valve towards the top, at pressures compatible with hydraulic fracturing pressure previously determined.

f) Treatment was stopped when 300 litres of grout was absorbed by the soil or after 10 minutes of no intake of grout.

g) Water loss tests were performed in specially located boreholes to document the benefit of the treatment, limiting the maximum pressure of the test at 1.0 kg/cm^2 not to fracture the soil.

At the beginning of the treatment special tests and experimental stretches of work were performed to supply the technical committee with data which, in turn, helped optimize its specifications. The special tests include:

- Carefully performed water loss tests in drained and in undrained conditions. Fig. 2 presents two of the referred tests, from which the pressure which causes hydraulic fracturing and the minor principal total stress σ_3 were interpreted. Following Bjerrum and Andersen (1972) discussions, the coefficient of earth pressure at rest Ko can be estimated, provided that σ_3 is assumed to be horizontal. This value was estimated to vary between $0.21 \leq Ko \leq 0.36$.

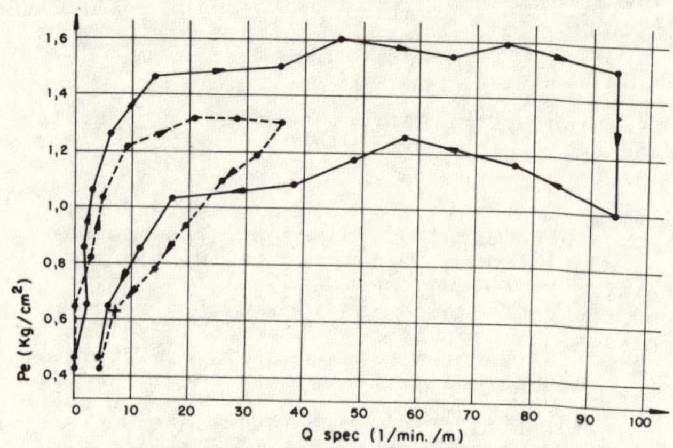

FIG. 2 - GROUTABILITY TEST - HYDRAULIC FRACTURING OF THE RESIDUAL SOIL

- Inspection trenches specially dug in treated areas showed valid the assumption of the orientation of σ_3.

Once the value of Ko is known, many theories exist which enable the pressure of hydraulic fracturing of a medium to be estimated (Bjerrum et al, 1972 ; Kennard, 1970; Haisom-Fauhurst, 1967; Morgenstern and Vaughan, 1963; etc.). The data collected allow the following relation to be established between the formulation proposed by Bjerrum in:

$$\Delta\mu\,obs = 7.82 + 1.02\ \Delta\mu Bjerrum\ (t/m^2)$$

- Special tests were also performed to try to collect data of pressure head losses in the grouting system.

4 EVALUATION OF THE TREATMENT

The evaluation of the improval of the residual soil of the foundation of Balbina dam with respect to its permeability was done through water loss tests carefully done after the treatment.

In Fig. 3 some test results are presented in relation to the information collected before grouting. The benefit of the treatment is obvious. In the mentioned figure, the interpretation of the water loss tests following Babouchkine's (1965) formulation is presented.

5 FILLING OF THE RESERVOIR

Filling of the reservoir started in mid 1987 and has been monitored continuously. Although the rainy season has not been able to fill the reservoir as expected, partial data is already available, and it indicates the benefit of the treatment.

Two different cross sections of the embankment were used during construction in order to separate the treatment of the foundation from the earthmoving and compacting, which have to be done in a very defined time of the year due to the heavy rains. For this matter in some areas the grout buffer was positioned in the upstream of the dam and linked to the impervious core of the dam by a compacted berm, while in some other areas the grout buffer was positioned just underneath the core of the dam.

Figures 4 to 7 present a cross section of the dam in the described situations with plots of the elevation of the compacted fill, of the water level in the reservoir and measured pore pressures in hydraulic piezometers.

As can be seen in the figures, operational water levels are to reach elevation 50.0 m so more data need to be collected and will be published by the authors as soon as available.

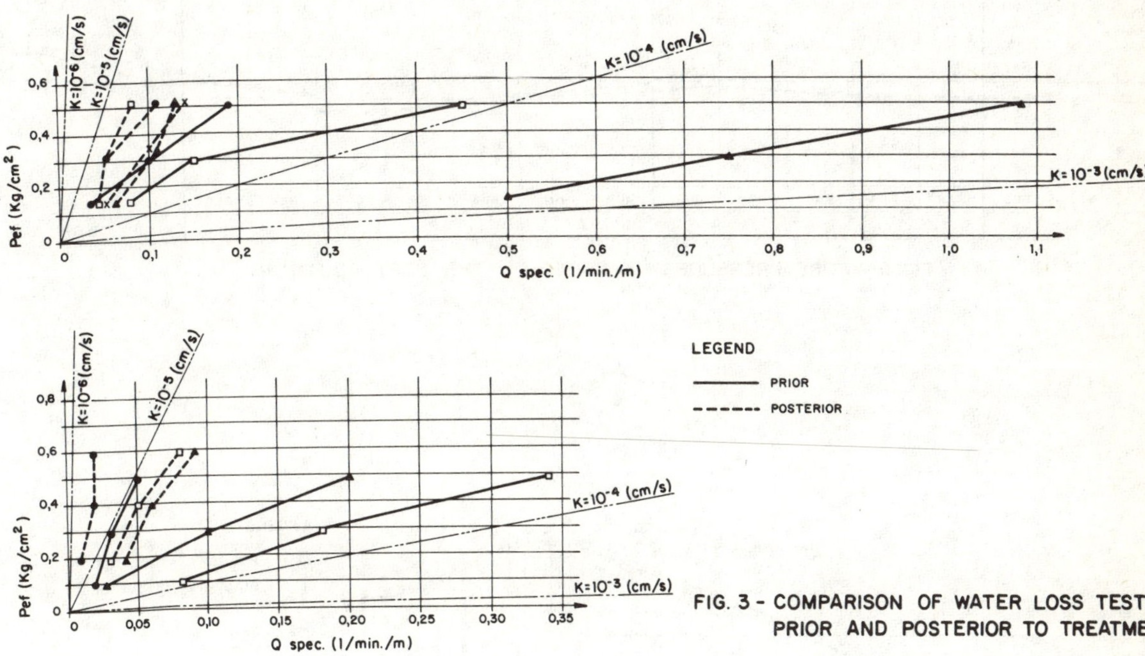

FIG. 3 - COMPARISON OF WATER LOSS TESTS PRIOR AND POSTERIOR TO TREATMENT

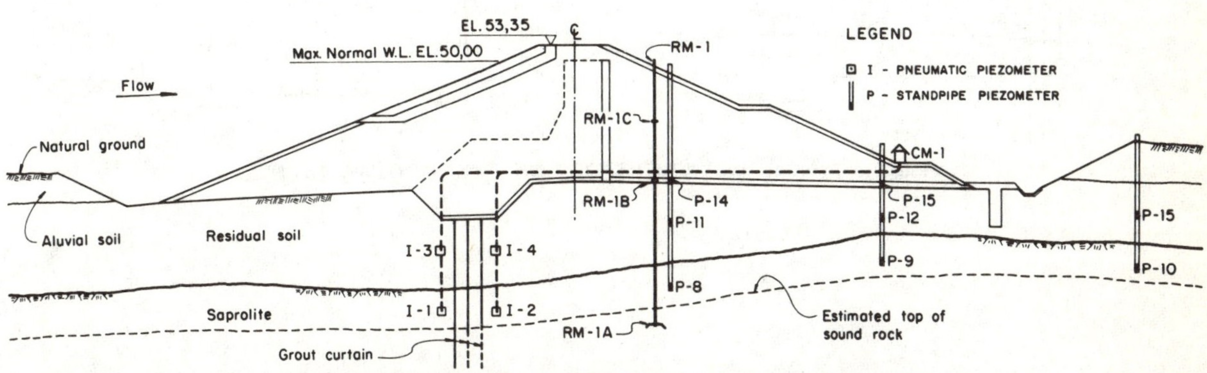

FIG. 4 - TYPICAL INSTRUMENTATION ON THE LEFT ABUTMENT

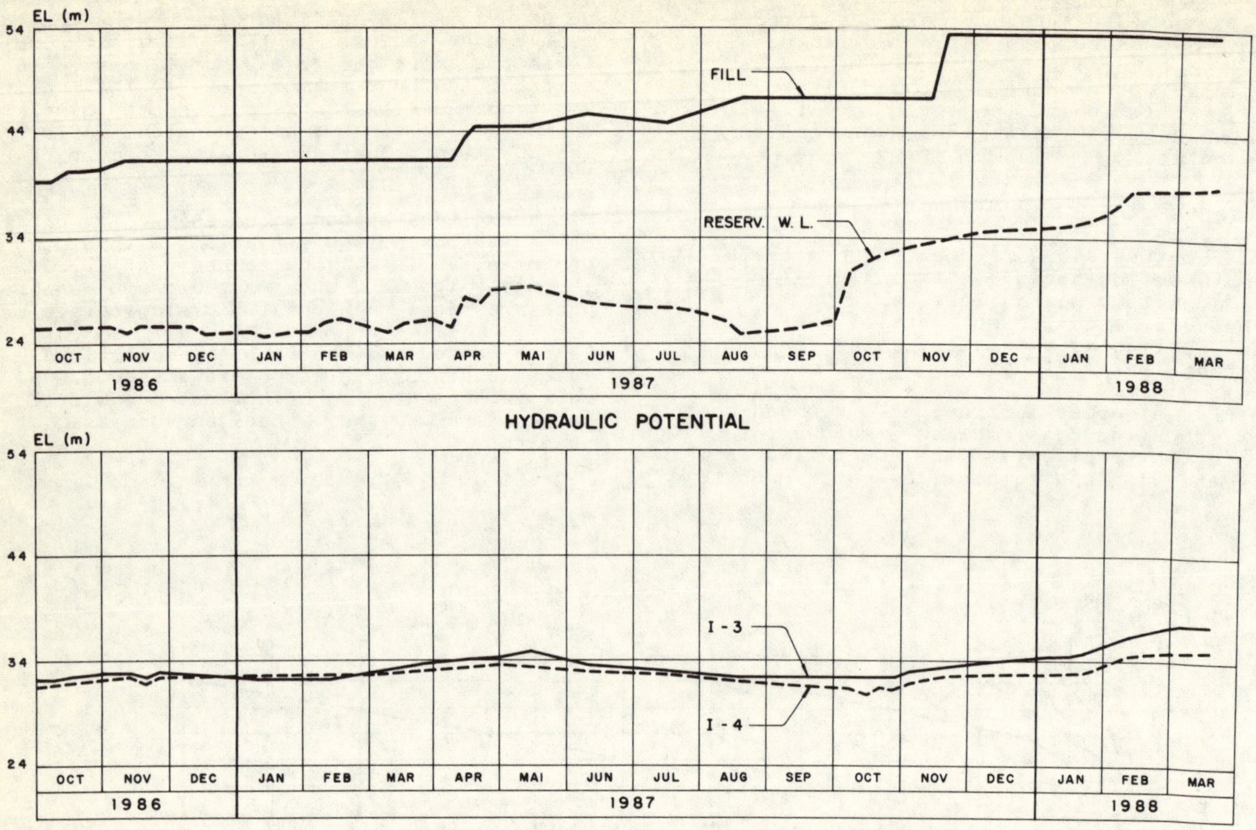

FIG. 5 - PORE PRESSURES MEASURED ON THE LEFT ABUTMENT

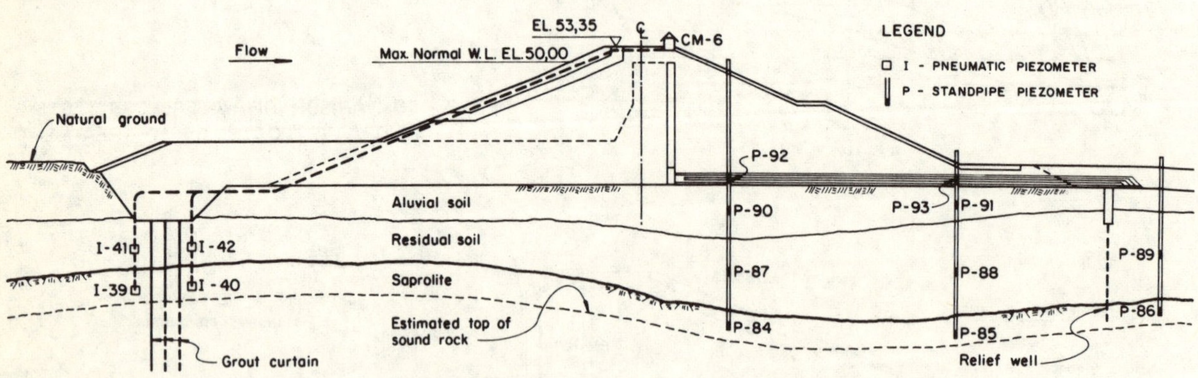

FIG. 6 - TYPICAL INSTRUMENTATION ON THE RIGHT ABUTMENT

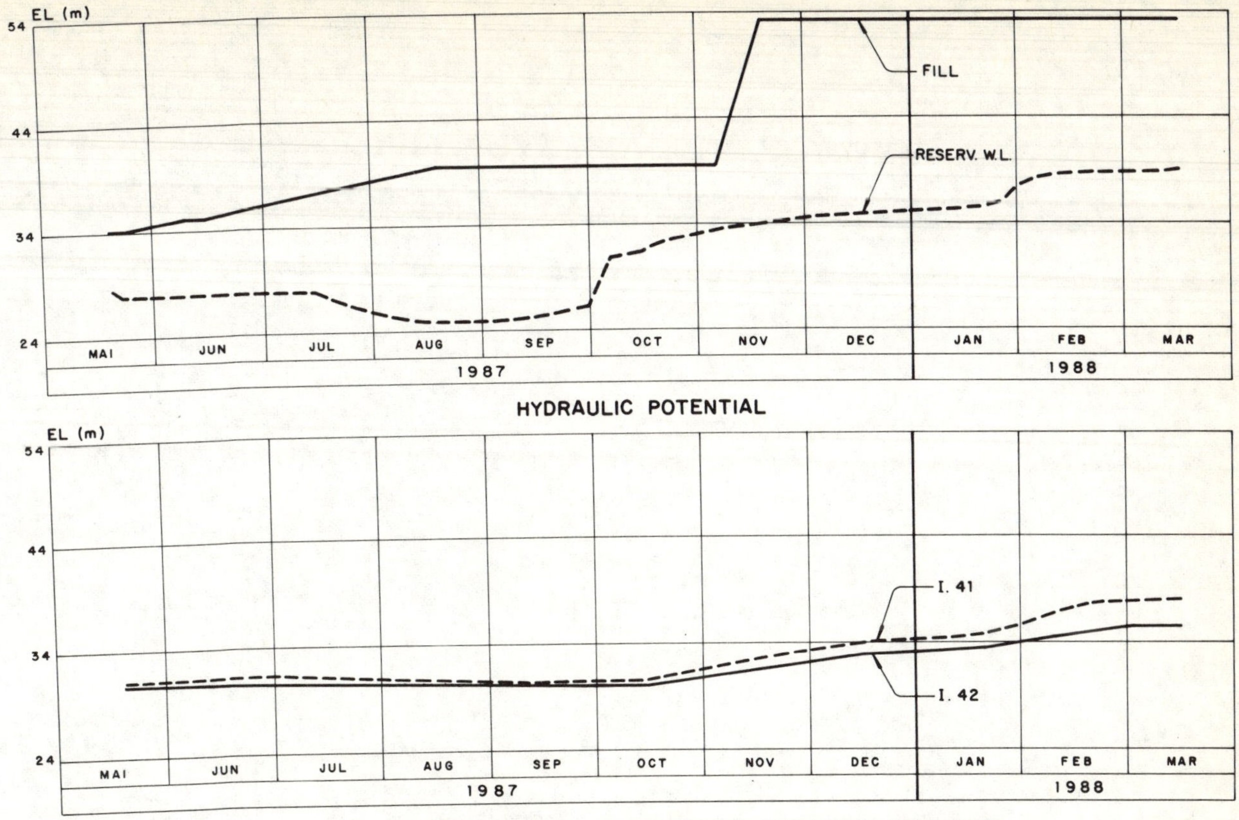

FIG. 7 - PORE PRESSURES MEASURED ON THE RIGHT ABUTMENT

6 CONCLUSIONS

The data collected so far as well as situ inspections demonstrate the validity of the use of the discussed treatment to homogenize and diminish the permeability of a residual soil with a net of open canaliculae. It is our belief that this type of solution should be pursued in equivalent situations, as well as developed for other geological environments in which this concept of treatment may be thought.

7 ACKNOWLEDGEMENTS

The authors wish to thank Eletronorte (Owner) and Construtora Andrade Gutierrez (Constructor) for all the support; Prof. Victor F.B. de Mello for his ever present orientation; and to all members of the technical team of Monasa-EngeRio involved in the design of the Balbina dam.

REFERENCES

Babouchkine in Bogomolov, G.V., 1965. Hydrogeologie. Ed. de la Paix, Moscow-URSS.

Bjerrum, L. and Andersen, K.H., 1972. In-situ measurements of lateral pressures in clay. Proc. Fifth European Conference SMFE, Madrid-Spain.

Bjerrum, L. et al, 1972. Hydraulic fracturing in field permeability testing. Geotechnique vol. 22.

de Mello, L.G.F.S., Franco Filho, J.M.M., and Alvise, C.R., 1987. O tratamento de solos residuais condicionados por canalículos: um caso de obra. In portuguese. Conferência Ibero-Americana sobre Aproveitamentos Hidráulicos, Lisbon-Portugal.

de Mello, V.F.B., 1977. Reflections on design decisions of practical significance to embankment dams. Geotechnique vol. 23.

Haimson, B. and Fairhurst, C., 1967. Initiation and extension of hydraulic fractures in rocks. Society of Petroleum Engineers Journal.

Kennard, R.M., 1970. The measurement of soil permeability in-situ by the constant head test. PhD thesis, University of London.

Morgenstern, N.R. and Vaughan, P.R., 1963. Some observations on allowable grouting pressures. Geotechnique.

Remy, J.P.P. et al, 1985. Choice of the foundation treatment of the Balbina Earth Dam. 15th International Conference on Large Dams, Lausanne-Switzerland.

Sathler, G. and Camargo, F.P., 1985. Tubular cavities in the residual soil of the foundation of Balbina Earth Dam - Amazon - Brazil. 15th International Conference on Large Dams.

Embankment test in Mukakuning Dam Project

D.DJARWADI, PT Bangun Tjipta Sarana, Indonesia

ABSTRACT: This paper deals with the embankment test of tropical soil as embankment material for dam construction. Field and laboratory tests are undertaken on the compacted embankment materials in order to ensure the soil characteristics after compaction meets the design parameters and specification. Three types of compaction machine are considered in the embankment test. A series of embankment test are carried out in order to determine the lifting tickness, number of passes, and type of compaction machine at economical cost to meet the required specification.

1 INTRODUCTION

Mukakuning Dam Project on the Mukakuning river, at the centre of Batam Island of Indonesia, some 30 miles from Singapore, is constructed for the purpose of supplying drinking water for new town of Batam Centre, at 15 km downstream of the dam site. The works were started on October 1985 and are due to completed on April 1988.

The project mainly consists of dam, spillway, waterway, intake tower and bottom outlet facility. The dam has 30 metre in height with a crest length of 380 metre, and the mean slopes upstream and downstream is 3H/1V. The dam is a zoned earthfill dam with total voluma are 750,000 cubic metres. Fig 1 shows the typical cross section of the dam and types material used in construction.

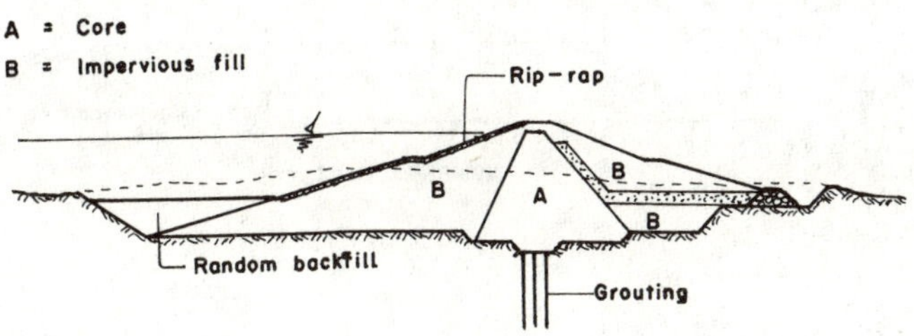

Fig.1 Typical cross section of dam

Impervious core fill (material A) with it's volume 150,000 cubic metres is belonging to the CH group in the Unified Soil Classification System, while impervious fill (material B) in upstream and downstream of core is belonging to the CL group in the same system. These types of soil can be found in the borrow pit surroundings the dam site within 5 km in distances.

2 TECHNICAL SPECIFICATIONS

The maximum allowable thicness of layer when placing the fills are 30 cm, and the maximum permissible grain size is 52.4 mm. The minimum degree of compaction required is 95% of maximum compaction determined by the Modified Proctor Compaction Test reference to AASHTO T 180-74 method C.

When placing embankment materials in dam, the moisture contents shall be within ± 2% of the optimum moisture content determined by Modified Proctor Compaction Test, AASHTO T 180-74 method C. Permeability of the embankment materials after compaction shall not less than $1X10^{-7}$ metre/sec. The gradation of the embankment materials suggested fall within the limit shown in table I.

Table I. Gradation of embankment materials

Grain size (mm)	Procentage Passing by weight.	
	Material A	Material B
25,40	100	100
10	90 - 100	90 - 100
5	85 - 98	85 - 98
1	83 - 96	80 - 95
0.50	80 - 95	75 - 90
0.1	70 - 90	60 - 85
0.02	55 - 88	40 - 80
0.002	20 - 45	10 - 40

During embankment test, field laboratory tests should be taken as directed in the specification (Batam Island Development Authority.1983). References of the tests are following to AASHTO (American Association of State Highway and Transportation Officials 1974) and/or ASTM (American Society for Testing Materials.1986) standards.

Moisture contents in the embankment test, and inspection before embankment in the construction of dam is determined using Speddy Moisture Tester which the values obtained by this test should be added by 2% to meet the moisture content determined by Moisture Content Test reference to ASTM D 2216-71. The addition of 2% are utilized based on the result of previous comparison tests between Speddy Moisture Tests and Moisture Content Test determined by ASTM D 2216-71. The description of tests and references shown in table II.

Table II. References for laboratory tests

DESCRIPTION OF TEST	DESTINATION	
	ASTM	ASSHTO
MOISTURE CONTENT	D 2216 - 71	
SPECIFIC GRAVITY	D 854 - 74	T100 - 74
LIQUID LIMIT	D 423 - 74	T89 - 74
PLASTIC LIMIT	D 424 - 74	T90 - 74
GRADATION	C 136 - 46	T27 - 74
HYDROMETER	D422 72	T88 - 72
THIN WALLED TUBE SAMPLING	D 1587	T207-74
MODIFIED PROCTOR COMPACTION	D1557 - 66	T180 - 74
CONSOLIDATION TEST	D2435-70	T216-74
TRIAXIAL TEST	D 2664-67	T226-68
PERMEABILITY TEST	D 2434-68	T215-70
DIRECT SHEAR TEST	D3080-72	T236-72
FIELD DENSITY (SAND CONE)	D 1556 - 6a	T 191 - 61

Specification also recomended a pneumatic tyred roller with operating weight of 15 tons, vibratory padfoot roller of 11 tons and smooth drum roller of 11 tons in operation weight, travelling at speed under 10 km/hr for construction the dam. The definitive compaction machine will be decided based on the embankment test result.

3 PRELIMINARY ATTEMPS

Borrow area investigation report as given in the Tender Document gives an indication of the soil characteristics. A series of 5.00 metres depth pits in 20.00 metres in distances are carried out in order to study the soil profiles at a greater depth for soil identification and selected samples for determineation of optimum moisture content, compaction, and sieve analyses.

Laboratory tests for triaxial, direct shear, consolidation, and permeability are also carried out for comparison with the design values and embankment tests result. The gradation boundary curve obtained from laboratory tests slightly different with those described in the specification for material A and material B as shown in fig 2.

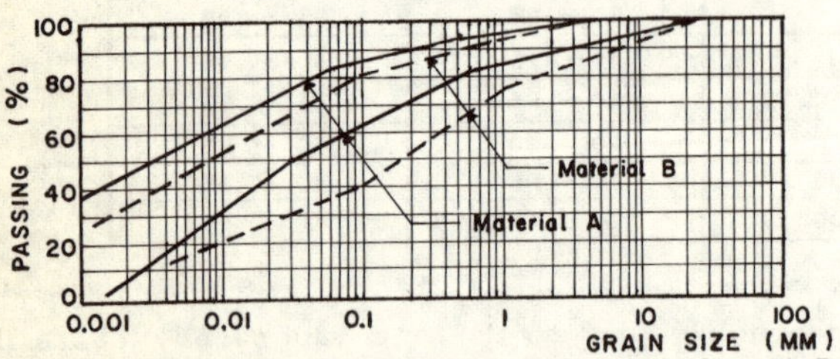

Fig 2. Gradation curve for material A and B

Specific gravity of embankment material varied from 27.5 kN/m^3 to 29.0 kN/m^3. The maximum dry density determined by Modified Proctor Compaction Test according to AASHTO T 180-74 method C were in the range of 17.0 kN/m^3 to 17.8 kN/m^3 for material A and 18.5 kN/m^3 to 19.3 kN/m^3 for material B, while the corresponding optimum moisture contents were 18.00% to 20.00% for material A, and 31.5% to 16.00% for material B.

Atterberg tests shows that plasticity index of material A and B are in the range of 28 - 35 and 19 - 25 respectively. It was confirmed that according to Mitchell and Sitar (1982),embankment material are classified as lateritic soil.

Fig 3 shows the position of embankment materials in Casagrande Plasticity Chart.

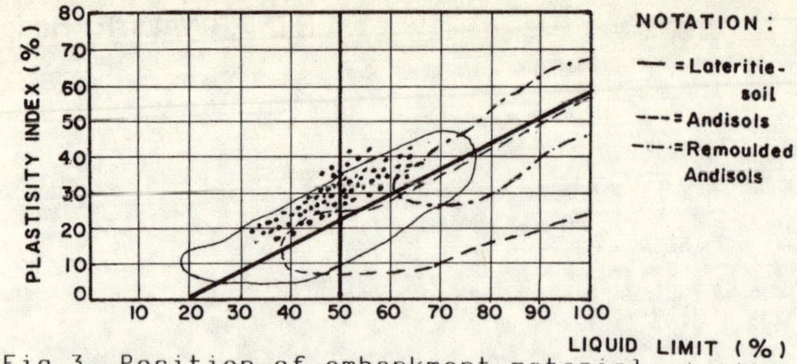

Fig 3. Position of embankment materials in the Casagrande Plasticity Chart after Mitchell & Sitar 1982

The natural humidity of soil was 25% to 32% in the condition covered by vegetation. In order to lowering the moisture contents, clearing, grubbing and stripping of vegetation and organic soil at 50 cm depth is carried out earlier before stock pilling was started. In site clearing borrow area was carried out 1 month before stock-pilling.

Since a 11 tons padfoot vibratory roller already mobilized in the job site and a 15 tons pneumatic tyred roller is available on the rental basis, preliminary embankment test is carried out using 2 machines for comparison and decision the definitive machine for construction. Compaction by 2 types of compaction machice are undertaken on the soil which has similar characteristic in 30 cm of thickness by 6, 8 and 12 passes. The compaction degree shows that field density after compaction has a little difference as shown in fig 4.

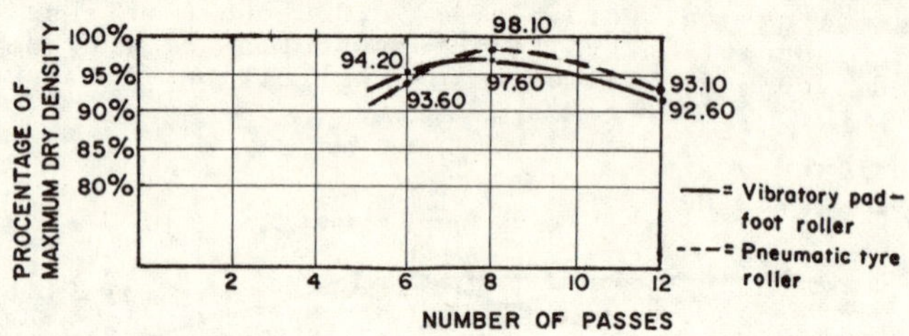

Fig 4. Compaction result of preliminary test

Based on the preliminary embankment test result, a vibratory padfoot roller with operating weight of 11 tons compacted of 30 cm in thickness of layer is adopted for compaction in construction Mukakuning Dam Project.

4 TEST PROCEDURE

Based on the preliminary embankment test, the detail compaction test will using a padfoot vibratory compaction machine. Embankment test are carried out on levelled ground of 30 X 25 square metres. devided into 5 strips of 6 X 52 square metres. A spreading layer of 30 cm were then placed and compacted at 4, 6, 8, 12 and 16 passes respectively. Another layer of the same thickness were then spread on the previous one and compacted in the same way. Embankment test stopped after reached 4 layers, and samples were taken from the last layer at the depth 15 cm from the surface. Sampling was taken using thin walled tube reference to AASHTO T 207-74. Number of samples taken for laboratory test and test description are shown in table III.

Table III. Test description and number of sample

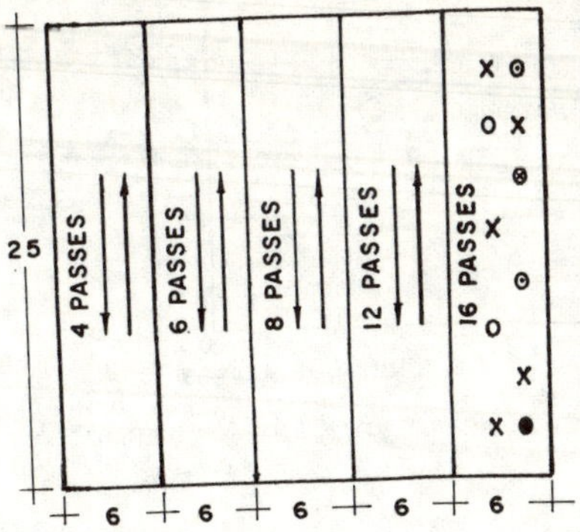

DESCRIPTION	CODE	NUMBER OF SAMPLE
Field density test	X	5
Gradation test	O	1
Permeability test	●	1
Triaxial test	⊙	2
Direct shear test	⊗	1
Consolidation test	⊕	1

5 TESTS RESULT

Although specification stated that the moisture content at works shall fall within + 2% from the optimum moisture content, and also before test the moisture content has been checked, for large volume it was still possible that a small part of soil has moisture contents more or less than permissible. Fig 5 shows the relation of dry density and moisture content in various condition of compaction obtained from embankment test.

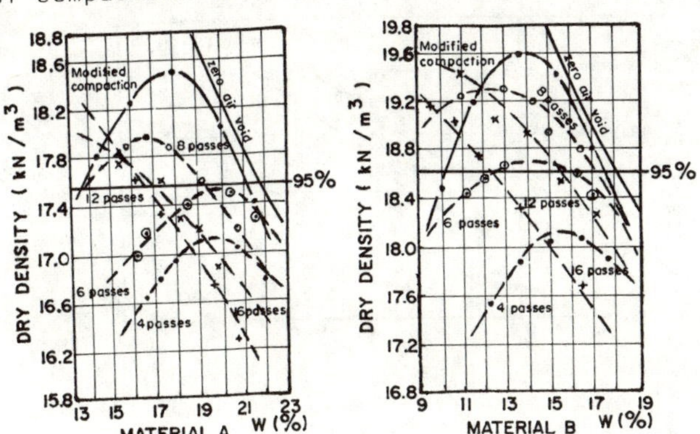

Fig 5. Compaction test values in various moisture contents

It shown that in the wet side compaction executed in the moisture contents exceed from the specification will produce the dry density less than specification at 95% maximum dry density, but on the dry side it can be considered to add number of passes or watering the embankment material to achieve the degree of compaction as specified, especially on the long dry season during construction, that moisture contents of embankment materials ussually drop untill 10% to 15%. From economical view that is reasonable to add the number of passes than to watering the embankment materials, but extra laboratory tests as permeability test, consolidation test,shear test, triaxial test and disperssivity test should be taken at the condition of moisture content below optimum moisture content.

In the permeability tests, it shown that the values are correspond to the compaction energy given to the soil during the embankment test. Fig 6 shows the relation of dry density, number of passes and permeability of the embankment materials.

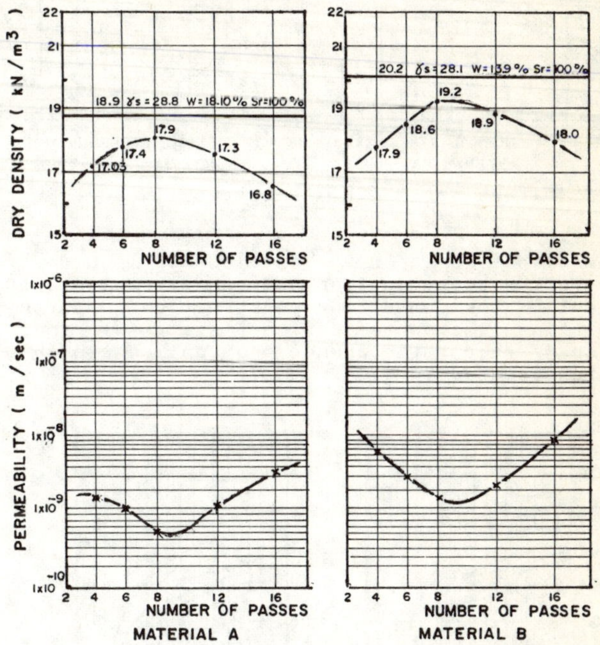

Fig 6. Permeability of embankment material related to compaction degree

In triaxial tests, the consolidated undrained on partly saturated and undrained on partly saturated tests are adopted (Bishop & Henkel.1962), while direct shear tests are undertaken in condition of consolidated undrained (Kezdi.1979). Figs 7 and 8 shows the result in relation of shear stress and normal stress for triaxial (CU) tests and direct shear (CU) tests condition respectively.

Notation: 4 PASSES, 6 PASSES, 8 PASSES, 12 PASSES, 16 PASSES

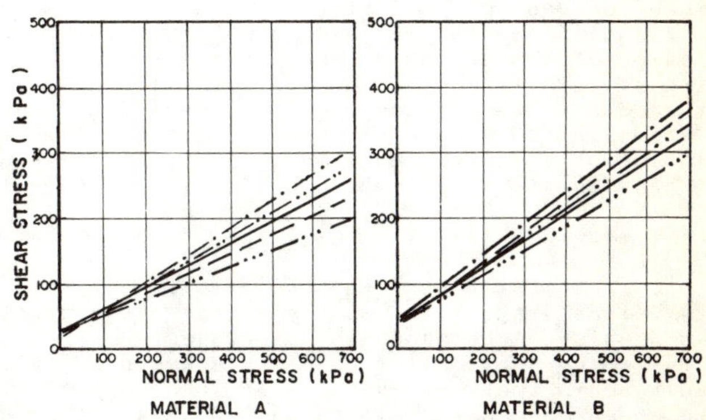

Fig 7. Triaxial (CU) tests result

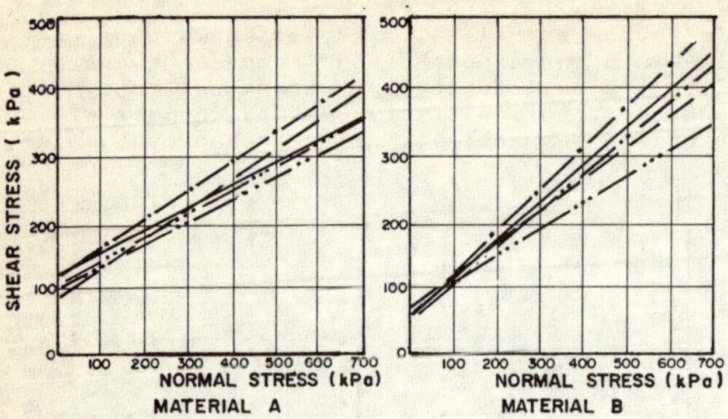

Fig 8. Direct shear (CU) tests result

Consolidation tests are carried out in order to anakyse the consolidation and consolidation settlement of the dam during construction and after completion. The consolidation tests result in relation of void ratio change and applied pressure are represented in fig 9.

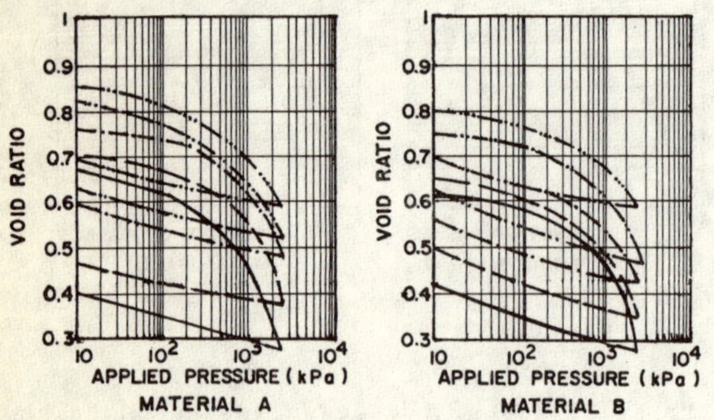

Fig 9. Consolidation tests result

6 DISCUSSION ON THE TESTS RESULT

6.1 Compaction

Fig 6 shows that the maximum dry density in relation with the optimum moisture content are achieved on the compaction by 8 passes. Compaction degree are increasing when embankment materials compacted by 4, 6 and 8 passes. It is believed that the increasing of compaction degree are due to volume change (compression) in the soil mass during compaction. But after reached it's peak, compaction degree are decreasing when compacted by 12 and 16 passes. It is supposed that the decreasing of compaction degree are caused by increasing of void ratio in the soil mass. The evidences appear in the series of consolidation tests as shown in fig 9 support the supposition. The relation of volume change and compaction degree in the embankment test shown in fig 10.

6.2 Permeability

According to the tests result in fig 6, permeability values are satisfy mostly at the above specified K value, eventhough it was compacted by 4 passes which has smaller compaction energy. Tests result shows that permeability are decreasing when embankment materials compacted by 4, 6 and 8 passes, but increasing when compacted more than 8 passes, as shown in compacted by 12 and 16 passes. Material A which belongs to the CH group in the CH group in the Unified Soil Classification System has smaller permeability values than material B which belongs to the CL group in the same classification.

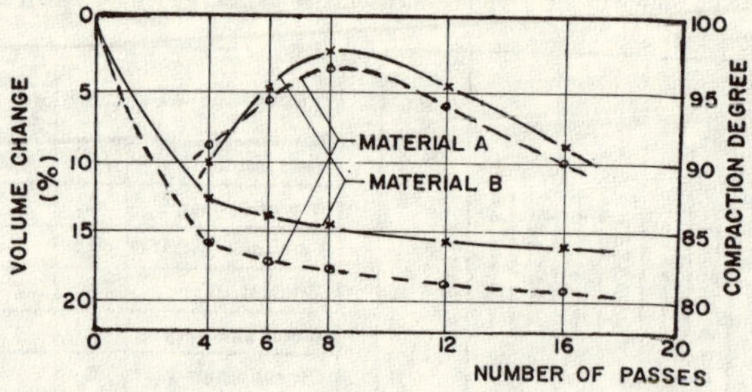

Fig 10. Relation of volume change and compaction degree

6.3 Internal anle of friction and cohesion

Triaxial tests and direct shear tests result shows a phenomenon that if dry density of embankment materials increase, angle of internal friction become larger, and cohesion are decrease. Fig 11 shows the relation of dry density and internal angle of friction obtained from direct shear tests in the laboratory and embankment tests The figure represent more than 100 tests result (Djarwadi.1986).

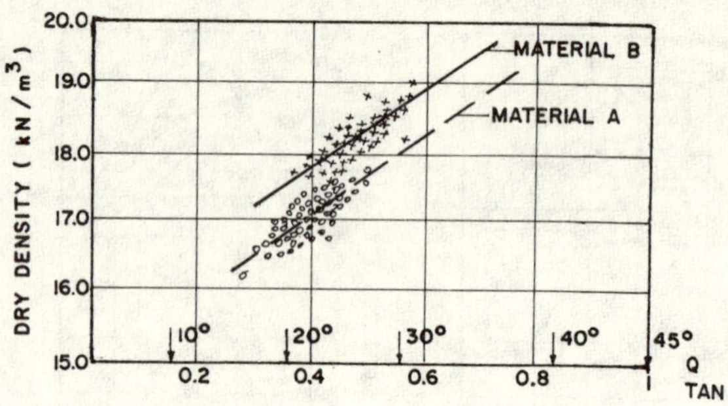

Fig 11. Relation dry density and angle of internal friction embankment materials

Based on the tests result, the internal angle of friction material A and B are defined at 18° and 24° respectively, while the cohesion for material A and B are defined at 30 kPa and 20 kPa respectively. These values are greater than indicated in the design calculation which given by the Engineer. Using the internal angle of friction and cohesion obtained from embankment test for checking the stability of the dam, it is confirmed that the mean slopes of 3H/1V for both upstream and downstream gives the Factor of Safety more than 1.90 in various condition, during construction, after completion and in operation.

6.4 Consolidation

Tests result indicated that the different energy

compaction given to the embankment materials will create different initial void ratio (e_o) in the consolidation tests, where the larger energy compaction will create bigger initial void ratio. It was confirmed that different void ratio before and after tests has a smaller values for compaction which have larger energy. This can be concluded that Compression Index (C_c) is decrease in corresponds the increasing compaction energy.

Using Compression Index obtained from consolidation tests taken from embankment tests by 8 passes, the total settlement of the dam is smaller than indicated in the design.

Considering the tests result from embankment tests, the compaction using vibratory padfoot roller with 11 tons opertion weight, compacted 30 cm thickness of embankment materials in 8 passes at the speed 8 km/hr is utilized in the construction Mukakuning Dam Project.

7 CONCLUSION

Embankment test which ussually undertaken before construction the main work is very important, since the results will reflecting the performances of the main work. From the test results described above the following may be pointed out:

1. Test results reflected that the lateritic soil in tropical region are satisfactory enough for embankment material in earthfill dams.
2. Compaction embankment material of 30 cm in thickness by 8 passes of 11 tons vibratory padfoot roller in operation weight can be utilized in construction.
3. In construction earthfill dam using lateritic tropical soil, the specification of compaction at least 95% of the Modified Proctor Compaction Test determined by AASHTO T 180-74 is easy achieved.
4. From the previous experience in construction of earthfill dams, and embankment test in Mukakuning Dam Project, a vibratory padfoot roller can be used effectively for compaction of clay in earthfill dams.

REFERENCES

AASHTO.1974. Standard Specifications for Transportation Materials and Method of Sampling and Testing, Part II. Method of Sampling and Testing. Washington D.C: AASHTO.

ASTM. 1986. Annual Book of ASTM Standards.Volume 04.08/Soil and Rock, Building Stones.Philadelphia: ASTM.

Batam Island Development Authority.1983. Technical Specifications for Construction Mukakuning Dam Project (in Indonesia). Jakarta: BIDA.

Bishop,A.W. and Henkel,D.J. 1962. The measurement of soil properties in the triaxial test.London: Edward Arnold.

Djarwad,D. 1986. Report on embankment test of Mukakuning dam project (in Indonesia) Batam: unpublished paper.

Kezdi,A. 1979. Handbook of Soil Mechanics,Vol 2 Soil testing, p 137-144.Amsterdam: Elsevier.

Mitchell,J.K. and Sitar,N. 1982. Engineering Properties of tropical residual soils, Proceedings Conference on Engineering and Construction of tropical and residual soils, ASCE: 30-57.

The use of non-traditional tropical and residual materials for pavement construction A review

M.D.GIDIGASU, Council for Scientific and Industrial Research (CSIR), Building and Road Research Institute, Kumasi, Ghana

ABSTRACT. Experiences in the tropics over the years have shown that paving materials acceptance specifications developed for temperate zone materials and environments sometimes offer limited application for pavement construction in the tropics.

This is so because materials for which the standard specifications were tailored are not always locally available in the tropics.

The alternative line of action is to define how best to utilise the available local materials to produce good motorable pavements under specific climatic conditions.

This paper has attempted to review some useful contributions to the testing and evaluation of typical tropical natural and lateritic rock aggregates, lateritic gravels and saprolitic soils for pavement construction in the tropics. The use of engineering geological and pedological concepts and the need for knowledge of local climatic indicators are considered important in dealing with these materials in tropical highway geotechnical engineering.

The literature study has emphasized the need to develop local acceptance specifications and construction control procedures and execution of work in relation to the environmental factors in order to prolong the life of all weather tropical road pavements.

Some experiences gained from laboratory testing and construction control and field pavement performance studies are also briefly discussed. For example, the variability of properties of some subgrade, sub-base and base lateritic soils necessitated statistical approach to defining significant paving materials selection properties.

The study has also shown that some paving materials that would have been rejected based upon standard laboratory classification and identification tests results alone, may through proper control of construction and execution of work give satisfactory performance of pavements constructed with them in specific climatic and drainage environments.

Apparently, the basis for rejecting or accepting many tropical lateritic and saprolitic paving materials should not be the results of laboratory tests alone but also the results of construction control and field performance studies of new and existing pavements.

The paper is a contribution to developments in tropical highway geotechnical engineering.

1. INTRODUCTION

The development of specifications for temperate zone paving materials has been the results of tedious and long term laboratory and field studies. The process has been a combination of theoretical and practical definition of optimum grading characteristics of materials that would yield the highest compaction density (e.g. Fuller and Thompson, 1907; Zemour and Durrier, 1966) and the relation between the fines, gravel contents and plasticity and the desired grading curves (e.g. Dunn, 1966).

Another aspect of selecting paving aggregates which has received serious attention is the evaluation of the influence of deleterious constituents on the performance of natural gravel pavements (e.g. Venters and Lewis,1954). The strength and breakage behaviour of aggregates during construction and under traffic loads have also been extensively studied both in the laboratory and under pavement construction conditions (e.g. Shelburne,1939, 1941; Shergold, 1948; Shergold and Hosking, 1963; Melville,1948; Road Research Laboratory,1952; Dunn,1966; Day, 1962; Farrah and Thenoz, 1960). The contributions of geological, petrographical, physical, chemical and mineralogical factors on the laboratory and field characterisation of paving gravels and aggregates has also received serious studies (e.g. Hartley, 1974; Lee and Kennedy, 1975; Reed,1967; Scott, 1955; Wylde, 1975, 1976). The results of these and other studies have resulted in the formulation of specifications for gravel and aggregate paving materials in different temperate zone countries (e.g. Zemour and Durrier,1966).

The development of paving materials specifications for tropical materials has not resulted from systematic European and North American type of long term laboratory and field performance studies. In fact the temperate zone paving materials acceptance specifications were in most cases transferred without local assessment for application in many tropical and sub-tropical countries. For example, the gravel acceptance specifications used until a few years ago in Ghana was the one adopted in Great Britain (e.g. Road Research Laboratory, 1952).

The use of non-traditional tropical lateritic and saprolitic materials in low cost pavement construction has posed many interesting problems. While the theme of the 2nd International Conference on Geomechanics in tropical soils, following the first one held in Brazil is to be confined to lateritic and saprolitic soils some light is thrown on the difficulties involved in utilising other equally abundant and unpredictable tropical and residual paving materials. For example, collapsing residual and transported materials constitute problem paving materials in different parts of the tropical areas (e.g. Knight and Dehlen, 1963). Similarly, the salt bearing soils are problem roading materials in the Mediterranean areas and extensive studies on these materials have resulted in developing some useful guidelines relating to their utilization (e.g. Fookes and French, 1977). Failures resulting from the use of natural aggregates containing soluble salts has also been reported by Blight (Blight, 1976). Pavement performance on expansive soils has also been a source of concern in many parts of the tropics. For example, cases of heave of pavements on this soils have been extensively reported (e.g. Williams,1965). As regards lateritic and saprolitic paving materials which form the theme of the conference, a lot of published information is available scattered in various sources. Attempt is made to summarise some relevant information and to highlight developments made in pavement construction practice in tropical and sub-tropical regions using these materials.

For the purpose of this paper the terms "lateritic" and "saprolitic" soils are the same as those defined in the progress report of the International Technical Committee on Tropical Soils of ISSMFE (Technical Report, 1985) and by Gidigasu (1976).

To enable us appreciate the nature and geotechnical behaviour of lateritic and saprolitic soils the decomposing rock should also be considered to complete the chain of these materials which range from decomposing rock, through saprolitic soils (finegrained soils retaining parent rock structure), lateritic gravels to lateritic rocks (or crusts). Apparently, the discussion would cover all the four major stages of the reddish tropically weathered materials. This should be so because it has been shown (e.g. Little,1969; Lohnes et al, 1971, 1976; Gidigasu,1974,1976; Brand and Philipson,1985) that lateritic and saprolitic materials constitute a chain of materials ranging from decomposing rock to lateritic (pedogenic) rock. These materials differ from one another in many respects; compositionally (physically, chemically, structurally and mineralogically) and methods of testing and evaluating each group or grade of materials have been shown to be different in many respects (e.g. Gidigasu, 1976). Starting from the decomposing natural rock aggregates, considerable amount of failures have been reported regarding their use as paving materials. Available information indicates that the origin of rocks, compositional factors of the soils and environmental factors are generally responsible for most failures of pavements (e.g. Scott, 1966; Wylde,1976; Weinert, 1964). Failure or stress development in the tropical low cost pavements is a major problem and constitutes a major constraint to the development of good network of low to medium trafficked roads in many tropical countries in the face of shortage of financial resources to build modern pavements, though in most cases traffic weight might justify this alternative course. Maintenance of these pavements are generally high by developing countries standards.

2. ORIGIN AND NATURE OF TROPICAL LATERITIC AND SAPROLITIC PAVING MATERIALS

The term non-traditional materials is used in a broad sense to include tropical and residual soils whose geotechnical behaviour deviate from the expectations of the traditional temperate zone paving materials for which the standard acceptance specifications have been tailored. In West Africa the only materials that are close to these materials are the weathered quartz rich, feldspar poor igneous rocks and transported stream quartzitic gravels. The origin and favourable highway geotechnical characteristics of these well-graded, durable, low plasticity nature of these materials make them to almost conform to the ideal temperate zone paving gravels. The detail highway geotechnical properties of these materials have been discussed elsewhere (e.g. Clare and Beaven, 1962; Arulanandan,1969; Gidigasu,1980). All other tropical materials of doubtful highway geotechnical behaviour are grouped here as non-traditional paving materials.

To be able to make sense out of a plethora of information and conflicting data on the reddish tropically weathered paving materials one need to briefly look at the significant factors which control the processes of their formation and the key environmental (especially climatic) indicators that shape the evolution of the materials to the state where they are about to be won and used for pavement construction. In a broad sense, the influence of environmental factors including geomorphology, climate, drainage conditions and how the particular mode of soil forming process determine the compositional factors such as physical, physico-chemical, chemical and mineralogical composition have been discussed on macro geographical levels by various authors (e.g. Bunting, 1960; Kellog, 1941; Peltier,1950; Strakhov, 1967; Clare, 1960; Clare and Beaven, 1962; Loughnan,1969; Gidigasu,1976). Environmental controls apparently influence the trend of weathering, evolution and nature of construction materials formed and available under varied tropical conditions. On the micro geographical scale temperature profiles, (e.g. Richards, 1969); moisture profiles (e.g. Aitchison, 1952) and variation of soil strength with depth (e.g. Aitchison, 1952) are significant interdependent factors; and these considerably influence the nature of the materials and hence the mode of their selection, evaluation, pavement design, construction and subsequent pavement performance patterns. For example, within Ghana basic geotechnical characteristics established during two-year studies in the various climatic zones have established that pavement engineering technology have to vary from one climatic zone to the other; this has been confirmed by subsequent studies (e.g. Gidigasu and Mate Korley, 1984). The formation of lateritic and saprolitic soils in West African condition generally leads to the leaching away of silica and other basic constituents and the enrichment of the residual materials either predominantly by alumina (Al_2O_3) or iron oxide (Fe_2O_3) or by both in varying proportions. For example, the chemical, mineralogical and geotechnical implications of the lateritisation processes reported from Ghana and Puerto Rico have been discussed by Lognes et al, (1976) and Gidigasu (1976).

In the light of the complexity, especially in terms of variability of compositional factors of lateritic and saprolitic soils a very comprehensive programme is generally required for evaluating and utilising them in earthworks engineering if some understanding of their behaviour in engineered structures is to be forthcoming. For example, a flow chart for compositional analysis of these materials (Mitchell, 1976) seems very applicable based upon available literature on the problems of winning, testing and utilising these unusual soils in tropical geotechnical engineering (e.g. Gidigasu, 1976).

3. GEOTECHNICAL CHARACTERISATION OF FINEGRAINED LATERITIC AND SAPROLITIC SOILS

The success of laboratory and field geotechnical evaluation of finegrained lateritic and saprolitic soils begins with some appreciation of the engineering geology of the soils. For example, Loubser (1967) threw considerable light on the origin of schistose rocks, shales and phyllite and through this one could speculate the potential laboratory testing, construction and performance problems that one may encounter in dealing with these materials as subgrade formations and as paving materials. The main problem in evaluating finegrained lateritic and saprolitic soils lies in the difficulty of getting reproducible test results using traditional pretest sample preparations and laboratory testing procedures. For example, pretest preparations and testing procedures may considerably influence the particle size distribution, Atterberg limits and compaction characteristics of lateritic and saprolitic soils (e.g. Hirashima,1948; Newill,1961; Sherwood, 1967; Townsend et al, 1969, 1971; Gidigasu,1976, 1980a, 1980c, 1980d, 1987; Gidigasu and Yeboa, 1972). However, through careful selection of pretest sample preparation methods and testing procedures it has been possible to classify typical lateritic and saprolitic soils over various parent rocks including phyllite and mica rich granite in Ghana (e.g. Gidigasu, 1980d, 1980f). The process of profile development also influences the morphology of finegrained lateritic and saprolitic soils. The three zones of lateritic soil profile, nemely the lateritic, mottled and pallid zones (e.g. Loughnan,1969; Gidigasu,1976) have characteristic differences in terms of texture, structure, chemical and mineralogical composition and geotechnical properties (e.g. Loughnan,1969; Gidigasu,1976, 1980d). For example, the compaction characteristics also do vary with depth in typical residual soil profiles even in the same soil series in the same climatic zone (e.g. Gidigasu, 1980d). The variation of bearing strength expressed in terms of the CBR and its variation under pavement with time is also a common phenomenon (e.g. Arulanandan,1969). The swell characteristics are also influenced by the morphological variations and the degree of compaction and moulding moisture content (e.g. Gidigasu,1980d). One notable characteristic of finegrained and even gravelly lateritic and saprolitic soils is that excessive compaction tends to destroy the soil structure, and induce lower bearing strength (e.g. Hammond,1970; Gidigasu,1980a). The structural sensitivity of these soils has called for studies aimed at evaluating the possibility of stabilising them to reduce swell, increase bearing strength (e.g. Gidigasu and Mate Korley, 1982) as well as to control their response to soaking effects. The main conclusion to be drawn from the foregoing is that these materials have to be handled with open mindedness using a broad spectrum of traditional and locally developed significant testing procedures that enable engineers to predict the behaviour of the materials as subgrades and as materials for the structural pavement layers.

Increased studies aimed at correlating results of field performance and laboratory studies are urgently required to promote judicious use of these materials in tropical pavement construction.

4. GRAVELLY MATERIALS AND DECOMPOSING ROCKS

Genetically many types of paving coarse materials abound in the tropics. The genetic characteristics of lateritic gravels and stones identified in Ghana are discussed elsewhere (e.g. Gidigasu, 1976). Identification problems noted with finegrained lateritic and saprolitic soils are also common to fines in the coarse materials. The degree of desiccation, clay mineralogy and the cementing effects of salts (Al_2O_3 or Fe_2O_3) have significant influence on the grading, Atterberg limits and strength of some lateritic soils (Lohnes et al, 1971, 1976). For detailed discussion on this, one could refer to other works relating to the problem (e.g. De Graft-Johnson, Bhatia and Hammond,1972; De Graft-Johnson, Bhatia and Gidigasu, 1969; Gidigasu, 1976; Millard, 1962; Nanda and Krishnamachari, 1952; Philip, 1952). Careful choice of pretesting preparation of samples and testing procedures are again required to obtain reproducible results. Lateritic gravels identified in Ghana have been grouped into four classes with specific ranges of geotechnical properties (DeGraft-Jn., 1972). Typical grading curves of lateritic gravels from some African countries have been reported elsewhere (Gidigasu, 1976). Most of the standard aggregate test are applicable to most tropical decomposing rocks, soft aggregates, lateritic gravels, and crushed lateritic stones (e.g. De Graft-Johnson et al,1972; Gidigasu,1976). For example, the Los Angeles Abrasion test (Road Research Lab.,1952) and the Aggregate impact value (Brit.Std.,1961) have found useful application in evaluating lateritic gravels and lateritic stone aggregates for pavement construction purposes (e.g. De Graft-Johnson, Bhatia and Hammond,1972). There are, however, cases where these tests are unable to provide good prediction of their behaviour in pavements. Sometimes climatic conditions and rate of chemical weathering undermine the usefulness of these tests. Consequently, attempts have been made to evolve new and non-traditional test procedures which are more predictive of their field behaviour. For example, the so-called modified Aggregate impact test, ten per cent fines test, drying and wetting test, acidity soundness tests have been found very useful (e.g. Tubey and Beaven, 1966; De Graft-Johnson et al, 1972; Hosking and Tubey, 1969; Netterberg,1971). The most significant contribution to the study of doubtful tropical and sub-tropical aggregates have been made in Australia by Wylde (1975, 1976), and in South Africa by Weinert (1961, 1964, 1965, 1968; Weinert and Clauss, 1962. 1967). Other areas of significant contributions have been shrinkage, specific surface tests (e.g. Roper, 1950), Methylene absorption test (e.g. Davidson, 1972), Washington degradation test (e.g. Davidson, 1972) and secondary mineralogical studies (e.g. Weinert, 1964; Scott, 1955) as well as petrographical and mineralogical tests (e.g. Wylde, 1976). Typical results of factors affecting the compaction results are also reported elsewhere (e.g. Gidigasu, 1976; De Graft-Johnson et al, 1972). The genetic variability and influence of compositional factors have also been shown to lead to correlations between properties for some soil deposits and no correlations reported for other similar deposits (e.g. Gidigasu and Mate Korley, 1984); climatic condition of the formation of the soils has been found to influence correlations between these significant highway geotechnical properties (e.g. Gidigasu and Mate Korley, 1984). Consequently, it would be appropriate to emphasize the need to introduce climatic indicators in evaluating paving materials in the tropics. For example, Weinert (1964) proposed the so-called N-climatic factor for separating areas of predominantly physical weathering from those of predominantly chemical weathering of aggregates in pavements. Using this factor and the secondary mineral counts in chemically decomposing rocks it has been possible to utilize some of the rapidly weatherable basic igneous rocks in South Africa for pavement construction (e.g. Weinert, 1974).

5. FIELD COMPACTION PROBLEMS

One major problem in utilising lateritic and saprolitic soils, gravels, etc. is the poor reproducibility of laboratory test results under field conditions. For most materials the standard methods of plant selection and field compaction control methods yield good results. For others there has been the need to replace the standard procedures by say combining different compaction efforts. In some cases, compaction tends to damage the macro structure of the materials and hence reduce the bearing strength considerably (e.g. Hammond, 1970; Gidigasu,1980e). Construction of experimental test sections appear to be a promising method of dealing with doubtful and unpredictable lateritic and saprolitic paving materials. The significant difference between typical laboratory and field compaction test results as obtained in Ghana is given in Fig.1. It would appear that blanket application of temperate zone plant selection procedures in relation to the textural classification of the materials may bring a lot of field construction problems under tropical materials and climatic conditions. Each materials has to undergo some experimental field compaction test section studies especially before using them on a long road project to avoid expensive future maintenance problems, and premature pavement failures.

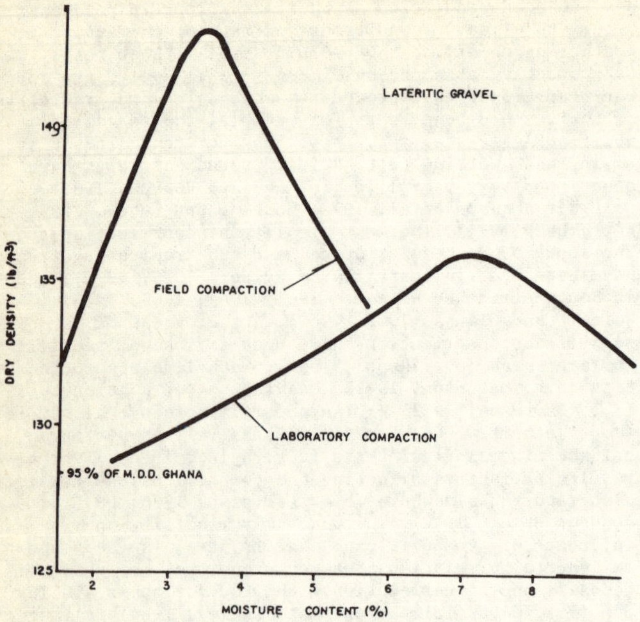

Fig.1. Relation between laboratory and field compaction curves for a typical lateritic gravel from Ghana.

6. DEFINITION OF REPRESENTATIVE SOIL PARAMETERS FOR PAVEMENT CONSTRUCTION

From the foregoing, variability of tropical lateritic and saprolitic soils poses a serious problem, calling for the use of statistical methods for defining representative soil strength and other significant properties relevant to pavement design and construction. The main problem relates to the finegrained lateritic and saprolitic soils but in some cases the crushing of the gravels under rollers during field compaction also lead to problems even though the gravels might have passed standard aggregate strength tests (e.g. Gidigasu, 1976). Simple histograms have been used to define basic soil properties for both laboratory and construction control test results (e.g. Gidigasu, 1980a).

7. SPECIFICATIONS FOR LATERITIC AND SAPROLITIC PAVING MATERIALS

Some studies relating to the behaviour of lateritic and saprolitic materials in pavements have been reported (e.g. Remillon, 1955; Philip, 1952; Nanda and Krishnamachari, 1958; Ackroyd, 1963, 1967; Arulanandan, 1969; De Graft-Johnson et al, 1972; Gidigasu, 1975). For example, specifications of lateritic gravels and aggregates for pavement construction in some countries have been discussed elsewhere (e.g. Tables 1 and 2; and Figs. 2 and 3). Correlations have been reported between the environmental factors (including the rock type)and the performance of roads built with lateritic gravels and crushed aggregates (e.g. Remillon, 1955; Nanda and Krishnamachari, 1958; Gidigasu, 1975; and Weinert, 1974). Based upon the results of laboratory studies and field performance evaluations of existing pavements in many African countries optimum grading and other highway geotechnical properties of acceptable lateritic gravels for a number of African countries have been proposed (e.g. Table 3).

Table 1 - Relation between the chemical composition of Nigerian pisoliths and probable field performance in road pavement (after Ackroyd, 1967)

Chemical composition			Nature of the material	Probable rating
SiO_2	Fe_2O_3	Al_2O_3		
8.4 15.5 19.1	65.1 53.9 50.0	17.4 19.7 27.3	Hard concretionary gravels	Good base and surfacing materials
39.3 48.1	30.4 25.4	20.2 16.0	Weak concretionary gravels	Sub-base materials and fills
38.1 41.2 40.8	28.5 28.6 5.8	20.0 16.8 41.6	Nodules in clay matrix	Groundwater laterites, sub-base and subgrade in materials

Table 2 - Recommended criterion for the rating of Ghanaian lateritic rock aggregates and pisoliths for pavement construction (after De Graft-Johnson et al., 1972

Specific gravity	Water absorption after 24 h soaking(%)	Aggregate impact value (%)	Los Angeles abrasion value (%)	Rating based on probable in-situ behaviour
> 2.85	< 4	<30	< 40	excellent
2.85-2.75	4 - 6	30-40	40-50	good
2.75-2.58	6 - 8	40-50	50-60	fair
< 2.58	>8	>50	>60	poor

There is need to intensify studies on development of laboratory pretest preparation methods and testing pro cedures. The results may be modification of known standard tests or development of entirely new ones. Construction studies and documentation and dissemination of information on field performance of these materials is of paramount importance. It is expected that research projects will be geared towards correlating laboratory and field performance studies which would form the basis of developing reliable acceptance specifications for lateritic and saprolitic soils occurring in various parts of the tropical and sub-tropical environments.

8. SUMMARY OF CONCLUSIONS

The development of temperate zone paving materials acceptance specifications and field construction control procedures has been the result of long term field recognition, laboratory standardisation of testing procedures of specific materials, experimental field construction control studies and performance evaluation of pavement constructed with the various genetic materials over a number of years. Wholesale transfer of these specifications to tropical materials and environments has failed to give desired results because the materials for which the standard European and North American specifications were tailored are just now available in some parts of the tropics and the dynamic influence of climatic factors on materials in tropical pavements seem to defy the anticipated satisfactory performance of most tropical roads. The standard specifications have found useful application for some tropical materials and environments. However, new tailoring of specifications has to be done for some non-traditional tropical soft and decomposing rocks, finegrained lateritic and saprolitic soils as well as for lateritic stone aggregates and gravels.

Attempts made to date to satisfy this need has been reviewed to the extent that the allowable length of the paper could be satisfied.

Considerable contributions have been summarised and more work is required to evolve reliable acceptance and pavement construction control specifications. New testing procedures appear to be necessary where standard ones fail to produce reproducible test results. Indeed, basic studies into the micro and macro structures, petrography, geological origin, mode of formation and trends of evolution of the materials are all significant factors that need to be considered to achieve satisfactory results. Engineering geology constitutes a useful tool for making the necessary impact in developing these specifications. Apart from the lateritic and saprolitic materials, there are a whole lot of other non-traditional pedogenic materials and soils of great potential for base, sub-base and subgrade construction. Indeed, certain environments are dominated by potential paving materials other than lateritic and saprolitic materials and the sooner these materials are also addressed through the promotion of such international co-operative efforts the better. For example, the black clays and expansive soils

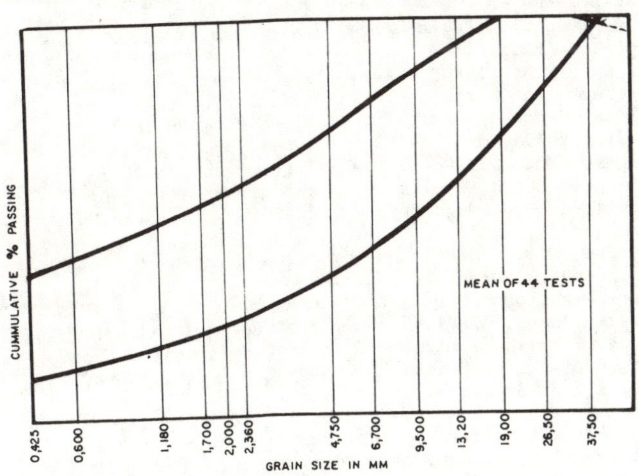

Fig.2. Lateritic base course materials specification for South African conditions (see Netterberg, 1976

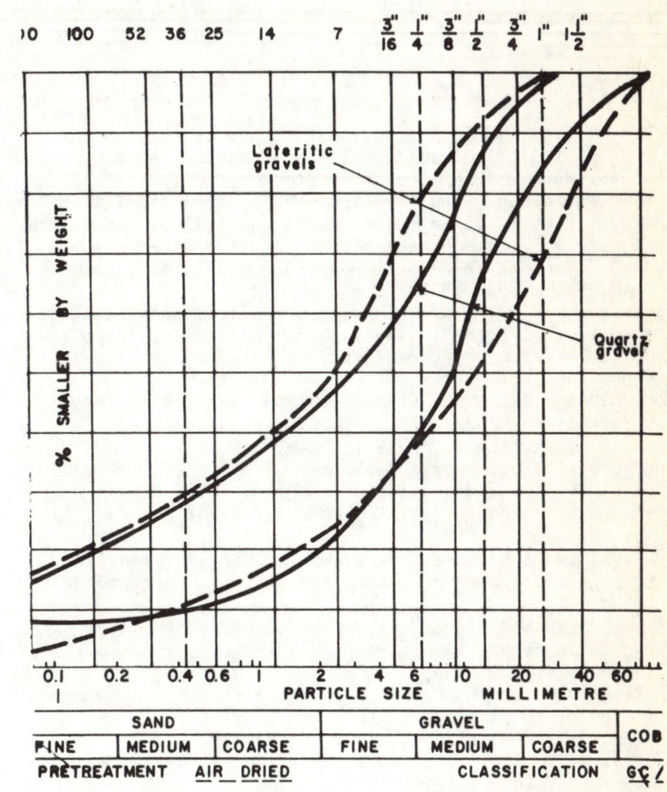

Fig.3. Grading specification of lateritic and quartzitic gravels from Ghana.

Table 3 - Typical lateritic base materials specifications in some African countries

Country	Material	GRDATION (Passing) 1½	3/4	3/8	3/16	No.10	No.40	No.200	Soaked C.B.R.	Liquid Limit (%)	Plasticity Index (%)
Malawi	Gravel	100	60-90	45-75	30-160	20-50	10-30	5-15	> 85	30	6
Niger	Gravel	100	-	-	-	-	-	<25	80		12
Nigeria	Gravel	100	80-100	55-80	40-60	30-50	-	5-15	> 80	25	10
Mali	Gravel	A.A.S.H.O.	SPECIFICATION					-	> 50	-	6-16
Gambia	Gravel	80	60-100	25-85	28-62	25-50	22-44	13-28	95% AASHO MOD	20-37	13-22
Zambia	Gravel	100	-	-	-	-	-	15-20	120	30	6
Uganda	Gravel	100	85-100	65-100	54-100	43-90	30-37	19-38	-	37-48	16-25

have received a fair share of these efforts; there are other materials especially the pedogenic materials such as calcretes which has been single handedly studied over two decades by Netterberg in South Africa.

Exchange of information on the problems of recognition of various genetic materials, testing, construction control and prediction of probable field performance of pavements in different environments will help solve this thorny problem of ensuring satisfactory pavement engineering practice in tropical and sub-tropical environments.

The paper is a contribution to developments in tropical highway geotechnical engineering.

9. REFERENCES

Ackroyd, L.W., 1963. The correlation between engineering and pedological classification systems in Western Nigeria and its implications. Proc.Reg.Conf.Africa Soil Mech.Found.Eng., 3rd, Salisbury,Vol.1, pp: 115-118.

Ackroyd, L.W.,1967. Formation and properties of concretionary and non-concretionary soils in Western Nigeria. Proc.Reg.Conf.Africa Soil Mech.Found.Eng., 4th, Cape Town, Vol.1, pp: 47-52.

Aitchison,G.D., 1952. The physical condition of the soil as a modifying factor in the measurement and interpretation of shear strength. Proc.Austr.New Zeal.Conf.Soil Mech. Found.Eng,,Parkville, Australia, pp: 15-29.

Arulanandan, K., 1969. Classification, engineering properties and behaviour of laterites. Spec.Sess.Lateritic Soils; Proc.Intern.Conf.Soil Mech.Found.Eng., 7th, Mexico, Vol.2, pp:181-190.

Blight,G.E.,1976. Migration of subgrade salts damage thin pavements. Proc.Am.Soc. Civ.Engrs.,Vol.102,No. TE4, pp:779-791.

British Standards Institution (BSI), 1961. Sampling and testing of aggregates, sands and fillers. Brit.Std.812, London, 84pp.

British Standards Institution (BSI), 1975. Methods of test for soils for civil engineering purposes. Brit.Std.1377, London, 143pp.

Buckland,A.H., 1960. Roading aggregate quality - the degradation of roading aggregates. Proc.New Zealand Roading Symposium, Wellington, Vol.2, pp:692-715.

Bunting,B.T., 1965. The geography of soils. Aldine Publ. Company, Chicago, 1st edition, 313pp.

Clare, K.E., 1960. Road making gravels and soils in Central Africa. Road Res.Lab (U.K.), Overseas Bull.12, 43pp.

Clare, K.E. and Beaven,P.J., 1962. Soils and other road making materials in Nigeria. Road Res.Lab. (U.K.), Technical Paper 57: 48pp.

Committee on Tropical Soils of the ISSMFE (1985). Peculiarities of geotechnical behaviour of tropical lateritic and saprolitic soils. Progress Report (1982-1985). Intern.Conf. on Geomechanics in tropical lateritic and saprolitic soils. 1st Sao Paulo, Brazil, Progress Report, 449pp.

Davidson,W.H., 1972. The influence of constitution on the engineering properties of crushed volcanic breccia. Proc.Conf.Austr.Road Res.Board, Melbourne, Vol.6 (5); 71-90.

Day, H.L., 1962. A progress report on studies of degradation of aggregate bases. Highw.Res.Board Bull.344, pp:8-16.

De Graft-Johnson, J.W.S., Bhatia, H.S. and Gidigasu,M.D., 1969. Engineering characteristics of laterite gravels of Ghana. Spec.Session Lateritic Soils. Proc.Intern. Conf.Soil Mech.Found.Eng., 7th, Mexico, Vol.1, pp:117-128.

De Graft-Johnson,J.W.S., Bhatia, H.S. and Hammond, A.A., 1972. Lateritic gravel evaluation for road construction. J.Soil Mech.Found.Eng., Div. Am.Soc. Civ.Engrs.,Vol.98 (SM11), pp:1245-1265).

De Graft-Johnson, J.W.S., Bhatia,H.S. and Yeboa,S.L.,1972. Influence of geology and physical properties on strength characteristics of lateritic gravels for road pavements. Highw.Res.Board, Record 405, pp:87-104.

Dunn,C.S., 1966a. Specification for crushed stone road base and sub-base aggregates. J.Brit.Granite and Whitestone Federation, Vol.6, No.2, pp:7-22.

Dunn,C.S.,1966b. The stability of aggregates used in road sub-base; Roads and Road Construction, Vol.44, No.519, pp:1-12.

Fookes,P.G. and French,W.J., 1977. Soluble salt damage to surfaced roads in the Middle East. The Highway Engineer, Vol.XXIV No.12, pp:10-20.

Frost,R.J., 1967. Review of specifications for gravel paving materials for low cost roads. Proc.South East Asia Reg.Conf. Soil Engineering, Bangkok, pp:451-460.

Fuller,W.B. and Thompson,S.E., 1907. The laws of proportioning concrete. Transactions Am.Soc.Civ.Engrs.,Vol.59, pp:67-143.

Gidigasu, M.D., 1975a. Behaviour of lateritic soils in roads. Ghana Engineer, Vol.7(1): 52-78.

Gidigasu, M.D., 1980a. Geotechnical evaluation of residual gravels in pavement construction. Engineering Geology, Amsterdam, Vol.15, pp:173-194.

Gidigasu, M.D.,1980b. Some contributions to tropical soils engineering in Ghana. Proc.Reg.Conf.Africa Soil Mech. and Foundation Engineering, 7th, Accra, Vol.2, pp: 509-626.

Gidigasu, M.D., 1980c. Variability of particle size distribution of a red mud with different pretest preparations. Proc.Reg.Conf.Africa Soil Mech.Found.Eng., 7th, Accra, Vol.1, pp:171-178.

Gidigasu, M.D., 1980d. Variability of geotechnical properties of subgrade soils in a residual profile over phyllite. Proc.Reg.Conf. Africa Soil Mech.Found.Eng., 7th, Accra, Vol.1, pp:95-104.

Gidigasu, M.D., 1980e. Some problems relating to compaction control of problem soils. Proc.Reg.Conf.Africa Soil Mech.Found.Eng., 7th, Accra, Vol.2, pp:785-796.

Gidigasu, M.D., 1980f. Highway geotechnical characterisation of residual micaceous soils over granite. Proc. Reg.Conf.Africa Soil Mech.Found. Eng., 7th, Accra, Vol.2, pp:831-844.

Gidigasu, M.D., 1982/83. Development of acceptance specifications for tropical gravel paving materials. Engineering geology, Amsterdam, Vol.19, pp:213-240.

Gidigasu, M.D. and Mate Korley, E., 1984. Tropical gravel paving materials specifications in relation to the environment. Proc.Reg.Conf.Africa Soil Mech.Found Eng., 8th, Harare, Vol.1, pp:267-273.

Gidigasu, M.D. and Yeboa, S.L., 1972. Significance of pretesting preparations in evaluating index properties of lateritic materials. Highway Res.Board,Washington, Record 405, pp:105-116.

Hartley,A., 1974. A review of the geological factors influencing the mechanical properties of road surface aggregates. Q.Jl.Eng. Geol., London,Vol.7, pp:69-100.

Hirashima, K.B., 1948. Highway experiences with thixotropic volcanic clay. Proc.Highw.Res. Board, Vol.28; pp.481-496.

Hosking,J.R. and Tubey,L.W., 1969. Research on low-grade aggregates. Road Research Laboratory (U.K.), Report LR 293, 30pp.

Kellog,C.G., 1941. Climate and soil. Year Book of Agriculture, United States Dept. of Agric.,pp:262-291.

Knight,K. and Dehlen,G.L., 1963. The failure of a road constructed on a collapsing soil. Proc.Reg.Conf.Africa Soil Mechanics and Found.Eng., 3rd, Salisbury, pp.31-34.

Lee, G. and Kennedy,C.K., 1975. Quality, shape and degradation of aggregates. Q.Jl. Eng.Geol., London, Vol.8, pp:193-209.

Little,A.L.,1969. The engineering classification of residual tropical soils. Proc.Spec.Session,Lateritic soils. Proc.Intern.Conf.Soil Mech.Found.Eng., 7th, Mexico, Vol.1, pp:1-10.

Lohnes,R.A., Fish,R.O. and Demirel,T., 1971. Geotechnical properties of selected Puerto Rican soils in relation to climate and parent rock. Geol.Soc. Amer., Bull.2, pp.2617-2624.
Lohnes,R.A., Demirel,T. and Tuncer,E.R., 1976. Strength and structure of laterites and lateritic soils. Engineering Res.Institute, Iowa State University, Iowa, Report ISU-ERI-Ames-77021; 83pp.
Loughnan,F.C., 1969. Chemical weathering of silicate minerals. Elsevier, New York, N.Y., 154pp.
Melville,P.L., 1948. Weathering studies of some aggregates. Proc.Highw.Res.Board, Bulletin 28, pp: 238-248.
Millard,R.S., 1962. Road Building in the tropics. Journal applied chemistry, Vol.12, pp:342-357.
Minor,C.E., 1960. Degradation of aggregates. American Society for Testing and Materials (ASTM), Special Technical Publication No.277, pp:109-121.
Nanda,R.L. and Krishnamachari, R., 1958. Study of soft aggregates from different parts of India with a view to their use in road construction. I. laterites. Central Road Res.Inst., India, Road Res. Paper 15, 32pp.
Netterberg, F., 1971. Calcrete in road construction. National Institute for Road Research, Pretoria, Bulletin 10, 73pp.
Netterberg, F., 1975. Pedogenic materials. Specialty Session D., Proc.Reg.Conf.Africa Soil Mech.Found.Eng., 6th, Cape Town, Vol.2, pp:193-212.
Netterberg, F., 1985. Pedocretes; Engineering geology of South Africa, Pretoria, Vol.4, pp:286-306.
Netterberg, F. and Caiger,J.H., 1983. A geotechnical classification of calcretes and other pedocretes. Proc. Residual deposits, surface related weathering processes and materials. The Geological Society, London, pp:235-243.
Newill,D., 1961. A laboratory investigation of two red clays from Kenya. Geotechnique, 11:303-318.
Peltier, L., 1950. The geographic cycle in periglacial regions as it is related to climatic geomorphology. Ann. Assoc. Geol., 40: 214-240.
Philip, 1952. Discussion on: Laterites by J. Drefus, Rev.Gen.Routes et Aerodromes, 22(245):100-101.
Reed,J.J., 1967. Application of recent petrological advances to selection of road aggregates. Proc. New Zealand Roading Symposium, Vol.2, pp:716-742.
Remillon,A., 1955. Stabilisation of laterites. Highway Res.Board,Washington, Bulletin 108, pp:96-101.
Richards, B.G., 1967. Moisture flow and equilibria in unsaturated soils for shallow foundations. ASTM Special Tech.Publ.No.47: 4-34.
Richards, B.G., 1969. Pavement temperatures and their engineering significance in Australia. Highw. Res. Board, Special Report 103, pp:254-266.
Road Research Laboratory, (RRL), 1952. Soil mechanics for road engineers, Her Majesty's Stationery Office, London, 541pp.
Roper, H., 1959. Study of shrinking aggregates. National Build.Res. Inst., Pretoria, Special Tech. Report No.502, 136pp.
Russan, K. and Coleman, J.D., 1961. Effect of climatic factors on subgrade conditions. Geotechnique, London, Vol.11, No.1, pp:22-28.
Scott,L.E., 1955. Secondary minerals in rocks as a cause of pavement and base failures. Proc.Highw. Res. Board, Vol.34, pp:412-417.
Shelburne, T.E., 1939. Degradation of aggregates under road rollers, Proc. ASTM, Vol.39, pp:950-970.
Shelburne, T.E., 1941. Crushing resistance of surface treatment aggregates. Road Abstract 8, 1976.
Shergold, F.A., 1948. A review of available information on the significance of roadstone tests. Road Res. Laboratory, U.K., Technical Paper 10, 16pp.
Shergold, F.A. and Hosking, J.R., 1962. Results of dry and wet sieve analysis and plasticity tests of wet-mix aggregates. Road Res.Lab.(U.K.), Note208, 4pp.
Shergold, F.A. and Hosking,J.R., 1963. Investigation of test procedures for argillaceous and gritty rocks in relation to breakdown under traffic. Roads and Road Construction, Vol.41, No.492, pp:376-378.
Sherwood, P.T., 1967. Classification tests on African red clay and Keuper marl. Q.Jl. Eng.Geol.,London, 1(1), 47-55.
Strakhov, N.M., 1967. Principles of Lithogenesis. Oliver and Boyd, London, Vol.1, 245pp.
Tanner, J.S., 1963. A review of the methods of pavement design used in overseas territories and the performance of roads constructed by the various design methods. Proc. Reg. Conf.Africa Soil Mech.Found.Eng., 3rd, Salisbury, Vol.1, pp: 93-100.
Townsend, F.C., Manke, G.P., and Parcher,J.V., 1971. The influence of sesquioxides on laterite soil properties. Highw. Res.Board, Record 374, pp:80-92.
Townsend, F.C., Manke, G.P. and Parcher, J.V., 1969. The effect of remoulding on the properties of a laterite soil. Highw. Res.Board, Record 284, pp:76-84.
Thornthwaite, C.W., 1948. An approach to a rational classification of climate. Geol.Review, Vol.38, pp.55-94.
Tubey, L.W., and Beaven, P.J., 1966. A study of the petrology of some soft limestones from Jamaica in relation to their engineering properties. Transp.Road Res.Lab., (U.K.)., Report LR.21, 37pp.
Venters, E. and Lewis, D.W., 1954. Deleterious constituents in Indiana gravels. Highw. Res. Board, Bulletin 94, pp:1-10.
Weinert, H.H., 1964. Basic igneous rocks in road foundations. National Institute for Road Research, Pretoria, Research Report No.218, 47pp.
Weinert, H.H., 1974. A climatic index of weathering and its application in road construction. Geotechnique, London, Vol.24, No.4, pp:475-488.
Weinert,H.H. and Clauss, K.A., 1967. Soluble salts in road foundations. Proc.Reg.Conf.Africa Soil Mech. Found.Eng., 4th, Cape Town, Vol.1, pp:213-218.
Weinert, H.H., Clauss, K.A., Winterbach,D.J. and Merwe, van der, 1960. Determination of the soundness of weathered basic igneous rocks (dolerites) used in road foundations. National Institute for Road Research, Pretoria, 21pp. (unpublished report).
Williams, A.A.B., 1965. The deformation of roads resulting from moisture changes in expansive soils in South Africa. Moisture equilibria and moisture changes in soils beneath covered areas. Butterworths, Australia, pp:143-155.
Williams, A.A.B. and Simons, N.E., 1963. The heaving of roads on expansive clays with suggested remedies based on field moisture conditions. Proc.Reg.Conf.Africa Soil Mech.Found.Eng., 3rd, Salisbury, pp: 47-51.
Wylde, L.J., 1975. Report on the first Austr. Road Res. Board Workshop on crushed rock and aggregates. Aust. Road Res. Board, Melbourne, Report No.53, 49pp.
Wylde, L.J., 1976a. Degradation of road aggregates. Journal of Aust.Road Res.Board, Vol.6(1), pp:22-29.
Wylde,L.J., 1976b. In service deterioration of road aggregates. Australian Road Research Board, Internal Report A/R 172-2, 21pp.
Zemour, P. and Durrier, J., 1966. Granularite des assizes stabilise mecaniqueinent. Bull. de Liaiason des Laboratoire Routier, Paris, 18 Mars-Avril, 1-1 to 1-48.

Naturally occurring gravels for road construction

D.W.HIGHT, Geotechnical Consulting Group, London, UK
D.G.TOLL, Imperial College, London, UK
H.GRACE, Henry Grace and Partners, Weybridge, Surrey, UK

ABSTRACT: Many naturally occurring gravels fall outside the limits set by current specifications for road base construction materials, but, notwithstanding this, they are often used as road bases in rural, low volume roads in developing countries. Where they have been used, there have been many cases where they have performed satisfactorily, even though they do not comply with local specifications. A detailed study of the behaviour of compacted naturally occurring gravels, in particular a lateritic gravel from Kenya, has given some insight into the important features of these materials. The good performance of the materials is attributed to their well-graded nature, and especially to the presence of a small amount of clay, well-distributed through the soil. When adequately compacted, they exhibit high strength and stiffness, even when saturated, and have low permeability.

1. INTRODUCTION

Specifications used for road construction in the developing countries were originally based on experience in Europe and North America, with little or no adjustment being made for the differences in climatic conditions and soil types. These specifications have been found to be too stringent for tropical climates, and progressive relaxation of the standards has taken place over subsequent years. Even so, many locally available, naturally occurring clayey gravels and clayey sands, which have been shown to perform well as road bases in low volume roads, are still excluded.

The economic implications of using local natural gravels, rather than having to resort to crushed stone construction, are enormous. Good quality rock is often difficult to locate in tropical regions due to the extensive weathering, which is a feature of the climate. With the high costs of transporting rock, often over large distances, combined with the costs of crushing, many road construction projects, particularly low volume rural roads, would not be viable. However, in many cases, gravel can be obtained from sites adjacent to the road, and, when used as-dug, both transport and processing costs are reduced considerably. Studies of alternative forms of construction in Kenya have shown that construction costs can be reduced by up to 60% by using local materials (Grace 1983).

Where natural gravels have been used in the construction of such roads, these generally are unsurfaced roads, i.e. having no bituminous surfacing or wearing course. Unsurfaced roads present problems, becoming very dusty during dry weather and slippery or even impassable in wet weather. Erosion by both traffic and water removes large quantities of the base material, often requiring complete reconstruction after only five or so years. When the alternative of surfaced roads are constructed, these are built to a higher standard; application of conventional specifications leads to the exclusion of the naturally occurring gravels. There is a clear need for an intermediate type of construction, where surfaced roads are built using local materials to provide cheap rural roads, suitable for carrying the low traffic volumes and which allow year round, all-weather access.

A fundamental understanding of why some naturally occurring materials perform well in road construction, could lead to their adoption, and help in establishing criteria for their selection and treatment. A laboratory investigation has been carried out into natural lateritic and quartzitic gravels that have been used successfully as road bases in low volume surfaced roads in Kenya and Malawi. This has lead to the identification of the key factors responsible for their good performance.

2. GRADING

Grading envelopes for the materials which have been used in road base construction in both Kenya and Malawi are shown on Fig. 1. More complete gradings, including clay fraction determinations, are shown for some of the gravels. Also shown is the Fuller grading (Fuller and Thompson, 1907) for a maximum particle size of 10 mm.

Although covering a broad spectrum of gradings, the form of the grading curves tends to be similar. They are all well-graded, containing particle sizes from gravel down to clay. The grading curves show some similarities with the Fuller Grading. There

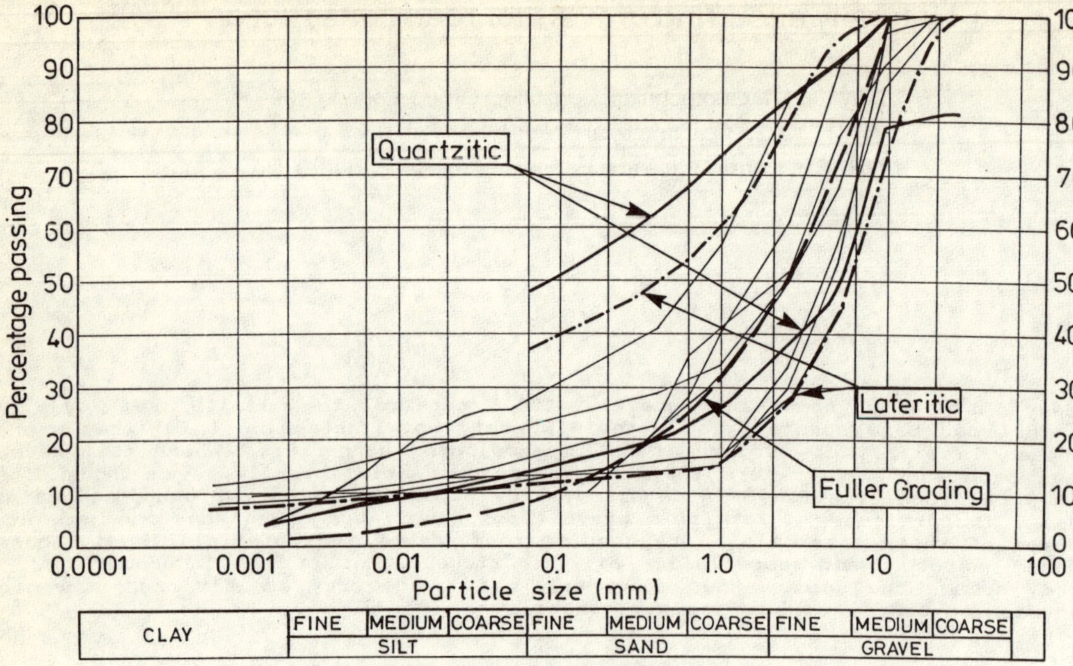

Fig. 1 Grading curves for naturally occurring gravels used in road base construction in Kenya and Malawi

is a sharp change in particle sizes in the coarse fraction, with a levelling off in the finer fractions. However, the materials all contain a greater amount of fines than the 'ideal' case. Corrections for changes in specific gravity for different particle sizes (Charman, 1988) increase this difference, although not significantly.

The percentage passing the 63 μm sieve ranges between 12 and 45. Many specifications limit this percentage to a maximum of 15 or 20, although there have been relaxations in some countries to allow up to 40%. Clay forms a significant proportion of the fines, with clay percentages of 8-13% in the lateritic gravels and 5-6% in the quartzitic gravels. This means that in the lateritic materials 53 to 71% of the fines (<63 μm) consist of clay (<2 μm), whereas in the quartzitic gravels 23-36% of the fines are clay.

The lateritic gravels are susceptible to breakdown during compaction. Under laboratory compaction the fines content increased from 15% to 32% and the clay fraction from 8 to 16%. The grading became much smoother. However, this large amount of breakdown was not observed in samples taken from the road bases. There was an increase in fines and slight smoothing of the grading, but field compaction caused less breakdown than laboratory compaction.

Materials having the type of grading shown are easy to compact (Yoder, 1957). The breakdown of particles during compaction further enhances the compactability.

3. ATTERBERG LIMITS AND CLAY ACTIVITY

The method of preparation has significant effects on the results of the Atterberg limit tests. A lateritic gravel having a plasticity index (PI) of greater than 30% in an undried state gave an apparent PI of 15 to 25% after air drying, reducing to 10 to 15% when oven dried. The use of Atterberg limit tests as a means of classification has little use unless standardisation of the method of preparation is achieved. Although the true PI in the undried state is of most use when assessing the activity of the clay fraction, it is more difficult to determine. If the PI determined on dried samples was adopted as a standard, it would need to be recognised that this would not be a true measure of the plasticity of the material, and materials susceptible to drying would then need to be treated as special cases.

From the undried PI values it is possible to determine the activity of the clay minerals present. The values of activity generally fell between 0.6 and 0.9, i.e. inactive to normal activity (Skempton, 1953). These values are consistent with the mineralogical determinations, which indicated the clay minerals to be kaolinite and illite, both low activity minerals.

4. FABRIC

The fabric of compacted soils plays a vital rôle in determining their behaviour. Samples compacted dry of optimum develop an aggregated fabric (Brackley, 1973). The soil within the aggregations or 'packets' is held together by suction, and the packets act as individual larger particles. The soil will, therefore, behave in a 'coarser' fashion than would be justified by the grading. As the moisture content at compaction nears optimum, the packets become less dominant, and wet of optimum a more uniform fabric is produced.

Even in material compacted wet, when aggregations are not present, the arrangement of the different particle sizes is important. Toll et al (1987) describe studies of the fabric of lateritic and quartzitic gravels, compacted in the field and in the laboratory. These studies showed there was little direct contact between the coarse particles. Instead the fines formed a matrix which surrounded and separated the granular fraction. In the lateritic gravels the granular fraction was made up of iron oxide-cemented kaolinite, whereas quartz formed the coarse particles in the quartzitic gravels. Apart from this, the fabrics were very similar.

The clay fraction was widely distributed throughout the soil, surrounding the silt and sand particles within the matrix. This wide distribution allows the clay to play a dominant rôle in controlling the behaviour. For a soil which has clay particles uniformly distributed throughout the fabric, the clay will be able to act as a binder, which holds the granular material together. It will also restrict the permeability of the soil. If the clay particles are not arranged in this way, but are present as aggregations or pockets in a predominantly granular framework, the clay will neither provide the binding action nor significantly reduce the permeability. Therefore, not only is the amount of clay important, but also how it is arranged within the soil.

The range of pore sizes in the compacted materials was wide. The larger spaces between the coarse particles were generally filled, although occasional voids larger than 0.5 mm were present. In the matrix material there were frequent voids in the 10-20 micron range, but there were also many less than 1 micron.

The microscope studies showed a fairly random fabric with little orientation or anisotropy, at least at the macro-fabric level. This is important in materials which are subjected to wheel loading where stresses rotate through 180°.

5. SUCTION

Suction measurements in compacted samples from both triaxial tests and from specially instrumented CBR tests (Toll, 1988) are plotted as points in Fig. 2, for a lateritic gravel from Kenya. Figure 2 also shows suction paths for the same material, compacted at different initial moisture contents and allowed to wet and dry from that initial state. The suctions are plotted against moisture content, rather than degree of saturation, as it was found that changes in density had little effect on the suction. This can be explained by the fact that suction is largely controlled by the clay fraction.

Overall changes in density of a material such as the lateritic gravel, affect only the macro fabric with little effect on the clay fraction. The bulk degree of saturation, which is influenced mainly by the macro-fabric, will, therefore, be a poor indicator of suction.

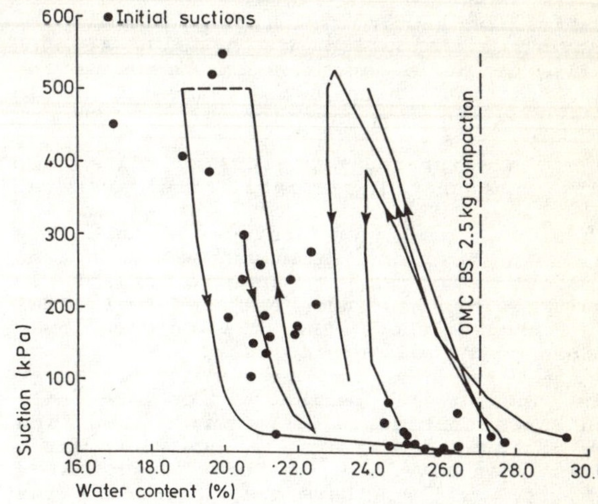

Fig. 2: Suction measurements in a compacted lateritic gravel

Significantly different suctions were measured in samples compacted at the same moisture content, although a general trend of increasing suction with decreasing moisture content can be seen (Fig. 2). The wetting and drying curves do not follow this trend. The wetting curve for an initially dry sample follows a steeper path, with moisture contents tending to follow the lower bound of the compacted samples. The curve for drying from a wet state shows much higher suctions being achieved at a particular moisture content than in the samples compacted at the same value of moisture content. This means that a wide range of suctions can exist in the same material at the same moisture content. Therefore, the suctions existing in the field and those in samples prepared at the same moisture content in the laboratory can be very different by as much as hundreds of kilopascals. Even comparisons between two field conditions at the same moisture content but at different points in a wetting/drying cycle cannot be made. Moisture content alone, without a measure of suction, is a poor indicator of field behaviour.

It can be seen that at moisture contents near optimum (BS 2.5 kg rammer) the suctions are low (Fig. 2), but they become very significant in samples compacted dry of optimum, reaching values of 400 kPa at moisture contents below 75% of optimum. These high suctions might appear surprising in a predominantly granular material containing only 8% clay. However, the wide distribution of the clay fraction, resulting in many fine pores less than 1 μm in diameter, explain how this can occur. The importance of a small amount of clay is again emphasised, and the effects of suction should not be overlooked in granular materials if they have clay present.

6. PERMEABILITY AND INFILTRATION

Saturated permeabilities measured in the compacted lateritic gravel ranged from 2.5 x 10^{-7} to 4.5 x 10^{-8} cm/sec. These permeabilities are extremely low for a predominantly granular material, being typical of values measured in clays. However, the observations on fabric explain why this is so. The uniform distribution of the clay through the soil, forming a matrix which surrounds the coarser fraction, means the permeability is controlled by the clay fraction.

Although the saturated permeability values are low, the unsaturated permeabilities to water are even lower, by three to four orders of magnitude. The unsaturated permeability varies with suction, although at suctions above about 200 kPa it appears relatively constant. This is due to the fact that suction changes above 200 kPa produce only small changes in moisture content.

The low unsaturated permeabilities are only relevant when considering flow under a suction gradient, where there are no positive water pressures involved. The unsaturated permeability will, therefore, control the process of moisture equalisation and redistribution, but not that of infiltration. In the case of infiltration, a wetting front moves through the soil, and behind the front the soil approaches saturation. Flow into the soil will, therefore, be governed by saturated permeabilities, although the rate of progress of a wetting front will depend on the suction in the soil ahead of the front.

The change in moisture content needed to reduce the suction to low values is very small, as can be seen by the steep wetting curves shown in Fig. 2, although greater changes in moisture content are needed to totally reduce the suction to zero. Even though permeabilities are low, the time taken for suctions to decrease to zero is short. Infiltration tests performed in the laboratory show that if small positive pressures are applied, the suctions throughout a 100 mm thick layer reduce to zero in one to two hours (Toll, 1988). This means that suction cannot be relied upon to maintain the performance of an unsufaced road during or immediately after a rain storm.

7. STRENGTH

Stress-strain curves and stress paths are shown in Fig. 3 for a series of undrained triaxial compression tests carried out on compacted samples of a lateritic gravel, which were saturated before testing. Although the six tests follow different stress paths, they all end on a unique stress ratio line, the critical state line. The three loose samples (Tests 1, 2 and 3) climb up to, and then continue to follow, this constant stress ratio line. The denser samples (4, 5 and 6) cross over it to higher stress ratios but then return to it as the strain increases. The maximum stress ratios are plotted as equivalent angles of friction against dry density in Fig. 4.

Some typical results of unsaturated tests are shown in Fig. 5. In all the four tests shown, the samples were sheared with σ_3 held constant at 50 kPa above the pore air pressure. The total stress path is shown on Fig. 5. The stress paths are plotted as $(\sigma_1 - \sigma_3)/2$ against $(\sigma_1 + \sigma_3)/2 - u_w$, where u_w is the pore water pressure. These cannot be treated as effective stress paths because the soil is unsaturated. The samples were sheared under constant water content conditions, with no drainage of water allowed. However, it can be seen that the stress paths are roughly parallel to the total stress path, indicating the soil is behaving in a drained fashion due to the unsaturated conditions. The stress-strain curves in Fig. 5 show the unsaturated behaviour to be much stiffer, and also more brittle than that shown by the saturated undrained tests.

The lateritic gravel, although consisting of relatively weakly cemented particles, which are themselves separated by a matrix of fines, shows itself to have a high angle of friction. Even in loose samples, ϕ' was 40° and, in the denser samples, maximum values of ϕ'exceeded

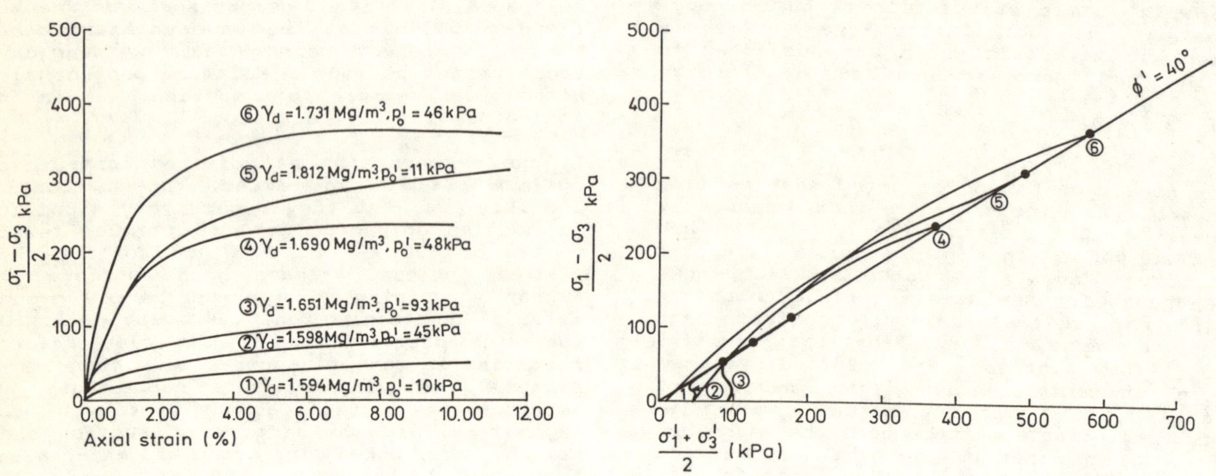

Fig. 3 Undrained triaxial test results on saturated samples of a lateritic gravel

60°. The material was found to be highly dilatant, this being a reflection of the grading and density. In the material compacted dry of optimum the dilatancy is even further enhanced by the presence of soil packets. The aggregated fabric results in the soil having a looser packing, yet it behaves in a more dilatant fashion than samples at the same density and stress level, but compacted wet of optimum. This is because the aggregations of particles act as single large particles, and the soil thus behaves in a 'coarser' way.

It might seem that compacting dry of optimum is preferable, as the tendency to dilate is increased. However, there is a danger in this, since the 'packet fabric' can be destroyed by shear, particularly in saturated conditions. If the material was subjected to cycles of loading when saturated, the fabric could be broken down, removing the tendency to dilate which imparts the high strength. Therefore, densification is the best method of increasing strength, and compaction to the maximum density which can be achieved with the available compactive effort is preferable.

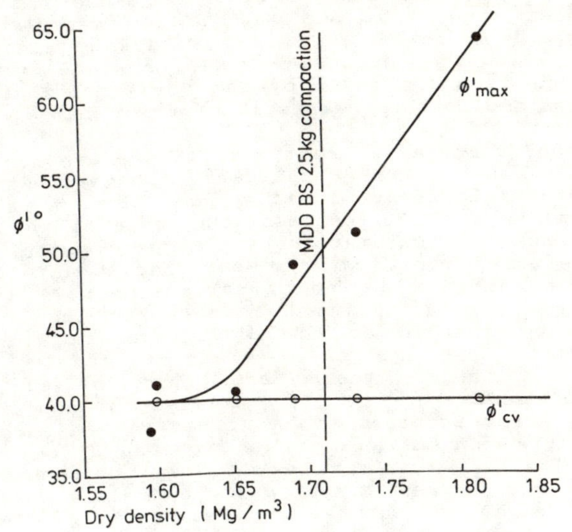

Fig. 4 Variation of angle of friction with density for a compacted lateritic gravel

Although the lateritic gravel has a high angle of friction, this is not enough in itself to impart high strength. The strength will depend both on the suction and on the total stresses acting within the road base. The contribution that suction makes to strength is affected by the degree of saturation. At low moisture contents the suction may be very high, yet the contribution to strength may not be large due to a low degree of saturation. However, in broad terms a high suction will generally produce a high strength.

It is important to recognise that changing the moisture content after compaction will produce different changes in strength to those produced by changing the compaction moisture content. For a soil which has been compacted wet and allowed to dry back the suction will be considerably higher than in a soil which is compacted at the final value of moisture content (see Section 5). The strength of a soil which has dried back will, therefore, be greater than in a soil compacted at the final moisture content. This is another argument in favour of compacting at optimum moisture content, if drying is allowed afterwards. Also, it is easy to see that there will be a discrepancy between tests carried out in the laboratory in which the soil is compacted at a particular moisture content and measurements made in the field after wetting or drying has taken place, even when the comparison is made at the same moisture content. Such discrepancies between field measurements and laboratory test results have been observed (Grace et al, 1985).

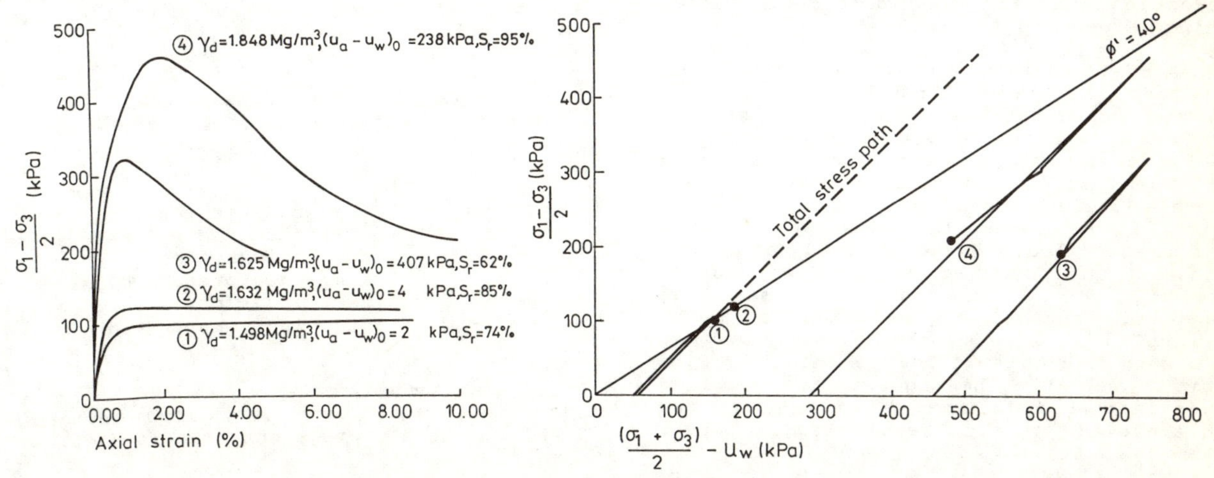

Fig. 5 Triaxial test results on unsaturated samples of a lateritic gravel

In order to assess the most critical time in the history of a road base, we can consider the different processes involved. Although this is done in a very simplistic manner, it gives some insight into the important areas of performance.

After compaction, the road base is likely to exist in a state with negligible vertical stress, but with some horizontal stress locked in by compaction. The values of suction may be quite significant, particularly if the base is allowed to dry. As the base dries out, shrinkage will take place and cracking is likely to occur, causing the horizontal total stresses to drop to zero. In this case, the strength of the soil will be controlled entirely by suction.

If the base then starts to wet up, either due to infiltration or by moisture equalisation after surfacing, the suction will reduce. As the suction drops the soil will swell, tending to close any cracks formed during drying out, and allowing horizontal stresses to be reimposed. Therefore, after wetting up, the state of stress could be such that there would be no contribution to strength from suction and only the lateral stress would maintain the stability of the base.

If the base has remained unsurfaced during these processes, the cracks may become infilled with dust and fine material. In this case, the stresses reimposed after wetting could be higher than the initial values. Continued cycles of drying and wetting, combined with infilling of the cracks, could result in a gradual build up of horizontal stress in the base. If, on the other hand, the base was surfaced, any cracks could remain unfilled and the horizontal stresses would not be fully reimposed as the base wets up. The stresses would only be maintained under a surfaced road if the base was not allowed to dry out before surfacing. The prevention of drying before surfacing would be diffcult in practice. Therefore, the alternative of leaving the base unsurfaced for a season of drying and wetting while open to traffic is preferable. This also has the advantage that poorly compacted areas become readily identifiable and can be corrected before surfacing (Grace and Toll, 1987).

With seasonal changes in suction, the soil could move back and forth between two extremes; one extreme, where strength is controlled by suction only and total stresses become zero due to cracking, and the other extreme where suctions reduce to zero and only the total lateral stress is acting. In the first condition the suction is high and the soil is unsaturated. Since it is unsaturated, any change in pore water pressure during loading will be small, and the stress path followed will approximate to a drained path. Some examples of stress paths for unsaturated samples with a high suction are shown in Fig. 5 (Tests 3 and 4). The high suction combined with little change in pore water pressure and a high angle of friction results in high peak strength.

For a soil at the other extreme, there could be differences in behaviour depending on the degree of saturation. If this state has been reached (i.e. the suction has reduced to zero) by the material approaching full saturation, the behaviour will be undrained under wheel loading (because of the low permeability) and the tendency to dilate under shear will produce negative pore water pressures. The stress paths followed will be similar to those shown in Fig. 3. Therefore, if the material is close to saturation, the material will have high strength even when starting from a low initial effective stress.

If the soil arrives at a zero suction condition without approaching full saturation, this would be the most critical condition. The stress paths followed would then be like those for Tests 1 and 2 on Fig. 5. In this case, the behaviour would be close to a drained condition (due to the low degree of saturation) but with a low initial effective stress. The strength would now be very dependent on the value of the total horizontal stress and on the peak angle of friction.

In an unsurfaced road, suctions are only likely to reduce completely to zero when subjected to direct infiltration during a rain storm. In this process the material will approach full saturation by the progression of a wetting front, so the combination of low suction and low degree of saturation will not occur. In a surfaced road the values of suction achieved at the equilibrium moisture condition are unlikely to reduce entirely to zero, although there are not sufficient measurements of field suctions to confirm this. For suctions to reduce to zero a large moisture content change is required (Fig. 2) and full saturation would be approached. However, suctions may drop to low values while the degree of saturation remains low. The rôle of the total stress would then become extremely important. Provided the material was well-compacted, with a reasonably high compaction stress locked in, it could still perform satisfactorily. However, if the density was low and only small values of total stress were maintained, failure could result. If the base was left exposed to the weather and to traffic for a season before surfacing, to allow the stresses to build up, it would help if this situation did arise.

In conclusion, the peak strength in the different conditions can be ranked in the following order:-

Low Strength- Loose, fully saturated
Loose, unsaturated, zero suction
Dense, unsaturated, zero suction
Loose, unsaturated, high suction
Dense, unsaturated, high suction
High Strength-Dense, fully saturated

This means that a loose material, if fully saturated, will have very poor performance. Even if not saturated, but at zero suction, it will still have low strength. It will only perform well if high suctions are present.

On the other hand, a dense material will still perform well even if fully saturated, and the most critical time for a dense road base will be if suctions reduce while the degree of saturation remains low.

8. STIFFNESS

From the previous section it can be seen that, in the majority of conditions, provided the material is adequately compacted, the strength is sufficiently high that shear failure within the road base is unlikely to occur. The performance of the base will depend on the deformations which occur at stress levels below peak strength. Stiffness is, therefore, a more important parameter than strength in these conditions.

Recently developed strain measuring devices were used in the study and these allowed true measurements of strain to better than 0.01% (Burland and Symes, 1982). Before the development of sample-mounted devices, all measurements of stiffness were subject to errors in the external strain measurements, due to system compliance and sample bedding. Only limited data is, therefore, available on true stiffness values.

The test results showed that the stress-strain behaviour was distincly non-linear, with much stiffer behaviour at low strains. Therefore, values of stiffness need to be quoted for particular strain levels. It was also seen that stiffness increased as a function of mean effective stress. By normalising the stiffness by the effective stress, it is then possible to examine the variations in stiffness due to density. Figure 6 shows the undrained stiffness measured in triaxial tests on saturated samples, normalised by the initial effective stress (p_o'), and plotted against compaction density. Below a threshold value of density (below MDD for BS 2.5 kg compaction) the density did not affect the stiffness, but above this threshold, large increases in stiffness were produced by increasing the density.

The normalised stiffnesses, Eu/p_o', measured in the lateritic gravel at the lower densities (below MDD for BS 2.5 kg compaction) were 2000 to 3000 at a strain level of 0.01%. These compare well with values measured in other soils (c.f. Jardine et al, 1984), particularly if the comparison is made with lightly overconsolidated behaviour. However, the values measured in denser samples (at or above MDD for BS 4.5 kg compaction) were over 5000. This represents extremely stiff behaviour, considerably higher than anything measured previously in uncemented materials, even when heavily overconsolidated. However, it is not yet known whether this is a special property of the lateritic gravel or simply a property of densely compacted materials in general, as there is no comparable data on other materials when densely compacted.

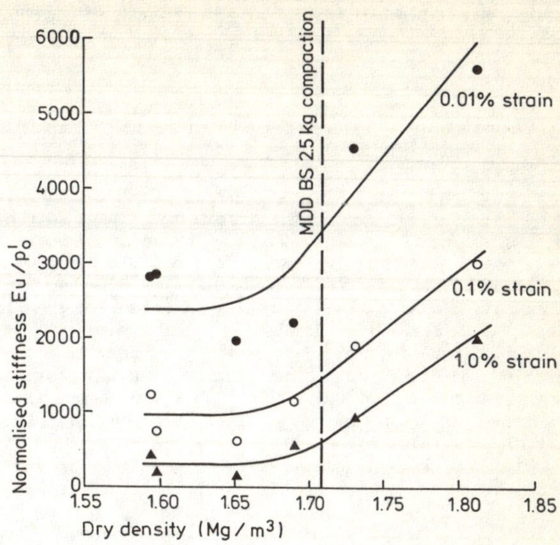

Fig. 6 Variation of undrained stiffness with density for a compacted lateritic gravel

As with strength, although density is important, the stiffness is finally determined by the effective stress. For the case of a densely compacted material, when the suction is high, the stiffness will be high, but when the effective stresses have dropped to lower values (with only the total horizontal stresses acting) the stiffness will be much reduced. However, the results of tests at effective stresses of only 10 kPa showed that for a well compacted material the stiffness was still adequate. Only 0.1% strain was developed under the application of a shear stress of 20 kPa. Such strains should not be large enough to produce significant deformations of the road base.

9. CONCLUSIONS

The successful performance of non-standard materials as road bases in low volume roads can be attributed to the following characteristics:-

- A wide Fuller-like grading, which produces dilatant behaviour under shear for most densities. This also gives good compactability.

- The presence of sufficient clay to allow high suctions to develop and also produce low permeability. It seems from the materials which have performed satisfactorily that a clay fraction between 5% and 15% is enough to achieve this. The distribution of the clay is also important in this respect.

- The low activity of the clay minerals, which restricts shrinkage and swelling.

- A high friction angle, even though there is little direct contact between the coarse grains. This is due to the overall grading and the mineralogy.

High stiffness when adequately compacted, even at low suctions.

ACKNOWLEDGEMENTS

The laboratory investigation at Imperial College was funded by the Science and Engineering Research Council. Field investigations in Kenya were funded by the Leverhulme Trust and the Overseas Development Administration. The Trial Construction in Malawi was funded by the Malawi Government. The University of Nairobi, The Kenya Government Materials Testing Laboratory, The Malawi Government Ministry of Works and Supplies and Scott Wilson Kirkpatrick and Partners carried out the field testing under the direction of Mr Grace. The Transport and Road Research Laboratory have contributed to the study as advisors.

REFERENCES

Brackley, I.J.A. (1973), A model of unsaturated clay structure and its application to swell behaviour, VIth Reg. Conf. Africa SMFE, Durban.

Burland, J.B. and Symes, M.J. (1982), A simple axial displacement gauge for use in the triaxial apparatus, Geotechnique 31, 1, 62-65.

Charman, J.H. (1988), Laterite in road pavements, CIRIA Special Publication 47.

Fuller, W.B. and Thompson, S.E. (1907), The laws of proportioning concrete, Trans. Amer. Soc. Civ. Engrs. 1907, 59.

Grace, H. (1983), Kenya low volume road study, Report on Phase I.

Grace, H., Hight, D.W. and Toll, D.G. (1985), Some recent work on the use of locally occurring materials for bases of bituminous surfaced low volume roads, 1st Int. Conf. Geomechanics of Tropical, Lateritic and Saprolitic Soils, Brasilia.

Grace, H. and Toll, D.G. (1987), Recent investigations into the use of plastic laterites as bases for bituminous surfaced low volume roads, IVth Int. Cont. on Low Volume Roads, Transportation Research Record 1106, Vol 1, 80-88.

Jardine, R.J., Symes, M.J. and Burland, J.B. (1984), The measurement of soil stiffness in the triaxial apparatus, Geotechnique 34, 3, 323-340.

Skempton, A.. (1953), The colloidal activity of clays, Proc. 3rd Int. Conf. Soil Mech., Zurich, 1, 57-61.

Toll, D.G. (1988), The behaviour of unsaturated compacted naturally occurring gravel, Ph.D. thesis, University of London.

Toll, D.G., Hight, D.W. and Shaw, H.F. (1987), The rôle of soil fabric in determining the engineering behaviour of compacted lateritic and quartzitic gravels, Zeitshrift fur Geomorphologie, June 1987, 133-144.

Yoder, E.J. (1957). Principles of Soil Stabilisation, Proc. 43rd Annual Purdue Road School, 92, 41(3).

Ground improvement using soil cement technique

A.C.M.MARQUES, Civil Engineer, Themag Engenharia Ltda, Brazil
N.C.T.FRANÇOSO, Civil Engineers of Themag Engenharia Ltda, Brazil
P.R.TANOUYE, Civil Engineers of Themag Engenharia Ltda, Brazil

ABSTRACT

The basic characteristics of South-Southeastern region of Brazil related to the most important problems faced presently in connection with the construction and maintenance of roads are presented. Special attention is given to the relationship durability/cost and to the methodologies proposed to solve the chronic lack of resources.

The evolution of the solutions for ground improvement of mountainous sites is briefly shown. Emphasis is given to applications using soil-cement.

The consolidation of an embankment at km 66, SP-55 highway is shown in detail, with the application of soil cement as soil-mass stiffening element. A qualitative comparison is made with the other techniques for soil stiffening used in Brazil.

The need for the development of guidelines and norms more suitable for tropical soil as well as for the dynamic process of design, construction and maintenance of roads is demonstrated, by means of manuals and regional geotechnical charts.

INTRODUCTION

The choice of solutions for ground improvement is everpresent in connection with growth and development of the Brazilian highway network. The solutions chosen should take into consideration not only the aspects for technical and economic feasability, but also the ease and time required for execution and further maintenance and repairs.

In this sense, it is very important to establish a data basis containing information on relief, pedology, hydrology, geology, geotechnical and environmental aspects. This may help to correlate different types of occurrences (slides, failures, erosion processes, etc.) with particular site characteristics, aiming at optimizing the solution adopted.

THE BRAZILIAN HIGHWAY AND RAILWAY NETWOORK

The great development of the Brazilian highway network occurred as of 1956, upon the implementation of auto-industry in the country.
At present, the total lenght is approximately 1.5 million km. Only 8% of those are paved. Whereas highways experienced spectacular growth, the railway network has remained stagnate for the past 4 decades. The total lenght today is not more than 30,000 km.

Figure 1 shows the main highways and railways of the south-southeastern region. That is the region of highest interest, for having the most developed network, associated with the largest number of occurrences affecting its infrastructure.

BASIC CONDITIONING ASPECTS

Relief Aspects

The main relief features in this region are the Atlantic Plateau and the Coast Flatlands.

The Atlantic Plateau consists of ancient ground (see Figure 2), corresponding to the Eastern portion of the Brazilian Plateau, ranging from the Northeast to Rio Grande do Sul. The Southeast highlands are located there. Serras do Mar, da Mantiqueira, do Espinhaço and the Chapadas (Plateaus) Diamantina and Caparaõ stand out.

Those lands had high elevation in the geological past. They suffered intense erosion and decomposition processes along geological eras.

Geology and Pedology Aspects

From the theree geological provinces existing in continental masses, only two are present in Brazilian territory: ancient masses and shields and the sedimentary basins. Recent foldings of later geological eras are not therefore present.

Ancient shields or masses were predominantly formed during the Precambrian, 36% of the total area of the Brazilian territory date from that period. Rocks from that period form in Brazil the crystalline basement. Granites and gneisses are the most predominant. Micashists, marbles, sienites, quartzites and others also occur.

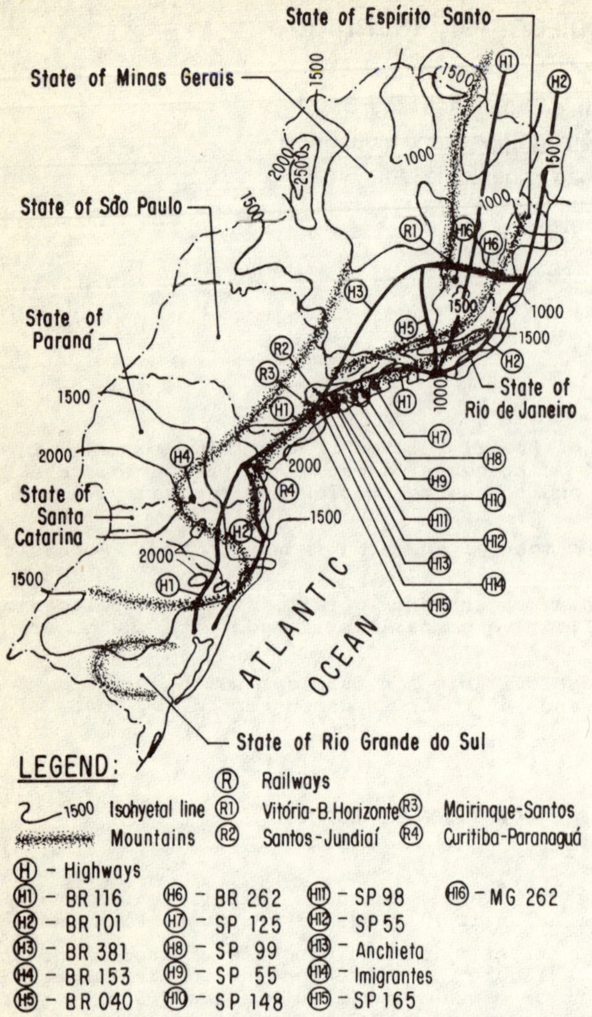

FIG. 1 - MAP OF SOUTH-SOUTHEST BRAZIL, WITH INDICATIONS OF ISOHYETAL LINES, MOST IMPORTANT MOUNTAINS, HIGHWAYS AND RAILWAYS

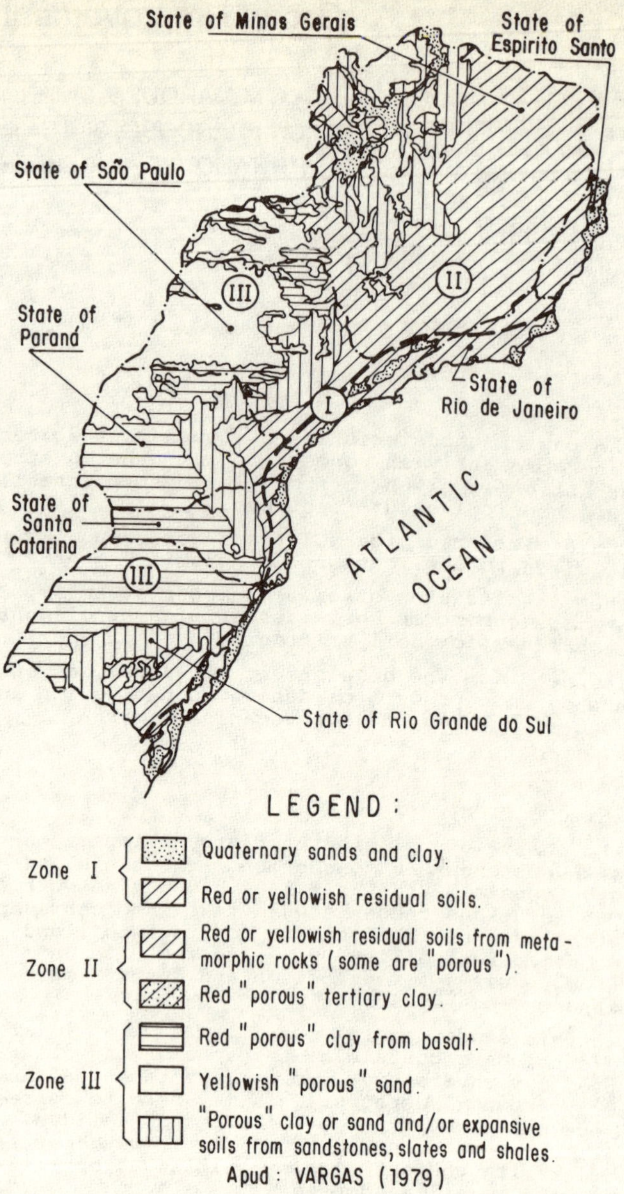

FIG. 2 - GEOLOGICAL MAP OF SOUTH-SOUTHEAST BRAZIL

Sedimentary basins cover most of the Brazilian territory. The large extension ones are the most important: Amazonas Basin, Mid-North Basin, Paraná Basin, São Francisco Basin, and Mato-Grosso Marshlands Basin. The small extension ones are: Recôncavo Baiano, Coastal Basins and small basins on the plateau.
A simplified geological map of the south-southeastern region is shown in Figure 2 (Ref. 13).

Pedological surveys carried out in the Brazilian territory show distinct degrees of abstraction, corresponding to exploratory reconnaissances and detailed surveys (Ref. 7).

Amongst others, the following soil classes mainly occur in the South-Southeastern region:

- red latosols: (formed from crystalline rocks and sandstone);

- dark red latosols: (developed from sandy-clayey soils);

- rhodic ferralsols: (porous and well drained soils);

- eutric nitosols: (erosive soils);

- yellow-red podzolic: (more erodible soils);

- vertisols: (expansive soils).

Hydrological Aspects

In a general way, the rain pattern in the South-Southeastern region is characterized by the existence of a very wet season in the November-March period and a dry season from May to September.

At Serra do Mar and Serra da Mantiqueira there is a local trend for increase of precipitation because both of them are located along the trajectories of polar weather fronts and tropical instability lines, besides being parallel to the shoreline. The increase is due to orographic rains caused by the intensification of turbulence and dynamic air rising due to the mountain obstacles.

From the exposed, this region may be considered critical from the standpoint of pluviosity. The intensity of the precipitation is directly related to the occurrance of landslides and mass failures. This fact is confirmed by the large number of events, recorded in the practice of consolidation of cuts and fills in the mountain areas. Yearly isohyiets for the South-Southeastern region are shown in Figure 2.

MAIN PROBLEMS FOR CONSTRUCTION AND MAINTENANCE

Problems of lack of funds for highway and railway maintenance and conservation have been faced in Brazil for about three decades, mostly with respect to the infrastructure. Also in relation to the construction of new highways, in spite of the large amount of the investments, the occurrence of slope stability problems (swelling soils, erodible silts, etc.) has been observed with considerable frequency, after short periods of utilization.

Mainly in the mountain areas, where some characteristic types of tropical soils have been associated to the intense pluviometric precipitation occurrences, the evidence of phenomena damaging highway infrastructure has reached preoccupying levels. A number of occurrences has been systematically recorded every year, some of them catastrophic in terms of both human lives and material losses. Among the most recent ones, the slides occurred in January and February 1988 at BR-101 highway (section between Ubatuba and Paraty), SP-55 highway (sections between Ubatuba and Caraguatatuba; São Sebastião and Bertioga) and in the cities of Rio de Janeiro and Petrópolis can be mentioned.

The nature of these phenomena caused by the characteristics of South-Southeastern region, can be summarized according to what was proposed by Couto (Ref.1):

- Group I - Instability inducing processes (problems of erosion and liquefaction of soils);
- Group II - Surface instabilities;
- Group III - Internal instabilities;
- Group IV - Large mass instabilities;
- Group V - Instabilities due to local geological peculiarities.

Another classification of the events can be associated, according to suggestions by Vargas (Ref. 14):

- Plastic and viscous movements:
  . surface layers creeping
  . talus slide
- Slides along failure surfaces
  . planar (thin mantle on rock surface)
  . rotary (residual soils and saprolites)
- Rock mass structural slides
  . slabs or wedges (saturated masses)
  . block falls
- Avalanches
  . mud flows (erosion or liquefaction of surface layers)
  . rock block flows (hydraulic demolition of fractured and decomposed masses).

All the premisses for the application of analysis methodology that allows for good characterization and hierarchizing of the problems can be established based on the information available in the data basis (Ref.2). The development of this subject is underway in conjunction with the São Paulo State Highway Department - DER. However, in order to keep highway maintenance on a rational basis, economic solutions have been adopted and distributed on a more judicious manner. For that, simplified decision methods (Refs. 4 and 6) and simple techniques (Ref.11) using new materials and equipment available have been used. The basic premise is to guarantee minimum safety levels to the whole road stretch, instead of isolated treatment of the problem sites. Figure 3, for instance, depicts the idea of reliability factors along a determined road section as a function of the safety level that is aimed to be reached with the available funds.

The subject of cost brings special complications to the selection of solutions. Due to frequent changes in the economy of the country, price fluctuations favor different materials at different times, and discourage some specialized services. So, costs practiced during a certain period may differ substantially from those of other period. Significant deviations from the world average often occur. Furthermore, the adaptation of the agencies responsible for the highways has been done very slowly, with the transposition of temperate countries techniques to technology applied to tropical soils. This fact, that seems obvious, is a significant obstacle for the adoption of modern technology.

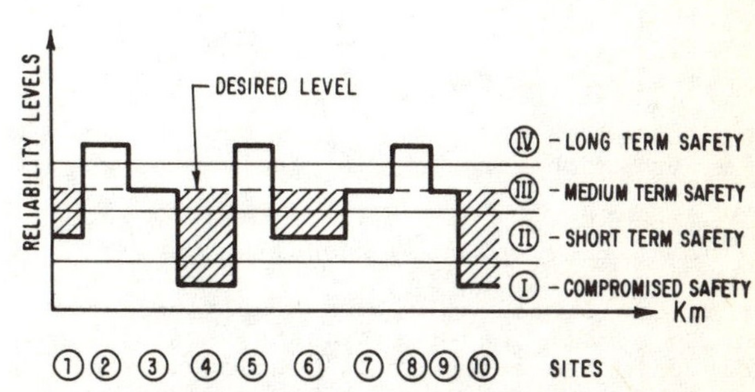

FIG. 3 - PHILOSOPHY OF TREATMENT OF THE SITES

The simple addition of a new item to the official price schedules, as well as its technical specification usually take quite some time. The progress of technology adapted to the tropical peculiarities is thus delayed.

SOIL CEMENT IN BRAZIL: EXPERIENCE AND UTILIZATION

The use of Portland cement for soil stabilization started in the United States in 1935 after research carried out by the Portland Cement Association (PCA).

The first use of soil cement in Brazil occurred after the first research works by the Brazilian Portland Cement Association - ABCP in 1941. The pionner landmark was the application for paving Petrolina airport, Pernambuco State.

Applications of soil cement have experienced substantial broadening from its initial use for pavements only. This has been more significant in the past 20 years, and the new applications include: stabilization by means of soil cement mud grouting (Anchieta highway, connecting São Paulo to Santos), slope protection and dam lining (Rosana Hydroelectric Scheme) manufacturing of blocks and bricks for popular housing, utilization as earth retaining structure and slope protection (Juiz de Fora City slopes, Mairinque-Santos railway and D. Pedro I highway) beside several uses for pavements.

The example presented in this paper is another variation of use of soil cement. In this case, a reinforced soil mass was used (conventional fill and layers of compacted soil cement) conveniently zoned in such a way to guarantee also surface protection for the slope.

CONSOLIDATION SOLUTION: TYPES AND EVOLUTION

In spite of the problems related to cost and the adoption of criteria for tropical soils, the most recent techniques for slope treatment have been used (Ref.15), such as:

- Anchored bulkheads
  . anchor rods and rack bolts, slab with reinforced concrete or shotcrete with anchoring beam.

- Pile walls with embedded lengths ("Pranchada")
  . (steel of concrete piles combined or not with the rods)

- "Pali-Radice"
  . (micro-piles, "presso-ancoragem", grouted piles, CCP, etc.)

- Reinforced Earth
  . (micro-anchoring, reinforced earth, etc.)

- Enveloped Soil
  . (geotextile, plastic meshes, etc.).

However, in Brazil solutions involving cement grouting usually have high costs, mostly if compared to the costs of conventional earth moving and to the funds available for highway maintenance and rehabilitation.

The use of the classical solutions can not be considered for the mountain regions due to inadequate topographical, conditions. absence of borrow areas, environmental restraints and high leisure property density.

Traditionally many gravity walls were constructed for the solution of these problems, due to the abundance of granite and gneiss rocks and cheap labor (e.g. São Paulo-Santos Railway, FEPASA São Paulo State Railway Co. constructed in 1927 and Santos-Jundiaí railway, RFFSA, Federal Railway Co., constructed in 1886). Reinforced concrete walls associated to anchored bulkheads (e.g. Anchieta highway, between São Paulo and Santos, and the rehabilitation of SP-99 highway, from São José dos Campos to Caraguatatuba, after the catastrophe of 1967). This way, anchored retaining structures predominated from 1965 to 1978 for the solution of problems of cut and fill mixed cross section at mountain or wavy regions (e.g. railway connection Vitória-Minas, CVRD, iron ore). Due to amplitude of the problems, cost increase and the economic condition of the country, cheaper methods were needed, but keeping the same efficiency. This trend, associated to the development of new materials and new worldwide techniques, allowed for the construction of works suited to the Brazilian conditions.

So, there was a natural shift from anchored bulk heads to solutions of embedded pile walls with or without tie rods. Costs were reduced, by decreasing the number of tie rods and eliminating eventual foundation treatment.

The evolution of the application of these solutions is depicted in the sequence shown in Figure 4.

Solutions B, C and D illustrate the preoccupation of optimizing the application of the most economical solution (earth moving) with the safest and quickest solution (anchoring). Table I summarizes a comparison of costs valid for stabilization works at SP-55.

These typical procedures for mountain regions also apply to all other cases (soft soils, cuts in fissured soils, etc.) present in Brazil.

THE FILL AT KM 66+00, SP-55 CARAGUATATUBA-UBATUBA

Figures 5 and 6 give an idea of the extension, the type and the seriouness of the problem. Material of the slide reached the Domingas Dias beach, taking along wooden bridges and the pavement of the streets of an existing condominium.

Figure 7 shows the surface geological mapping and the typical geological-geotechnical cross section. The local geology indicates the presence of rocks of the crystalline basement (migmatites) with intrusion of a diabase dyke attitude N70^{o} E/80° SE). The cross section shows a 7,0 m thick fill composed of reddish brown sandy clay overlaying residual soil composed of reddish-yellow silty sand with small thickness overlaying rock outcrops.

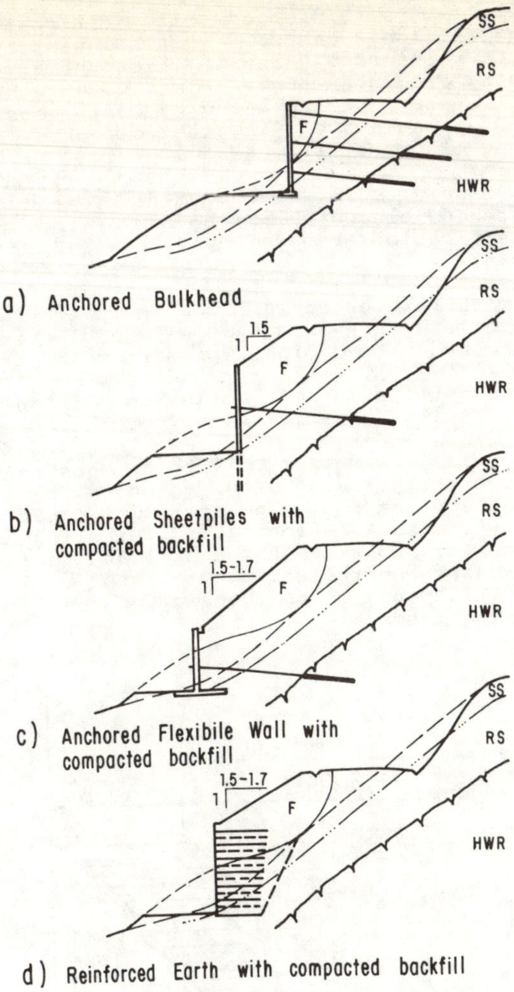

FIG. 4 - EVOLUTION OF THE SOLUTIONS OF GROUND IMPROVEMENT OF THE WORKS EXECUTED ON MOUNTAIN REGIONS

FIG. 5 - AIR VIEW OF THE SLIDE

TABLE I - COST RATIO FOR EARTH RETAINING SOLUTIONS IN MOUNTAIN REGIONS

SOLUTION	COST RATIO
A - ANCHORED BULKHEAD	1,71
B - ANCHORED SHEETPILES WITH COMPACTED BACKFILL	1,52
C - ANCHORED FLEXIBLE WALL WITH COMPACTED BACKFILL	1,16
D - REINFORCED EARTH WITH COMPACTED BACKFILL	1,00

NOTE: COST RATIO REFERED TO THE CHEAPEST SOLUTION

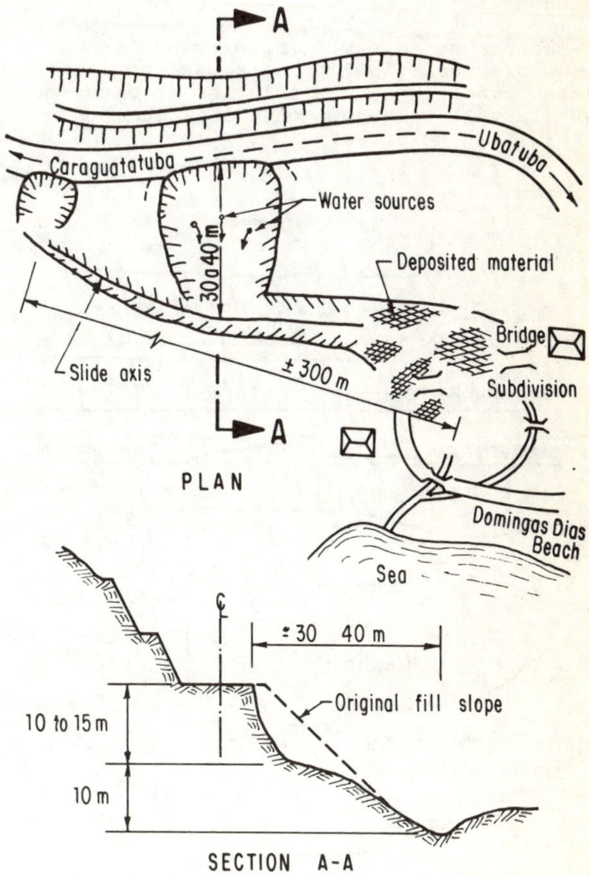

FIG. 6 - SKETCH OF THE SLIDE - PLAN VIEW AND CROSS SECTION

Records of the intense rains of January 23 and 24, 1985 in the pluviometers of Caputera and Ubatuba, located within a radius of 20 km from the site show respectively 109.4 mm and 123.8 mm in 24 hours, causing a phenomenon of avalanche.

The movement of the debris (soil, blocks, trees) remained confined by the diabase dyke which directed the flow of water to the site reaching the Domingas Dias beach after about 300 m. The phenomenon started after a loud noise, according to local people.

The solutions proposed and analyzed for the site were:

- reinforced earth with compacted backfill;

- anchored bulkhead to the highway level;

- partial anchored bulkhed and compacted backfill;

- anchored flexible wall and backfill with soil cement; and

- viaduct.

For reasons of safety, time of execution, borrow area availability and construction related conditions, the solution of anchored flexible wall with compacted soil cement backfill was selected.

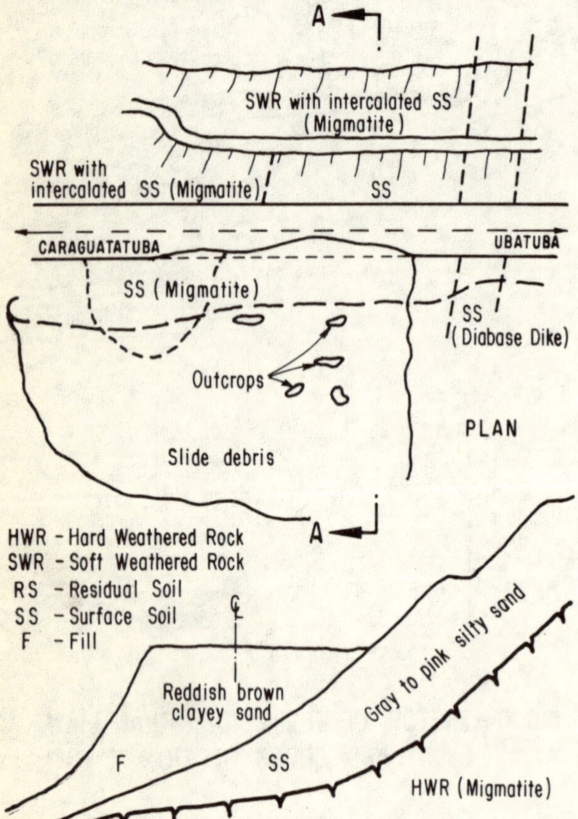

FIG. 7 - GEOLOGICAL-GEOTECHNICAL MAPPING AND TYPICAL CROSS SECTION

The backfill is a reinforced soil, stiffened with layers of soil cement. This solution is similar to reinforced earth in which the reinforcing elements (steel tapes, wire mesh, geotextile, etc.) are replaced by continuous zones of soil cement playing the role of stiffening elements (Fig. 8).

The technique of soil reinforcement by means of "stiffening elements" may be summarized by the classification proposed by Mitchell (Ref.8), as follows:

. Walls with reinforced soil, earth reinforced with steel tapes, VSL masses with a plastic mesh, masses with gabion type wire mesh, etc.

. Reinforced fills on soft soils application of geotextiles, plastic meshes, etc.

. Nailed soils: in Brazil only these ones are not yet systematically used, in spite of many cases of application. In Mitchell's paper he found by means of three practical applications that the ratio between horizontal and vertical stresses and the stiffness of the mass plus wall depends on the density of the stiffening materials.

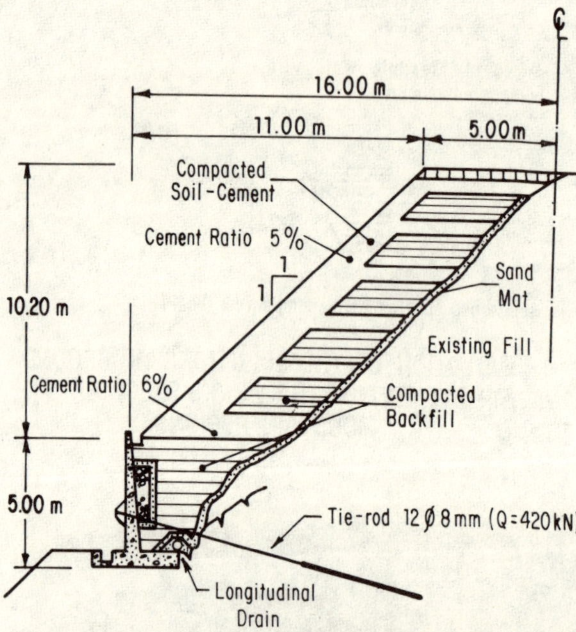

SOIL PROPERTIES			
Material	γ (KN/m^3)	c' (KPa)	φ (°)
Compacted Soil Cement	18	100	35
Compacted Backfill	18	15	30
Existing Fill	18	10	30

FIG. 8 - TYPICAL CROSS SECTION OF THE SOLUTION

Without a specific mechanism for the adopted solution, limit equilibrium global stability

TABLE II - RESULTS OF UNCONFINED COMPRESSION AND DURABILITY TESTS (12 CYCLES)

CEMENT RATIO (%)	SPECIMEN	SPECIMEN MOLDING DATA			UNCONFINED COMPRESSIVE STRENGHT(KPa)
		ρ(kg/m)	w (%)	C D (%)	
4	1	1735	15.5	97	1800
	2	1737	15.5	97	1760
	3	1734	15.5	97	1660
6	1	1728	16.0	97	2130
	2	1727	16.4	97	2330
	3	1741	16.4	98	2480

CEMENT RATIO (%)	SPECIMEN	SPECIMEN MOLDING DATA			LOSS OF WEIGHT (%)
		ρ(kg/m³)	W (%)	C D (%)	
4	1	1731	15.5	97	17.7
	2	1736	15.5	97	18.6
6	1	1722	16.4	97	9.4
	2	1728	16.0	97	7.6

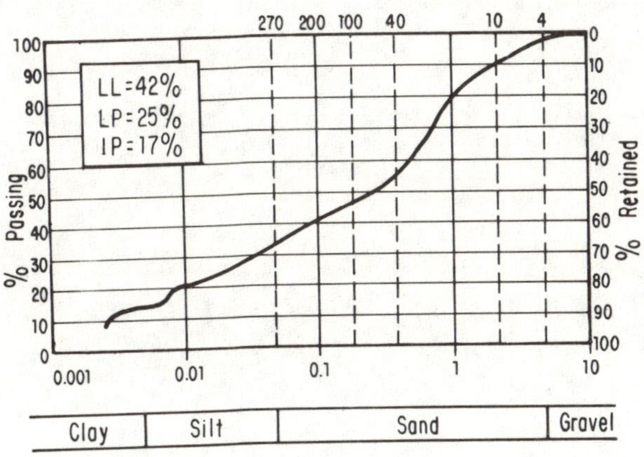

FIG. 9 - GRAIN SIZE DISTRIBUTION OF THE MIX (Borrow area of SP-55 Km 68)

FIG. 10 - SIDE VIEW OF THE EXECUTED WORK DETAIL OF THE SOIL CEMENT MASS

was evaluated and local stability between layers of soil cement. Soil cement mix was determined based on criteria recommended by the PCA, Burec and ABCP (Refs. 3 and 9). Strength and durability tests were carried out on the material from the borrow area of highway SP-55 km 68, with 4% and 6% cement ratios (in weight), as shown on table II. The grain-size distribution of the soil is shown in Fig. 9.

Because the role of soil cement here is different from those of pavement applications and dam slope protection, the cement weight ratio was specified as 5% for the dry soil in the main section of the fill and 6% at the base of the fill next to the transition of the anchored wall top (see Fig. 8).

Figure 10 shows the site as finished and released for traffic since February 1986. Performance has been satisfactory so far.

This solution has provided information for other proposals of sites where large volumes of fill would be replaced by smaller volumes of cement stiffened soil (with or without wire mesh working as tension reinforcement) adopting nearly vertical slopes (ca. 2V : 1H).

CONCLUSIONS

In spite of disagreements, most of the authors recognize that the procedures developed and used for temperate region soils do not apply for tropical soils. In practical terms, the importance of geological-geotechnical cartography including pedology and geomorphology is evident, as well as some other aspects of the geological science. An understanding of the inter-relationship and applications, fitting in this case the large dimensions of the country and the lack of funds is thus favored.

Geotechnical charts and highway manuals are being prepared including a large record (survey of critical points) for the São Paulo State Highway Department, aiming at the establishment of methodology for control maintenance of our highway network.

Also included are methodological arrangement of decicion taking process and hierarchizing as a function of different parameters. The adoption of standardized solutions feasible both technically and economically is permitted.

It is worthwile to stress the importance and the need for the sequence of studies in order to optimize the criteria for dimensioning soil cement mix determinations for stabilization works. Those criteria available today are mainly applied for pavement works.

ACKNOWLEDGMENT

This research has had support from DER the State Highway Department through its Plan for Adaptation of New Consolidation Technologies for Tropical soils.

REFERENCES

1. Couto, A.M.C. (1986). "Follow-up and Observation of Different Slope Retaining Proc-

esses", (in Portuguese) Anais do I Seminário de Altos Estudos sobre Contenção de Encostas, Petrópolis.

2. Françoso, N.C.T. et al (1982). "New Technology for Highway Infrastructure Maintenance", (in Portuguese) Anais do 7º Congresso Brasileiro de Mecânica dos Solos e Engenharia de Fundações, Recife.

3. Françoso, N.C.T. et al (1983). "Critical Analysis of Methods for Soil Cement Mix Determination for Earth Dam Slope Protection", (in Portuguese) Anais do XV Seminário Nacional de Grandes Barragens, Rio de Janeiro.

4. Françoso, N.C.T. e Ferreira, L.C. (1985). "Highway Maintenance in Brazil", (in Portuguese) Revista de Engenharia Mackenzie, Março/Abril, São Paulo.

5. Gomes, J.M. (1986). "The Use of Soil Cement for Earth Retaining Works", (in Portuguese) Boletim Técnico da Associação Brasileira de Cimento Portland, São Paulo.

6. Hachich, W.C. e Luz, P.A.C. (1988). "Highway Infrastructure Maintenance: Hierarchization and Reliability", (in Portuguese) Anais EPUSP, São Paulo, Vol.IB, pp 1083-1114.

7. Larach, J.O. et al (1981). "Pedological Surveys in Brazil Carried out by SNLCS - EMBRAPA", (in Portuguese) Anais do Simpósio Brasileiro de Solos Tropicais em Engenharia, Rio de Janeiro, pp 13-29.

8. Mitchell, J.K. "Reinforcement for Earth Work Construction and Ground Stabilization", Anais do VIII PCSMFE, Cartagena, Vol.I, pp 349-380.

9. Pinto, C.S. (1980). "Evolution of Laboratory Research About Soil Cement", (in Portuguese) Boletim Técnico da Associação Brasileira de Cimento Portland, São Paulo.

10. Pitta, M.R. (1983). "Soil Cement for Lining of Earth Dams, Dikes and Reservoirs", (in Portuguese) Boletim Técnico da Associação Brasileira de Cimento Portland, São Paulo.

11. Tanouye, P.R. e Katsutani, L.T. (1982). "Application of Simple Techniques for the Consolidation of Highway and Railway Infrastructure", (in Portuguese) Anais do 7º Congresso Brasileiro de Mecânica dos Solos e Engenharia de Fundações, Recife.

12. Vargas, E.J. et al (1986). "A Study of the Relationship Between the Stability of Slopes in Residual Soils and Rain Intensity", Proc. of International Symposium on Environmental Geotechnology, Allemtown, Vol.I, pp 491-500.

13. Vargas, M. (1984). "Maintenance and Rehabilitation Plan for Highways in Mountain Areas of the State of São Paulo" (in Portuguese) Anais do I Seminário Regional de Estradas Vicinais, Ibiúna.

14. Vargas, M. (1985). "The Concept of Tropical Soils, Proc. of I International Conference on Geomechanics in Tropical Lateritic and Saprolitic Soils, Brasília, Vol.3, pp 101-134.

15. Vargas, M. (1986). "Slope Stabilization in Brazil National Technique", (in Portuguese) Anais do I Seminário de Altos Estudos sobre Contenção de Encostas, Petrópolis.

16. Vieira, P.H. (1986). "Micro-anchoring", (in Portuguese) Anais do 1º Seminário de Altos Estudos sobre Contenção de Encostas, Petrópolis.

17. Wolle, C.M. (1988). "Fills Reinforced with Geotextile", (in Portuguese) Boletim Técnico D.E.R., São Paulo, nº 1, pp 15-22.

Engineering behaviour of laterites in road constructions

G.T.H.SWEERE, Delft University of Technology, Netherlands
P.J.GALJAARD, Delft University of Technology, Netherlands
H.TJONG TJIN JOE, Delft University of Technology, Netherlands

ABSTRACT: During the reconstruction of the road Moengo-Albina in Suriname, major problems were encountered with the laterite materials used. Heavy rainfall combined with trafficking led to complete failures of sections of the road under construction, whilst other sections of the road sustained the influence of weather and traffic without problems. In a back-analysis of the problems encountered, the composition and mechanical properties of the lateritic materials used were studied in detail. The in-situ performance and the results of cyclic loading triaxial tests could be clearly traced back to the composition of the materials. Soil suction showed to play a major role in the performance of the unsaturated road building materials.

1 INTRODUCTION

In the tropics, laterites are used extensively as a road building material. They are applied both in unbound base and sub-base layers and in surface layers. Laterites are formed from parent rock by an intense process of accumulation of iron, and leaching of other constituents of the parent material like quartz |Sweere, 1982|. The high iron-content causes the characteristic red colour of laterites.

In a process of redistribution and subsequent crystallization of the iron in the weathering product, materials of various particle sizes are formed. Concretions of iron are formed, resulting in a hard, gravel-sized material. The kaolinitic clay resulting from the weathering proces is stained red by the abundant iron, and clay particles are cemented together to form larger particles. Often, the sand-sized fraction is virtually absent, resulting in so-called "gap-graded" laterites.

Laterites can perform very well as a road building material. In case of gap-graded laterites, an almost ideal base course material can be obtained, provided the proper mixing ratio of gravel to clay sized particles is present in the material. After compaction, the pavement layer will then consist of a skeleton of gravel-sized particles, the voids of which are filled with clay- and silt-sized material. If too much fine material is present in a gap-graded laterite, the pavement layer after compaction will consist of a matrix of fine material, with coarse particles "floating" in it. This material will behave properly in the dry condition, but may readily loose its strength on wetting |Sweere and Galjaard, 1984|.

Continuously graded laterites also occur. Again, the particle size distribution of the material is one of the predominant factors with respect to performance of the material in the pavement construction. Continuously graded laterites containing a sufficient amount of sand- and gravel-sized material will perform well in the pavement layer, whilst materials containing a surplus of clayey material will perform poorly when the pavement layer is wetted.

Laterites can be selected for road building purposes using the relationship between particle size distribution and in-pavement performance. Such a selection however is only possible when sufficient good quality material is available within reasonable hauling distances from the construction site. This is often not the case, and the pavement engineer then has to work with poorer quality materials.

This paper presents a case study of the problems that were encountered on a project where poor quality laterites had to be used for reasons of availability and hauling distances. During the reconstruction of the road Moengo to Albina in Suriname, South America, laterites from various horizons in seven borrow-pits were applied in the base course. During the reconstruction, the road had to be kept open to normal commercial traffic, which involved among others 35-ton loading capacity mining trucks. The construction of the base course was followed by the wettest rainy season in ten years. Major sections of the base course completely failed under the combined influence of rain and traffic. Other sections, however, performed remarkably well under the same circumstances.

In a back-analysis of the problems encountered, a chemical and physical analysis of the laterites used was performed. The in-service behaviour of the laterites could be traced back to the composition of the materials. Cyclic loading triaxial tests were performed to study the mechanical behaviour of the laterites in more detail. This behaviour again could be clearly traced back to the composition of the materials investigated.

2 THE ROAD MOENGO-ALBINA

The towns of Moengo and Albina are situated in the North-Eastern part of Suriname in South America. The reconstruction project dealt with in this paper involved raising the level of the existing road with 0.5 - 7 m. Laterites were used as base and sub-base material, covered later with two layers of asphalt concrete of 10 cm total thickness. Table I shows some construction data of this project.

Table I. Construction data

road length	42.4 km
earth moving	760 000 m^3
road fill	400 000 m^3
laterites used	270 000 m^3
asphalt concrete	61 200 tons
thickness of binder layer	6.5 cm
thickness of surface layer	3.5 cm

With the exception of three sections totalling 7 km the

existing road was within the construction area of the new road. The continuous passage of the normal commercial traffic was a main condition, which implied that during construction the materials used as road fill had to serve as a road pavement.

The road fill materials were obtained from seven borrowpits located alongside the construction area. Starting at Moengo, these borrowpits were indicated in alphabetical order: borrowpit A (km 2), B (km 6), C (km 11), D (km 18), F (km 23), H (km 26) and I (km 42). Between km 28 and km 41 the surrounding sand was used as road fill, covered by a 10 cm thick laterite layer.

The lateritic material used was obtained from various horizons in the borrowpit profile, as shown in figure 1.

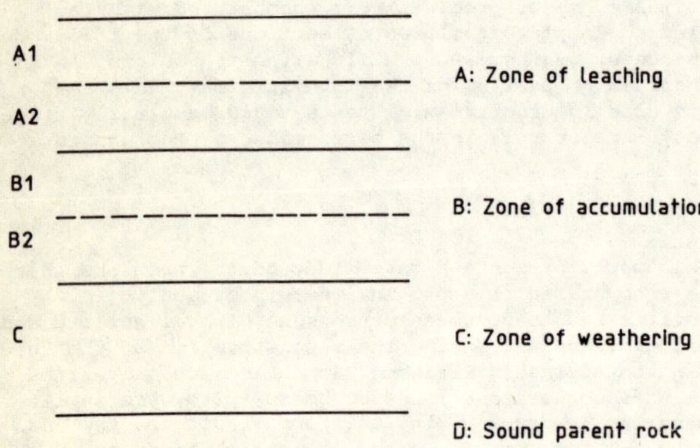

Figure 1. Schematic representation of soil profile.

The lateritic materials used will be coded in this paper using three characters: the first character indicating the borrowpit, and the second and third character indicating the horizon from which the material was obtained. H A2, for example, thus is a material from borrowpit H, horizon A2.

3 PROBLEMS ENCOUNTERED DURING CONSTRUCTION

The selection of lateritic materials for the road fill, the sub-base and the base of the road Moengo-Albina was mainly based on engineering judgement and experience. Standard, temperate-zone developed specifications could not be used, since materials meeting these specifications simply were not available. These standard specifications are often too rigid for application on low-volume road projects in the tropics: lateritic materials not meeting standard specifications have been used successfully in road building projects throughout the tropics (e.g. Grace et al., 1985).

The main complicating factor during the construction of the road however was that the normal commercial traffic had to use the road under construction. This commercial traffic amongst others involved 15 ton loading capacity trucks transporting agricultural products and 35 ton loading capacity mining trucks. On top of the commercial traffic, 20 on-the-job trucks of 8 ton loading capacity travelled the road under construction.

During the dry season of the construction year no serious problems were encountered in spite of the heavy traffic loading. However, in the following wet season rainfall was the heaviest in the last ten years with monthly averages of over 500 mm. Whole sections of the road under construction failed under the combined influence of rainfall and traffic, and had to be rebuilt completely. In other sections of the road, however, the lateritic materials used showed adequate strength to carry the heavy traffic loading.

Following the major problems encountered, the decision was taken to learn as much as possible from these costly mistakes, and perform a detailed back-analysis of the road failures. This analysis consisted of two parts:

- A detailed compositional analysis of the laterites used in the construction of the road Moengo-Albina, involving particle size analysis, chemical analysis etc.
- A detailed mechanical analysis of selected samples, using cyclic loading triaxial tests.

The compositional analysis was performed in Suriname, and the mechanical analysis was carried out at Delft University of Technology in the Netherlands.

4 COMPOSITIONAL ANALYSIS

For the compositional analysis, samples were taken from each of the horizons in all the seven borrowpits. A selection of these materials was later investigated in the cyclic loading triaxial tests. To limit the length of this paper, only these materials will be dealt with here. The full details of the compositional analysis were reported elsewhere |Tjong Tjin Joe, 1982|. Table II shows the materials dealt with in this paper, together with a qualification of the behaviour in the pavement construction.

Table II. Materials tested

borrowpit	horizon	material code	performance
H	A2	HA2	poor
A	B1	AB1	good
B	B1	BB1	good
I	B1	IB1	good
B	B2	BB2	poor
A	C	AC	poor

Table III shows the particle size distributions of the materials tested.

Table III. Particle size distributions

material	percentage passing square opening |mm|							
	0.075	0.150	0.300	0.600	2.36	4.75	9.5	19	25
HA2	41	45	53	68	97	100			
AB1	24	25	29	35	50	55	65	87	92
BB1	29	30	32	34	55	57	63	85	91
IB1	31	34	41	49	76	84	89	98	100
BB2	50	50	53	58	93	95	96	100	
AC	54	58	66	75	87	92	96	100	

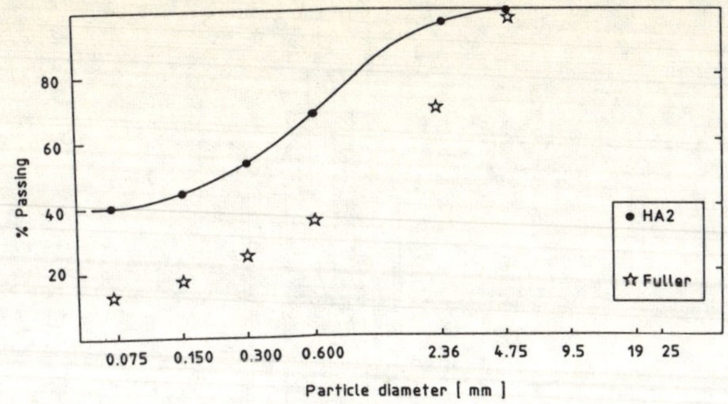

Figure 2. Particle size distribution of material HA2.

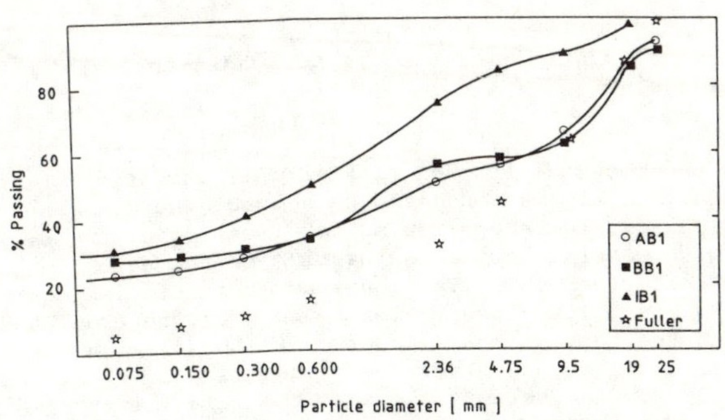

Figure 3. Particle size distribution of materials AB1, BB1 and IB1.

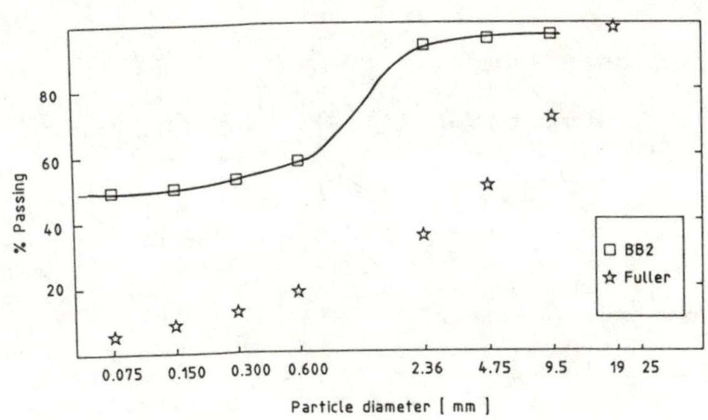

Figure 4. Particle size distribution of material BB2.

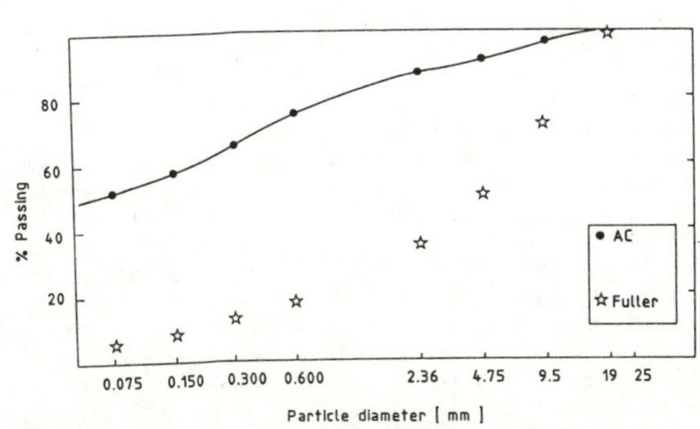

Figure 5. Particle size distribution of material AC.

Figures 2 through 5 show a graphical representation of these particle size distributions. In these figures, the materials are grouped together according to the horizon from which they were obtained.

As can be seen from figures 2 through 5, all the materials tested have continuous gradings. Fuller and Thompson (1907) have shown that for a particulate medium to have a maximum number of intergrain contact points, the material should have a particle size distribution represented by the formula:

$$p = 100 \sqrt{\frac{d}{d_{max}}}$$

where p = percentage smaller than particle diameter d
d_{max} = maximum particle diameter in the material.

The optimal grading curves calculated from the maximum particle diameter of each group have also been plotted in figures 2 through 5. As can be seen from figures 2, 4 and 5, the materials from the A2, B2 and C horizons have particle size distributions that deviate substantially from the Fuller curve. These materials show a substantial shortage of coarse particles and a clear surplus of fine material. Although these materials do contain some gravel sized particles, the percentage of these particles is quite low. Practically speaking, these materials are therefore mixtures of sand-sized particles and fines.

Such materials might be used as road fill, if they are applied in dry weather and paved before the start of the rainy season. For practical reasons, the latter condition was not met during the construction of the road Moengo-Albina, and the materials from the A2, B2 and C horizons softened dramatically under the influence of rainfall and traffic, and ultimately had to be replaced by other materials.

Figure 3 shows that the materials from the B1 horizon (AB1, BB1 and IB1) contain a substantial percentage of gravel-sized particles. The materials AB1 and BB1 have particle size distributions that lie relatively close to the Fuller-curve. These materials performed well in the road construction, and sustained the combined action of rainfall and traffic without problems.

The particle size distribution of material IB1 deviates substantially from the Fuller-curve: as the materials from the other horizons, this material too shows a shortage of gravel sized particles and a surplus of fines. The percentage of gravel sized particles, however, is substantially higher than for the materials from the other horizons. This is believed to be one of the reasons that material IB1 performed well in the road construction. As will be shown later, material IB1 furthermore has a much higher iron content than the materials from the other horizons. Alternate wetting and drying cycles may therefore have led to an increase in strength of this material after placement in the road construction. Grace et al. |1985| reported such increases in strength under influence of trafficking superimposed on cycles of wetting and drying.

Table IV. Chemical composition

material	L.O.I. %	SiO_2 %	Fe_2O_3 %	TiO_2 %	Al_2O_3 %
HA2	8.8	64.5	8.3	1.0	17.4
BB1	14.6	14.8	43.8	1.45	25.4
IB1	4.7	73.5	19.0	0.6	2.2
BB2	12.9	43.8	8.9	1.25	33.2

Table IV shows the chemical composition of the materials HA2, BB1, IB1 and BB2. For lack of sufficient testing capacity, the materials AB1 and AC were omitted from the detailed chemical analysis, AB1 because it was judged to be almost an exact copy of BB1, and AC because on sight it was judged to be a kaolinitic clay with some coarse particles in it.

The specific gravity of the laterites was determined using the pycnometer-method |Head, 1980|. Table V shows the results of this determination.

Table V. Specific gravity

material	specific gravity
HA2	2.82
AB1	2.85
BB1	3.08
IB1	2.90
BB2	2.84
AC	2.80

The particle size distribution of the materials (see Table III), their chemical composition (see Table IV) and their specific gravity (see Table V) are - of course - closely interrelated: a high percentage of iron leads to an increase in coarse graded particles and to an increase in specific gravity. The iron content therefore is clearly indicative of the material behaviour in the pavement construction: a higher iron content leads to a better in-situ performance.

5 CYCLIC LOADING TRIAXIAL TESTS

To make a detailed study of the mechanical properties of the materials used in the construction of the Moengo-Albina road, samples of a limited number of selected materials were investigated using cyclic loading triaxial tests. Since equipment for these rather complicated tests was not available in Suriname, the tests were carried out at Delft University of Technology in the Netherlands. The materials to be tested were selected in such a way, that from each of the horizons from which materials were used in the road construction at least one material was tested. From the B1-horizon, from which the materials that performed well in the road construction, three materials were investigated (borrowpits A, B and I). From the poor-performance materials, one material from the A2 horizon was tested (borrowpit H), one from the B2-horizon (borrowpit B) and one from the C-horizon (borrowpit A).

Figure 6 shows the schematics of the cyclic loading triaxial apparatus used in this investigation. A cylindrical specimen of 100 mm diameter and 200 mm height is placed on the pedestal of the triaxial cell. A rubber membrane encloses the specimen, with an airtight connection to pedestal and top cap. A constant confining stress is applied to the specimen through air pressure in the triaxial cell. Axial loading of the specimen is performed through an MTS servo-hydraulic actuator, connected to the loading ram of the triaxial cell. A load cell incorporated in the loading ram measures the applied load. Axial deformation of the specimen is measured by an LVDT (Linear Variable Displacement Transducer), connected to the loading ram outside the triaxial cell. Radial deformation of the specimen is measured by three non-contacting sensors at midheight of the specimen.

The soil suction in the triaxial specimen is measured throughout the test by means of a so-called "tensiometer". A high air-entry ceramic tip protrudes from the pedestal of the triaxial cell into the specimen. The ceramic tip is connected by a double hose to the tensiometer, thereby allowing for flushing of the ceramic tip with water to obtain saturation of the ceramic tip/ tensiometer assembly. The tensiometer itself is required with a low volume-change electronic pressure transducer. During the test, the soil suction in the triaxial specimen will cause a very small quantity of water to travel from the ceramic tip into the triaxial specimen, until equilibrium is reached between the negative pore water pressure in the specimen and the resulting negative pressure in the tensiometer. The latter pressure can be read from the read-out unit connected to the electronic pressure transducer.

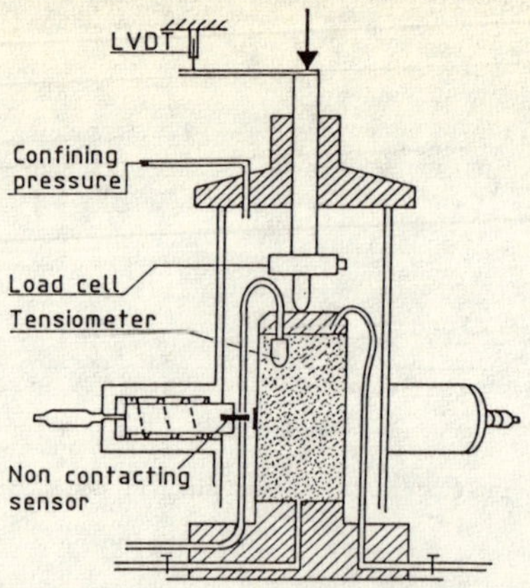

Figure 6. Principle of the cyclic loading triaxial apparatus.

6 SPECIMEN PREPARATION

Samples of the materials HA2, AB1, BB1, IB1, BB2 and AC were transported from Suriname to Delft for triaxial testing. The materials from the B1-horizon (AB1, BB1 and IB1) contained small amounts of particles with a diameter d > 19 mm (see Table III). In view of the limited dimensions of the triaxial specimen, the particles with d > 19 mm were crushed to d < 19 mm, and then re-added to the sample material.

The sample material was then brought to the moisture content required for compaction, and subsequently compacted in a 100 mm diameter steel mould. Compaction was done in five layers, using a full-faced compaction drop-hammer. Per layer, ten blows of the hammer were applied, resulting in a compaction energy equal to that of AASHTO T99-74 standard Proctor compaction level |AASHTO, 1978|. To obtain a smooth transition between the successive compaction layers, each layer was duly scratched after compaction, prior to admission of the material of the following layer. After completion of the compaction procedure, the specimen was extruded from the steel mould and placed in the triaxial cell.

Since it was the aim of the triaxial testing to back-analyse the problems encountered in the road construction on wetting of the materials, the triaxial tests were carried out at two material conditions, namely a relatively dry condition and a very wet condition. After compaction, the triaxial specimens were first brought to a specified degree of saturation of S_r = 70%, either by drying or wetting of the specimen, depending on the initial degree of saturation after compaction. The degree of saturation of the triaxial specimen during drying or wetting was determined by continuously weighing of the specimen/triaxial pedestal assembly. The amount of water lost or taken in by the specimen could thus be measured, and the degree of saturation S_r could be calculated from the known initial moisture content, the moisture loss or intake and the specific gravity of the material at hand.

After the required degree of saturation S_r = 70% for the "dry" condition was reached, the triaxial specimen was tested for resilient properties. After completion of the tests, the specimen was then allowed to take in water under a hydraulic head of 150 kPa for a period of approximately one week, again with continuous weighing of the specimen/pedestal assembly. After this wetting-procedure, the triaxial specimens were again tested for resilient properties. Table VI shows the specimen characteristics after compaction, and in the "dry" and the "wet" testing conditions.

Table VI. Specimen characteristics.

material	compaction			dry			wet		
	ρ_d |kg/m³|	w |%|	S_r |%|	w |%|	S_r |%|	pF |-|	w |%|	S_r |%|	pF |-|
HA2	1694	21.3	90.4	16.5	70.0	>2.9	21.6	91.8	2.3
AB1	1890	10.2	57.0	12.5	70.0	1.9	15.5	87.1	1.8
BB1	1851	11.9	55.3	15.1	70.0	1.7	18.8	87.3	1.8
IB1	1599	24.4	86.9	19.6	70.0	>2.9	25.1	89.5	1.9
BB2	1860	17.1	91.1	13.2	70.0	>2.9	17.3	92.2	2.2
AC	1560	25.7	90.6	19.8	70.0	>2.9	25.1	88.4	2.2

Also shown in table VI is the soil suction in the triaxial specimens, as measured with the tensiometer. The suction is expressed in the standard pF-units, where pF is the logarithm of the negative pore water pressure in cm H_2O. As can be seen from table VI, the materials from the A2, B2 and C horizons in the dry condition have a suction of pF > 2.9 (due to the tensiometer principle suction values of pF > 2.9 cannot be measured with the tensiometer). These high suction values are attributable to the large amount of fines in these materials. The coarser graded materials from the B1-horizon (AB1 and BB1) show modest suction values in the dry condition, whereas the more finely grained material IB1 from the same horizon again has a suction of pF > 2.9 in the dry condition.

After wetting, all specimens have suction values within the measuring range of the tensiometer. The materials from the A2, B2 and C horizons and material IB1 show a substantial decrease in suction values, whereas the suction in materials AB1 and BB1 are practically constant. The measured small increase in suction upon wetting of material IB1 indicates the accuracy of this way of measuring soil suction, since in reality suction of course cannot increase upon wetting.

7 CYCLIC LOADING TRIAXIAL TEST PROCEDURE

The principle of the cyclic loading triaxial test is to simulate as closely as possible the stress conditions occurring in pavement constructions, and to measure the resulting deformation. Figure 7 shows the stresses imposed on the triaxial specimen in the cyclic loading triaxial tests. The constant confining stress σ_3 simulates the overburden stress in the pavement construction, whilst the cyclic deviator stress σ_d simulates the axial stress caused by traffic loading.

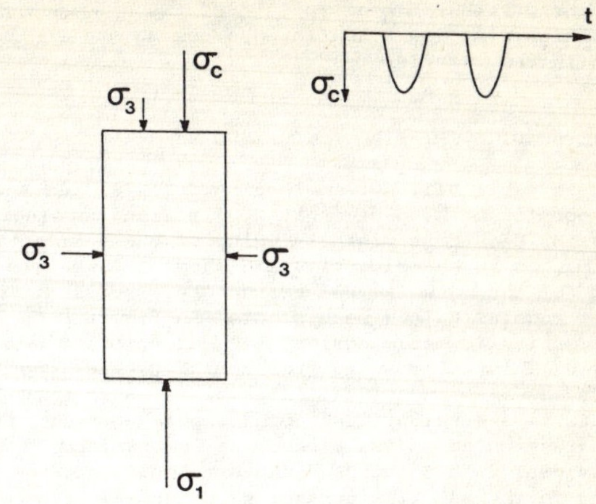

Figure 7. Stresses imposed on specimen in cyclic loading triaxial test.

Cyclic loading triaxial test results are usually interpreted using the resilient modulus M_r, which is calculated according to:

$$M_r = \frac{\sigma_d}{\varepsilon_{a,r}} \tag{1}$$

where M_r = resilient modulus |MPa|
σ_d = cyclic deviator stress |kPa|
$\varepsilon_{a,r}$ = axial resilient deformation |10^{-3}|

For soils and unbound granular materials, M_r is stress-dependent. Therefore, tests have to be run at a number of different combinations of stresses to obtain the relationship between resilient modulus and stress level. For granular materials, this relationship is often expressed using the well-known K-θ model:

$$M_r = K_1 \left(\frac{\theta}{\theta_0}\right)^{K_2} \tag{2}$$

where θ = sum of principal stresses |kPa|
θ_0 = 1 kPa = reference stress |kPa|
K_1 = material parameter |MPa|
K_2 = material parameter |-|

The reference stress θ_0 in equation (2) is used for mathematical reasons only, i.e. rendering the stress parameter in the equation dimensionless.

For the research described herein, cyclic loading triaxial tests were carried out at five levels of the confining stress σ_3, and at three to five levels of the deviator stress σ_d. Table VII shows the stresses applied.

Table VII. Stresses applied in cyclic loading triaxial tests.

σ_3 |kPa|	σ_d |kPa|				
10	10	20	30	40	50
20	20	40	60	80	
30	30	60	90	120	
40	40	80	120	160	
50	50	100	150		

As can be seen from this table, the higher ratios of deviator stress over confining stress were only used at the lower levels of confining stress. This was done in view of the stress ratio at failure of the specimens, which is higher at lower confining stresses due to the cohesion in the material tested.

8 CYCLIC LOADING TRIAXIAL TEST RESULTS

As shown in table VII, cyclic loading triaxial tests were carried out for each specimen at 20 different combinations of stresses. Each test yields a value of the resilient modulus M_r. As an example, figure 8 shows the results obtained for the material HA2 in the dry condition. The resilient modulus M_r has been plotted in figure 8 on a double-logarithmic scale against the sum of principal stresses θ. The test results from the 20 cyclic loading tests are shown as individual data points. Also shown in the figure is a straight line according to equation (2). Using regression analysis, a best fit is obtained between the individual data points and the straight line, resulting in the material parameters K_1 and K_2. The multiple correlation coefficient squared R^2 is a measure of the fit between the straight line and the individual data points.

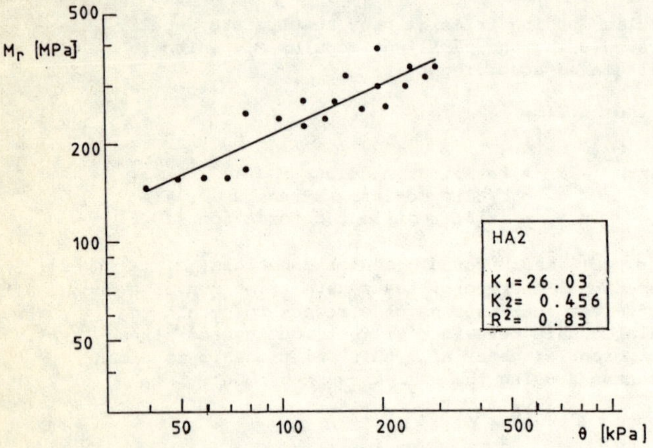

Figure 8. Cyclic loading triaxial test results for material HA2, dry condition.

Table VIII summarizes the results of the cyclic loading triaxial tests on the six materials, tested in the dry and the wet condition.

Table VIII. Cyclic loading triaxial test results

Material	dry			wet		
	K_1 |MPa|	K_2 |-|	R^2 |-|	K_1 |MPa|	K_2 |-|	R^2 |-|
HA2	26.0	0.46	0.83	60.8	0.03	0.04
AB1	15.6	0.51	0.71	15.9	0.50	0.71
BB1	92.6	0.15	0.14	28.9	0.36	0.52
IB1	76.1	0.26	0.69	17.4	0.22	0.25
BB2	18.8	0.53	0.81	15.7	0.32	0.42
AC	16.7	0.45	0.72	135.3	-0.24	0.37

Figure 9 shows the M_r-θ relationships for the materials from the A2, B2 and C horizons, for the dry and wet condition. All three materials show a substantial drop in M_r upon wetting. Both in the dry and the wet condition, materials HA2 and BB2 show a higher resilient modulus than material AC. After wetting, material AC shows a typical soil-like behaviour (decreasing stiffness with increasing stress).

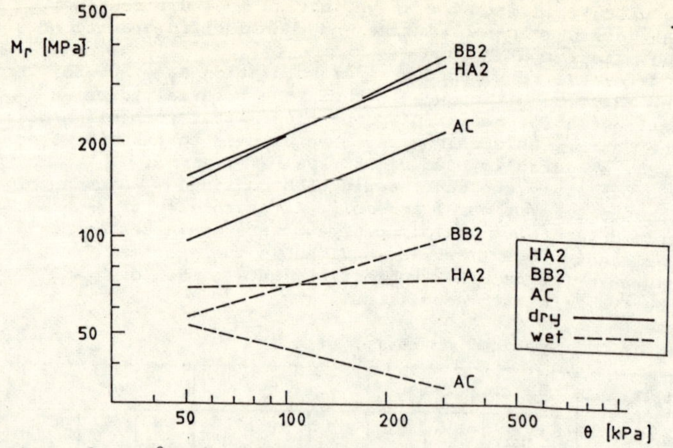

Figure 9. M_r-θ relationships for materials HA2, BB2 and AC, for the dry and wet condition.

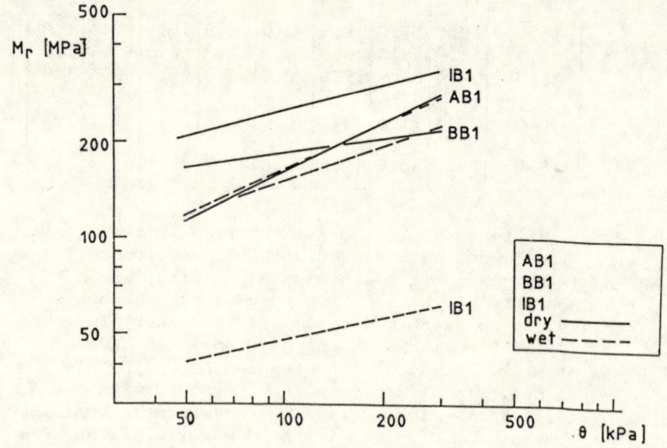

Figure 10. M_r-θ relationships for materials AB1, BB1 and IB1.

Figure 10 shows the M_r-θ relationships for the materials from the B1 horizon, for the dry and the wet condition. For material AB1, the resilient modulus remains constant upon wetting, whereas for material BB1 the resilient modulus shows a decrease in the low stress region and remains constant in the high stress region. The behaviour of material IB1 is quite distinct from that of AB1 and BB1; this material shows a drastic drop in resilient modulus upon wetting, as did the materials from the other horizons.

9 DISCUSSION

In the road Moengo-Albina, the materials from the A2, B2 and C horizon performed properly in the dry condition. Upon wetting during the rainy season, however, whole sections of the road built with these materials failed completely under the influence of traffic. The results from the cyclic loading triaxial tests are consistent with this in-situ behaviour: in the dry condition, resilient moduli for these materials were of the same order of magnitude as those for the materials from the B1 horizon. Upon wetting a substantial drop in resilient moduli was found in the cyclic loading triaxial tests.

The materials from the B1 horizon performed well in the road pavement, both in the dry and the wet condition. For the materials AB1 and BB1, cyclic loading triaxial test results are consistent with this in-situ behaviour: high values for resilient moduli were found for the dry condition, and were retained upon wetting of the triaxial specimens. For the material IB1, in-situ and laboratory performance are not consistent: material IB1 performed well in-situ both in the dry and the wet condition, whereas a substantial reduction in resilient moduli was found upon wetting of the specimen.

As discussed earlier, the materials from the A2, B2 and C horizons are poorly graded, with a shortage of coarse graded particles and a surplus of fines (see figures 2 through 5). Thanks to the abundance of fines, these materials have a high soil suction in the dry condition (see table VI). Soil suction is well known to increase the strength of unsaturated soils considerably |Fredlund and Rahardjo, 1985|. The good in-situ performance in the dry condition and the resilient moduli in that condition are attributable to soil suction. Upon wetting, the soil suction in these materials is reduced substantially, resulting in a drastic drop in resilient moduli and failure in the road construction.

The materials AB1 and BB1 have far better gradings than the other materials. They contain a larger amount of coarse graded particles, and far less fines. The soil suction in these materials in the dry condition is consequenty lower, and is virtually constant upon wetting. These better graded materials obtain their strength and stiffness from aggregate interlock, and not from soil suction. Therefore, upon wetting strength and stiffness remain intact for these materials.

Material IB1 has a poorer grading than materials AB1 and BB1, in the sense that IB1 shows a shortage of coarse graded particles. In the dry condition, this material has a high soil suction, and consequently high resilient moduli. The suction is reduced substantially upon wetting, with a resulting drop in resilient moduli. So far, the performance of material IB1 is parallel to that of the materials from the A2, B2 and C horizons. In-situ, however, material IB1 performed well both in the dry ànd the wet condition. This good performance in the wet condition of this high iron-content material must be attributable to an increase in strength under the combined influence of trafficking and alternate wetting and drying, as reported by Grace et al. |1985|.

10 CONCLUSIONS

1. The good performance in the dry condition of the poorly graded lateritic materials investigated is attributable to high values of soil suction. Upon wetting, the suction is reduced substantially, resulting in a drastic drop in bearing capacity.

2. The good performance of the better graded lateritic materials is attributable to aggregate interlock, which is not affected by water. Therefore, these materials performed well in the dry ànd the wet condition.

ACKNOWLEDGEMENT

These investigations were supported in part by the Netherlands Technology Foundation (STW).

REFERENCES

American Association of State Highway and Transportation Officials (1978), "Standard specifications for transportation materials and methods for sampling and testing", Part 2, "Methods for sampling and testing", AASHTO, Washington D.C.

Fredlund, D.G. and B.H. Rahardjo (1985), "Theoretical context for understanding unsaturated residual soil behaviour", Proceedings, First International Conference on Geomechanics in Tropical Lateritic and Saprolitic Soils "Tropica LS '85", Brasilia, Volume 1, pp. 295-306.

Fuller, W.B. and S.B. Thompson (1907), "The laws of proportioning concrete", A.S.C.E. Transcripts, Vol. 59, Paper No. 1053.

Grace, H., D.W. Higth and D.G. Toll (1985), "Some recent work on the use of locally occurring materials as bases for bituminous surfaced low volume roads", Proceedings, First International Conference on Geomechanics in Tropical Lateritic and Saprolitic Soils "Tropica LS '85", Volume 2, Brasilia, pp. 295-307.

Head, K.H. (1980), "Manual of Soil Laboratory Testing", Volume 1, "Soil Classification and Compaction Tests", Pentech Press, Plymouth.

Sweere, G.T.H. (1982), "Genesis of laterites and its engineering implications", Proceedings, FURORIS-conference, University of Suriname, Paramaribo, pp. 61-68.

Sweere, G.T.H. and P.J. Galjaard (1984), "Compaction and strength characteristics of laterites as related to soil composition", Proceedings, International Conference on Roads and Development, Paris, pp. 835-841.

Tjong Tjin Joe, J.F.H. (1982), "Geological aspects of laterite soils used in the construction of the road Moengo-Albina", Proceedings, FURORIS-conference, University of Suriname, Paramaribo, pp. 69-84.

New considerations about typical tropical soils

E.VERTAMATTI, Instituto Technologico de Aeronautic, Brazil

ABSTRACT : Taking into account the importance of soil properties determination for engineering purposes, in special to the pavement constructions, and looking for new understanding of geotechnical peculiarities of typical tropical soils, a synthesis about a treatment of sedimentary, plinthitic and concretioned lateritic soils is presented considering the Brazilian Amazon soils gathered in the airfields of that region as pattern group. The MCT and Resilient methodologies were employed for the analysis over materials with real granulometry, using appropriate equipment for this. Accelerated changes in the properties of plinthitic soils are also considered. The results obtained permit the establishment of new geotechnical behavior standards adapted for the adequate selection and use of that typical tropical soils in pavement layers.

1 INTRODUCTION

In recent years it has been noted an increase in the number of worldwide researches about tropical soils technology in civil engineering, all of them contributing to develop more adequate criteria for use in tropical countries. This is due to the fact that tropical soils have shown different behaviors when compared to soils of other origins and that they are not conveniently represented by classical soil classification systems like HRB and USCS. In this case,many attempts (using objective or subjective factors) were elaborated to identify the intrinsic characteristcs of tropical soils and to foresee their geotechnical properties (TropicaLS'85). One of these, developed in Brazil for road purposes and denominated MCT - Miniature Compacted Tropical (Nogami&Villibor 1981,1985), permit the consideration of lateritic or non lateritic (saprolitic) behavior and the properties of different soils in relation to their genetic characteristics. Another criterium, also developed in Brazil and originated by studies of resilient deformations of flexible pavements (Preussler 1983, Medina 1985), permit to estimate resilient properties of materiais based on granulometric aspects. Both criteria were successfully employed in the study of many Brazilian soils.

Considering that the knowledge of tropical soils increase in proportion to the treatment of pedological formations not yet explored and that the Brazilian Amazon tropical soils were very little analyzed, a specific research was developed at the Aeronautical Institute of Technology in Brazil by the author (Vertamatti 1988) about the peculiarities of these soils.

For the research samples gathered in regional airfields were used to determine their unknown MCT and Resilient characteristics, with the purpose of analyzing the rational employment of local materials in pavement construction.

The importance of the study of Amazon soils is that this region covers about sixty-two percent of the Brazilian land, which corresponds to five thousand square kilometers,where there is a large occurence of fine sedimentary, plinthitic and concretioned soils, as can be noted in other tropical countries (Africa, for exemple). The airport reality was considered due to the fact that it is a very important transport system covering all the region over diverses pedological an geological units, where more than one hundred and twenty airfiels (see Figure 1) have been constructed in the last thirty years by the COMARA (The Federal Agency for airfield construction in the Amazon headed by the Ministry of Aeronautics).

Figure 1. Amazon airfield planning network.

2 PECULIARITIES OF TYPICAL AMAZON TROPICAL SOILS

In view of local climatical (Afi, Ami and Awi in the Köeppen climate classification system), geomorphological (predominance of low planalts and montain zones), geological (Pré-cambriane and Çenozoic formations) and fluvial (large abundance of rivers) conditions, there were proper conditions for the formation of typical amazon tropical soils which can be grouped in three peculiar cathegories : fine sedimentary, plinthitic and concretioned (laterite) soils.

A statistical treatment of geotechnical data available in constructed airfields (Vertamatti 1986) showed that these types of soils have distinct granulometrical bands but practically do not have coarse and medium sand fractions and fine sand and fine fractions with ample values. Because of this, it was possible to use the sieve number 10 (mesch of 2 mm) as a new class separator, since the fine sedimentary soils pass integrally through sieve 10; the concretioned soils have more than 50% of material retained in sieve 10 (only iron nodulus or with rounded quartz gravel in the case of fluvial terraces) and the plinthitic soils have low or null percentage of iron nodulus retained in sieve 10. These soils also have predominance of CBR varying from 3 to 150, maximum dry density from 16,7 KN/m³ to 22,1 KN/m³, optimum moisture from 5% to 18%, low to medium plasticity and silt fraction percentage in the fine fraction from 10% to 50%. These aspects added to mineralogical analysis of sand fraction (predominance of quartz and iron particles in some cases) permit to ponder that silt fractions are of quartzific nature. The clay fraction is formed by kaulinite and iron and alumine oxides-hidróxides respectively of 4% and 17%. The characteristics above give a peculiar behavior to Amazon soils.

The plinthitic soils are materials which have iron segregation in red mosqueate forms and occur in plane topography near to the water level fluctuation zones.

In relation to the concretioned soils, although they have been considered as a singular material up to now, it was verified that they present a great variety of granulometry (50% to 90% retained in sieve 10) and physical characteristics of concretions (mature or hidromorfic nature, smooth to very rough superficial texture, lengthed to rounded form, massive to conglomerate structure, brittle or not by high finger pressure or absence of cavities, null to striking aderence of fines, absortion from 2% to 23%, and real density from 22,5 KN/m³ to 36,3 KN/m³, for example).

An interesting aspect to be noticed is that there is a cicle of granulometric evolution from plinthitic to concretioned soils, as presented in Figure 2. Using this figure and granulometric data of the soil it is possible to foresee a qualitative evolutional stage of concretioned soils and their peculiarities. This fact can be observed in the Amazon region due to the predominance of these soils in differents topography and drenability conditions.

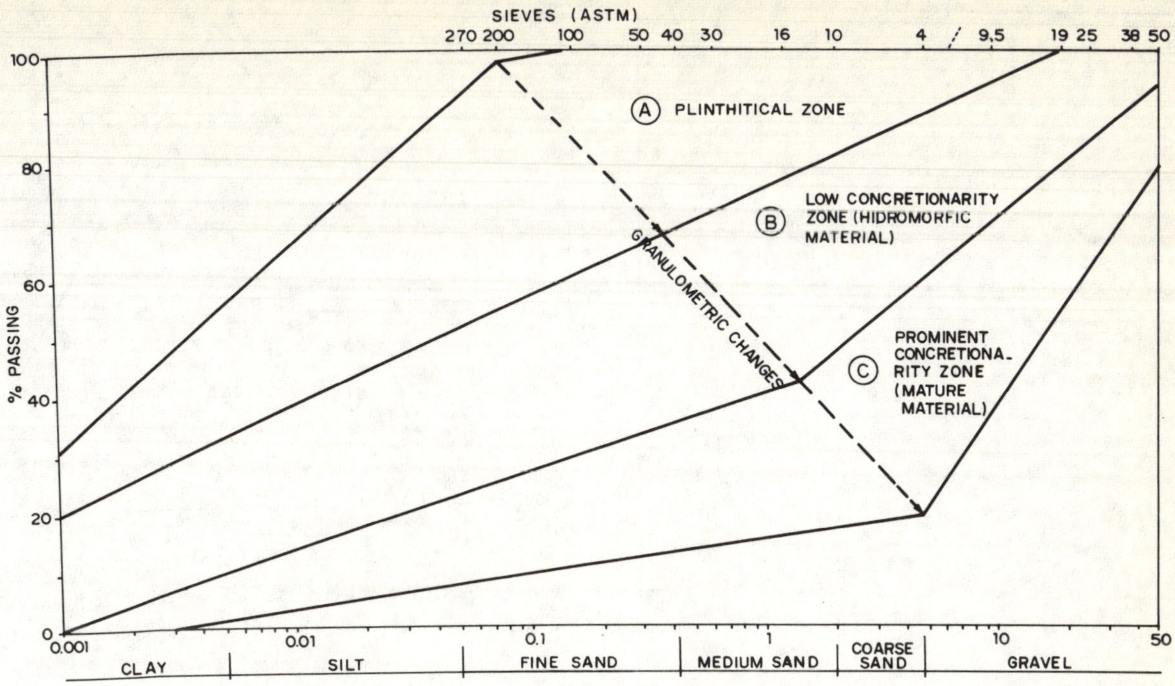

Figure 2. Cicle of granulometric evolution from plinthitic to concretioned soils.

One attempt to associate CBR (modified energy) to IG (Group Index) is presented in Figure 3, where the absence of correlation among those parameters in Amazon tropical soils can be noted, which is exactly the opposite of what happens in temperate soils (exemplified by AASHTO 1962, at standard energy).

3 THE PEDOLOGICAL CONDITIONING OF AIRFIELD CONSTRUCTION

The Amazon reality of materials produce a remarkable pedological conditioning of airfield construction. Most of the airfields were constructed using concretioned soils as base and/or subbase layers while soil stabilization (using cement or lime) was employed in some places where materials with adequate bearing capacity were not found. It has been registred that many pavements constructed with concretioned soils had a good geotechnical performance (high bearing capacity, easy drenability and good erosion resistence) while the use of stabilized soils have laid to expensive solutions which were not always of long durability.

In this last case the aggregates for asphalt concrete needed to be transported by raft of long distances (more than one thousand kilometers in some cases) in view of the lack of rocky outcrop.

Considering the types of pavement solutions employed and the Amazon soil formations (Embrapa 1981), it was possible to identify macro-zones with distinct geotechnical qualifications, which are illustrated in Figure 4. It is important to point out that exactly at the Solimões podzolic bazin where the soil stabilization has been adopted, there is a large occurence of plinthitic soils not yet considered as construction material.

4 THE MCT AND RESILIENT TREATMENT

The MCT criterium is based on the analysis of total compaction curves and lost of soaking relatives to samples prepared in different moisture conditions (with 5 cm diameters by 5 cm height) only considering materials or fractions passing through sieve 10. The Resilient treatment was made considering samples with 5,

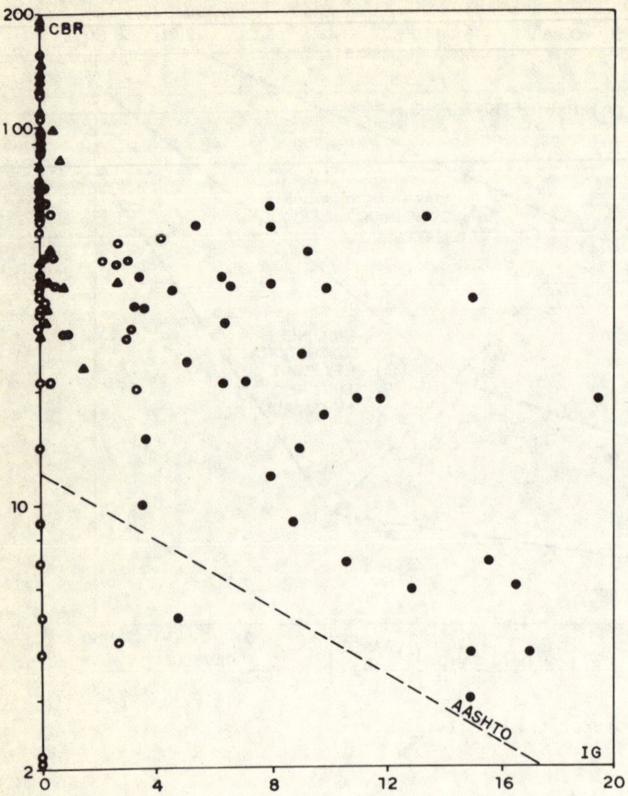

Figure 3. Association between the CBR and the IG parameters relative to fine (●), sandy (o) and concretioned (Δ) soils when compared with the AASTHO correlation example.

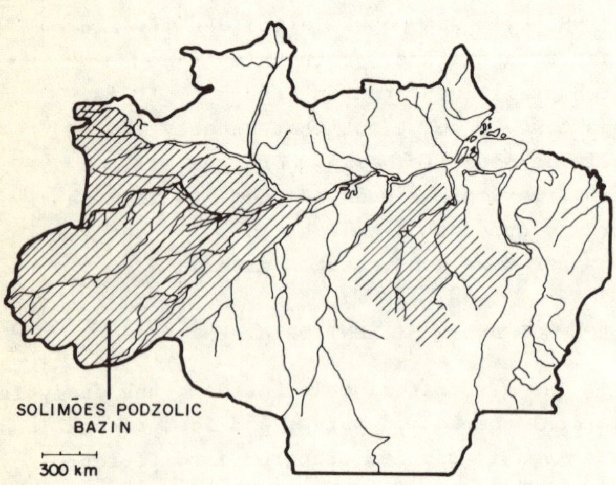

Figure 4. Amazon macro-zones with distinct geotechnical qualifications : employing of soil stabilization (hatched) or concretioned soils ( ).

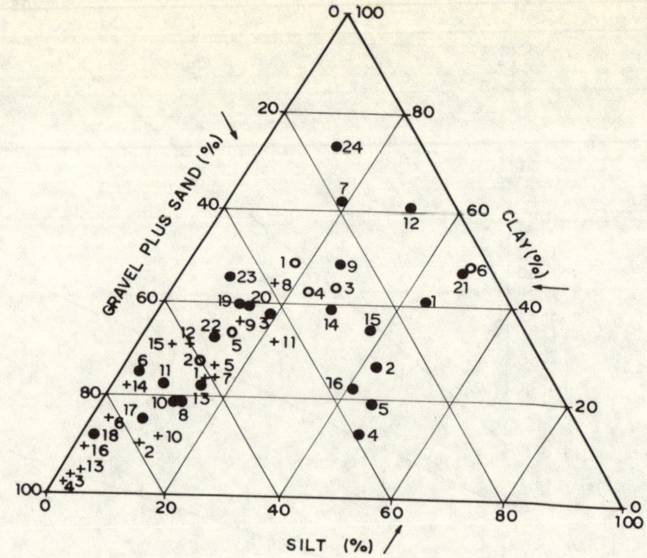

Figure 5. Granulometric Trilinear distribution of fine sedimentary (●), plinthitic (o) and concretioned (+) soils.

10 or 15 cm diameters (relation height-diameter of 2:1) employing a proper equipment developed for this (monitored by microcomputer). Some adaptations of this criteria were made (Vertamatti 1988) in order to permit the studies of the Amazon soils.

For the analysis, twenty-four fine sedimentary, six plinthitic and sixteen concretioned soil samples were gathered in airfields situated in different regions and soil formations, and corresponding to sub-grade, sub-base or base materials. The granulometric trilinear distribution of these samples is presented in Figure 5.

4.1 Fine sedimentary and plinthitic soils

The final results of fine sedimentary and plinthitic soils relative to the MCT treatment showed that these soils presented a transient behavior between lateritic an saprolitic soils and that the silt samples did not disclosed a saprolitic character (more typical of kaulinitic or micaceous silts). This fact reinforces the pondered quartzific nature of Amazon sedimentary silt soils. Sample number 20 corresponds to the soil denominated "albine lateritic soil" which has predominance of alumine compounds and geotechnical behavior similar to lateritic soils rich in iron oxides. The transient soils can be a mixture of different materials in partial stages of pedological evolution due to their recent geological deposition.

In respect to the resilient results the soils presented resilient modulus between the categories III to I. This variation band is exactly the one which needs a more criterious analysis to define the soil aplicability in pavement layers. The characteristics of fine Amazon soils is not sufficient to attribute high resilient modulus as occur in very evolutioned lateritic fine sand soils, but they have a better resilient quality when compared to micaceous soils and superior to that of temperate soils as exemplified by the AASHTO (1986).

From the obtained results it was possible to establish an association between the MCT and the resilient parameters (in abacus form) to estimate soil engineering properties, firstly tested by Vertamatti (1985).

Figure 6 shyntetizes the results attained.

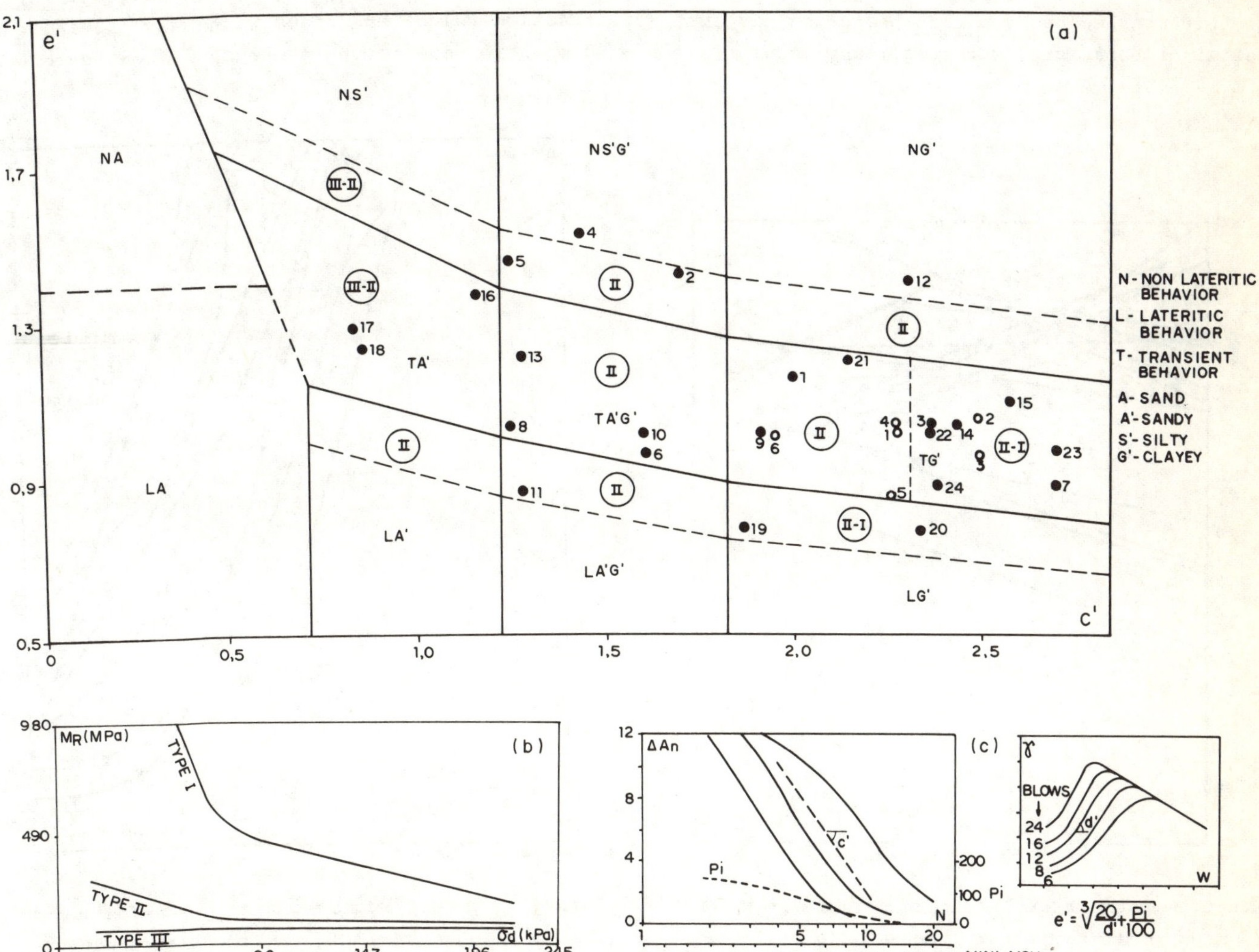

Figure 6. MCT and Resilient results attained (a) relative to fine sedimentary (●) and plinthitic (o) soils, with the resilient (b) and MCT (c) parameters definition.

4.2 Concretioned soils

Concretioned soils occur in many tropical countries and it has been investigated a lot in the last years (TropicaLS' 85) mainly to define specifications of base and sub-base materials using elements like granulometry, plasticity and abrasion. But, there is little knowledge of resilient properties in these soils.

Figure 7 presents resilient curves of concretioned soils considering variations of confinant stress in samples molded at modified energy with 10 or 15 cm diameter according to their granulometry. Due to different values of water absortion, the moistures are determined using natural portions of the samples.

It can be observed that there are distinct resilient behaviors, with resilient modulus varing from 49 MPa to 980 MPa according to the materials and confinant stresses. Due to this, it can be pondered that not only the granulometry but also factors like physical peculiarities of the concretions, genetical aspects, relative percentage of each fraction present and laterization degree of the fine fraction have a high influence to define their behaviour.

In the optimum condition of compaction samples were molded at the modified energy recording variations of height. In Figure 8 it can be noted that the compaction efficiency is very variable in these soils. There were cases of structural wedging for low energy (high energy level was not convenient); cases where final compaction was not reached (elevate fine fraction value) and cases with easy initial compaction due to the presence of concretions

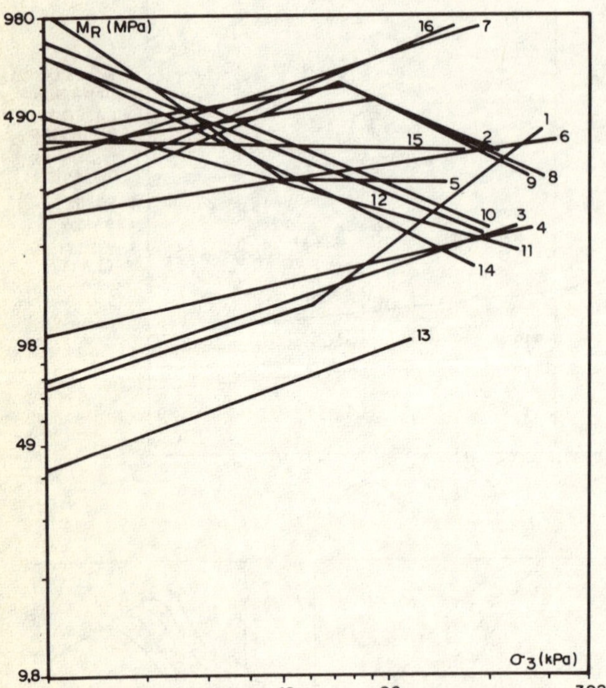

Figure 7. Resilient Modulus (M_R) of concretioned soils relative to confinant stress (σ_3) variations.

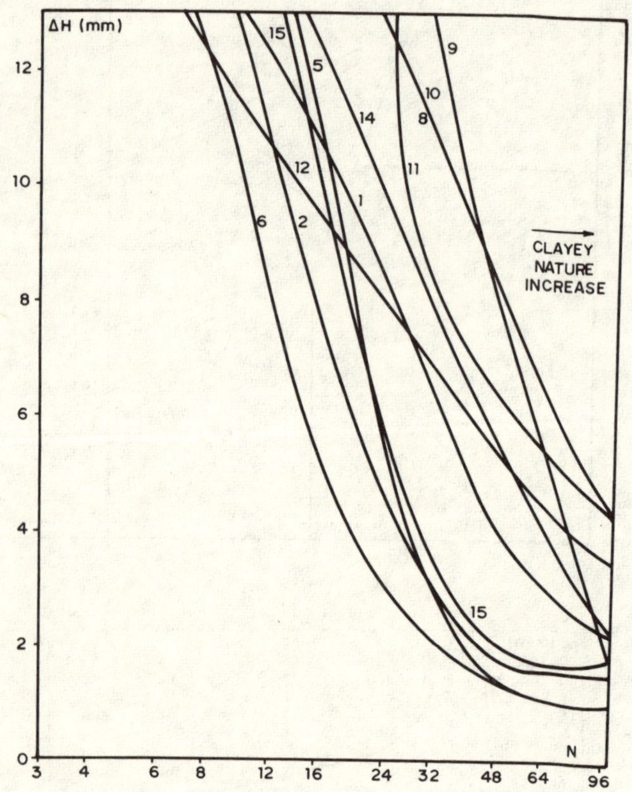

Figure 8. Curves of height reduction (ΔH) in function of the applyed blow number (N), relative to compacted samples with 10 cm diameter.

but with reduction of efficiency for higher levels of energy (due to the fine fraction and breaking of the concretions). Thus, the behavior and structural arrangement of these soils depend on the capacity of wedging and resistance of their gravel fraction. These facts cannot be observed only by classical procedures.

5 CHANGES IN PROPERTIES OF PLINTHITIC SOILS

The plinthitic samples were subjected to different cicles of wetting-drying (wetting by saturation or sprinkling of water and air drying) during five months with the aim of verifying changes in geotechnical properties when compared to the same soils in natural conditions. Using the MCT and the Resilient criteria, it was possible to verify effective changes in these soils (see Figure 9) where the samples showed an increase in lateritic behavior aspects and in resilient modulus values.

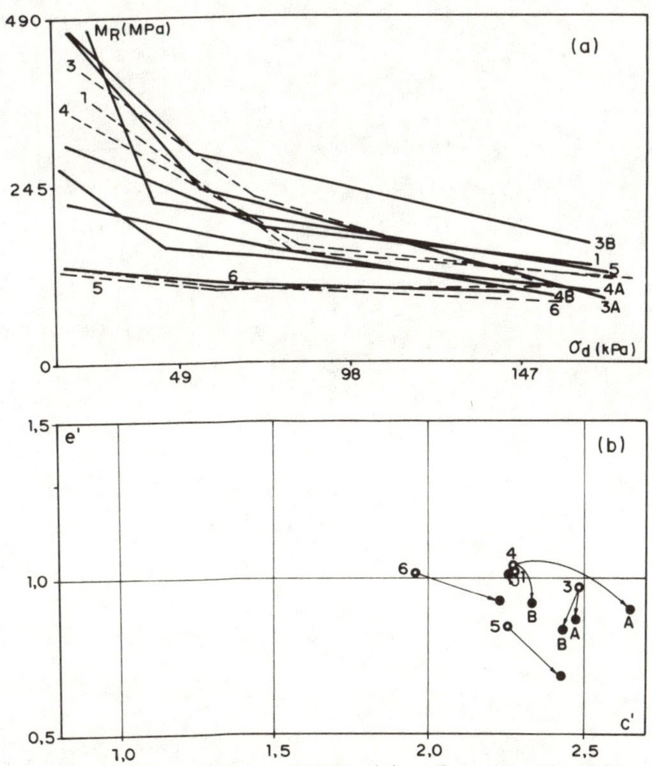

Figure 9. Changes in resilient (a) and in genetic (b) behavior of plinthitic soils due to different wetting-drying cicles (A,B) when compared to the natural conditions (---,o).

This phenomenon is not the same that occurs when a sample molded in a CBR cylinder is maintained drying before soaking, where an increase in CBR values due to drying can occur. The previous drying process does not constitute itself an agent of self-stabilization,but, on the other hand, the mutation of plinthitic soils in decorrence of slowly alternated cicles of wetting-drying produce an improvement in engineering properties. The degree in which this effect occurs depends of the iron segregation level in the mosqueate nucleus and of the possibilities of oxi-reduction reations.

6 CONCLUSIONS

Through this research it was possible to analyze typical tropical soils using new criteria not yet considered. The Amazon soils treatment using the MCT and the Resilient criteria allowed to identify the peculiarities of sedimentary, plinthitic and concretioned soils in the tropical environment which do not follow some of the tendencies established by materials of temperate climate. The behavior standards (like MCT and Resilient correlations) permit to estimate the genetic quality and geotechnical properties of the soils conccuring for the more adequate selection of materials in the planning of new works and analysis of existing pavements.

A new standard behavior was defined - the transient soils-relative to sedimentary tropical formations and their quartzific nature of silt fraction that have an influence on their behaviors. Generally, the fine soils constitute good materials as subgrade and with possibilities of use in pavement layers.

The beneficial mutational sensibility of plinthitic soils was demonstred. Although they are treated only in laboratory, they can constitute a new construction material in certain tropical regions if adequatly selectioned.

The concretioned soils showed variable characteristics and resilient behavior. Some of them are in the limit of behavior dictated by the gravel fraction or fine sand and fine fractions.In function of their peculiarities and existing stresses they can be good or inadequate materials for use as base or sub-base of pavements. Also the efficiency of compactation is very variable. These aspects show the necessity of selecting these soils adequatly to warrant the functionality and durability of pavements.

REFERENCES

American Association of State Highway and Transportation Officials (1962). AASHTO guide for design of pavement structures. Washington D. C., p. E. 4.

——(1986). AASHTO guide for design of pavement structures. Washington D.C., C. 4-II : 71-87.

Empresa Brasileira de Pesquisa Agropecuária (1981). Mapa de solos do Brasil E 1:5.000.000. SNLCS, Rio de Janeiro.

Medina, J, (1985). Classification of Tropical soils according to resilient characteristics for pavement design. First International Conference on Geomechanics in Tropical Lateritic and Saprolitic Soils. ISSMFE, Brasília, 2:321-326.

Nogami, J.S. & Villibor, D.F. (1981). Uma nova classificação de solos para finalidades rodoviárias. Simpósio Brasileiro de Solos Tropicais em Engenharia, ABMS, Rio de Janeiro, p. 30-41.

—— (1985). Additional considerations about a new geotechnical classification for tropical soils. First International Conference on Geomechanics in Tropical Lateritic and Saprolitic Soils. ISSMFE, Brasília,1:165-174.

Preussler, E.S. (1983). Estudo da deformação resiliente de pavimentos flexíveis e aplicação ao projeto de camadas de reforço. Tese de Doutorado, COPPE/UFRJ, Rio de Janeiro, 252 p.

TropicaLS'85(1985). First International Conference on Geomechanics in Tropical Lateritic and Saprolitic Soils. Progress Report. ISSMFE, Brasília.

Vertamatti, E. (1985). Potential use of tropical soils from the Amazon in airfield pavements. First International Conference on Geomechanics in Tropical Lateritic and Saprolitic Soils. ISSMFE, Brasília, 2:361-368.

——(1986). Propriedades básicas de solos superficiais da Região Amazônica. VIII Congresso Brasileiro de Mecânica dos Solos e Engenharia de Fundações. ABMS, Porto Alegre, 1:95-106.

——(1988). Contribuição ao conhecimento geotécnico de solos amazônicos com base na investigação de aeroportos e metodologias MCT e Resiliente. Tese de Doutorado, ITA, São José dos Campos - São Paulo, 276 p.

ACKNOWLEDGEMENTS

The autor wants to thank Prof. Job S. Nogami and Prof. Ernesto S. Preussler for their valuable help on the discussion of results. Also to thank Fundação de Amparo à Pesquisa no Estado de São Paulo, Diretoria de Engenharia de Aeronáutica and Comissão de Aeroportos da Região Amazonica for their decisive support on the research.

Natural and lime stabilised properties of Bandung clay

J.S.YOUNGER, Frank Graham International Ltd, Indonesia

ABSTRACT:

A 30m deep deposit of volcanic clay is located on the southern boundary of Bandung, West Java, Indonesia. X-ray diffraction testing indicated poor crystallinity with halloysite and some albite being mainly found along with a significant proportion of diatomaceous silica at some levels of testing. Unweathered layers of the deposit have very high but variable void ratios, sometimes in excess of 6, with corresponding high moisture contents. A variable but sometimes significant content of organic matter is also present. Apart from the surface weathered layers, investigations have indicated the presence of two other buried weathered layers. Interest in the deposit has increased since a major highway is located across one section of the deposit, the highly compressible characteristics of which pose construction problems and the necessity of countermeasures. The natural properties of the soil, and those following treatment with lime in the laboratory are described. Indications of likely settlement behaviour under embankment loading are also given.

1. INTRODUCTION

Bandung Clay, a Quaternary deposit of volcanic clay, lies on the southern boundary of the city of Bandung, the capital of West Java, Indonesia. The deposit extends over an area of about 50 sq. km and reaches a depth in excess of 30m for most of this area. The source of the original material for the soil is the volcano Tangkuban Perahu, located some 25km to the north. There is evidence of the formation of diatomaceous silica in quiescent periods. The fresh water lacustrine nature of the deposit is in contrast to the majority of Indonesian soft soils, which tend to be coastal or offshore.

Interest in the deposit has increased recently because of the construction of the Padalarang to Cileunyi (Panci) Toll Road, a 35km highway which will bypass the city of Bandung to the south and west. Eight kilometres of this highway passes directly over the soft deposit, and this has led to investigation into the consolidation and stability characteristics of the material. The pertinent properties of the soil for settlement and stability analysis form the basis of this paper. Results of laboratory tests where quicklime has been added are also reported.

2. SOIL TYPE AND BASIC PROPERTIES

X-ray diffraction testing at various depths has shown the presence of halloysite and some albite. A draft Geological Society Report on Tropical Residual Soils (1988) would classify the soil as being a fersiallitic andosol. This is a soil with volcanic affinities either directly developed or derived from outside the areas. However, a sample tested with the electron microscope from a depth of 6m (see Figure 1) showed clearly the large presence of diatomaceous silica, with the clay fragments as metahalloysite/kaolinite. This indicates near surface quiet periods and periods of weathering and dessication. The presence of diatoms has also been found at other depths.

Figure 1. Electron micrograph of soil from 6.0m depth

The soil is dark coloured, mainly amorphous, with enormous water-holding capacity exceeding 100% and reaching 200%.

The distribution of water content with depth as found at several locations along the proposed route of the Panci highway is presented in Figure 2, which also shows profiles of the variations in void ratio and maximum past (or critical) pressure with depth (Younger 1986). It can be seen that the distribution of these characteristics indicates possibly three different periods of geological deposition, i.e. there would appear to be two earlier periods when conditions, including a lowered water table, may have allowed the partial development of residually weathered profiles. In such an environment the development of chemical bonding is considered likely.

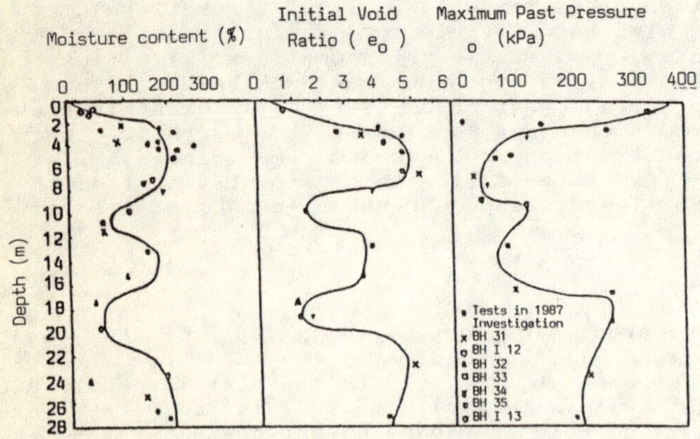

Figure 2. Profiles of natural moisture content, initial void ratio and maximum past pressure for boreholes between Sta 40 + 700 and Sta 43 + 000 (After Younger 1986)

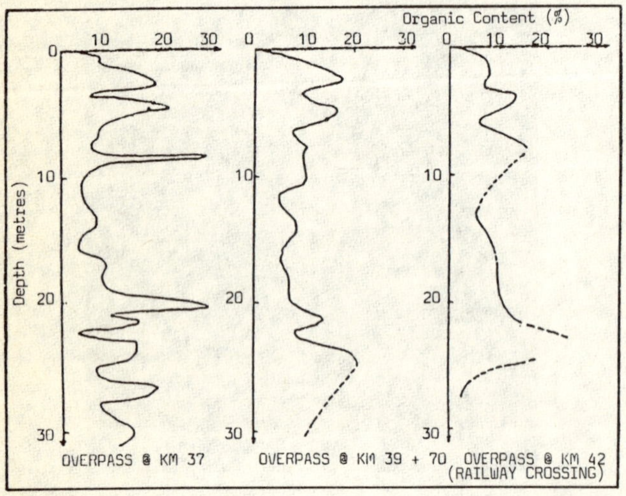

Figure 3. Profile with Depth of Organic Content for Boreholes Between Sta 37+000 and Sta 42+000 (After Younger 1986)

Void ratios for the soil in the non-weathered layers have been found to be in the region of 4-6, and variable. Rouse et al. (1986) reported similar high values for the allophane soils of Dominica, West indies. Figure 3 shows a tentative distribution of organic content with depth at three locations along the proposed route alignment. The presence of organics in the formation of andosols can be expected. Areas of higher organic content may reflect periods where vegetation may have been allowed to take hold on the deposit.

Figure 4 shows, for Bandung Clay, the relationship between Plasticity Index and Liquid Limit on the extended Casagrande plasticity chart. The results tend to conform to the pattern reported by Rouse et al. (1986) for freshwater depositions which included data presented by Wesley (1973) for Indonesian latosols and andosols. Those results which exhibit relatively lower PI values could indicate more marked presence of the weathering process. The plasticity values of current surface crustal or buried crustal samples were found to be significantly less.

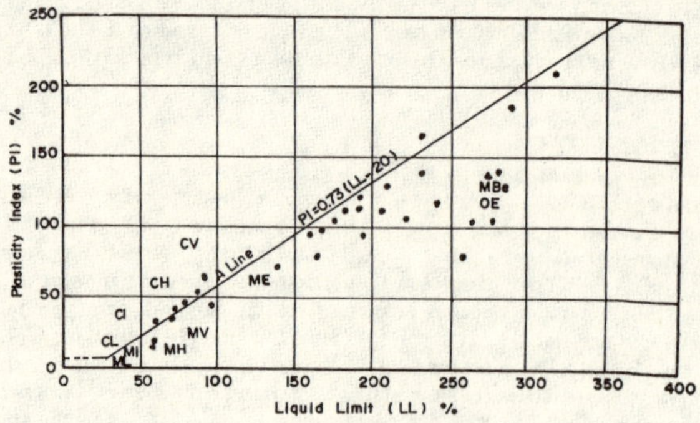

Figure 4. Extended Casagrande Plasticity Chart (Farrington, 1980) Illustrating Classification of Natural Soil at Time of Sampling.

The method of testing can have major influence on both particle size analysis and plasticity index tests. Warkentin (1974) and the draft Geological Society Report (1988) advise against use of grain size as a means of predicting behaviour of tropical clay soils under stress; plasticity index testing may be used with caution, taking note of the influence of testing. Wesley (1973), however, found that results of plasticity index tests for allophane and halloysitic clay soils in Java were little affected by soil-water mixing periods provided the paste was uniform before testing.

Despite high plasticity indices, high in situ moisture contents and low unit weights, tropical clay soils, particularly allophanes and halloysites, are often remarkably stable, observations noted by Wallace (1973) and Wesley (1973), a factor which Vaughan et al. (1988) attributes to the development of chemical bonding during the weathering process. This internal stability will have been a large factor in the generally successful recovery of open-drive samples, albeit with a measure of disturbance, during the original site investigation, where standard wash-boring methods were employed. Comment on this issue for site investigation practice in Indonesia has been made by Cook & Younger (1987) and Younger (1988).

3. CONSOLIDATION CHARACTERISTICS

(i) General

An extensive number of conventional oedometer consolidation tests were carried out during the original site investigation (Lyon/Indec 1983). Further tests have been carried out to examine consolidation properties of the main upper soft layer both in the natural state (Setionegoro 1986. Ridwan 1988) and following treatment with lime (Simanjuntak 1986. Helmi 1987). The main findings of these studies are outlined below.

(ii) Consolidation Parameters, e, c', Cc and C_α for Natural Soil

With reference to Figure 2, it can be seen that the void ratios of the three main soft layers are very high, in the range, generally, of 4-6. As stated above, similar values were also reported for Dominican allophanes (Rouse et al. 1986). In comparison with void ratios for most other reported clays these values are much higher, as shown in Figure 5, the one notable exception being Mexico City clays (Zeevaert 1973), which are also of volcanic origin. The void ratios in the weathered horizon levels and at the current surface fall in to more normal patterns, i.e. between 1 and 2. These buried horizons indicate changes in the geological environment. Although chemical bonding in the weathering state has been likely (Vaughan et al. 1988) the development of this bonding will have followed a reduction in the clay structure resulting from water loss in the normal air drying process. As indicated also in Figure 4, the changes in plasticity characteristics from air drying will have been effected before the chemical bonding stage.

Figure 2 showed the profile of the maximum past or critical pressure, σ_c', based on analysis of oedometer test results on samples recovered from several locations. Overconsolidation is quite marked in the soil depths below 17m, and evident in the horizon at 10m as well for the uppermost 3m. The depths between 3m and about 8m, and again between 11 and 17m, are close to the normally consolidated situation, although the conventional method of sampling and small specimen testing could be expected to adversely influence the true in situ critical pressure. A few tests carried out on samples recovered at depths between 4.2m and 5.2m using test specimens of increasing thickness showed critical pressures of 54, 82 and 110 KPa for thicknesses of 20 mm, 40 mm and 60 mm respectively. The Geological Society's Report on Tropical Residual Soils (1988) states that yield stresses determined in the laboratory are probably underestimated as a result of the effects of sample disturbance, chemical bonding being particularly susceptible to this. The standard oedometer, in particular, substantially underestimates yield in the field (Prusza et al. 1983. Dib 1985). The argument for piston sampling and larger diameter testing (Rowe & Barden 1966) is clearly indicated, since the accuracy of determination of σ_c' will affect significantly settlement analysis. More recent work by Holtz et al. (1986), shows clearly again the influence of sample disturbance on the determination of σ_c', in terms of sample size and method of sampling.

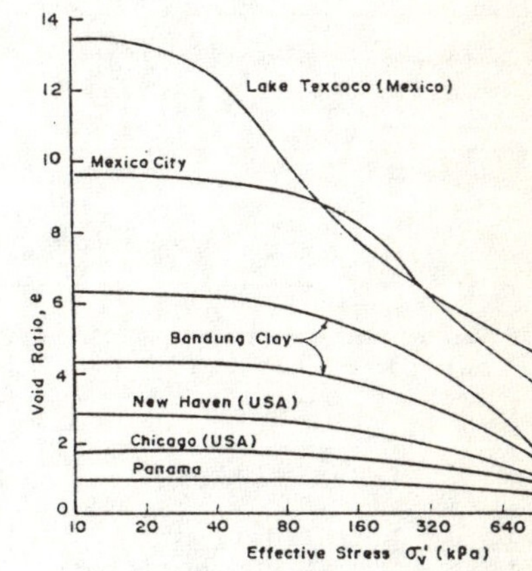

Figure 5. Compressibility Curves for Various Localities (After Zeevaert 1973)

Very significant values of coefficient of compression, C_c, have been found for the "normally" consolidated layers, generally of the order of 2.5 to 3.2 and sometimes greater. For weathered layers the values are much less, between 0.5 and 1.0, depending on depth and degree of change that has taken place.

Figure 3 showed that there was a steady but variable distribution of organics with depth in the deposit. Significant values of C_α, the coefficient of secondary compression have been found in conventional oedometer testing, particularly when applied effective stresses were greater than the critical pressure. Typical values are shown in Table 1. However, it was found that C_α values significantly reduced for loads beyond the critical pressure, if time of testing per load increment was increased for more than one day (Helmi 1987). A 30-50% reduction in computed C_α values was obtained provided a load increment was maintained for at least 6-10 days. For applied stress levels below the critical pressure, the C_α values were much lower than where σ_c' had been passed, although a small increase in C_α was evident with time, suggesting some structural reorientation taking place in the test specimens.

Table 1. Values of C_α with load increment for samples from km 42 + 096 on the Padalarang - Cileunyi Toll Road alignment

Depth m	Nat. M.C. %	Org. Cont. WT - %	Load increment kPa	C_α
3.65	131	13.28	0.262	0.005
			0.524	0.0064
			1.048	0.011
			2.096	0.018*
			4.192	0.153
8.00	193		0.262	0.0012
			0.524	0.0094
			1.048	0.0162
			2.096	0.016*
			4.192	0.0053
13.75	54		0.252	negl.
			0.524	negl.
			1.048	0.0008
			2.096	0.0024*
			4.192	0.0045
			8.384	0.0083
30.25	167		0.262	0.001
			0.524	0.0018
			1.048	0.0027*
			2.096	0.0073
			4.192	0.026
			8.384	0.033

* *Values appropriate to loadings under expected high fills*

(iii) Rate of Consolidation

Both vertical and horizontal rates of consolidation have been examined in the conventional oedometer. Figures 6 and 7 from Ridwan (1988) reflect the ranges of values obtained. It can be seen that, at applied stresses below the critical value, C_v and C_h values are high, particularly the latter. The draft Geological Society Report (1988) states that values in natural residual soils would be much higher than the grading of the soil would suggest. The horizontal coefficient permeability k_h is also significantly higher than the vertical, k_v for the Bandung soil, except at very low applied stress levels.

Once the critical pressure, σ_c', is reached and passed there is a significant drop in both C_v and C_h values (see also Medina & Liu 1982), indicating a marked reduction in the macro-pore structure. For the Bandung Clay, a considerable reduction in rate of pore pressure dissipation for any surface applied loads could thus be expected once the critical pressure had been surpassed for any given situation.

An ongoing monitored trial embankment loading is expected to provide some

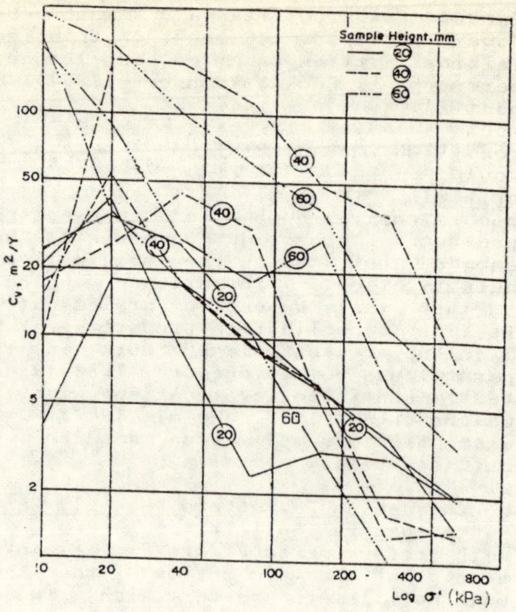

Figure 6. Coefficient of Radial Consolidation, C_{VR} c Log Effective Pressure, kPa

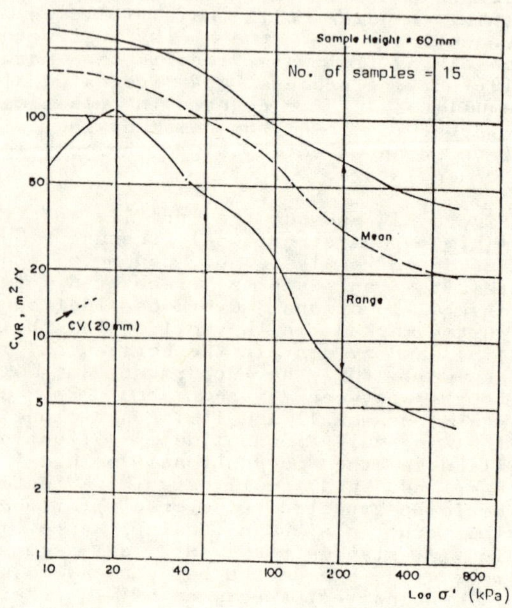

Figure 7. Coefficient of Consolidation, C_v, Log Effective pressure, σ'

further information in due course. However, the consequence of the high initial in-situ coefficient of permeability would mean that loading at normal engineering rates of construction would not likely result in undrained conditions, and excess pore pressures would not be generated until the critical stress at any given depth had been passed. No increase in pore pressures has been observed for trial embankment heights constructed to heights between 2.0 and 2.5m.

It is well established, however, that direct use of rates of consolidation parameters from conventional oedometer testing often leads to considerable under-estimations of the rates of dissipation of pore pressure and related rates of settlement (Rowe 1968 & 1972), although Mitchell (1976) has reported on a number of anomalous situations. Murray (1971) and Symons & Murray (1975) on carefully monitored sites found that closer evaluation of field performance required the taking of in situ permeability tests (coupled with conventional laboratory m_v values) in order to obtain C_v values closer to the in situ condition. Conventional oedometer determinations (vertical drainage only) led to very significant underestimations of rates of settlement. In the present study, the laboratory tests with radial drainage should provide predictions closer to the expected field situation.

(iv) Settlement analysis

Based on an analysis of consolidation data obtained for Sta 42 + 096, where a high fill area to provide a railway crossing is envisaged, a multi-layer settlement analysis was carried out. Boussinesq stress distribution conditions were assumed (Poulus & Davis 1974) based on the shape of the embankment. The relationship between embankment height and settlement, both overall and separately for the main individual soft layers, is shown in Figure 8. Predictions indicate that embankment heights below about 2.5m would produce stress levels under a yield condition. Beyond this height settlements would become significant.

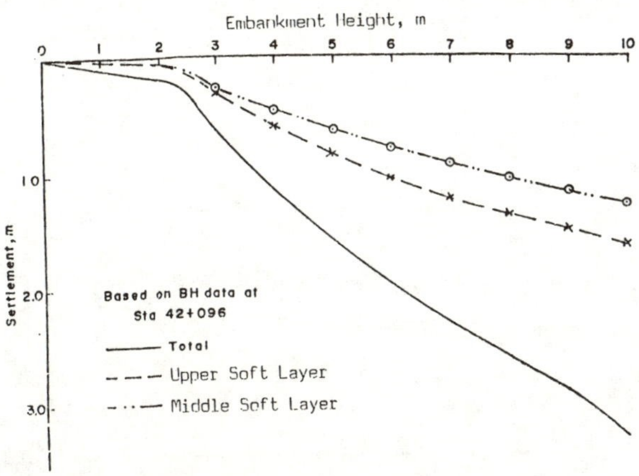

Figure 8. Embankment Height, m v Settlement, m

(v) Influence of Lime

Tests were carried out either introducing quicklime mixed totally throughout the laboratory specimens or in a central column. In the latter case, the height of the specimens was sometimes extended. As given in Figure 9, Simanjuntak (1986) showed that the introduction of lime columns for standard consolidation specimens led to improvements in the determined critical pressure for samples recovered from the upper main soft layer; results also being affected by column diameter. A 60mm diameter specimen indicated that the total specimen had been mixed with lime. Figure 10a shows the influence of lime content on C_c values. For the higher lime contents tried, 12% and 15%, conventionally measured C_v values were also increased at low applied effective stress values by a factor of 2 to 4.

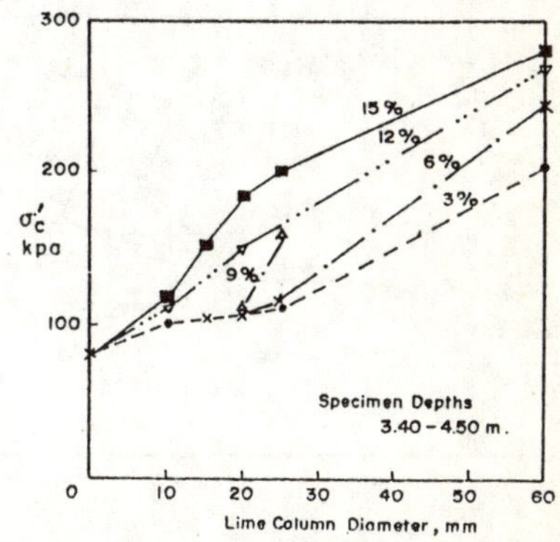

Figure 9. Maximum Part Pressure v Column Diameter for Various Percentages of Lime

Figure 11, after Helmi (1987), shows that there is significant dispersion of Ca ions in the clay following insertion of lime columns (10% lime content) measured in terms of change in pH value, with the untreated soil outside the column increasing from a value of 3 to 4 to a value between 6 and 7. It is expected that this will have taken place at the beginning, and under low applied stress levels when the macro-structure of the specimens was still large. C_c values also changed significantly, as shown in Figure 10b, and similar increases in C_v as reported by Simanjuntak (1986) were also found. Figure 12b shows also the influence of specimen height on C_c values; less reduction being observed where the column was longer. However, a more definitive appraisal requires use of larger cells and model foundations as used for a study on Bangkok clay (Younger et al. 1987).

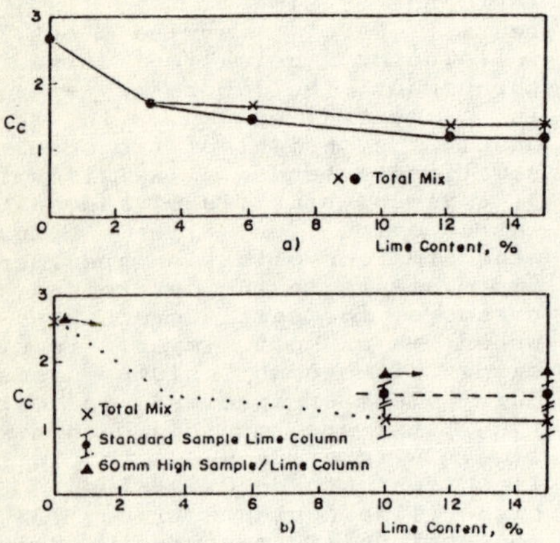

Figure 10. Compression Index, C_c v Lime Content %

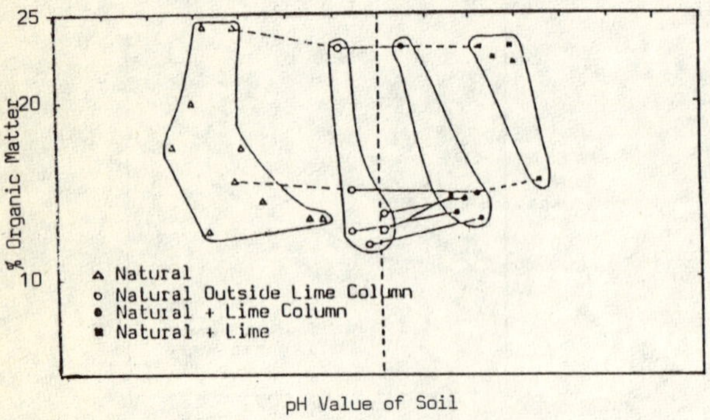

Figure 11. Relationship between Organic Matter and pH Values after Consolidation Tests.

4. UNDRAINED STRENGTH CHARACTERISTICS

The results reported here, as shown in Figure 12, on in situ vane tests to a depth of 30m, and laboratory undrained triaxial tests on samples recovered within 6m depth indicate the undrained strength characteristics of weathered and "natural" layers. Towards the surface, peak strengths in vane testing of 100-200 kPa were recorded, and from 3m to 6m, 40 - 100 kPa. Corresponding undrained triaxial tests on "undisturbed" samples recovered in conventional open-drive sampling gave values in the range 14-49 kPa, indicating that a fair measure of disturbance had occurred during sampling. Sensitivity ratios from vane testing indicated a wide range from moderately sensitive to very sensitive conditions.

In preparation of lime mixes, mixing led to a loss in strength of the natural material, with strength regain at 28 days reflecting largely only a value close to the likely undisturbed value before mixing. The very high natural moisture contents will have inhibited the lime from being more effective in terms of strength gain. Some results are shown in Figure 13. In practice, the in situ high moisture contents would indicate the requirement of higher than normal lime content, and this would probably mitigate against the economic feasibility of using the lime column method in this instance, taking note also that treatment would be required to about 18m.

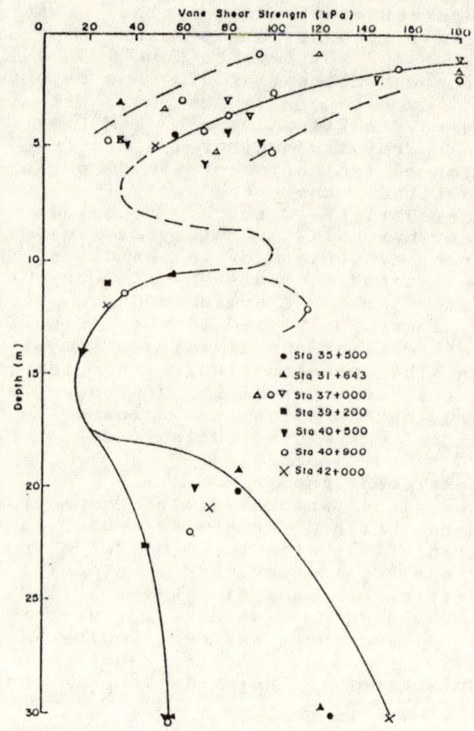

Figure 12. Vane Shear Strength with Depth.

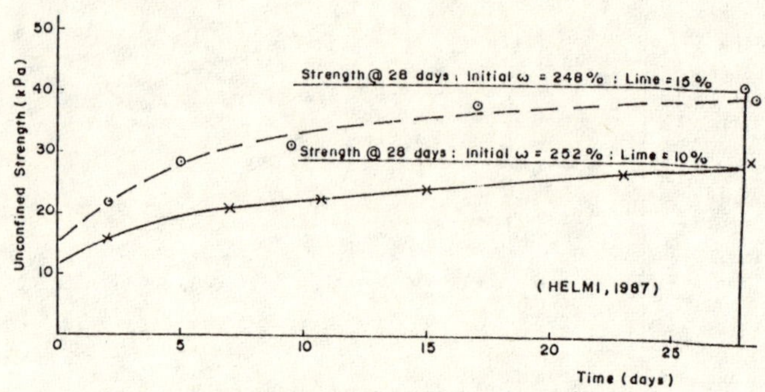

Figure 13. Unconfined Strength Gain with Time on Lime Treated Samples.

5. CONCLUSIONS

The strength and consolidation characteristics of Bandung Clay, a high moisture content volcanic material, have been described. The soils found have been variable, with halloysite, diatomaceous silica and albite being the main minerals identified.

Void ratios in the range of 4 to 6 or more were found in unweathered soils with corresponding high moisture contents. In oedometer testing consolidation settlement was small for applied stresses less than the measured critical pressure, but was very significant for higher applied stress levels, reflecting a breakdown in the macro-fabric. Sampling disturbance is expected to have reduced the likely in situ critical pressure. C_c values were in the range 2.5 - 3.5. For weathered material, void ratios and C_c values were significantly less, in line with more usual soft soils.

The observed significant secondary compression component past the critical pressure probably also reflected some clay fabric restructuring as well as the presence of organics. It was found that sustained application of a particular load beyond the conventional 24 hour loading period led to a reduction in the value of secondary compression of the order of 30-50%; a steady state value was obtained after 6 - 10 days.

High rates of consolidation were recorded for test specimens where the applied vertical stress in a standard oedometer was less than the critical pressure or yield value of the soil. For applied stresses beyond this point, the values of C_v and C_h decreased markedly, coming in line with those for more normal clay types, and reflecting the breakdown of the macro-fabric.

In the field, settlement analysis for highway loading, based on site investigation data, indicated that a critical height of embankment to initiate large consolidation settlement was about 2.5m.

In examining undrained strength both in situ vane and laboratory triaxial tests were carried out, the latter indicating a good measure of sampling disturbance. Strengths of unweathered material were in the very soft to soft category, but higher values of from 60-180 kPa were recorded in weathered layers, strength depending on the degree of weathering, mineral type, and likely chemical bonding.

Consolidation and strength tests using quicklime indicated some improvements in relevant properties, but the high initial moisture contents of unweathered material reduced the normally marked improvements associated with use of lime in clay soils. pH tests on laboratory samples with lime columns indicated that there was noticeable dispersion of Ca ions, significant changes in pH being found. It is expected that the dispersion will have occurred at early stages of testing below the yield stress, where there were high permeability values in the soil.

6. ACKNOWLEDGEMENTS

Use of field strength data from the Panci project has been afforded by P.T. Jasa Marga (Project Manager, Ir. Soelastadi) and by the consulting engineers Messrs Ammann & Whitney and Wiratman & Associates. Their interest in this study is appreciated. Discussion on the geology of volcanic soils with Mr. J.R. Cook, a member of the Geological Society Working Party now attached to the Institute of Road Engineering, Bandung, Indonesia, through T.P.O'Sullivan & Partners, is also appreciated. The work on X-ray diffraction testing was undertaken by Dr. A. Poole, Queen Mary College London and Mr. Ross Freeman, Sietronics, Canberra, Australia. The electron microscope photograph was supplied by Dr. A. Poole.

REFERENCES

Cook, J.R. & J.S. Younger (1987). Ground investigations for roads in the region. 3rd Indonesian Conference on Road Engineering, Bandung 1 : 95 : 120.

Dib, P.S. 1985. Compressibility characteristics of tropical soils making up the foundations of the Tucuru Dam in Amazonia, Brazil. Proc. 1st. Int. Conf. on Geomechanics in Tropical Lateritic & Saprolitic Soils, Brasilia, 2: 131-142.

Farrington, P. 1980. Earthworks and Foundations on recently deposited organic soils in the Lagos area. Proc. 7th Reg. Conf. for Africa on Soil Mech. and Found. Eng., Accra, Ghana, 235 - 259.

Geological Society 1988. Tropical residual soils. Draft Working Party Report of the Engineering Group of the Geological Society of London.

Helmi, A. 1987. Influence of lime on the secondary compression characteristics of Bandung clay. Thesis in partial fulfilment of requirements for M.Sc. Degree, Inst. of Technology Bandung, Indonesia.

Holtz, R.D., M.B. Jamiolknowski, & R. Lancellota 1986. Lessons from oedometer tests on high quality samples. J. Geot. Eng. Div., ASCE, 112.

Lyon/Indec 1983. Geotechnical Report Section B of Bandung bypass, Padalarang - Cileunyi Toll Highway, West Java, Indonesia (unpublished).

Medina, J. & B.S. Liu (1982). The influence of a collapsible foundation on the design of the Guri Embankment Dams. Trans. 14th Int. Congress on Large Dams, Rio de Janeiro, 2: 241 - 260.

Mitchell, J.K. 1986. Practical problems from surprising soil behaviour. 20th Terzaghi lecture, ASCE 1984 Annual Convention, J. Geot. Eng. Div., ASCE, 112.

Murray, R.T. 1971. Embankments constructed on soft foundations: settlement study at Avonmouth. DoE, RRL report LR 419, Crowthorne, U.K.

Poulous, H.G. & E.H. Davis 1974. Elastic Solutions for Soil and Rock Mechanics. John Wiley & Sons Inc. New York.

Prusza, Z., K. De Fries & R. Schmitt 1983. The compressibility of undisturbed and compacted residual soils at Guri. Proc. 7th Panamerican Conf. on Soil Mech. & Found. Eng., Vancouver, 1, 185-198.

Ridwan, M. 1988. Study of model sand columns in Bandung clay. Thesis in partial fulfilment of requirements for M.Sc. Degree, Inst. of Technology, Bandung, Indonesia.

Rouse, W.C., A.J. Reading & R.P.D. Walsh 1986. Volcanic soil properties in Dominica, West Indies. J. Eng. Geol. 23: 1-28.

Rowe, P.W. 1968. The influence of geological features of clay deposits on the design and performance of sand drains. Proc. Inst. of Civil Eng., London, Paper 70585.

Rowe, P.W. 1972. The relevance of soil fabric to site investigation practice. Geotechnique. 22: 195-300.

Rowe, P.W. & L. Barden 1966. A new consolidation cell, Geotechnique: 16: 162-170.

Setionegoro, N. 1986. Consolidation properties of Bandung clay. Thesis in partial fulfilment of requirements for M.Sc. Degree, Inst. of Technology Bandung, Indonesia.

Simanjuntak, A. 1986. A laboratory study of lime - soil columns in Bandung clay. Thesis in partial fulfilment of requirements for M.Sc. Degree, Inst. of Technology, Bandung, Indonesia.

Symons, I.F. & R.T. Murray 1975. Embankments on soft foundations: settlement & stability at Over causeway by-pass. DoE, TRRL Lab. Report 675, Crowthorne, U.K.

Vaughan, P.R., M. Maccarini & S.M. Mokhtar 1988. Indexing the engineering properties of residual soil. Quar. Jour. of Engineering Geology. 21.

Wallace, K.B. 1973. Structural behaviour of residual soils of the continually wet highlands of Papua New Guinea. Geotechnique. 23: 203-218.

Warkentin, B.P. 1974. Physical properties of Caribbean clay soils. Trop. Agric. (Trinidad), 51(2): 279-287.

Wesley, L.D. 1973. Some basic engineering properties of halloysites and allophane clays in Java, Indonesia. Geotechnique. 23: 471 - 494.

Younger, J.S. 1986. Construction over soft ground. Proc. of Pasca Sarjana Workshop on Developments in Highways & Transportation Engineering, Bandung, December.

Younger, J.S., A.S. Balasubramaniam, Lee Hong-Jen & M. Anshinnali 1987. A lime column model study in Bangkok clay. Proc. Eight Asian Regional Conf. on Soil Mechanics and Foundation Engineering, Kyoto, japan: 221-224.

Younger, J.S. 1988. Construction over Soft Ground with special emphasis on use of vertical drainage. Indonesian Soils Association Seminar, Medan, Indonesia, April.

Zeevaert, L. 1973. Foundation engineering for difficult subsoil conditions. Litton Educational Publishing Inc.